Selected Papers on

QUANTUM ELECTRODYNAMICS

Edited by Julian Schwinger
Professor of Physics,
Harvard University

DOVER PUBLICATIONS, INC., NEW YORK

Published in Canada by General Publishing Company, Ltd., 30 Lesmill Road, Don Mills, Toronto, Ontario.

Published in the United Kingdom by Constable and Company, Ltd., 10 Orange Street, London WC 2.

This Dover edition, first published in 1958, is a selection of papers published for the first time in collected form. The editor and publisher are indebted to the original authors, journals and the Columbia University Library for assistance and permission to reproduce these papers.

Standard Book Number: 486-60444-6
Library of Congress Catalog Card Number: 58-8521

Manufactured in the United States of America
Dover Publications, Inc.
180 Varick Street
New York, N.Y. 10014

CONTENTS

Preface vii

PAPERS **PAGE**

1 P.A.M. Dirac THE QUANTUM THEORY OF THE EMISSION AND ABSORPTION OF RADIATION
Proceedings of the Royal Society of London, Series A, Vol. 114, p. 243 (1927) 1

2 Enrico Fermi SOPRA L'ELLETTRODINAMICA QUANTISTICA
Atti della Reale Accademia Nazionale dei Lincei, Vol. 12, p. 431 (1930) 24

3 P.A.M. Dirac, V. A. Fock, and Boris Podolsky ON QUANTUM ELECTRODYNAMICS
Physikalische Zeitschrift der Sowjetunion, Band 2, Heft 6 (1932) 29

4 P. Jordan and E. Wigner ÜBER DAS PAULISCHE ÄQUIVALENZVERBOT
Zeitschrift für Physik, Vol. 47, p. 631 (1928) 41

5 W. Heisenberg ÜBER DIE MIT DER ENTSTEHUNG VON MATERIE AUS STRAHLUNG VERKNÜPFTEN LADUNGSSCHWANKUNGEN
Sachsiche Akademie der Wissenschaften, Vol. 86, p. 317 (1934) 62

6 V. S. Weisskopf ON THE SELF-ENERGY AND THE ELECTROMAGNETIC FIELD OF THE ELECTRON
Physical Review, Vol. 56, p. 72 (1939) 68

7 P.A.M. Dirac THEORIE DU POSITRON
Rapport du 7e Conseil Solvay de Physique, Structure et Proprietés des Noyaux Atomiques, p. 203 (1934) 82

8 V. S. Weisskopf ÜBER DIE ELEKTRODYNAMIK DES VAKUUMS AUF GRUND DER QUANTENTHEORIE DES ELEKTRONS *Kongelige Danske Videnskabernes Selskab, Mathematisk-fysiske Meddelelser XIV,* No. 6 (1936) 92

9 F. Bloch and A. Nordsieck NOTES ON THE RADIATION FIELD OF THE ELECTRON *Physical Review,* Vol. 52, p. 54 (1937) 129

10 H. M. Foley and P. Kusch ON THE INTRINSIC MOMENT OF THE ELECTRON *Physical Review,* Vol. 73, p. 412 (1948) 135

11 Willis E. Lamb, Jr. and Robert C. Retherford FINE STRUCTURE OF THE HYDROGEN ATOM BY A MICROWAVE METHOD *Physical Review,* Vol. 72, p. 241 (1947) 136

12 H. A. Bethe THE ELECTROMAGNETIC SHIFT OF ENERGY LEVELS *Physical Review,* Vol. 72, p. 339 (1947) 139

13 Julian Schwinger ON QUANTUM-ELECTRODYNAMICS AND THE MAGNETIC MOMENT OF THE ELECTRON *Physical Review,* Vol. 73, p. 416 (1948) 142

14 Julian Schwinger ON RADIATIVE CORRECTIONS TO ELECTRON SCATTERING *Physical Review,* Vol. 75, p. 898 (1949) 143

15 J. R. Oppenheimer ELECTRON THEORY *Rapports du 8e Conseil de Physique, Solvay,* p. 269 (1950) 145

16 S. Tomonaga ON A RELATIVISTICALLY INVARIANT FORMULATION OF THE QUANTUM THEORY OF WAVE FIELDS *Progress of Theoretical Physics,* Vol. I, p. 27 (1946) 156

17 Julian Schwinger QUANTUM ELECTRODYNAMICS, III: THE ELECTROMAGNETIC PROPERTIES OF THE ELECTRON—RADIATIVE CORRECTIONS TO SCATTERING *Physical Review,* Vol. 76, p. 790 (1949) 169

18 S. Tomonaga ON INFINITE FIELD REACTIONS IN QUANTUM FIELD THEORY *Physical Review,* Vol. 74, p. 224 (1948) 197

19 W. Pauli and F. Villars ON THE INVARIANT REGULARIZATION IN RELATIVISTIC QUANTUM THEORY *Reviews of Modern Physics,* Vol. 21, p. 434 (1949) 198

20 Julian Schwinger ON GAUGE INVARIANCE AND VACUUM POLARIZATION *Physical Review,* Vol. 82, p. 664 (1951) 209

21 R. P. Feynman THE THEORY OF POSITRONS *Physical Review,* Vol. 76, p. 749 (1949) 225

*This paper should properly appear following paper 12, but for reasons beyond my editorial control, it appears as the last paper. JS

PREFACE

ANY SELECTION of important contributions from the extensive literature of quantum electrodynamics necessarily reflects a particular viewpoint concerning the significance of those works, both historically and in their implications for the future progress of the subject. The following brief commentary is intended to indicate that viewpoint, and to supply a setting for the individual papers. The latter are referred to by consecutive numbers, and appear in the same order, which does not always correspond to the historical one. A few papers, which were omitted only because of limitations on the size of the volume, are mentioned explicitly in the text.

The development of quantum mechanics in the years 1925 and 1926 had produced rules for the description of systems of microscopic particles, which involved promoting the fundamental dynamical variables of a corresponding classical system into operators with specified commutators. By this means, a system, described initially in classical particle language, acquires characteristics associated with the complementary classical wave picture. It was also known that electromagnetic radiation contained in an enclosure, when considered as a classical dynamical system, was equivalent energetically to a denumerably infinite number of harmonic oscillators. With the application of the quantization process to these fictitious oscillators the classical radiation field assumed characteristics describable in the complementary classical particle language. The ensuing theory of light quantum emission and absorption by atomic systems [1] marked the beginning of quantum electrodynamics, as the theory of the quantum dynamical system formed by the electromagnetic field in interaction with charged particles (in a narrower sense, the lightest charged particles). The quantization procedure could be transferred from the variables of the fictitious oscillators to the components of the field in three-dimensional space, based upon the classical analogy between a field specified within small spatial cells, and equivalent particle systems. When it was attempted to quantize the complete electromagnetic field [W. Heisenberg and W. Pauli, Zeits. f. Physik 56, 1 (1929)], rather than the radiation field that remains after the Coulomb interaction is separated, difficulties were encountered that stem from the gauge ambiguity of the potentials that appear in the Lagrangian formulation of the Maxwell equations. The only real dynamical degrees of freedom are those of the radiation part of the field. Yet one can employ additional degrees of freedom which are suppressed finally by imposing a consistent restriction on the admissible states of the system [2]. To make more evident the relativistic invariance of the scheme, other equivalent forms were given to the theory by introducing different time coordinates for each of a fixed number of charged particles coupled to the electromagnetic field [3]. This formal period of quantization of the electromagnetic field was terminated by a critical analysis of the limitations in the accuracy of simultaneous measurements of two field strengths, produced by the known quantum restrictions on the simultaneous measurability of properties of material test bodies [N. Bohr and L. Rosenfeld, Kgl. Danske Vid. Sels., Math.—fys. Medd. 12, No. 8 (1933)]. The complete agreement of these considerations with the formal implications of the operator commutation relations indicated the necessity and consistency of applying the quantum mechanical description to all dynamical systems. The synthesis of the complementary classical particle and field languages in the concept of the quantized field, as exemplified in the treatment of the electromagnetic field, was found to be of general applicability to systems formed by arbitrary numbers of identical particles, although the rules of field quantization derived by analogy from those of particle mechanics were too restrictive, yielding only systems obeying the Bose-Einstein statistics. The replacement of commutators by anti-commutators was necessary to describe particles, like the electron, that obey the Fermi-Dirac statistics [4]. In the latter situation there is no realizable physical limit for which the system behaves as a classical field.

But, from the origin of quantum electrodynamics in the classical theory of point charges came a legacy of difficulties. The coupling of an electron with the electromagnetic field implied an infinite energy displacement, and, indeed, an infinite shift of all spectral lines emitted by an atomic system [J. R. Oppenheimer, Phys. Rev. 35,461 (1930)]; in the reaction of the electromagnetic field stimulated by the presence of the electron, arbitrarily short wave lengths play a disproportionate and divergent role. The phenomenon of electron-positron pair creation, which finds a natural place in the relativistic electron field theory, contributes to this situation in virtue of the fluctuating densities of charge and current that occur even in the vacuum state [5] as the matter-field counterpart of the fluctuations in electric and magnetic field strengths. In computing the energy of a single electron relative to that of the vacuum state, it is of significance that the presence of the electron tends to suppress the charge-current fluctuations induced by the fluctuating electromagnetic field. The resulting electron energy, while still divergent in its dependence upon the contributions of arbitrarily short wave lengths, exhibits only a logarithmic infinity [6]; the combination of quantum and relativistic effects has destroyed all correspondence with the classical theory and its strongly structure-dependent electromagnetic mass. The existence of current fluctuations in the vacuum has other implications, since the introduction of an electromagnetic field induces currents that tend to modify the initial field; the "vacuum" acts as a polarizable medium [7]. New non-linear electromagnetic phenomena appear, such as the scattering of one light beam by another, or by an electrostatic field. But, in the calculation of the current induced by weak fields, there occurred terms that depended divergently upon the contributions of high-energy electron-positron pairs. These were generally considered to be completely without physical significance, although it was noticed [8] that the contribution to the induced charge density that is proportional to the inducing density, with a logarithmically divergent coefficient, would result in an effective reduction of all densities by a constant factor which is not observable separately under ordinary circumstances. In contrast with the divergences at infinitely high energies, another kind of divergent situation was encountered in calculating the total probability that a photon be emitted in a collision of a charged particle. Here, however, the deficiency was evidently in the approximate method of calculation; in any deflection of a charged particle it is certain that "zero" frequency quanta shall be emitted, which fact must be taken into account if meaningful questions are to be asked. The concentration on photons of very low energy permitted a sufficiently accurate treatment to be developed [9], in which it was recognized that the correct quantum description of a freely moving charged particle includes an electromagnetic field that accompanies the particle, as in the classical picture. It also began to be appreciated that the quantum treatment of radiation processes was inconsistent in its identification of the mass of the electron, when decoupled from the electromagnetic field, with the experimentally observed mass. Part of the effect of the electromagnetic coupling is to generate the field that accompanies the charge, and which reacts on it to produce an electromagnetic mass. This is familiar classically, where the sum of the two mass contributions appears as the effective electron mass in an equation of motion which, under ordinary conditions, no longer refers to the detailed structure of the electron. Hence, it was concluded that a classical theory of the latter type should be the correspondence basis for a quantum electrodynamics [H. A. Kramers, Quantentheorie des Elektrons und der Strahlung, Leipzig, 1938].

Further progress came only with the spur of experimental discovery. Exploiting the wartime development of electronic and microwave techniques, delicate measurements disclosed that the electron possessed an intrinsic magnetic moment slightly greater than that predicted by the relativistic quantum theory of a single particle [10], while another prediction of the latter theory concerning the degeneracy of states in the excited levels of hydrogen was contradicted by observing a separation of the states [11]. (Historically, the experimental stimulus came entirely from the latter measurement; the evidence on magnetic anomalies received its proper interpretation only in consequence of the theoretical prediction of an additional spin magnetic moment.) If these new electron properties were to be understood as electrodynamic effects, the theory had to be recast in a usable form. The parameters of mass and charge associated with the electron in the formalism of electrodynamics are not the quantities measured under ordinary conditions. A free electron is accompanied by an electromagnetic field which effectively alters the inertia of the system, and an electromagnetic field is accompanied by a current of electron-positron pairs which effectively alters the strength of the field and of all charges. Hence a process of renormalization must be carried out, in which the initial parameters are eliminated in favor of those with immediate physical significance. The simplest approximate method of accomplishing this is to compute the electrodynamic corrections to some property and then subtract the effect of the mass and charge redefinitions. While this is a possible nonrelativistic procedure [12], it is not a satisfactory basis for relativistic calculations where the difference of two individually divergent terms is generally ambiguous. It was necessary to subject the conventional Hamiltonian electrodynamics to a transformation designed to introduce the proper description of single electron and photon states, so that the interactions among these particles would be characterized from the beginning by experimental parameters. As the result of this calculation [13], performed to the first significant order of approximation in the electromagnetic coupling, the electron acquired new electrodynamic properties, which were completely finite. These included an energy displacement in an external magnetic field corresponding to an additional spin magnetic moment, and a displacement of energy levels in a Coulomb field. Both predictions were in good accord with experiment, and later refinements in experiment and theory have only emphasized that agreement. However, the Coulomb calculation disclosed a serious flaw; the additional spin interaction that appeared in an electrostatic field was not that expected from the relativistic transformation properties of the supplementary spin magnetic moment, and had to be artificially corrected [14, footnote 5], [15]. Thus, a complete revision in the computational techniques of the relativistic theory could not be avoided. The electrodynamic formalism is invariant under Lorentz transformations and gauge transformations, and the concept of renormalization is in accord with these requirements. Yet, in virtue of the divergences inherent in the theory, the use of a particular coordinate system or gauge in the course of computation could result in a loss of covariance. A version of the theory was needed that manifested covariance at every stage of the calculation. The basis of such a formulation was found in the distinction between the elementary properties of the individual uncoupled fields, and the effects produced by the interaction between them [16], [J. Schwinger, Phys. Rev. 74, 1439 (1948)]. The application of these methods to the problems of vacuum polarization, electron mass, and the electromagnetic properties of single electrons now gave finite, covariant results which justified and extended the earlier calculations [17]. Thus, to the first approximation at least, the use of a covariant renormalization technique had produced a theory that was devoid of divergences and in agreement with experience, all high energy difficulties being isolated in the renormalization constants. Yet, in one aspect of these calculations, the preservation of gauge invariance, the utmost caution was required [18], and the need was felt for less delicate methods of evaluation. Extreme care would not be necessary if, by some device, the various divergent integrals could be rendered convergent while maintaining their general covariant features. This can be accomplished by substituting, for the mass of the particle, a suitably weighted spectrum of masses, where all auxiliary masses eventually tend to infinity [19]. Such a procedure has no meaning in terms of physically realizable particles.

It is best understood, and replaced, by a description of the electron with the aid of an invariant proper-time parameter. Divergences appear only when one integrates over this parameter, and gauge invariant, Lorentz invariant results are automatically guaranteed merely by reserving this integration to the end of the calculation [20].

Throughout these developments the basic view of electromagnetism was that originated by Maxwell and Lorentz—the interaction between charges is propagated through the field by local action. In its quantum mechanical transcription it leads to formalisms in which charged particles and field appear on the same footing dynamically. But another approach is also familiar classically; the field produced by arbitrarily moving charges can be evaluated, and the dynamical problem reformulated as the purely mechanical one of particles interacting with each other, and themselves, through a propagated action at a distance. The transference of this line of thought into quantum language [21], [22], [23] was accompanied by another shift in emphasis relative to the previously described work. In the latter, the effect on the particles of the coupling with the electromagnetic field was expressed by additional energy terms which could then be used to evaluate energy displacements in bound states, or to compute corrections to scattering cross-sections. Now the fundamental viewpoint was that of scattering, and in its approximate versions led to a detailed space-time description of the various interaction mechanisms. The two approaches are equivalent; the formal integration of the differential equations of one method supplying the starting point of the other [24]. But if one excludes the consideration of bound states, it is possible to expand the elements of a scattering matrix in powers of the coupling constant, and examine the effect of charge and mass renormalization, term by term, to indefinitely high powers. It appeared that, for any process, the coefficient of each power in the renormalized coupling constant was completely finite [25]. This highly satisfactory result did not mean, however, that the act of renormalization had, in itself, produced a more correct theory. The convergence of the power series is not established, and the series doubtless has the significance of an asymptotic expansion. Yet, for practical purposes, in which the smallness of the coupling parameter is relevant, this analysis gave assurance that calculations of arbitrary precision could be performed.

The evolutionary process by which relativistic field theory was escaping from the confines of its non-relativistic heritage culminated in a complete reconstruction of the foundations of quantum dynamics. The quantum mechanics of particles had been expressed as a set of operator prescriptions superimposed upon the structure of classical mechanics in Hamiltonian form. When extended to relativistic fields, this approach had the disadvantage of producing an unnecessarily great asymmetry between time and space, and of placing the existence of Fermi-Dirac fields on a purely empirical basis. But the Hamiltonian form is not the natural starting point of classical dynamics. Rather, this is supplied by Hamilton's action principle, and action is a relativistic invariant. Could quantum dynamics be developed independently from an action principle, which, being freed from the limitations of the correspondence principle, might automatically produce two distinct types of dynamical variables? The correspondence relation between classical action, and the quantum mechanical description of time development by a transformation function, had long been known [26]. It had also been observed that, for infinitesimal time intervals and sufficiently simple systems, this asymptotic connection becomes sharpened into an identity of the phase of the transformation function with the classically evaluated action [27]. The general quantum dynamical principle was found in a differential characterization of transformation functions, involving the variation of an action operator [28]. When the action operator is chosen to produce first order differential equations of motion, or field equations, it indeed predicts the existence of two types of dynamical variables, with operator properties described by commutators and anti-commutators, respectively [29]. Furthermore, the connection between the statistics and the spin of the particles is inferred from invariance requirements, which strengthens the previous arguments based upon properties of non-interacting particles [30]. The practical utility of this quantum dynamical principle stems from its very nature; it supplies differential equations for the construction of the transformation functions that contain all the dynamical properties of the system. It leads in particular to a concise expression of quantum electrodynamics in the form of coupled differential equations for electron and photon propagation functions [31]. Such functions enjoy the advantages of space-time pictorializability, combined with general applicability to bound systems or scattering situations. Among these applications has been a treatment of that most electrodynamic of systems—positronium, the metastable atom formed by a positron and an electron. The agreement between theory and experiment on the finer details of this system is another quantitative triumph of quantum electrodynamics [32].

The post-war developments of quantum electrodynamics have been largely dominated by questions of formalism and technique, and do not contain any fundamental improvement in the physical foundations of the theory. Such a situation is not new in the history of physics; it took the labors of more than a century to develop the methods that express fully the mechanical principles laid down by Newton. But, we may ask, is there a fatal fault in the structure of field theory? Could it not be that the divergences—apparent symptoms of malignancy—are only spurious byproducts of an invalid expansion in powers of the coupling constant and that renormalization, which can change no physical implication of the theory, simply rectifies this mathematical error? This hope disappears on recognizing that the observational basis of quantum electrodynamics is self-contradictory. The fundamental dynamical variables of the electron-positron field, for example, have meaning only as symbols of the localized creation and annihilation of charged particles, to which are ascribed a definite mass without reference to the electromagnetic field. Accordingly it should be possible, in principle, to confirm these properties by measurements, which, if they are to be uninfluenced by the coupling of the particles to the electromagnetic field, must be performed instantaneously. But there appears to be nothing in the formalism to set a standard for arbitrarily short times and, indeed, the assumption that over sufficiently small intervals the two fields behave as though free from interaction is contradicted by evaluating the supposedly small effect of the coupling. Thus, although the starting point of the theory is the independent assignment of properties to the two fields, they can never be disengaged to give those properties immediate observational significance. It seems that we have reached the limits of the quantum theory of measurement, which asserts the possibility of instantaneous observations, without reference to specific agencies. The localization of charge with indefinite precision requires for its realization a coupling with the electromagnetic field that can attain arbitrarily large magnitudes. The resulting appearance of divergences, and contradictions, serves to deny the basic measurement hypothesis. We conclude that a convergent theory cannot be formulated consistently within the framework of present space-time concepts. To limit the magnitude of interactions while retaining the customary coördinate description is contradictory, since no mechanism is provided for precisely localized measurements.

In attempting to account for the properties of electron and positron, it has been natural to use the simplified form of quantum electrodynamics in which only these charged particles are considered. Despite the apparent validity of the basic assumption that the electron-positron field experiences no appreciable interaction with fields other than electromagnetic, this physically incomplete theory suffers from a fundamental limitation. It can never explain the observed value of the dimensionless coupling constant measuring the electron charge. Indeed, since charge renormalization is a property of the electromagnetic field, and the latter is influenced by the behavior of every kind of fundamental particle with direct or indirect electromagnetic coupling, a full understanding of the electron charge can exist only when the theory of elementary particles has come to a stage of perfection that is presently unimaginable. It is not likely that future developments will change drastically the practical results of the electron theory, which gives contemporary quantum electrodynamics a certain enduring value. Yet the real significance of the work of the past decade lies in the recognition of the ultimate problems facing electrodynamics, the problems of conceptual consistency and of physical completeness. No final solution can be anticipated until physical science has met the heroic challenge to comprehend the structure of the sub-microscopic world that nuclear exploration has revealed.

JULIAN SCHWINGER

Cambridge, Mass.
1956

QUANTUM ELECTRODYNAMICS

The Quantum Theory of the Emission and Absorption of Radiation.

By P. A. M. Dirac, St. John's College, Cambridge, and Institute for Theoretical Physics, Copenhagen.

(Communicated by N. Bohr, For. Mem. R.S.—Received February 2, 1927.)

§ 1. *Introduction and Summary.*

The new quantum theory, based on the assumption that the dynamical variables do not obey the commutative law of multiplication, has by now been developed sufficiently to form a fairly complete theory of dynamics. One can treat mathematically the problem of any dynamical system composed of a number of particles with instantaneous forces acting between them, provided it is describable by a Hamiltonian function, and one can interpret the mathematics physically by a quite definite general method. On the other hand, hardly anything has been done up to the present on quantum electrodynamics. The questions of the correct treatment of a system in which the forces are propagated with the velocity of light instead of instantaneously, of the production of an electromagnetic field by a moving electron, and of the reaction of this field on the electron have not yet been touched. In addition, there is a serious difficulty in making the theory satisfy all the requirements of the restricted principle of relativity, since a Hamiltonian function can no longer be used. This relativity question is, of course, connected with the previous ones, and it will be impossible to answer any one question completely without at the same time answering them all. However, it appears to be possible to build up a fairly satisfactory theory of the emission of radiation and of the reaction of the radiation field on the emitting system on the basis of a kinematics and dynamics which are not strictly relativistic. This is the main object of the present paper. The theory is non-relativistic only on account of the time being counted throughout as a c-number, instead of being treated symmetrically with the space co-ordinates. The relativity variation of mass with velocity is taken into account without difficulty.

The underlying ideas of the theory are very simple. Consider an atom interacting with a field of radiation, which we may suppose for definiteness to be confined in an enclosure so as to have only a discrete set of degrees of freedom. Resolving the radiation into its Fourier components, we can consider the energy and phase of each of the components to be dynamical variables describing the radiation field. Thus if E_r is the energy of a component labelled r and θ_r is the corresponding phase (defined as the time since the wave was in a standard phase), we can suppose each E_r and θ_r to form a pair of canonically conjugate variables. In the absence of any interaction between the field and the atom, the whole system of field plus atom will be describable by the Hamiltonian

$$H = \Sigma_r E_r + H_0 \tag{1}$$

equal to the total energy, H_0 being the Hamiltonian for the atom alone, since the variables E_r, θ_r obviously satisfy their canonical equations of motion

$$\dot{E}_r = -\frac{\partial H}{\partial \theta_r} = 0, \quad \dot{\theta}_r = \frac{\partial H}{\partial E_r} = 1.$$

When there is interaction between the field and the atom, it could be taken into account on the classical theory by the addition of an interaction term to the Hamiltonian (1), which would be a function of the variables of the atom and of the variables E_r, θ_r that describe the field. This interaction term would give the effect of the radiation on the atom, and also the reaction of the atom on the radiation field.

In order that an analogous method may be used on the quantum theory, it is necessary to assume that the variables E_r, θ_r are q-numbers satisfying the standard quantum conditions $\theta_r E_r - E_r \theta_r = ih$, etc., where h is $(2\pi)^{-1}$ times the usual Planck's constant, like the other dynamical variables of the problem. This assumption immediately gives light-quantum properties to the radiation.* For if ν_r is the frequency of the component r, $2\pi\nu_r\theta_r$ is an angle variable, so that its canonical conjugate $E_r/2\pi\nu_r$ can only assume a discrete set of values differing by multiples of h, which means that E_r can change only by integral multiples of the quantum $(2\pi h)\,\nu_r$. If we now add an interaction term (taken over from the clasical theory) to the Hamiltonian (1), the problem can be solved according to the rules of quantum mechanics, and we would expect to obtain the correct results for the action of the radiation and the atom on one another. It will be shown that we actually get the correct laws for the emission and absorption of radiation, and the correct values for Einstein's A's and B's. In the author's previous theory,† where the energies and phases of the components of radiation were c-numbers, only the B's could be obtained, and the reaction of the atom on the radiation could not be taken into account.

It will also be shown that the Hamiltonian which describes the interaction of the atom and the electromagnetic waves can be made identical with the Hamiltonian for the problem of the interaction of the atom with an assembly of particles moving with the velocity of light and satisfying the Einstein-Bose statistics, by a suitable choice of the interaction energy for the particles. The number of particles having any specified direction of motion and energy, which can be used as a dynamical variable in the Hamiltonian for the particles, is equal to the number of quanta of energy in the corresponding wave in the Hamiltonian for the waves. There is thus a complete harmony between the wave and light-quantum descriptions of the interaction. We shall actually build up the theory from the light-quantum point of view, and show that the Hamiltonian transforms naturally into a form which resembles that for the waves.

The mathematical development of the theory has been made possible by the author's general transformation theory of the quantum matrices.‡ Owing to the fact that we count the time as a c-number, we are allowed to use the notion of the value of any dynamical variable at any instant of time. This value is

* Similar assumptions have been used by Born and Jordan [' Z. f. Physik,' vol. 34, p. 886 (1925)] for the purpose of taking over the classical formula for the emission of radiation by a dipole into the quantum theory, and by Born, Heisenberg and Jordan [' Z. f. Physik,' vol. 35, p. 606 (1925)] for calculating the energy fluctuations in a field of black-body radiation.

† ' Roy. Soc. Proc.,' A, vol. 112, p. 661, § 5 (1926). This is quoted later by, *loc. cit.*, I.

‡ ' Roy. Soc. Proc.,' A, vol. 113, p. 621 (1927). This is quoted later by *loc. cit.*, II. An essentially equivalent theory has been obtained independently by Jordan [' Z. f. Physik,' vol. 40, p. 809 (1927)]. See also, F. London, ' Z. f. Physik,' vol. 40, p. 193 (1926).

a q-number, capable of being represented by a generalised " matrix " according to many different matrix schemes, some of which may have continuous ranges of rows and columns, and may require the matrix elements to involve certain kinds of infinities (of the type given by the δ functions*). A matrix scheme can be found in which any desired set of constants of integration of the dynamical system that commute are represented by diagonal matrices, or in which a set of variables that commute are represented by matrices that are diagonal at a specified time.† The values of the diagonal elements of a diagonal matrix representing any q-number are the characteristic values of that q-number. A Cartesian co-ordinate or momentum will in general have all characteristic values from $-\infty$ to $+\infty$, while an action variable has only a discrete set of characteristic values. (We shall make it a rule to use unprimed letters to denote the dynamical variables or q-numbers, and the same letters primed or multiply primed to denote their characteristic values. Transformation functions or eigenfunctions are functions of the characteristic values and not of the q-numbers themselves, so they should always be written in terms of primed variables.)

If $f(\xi, \eta)$ is any function of the canonical variables ξ_k, η_k, the matrix representing f at any time t in the matrix scheme in which the ξ_k at time t are diagonal matrices may be written down without any trouble, since the matrices representing the ξ_k and η_k themselves at time t are known, namely,

$$\left.\begin{aligned} \xi_k(\xi'\xi'') &= \xi_k'\,\delta(\xi'\xi''), \\ \eta_k(\xi'\xi'') &= -ih\,\delta(\xi_1'-\xi_1'')\ldots\delta(\xi_{k-1}'-\xi_{k-1}'')\,\delta'(\xi_k'-\xi_k'')\,\delta(\xi_{k+1}'-\xi_{k+1}'')\ldots \end{aligned}\right\} . \tag{2}$$

Thus if the Hamiltonian H is given as a function of the ξ_k and η_k, we can at once write down the matrix $H(\xi'\,\xi'')$. We can then obtain the transformation function, (ξ'/α') say, which transforms to a matrix scheme (α) in which the Hamiltonian is a diagonal matrix, as (ξ'/α') must satisfy the integral equation

$$\int H(\xi'\xi'')\,d\xi''\,(\xi''/\alpha') = W(\alpha')\,.\,(\xi'/\alpha'), \tag{3}$$

of which the characteristic values $W(\alpha')$ are the energy levels. This equation is just Schrödinger's wave equation for the eigenfunctions (ξ'/α'), which becomes an ordinary differential equation when H is a simple algebraic function of the

* *Loc. cit.* II, § 2.

† One can have a matrix scheme in which a set of variables that commute are at all times represented by diagonal matrices if one will sacrifice the condition that the matrices must satisfy the equations of motion. The transformation function from such a scheme to one in which the equations of motion are satisfied will involve the time explicitly. See p. 628 in *loc. cit.*, II.

ξ_k and η_k on account of the special equations (2) for the matrices representing ξ_k and η_k. Equation (3) may be written in the more general form

$$\int \mathrm{H}\,(\xi'\xi'')\,d\xi''\,(\xi''/\alpha') = ih\,\partial\,(\xi'/\alpha')/\partial t, \tag{3'}$$

in which it can be applied to systems for which the Hamiltonian involves the time explicitly.

One may have a dynamical system specified by a Hamiltonian H which cannot be expressed as an algebraic function of any set of canonical variables, but which can all the same be represented by a matrix $\mathrm{H}(\xi'\xi'')$. Such a problem can still be solved by the present method, since one can still use equation (3) to obtain the energy levels and eigenfunctions. We shall find that the Hamiltonian which describes the interaction of a light-quantum and an atomic system is of this more general type, so that the interaction can be treated mathematically, although one cannot talk about an interaction potential energy in the usual sense.

It should be observed that there is a difference between a light-wave and the de Broglie or Schrödinger wave associated with the light-quanta. Firstly, the light-wave is always real, while the de Broglie wave associated with a light-quantum moving in a definite direction must be taken to involve an imaginary exponential. A more important difference is that their intensities are to be interpreted in different ways. The number of light-quanta per unit volume associated with a monochromatic light-wave equals the energy per unit volume of the wave divided by the energy $(2\pi h)\nu$ of a single light-quantum. On the other hand a monochromatic de Broglie wave of amplitude a (multiplied into the imaginary exponential factor) must be interpreted as representing a^2 light-quanta per unit volume for all frequencies. This is a special case of the general rule for interpreting the matrix analysis,* according to which, if (ξ'/α') or $\psi_{\alpha'}(\xi_k')$ is the eigenfunction in the variables ξ_k of the state α' of an atomic system (or simple particle), $|\psi_{\alpha'}(\xi_k')|^2$ is the probability of each ξ_k having the value ξ_k', [or $|\psi_{\alpha'}(\xi_k')|^2\,d\xi_1'\,d\xi_2' \ldots$ is the probability of each ξ_k lying between the values ξ_k' and $\xi_k' + d\xi_k'$, when the ξ_k have continuous ranges of characteristic values] on the assumption that all phases of the system are equally probable. The wave whose intensity is to be interpreted in the first of these two ways appears in the theory only when one is dealing with an assembly of the associated particles satisfying the Einstein-Bose statistics. There is thus no such wave associated with electrons.

* *Loc. cit.*, II, §§ 6, 7.

§ 2. *The Perturbation of an Assembly of Independent Systems.*

We shall now consider the transitions produced in an atomic system by an arbitrary perturbation. The method we shall adopt will be that previously given by the author,† which leads in a simple way to equations which determine the probability of the system being in any stationary state of the unperturbed system at any time.‡ This, of course, gives immediately the probable number of systems in that state at that time for an assembly of the systems that are independent of one another and are all perturbed in the same way. The object of the present section is to show that the equations for the rates of change of these probable numbers can be put in the Hamiltonian form in a simple manner, which will enable further developments in the theory to be made.

Let H_0 be the Hamiltonian for the unperturbed system and V the perturbing energy, which can be an arbitrary function of the dynamical variables and may or may not involve the time explicitly, so that the Hamiltonian for the perturbed system is $H = H_0 + V$. The eigenfunctions for the perturbed system must satisfy the wave equation

$$ih\, \partial\psi/\partial t = (H_0 + V)\, \psi,$$

where $(H_0 + V)$ is an operator. If $\psi = \Sigma_r a_r \psi_r$ is the solution of this equation that satisfies the proper initial conditions, where the ψ_r's are the eigenfunctions for the unperturbed system, each associated with one stationary state labelled by the suffix r, and the a_r's are functions of the time only, then $|a_r|^2$ is the probability of the system being in the state r at any time. The a_r's must be normalised initially, and will then always remain normalised. The theory will apply directly to an assembly of N similar independent systems if we multiply each of these a_r's by $N^{\frac{1}{2}}$ so as to make $\Sigma_r |a_r|^2 = N$. We shall now have that $|a_r|^2$ is the probable number of systems in the state r.

The equation that determines the rate of change of the a_r's is§

$$ih\dot{a}_r = \Sigma_s V_{rs} a_s, \tag{4}$$

where the V_{rs}'s are the elements of the matrix representing V. The conjugate imaginary equation is

$$-ih\dot{a}_r^* = \Sigma_s V_{rs}^* a_s^* = \Sigma_s a_s^* V_{sr}. \tag{4'}$$

† *Loc. cit.* I.

‡ The theory has recently been extended by Born ['Z. f. Physik,' vol. 40, p. 167 (1926)] so as to take into account the adiabatic changes in the stationary states that may be produced by the perturbation as well as the transitions. This extension is not used in the present paper.

§ *Loc. cit.*, I, equation (25).

If we regard a_r and $ih\,a_r^*$ as canonical conjugates, equations (4) and (4′) take the Hamiltonian form with the Hamiltonian function $F_1 = \Sigma_{rs} a_r^* V_{rs} a_s$, namely,

$$\frac{da_r}{dt} = \frac{1}{ih}\frac{\partial F_1}{\partial a_r^*}, \quad ih\frac{da_r^*}{dt} = -\frac{\partial F_1}{\partial a_r}.$$

We can transform to the canonical variables N_r, ϕ_r by the contact transformation

$$a_r = N_r^{\frac{1}{2}} e^{-i\phi_r/h}, \quad a_r^* = N_r^{\frac{1}{2}} e^{i\phi_r/h}.$$

This transformation makes the new variables N_r and ϕ_r real, N_r being equal to $a_r a_r^* = |a_r|^2$, the probable number of systems in the state r, and ϕ_r/h being the phase of the eigenfunction that represents them. The Hamiltonian F_1 now becomes

$$F_1 = \Sigma_{rs} V_{rs} N_r^{\frac{1}{2}} N_s^{\frac{1}{2}} e^{i(\phi_r - \phi_s)/h},$$

and the equations that determine the rate at which transitions occur have the canonical form

$$\dot{N}_r = -\frac{\partial F_1}{\partial \phi_r}, \quad \dot{\phi}_r = \frac{\partial F_1}{\partial N_r}.$$

A more convenient way of putting the transition equations in the Hamiltonian form may be obtained with the help of the quantities

$$b_r = a_r\, e^{-iW_r t/h}, \quad b_r^* = a_r^*\, e^{iW_r t/h},$$

W_r being the energy of the state r. We have $|b_r|^2$ equal to $|a_r|^2$, the probable number of systems in the state r. For $\dot{b}_r$ we find

$$\begin{aligned} ih\,\dot{b}_r &= W_r b_r + ih\,\dot{a}_r\, e^{-iW_r t/h} \\ &= W_r b_r + \Sigma_s V_{rs} b_s e^{i(W_s - W_r)t/h} \end{aligned}$$

with the help of (4). If we put $V_{rs} = v_{rs} e^{i(W_r - W_s)t/h}$, so that v_{rs} is a constant when V does not involve the time explicitly, this reduces to

$$\begin{aligned} ih\,\dot{b}_r &= W_r b_r + \Sigma_s v_{rs} b_s \\ &= \Sigma_s H_{rs} b_s, \end{aligned} \tag{5}$$

where $H_{rs} = W_r \delta_{rs} + v_{rs}$, which is a matrix element of the total Hamiltonian $H = H_0 + V$ with the time factor $e^{i(W_r - W_s)t/h}$ removed, so that H_{rs} is a constant when H does not involve the time explicitly. Equation (5) is of the same form as equation (4), and may be put in the Hamiltonian form in the same way.

It should be noticed that equation (5) is obtained directly if one writes down the Schrödinger equation in a set of variables that specify the stationary states of the unperturbed system. If these variables are ξ_h, and if $H(\xi'\xi'')$ denotes a matrix element of the total Hamiltonian H in the (ξ) scheme, this Schrödinger equation would be

$$ih\, \partial\psi(\xi')/\partial t = \Sigma_{\xi''} H(\xi'\xi'')\, \psi(\xi''), \tag{6}$$

like equation (3′). This differs from the previous equation (5) only in the notation, a single suffix r being there used to denote a stationary state instead of a set of numerical values ξ_k' for the variables ξ_k, and b_r being used instead of $\psi(\xi')$. Equation (6), and therefore also equation (5), can still be used when the Hamiltonian is of the more general type which cannot be expressed as an algebraic function of a set of canonial variables, but can still be represented by a matrix $H(\xi'\xi'')$ or H_{rs}.

We now take b_r and $ih\, b_r^*$ to be canonically conjugate variables instead of a_r and $ih\, a_r^*$. The equation (5) and its conjugate imaginary equation will now take the Hamiltonian form with the Hamiltonian function

$$F = \Sigma_{rs} b_r^* H_{rs} b_s. \tag{7}$$

Proceeding as before, we make the contact transformation

$$b_r = N_r^{\frac{1}{2}} e^{-i\theta_r/h}, \qquad b_r^* = N_r^{\frac{1}{2}} e^{i\theta_r/h}, \tag{8}$$

to the new canonical variables N_r, θ_r, where N_r is, as before, the probable number of systems in the state r, and θ_r is a new phase. The Hamiltonian F will now become

$$F = \Sigma_{rs} H_{rs} N_r^{\frac{1}{2}} N_s^{\frac{1}{2}} e^{i(\theta_r - \theta_s)/h},$$

and the equations for the rates of change of N_r and θ_r will take the canonical form

$$\dot{N}_r = -\frac{\partial F}{\partial \theta_r}, \qquad \dot{\theta}_r = \frac{\partial F}{\partial N_r}.$$

The Hamiltonian may be written

$$F = \Sigma_r W_r N_r + \Sigma_{rs} v_{rs} N_r^{\frac{1}{2}} N_s^{\frac{1}{2}} e^{i(\theta_r - \theta_s)/h}. \tag{9}$$

The first term $\Sigma_r W_r N_r$ is the total proper energy of the assembly, and the second may be regarded as the additional energy due to the perturbation. If the perturbation is zero, the phases θ_r would increase linearly with the time, while the previous phases ϕ_r would in this case be constants.

§3. *The Perturbation of an Assembly satisfying the Einstein-Bose Statistics.*

According to the preceding section we can describe the effect of a perturbation on an assembly of independent systems by means of canonical variables and Hamiltonian equations of motion. The development of the theory which naturally suggests itself is to make these canonical variables q-numbers satisfying the usual quantum conditions instead of c-numbers, so that their Hamiltonian equations of motion become true quantum equations. The Hamiltonian function will now provide a Schrödinger wave equation, which must be solved and interpreted in the usual manner. The interpretation will give not merely the probable number of systems in any state, but the probability of any given distribution of the systems among the various states, this probability being, in fact, equal to the square of the modulus of the normalised solution of the wave equation that satisfies the appropriate initial conditions. We could, of course, calculate directly from elementary considerations the probability of any given distribution when the systems are independent, as we know the probability of each system being in any particular state. We shall find that the probability calculated directly in this way does not agree with that obtained from the wave equation except in the special case when there is only one system in the assembly. In the general case it will be shown that the wave equation leads to the correct value for the probability of any given distribution when the systems obey the Einstein-Bose statistics instead of being independent.

We assume the variables b_r, ihb_r^* of § 2 to be canonical q-numbers satisfying the quantum conditions

$$b_r \,.\, ih\, b_r^* - ih\, b_r^* \,.\, b_r = ih$$

or

$$b_r b_r^* - b_r^* b_r = 1,$$

and

$$b_r b_s - b_s b_r = 0, \qquad b_r^* b_s^* - b_s^* b_r^* = 0,$$

$$b_r b_s^* - b_s^* b_r = 0 \qquad (s \neq r).$$

The transformation equations (8) must now be written in the quantum form

$$\left. \begin{aligned} b_r &= (\mathrm{N}_r + 1)^{\frac{1}{2}}\, e^{-i\theta_r/h} = e^{-i\theta_r/h}\mathrm{N}_r^{\frac{1}{2}} \\ b_r^* &= \mathrm{N}_r^{\frac{1}{2}} e^{i\theta_r/h} = e^{i\theta_r/h}(\mathrm{N}_r + 1)^{\frac{1}{2}}, \end{aligned} \right\} \tag{10}$$

in order that the N_r, θ_r may also be canonical variables. These equations show that the N_r can have only integral characteristic values not less than zero,† which provides us with a justification for the assumption that the variables are q-numbers in the way we have chosen. The numbers of systems in the different states are now ordinary quantum numbers.

† See § 8 of the author's paper 'Roy. Soc. Proc.,' A, vol. 111, p. 281 (1926). What are there called the c-number values that a q-number can take are here given the more precise name of the characteristic values of that q-number.

The Hamiltonian (7) now becomes

$$F = \Sigma_{rs} b_r^* H_{rs} b_s = \Sigma_{rs} N_r^{\frac{1}{2}} e^{i\theta_r/h} H_{rs} (N_s + 1)^{\frac{1}{2}} e^{-i\theta_s/h}$$
$$= \Sigma_{rs} H_{rs} N_r^{\frac{1}{2}} (N_s + 1 - \delta_{rs})^{\frac{1}{2}} e^{i(\theta_r - \theta_s)/h} \tag{11}$$

in which the H_{rs} are still c-numbers. We may write this F in the form corresponding to (9)

$$F = \Sigma_r W_r N_r + \Sigma_{rs} v_{rs} N_r^{\frac{1}{2}} (N_s + 1 - \delta_{rs})^{\frac{1}{2}} e^{i(\theta_r - \theta_s)/h} \tag{11'}$$

in which it is again composed of a proper energy term $\Sigma_r W_r N_r$ and an interaction energy term.

The wave equation written in terms of the variables N_r is†

$$ih \frac{\partial}{\partial t} \psi (N_1', N_2', N_3' \ldots) = F\psi (N_1', N_2', N_3' \ldots), \tag{12}$$

where F is an operator, each θ_r occurring in F being interpreted to mean $ih\,\partial/\partial N_r'$. If we apply the operator $e^{\pm i\theta_r/h}$ to any function $f(N_1', N_2', \ldots N_r', \ldots)$ of the variables $N_1', N_2', \ldots$ the result is

$$e^{\pm i\theta_r/h} f(N_1', N_2', \ldots N_r', \ldots) = e^{\mp \partial/\partial N_r'} f(N_1', N_2', \ldots N_r' \ldots)$$
$$= f(N_1', N_2', \ldots N_r' \mp 1, \ldots).$$

If we use this rule in equation (12) and use the expression (11) for F we obtain‡

$$ih \frac{\partial}{\partial t} \psi(N_1', N_2', N_3' \ldots)$$
$$= \Sigma_{rs} H_{rs} N_r'^{\frac{1}{2}} (N_s' + 1 - \delta_{rs})^{\frac{1}{2}} \psi(N_1', N_2' \ldots N_r' - 1, \ldots N_s' + 1, \ldots). \tag{13}$$

We see from the right-hand side of this equation that in the matrix representing F, the term in F involving $e^{i(\theta_r - \theta_s)/h}$ will contribute only to those matrix elements that refer to transitions in which N_r decreases by unity and N_s increases by unity, *i.e.*, to matrix elements of the type $F(N_1', N_2' \ldots N_r' \ldots N_s'; N_1', N_2' \ldots N_r' - 1 \ldots N_s' + 1 \ldots)$. If we find a solution $\psi(N_1', N_2' \ldots)$ of equation (13) that is normalised [*i.e.*, one for which $\Sigma_{N_1', N_2' \ldots} |\psi(N_1', N_2' \ldots)|^2 = 1$] and that satisfies the proper initial conditions, then $|\psi(N_1', N_2' \ldots)|^2$ will be the probability of that distribution in which N_1' systems are in state 1, N_2' in state 2, ... at any time.

Consider first the case when there is only one system in the assembly. The probability of its being in the state q is determined by the eigenfunction

† We are supposing for definiteness that the label r of the stationary states takes the values 1, 2, 3, ...

‡ When $s = r$, $\psi(N_1', N_2' \ldots N_r' - 1 \ldots N_s' + 1)$ is to be taken to mean $\psi(N_1' N_2' \ldots N_r' \ldots)$.

$\psi(N_1', N_2', \ldots)$ in which all the N′'s are put equal to zero except N_q', which is put equal to unity. This eigenfunction we shall denote by $\psi\{q\}$. When it is substituted in the left-hand side of (13), all the terms in the summation on the right-hand side vanish except those for which $r = q$, and we are left with

$$ih\,\frac{\partial}{\partial t}\psi\{q\} = \Sigma_r H_{qs}\psi\{s\},$$

which is the same equation as (5) with $\psi\{q\}$ playing the part of b_q. This establishes the fact that the present theory is equivalent to that of the preceding section when there is only one system in the assembly.

Now take the general case of an arbitrary number of systems in the assembly, and assume that they obey the Einstein-Bose statistical mechanics. This requires that, in the ordinary treatment of the problem, only those eigenfunctions that are symmetrical between all the systems must be taken into account, these eigenfunctions being by themselves sufficient to give a complete quantum solution of the problem.† We shall now obtain the equation for the rate of change of one of these symmetrical eigenfunctions, and show that it is identical with equation (13).

If we label each system with a number n, then the Hamiltonian for the assembly will be $H_A = \Sigma_n H(n)$, where $H(n)$ is the H of § 2 (equal to $H_0 + V$) expressed in terms of the variables of the nth system. A stationary state of the assembly is defined by the numbers $r_1, r_2 \ldots r_n \ldots$ which are the labels of the stationary states in which the separate systems lie. The Schrödinger equation for the assembly in a set of variables that specify the stationary states will be of the form (6) [with H_A instead of H], and we can write it in the notation of equation (5) thus :—

$$ih\dot{b}(r_1r_2\ldots) = \Sigma_{s_1, s_2\ldots} H_A(r_1r_2\ldots;\ s_1s_2\ldots)\, b(s_1s_2\ldots), \tag{14}$$

where $H_A(r_1r_2\ldots;\ s_1s_2\ldots)$ is the general matrix element of H_A [with the time factor removed]. This matrix element vanishes when more than one s_n differs from the corresponding r_n; equals $H_{r_ms_m}$ when s_m differs from r_m and every other s_n equals r_n; and equals $\Sigma_n H_{r_nr_n}$ when every s_n equals r_n. Substituting these values in (14), we obtain

$$ih\dot{b}(r_1r_2\ldots) = \Sigma_m\Sigma_{s_m\neq r_m} H_{r_ms_m} b(r_1r_2\ldots r_{m-1}s_mr_{m+1}\ldots) + \Sigma_n H_{r_nr_n} b(r_1r_2\ldots). \tag{15}$$

We must now restrict $b(r_1r_2\ldots)$ to be a symmetrical function of the variables $r_1, r_2\ldots$ in order to obtain the Einstein-Bose statistics. This is permissible since if $b(r_1r_2\ldots)$ is symmetrical at any time, then equation (15) shows that

† *Loc. cit.*, I, § 3.

$\dot{b}(r_1 r_2 \ldots)$ is also symmetrical at that time, so that $b\ (r_1 r_2 \ldots)$ will remain symmetrical.

Let N_r denote the number of systems in the state r. Then a stationary state of the assembly describable by a symmetrical eigenfunction may be specified by the numbers $N_1, N_2 \ldots N_r \ldots$ just as well as by the numbers $r_1, r_2 \ldots r_n \ldots$, and we shall be able to transform equation (15) to the variables $N_1, N_2 \ldots$. We cannot actually take the new eigenfunction $b\ (N_1, N_2 \ldots)$ equal to the previous one $b\ (r_1 r_2 \ldots)$, but must take one to be a numerical multiple of the other in order that each may be correctly normalised with respect to its respective variables. We must have, in fact,

$$\Sigma_{r_1, r_2 \ldots} |b(r_1 r_2 \ldots)|^2 = 1 = \Sigma_{N_1, N_2 \ldots} |b(N_1, N_2 \ldots)|^2,$$

and hence we must take $|b(N_1, N_2 \ldots)|^2$ equal to the sum of $|b(r_1 r_2 \ldots)|^2$ for all values of the numbers $r_1, r_2 \ldots$ such that there are N_1 of them equal to 1, N_2 equal to 2, etc. There are $N!/N_1!\,N_2!\ldots$ terms in this sum, where $N = \Sigma_r N_r$ is the total number of systems, and they are all equal, since $b(r_1 r_2 \ldots)$ is a symmetrical function of its variables $r_1, r_2 \ldots$. Hence we must have

$$b(N_1, N_2 \ldots) = (N!/N_1!\,N_2!\ldots)^{\frac{1}{2}}\, b(r_1 r_2 \ldots).$$

If we make this substitution in equation (15), the left-hand side will become $ih\,(N_1!\,N_2!\ldots/N!)^{\frac{1}{2}}\,\dot{b}\,(N_1, N_2 \ldots)$. The term $H_{r_m s_m} b\,(r_1 r_2 \ldots r_{m-1} s_m r_{m+1} \ldots)$ in the first summation on the right-hand side will become

$$[N_1!\,N_2!\ldots(N_r-1)!\ldots(N_s+1)!\ldots/N!]^{\frac{1}{2}}\, H_{rs} b\,(N_1, N_2 \ldots N_r-1 \ldots N_s+1 \ldots), \quad (16)$$

where we have written r for r_m and s for s_m. This term must be summed for all values of s except r, and must then be summed for r taking each of the values $r_1, r_2 \ldots$. Thus each term (16) gets repeated by the summation process until it occurs a total of N_r times, so that it contributes

$$N_r [N_1!\,N_2!\ldots(N_r-1)!\ldots(N_s+1)!\ldots/N!]^{\frac{1}{2}}\, H_{rs} b\,(N_1, N_2 \ldots N_r-1 \ldots N_s+1 \ldots)$$
$$= N_r^{\frac{1}{2}} (N_s+1)^{\frac{1}{2}} (N_1!\,N_2!\ldots/N!)^{\frac{1}{2}}\, H_{rs} b\,(N_1, N_2 \ldots N_r-1 \ldots N_s+1 \ldots)$$

to the right-hand side of (15). Finally, the term $\Sigma_n H_{r_n r_n} b\,(r_1, r_2 \ldots)$ becomes

$$\Sigma_r N_r H_{rr} \,.\, b\,(r_1 r_2 \ldots) = \Sigma_r N_r H_{rr} \,.\, (N_1!\,N_2!\ldots/N!)^{\frac{1}{2}}\, b\,(N_1, N_2 \ldots).$$

Hence equation (15) becomes, with the removal of the factor $(N_1!\,N_2!\ldots/N!)^{\frac{1}{2}}$,

$$ih\dot{b}\,(N_1, N_2 \ldots) = \Sigma_r \Sigma_{s \neq r} N_r^{\frac{1}{2}} (N_s+1)^{\frac{1}{2}} H_{rs} b\,(N_1, N_2 \ldots N_r-1 \ldots N_s+1 \ldots)$$
$$+ \Sigma_r N_r H_{rr} b\,(N_1, N_2 \ldots), \quad (17)$$

which is identical with (13) [except for the fact that in (17) the primes have been omitted from the N's, which is permissible when we do not require to refer to the N's as q-numbers]. We have thus established that the Hamiltonian (11) describes the effect of a perturbation on an assembly satisfying the Einstein-Bose statistics.

§ 4. *The Reaction of the Assembly on the Perturbing System.*

Up to the present we have considered only perturbations that can be represented by a perturbing energy V added to the Hamiltonian of the perturbed system, V being a function only of the dynamical variables of that system and perhaps of the time. The theory may readily be extended to the case when the perturbation consists of interaction with a perturbing dynamical system, the reaction of the perturbed system on the perturbing system being taken into account. (The distinction between the perturbing system and the perturbed system is, of course, not real, but it will be kept up for convenience.)

We now consider a perturbing system, described, say, by the canonical variables J_k, ω_k, the J's being its first integrals when it is alone, interacting with an assembly of perturbed systems with no mutual interaction, that satisfy the Einstein-Bose statistics. The total Hamiltonian will be of the form

$$H_T = H_P(J) + \Sigma_n H(n),$$

where H_P is the Hamiltonian of the perturbing system (a function of the J's only) and H (n) is equal to the proper energy $H_0(n)$ plus the perturbation energy $V(n)$ of the nth system of the assembly. H (n) is a function only of the variables of the nth system of the assembly and of the J's and w's, and does not involve the time explicitly.

The Schrödinger equation corresponding to equation (14) is now

$$ih\dot{b}(J', r_1r_2\ldots) = \Sigma_{J''}\Sigma_{s_1, s_2}\ldots H_T(J', r_1r_2\ldots; J'', s_1s_2\ldots)\, b(J'', s_1s_2\ldots),$$

in which the eigenfunction b involves the additional variables J_k'. The matrix element $H_T(J', r_1r_2 \ldots; J'', s_1s_2 \ldots)$ is now always a constant. As before, it vanishes when more than one s_n differs from the corresponding r_n. When s_m differs from r_m and every other s_n equals r_n, it reduces to $H(J'r_m; J''s_m)$, which is the $(J'r_m; J''s_m)$ matrix element (with the time factor removed) of $H = H_0 + V$, the proper energy plus the perturbation energy of a single system of the assembly; while when every s_n equals r_n, it has the value $H_P(J')\,\delta_{J'J''} + \Sigma_n H(J'r_n; J''r_n)$. If, as before, we restrict the eigenfunctions to be symmetrical in the variables r_1, r_2 ..., we can again transform to the variables N_1, N_2 ..., which will lead, as before, to the result

$$i h \dot{b}(J', N_1', N_2' \ldots) = H_P(J')\, b(J', N_1', N_2' \ldots)$$
$$+ \Sigma_{J''} \Sigma_{r,s} N_r'^{\frac{1}{2}} (N_s' + 1 - \delta_{rs})^{\frac{1}{2}} H(J'r; J''s)\, b(J'', N_1', N_2' \ldots N_r' - 1 \ldots N_s' + 1 \ldots) \quad (18)$$

This is the Schrödinger equation corresponding to the Hamiltonian function

$$F = H_P(J) + \Sigma_{r,s} H_{rs} N_r^{\frac{1}{2}} (N_s + 1 - \delta_{rs})^{\frac{1}{2}} e^{i(\theta_r - \theta_s)/h}, \quad (19)$$

in which H_{rs} is now a function of the J's and w's, being such that when represented by a matrix in the (J) scheme its (J′ J″) element is $H(J'r; J''s)$. (It should be noticed that H_{rs} still commutes with the N's and θ's.)

Thus the interaction of a perturbing system and an assembly satisfying the Einstein-Bose statistics can be described by a Hamiltonian of the form (19). We can put it in the form corresponding to (11′) by observing that the matrix element $H(J'r; J''s)$ is composed of the sum of two parts, a part that comes from the proper energy H_0, which equals W_r when $J_k'' = J_k'$ and $s = r$ and vanishes otherwise, and a part that comes from the interaction energy V, which may be denoted by $v(J'r; J''s)$. Thus we shall have

$$H_{rs} = W_r \delta_{rs} + v_{rs},$$

where v_{rs} is that function of the J's and w's which is represented by the matrix whose (J′ J″) element is $v(J'r; J''s)$, and so (19) becomes

$$F = H_P(J) + \Sigma_r W_r N_r + \Sigma_{r,s} v_{rs} N_r^{\frac{1}{2}} (N_s + 1 - \delta_{rs})^{\frac{1}{2}} e^{i(\theta_r - \theta_s)/h}. \quad (20)$$

The Hamiltonian is thus the sum of the proper energy of the perturbing system $H_P(J)$, the proper energy of the perturbed systems $\Sigma_r W_r N_r$ and the perturbation energy $\Sigma_{r,s} v_{rs} N_r^{\frac{1}{2}} (N_s + 1 - \delta_{rs})^{\frac{1}{2}} e^{i(\theta_r - \theta_s)/h}$.

§ 5. *Theory of Transitions in a System from One State to Others of the Same Energy.*

Before applying the results of the preceding sections to light-quanta, we shall consider the solution of the problem presented by a Hamiltonian of the type (19). The essential feature of the problem is that it refers to a dynamical system which can, under the influence of a perturbation energy which does not involve the time explicitly, make transitions from one state to others of the same energy. The problem of collisions between an atomic system and an electron, which has been treated by Born,* is a special case of this type. Born's method is to find a *periodic* solution of the wave equation which consists, in so far as it involves the co-ordinates of the colliding electron, of plane waves,

* Born, 'Z. f. Physik,' vol. 38, p. 803 (1926).

representing the incident electron, approaching the atomic system, which are scattered or diffracted in all directions. The square of the amplitude of the waves scattered in any direction with any frequency is then assumed by Born to be the probability of the electron being scattered in that direction with the corresponding energy.

This method does not appear to be capable of extension in any simple manner to the general problem of systems that make transitions from one state to others of the same energy. Also there is at present no very direct and certain way of interpreting a periodic solution of a wave equation to apply to a non-periodic physical phenomenon such as a collision. (The more definite method that will now be given shows that Born's assumption is not quite right, it being necessary to multiply the square of the amplitude by a certain factor.)

An alternative method of solving a collision problem is to find a *non-periodic* solution of the wave equation which consists initially simply of plane waves moving over the whole of space in the necessary direction with the necessary frequency to represent the incident electron. In course of time waves moving in other directions must appear in order that the wave equation may remain satisfied. The probability of the electron being scattered in any direction with any energy will then be determined by the rate of growth of the corresponding harmonic component of these waves. The way the mathematics is to be interpreted is by this method quite definite, being the same as that of the beginning of § 2.

We shall apply this method to the general problem of a system which makes transitions from one state to others of the same energy under the action of a perturbation. Let H_0 be the Hamiltonian of the unperturbed system and V the perturbing energy, which must not involve the time explicitly. If we take the case of a continuous range of stationary states, specified by the first integrals, α_k say, of the unperturbed motion, then, following the method of § 2, we obtain

$$ih\,\dot{a}(\alpha') = \int V(\alpha'\alpha'')\,d\alpha'' \,.\, a(\alpha''), \tag{21}$$

corresponding to equation (4). The probability of the system being in a state for which each α_k lies between α_k' and $\alpha_k' + d\alpha_k'$ at any time is $|a(\alpha')|^2 d\alpha_1' \,.\, d\alpha_2' \ldots$ when $a(\alpha')$ is properly normalised and satisfies the proper initial conditions. If initially the system is in the state α^0, we must take the initial value of $a(\alpha')$ to be of the form $a^0 \,.\, \delta(\alpha' - \alpha^0)$. We shall keep a^0 arbitrary, as it would be inconvenient to normalise $a(\alpha')$ in the present case. For a first approximation we may substitute for $a(\alpha'')$ in the right-hand side of (21) its initial value. This gives

$$ih\,\dot{a}(\alpha') = a^0 \mathrm{V}(\alpha'\alpha^0) = \alpha^0 v(\alpha'\alpha^0)\, e^{i[\mathrm{W}(\alpha') - \mathrm{W}(\alpha^0)]t/h},$$

where $v(\alpha'\alpha^0)$ is a constant and $\mathrm{W}(\alpha')$ is the energy of the state α'. Hence

$$ih\,a(\alpha') = a^0\,\delta(\alpha' - \alpha^0) + a^0 v(\alpha'\alpha^0)\,\frac{e^{i[\mathrm{W}(\alpha') - \mathrm{W}(\alpha^0)]t/h} - 1}{i[\mathrm{W}(\alpha') - \mathrm{W}(\alpha^0)]/h}. \qquad (22)$$

For values of the α_k' such that $\mathrm{W}(\alpha')$ differs appreciably from $\mathrm{W}(\alpha^0)$, $a(\alpha')$ is a periodic function of the time whose amplitude is small when the perturbing energy V is small, so that the eigenfunctions corresponding to these stationary states are not excited to any appreciable extent. On the other hand, for values of the α_k' such that $\mathrm{W}(\alpha') = \mathrm{W}(\alpha^0)$ and $\alpha_k' \neq \alpha_k^0$ for some k, $a(\alpha')$ increases uniformly with respect to the time, so that the probability of the system being in the state α' at any time increases proportionally with the square of the time. Physically, the probability of the system being in a state with exactly the same proper energy as the initial proper energy $\mathrm{W}(\alpha^0)$ is of no importance, being infinitesimal. We are interested only in the integral of the probability through a small range of proper energy values about the initial proper energy, which, as we shall find, increases linearly with the time, in agreement with the ordinary probability laws.

We transform from the variables $\alpha_1, \alpha_2 \ldots \alpha_u$ to a set of variables that are arbitrary independent functions of the α's such that one of them is the proper energy W, say, the variables W, $\gamma_1, \gamma_2, \ldots \gamma_{u-1}$. The probability at any time of the system lying in a stationary state for which each γ_k lies between γ_k' and $\gamma_k' + d\gamma_k'$ is now (apart from the normalising factor) equal to

$$d\gamma_1' \,.\, d\gamma_2' \ldots d\gamma_{u-1}' \int |a(\alpha')|^2 \frac{\partial(\alpha_1', \alpha_2' \ldots \alpha_u')}{\partial(\mathrm{W}', \gamma_1' \ldots \gamma_{u-1}')}\, d\mathrm{W}'. \qquad (23)$$

For a time that is large compared with the periods of the system we shall find that practically the whole of the integral in (23) is contributed by values of W′ very close to $\mathrm{W}^0 = \mathrm{W}(\alpha^0)$. Put

$$a(\alpha') = a(\mathrm{W}', \gamma') \quad \text{and} \quad \partial(\alpha_1', \alpha_2' \ldots \alpha_u')/\partial(\mathrm{W}', \gamma_1' \ldots \gamma_{u-1}') = \mathrm{J}(\mathrm{W}', \gamma').$$

Then for the integral in (23) we find, with the help of (22) (provided $\gamma_k' \neq \gamma_k^0$ for some k)

$$\int |a(\mathrm{W}', \gamma')|^2\, \mathrm{J}(\mathrm{W}', \gamma')\, d\mathrm{W}'$$

$$= |a^0|^2 \int |v(\mathrm{W}', \gamma'; \mathrm{W}^0, \gamma^0)|^2\, \mathrm{J}(\mathrm{W}', \gamma')\, \frac{[e^{i(\mathrm{W}'-\mathrm{W}^0)t/h} - 1][e^{-i(\mathrm{W}'-\mathrm{W}^0)t/h} - 1]}{(\mathrm{W}' - \mathrm{W}^0)^2}\, d\mathrm{W}'$$

$$= 2|a^0|^2 \int |v(\mathrm{W}', \gamma'; \mathrm{W}^0, \gamma^0)|^2\, \mathrm{J}(\mathrm{W}', \gamma')[1 - \cos(\mathrm{W}' - \mathrm{W}^0)t/h]/(\mathrm{W}' - \mathrm{W}^0)^2 \,.\, d\mathrm{W}'$$

$$= 2|a^0|^2\, t/h \,.\, \int |v(\mathrm{W}^0 + hx/t, \gamma'; \mathrm{W}^0, \gamma^0)|^2\, \mathrm{J}(\mathrm{W}^0 + hx/t, \gamma')(1 - \cos x)/x^2 \,.\, dx,$$

if one makes the substitution $(W'-W^0)t/h = x$. For large values of t this reduces to

$$2\,|a^0|^2 t/h \,.\, |v(W^0, \gamma'\,;\; W^0, \gamma^0)|^2 J(W^0, \gamma') \int_{-\infty}^{\infty} (1-\cos x)/x^2 \,.\, dx$$

$$= 2\pi\,|a^0|^2 t/h \,.\, |v(W^0, \gamma'\,;\; W^0, \gamma^0)|^2 J(W^0, \gamma').$$

The probability per unit time of a transition to a state for which each γ_k lies between γ_k' and $\gamma_k' + d\gamma_k'$ is thus (apart from the normalising factor)

$$2\pi\,|a^0|^2/h \,.\, |v(W^0, \gamma'\,;\; W^0, \gamma^0)|^2 J(W^0, \gamma')\, d\gamma_1' \,.\, d\gamma_2' \ldots d\gamma_{u-1}', \tag{24}$$

which is proportional to the square of the matrix element associated with that transition of the perturbing energy.

To apply this result to a simple collision problem, we take the α's to be the components of momentum p_x, p_y, p_z of the colliding electron and the γ's to be θ and ϕ, the angles which determine its direction of motion. If, taking the relativity change of mass with velocity into account, we let P denote the resultant momentum, equal to $(p_x^2 + p_y^2 + p_z^2)^{\frac{1}{2}}$, and E the energy, equal to $(m^2c^4 + P^2c^2)^{\frac{1}{2}}$, of the electron, m being its rest-mass, we find for the Jacobian

$$J = \frac{\partial(p_x, p_y, p_z)}{\partial(E, \theta, \phi)} = \frac{EP}{c^2}\sin\theta.$$

Thus the $J(W^0, \gamma')$ of the expression (24) has the value

$$J(W^0, \gamma') = E'P' \sin\theta'/c^2, \tag{25}$$

where E′ and P′ refer to that value for the energy of the scattered electron which makes the total energy equal the initial energy W^0 (*i.e.*, to that value required by the conservation of energy).

We must now interpret the initial value of $a(\alpha')$, namely, $a^0\,\delta(\alpha' - \alpha^0)$, which we did not normalise. According to § 2 the wave function in terms of the variables α_k is $b(\alpha') = a(\alpha')\,e^{-iW't/h}$, so that its initial value is

$$a^0\,\delta(\alpha' - \alpha^0)\,e^{-iW't/h} = a^0\,\delta(p_x' - p_x^0)\,\delta(p_y' - p_y^0)\,\delta(p_z' - p_z^0)\,e^{-iW't/h}.$$

If we use the transformation function*

$$(x'/p') = (2\pi h)^{-3/2} e^{i\Sigma_{xyz} p_x' x'/h},$$

and the transformation rule

$$\psi(x') = \int (x'/p')\,\psi(p')\,dp_x'\,dp_y'\,dp_z',$$

we obtain for the initial wave function in the co-ordinates x, y, z the value

$$a^0\,(2\pi h)^{-3/2}\,e^{i\Sigma_{xyz} p_x^0 x'/h}\,e^{-iW't/h}.$$

* The symbol x is used for brevity to denote x, y, z.

This corresponds to an initial distribution of $|a^0|^2 (2\pi h)^{-3}$ electrons per unit volume. Since their velocity is P^0c^2/E^0, the number per unit time striking a unit surface at right-angles to their direction of motion is $|a^0|^2 P^0c^2/(2\pi h)^3 E^0$. Dividing this into the expression (24) we obtain, with the help of (25),

$$4\pi^2 (2\pi h)^2 \frac{E'E^0}{c^4} |v(p'; p^0)|^2 \frac{P'}{P^0} \sin\theta' \, d\theta' \, d\phi'. \tag{26}$$

This is the effective area that must be hit by an electron in order that it shall be scattered in the solid angle $\sin\theta' \, d\theta' \, d\phi'$ with the energy E'. This result differs by the factor $(2\pi h)^2/2mE' \, . \, P'/P^0$ from Born's.* The necessity for the factor P'/P^0 in (26) could have been predicted from the principle of detailed balancing, as the factor $|v(p'; p^0)|^2$ is symmetrical between the direct and reverse processes.†

§ 6. *Application to Light-Quanta.*

We shall now apply the theory of § 4 to the case when the systems of the assembly are light-quanta, the theory being applicable to this case since light-quanta obey the Einstein-Bose statistics and have no mutual interaction. A light-quantum is in a stationary state when it is moving with constant momentum in a straight line. Thus a stationary state r is fixed by the three components of momentum of the light-quantum and a variable that specifies its state of polarisation. We shall work on the assumption that there are a finite number of these stationary states, lying very close to one another, as it would be inconvenient to use continuous ranges. The interaction of the light-quanta with an atomic system will be described by a Hamiltonian of the form (20), in which $H_P(J)$ is the Hamiltonian for the atomic system alone, and the coefficients v_{rs} are for the present unknown. We shall show that this form for the Hamiltonian, with the v_{rs} arbitrary, leads to Einstein's laws for the emission and absorption of radiation.

The light-quantum has the peculiarity that it apparently ceases to exist when it is in one of its stationary states, namely, the zero state, in which its momentum, and therefore also its energy, are zero. When a light-quantum is absorbed it can be considered to jump into this zero state, and when one is emitted it can be considered to jump from the zero state to one in which it is

* In a more recent paper ('Nachr. Gesell. d. Wiss.,' Gottingen, p. 146 (1926)) Born has obtained a result in agreement with that of the present paper for non-relativity mechanics, by using an interpretation of the analysis based on the conservation theorems. I am indebted to Prof. N. Bohr for seeing an advance copy of this work.

† See Klein and Rosseland, 'Z. f. Physik,' vol. 4, p. 46, equation (4) (1921).

physically in evidence, so that it appears to have been created. Since there is no limit to the number of light-quanta that may be created in this way, we must suppose that there are an infinite number of light-quanta in the zero state, so that the N_0 of the Hamiltonian (20) is infinite. We must now have θ_0, the variable canonically conjugate to N_0, a constant, since

$$\dot{\theta}_0 = \partial F/\partial N_0 = W_0 + \text{terms involving } N_0^{-\frac{1}{2}} \text{ or } (N_0+1)^{-\frac{1}{2}}$$

and W_0 is zero. In order that the Hamiltonian (20) may remain finite it is necessary for the coefficients v_{r0}, v_{0r} to be infinitely small. We shall suppose that they are infinitely small in such a way as to make $v_{r0}N_0^{\frac{1}{2}}$ and $v_{0r}N_0^{\frac{1}{2}}$ finite, in order that the transition probability coefficients may be finite. Thus we put

$$v_{r0}(N_0+1)^{\frac{1}{2}}e^{-i\theta_0/h} = v_r, \quad v_{0r}N_0^{\frac{1}{2}}e^{i\theta_0/h} = v_r^*,$$

where v_r and v_r^* are finite and conjugate imaginaries. We may consider the v_r and v_r^* to be functions only of the J's and w's of the atomic system, since their factors $(N_0+1)^{\frac{1}{2}}e^{-i\theta_0/h}$ and $N_0^{\frac{1}{2}}e^{i\theta_0/h}$ are practically constants, the rate of change of N_0 being very small compared with N_0. The Hamiltonian (20) now becomes

$$F = H_P(J) + \Sigma_r W_r N_r + \Sigma_{r\neq 0}[v_r N_r^{\frac{1}{2}} e^{i\theta_r/h} + v_r^*(N_r+1)^{\frac{1}{2}}e^{-i\theta_r/h}]$$
$$+ \Sigma_{r\neq 0}\Sigma_{s\neq 0} v_{rs} N_r^{\frac{1}{2}}(N_s+1-\delta_{rs})^{\frac{1}{2}} e^{i(\theta_r-\theta_s)/h}. \qquad (27)$$

The probability of a transition in which a light-quantum in the state r is absorbed is proportional to the square of the modulus of that matrix element of the Hamiltonian which refers to this transition. This matrix element must come from the term $v_r N_r^{\frac{1}{2}} e^{i\theta_r/h}$ in the Hamiltonian, and must therefore be proportional to $N_r'^{\frac{1}{2}}$ where N_r' is the number of light-quanta in state r before the process. The probability of the absorption process is thus proportional to N_r'. In the same way the probability of a light-quantum in state r being emitted is proportional to $(N_r'+1)$, and the probability of a light-quantum in state r being scattered into state s is proportional to $N_r'(N_s'+1)$. Radiative processes of the more general type considered by Einstein and Ehrenfest,† in which more than one light-quantum take part simultaneously, are not allowed on the present theory.

To establish a connection between the number of light-quanta per stationary state and the intensity of the radiation, we consider an enclosure of finite volume, A say, containing the radiation. The number of stationary states for light-quanta of a given type of polarisation whose frequency lies in the

† 'Z. f. Physik,' vol. 19, p. 301 (1923).

range ν_r to $\nu_r + d\nu_r$ and whose direction of motion lies in the solid angle $d\omega_r$ about the direction of motion for state r will now be $A\nu_r^2 d\nu_r d\omega_r/c^3$. The energy of the light-quanta in these stationary states is thus $N_r' \,.\, 2\pi h\nu_r \,.\, A\nu_r^2 d\nu_r d\omega_r/c^3$. This must equal $Ac^{-1}I_r d\nu_r d\omega_r$, where I_r is the intensity per unit frequency range of the radiation about the state r. Hence

$$I_r = N_r' (2\pi h)\nu_r^3/c^2, \tag{28}$$

so that N_r' is proportional to I_r and $(N_r' + 1)$ is proportional to $I_r + (2\pi h)\nu_r^3/c^2$. We thus obtain that the probability of an absorption process is proportional to I_r, the incident intensity per unit frequency range, and that of an emission process is proportional to $I_r + (2\pi h)\nu_r^3/c^2$, which are just Einstein's laws.* In the same way the probability of a process in which a light-quantum is scattered from a state r to a state s is proportional to $I_r [I_s + (2\pi h)\nu_r^3/c^2]$, which is Pauli's law for the scattering of radiation by an electron.†

§7. *The Probability Coefficients for Emission and Absorption.*

We shall now consider the interaction of an atom and radiation from the wave point of view. We resolve the radiation into its Fourier components, and suppose that their number is very large but finite. Let each component be labelled by a suffix r, and suppose there are σ_r components associated with the radiation of a definite type of polarisation per unit solid angle per unit frequency range about the component r. Each component r can be described by a vector potential κ_r chosen so as to make the scalar potential zero. The perturbation term to be added to the Hamiltonian will now be, according to the classical theory with neglect of relativity mechanics, $c^{-1}\Sigma_r \kappa_r \dot{X}_r$, where X_r is the component of the total polarisation of the atom in the direction of κ_r, which is the direction of the electric vector of the component r.

We can, as explained in §1, suppose the field to be described by the canonical variables N_r, θ_r, of which N_r is the number of quanta of energy of the component r, and θ_r is its canonically conjugate phase, equal to $2\pi h\nu_r$ times the θ_r of §1. We shall now have $\kappa_r = a_r \cos \theta_r/h$, where a_r is the amplitude of κ_r, which can be connected with N_r as follows:—The flow of energy per unit area per unit time for the component r is $\frac{1}{2}\pi c^{-1} a_r^2 \nu_r^2$. Hence the intensity

* The ratio of stimulated to spontaneous emission in the present theory is just twice its value in Einstein's. This is because in the present theory either polarised component of the incident radiation can stimulate only radiation polarised in the same way, while in Einstein's the two polarised components are treated together. This remark applies also to the scattering process.

† Pauli, 'Z. f. Physik,' vol. 18, p. 272 (1923).

per unit frequency range of the radiation in the neighbourhood of the component r is $I_r = \frac{1}{2}\pi c^{-1} a_r^2 \nu_r^2 \sigma_r$. Comparing this with equation (28), we obtain $a_r = 2(h\nu_r/c\sigma_r)^{\frac{1}{2}} N_r^{\frac{1}{2}}$, and hence

$$\kappa_r = 2(h\nu_r/c\sigma_r)^{\frac{1}{2}} N_r^{\frac{1}{2}} \cos \theta_r/h.$$

The Hamiltonian for the whole system of atom plus radiation would now be, according to the classical theory,

$$F = H_P(J) + \Sigma_r (2\pi h\nu_r) N_r + 2c^{-1}\Sigma_r (h\nu_r/c\sigma_r)^{\frac{1}{2}} \dot{X}_r N_r^{\frac{1}{2}} \cos \theta_r/h, \tag{29}$$

where $H_P(J)$ is the Hamiltonian for the atom alone. On the quantum theory we must make the variables N_r and θ_r canonical q-numbers like the variables J_k, w_k that describe the atom. We must now replace the $N_r^{\frac{1}{2}} \cos \theta_r/h$ in (29) by the real q-number

$$\tfrac{1}{2}\{N_r^{\frac{1}{2}} e^{i\theta r/h} + e^{-i\theta r/h} N_r^{\frac{1}{2}}\} = \tfrac{1}{2}\{N_r^{\frac{1}{2}} e^{i\theta r/h} + (N_r + 1)^{\frac{1}{2}} e^{-i\theta r/h}\}$$

so that the Hamiltonian (29) becomes

$$F = H_P(J) + \Sigma_r (2\pi h\nu_r) N_r + h^{\frac{1}{2}} c^{-\frac{3}{2}} \Sigma_r (\nu_r/\sigma_r)^{\frac{1}{2}} \dot{X}_r \{N_r^{\frac{1}{2}} e^{i\theta r/h} + (N_r + 1)^{\frac{1}{2}} e^{-i\theta r/h}\}. \tag{30}$$

This is of the form (27), with

$$v_r = v_r^* = h^{\frac{1}{2}} c^{-\frac{3}{2}} (\nu_r/\sigma_r)^{\frac{1}{2}} \dot{X}_r \tag{31}$$

and $$v_{rs} = 0 \qquad (r, s \neq 0).$$

The wave point of view is thus consistent with the light-quantum point of view and gives values for the unknown interaction coefficient v_{rs} in the light-quantum theory. These values are not such as would enable one to express the interaction energy as an algebraic function of canonical variables. Since the wave theory gives $v_{rs} = 0$ for $r, s \neq 0$, it would seem to show that there are no direct scattering processes, but this may be due to an incompleteness in the present wave theory.

We shall now show that the Hamiltonian (30) leads to the correct expressions for Einstein's A's and B's. We must first modify slightly the analysis of § 5 so as to apply to the case when the system has a large number of discrete stationary states instead of a continuous range. Instead of equation (21) we shall now have

$$ih\,\dot{a}(\alpha') = \Sigma_{\alpha''} V(\alpha'\alpha'')\, a(\alpha'').$$

If the system is initially in the state α^0, we must take the initial value of $a(\alpha')$ to be $\delta_{\alpha'\alpha^0}$, which is now correctly normalised. This gives for a first approximation

$$ih\,\dot{a}(\alpha') = V(\alpha'\alpha^0) = v(\alpha'\alpha^0)\, e^{i[W(\alpha') - W(\alpha^0)]t/h},$$

which leads to

$$ih\, a(\alpha') = \delta_{\alpha'\alpha^0} + v(\alpha'\alpha^0)\, \frac{e^{i[W(\alpha') - W(\alpha^0)]t/h} - 1}{i[W(\alpha') - W(\alpha^0)]/h},$$

corresponding to (22). If, as before, we transform to the variables W, γ_1, $\gamma_2 \cdots \gamma_{u-1}$, we obtain (when $\gamma' \neq \gamma^0$)

$$a(\mathrm{W}'\gamma') = v(\mathrm{W}', \gamma'\,;\ \mathrm{W}^0, \gamma^0)\,[1 - e^{i(\mathrm{W}'-\mathrm{W}^0)t/h}]/(\mathrm{W}' - \mathrm{W}^0).$$

The probability of the system being in a state for which each γ_k equals γ_k' is $\Sigma_{\mathrm{W}'}\,|\,a(\mathrm{W}'\,\gamma')|^2$. If the stationary states lie close together and if the time t is not too great, we can replace this sum by the integral $(\Delta\mathrm{W})^{-1}\int|\,a(\mathrm{W}'\gamma')\,|^2\,d\mathrm{W}'$, where $\Delta\mathrm{W}$ is the separation between the energy levels. Evaluating this integral as before, we obtain for the probability per unit time of a transition to a state for which each $\gamma_k = \gamma_k'$

$$2\pi/h\Delta\mathrm{W}\,.\,|\,v(\mathrm{W}^0, \gamma'\,;\ \mathrm{W}^0, \gamma^0)\,|^2. \qquad (32)$$

In applying this result we can take the γ's to be any set of variables that are independent of the total proper energy W and that together with W define a stationary state.

We now return to the problem defined by the Hamiltonian (30) and consider an absorption process in which the atom jumps from the state J^0 to the state J' with the absorption of a light-quantum from state r. We take the variables γ' to be the variables J' of the atom together with variables that define the direction of motion and state of polarisation of the absorbed quantum, but not its energy. The matrix element $v(\mathrm{W}^0, \gamma'\,;\ \mathrm{W}^0, \gamma^0)$ is now

$$h^{1/2}c^{-3/2}(\nu_r/\sigma_r)^{1/2}\,\dot{\mathrm{X}}_r(\mathrm{J}^0\mathrm{J}')\mathrm{N}_r^{\,0},$$

where $\dot{\mathrm{X}}_r(\mathrm{J}^0\mathrm{J}')$ is the ordinary $(\mathrm{J}^0\mathrm{J}')$ matrix element of $\dot{\mathrm{X}}_r$. Hence from (32) the probability per unit time of the absorption process is

$$\frac{2\pi}{h\Delta\mathrm{W}}\cdot\frac{h\nu_r}{c^3\sigma_r}\,|\,\dot{\mathrm{X}}_r(\mathrm{J}^0\mathrm{J}')\,|^2\mathrm{N}_r^{\,0}.$$

To obtain the probability for the process when the light-quantum comes from any direction in a solid angle $d\omega$, we must multiply this expression by the number of possible directions for the light-quantum in the solid angle $d\omega$, which is $d\omega\,\sigma_r\Delta\mathrm{W}/2\pi h$. This gives

$$d\omega\,\frac{\nu_r}{hc^3}\,|\,\dot{\mathrm{X}}_r(\mathrm{J}_0\mathrm{J}')\,|^2\,\mathrm{N}_r^{\,0} = d\omega\,\frac{1}{2\pi h^2 c\nu_r^{\,2}}\,|\,\dot{\mathrm{X}}_r(\mathrm{J}^0\mathrm{J}')\,|^2\,\mathrm{I}_r$$

with the help of (28). Hence the probability coefficient for the absorption process is $1/2\pi h^2 c\nu_r^{\,2}\,.\,|\,\dot{\mathrm{X}}_r(\mathrm{J}^0\mathrm{J}')\,|^2$, in agreement with the usual value for Einstein's absorption coefficient in the matrix mechanics. The agreement for the emission coefficients may be verified in the same manner.

The present theory, since it gives a proper account of spontaneous emission, must presumably give the effect of radiation reaction on the emitting system, and enable one to calculate the natural breadths of spectral lines, if one can overcome the mathematical difficulties involved in the general solution of the wave problem corresponding to the Hamiltonian (30). Also the theory enables one to understand how it comes about that there is no violation of the law of the conservation of energy when, say, a photo-electron is emitted from an atom under the action of extremely weak incident radiation. The energy of interaction of the atom and the radiation is a q-number that does not commute with the first integrals of the motion of the atom alone or with the intensity of the radiation. Thus one cannot specify this energy by a c-number at the same time that one specifies the stationary state of the atom and the intensity of the radiation by c-numbers. In particular, one cannot say that the interaction energy tends to zero as the intensity of the incident radiation tends to zero. There is thus always an unspecifiable amount of interaction energy which can supply the energy for the photo-electron.

I would like to express my thanks to Prof. Niels Bohr for his interest in this work and for much friendly discussion about it.

Summary.

The problem is treated of an assembly of similar systems satisfying the Einstein-Bose statistical mechanics, which interact with another different system, a Hamiltonian function being obtained to describe the motion. The theory is applied to the interaction of an assembly of light-quanta with an ordinary atom, and it is shown that it gives Einstein's laws for the emission and absorption of radiation.

The interaction of an atom with electromagnetic waves is then considered, and it is shown that if one takes the energies and phases of the waves to be q-numbers satisfying the proper quantum conditions instead of c-numbers, the Hamiltonian function takes the same form as in the light-quantum treatment. The theory leads to the correct expressions for Einstein's A's and B's.

RENDICONTI DELLA R. ACCADEMIA NAZIONALE DEI LINCEI
Classe di Scienze fisiche, matematiche e naturali.
Estratto dal vol. XII, serie 6ª, 2° sem., fasc. 9. – Roma, novembre 1930-VIII.

Fisica. — *Sopra l'elettrodinamica quantistica.* Nota II [1] di E. FERMI, presentata dal Socio O. M. CORBINO.

In una Nota pubblicata recentemente in questi «Rendiconti» [2] ho scritto in forma quantistica le equazioni dell'elettrodinamica; cioè le equazioni del sistema costituito dal campo elettromagnetico e da un numero qualunque di cariche elettriche puntiformi. Le equazioni scritte allora si riferivano al caso non relativistico; presupponevano cioè che la velocità delle cariche non fossero molto elevate. Esse possono tuttavia senza alcuna difficoltà essere scritte in forma relativistica, basandosi sopra la teoria di Dirac dell'elettrone rotante. È noto che recentemente anche W. Heisenberg e W. Pauli [3] hanno trattato il problema dell'elettrodinamica quantistica. Siccome però i metodi seguiti da questi autori sono essenzialmente diversi dai miei, credo non inutile pubblicare anche i miei risultati.

La forma definitiva in cui verranno espressi i risultati di questo lavoro è particolarmente semplice. Troveremo infatti che la Hamiltoniana che, nel senso del principio di corrispondenza, rappresenta la naturale traduzione quantistica dell'elettrodinamica classica, si ottiene semplicemente aggiungendo alla Hamiltoniana della teoria dell' irradiazione di Dirac un termine che rappresenta l'energia elettrostatica del sistema di cariche elettriche; per modo che, nella presente forma, l'elettrodinamica quantistica viene a non essere in alcun modo più complicata della teoria di Dirac dell'irradiazione. Questa semplificazione si può raggiungere come vedremo mediante una opportuna espressione della condizione

$$\frac{1}{c}\frac{\partial V}{\partial t} + \operatorname{div} U = 0 \tag{1}$$

che lega tra di loro i potenziali scalare e vettore e che, anche nella teoria di Heisenberg e Pauli costituisce uno degli elementi più caratteristici dell'elettrodinamica quantistica.

Nella Nota I abbiamo trovata l'espressione (21) che rappresenta l'Hamiltoniana del nostro sistema. Se, invece della meccanica classica, vogliamo rappresentare il moto dei punti per mezzo della Hamiltoniana di Dirac,

(1) Pervenuta all'Accademia il 29 settembre 1930.
(2) E. FERMI, «Rend. Lincei», *9*, 881, 1929. Citata nel seguito con I.
(3) W. HEISENBERG und W. PAULI, «Zs. f. Phys.», *56*, 1, 1929; *59*, 160, 1930.

possiamo verificare facilmente che, al posto della (21) I dobbiamo usare l'Hamiltoniana seguente:

$$(2)\quad H = -c\sum_i \gamma_i \times p_i - \sum_i \delta_i m_i c^2 + \sum_i e_i c\sqrt{\frac{8\pi}{\Omega}}\sum_s Q_s \cos\Gamma_{si} +$$

$$+\sum_s e_i c\sqrt{\frac{8\pi}{\Omega}}\sum_s \gamma_i \times (\alpha_s \chi_s + A_{s1} w_{s1} + A_{s2} w_{s2}) \sin\Gamma_{si} +$$

$$+\sum_s\left[\frac{1}{2}(\omega_{s1}^2 + \omega_{s2}^2 + \tilde{\omega}_s^2 - P_s^2) + 2\pi^2\nu_s^2 (w_{s1}^2 + w_{s2}^2 + \chi_s^2 - Q_s^2)\right].$$

Le notazioni sono quelle della Nota I; δ_i e γ_i rappresentano un q — scalare e un q — vettore tali che $\delta_i, \gamma_{ix}, \gamma_{iy}, \gamma_{iz}$ sono i quattro operatori, rappresentabili con matrici del quarto ordine, che intervengono nella Hamiltoniana relativa all'i.esimo punto materiale; naturalmente le γ e la δ relative a uno dei punti sono permutabili con le γ e la δ relative a un altro dei punti.

Osserviamo in particolare che dall'Hamiltoniana (2) risulta

$$(3)\quad \begin{cases} \dot{Q}_s = \dfrac{\partial H}{\partial P_s} = -P_s \;; & \dot{P}_s = -\dfrac{\partial H}{\partial Q_s} = 4\pi^2\nu_s^2 Q_s - c\sqrt{\dfrac{8\pi}{\Omega}}\displaystyle\sum_i e_i \cos\Gamma_{si} \\ \dot{\chi}_s = \dfrac{\partial H}{\partial \tilde{\omega}_s} = \tilde{\omega}_s \;; & \dot{\tilde{\omega}}_s = -\dfrac{\partial H}{\partial \chi_s} = -4\pi^2\nu_s^2 \chi_s - c\sqrt{\dfrac{8\pi}{\Omega}}\displaystyle\sum_i e_i \gamma_i \times \alpha_s \sin\Gamma_{si} \end{cases}$$

e

$$\dot{X}_i = \frac{\partial H}{\partial P_i} = -c\gamma_i$$

da cui

$$\frac{d}{dt}\cos\Gamma_{si} = 2\pi\nu_s \gamma_i \times \alpha_s \sin\Gamma_{si}.$$

Da questa equazione e dalle (3) risulta subito che anche dalla nuova Hamiltoniana (2) deriva l'equazione (18) I; e quindi che se l'espressione (19) I si annulla insieme alla sua derivata prima all'istante zero, essa resta sempre nulla in virtù delle equazioni differenziali. E resta quindi verificata la condizione (1), equivalente alla (19) I. Per mezzo delle (3), la (19) I si può scrivere

$$(4)\qquad 2\pi\nu_s \chi_s - P_s = 0$$

e la sua derivata, a meno di un fattore costante

$$(5)\qquad \tilde{\omega}_s - 2\pi\nu_s Q_s + \frac{c}{2\pi\nu_s}\sqrt{\frac{8\pi}{\Omega}}\sum_i e_i \cos\Gamma_{si} = 0.$$

In una interpretazione classica potremmo dunque dire che l'elettrodinamica ordinaria si ottiene integrando le equazioni canoniche dedotte dalla (2) e imponendo (4) e (5) come condizioni iniziali; ciò basta, poichè si è detto che se (4) e (5) sono verificate all'istante zero, esse lo sono anche automaticamente a un istante qualsiasi.

Per tradurre tutto questo nel linguaggio della meccanica quantistica, osserviamo che, affinchè le due grandezze (4) e (5) possano avere simultaneamente il valore zero, è necessario che esse siano commutabili, poichè altrimenti il fatto che una delle due grandezze ha un valore determinato renderebbe di necessità indeterminato il valore dell'altra. Ora si verifica facilmente, in base alle ordinarie regole di commutazione, che i primi membri di (4) e (5) sono effettivamente commutabili; si può quindi anche quantisticamente attribuire ad essi allo stesso istante il valore determinato zero.

Al procedimento classico di integrazione delle equazioni canoniche con valori arbitrari delle costanti di integrazione, corrisponde, nella meccanica ondulatoria l'integrazione dell'equazione di Schroedinger corrispondente alla Hamiltoniana (2), scegliendo arbitrariamente la funzione che rappresenta lo scalare di campo

$$\psi = \psi(t, x_i, \Phi_i, w_{s1}, w_{s2}, \chi_s, Q_s) \tag{6}$$

(σ_i rappresenta simbolicamente la coordinata interna « spincoordinate » dell'i.esimo corpuscolo) per il valore $t = 0$ del tempo. Se vogliamo invece soddisfare le condizioni (4) e (5) non possiamo più lasciare arbitraria questa funzione; resta invece determinato il modo secondo cui essa dipende dalle variabili χ_s e Q_s. Siccome infatti $\tilde{\omega}_s$, coniugata di χ_s, deve avere, secondo la (5) il valore

$$\tilde{\omega}_s = 2\pi\nu_s Q_s - \frac{c}{2\pi\nu_s}\sqrt{\frac{8\pi}{\Omega}}\sum_i e_i \cos\Gamma_{si}$$

risulta che ψ deve dipendere da χ_s nel fattore

$$e^{\frac{2\pi i}{h}\chi_s\left(2\pi\nu_s Q_s - \frac{c}{2\pi\nu_s}\sqrt{\frac{8\pi}{\Omega}}\sum_i e_i \cos\Gamma_{si}\right)}. \tag{7}$$

Dalla (4) segue in modo simile che Q_s deve intervenire soltanto nel fattore

$$e^{\frac{2\pi i}{h} 2\pi\nu_s Q_s \chi_s} \tag{8}$$

il quale è del resto già contenuto nel fattore (7). In conclusione la soluzione corrispondente alle condizioni (4) e (5) deve avere la forma:

$$\psi = \left[\prod_s e^{\frac{2\pi i}{h}\chi_s\left(2\pi\nu_s Q_s - \frac{c}{2\pi\nu_s}\sqrt{\frac{8\pi}{\Omega}}\sum_i e_i \cos\Gamma_{si}\right)}\right]\varphi(t, x_i, \sigma_i, w_{s1}, w_{s2}). \tag{9}$$

Dobbiamo ora dimostrare che effettivamente si può soddisfare l'equazione di Schroedinger per mezzo della posizione (9). L'equazione di Schroedinger dedotta dall'Hamiltoniana (2) è

$$(10) \qquad -\frac{h}{2\pi i}\frac{\partial\psi}{\partial t} = \mathrm{H}\psi$$

dove H è naturalmente interpretato come un operatore. Sostituendo nella (10) al posto di ψ l'espressione (9) si trova, con calcoli non difficili la seguente equazione a cui deve soddisfare φ:

$$(11) \qquad -\frac{h}{2\pi i}\frac{\partial\varphi}{\partial t} = \mathrm{R}\varphi$$

dove R rappresenta il seguente operatore

$$(12) \quad \mathrm{R} = -c\sum_i \gamma_i \times p_i - \sum_i \delta_i m_i c^2 + \sum_i e_i c\sqrt{\frac{8\pi}{\Omega}}\sum_s \gamma_i \times (\mathrm{A}_{s1} w_{s1} + \mathrm{A}_{s2} w_{s2}) \sin\Gamma_{si} +$$
$$+ \sum_s \left[\frac{1}{2}(\omega_{s1}^2 + \omega_{s2}^2) + 2\pi^2\nu_s^2(w_{s1}^2 + w_{s2}^2)\right] + \frac{c^2}{\pi\Omega}\sum_s \frac{1}{\nu_s^2}\left(\sum_i e_i \cos\Gamma_{si}\right)^2.$$

A prescindere dall'ultimo termine, R coincide con l'Hamiltoniana della teoria dell'irradiazione di Dirac, in cui si trascura il potenziale scalare e la componente longitudinale del potenziale vettore, considerando solo il campo determinato delle componenti trasversali del potenziale vettore, e cioè il solo campo di radiazione. Dobbiamo discutere il significato dell'ultimo termine. Per questo lo trasformiamo nel modo seguente:

$$\frac{c^2}{\pi\Omega}\sum_s \frac{1}{\nu_s^2}\left(\sum_i e_i \cos\Gamma_{si}\right)^2 = \frac{c^2}{\pi\Omega}\sum_{sij} e_i e_j \frac{\cos\Gamma_{si}\cos\Gamma_{sj}}{\nu_s^2} =$$
$$= \frac{c^2}{\pi\Omega}\sum_{ij} e_i e_j \sum_s \frac{\cos\Gamma_{si}\cos\Gamma_{sj}}{\nu_s^2}.$$

La somma rispetto ad s si può trasformare in un integrale e si trova, con calcoli privi di difficoltà

$$\sum_s \frac{\cos\Gamma_{si}\cos\Gamma_{sj}}{\nu_s^2} = \frac{\pi\Omega}{2c^2}\frac{1}{r_{ij}}$$

dove r_{ij} rappresenta la distanza tra i due punti i e j. Sostituendo troviamo

$$(13) \qquad \frac{c^2}{\pi\Omega}\sum_s \frac{1}{\nu_s^2}\left(\sum e_i \cos\Gamma_{si}\right)^2 = \frac{1}{2}\sum_{ij}\frac{e_i e_j}{r_{ij}}.$$

La (13) ci dà dunque semplicemente la ordinaria espressione dell'energia elettrostatica; come nell'elettrostatica classica, l'espressione (13) diventa infinita nel caso di cariche elettriche puntiformi. Questo inconveniente, più che dalla elettrodinamica, deriva dalla imperfetta conoscenza della struttuta dell'elettrone, e potrebbe p. es. venir eliminato considerando elettroni di raggio finito. Noi lo elimineremo formalmente, come si suol fare anche nell'elettrostatica classica, escludendo dalla somma (13) i termini per cui $i = j$ che rappresentano in certo modo una costante additiva infinitamente grande. Indicheremo ciò con un'apice al segno $\sum$. Per mezzo della (13), la (12) diventa:

$$(14)\quad R = -c\sum_i \gamma_i \times p_i - \sum_i \delta_i m_i c^2 + \sum_i e_i c \sqrt{\frac{8\pi}{\Omega}} \sum_s \gamma_i + (A_{s1} w_{s1} + A_{s2} w_{s2}) \sin \Gamma_{si} +$$
$$+ \sum_s \left[\frac{1}{2}(\omega_{s1}^2 + \omega_{s2}^2) + 2\pi^2 \nu_s^2 (w_{s1}^2 + w_{s2}^2)\right] + \frac{1}{2} \sum_{ij}{}' \frac{e_i e_j}{r_{ij}}.$$

Osserviamo infine che la funzione φ che abbiamo sostituito allo scalare di campo ψ per mezzo della (9) può in tutte le considerazioni sostituirsi ad esso, da cui differisce per un fattore complesso di modulo 1. La φ soddisfa all'equazione (11) che è del tipo di una equazione di Schroedinger in cui però si deve prendere come Hamiltoniana R invece di H Questa nuova Hamiltoniana R, come si legge immediatamente dalla (14) è costituita semplicemente aggiungendo l'ordinaria espressione dell'energia elettrostatica alla ordinaria Hamiltoniana dei termini di pura radiazione. Possiamo dunque concludere che, in questa forma, il problema di elettrodinamica quantistica non è in alcun modo più complicato di un ordinario problema di teoria della radiazione.

Naturalmente, come già abbiamo accennato, anche questa teoria conserva in se due difetti fondamentali che però più che di origine elettrodinamica, possono considerarsi derivanti dalla non completa conoscenza della struttura elettronica. Essi sono la possibilità che ha l'elettrone di Dirac di passare a livelli energetici con energia negativa ed il fatto che l'energia intrinseca a valore infinito se si ammette l'elettrone esattamente puntiforme.

ON QUANTUM ELECTRODYNAMICS.

By P. A. M. Dirac, V. A. Fock and Boris Podolsky.

(Received October 25, 1932.)

In the first part of this paper the equivalence of the new form of relativistic Quantum Mechanics[1] to that of Heisenberg and Pauli[2] is proved in a new way which has the advantage of showing their physical relation and serves to suggest further development considered in the second part.

Part I. Equivalence of Dirac's and Heisenberg-Pauli's Theories.

§ 1. Recently Rosenfeld showed[3] that the new form of relativistic Quantum Mechanics[1] is equivalent to that of Heisenberg and Pauli.[2] Rosenfeld's proof is, however, obscure and does not bring out some features of the relation of the two theories. To assist in the further development of the theory we give here a simplified proof of the equivalence.

Consider a system, with a Hamiltonian H, consisting of two parts A and B with their respective Hamiltonians H_a and H_b and the interaction V. We have

$$H = H_a + H_b + V, \tag{1}$$

where

$$H_a = H_a(p_a q_a T); \quad H_b = H_b(p_b q_b T);$$
$$V = V(p_a q_a p_b q_b T)$$

and T is the time for the entire system. The wave function for the entire system will satisfy the equation[4]

$$(H - ih\partial/\partial T)\psi(q_a q_b T) = 0 \tag{2}$$

and will be a function of the variables indicated.

[1] Dirac, Proc. Roy. Soc. A **136**, 453, 1932.

[2] Heisenberg and Pauli, ZS. f. Physik, **56**, 1, 1929 and **59**, 168, 1930.

[3] Rosenfeld, ZS. f. Physik **76**, 729, 1932.

[4] h is Planck's constant divided by 2π.

Now, upon performing the canonical transformation

$$\psi^* = e^{\frac{i}{h}H_b T}\psi, \tag{3}$$

by which dynamical variables, say F, transform as follows

$$F^* = e^{\frac{i}{h}H_b T} F e^{-\frac{i}{h}H_b T}, \tag{4}$$

Eq. (2) takes the form

$$(H_a^* + V^* - ih\partial/\partial T)\psi^* = 0. \tag{5}$$

Since H_a commutes with H_b, $H_a^* = H_a$. On the other hand, since the functional relation between variables is not disturbed by the canonical transformation (3), V^* is the same function of the transformed variables p^*, q^* as V is of p, q. But p_a and q_a commute with H_b so that $p_a^* = p_a$, $q_a^* = q_a$. Therefore

$$V^* = V(p_a q_a p_b^* q_b^*), \tag{6}$$

where

$$\left.\begin{aligned} q_b^* &= e^{\frac{i}{h}H_b T} q_b e^{-\frac{i}{h}H_b T} \\ p_b^* &= e^{\frac{i}{h}H_b T} p_b e^{-\frac{i}{h}H_b T} \end{aligned}\right\} \tag{7}$$

It will be shown in § 7, after suitable notation is developed, that Eqs. (7) are equivalent to

$$\left.\begin{aligned} \partial q_b^*/\partial t &= \frac{i}{h}(H_b q_b^* - q_b^* H_b) \\ \partial p_b^*/\partial t &= \frac{i}{h}(H_b p_b^* - p_b^* H_b) \end{aligned}\right\} \tag{8}$$

where t is the separate time of the part B.

These, however, are just the equations of motion for the part B alone, unperturbed by the presence of part A.

§ 2. Now let part B correspond to the field and part A to the particles present. Eqs. (8) must then be equivalent to Maxwell's equations for empty space. Eq. (2) is then the wave equation of Heisenberg-Pauli's theory, while Eq. (5), in which the perturbation is expressed in terms of potentials corresponding to empty space, is the wave equation of the new theory. Thus, this theory corresponds to treating separately a part of the system, which is in some problems more convenient.[1]

Now, H_a can be represented as a sum of the Hamiltonians for the separate particles. The interaction between the particles is not included in H_a for this is taken to be the result of interaction between the particles and the field. Similarly, V is the sum of interactions between the field and the particles. Thus, we may write

$$\left.\begin{aligned} H_a &= \sum_{s=1}^{n}(c\alpha_s \cdot p_s + m_s c^2 \alpha_s^{(4)}) = \sum_{s=1}^{n} H_s \\ &\text{and} \\ V^* &= \sum_{s=1}^{n} V_s^* = \sum_{s=1}^{n} \varepsilon_s\left[\Phi(r_s, T) - \alpha_s \cdot A(r_s, T)\right] \end{aligned}\right\} \quad (9)$$

where r_s are the coordinates of the s-th particle and n is the number of particles.

Eq. (5) takes the form

$$\left[\sum_{s=1}^{n}(H_s + V_s^*) - ih\partial/\partial T\right]\psi^*(r_s; J; T) = 0, \quad (10)$$

J stands for the variables describing the field. Besides the common time T and the field time t an individual time $t_s = t_1$, $t_2, \ldots t_n$ is introduced for each particle. Eq. (10) is satisfied by the common solution of the set of equations

$$\left.\begin{aligned} &(R_s - ih\partial/\partial t_s)\psi^* = 0, \\ &\text{where} \\ &R_s = c\alpha_s \cdot p_s + m_s c^2 \alpha_s^4 + \varepsilon_s[\Phi(r_s t_s) - \alpha_s \cdot A(r_s t_s)] \end{aligned}\right\} \quad (11)$$

and $\psi^* = \psi^*(r_1\, r_2 \ldots r_n;\ t_1\, t_2 \ldots t_n;\ J)$, when all the t's are put equal to the common time T.

Now, Eqs. (11) are the equations of Dirac's theory. They are obviously relativistically invariant and form a generalization of Eq. (10). This obvious relativistic invariance is achieved by the introduction of separate time for each particle.

§ 3. For further development we shall need some formulas of quantization of electromagnetic fields and shall use

[1] This is somewhat analogous to Frenkel's method of treating incomplete systems, see Frenkel, Sow. Phys. 1, 99, 1932.

for this purpose some formulas obtained by Fock and Podolsky.[1] Starting with the Lagrangian function

$$L=\frac{1}{2}(\mathfrak{E}^2-\mathfrak{H}^2)-\frac{1}{2}\left(\operatorname{div} A+\frac{1}{c}\dot{\Phi}\right)^2, \tag{12}$$

taking as coordinates $(Q_0\, Q_1\, Q_2\, Q_3)$ the potentials $(\Phi\, A_1\, A_2\, A_3)$, and retaining the usual relations

$$\mathfrak{E}=-\operatorname{grad}\Phi-\frac{1}{c}\dot{A};\quad \mathfrak{H}=\operatorname{curl} A, \tag{13}$$

one obtains

$$\left.\begin{aligned} (P_1\, P_2\, P_3)=P&=-\frac{1}{c}\mathfrak{E};\\ P_0&=-\frac{1}{c}\left(\operatorname{div} A+\frac{1}{c}\dot{\Phi}\right); \end{aligned}\right\} \tag{14}$$

and the Hamiltonian

$$H=\frac{c^2}{2}(P^2-P_0^2)+\frac{1}{2}\sum_{1,2,3}\left(\frac{\partial Q_1}{\partial x_2}-\frac{\partial Q_2}{\partial x_1}\right)^2$$

$$-cP_0\sum_{l=1}^{3}\frac{\partial Q_l}{\partial x_l}-cP\cdot\operatorname{grad} Q_0. \tag{15}$$

The equations of motion are[2]

$$\left.\begin{aligned} \dot{A}&=c^2 P-c\operatorname{grad}\Phi,\\ \dot{\Phi}&=-c^2 P_0-c\operatorname{div} A,\\ \dot{P}&=\Delta A-\operatorname{grad}\operatorname{div} A-c\operatorname{grad} P_0,\\ \dot{P}_0&=-c\operatorname{div} P, \end{aligned}\right\} \tag{16}$$

On elimination of P and P_0, Eqs. (16) give the D'Alembert equations for the potentials Φ and A. To obtain Maxwell's equation for empty space one must set $P_0=0$. The quantization rules are expressed in terms of the amplitudes of the Fourier's integral. Thus, for every $F=F(xyzt)$,

[1] Fock and Podolsky, Sow. Phys. **1**, 801, 1932, later quoted as l. c. For other treatments see Jordan and Pauli, ZS. f. Physik, **47**, 151, 1928 or Fermi, Rend. Lincei, **9**, 881, 1929. The Lagrangian (12) differs from that of Fermi only by a four-dimensional divergence.

[2] A dot over a field quantity will be used to designate a derivative with respect to the field time t.

amplitudes $F(k)$ and $F^+(k)$ are introduced by the equation

$$F = \left(\frac{1}{2\pi}\right)^{3/2} \int \left\{F(k)\, e^{-ic|k|t+ik\cdot r} + F^+(k)\, e^{+ic|k|t-ik\cdot r}\right\} dk \qquad (17)$$

where $r = (xyz)$ is the position vector, $k = (k_x k_y k_z)$ is the wave vector having the magnitude $|k| = 2\pi/\lambda$, $dk = dk_x dk_y dk_z$, the integration being performed for each component of k from $-\infty$ to ∞. In terms of the amplitudes equations of motion can be written

$$\left.\begin{aligned} P(k) &= \frac{i}{c}\,[k\,\Phi(k) - |k|\,A(k)] = -\frac{1}{c}\,\mathfrak{E}(k) \\ P_0(k) &= \frac{i}{c}\,[\,|k|\,\Phi(k) - k\cdot A(k)] \end{aligned}\right\} \qquad (18)$$

the other two equations being algebraic consequences of these.

The commutation rules for the potentials are

$$\left.\begin{aligned} \Phi^+(k)\,\Phi(k') - \Phi(k')\,\Phi^+(k) &= \frac{ch}{2\,|k|}\,\delta(k-k') \\ A_l^+(k)\,A_m(k') - A_m(k')\,A_l^+(k) &= -\frac{ck}{2\,|k|}\,\delta_{lm}\,\delta(k-k') \end{aligned}\right\} \qquad (19)$$

all other combinations of amplitudes commuting.

Part II. The Maxwellian Case.

§ 4. For the Maxwellian case the following additional considerations are necessary. In obtaining the field variables, besides the regular equations of motion of the electromagnetic field one must use the additional condition $P_0 = 0$, or $-cP_0 = \operatorname{div} A + \dot{\Phi}/c = 0$. This condition cannot be regarded as a quantum mechanical equation, but rather as a condition on permissible ψ functions. This can be seen, for example, from the fact that, when regarded as a quantum mechanical equation, $\operatorname{div} A + \dot{\Phi}/c = 0$ contradicts the commutation rules. Thus, only those ψ's should be regarded as physically permissible which satisfy the condition

$$-cP_0\psi = \left(\operatorname{div} A + \frac{1}{c}\dot{\Phi}\right)\psi = 0. \qquad (20)$$

Condition (20), expressed in terms of amplitudes by the use of Eq. (18), takes the form

$$\left.\begin{aligned} i\,[k\cdot A(k)-|k|\,\Phi(k)]\,\psi=0 \\ \text{and} \qquad -i\,[k\cdot A^{+}(k)-|k|\,\Phi^{+}(k)]\,\psi=0. \end{aligned}\right\} \qquad (20')$$

To these must, of course, be added the wave equation

$$(H_b - ih\,\partial/\partial t)\,\psi=0, \qquad (21)$$

where H_b is the Hamiltonian for the field

$$H_b = 2\int\{\,A^{+}(k)\cdot A(k)-\Phi^{+}(k)\,\Phi(k)\,\}\,|\,k\,|^2\,dk, \qquad (22)$$

as in l. c.

If a number of equations $A\psi=0$, $B\psi=0$, etc., are simultaneously satisfied, then $AB\psi=0$, $BA\psi=0$, etc.; and therefore $(AB-BA)\,\psi=0$, etc. All such new equations must be consequences of the old, i. e. must not give any new conditions on ψ. This may be regarded as a test of consistency of the original equations. Applying this to our Eqs. (20′) and (21) we have

$$\begin{aligned} &P_0(k)\,P_0^{+}(k')-P_0^{+}(k')\,P_0(k) \\ &= c^2\,[k\cdot A(k)\,k'\cdot A^{+}(k')-k'\cdot A^{+}(k')\,k\cdot A(k)] \\ &\qquad + c^2\,|k|\,|k'|\,[\Phi(k)\,\Phi^{+}(k')-\Phi^{+}(k')\,\Phi(k)] \end{aligned} \qquad (23)$$

since A's commute with Φ's. Applying now the commutation rules of Eq. (19), we obtain

$$\begin{aligned} &P_0(k)\;P_0^{+}(k')-P_0^{+}(k')\;P_0(k) \\ &= \frac{c^3 h}{2\,|k|}\Big(\sum_{l,m} k_l k_m \delta_{lm}-|k|^2\Big)\,\delta(k-k')=0. \end{aligned} \qquad (24)$$

Eq. (24) is satisfied in consequence of quantum-mechanical equations, hence

$$[P_0(k)\,P_0^{+}(k')-P_0^{+}(k')\,P_0(k)]\,\psi=0$$

is not a condition on ψ. Thus, conditions (20′) are consistent. Since $P_0(k)$ and $P_0^{+}(k)$ commute with $\partial/\partial t$, to test the consistence of condition (20) with (21) one must test the condition

$$(H_b P_0-P_0 H_b)\,\psi=0 \qquad (25)$$

However, since $\dot{P}_0=(i/h)\,(H_b P_0-P_0 H_b)$, Eq. (25) takes the form $\dot{P}_0\psi=0$, or in F o u r i e r's components

$$\dot{P}_0(k)\,\psi=-ic\,|k|\,P_0(k)\,\psi=0$$

and

$$\dot{P}_0^+(k)\,\psi = ic\,|k|\,P_0^+(k)\,\psi = 0.$$

But these are just the conditions (20′). Thus, conditions (20) and (21) are consistent.

§ 5. The extra condition of Eq. (20) is not an equation of motion, but is a "constraint" imposed on the initial coordinates and velocities, which the equations of motion then preserve for all time. The existence of this constraint for the Maxwellian case is the reason for the additional considerations, mentioned at the beginning of § 4. It turns out that we must modify this constraint when particles are present, in order to get something which the equations of motion will preserve for all time.

The conditions (20′) as they stand, when applied to ψ, are not consistent with Eqs. (11). It is, however, not difficult to see that they can be replaced by a somewhat different set of conditions[1]

$$C(k)\,\psi = 0 \quad \text{and} \quad C^+(k)\,\psi = 0, \tag{26′}$$

where

$$C(k) = i\,[k\cdot A(k) - |k|\,\Phi(k)] + \frac{i}{2\,(2\,\pi)^{3/2}\,|k|}\sum_{s=1}^{n} \varepsilon_s e^{ic\,|k|\,t_s - ik\cdot r_s}. \tag{27′}$$

Terms in $C(k)$ not contained in $-cP_0(k)$ are functions of the coordinates and the time for the particles. They commute with $H_b - ih\,\partial/\partial t$, with $P_0(k)$ and with each other. Therefore Eqs. (26′) are consistent with each other and with Eq. (21). It remains to show that Eqs. (26′) are consistent with Eqs. (11). In fact $C(k)$ and $C^+(k)$ commute with $R_s - ih\,\partial/\partial t_s$. We shall show this for $C(k)$.

Designating, in the usual way, $AB - BA$ as $[A, B]$, we see that it is sufficient to show that

$$\left[C(k),\; p_s - \frac{\varepsilon_s}{c}\,A(r_s\,t_s)\right] = 0 \tag{28}$$

and

$$[C(k),\; ih\,\partial/\partial t_s - \varepsilon_s\,\Phi(r_s\,t_s)] = 0. \tag{29}$$

[1] We shall drop the asterisk and in the following use ψ instead of ψ^*.

By considering the form of $C(k)$, these become respectively

$$[k\cdot A(k),\ A(r_s t_s)] - \frac{c}{2(2\pi)^{3/2}|k|} e^{ic|k|t_s}[e^{-ik\cdot r_s}, p_s] = 0. \quad (30)$$

and

$$[|k|\Phi(k),\ \Phi(r_s t_s)] + \frac{1}{2(2\pi)^{3/2}|k|} e^{-ik\cdot r_s}[e^{ic|k|t_s}; ih\partial/\partial t_s] = 0. \quad (31)$$

Now

$$[k\cdot A(k),\ A(r_s t_s)] = \left(\frac{1}{2\pi}\right)^{3/2} \int [k\cdot A(k),\ A^{+}(k')]\, e^{ic|k'|t_s - ik'\cdot r_s}\, dk',$$

by Eq. (17) and because $A(k)$ commutes with $A(k')$. Using the commutation formulas and performing the integration it becomes

$$\frac{chk}{2(2\pi)^{3/2}|k|} e^{ic|k|t_s - ik\cdot r_s}. \quad (32)$$

On the other hand

$$[e^{-ik\cdot r_s}, p_s] = hi\,\mathrm{grad}\, e^{-ik\cdot r_s} = hke^{-ik\cdot r_s} \quad (33)$$

Thus, Eq. (30) is satisfied. Similarly Eq. (31) is satisfied because

$$[|k|\,\Phi(k),\ \Phi(r_s t_s)] = \frac{-ch}{2(2\pi)^{3/2}} e^{ic|k|t_s - ik\cdot r_s} \quad (34)$$

and

$$[e^{ic|k|t_s},\ ih\partial/\partial t_s] = ch|k|\, e^{ic|k|t_s}. \quad (35)$$

Thus, conditions (26′) satisfy all the requirements of consistence. It can be shown that these requirements determine $C(k)$ uniquely up to an additive constant, if it is taken to have the form $i[k\cdot A(k) - |k|\,\Phi(k)] + f(r_s t_s)$.

§ 6. We shall now show that the introduction of separate time for the field and for each particle allows the use of the entire vacuum electrodynamics of § 3 and l. c., except for the change discussed in § 5. In fact, we shall show that Maxwell's equations of electrodynamics, in which enter current or charge densities, become conditions on ψ function.

For convenience we collect together our fundamental equations.

The equations of vacuum electrodynamics are

$$\mathfrak{E} = -\operatorname{grad}\Phi = \frac{1}{c}\operatorname{div} A; \quad \mathfrak{H} = \operatorname{curl} A \tag{13}$$

$$\Delta\Phi - \frac{1}{c^2}\ddot{\Phi} = 0; \qquad \Delta A - \frac{1}{c^2}\ddot{A} = 0. \tag{36}$$

The wave equations are

$$\left.\begin{aligned} &(R_s - ih\partial/\partial t_s)\psi = 0, \\ \text{where} \quad &R_s = c\alpha_s\cdot p_s + m_s c^2\alpha_s^{(4)} - \varepsilon_s\alpha_s\cdot A(r_s t_s) + \varepsilon_s\Phi(r_s t_s). \end{aligned}\right\} \tag{11}$$

The additional conditions on ψ function are

$$C(k)\psi = 0 \quad \text{and} \quad C^+(k)\psi = 0, \tag{26'}$$

where

$$C(k) = i[k\cdot A(k) - |k|\,\Phi(k)] + \frac{i}{2(2\pi)^{3/2}|k|}\sum_{s=1}^{n}\varepsilon_s e^{ic|k|t_s - ik\cdot r_s}. \tag{27'}$$

We transform the last two equations by passing from the amplitudes $C(k)$ and $C^+(k)$ to $C(r, t)$ by means of Eq. (17). Thus we obtain

$$C(r, t)\psi = 0 \tag{26}$$

and

$$C(r, t) = \operatorname{div} A + \frac{1}{c}\frac{\partial\Phi}{\partial t} - \sum_{s=1}^{n}\frac{\varepsilon_s}{4\pi}\Delta(X - X_s), \tag{27}$$

where X and X_s are four dimensional vectors $X = (x\,y\,z\,t)$, $X_s = (x_s y_s z_s t_s)$ and Δ is the so-called invariant delta function[1]

$$\Delta(X) = \frac{1}{|r|}[\delta(|r| + ct) - \delta(|r| - ct)]. \tag{37}$$

From Eqs. (13) follows immediately

$$\operatorname{div}\mathfrak{H} = 0 \quad \text{and} \quad \operatorname{curl}\mathfrak{E} + \frac{1}{c}\frac{\partial}{\partial t}\mathfrak{H} = 0 \tag{38}$$

so that these remain as quantum-mechanical equations. Using Eqs. (13) and (36) and condition (26) we obtain by direct calculation

$$\left(\operatorname{curl}\mathfrak{H} - \frac{1}{c}\frac{\partial\mathfrak{E}}{\partial t}\right)\psi = \operatorname{grad}\sum_{s=1}^{n}\frac{\varepsilon_s}{4\pi}\Delta(X - X_s)\psi \tag{39}$$

[1] See Jordan and Pauli, ZS. f. Physik **47**, 159, 1928.

and

$$(\operatorname{div} \mathfrak{E})\psi = -\frac{1}{c}\left(\frac{\partial}{\partial t}\sum_{s=1}^{n}\frac{\varepsilon_s}{4\pi}\Delta(X - X_s)\right)\psi. \tag{40}$$

Now, let us consider what becomes of these equations when we put $t = t_1 = t_2 = \ldots = t_n = T$, which is implied in Maxwell's equations and which we shall write for short as $t_s = T$.

For any quantity $f = f(t\,t_1\,t_2 \ldots t_n)$

$$\frac{\partial f(TTT\ldots T)}{\partial T} = \left[\left(\frac{\partial f}{\partial t}\right) + \left(\frac{\partial f}{\partial t_1}\right) + \ldots + \left(\frac{\partial f}{\partial t_n}\right)\right]_{t_s = T} \tag{41}$$

and for each of the n derivatives $\partial/\partial t_s$ we have an equation of motion

$$\frac{\partial f}{\partial t_s} = \frac{i}{h}(R_s f - f R_s). \tag{42}$$

If we put $f = A(r, t)$ or $f = \Phi(r, t)$, then, since both commute with R_s, $\partial f/\partial t_s = 0$ and we get

$$\frac{\partial A}{\partial t} = \frac{\partial A}{\partial T} \quad \text{and} \quad \frac{\partial \Phi}{\partial t} = \frac{\partial \Phi}{\partial T}. \tag{43}$$

It follows that

$$\mathfrak{E} = -\frac{1}{c}\frac{\partial A}{\partial T} - \operatorname{grad}\Phi; \quad \mathfrak{H} = \operatorname{curl} A, \tag{44}$$

so that the form of the connection between the field and the potentials is preserved. Remembering that for $t = t_s$ we have $\Delta(X - X_s) = 0$ and hence $\operatorname{grad}\Delta(X - X_s) = 0$, and using Eqs. (26), (39) and (40) we obtain

$$\left(\operatorname{div} A + \frac{1}{c}\frac{\partial \Phi}{\partial T}\right)\psi = 0, \tag{45}$$

$$\left(\operatorname{curl}\mathfrak{H} - \frac{1}{c}\frac{\partial \mathfrak{E}}{\partial t}\right)_{t_s = T}\psi = 0 \tag{46}$$

and

$$(\operatorname{div}\mathfrak{E})\psi = -\sum_{s=1}^{n}\frac{\varepsilon_s}{4\pi}\left[\frac{1}{c}\frac{\partial}{\partial t}\Delta(X - X_s)\right]_{t=t_s}\psi. \tag{47}$$

For further reduction of Eq. (46) we must use Eqs. (41) and (42), from which follows

$$\left(\frac{1}{c}\frac{\partial \mathfrak{E}}{\partial t}\right)_{t_s=T} = \frac{1}{c}\frac{\partial \mathfrak{E}}{\partial T} - \sum_{s=1}^{n} \frac{i}{ch}[R_s, \mathfrak{E}] \tag{48}$$

and $[R_s, \mathfrak{E}]$ is easily calculated, because the only term in R_s which does not commute with $\mathfrak{E}$ is $-\varepsilon_s \alpha_s \cdot A(r_s\, t_s)$, and $-\mathfrak{E}/c$ is the momentum conjugate to A. In this way we obtain

$$[R_s, \mathfrak{E}] = ich\,\varepsilon_s \alpha_s\, \delta(r - r_s). \tag{49}$$

For the reduction of Eq. (47) we need only remember[1] that

$$\left[\frac{1}{c}\frac{\partial}{\partial t}\Delta(X)\right]_{t=0} = -4\pi\,\delta(r). \tag{50}$$

Thus, Eqs. (46) and (47) become

$$\left(\operatorname{curl}\mathfrak{H} - \frac{1}{c}\frac{\partial \mathfrak{E}}{\partial T}\right)\psi = \sum_{s=1}^{n} \varepsilon_s \alpha_s\, \delta(r - r_s)\,\psi \tag{51}$$

and

$$(\operatorname{div}\mathfrak{E})\psi = \sum_{s=1}^{n} \varepsilon_s\, \delta(r - r_s)\,\psi, \tag{52}$$

which are just the remaining Maxwell's equations appearing as conditions on ψ. Eq. (52) is the additional condition of Heisenberg-Pauli's theory.

§ 7. We shall now derive Eq. (8) of § 1. For this we need to recall that the transformation (7) is a canonical transformation which preserves the form of the algebraic relations between the variables, as well as the equations of motion. These will be, in the exact notation now developed,

$$\frac{\partial q_b^*}{\partial T} = \frac{i}{h}[H^*, q_b^*]_{t_s=T}; \qquad \frac{\partial p_b^*}{\partial T} = \frac{i}{h}[H^*, p_b^*]_{t_s=T}. \tag{53}$$

As we have seen in the discussion following Eq. (5)

$$H^* = H_a + H_b + V^* \tag{54}$$

[1] Heisenberg und Pauli, ZS. f. Physik **56**, 34, 1929.

and since q_b and p_b commute with H_a, q_b^* und p_b^* commute with H_a^* and hence with H_a. Therefore Eqs. (53) become

$$\left.\begin{aligned} \frac{\partial q_b^*}{\partial T} &= \frac{i}{h}\{[H_b, q_b^*] + [V^*, q_b^*]\}_{t_s=T} \\ \frac{\partial p_b^*}{\partial T} &= \frac{i}{h}\{[H_b, p_b^*] + [V^*, p_b^*]\}_{t_s=T} \end{aligned}\right\} \tag{55}$$

On the other hand, we have from Eqs. (41) and (42)

$$\left.\begin{aligned} \frac{\partial q_b^*}{\partial T} &= \left\{\frac{\partial q_b^*}{\partial t} + \frac{i}{h}\sum_{s=1}^{n}[R_s, q_b^*]\right\}_{t_s=T} \\ &\text{and} \\ \frac{\partial p_b^*}{\partial T} &= \left\{\frac{\partial p_b^*}{\partial t} + \frac{i}{h}\sum_{s=1}^{n}[R_s, p_b^*]\right\}_{t_s=T} \end{aligned}\right\} \tag{56}$$

Now the only term in R_s which does not commute with p_b^* and q_b^* is V_s^* so that

$$[R_s, q_b^*] = [V_s^*, q_b^*] \quad \text{and} \quad [R_s, p_b^*] = [V_s^*, p_b^*]. \tag{57}$$

Since $\sum V_s^* = V^*$, Eqs. (56) become

$$\left.\begin{aligned} \frac{\partial q_b}{\partial T} &= \left\{\frac{\partial q_b^*}{\partial t} + \frac{i}{h}[V^*, q_b^*]\right\}_{t_s=T} \\ \frac{\partial p_b^*}{\partial T} &= \left\{\frac{\partial p_b^*}{\partial t} + \frac{i}{h}[V^*, q_b^*]\right\}_{t_s=T} \end{aligned}\right\} \tag{58}$$

Comparison of Eqs. (55) with (58) finally gives

$$\left.\begin{aligned} \left(\frac{\partial q_b^*}{\partial t}\right)_{t=T} &= \frac{i}{h}[H_b, q_b^*]_{t=T} \\ \left(\frac{\partial p_b^*}{\partial t}\right)_{t=T} &= \frac{i}{h}[H_b, p_b^*]_{t=T} \end{aligned}\right\} \tag{59}$$

which is, in the more exact notation, just Eqs. (8).

Cambridge, Leningrad and Kharkov.

Über das Paulische Äquivalenzverbot.

Von **P. Jordan** und **E. Wigner** in Göttingen.

(Eingegangen am 26. Januar 1928.)

Die Arbeit enthält eine Fortsetzung der kürzlich von einem der Verfasser vorgelegten Note „Zur Quantenmechanik der Gasentartung", deren Ergebnisse hier wesentlich erweitert werden. Es handelt sich darum, ein ideales oder nichtideales, dem Paulischen Äquivalenzverbot unterworfenes Gas zu beschreiben mit Begriffen, die keinen Bezug nehmen auf den abstrakten Koordinatenraum der Atomgesamtheit des Gases, sondern nur den gewöhnlichen dreidimensionalen Raum benutzen. Das wird ermöglicht durch die Darstellung des Gases vermittelst eines gequantelten dreidimensionalen Wellenfeldes, wobei die besonderen nichtkommutativen Multiplikationseigenschaften der Wellenamplitude gleichzeitig für die Existenz korpuskularer Gasatome und für die Gültigkeit des Paulischen Äquivalenzverbots verantwortlich sind. Die Einzelheiten der Theorie besitzen enge Analogien zu der entsprechenden Theorie für Einsteinsche ideale oder nichtideale Gase, wie sie von Dirac, Klein und Jordan ausgeführt wurde.

§ 1. Schon bei den ersten Untersuchungen zur systematischen Ausbildung der Matrizentheorie der Quantenmechanik ergaben sich Hinweise darauf, daß die bekannten Schwierigkeiten der Strahlungstheorie überwunden werden könnten, indem man nicht nur auf die materiellen Atome, sondern auch auf das Strahlungsfeld die quantenmechanischen Methoden anwendet*. In diesem Sinne sind durch mehrere Arbeiten** der letzten Zeit Fortschritte erzielt worden einerseits bezüglich einer quantenmechanischen Beschreibung des elektromagnetischen Feldes, andererseits bezüglich einer Formulierung der Quantenmechanik materieller Teilchen, welche die Wellendarstellung im abstrakten Koordinatenraum vermeidet zugunsten einer Darstellung durch quantenmechanische Wellen im gewöhnlichen dreidimensionalen Raume, und welche die Existenz materieller Teilchen in ähnlicher Weise zu erklären sucht, wie durch die Quantelung der elektromagnetischen Wellen die Existenz von Lichtquanten bzw. jeder durch die Annahme von Lichtquanten zu deutende physikalische Effekt erklärt wird.

Man verfährt bei dieser Beschreibung so, daß man diejenige, als q-Zahl aufzufassende Größe N_r, welche in korpuskulartheoretischer Um-

---

* M. Born, W. Heisenberg und P. Jordan, ZS. f. Phys. **35**, 557, 1926.

** P. A. M. Dirac, Proc. Roy. Soc. London (A) **114**, 243, 719, 1927; P. Jordan, ZS. f. Phys. **44**, 473, 1927 (im folgenden als A bezeichnet); P. Jordan und O. Klein, ZS. f. Phys. **45**, 751, 1927; P. Jordan, ZS. f. Phys. **45**, 766, 1927 (im folgenden als B bezeichnet); P. Jordan und W. Pauli jr., ZS. f. Phys. (im Erscheinen).

deutung die Anzahl von Atomen (etwa innerhalb eines Kastens) im r-ten Quantenzustande mißt, in zwei Faktoren

$$N_r = b_r^\dagger b_r \tag{1}$$

der Form

$$\left.\begin{aligned} b_r &= e^{-\frac{2\pi i}{h}\Theta_r} N_r^{1/2}, \\ b_r^\dagger &= N_r^{1/2} e^{\frac{2\pi i}{h}\Theta_r} \end{aligned}\right\} \tag{2}$$

zerlegt, wobei man fordert, daß N_r, Θ_r kanonisch konjugiert seien. Legt man nun als Definition kanonisch konjugierter Größen diejenige zugrunde, die von einem der Verfasser kürzlich vorgeschlagen wurde*, so erhält man die Möglichkeit, in dieser Form nicht nur die Einsteinsche Statistik darzustellen, bei der die Eigenwerte N_r' von N_r durch

$$N_r' = 0, 1, 2, 3, \ldots \tag{3}$$

gegeben sind, sondern auch die Paulische, bei der nur

$$N_r' = 0, 1 \tag{4}$$

in Frage kommt. Man erhält dann ferner sofort neben (2) weitere Gleichungen, und zwar im Einsteinschen Falle

$$\left.\begin{aligned} b_r &= (1 + N_r)^{1/2} e^{-\frac{2\pi i}{h}\Theta_r}, \\ b_r^\dagger &= e^{\frac{2\pi i}{h}\Theta_r} (1 + N_r)^{1/2}; \end{aligned}\right\} \tag{5'}$$

aber statt dessen im Paulischen Falle:

$$\left.\begin{aligned} b_r &= (1 - N_r)^{1/2} e^{-\frac{2\pi i}{h}\Theta_r}, \\ b_r^\dagger &= e^{\frac{2\pi i}{h}\Theta_r} (1 - N_r)^{1/2}, \end{aligned}\right\} \tag{5''}$$

wie in A gezeigt wurde.

Diese Formeln stützen bereits sehr die Überzeugung, daß diese Darstellungsweise des Paulischen Äquivalenzverbotes dem Wesen der Sache entspricht und in ihrer weiteren Verfolgung zu richtigen Ergebnissen führen wird. Die Formeln (5'), (5'') stehen nämlich in enger Beziehung einerseits zu den Problemen der stoßartigen Wechselwirkungen von Korpuskeln, und andererseits zu den Dichteschwankungen quantenmechanischer Gase.

---

* P. Jordan, ZS. f. Phys. **44**, 1, 1927.

§ 2. Was zunächst die Wechselwirkungen betrifft, so mag es erlaubt sein, aus einer früheren Note folgendes zu wiederholen*: In einem abgeschlossenen Kasten mögen (endlich oder unendlich viele) Arten verschiedener Teilchen (materielle oder Lichtquanten) vorhanden sein. Die Dichte der l-ten Teilchenart pro Zelle im Phasenraum sei $n^{(l)}(E)$, wo E die zu den betrachteten Zellen gehörige Energie bedeutet. Die Gesamtzahlen $N^{(l)}$ [Integrale der $n^{(l)}(E)$ über den Phasenraum] seien beliebig vielen ($j = 1, 2, \ldots$) linearen Nebenbedingungen

$$\sum_l C_l^j N^{(l)} = C^j = \text{const} \tag{6}$$

unterworfen (Beispiele dazu a. a. O.), wo die C_l^j ganze positive oder negative Zahlen (oder Null) sind. Dann wird im statistischen Gleichgewicht

$$n^{(l)}(E) = \frac{1}{e^{\sum_j C_l^j a_j(T) + \frac{E}{kT}} \pm 1}, \tag{7}$$

wobei das positive oder negative Vorzeichen in ± 1 zu wählen ist, je nachdem, ob die l-te Teilchenart dem Pauliprinzip oder der Einsteinstatistik gehorcht.

Als Wechselwirkungsprozesse sind nun natürlich nur solche zuzulassen, bei denen die Forderungen (6) nicht verletzt werden. Eine bestimmte Form eines solchen Elementaraktes ist zu beschreiben durch Angabe der Indizes l und der Geschwindigkeiten der vor dem Elementarakt vorhandenen und der nach dem Prozeß vorhandenen mitwirkenden Teilchen. Es seien $n_1^+, n_2^+, \ldots, n_r^+$ die zugehörigen Dichten $n^{(l)}(E)$ für diejenigen vor dem Prozeß vorhandenen Teilchen, welche der Einsteinstatistik folgen; und $n_1^-, n_2^-, \ldots, n_s^-$ die $n^{(l)}(E)$ für die vor dem Prozeß vorhandenen mitwirkenden Teilchen Paulischer Art. Entsprechend sollen sich $m_1^+, m_2^+, \ldots, m_\varrho^+$; $m_1^-, m_2^-, \ldots, m_\sigma^-$ auf die nach dem Prozeß übriggebliebenen bzw. neu erzeugten Teilchen beziehen. Dann muß aus statistisch-thermodynamischen Gründen die Wahrscheinlichkeit des Elementarakts proportional mit

$$\begin{aligned} &n_1^+ n_2^+ \ldots n_r^+ n_1^- n_2^- \ldots n_s^- (1 + m_1^+)(1 + m_2^+) \ldots \\ &\ldots (1 + m_\varrho^+)(1 - m_1^-)(1 - m_2^-) \ldots (1 - m_\sigma^-) \end{aligned} \tag{8}$$

angenommen werden; die des inversen Elementarakts entsprechend proportional mit

$$m_1^+ \ldots m_\varrho^+ m_1^- \ldots m_\sigma^- (1 + n_1^+) \ldots (1 + n_r^+)(1 - n_1^-) \ldots (1 - n_s^-). \tag{9}$$

* P. Jordan, ZS. f. Phys. **41**, 711, 1927. Anmerkung nach Abschluß der Arbeit: Dieselben Formeln sind kürzlich von Bothe, ZS. f. Phys. **46**, 327, 1928, erneut erörtert worden.

Bezüglich der Faktoren $(1 - m_1^\dagger)$ usw. bei den Einsteinschen Teilchen ist nun von Dirac bei der Untersuchung der Absorption und Emission von Licht durch Atome gezeigt worden, daß ihre Gestalt unmittelbar folgt aus der Gestalt der entsprechenden Faktoren in den Formeln (5'). Entsprechend wird die Form der Glieder $(1 - m_1^-)$ usw. in (8), (9) auf (5'') zurückzuführen sein.

Was andererseits die Schwankungserscheinungen anbetrifft, so ist in A darauf hingewiesen, daß das Schwankungsquadrat der Teilchendichte in einem Volumen, welches mit einem großen Volumen bezüglich der zu einem engen Frequenzintervall $\Delta\nu$ gehörigen Wellen kommuniziert, nach den bekannten Einsteinschen Formeln einen Wert proportional zu

$$n_r(1 + n_r) \tag{10'}$$

besitzt. (Für klassische Wellen wäre es proportional n_r^2). Die von Pauli berechnete analoge Größe bei einem Fermischen Gase ist aber proportional

$$n_r(1 - n_r); \tag{10''}$$

und in (10'), (10'') zeigt sich der Unterschied vom Einsteinschen und Paulischen Gase wieder in derselben Form wie in (5'), (5'').

Von dem Einsteinschen Gas bzw. dem Boseschen Wellenfeld besitzt man auf Grund der Arbeiten, die auf S. 631 genannt wurden, bereits eine weitgehende Kenntnis. Wir beschäftigen uns im folgenden damit, in ähnlicher Weise die in A begründete Theorie des Pauligases zu vertiefen.

§ 3. Wir wiederholen hier der Deutlichkeit halber einige in A gebrachte Formeln. Die Größen b_r, $b_r^\dagger$, N_r, Θ_r sind darstellbar durch die Matrizen

$$\left.\begin{aligned} b_r &= \begin{pmatrix} 0 & 1 \\ 0 & 0 \end{pmatrix}_r; \quad b_r^\dagger = \begin{pmatrix} 0 & 0 \\ 1 & 0 \end{pmatrix}_r; \\ N_r &= \begin{pmatrix} 0 & 0 \\ 0 & 1 \end{pmatrix}_r; \quad \Theta_r = \frac{h}{4}\begin{pmatrix} 0 & 1 \\ 1 & 0 \end{pmatrix}_r. \end{aligned}\right\} \tag{11}$$

Hierbei ist jeweils

$$\begin{pmatrix} \alpha_{11} & \alpha_{12} \\ \alpha_{21} & \alpha_{22} \end{pmatrix}_r \tag{12}$$

eine Matrix, deren Zeilen und Spalten bezeichnet werden durch eine Reihe von Indizes, deren jeder den Wert 0 oder 1 haben kann; und zwar ist (12) eine Diagonalmatrix in bezug auf den ersten bis $(r - 1)$-ten Index und in bezug auf den $(r + 1)$-ten und die folgenden Indizes.

Neben den schon in § 1 besprochenen Gleichungen gelten die Formeln:

$$\left.\begin{aligned} b_r^\dagger b_r &= N_r; \quad b_r b_r^\dagger = 1 - N_r; \\ N_r^2 &= N_r; \quad (b_r^\dagger)^2 = (b_r)^2 = 0; \\ e^{\frac{2\pi i}{h}\theta_r} &= i\,\frac{4}{h}\,\Theta_r. \end{aligned}\right\} \tag{13}$$

Man kann alle diese Größen ausdrücken durch drei Größen $k_1^{(r)}$, $k_2^{(r)}$, $k_3^{(r)}$, die den Multiplikationsregeln der Quaternionen folgen:

$$\left.\begin{aligned} & k_1^{(r)} k_2^{(r)} = -k_2^{(r)} k_1^{(r)} = k_3^{(r)}, \\ & \cdots\cdots\cdots\cdots\cdots \\ & \cdots\cdots\cdots\cdots\cdots \\ & (k_1^{(r)})^2 = (k_2^{(r)})^2 = (k_3^{(r)})^2 = -1, \end{aligned}\right\} \tag{14}$$

wobei durch die Punkte die beiden aus der angeschriebenen Gleichung durch zyklische Permutation der 1, 2, 3 hervorgehenden Gleichungen angedeutet sein sollen.

Nämlich:

$$\left.\begin{aligned} b_r &= -\frac{i\,k_2^{(r)} - k_3^{(r)}}{2}, \quad b_r^\dagger = -\frac{i\,k_2^{(r)} + k_3^{(r)}}{2}; \\ N_r &= -\frac{i\,k_1^{(r)} - 1}{2}; \quad \Theta_r = -\frac{h}{4}\, i\, k_2^{(r)}. \end{aligned}\right\} \tag{15}$$

Die Quaternionen $k_1^{(r)}$, $k_2^{(r)}$, $k_3^{(r)}$ werden dabei selbst dargestellt durch die Matrizen

$$k_1^{(r)} = \begin{pmatrix} -i & 0 \\ 0 & i \end{pmatrix}_r; \quad k_2^{(r)} = \begin{pmatrix} 0 & i \\ i & 0 \end{pmatrix}_r; \quad k_3^{(r)} = \begin{pmatrix} 0 & 1 \\ -1 & 0 \end{pmatrix}_r. \tag{16}$$

§ 4. Ebenso wie in B für die Bosesche Statistik ausgeführt wurde kann man auch bei der Fermischen Statistik die Heisenberg-Diracschen Determinantenformeln für die Herleitung der antisymmetrischen Schrödingerschen Eigenfunktionen des Gesamtsystems aus denen eines Einzelatoms übertragen auf beliebige Wahrscheinlichkeitsamplituden. Eine solche Amplitude sei für ein Einzelatom gegeben durch

$$\Phi_{\alpha p}^{\beta q} = \Phi_{\alpha p}(\beta', q'). \tag{17}$$

Um die Vorzeichen der aus ihr zu bildenden Determinanten eindeutig zu machen, legen wir für die Eigenwerte β' von β in beliebiger, aber ein für allemal bestimmter Weise eine Reihenfolge fest. Wir bezeichnen die so definierte Anordnung für zwei spezielle Eigenwerte β', β'' von β durch $\beta' < \beta''$ bzw. $\beta'' < \beta'$, ohne jedoch mit dem Zeichen $<$ notwendig die Bedeutung „kleiner als" zu verbinden. Genau so verfahren wir mit den Eigenwerten q' von q und allgemein mit den Eigenwerten Q' jeder meßbaren Größe Q beim Einzelatom. Danach kann man jeder Amplitude (17) für das Einzelatom in eindeutiger Weise eine antisymmetrische Amplitude $\Psi_{\alpha p}^{\beta q}$ für ein System von N energetisch ungekoppelten Teilchen zuordnen. Wir schreiben

$$\Psi_{\alpha p}^{\beta q} = \Psi_{\alpha p}(\beta^{(1)}, \beta^{(2)}, \ldots, \beta^{(N)};\; q^{(1)}, q^{(2)}, \ldots, q^{(N)}), \tag{18}$$

worin

$$\left.\begin{array}{c} \beta^{(1)} < \beta^{(2)} < \cdots < \beta^{(N)}, \\ q^{(1)} < q^{(2)} < \cdots < q^{(N)} \end{array}\right\} \tag{19}$$

sein soll; und dann:

$$\Psi_{\alpha p}^{\beta q} = \frac{1}{N!} \sum_{(n)} \varepsilon_n \prod_{k=1}^{N} \Phi_{\alpha p}(\beta^{(k)}, q^{(n_k)}), \tag{20}$$

wo die Summe über alle $N!$ Permutationen $n_1, n_2, \ldots, n_N$ der Zahlen 1, 2, …, N zu erstrecken ist, während ε_n gleich $+1$ ist für gerade Permutationen und -1 für ungerade.

Nach (20) verschwindet $\Psi_{\alpha p}^{\beta q}$, sobald zwei der Größen $\beta^{(k)}$ einander gleich werden. Das heißt physikalisch: Es kommt nicht vor, daß irgend eine nichtentartete Größe β bei zwei verschiedenen Teilchen des Systems gleichzeitig denselben Wert annimmt. Wählen wir für β insbesondere das System der Quantenzahlen, so gibt dieser Satz das Paulische Äquivalenzverbot in seiner ursprünglichen Fassung. Wir wollen im folgenden das gleichzeitige Bestehen dieses Satzes für alle Größen β als den eigentlichen Inhalt des Paulischen Äquivalenzverbots betrachten.

Wir beschäftigen uns übrigens im folgenden vorwiegend mit dem Falle, daß jede Größe β am Einzelatom nur endlich viele, sagen wir K, Eigenwerte hat. Nur gelegentlich werden wir näher hinweisen auf den Grenzübergang $K \to \infty$, der im allgemeinen keinerlei Schwierigkeit macht. Wir wollen die K Eigenwerte jeder Größe β numerieren mit $\beta'_1, \beta'_2, \ldots, \beta'_K$, und zwar so, daß die oben vorausgesetzte Anordnung der Eigenwerte gerade die Form

$$\beta'_1 < \beta'_1 < \cdots < \beta'_K \tag{21}$$

gewinnt.

§ 5. Die in solcher Weise definierten antisymmetrischen Amplituden sind nun in eindeutiger Weise darstellbar als Funktion von Argumenten

$$N'(\beta');\quad N'(q') \tag{22}$$

mit folgender Bedeutung: $N'(\beta')$ ist die Anzahl von Atomen, bei denen β den Wert β' hat; ist also β' ein diskreter Eigenwert, so ist nach dem allgemeinen Paulischen Äquivalenzverbot

$$N'(\beta') = 0 \text{ oder } 1. \tag{23}$$

Liegt dagegen β' in einem kontinuierlichen Eigenwertgebiet, so haben wir zu schreiben:

$$N'(\beta') = \sum_{k=1}^{N'} \delta(\beta' - \beta'_k), \tag{24}$$

wenn insgesamt N' Teilchen vorhanden sind; das Integral von $N'(\beta')$ über ein Teilstück des Eigenwertgebietes ist dann die Anzahl der Atome, bei denen die Werte von β in dieses Teilstück fallen.

Wir begnügen uns aber nicht mit der rein mathematischen Einführung der neuen Größen $N'(\beta')$, $N'(q')$, sondern gehen zu einer neuen physikalischen Theorie über, indem wir annehmen, das Gesamtgas sei ein System, das durch ein kanonisches System von q-Zahlgrößen

$$N(\beta'); \quad \Theta(\beta') \tag{25}$$

beschrieben werden kann, wobei die $N'(\beta')$ gerade die Eigenwerte von $N(\beta')$ darstellen. Dann sind die $N(\beta')$, $\Theta(\beta')$ in der in § 3 erläuterten Weise durch Matrizen darzustellen; den verschiedenen Eigenwerten β' entsprechen die verschiedenen Werte der in § 3 gebrauchten Indizes r, s. Insbesondere gilt für diskretes β' die Gleichung

$$N(\beta') \,.\, [1 - N(\beta')] = 0, \tag{26}$$

für nicht diskretes β' kann man statt dessen schreiben:

$$N(\beta') \,.\, [\delta(\beta' - \beta'') - N(\beta'')] = \begin{cases} N(\beta')\, N(\beta'') & \text{für} \quad \beta' \neq \beta'', \\ 0 & \text{für} \quad \beta' = \beta'' \end{cases} \tag{27}$$

Während nun die q-Zahlen $N(\beta')$ durch ihre physikalische Bedeutung völlig definiert sind, ist dasselbe natürlich nicht der Fall für die $\Theta(\beta')$, wenn wir von ihnen nur verlangen, daß sie kanonisch konjugiert zu den $N(\beta')$ seien. Man muß diese Nichteindeutigkeit natürlich beseitigen bzw. beschränken, wenn man eindeutige Relationen zwischen den Größen $N(\beta')$, $\Theta(\beta')$ und $N(q')$, $\Theta(q')$ erhalten will. Wir werden nun im weiteren Verlauf unserer Betrachtungen sehen: Man kann, nachdem für jede Größe β, q usw. in der oben besprochenen Weise eine Reihenfolge der Eigenwerte β' und q' usw. festgelegt ist, ein gewisses System von konjugierten Phasen $\Theta(\beta')$, $\Theta(q')$ usw. zu den $N(\beta')$, $N(q')$ usw. bestimmen derart, daß einfache und eindeutige Relationen zwischen den verschiedenen kanonischen Systemen $N(\beta')$, $\Theta(\beta')$; $N(q')$, $\Theta(q')$ usw. entstehen. Man hat dabei aber noch verschiedene Möglichkeiten für die Definition der $\Theta(\beta')$ zu den $N(\beta')$, und diese verschiedenen Möglichkeiten können eindeutig zugeordnet werden den verschiedenen konjugierten Impulsen α zu β. Wir bezeichnen deshalb die q-Zahlgrößen, deren Theorie wir im folgenden entwickeln wollen, endgültig mit

$$N(\beta'); \quad \Theta_\alpha(\beta') \tag{28}$$

bzw.

$$N(q'); \quad \Theta_p(q') \text{ usf.} \tag{29}$$

Die gebildeten Verhältnisse besitzen offenbar die denkbar größte Analogie zu den in B erörterten Verhältnissen im Boseschen Falle, soweit man überhaupt angesichts der tiefgehenden Verschiedenheit beider Fälle eine Analogie erwarten kann.

§ 6. Die zwei K Größen

$$N(\beta'), \quad \Theta_\alpha(\beta')$$

müssen als q-Zahlen gewisse Funktionen der q-Zahlen

$$N(q'), \quad \Theta_p(q')$$

sein; dieser funktionale Zusammenhang soll jetzt besprochen werden. Im Bose-Einsteinschen Falle galt einfach

$$\left.\begin{aligned} b_\alpha(\beta') &= \sum_{q'} \Phi_{\alpha p}(\beta', q')\, b_p(q'), \\ b_\alpha^\dagger(\beta') &= \sum_{q'} b_p^\dagger(q')\, \Phi_{p\alpha}(q', \beta'); \end{aligned}\right\} \tag{30}$$

aber diese Formeln gelten nicht für das Paulische Gas. Statt dessen gelten Formeln

$$\left.\begin{aligned} a_\alpha(\beta') &= \sum_{q'} \Phi_{\alpha p}(\beta', q')\, a_p(q'), \\ a_\alpha^\dagger(\beta') &= \sum_{q'} a_p^\dagger(q')\, \Phi_{p\alpha}(q', \beta'), \end{aligned}\right\} \tag{30a}$$

wenn wir die Größen a, $a^\dagger$ durch

$$\left.\begin{aligned} a_p(q') &= v(q') \cdot b_p(q'), \\ a_p^\dagger(q') &= b_p^\dagger(q') \cdot v(q'); \end{aligned}\right\} \tag{31}$$

$$v(q') = \prod_{q'' \leqq q'} \{1 - 2\,N(q'')\} \tag{32}$$

definieren. Hier ist also $v(q')$ das Produkt der Größen $1 - 2\,N(q'')$ für $q'' = q'$ und alle vor q' kommenden q''. Es ist also $v(q')$ eine Diagonalmatrix, deren Diagonalelemente sämtlich gleich $+1$ oder -1 sind; und es wird

$$[v(q')]^2 = 1. \tag{33}$$

Der vollständige mathematische Beweis für die Richtigkeit dieser Formeln (30a) wird sich in den §§ 8 und 9 ergeben. Hier wollen wir lediglich die Multiplikationseigenschaften der Größen a, $a^\dagger$ untersuchen und die Invarianz dieser Multiplikationseigenschaften gegen Transformationen (30 a) nachweisen.

Zunächst wird

$$\left.\begin{aligned} b_p(q')\,.\,v(q'') &= \begin{cases} -v(q'')\,.\,b(q') & \text{für } q' \leqq q'', \\ v(q'')\,.\,b(q') & \text{für } q' > q''; \end{cases} \\ b_p^\dagger(q')\,.\,v(q'') &= \begin{cases} -v(q'')\,.\,b^\dagger(q') & \text{für } q' \leqq q'', \\ v(q'')\,.\,b^\dagger(q') & \text{für } q' > q''. \end{cases} \end{aligned}\right\} \tag{34}$$

Der Beweis ergibt sich leicht daraus, daß z. B.

$$\begin{aligned} b_p^\dagger(q')\,.\,\{1 - 2N(q')\} &= N(q')\, e^{\frac{2\pi i}{h}\,\Theta_p(q')}\{1 - 2N(q')\} \\ &= -\{1 - 2N(q')\}\, b_p^\dagger(q') \end{aligned} \tag{35}$$

wird. Dann wird weiter

$$\left.\begin{aligned} a_p(q')\,a_p(q'') &= -a_p(q'')\,a_p(q'), \\ a_p^\dagger(q')\,a_p^\dagger(q'') &= -a_p^\dagger(q'')\,a_p^\dagger(q'). \end{aligned}\right\} \tag{36}$$

Man beweist z. B.

$$\begin{aligned} a_p(q')\,a_p(q'') &= v(q')\,b_p(q')\,v(q'')\,b_p(q'') \\ &= \begin{cases} -v(q')\,v(q'')\,b_p(q')\,b_p(q'') & \text{für } q' \leqq q'', \\ v(q')\,v(q'')\,b_p(q')\,b_p(q'') & \text{für } q' > q''; \end{cases} \end{aligned} \tag{37}$$

nun ist aber insbesondere für $q' = q''$:

$$b_p(q')\,b_p(q'') = [b_p(q')]^2 = 0, \tag{38}$$

also auch

$$[a_p(q')]^2 = 0, \tag{39}$$

womit für $q' = q''$ die Formel (36) schon bewiesen ist. Man sieht danach aus (37), daß das Produkt (36) in der Tat antisymmetrisch in q', q'' ist.

Ferner gilt

$$a_p^\dagger(q')\,a_p(q'') + a_p(q'')\,a_p^\dagger(q') = \delta(q' - q''). \tag{40}$$

Denn es wird der links stehende Ausdruck gleich

$$\left.\begin{aligned} & b_p^\dagger(q')\,v(q')\,v(q'')\,b_p(q'') + v(q'')\,b_p(q'')\,b_p^\dagger(q')\,v(q') \\ = & \begin{cases} v(q')\,v(q'')\,b_p^\dagger(q')\,b_p(q'') - v(q'')\,v(q')\,b_p(q'')\,b_p^\dagger(q') & \text{für } q' < q''; \\ v(q')\,v(q'')\,b_p^\dagger(q')\,b_p(q'') + v(q'')\,v(q')\,b_p(q'')\,b_p^\dagger(q') & \text{für } q' = q''; \\ -v(q')\,v(q'')\,b_p^\dagger(q')\,b_p(q'') + v(q'')\,v(q')\,b_p(q'')\,b_p^\dagger(q') & \text{für } q' > q''; \end{cases} \end{aligned}\right\} \tag{41}$$

also verschwindet er in der Tat für $q' \neq q''$ und wird für $q' = q''$ gleich

$$[v(q')]^2\,.\,[b_p^\dagger(q')\,b_p(q') + b_p(q')\,b_p^\dagger(q')] = 1. \tag{42}$$

Nunmehr zeigen wir, daß die Gleichungen (36) und (40) wirklich invariant sind gegen Transformationen (30 a). Es ergibt sich

$$a_\alpha(\beta')\,a_\alpha(\beta'') + a_\alpha(\beta'')\,a_\alpha(\beta')$$

$$= \sum_{q'q''} \Phi_{\alpha p}(\beta', q')\,\Phi_{\alpha p}(\beta'', q'') \cdot \{a_p(q')\,a_p(q'') + a_p(q'')\,a_p(q')\} = 0; \quad (43)$$

$$a_\alpha^\dagger(\beta')\,a_\alpha(\beta'') + a_\alpha(\beta'')\,a_\alpha^\dagger(\beta')$$

$$= \sum_{q'q''} \Phi_{p\alpha}(q', \beta')\,\Phi_{\alpha p}(\beta'', q'') \cdot \{a_p^\dagger(q')\,a_p(q'') + a_p(q'')\,a_p^\dagger(q')\} = \delta(\beta' - \beta''). \quad (44)$$

Während also die Größen b, $b^\dagger$ im Paulischen Falle ebenso wie im Einsteinschen Falle die Eigenschaft haben, daß $b(\beta')$ mit $b(\beta'')$ und $b^\dagger(\beta')$ für $\beta'' \neq \beta'$ vertauschbar ist, kommt den a, $a^\dagger$ diese Eigenschaft nicht mehr zu. Trotzdem besitzen die a, $a^\dagger$ des Fermischen Gases in gewisser Hinsicht eine engere Analogie zu den b, $b^\dagger$ des Einsteingases, als die b, $b^\dagger$ des Pauligases selbst; man sieht das besonders deutlich durch die Gegenüberstellung:

Bose-Einstein	Pauli
$b_\alpha(\beta')\,b_\alpha(\beta'') - b_\alpha(\beta'')\,b_\alpha(\beta') = 0;$	$a_\alpha(\beta')\,a_\alpha(\beta'') + a_\alpha(\beta'')\,a_\alpha(\beta') = 0;$
$b_\alpha^\dagger(\beta')\,b_\alpha(\beta'') - b_\alpha(\beta'')\,b_\alpha^\dagger(\beta') = \delta(\beta' - \beta'');$	$a_\alpha^\dagger(\beta')\,a_\alpha(\beta'') + a_\alpha(\beta'')\,a_\alpha^\dagger(\beta') = \delta(\beta' - \beta'');$
$b_\alpha^\dagger(\beta')\,b_\alpha(\beta') = N(\beta');$	$a_\alpha^\dagger(\beta')\,a_\alpha(\beta') = N(\beta');$
$b_\alpha(\beta') = \sum_{q'} \Phi_{\alpha p}(\beta', q')\,b_p(q');$	$a_\alpha(\beta') = \sum_{q'} \Phi_{\alpha p}(\beta', q')\,a_p(q').$

Wir haben diese Gleichungen abgeleitet, indem wir das Paulische Äquivalenzverbot von vornherein zugrunde legten. Es zeigt sich aber, daß umgekehrt diese Multiplikationseigenschaften der a, $a^\dagger$ bereits die möglichen Eigenwerte der $N(\beta')$ bestimmen und die Vertauschbarkeit (gleichzeitige Beobachtbarkeit) von $N(\beta')$ und $N(\beta'')$ nach sich ziehen. Infolgedessen können wir sagen, daß die Existenz korpuskularer Teilchen und die Gültigkeit des Paulischen Prinzips als eine Folgerung aus den quantenmechanischen Multiplikationseigenschaften der de Broglieschen Wellenamplituden aufgefaßt werden dürfen, da in den beiden Gleichungen

$$N(\beta')\,N(\beta'') - N(\beta'')\,N(\beta') = 0, \quad (45)$$

$$N'(\beta') = 0 \text{ oder } 1 \quad (46)$$

diese Tatsachen vollständig ausgedrückt sind. Die Gleichung (45) folgt sofort. Der Beweis, daß auch (46) aus den Multiplikationsregeln der $a(\beta')$, $a^\dagger(\beta')$ folgt, ergibt sich folgendermaßen:

Auf Grund von

$$a_\alpha^\dagger(\beta')\,a_\alpha(\beta') + a_\alpha(\beta')\,a_\alpha^\dagger(\beta') = 1 \tag{47}$$

gilt wegen

$$[a_\alpha(\beta')]^2 = 0$$

die Gleichung

$$a_\alpha^\dagger(\beta')\,a_\alpha(\beta')\,a_\alpha^\dagger(\beta')\,a_\alpha(\beta')$$

$$= a_\alpha^\dagger(\beta')\,.\,[1 - a_\alpha^\dagger(\beta')\,a_\alpha(\beta')]\,.\,a_\alpha(\beta') = a_\alpha^\dagger(\beta')\,a_\alpha(\beta'); \tag{48}$$

also wird

$$N(\beta')\,.\,[1 - N(\beta')] = a_\alpha^\dagger(\beta')\,a_\alpha(\beta')\,.\,[1 - a_\alpha^\dagger(\beta')\,a_\alpha(\beta')] = 0. \tag{49}$$

Es sei noch betont: Da $v(\beta')$ aus den $N(\beta'')$ allein zu bilden ist (nachdem eine Reihenfolge der Eigenwerte festgelegt wurde), so kann man vermittelst

$$\left.\begin{aligned} b_\alpha(\beta') &= v(\beta')\,a_\alpha(\beta'),\\ b_\alpha^\dagger(\beta') &= a_\alpha^\dagger(\beta')\,v(\beta') \end{aligned}\right\} \tag{50}$$

die b, $b^\dagger$ eindeutig durch die a, $a^\dagger$ definieren. Man kann also in der Tat die a, $a^\dagger$ als die ursprünglichen Größen der Theorie und alle anderen Größen als Funktionen der a, $a^\dagger$ betrachten.

Endlich sei hervorgehoben, daß die Gesamtzahl N der vorhandenen Teilchen gegenüber den betrachteten Transformationen invariant bleibt:

$$N = \sum_{\beta'} N(\beta') = \sum_{q'} N(q'). \tag{51}$$

Diese Invarianz ist offenbar nur ein anderer Ausdruck dafür, daß (30a) eine unitäre Transformation ist.

Zusatz bei der Korrektur. Es zeigt sich durch eine genauere Betrachtung, die im Anhang mitgeteilt wird, daß die Multiplikationsregeln der a, $a^\dagger$ nicht nur die Eigenwerte der $N(\beta')$ schon bestimmen, sondern überhaupt die Matrizen a, $a^\dagger$ bis auf eine kanonische Transformation der Matrizendarstellung festlegen.

§ 7. Für ein eindimensionales Kontinuum mit der Schwingungsgleichung

$$\frac{\partial^2\psi}{\partial x^2} = \frac{\partial^2\psi}{\partial t^2}; \qquad \psi = \psi(x, t) \tag{52}$$

und der Randbedingung

$$\psi(0, t) = \psi(l, t) = 0 \tag{53}$$

war in A versuchsweise die räumliche Teilchendichte der Wellenkorpuskeln definiert durch

$$N(x) = \psi^\dagger \psi, \tag{54}$$

$$\psi = \sum_{r=1}^{\infty} b_r \sin r \frac{\pi}{l} x, \tag{55}$$

wo $N_r = b_r^\dagger b_r$ die Anzahl von Teilchen im r-ten Quantenzustand der Translation bedeutet.

Wir haben jetzt (55) zu korrigieren, indem wir die b_r durch entsprechende a_r ersetzen:

$$\psi = \sum_{r=1}^{\infty} a_r \sin r \frac{\pi}{l} x \cdot \tag{56}$$

Die in A durchgeführte Berechnung der Dichteschwankungen kann aber sofort von (55) auf (56) übertragen werden und zeigt dann, daß (56), wie es sein muß, wirklich die richtige Formel liefert. Es wurde nämlich aus (55) erhalten, daß der fragliche quadratische Mittelwert $\overline{\Delta^2}$ proportional sei mit

$$\overline{b_r^\dagger b_r} \, . \, \overline{b_r b_r^\dagger} = \overline{N_r} \, . \, \overline{(1 - N_r)}, \tag{57}$$

wo die Querstriche die Mittelung über ein infinitesimales Frequenzgebiet im Anschluß an die Frequenz ν_r bedeuten, so daß sich die in § 2 erwähnte Formel

$$\overline{\Delta^2} = \text{const.}\, n_r (1 - n_r) \tag{58}$$

ergibt. Rechnet man nun entsprechend mit (56), so wird $\overline{\Delta^2}$ proportional mit

$$\overline{a_r^\dagger a_r} \, . \, \overline{a_r a_r^\dagger} = \overline{N_r} \, . \, \overline{(1 - N_r)}, \tag{59}$$

d. h. das Endergebnis bleibt ungeändert.

§ 8. Wir wollen nun den in § 6 angekündigten vollständigen Beweis für die Äquivalenz der Formeln (30 a) mit den Formeln der gewöhnlichen Darstellung im mehrdimensionalen Koordinatenraum antreten. Wir müssen uns in diesem Koordinatenraum auf solche Größen (Operatoren) beschränken, die symmetrisch in den gleichen Teilchen sind; außerdem aber beschränken wir uns in diesem Paragraphen auf Größen, die aus einer Summe bestehen, wo in jedem Summand nur ein Elektron vorkommt. Von dieser Art ist die Energie eines idealen Gases. Diese Operatoren haben also die Gestalt

$$V = V_1 + V_2 + \cdots + V_N, \tag{60}$$

wo die V_i immer dieselbe Größe darstellen, nur an verschiedenen (am 1-, 2-, …, N-ten) Teilchen gemessen.

Unsere Wellenfunktion dagegen wird [§ 5, Gleichung (22), (23)] von den $N'(\beta'_k)$ abhängen. Dies erscheint in der Tat vom Standpunkt der Quantenmechanik als der naturgemäße Ansatz, da ja ein „maximaler Versuch" — wegen der Gleichheit der Teilchen — immer nur bestimmen kann, wie viele Teilchen im Zustand $\beta_1, \beta_2, \ldots, \beta_K$ sind, während die Frage, in welchem Zustand ein bestimmtes Elektron ist, nicht entschieden werden kann. Im Sinne des Paulischen Äquivalenzverbotes haben wir in (23) den Wertbereich der $N'(\beta'_k)$ auf 0,1 beschränkt.

Wir nehmen nun noch, wie in § 4 bereits betont, der Einfachheit halber an (was in Wahrheit niemals erfüllt ist), daß ein Elektron nur endlich viele (sagen wir K) Zustände annehmen kann, die wir also mit $\beta_1, \beta_2, \ldots, \beta_K$ bezeichnen. Dann hat die im folgenden einzuführende Wellenfunktion $\Psi(N'(\beta'_1), N'(\beta'_2), \ldots, N'(\beta'_K))$ gerade K Argumente und ist für 2^K Wertsysteme der Argumente definiert. Der Grenzübergang $K \to \infty$ scheint keine wesentlichen Schwierigkeiten zu bieten.

Die nun folgenden Betrachtungen lassen sich am einfachsten ausführen, wenn man den Zustand eines einzelnen Elektrons im mehrdimensionalen Koordinatenraum mit einer Wellenfunktion beschreibt, deren Argument β' ist. Das bedeutet, daß beim Einzelelektron die Messung eben die Bestimmung der Größe β ist, deren Wertbereich also K Zahlen $\beta'_1, \beta'_2, \ldots, \beta'_K$ umfaßt.

Haben wir nun im mehrdimensionalen Koordinatenraum eine antisymmetrische Wellenfunktion von N' Elektronen

$$\psi(\beta'^1, \beta'^2, \ldots, \beta'^{N'}), \tag{61}$$

so bestimmen wir, daß wir diesen Zustand fortan in unserem neuen N-Raume durch die Wellenfunktion $\Psi(N'(\beta'_1), N'(\beta'_2), \ldots, N'(\beta'_K))$ beschreiben wollen. Dabei sei

$$\Psi(N'(\beta'_1), \ldots, N'(\beta'_K)) = \frac{1}{\sqrt{N'!}} \psi(\beta'^1, \ldots, \beta'^{N'}). \tag{62}$$

Diese Gleichung ist so zu verstehen, daß Ψ überall 0 ist, wo nicht genau N von den $K (K > N)$ Zahlen $N'(\beta'_1), \ldots, N'(\beta'_K)$ gleich 1, die übrigen gleich 0 sind. Um den Wert an diesen Stellen zu bestimmen, setzt man rechts für die $\beta'^1, \beta'^2, \ldots, \beta'^{N'}$ jene Werte ein, für die eben $N'(\beta'_i) = 1$ ist, und zwar für β'^1 den (im Sinne der in § 4 getroffenen Anordnung) ersten, für β'^2 den folgenden, ..., für $\beta'^{N'}$ den letzten.

Wir wollen an dieser Zuordnung einer Funktion

$$\Psi(N'(\beta'_1), N'(\beta'_2), \ldots, N'(\beta'_K))$$

im neuen N-Raume zu einer Funktion

$$\psi(\beta'^1, \beta'^2, \ldots, \beta'^{N'})$$

im mehrdimensionalen Koordinatenraume im folgenden immer (also auch, wenn ψ keine Wellenfunktion ist) festhalten. Es ist zu beachten, daß es für das Vorzeichen von Ψ wichtig ist, in ψ die $\beta'_1, \beta'_2, \ldots, \beta'_K$ in der einmal festgesetzten Reihenfolge

$$(\beta'_1 < \beta'_2 < \cdots < \beta'_K)$$

einzusetzen, da nur dadurch das Vorzeichen von Ψ eindeutig geregelt wird; und nur hierdurch ist die Zuordnung einer eindeutigen Funktion Ψ zu der Funktion ψ möglich.

Umgekehrt wird aber auch ψ durch Ψ eindeutig bestimmt*: an den Stellen, für welche $\beta'^1 < \cdots < \beta'^{N'}$ gilt, durch (62), überall sonst durch die Forderung der Antisymmetrie.

Die einzelnen Teile (z. B. V_1) des Operators V [Gleichung (60)] im mehrdimensionalen Koordinatenraum sind im wesentlichen hermiteische Matrizen von K Zeilen und K Spalten. In der ν-ten Zeile und μ-ten Spalte stehe $H_{\nu\mu}$. Dann ist der ganze Operator V mit der Matrix

$$H_{\nu_1 \nu_2 \ldots \nu_{N'};\, \mu_1 \mu_2 \ldots \mu_{N'}} = H_{\nu_1 \mu_1} \delta_{\nu_2 \mu_2} \delta_{\nu_3 \mu_3} \ldots \delta_{\nu_{N'} \mu_{N'}} + \delta_{\nu_1 \mu_1} H_{\nu_2 \mu_2} \delta_{\nu_3 \mu_3} \ldots \delta_{\nu_{N'} \mu_{N'}} + \cdots + \delta_{\nu_1 \mu_1} \delta_{\nu_2 \mu_2} \delta_{\nu_3 \mu_3} \ldots H_{\nu_{N'} \mu_{N'}} \tag{63}$$

identisch. Wir schreiben zur Abkürzung

$$V\psi(\beta'^1 \ldots \beta'^{N'}) = \overline{\psi}(\beta'^1 \ldots \beta'^{N'}), \tag{64}$$

dann ist also

$$\overline{\psi}(\beta'_{\nu_1} \ldots \beta'_{\nu_{N'}}) = \sum_{\mu_1, \mu_2, \ldots, \mu_{N'} = 1}^{K} H_{\nu_1 \ldots \nu_{N'};\, \mu_1 \ldots \mu_{N'}}\, \psi(\beta'_{\mu_1} \ldots \beta'_{\mu_{N'}}). \tag{65}$$

Aus diesem $\overline{\psi}(\beta'^1, \beta'^2, \ldots, \beta'^{N'})$ bilden wir — genau wie in (62) — ein $\overline{\Psi}(N'(\beta'_1), \ldots, N'(\beta'_K))$ durch

$$\overline{\Psi}(N'(\beta'_1), \ldots, N'(\beta'_K)) = \overline{\psi}(\beta'^1, \ldots, \beta'^{N'}) \frac{1}{\sqrt{N'!}}. \tag{62a}$$

Wir behaupten nun, daß

$$\overline{\Psi}(N'(\beta'_1), \ldots, N'(\beta'_K)) = \Omega\, \Psi(N'(\beta'_1), \ldots, N'(\beta'_K)), \tag{66}$$

wo der Operator Ω

$$\Omega = \sum_{\varkappa, \lambda = 1}^{K} H_{\varkappa\lambda} a_\varkappa^\dagger a_\lambda \tag{66a}$$

ist, mit den a aus (31).

---

* Wir setzen voraus, daß Ψ wieder überall verschwindet, wo nicht

$$N'(\beta'_1) + \cdots + N'(\beta'_K) = N'$$

ist.

Gleichung (65) ist zunächst sicher richtig an allen Stellen (für alle Wertsysteme der Argumente), wo die Anzahl der 1 im Argumentsystem von $\overline{\Psi}$ nicht eben gleich N' ist. Dann verschwindet nämlich die linke Seite wegen (62), (62a), und auch auf der rechten Seite stehen lauter Nullen, da $a_\varkappa^\dagger a_\lambda$ die Anzahl der 1 nicht ändert.

An den Stellen aber, wo etwa die

$$N'(\beta'_{i_1}),\ N'(\beta'_{i_2}),\ \ldots,\ N'(\beta'_{i_{N'}})$$

(also genau N') gleich 1 sind, ist $\sqrt{N'!}\ \overline{\Psi}$ gleich

$$\overline{\psi}(\beta'_{i_1}, \ldots, \beta'_{i_{N'}}),$$

also gleich

$$\sum_{\mu_1 \ldots \mu_{N'}=1}^{K} H_{i_1 \ldots i_{N'};\, \mu_1 \ldots \mu_{N'}}\, \psi(\beta'_{\mu_1}, \ldots, \beta'_{\mu_{N'}})$$

$$= \sum_{\mu_1=1}^{K} H_{i_1 \mu_1}\, \psi(\beta'_{\mu_1}, \beta'_{i_2}, \ldots, \beta'_{i_{N'}}) + \sum_{\mu_2=1}^{K} H_{i_2 \mu_2}\, \psi(\beta'_{i_1}, \beta'_{\mu_2}, \ldots, \beta'_{i_{N'}}) +$$

$$\cdots + \sum_{\mu_{N'}=1}^{K} H_{i_{N'} \mu_{N'}}\, \psi(\beta'_{i_1}, \beta'_{i_2}, \ldots, \beta'_{\mu_{N'}})$$

$$= \sum_{r=1}^{N'} \sum_{\mu=1}^{K} H_{i_r \mu}\, \psi(\beta'_{i_1}, \beta'_{i_2}, \ldots, \beta'_{i_{r-1}}, \beta'_{\mu}, \beta'_{i_{r+1}}, \ldots, \beta'_{i_{N'}}), \quad (67)$$

das zweite ersieht man aus (63). Unsere Absicht ist nun, die rechte Seite von (67) durch die Ψ auszudrücken. Zu diesem Zwecke bemerken wir, daß in (67) rechts die β'_i schon in der richtigen Reihenfolge stehen, nur das jeweils auftretende β'_μ ist an der falschen Stelle. Ist etwa*

$$\beta'_{i_{s-1}} < \beta'_\mu \leqq \beta'_{i_s},$$

so können wir die Reihenfolge der β' zur richtigen machen, indem wir β'_μ über die zwischenliegenden β'_i von der Stelle zwischen $\beta'_{i_{r-1}}$ und $\beta'_{i_{r+1}}$ an die Stelle zwischen $\beta'_{i_{s-1}}$ und β'_{i_s} verschieben. Dabei multipliziert sich die antisymmetrische Funktion ψ mit $(-1)^z$, wo z die Anzahl der zwischen den beiden angegebenen Stellen stehenden β_i ist.

Wir können (67) also auch schreiben

$$\overline{\psi}(\beta'_{i_1}, \ldots, \beta'_{i_{N'}}) = \sum_{r=1}^{N} \sum_{\mu=1}^{K} \pm H_{i_r \mu}\, \psi(\beta'_{i_1}, \ldots, \beta'_{\mu}, \ldots, \beta'_{i_{N'}}), \quad (67\text{a})$$

* Das Gleichheitszeichen kommt nicht in Frage, da dann das entsprechende ψ doch verschwindet.

wo natürlich noch das Vorzeichen $\pm$ von r und μ abhängt, aber schon

$$\beta'_{i_1} < \cdots < \beta'_\mu < \cdots < \beta'_{N'}$$

gilt. Wenn wir (62) beachten, können wir dies auch*

$$\overline{\Psi}(x_1 \ldots x_K) = \sum_{x_j=1} \sum_{x_l=0} \pm H_{jl} \Psi(x_1, \ldots, x_{j-1}, 0, x_{j+1}, \ldots, x_{l-1}, 1, x_l, \ldots, x_K)$$
$$+ \sum_{x_j=1} H_{jj} \Psi(x_1, \ldots, x_K) \qquad (68)$$

schreiben, wie man leicht überlegt. In den Summanden nämlich, in (67a), wo $i_r \neq \mu$ ist [erstes Glied in (68)], sind dieselben Argumente vorhanden wie an der linken Seite, es fehlt nur β'_{i_r}, was links vorhanden war ($x_j = 1$), dagegen ist β'_μ hinzugekommen, und wir können annehmen, daß es nicht da war ($x_l = 0$), da sonst ψ doch verschwinden würde. Ist $i_r = \mu$ [zweites Glied in (68)], so sind rechts in ψ dieselben Argumente wie links.

Das Vorzeichen in (68) bestimmt sich offenbar dadurch, daß man + oder — setzt, je nachdem zwischen der ausgeschriebenen 0 und 1 eine gerade oder ungerade Anzahl von 1 steht. (Über ebenso viele β'_i mußte man das entsprechende β'_μ hinüberschieben.) Dies ist aber die Anzahl der 1, die links von x_l stehen, vermindert um die Anzahl der 1, die links von x_j stehen.

Obwohl jetzt die Richtigkeit der vorangehenden Formeln schon klar ist, wollen wir diese Gedanken doch zu Ende führen. Wir wissen, daß der Wertbereich der Argumente von Ψ ingesamt 2^K Stellen umfaßt, indem für jedes der $x_k = N'(\beta'_k)$ entweder $+1$ oder 0 gesetzt werden kann. Ein darauf wirkender linearer Operator ist also eine Matrix mit 2^K Zeilen und ebensoviel Spalten. Wir bezeichnen jede Zeile oder Spalte mit K Indizes (den $x_1, \ldots, x_K$ entsprechend), die jeweils 1 oder 0 sein können.

Der Operator a_λ ist entsprechend § 3 und § 6 zu definieren (wir schreiben der besseren Übersicht halber die Indizes als Argumente) durch

$$a_\lambda(x_1, x_2, \ldots, x_K;\ y_1, y_2, \ldots, y_K)$$
$$= (-1)^{x_1 + x_2 + \cdots + x_{\lambda-1}} \delta_{x_1 y_1} \delta_{x_2 y_2} \cdots \delta_{x_{\lambda-1} y_{\lambda-1}} \delta_{x_\lambda 0} \delta_{y_\lambda 1} \delta_{x_{\lambda+1} y_{\lambda+1}} \cdots \delta_{x_K y_K}, \qquad (69)$$

und entsprechend ist dann $a^\dagger_\varkappa$

$$a^\dagger_\varkappa(x_1, \ldots, x_K;\ y_1, \ldots, y_K)$$
$$= (-1)^{x_1 + \cdots + x_{\varkappa-1}} \delta_{x_1 y_2} \cdots \delta_{x_{\varkappa-1} y_{\varkappa-1}} \delta_{x_\varkappa 1} \delta_{y_\varkappa 0} \cdots \delta_{x_K y_K}. \qquad (69\,a)$$

* Wir setzen aus Bequemlichkeitsgründen x_k für $N'(\beta'_k)$.

Mit Hilfe dieser Formeln kann (68) auch so geschrieben werden:

$$\overline{\Psi}(x_1, \ldots, x_K)$$

$$\sum_{\lambda, \varkappa=1}^{K} H_{\varkappa\lambda} \sum_{\substack{y_1 \ldots y_K = 0,1 \\ z_1 \ldots z_K = 0,1}} a_\varkappa^\dagger (x_1 \ldots x_K; y_1 \ldots y_K) \cdot a_\lambda (y_1 \ldots y_K; z_1 \ldots z_K) \Psi(z_1 \ldots z_K). \quad (70)$$

Wie man sich mit einiger Mühe überlegen kann, was aber schwer hingeschrieben werden kann. Damit ist (66) gewonnen.

§ 9. Wir haben also in § 8 folgendes gesehen: Jeder antisymmetrischen Funktion, welche definiert ist in den Koordinatenräumen mit allen Anzahlen $N' < K$ von Dimensionen, entspricht durch (62) eine Funktion im neuen Raume. Es entspricht dann dem Operator (60) $V = V_1 + V_2 + \cdots + V_N$ [mit der Matrix in (64) $H_{r_1 \ldots r_{N'}; \mu_1 \ldots \mu_{N'}}$] im Koordinatenraum der Operator Ω von (68) im neuen N'-Raum. Der Operator Ω ist dabei

$$\Omega = \sum_{\varkappa, \lambda=1}^{K} H_{\varkappa\lambda} a_\varkappa^\dagger a_\lambda \quad (66\text{a})$$

mit den $a_\varkappa^\dagger$, a_λ von (69), (69a).

Es folgt hieraus, daß einer Eigenfunktion von V eine Eigenfunktion von Ω entspricht. Wenn wir noch zeigen können, daß das innere Produkt zweier Funktionen im Koordinatenraum denselben Wert hat wie dasjenige der entsprechenden Funktionen im neuen N'-Raum, so sind wir mit dem Beweis fertig. Im Koordinatenraum ist

$$(\psi\,\varphi) = \sum_{\beta'^1, \ldots, \beta'^{N'} = \beta'_1}^{\beta'_K} \psi(\beta'^1 \ldots \beta'^{N'})\, \tilde{\varphi}(\beta'^1 \ldots \beta'^{N'}), \quad (71)$$

was wegen der Antisymmetrie

$$(\psi\,\varphi) = N'! \sum_{\substack{\beta'^1, \ldots, \beta'^{N'} = \beta'_1 \\ \beta'^1 < \; < \beta'^{N'}}}^{\beta'_K} \psi(\beta'^1 \ldots \beta'^{N'})\, \tilde{\varphi}(\beta'^1 \ldots \beta'^{N'}) \quad (72)$$

ergibt. Andererseits gilt im neuen Raume

$$(\Psi\,\Phi) = \sum_{x_1 \ldots x_K = 0,1} \Psi(x_1 \ldots x_K)\, \tilde{\Phi}(x_1 \ldots x_K), \quad (73)$$

was mit Rücksicht auf (62) eben (72) ist.

Wir möchten noch bemerken, daß die q-Zahlrelationen (36), (40) natürlich sofort aus der Formel (69) hervorgehen.

§ 10. Wir müssen schließlich Operatoren betrachten, die keine Zerlegung mehr in die Gestalt (60) gestatten. Von dieser Form ist die

Energie eines nichtidealen Gases. Wir beschränken uns dabei zunächst auf solche, die in Teile zerlegbar sind, die jeweils nur zwei, stets verschiedene, Teilchen enthalten. Die Erledigung dieser Aufgabe ergibt sich durch Analogisierung der entsprechenden Formeln für die Bose-Einsteinsche Statistik *. Der Operator V läßt sich dann schreiben:

$$V = \sum_{\substack{j,\,k=1 \\ j<k}}^{N'} V_{jk}, \tag{60b}$$

wobei V die Matrix

$$H_{\nu_1 \ldots \nu_{N'};\, \mu_1 \ldots \mu_{N'}}$$

$$= \sum_{\substack{j,\,k=1 \\ j<k}}^{N'} H_{\nu_j \nu_k;\, \mu_j \mu_k} \delta_{\nu_1 \mu_1} \cdots \delta_{\nu_{j-1} \mu_{j-1}} \delta_{\nu_{j+1} \mu_{j+1}} \cdots \delta_{\nu_{k-1} \mu_{k-1}} \delta_{\nu_{k+1} \mu_{k+1}} \cdots \delta_{\nu_{N'} \mu_{N'}} \tag{63b}$$

entspricht. Es ist dann

$$V\psi(\beta'^1, \ldots, \beta'^{N'}) = \overline{\psi}(\beta'^1, \ldots, \beta'^{N'}) \tag{64b}$$

mit Hilfe von (63b) [ebenso wie (65)] zu berechnen. Dann ist wieder die „richtige Reihenfolge" der β auf der rechten Seite herzustellen. Dabei können die β'_i stehengelassen bleiben, β'_μ und $\beta'_{\mu'}$ müssen über eine Anzahl von β'_i hinübergeschoben werden, wobei sich wieder das Vorzeichen ändern kann.

Beachten wir wieder (62), so können wir wieder links und rechts für $\overline{\psi}$ bzw. ψ einsetzen $\overline{\Psi}$ bzw. Ψ. Unter Beachtung von (69) findet man nun, daß dem Operator (60b) nunmehr der Operator

$$\frac{1}{2!} \sum_{\substack{\lambda_1,\,\lambda_2=1 \\ \varkappa_1,\,\varkappa_2=1}}^{K} H_{\varkappa_1 \varkappa_2;\, \lambda_1 \lambda_2}\, a^\dagger_{\varkappa_1} a^\dagger_{\varkappa_2} a_{\lambda_2} a_{\lambda_1} \tag{66b}$$

entspricht.

Es ist befriedigend, daß die Rückwirkung der Teilchen auf sich selbst wieder durch die nichtkommutativen Multiplikationseigenschaften der Wellenamplituden im dreidimensionalen Raume automatisch ausgeschlossen wird. Im Bose-Einsteinschen Falle wurde dieser Umstand deutlich gemacht durch Formel (40) der Arbeit von Jordan und Klein. Daß dieselbe Formel auch hier gilt, folgt aus der leicht beweisbaren Formel

$$a^\dagger_k a_k a^\dagger_l a_l - a^\dagger_k a^\dagger_l a_l a_k = \delta_{kl} a^\dagger_k a_k.$$

Wir wenden uns endlich zu dem Fall von Operatoren, die aus Summanden bestehen, welche jeweils in $n > 2$ Teilchen symmetrisch sind,

* P. Jordan und O. Klein, a. a. O.

während in der Summe alle diejenigen Summanden auszulassen sind, welche dasselbe Teilchen zweimal enthalten (was eine Wechselwirkung des Teilchens mit sich selbst bedeuten würde). Eine Verallgemeinerung der obigen Betrachtungen führt dann zu den folgenden Formeln, welche analog sind der in B angegebenen Verallgemeinerung* der Formel von Jordan und Klein:

$$\frac{1}{n!}\sum_{\lambda_1\ldots\lambda_n=1}^{K}\;\sum_{\varkappa_1\ldots\varkappa_2=1}^{K} H_{\varkappa_1\ldots\varkappa_n;\,\lambda_1\ldots\lambda_n}\, a_{\varkappa_1}^{\dagger}\ldots a_{\varkappa_n}^{\dagger}\, a_{\lambda_n}\ldots a_{\lambda_1}. \qquad (66\,\mathrm{c})$$

§ 11. Wir können endlich die erhaltenen Ergebnisse in einer etwas anderen Form so aussprechen: Es gibt bei einem Paulischen Mehrkörperproblem eine Wahrscheinlichkeitsamplitude, welche die Wahrscheinlichkeit dafür bestimmt, daß, nachdem für die meßbaren Größen $N(\beta_1')$, $N(\beta_2')$, …, $N(\beta_K')$ ein gewisses Wertsystem $N'(\beta_1')$, $N'(\beta_2')$, …, $N'(\beta_K')$ gemessen worden ist, für die anderen, entsprechend definierten Größen $N(q_1')$, $N(q_2')$, …, $N(q_K')$ die Werte $N'(q_1')$, $N'(q_2')$, …, $N'(q_K')$ gefunden werden.

Eine solche Amplitude wird erhalten durch

$$\Phi^{N(\beta'),\,N(q')}_{-\Theta_\alpha(\beta'),\,-\Theta_p(\gamma')} = 0 \quad \text{für} \quad \sum_{\beta'} N'(\beta') \neq \sum_{\gamma'} N'(\gamma')$$

und

$$\Phi^{N(\beta'),\,N(q')}_{-\Theta(\beta'),\,-\Theta_p(\gamma')} = N'!\,\Psi^{\beta q}_{\alpha p} \quad \text{für} \quad \sum_{\beta'} N'(\beta') = \sum_{\gamma'} N'(q'), \qquad (74)$$

wobei $\Phi^{\beta q}_{\alpha p}$, $\Psi^{\beta q}_{\alpha p}$ die in § 4 besprochenen Funktionen sind.

Die in A in der Form

$$\Big\{\sum_{rs} H_{rs}\, b_r^{\dagger}\, b_s - W\Big\}\,\Phi = 0$$

angegebene Funktionalgleichung für die zum Gesamtsystem gehörige Amplitude Φ lautet nunmehr bei unserer Vorzeichenbestimmung der in Φ enthaltenen Determinanten:

$$\Big\{\sum H_{rs}\, a_r^{\dagger}\, a_s - W\Big\}\,\Phi = 0;$$

diese Abänderung ist nötig, weil in A die Vorzeichenzweideutigkeit dieser Determinanten nicht ausreichend berücksichtigt wurde; die Matrizen a_r unterscheiden sich, wie wir wissen, von den b_r nur bezüglich der Vorzeichen ihrer verschiedenen Elemente. Es scheint sehr befriedigend, daß die für die Bildung des Energieausdrucks notwendige Einführung der

* Vgl. B, Formel (34).

Größen a, $a^\dagger$ gleichzeitig auch zu einfachen Multiplikationsgesetzen führte, wie wir in § 6 gesehen haben.

Es sei endlich hervorgehoben, daß die in § 6 erörterten Multiplikationsgesetze der gequantelten Amplituden $a_p(q')$ in Analogie zu den von Jordan und Pauli entwickelten relativistisch invarianten Multiplikationsregeln des ladungsfreien elektromagnetischen Feldes leicht relativistisch verallgemeinert werden können, so daß man die dem Paulischen Äquivalenzverbot entsprechende Quantelung der de Broglieschen Wellen in relativistisch invarianter Form erhält. Eine genauere Darlegung soll jedoch vorläufig zurückgestellt werden.

Zusatz bei der Korrektur: Zwischen den $2K$ Operatoren a_1, $a_2, \ldots, a_K$; $a_1^\dagger, a_2^\dagger, \ldots, a_K^\dagger$ bestehen die Relationen

$$\left.\begin{aligned} a_\varkappa a_\lambda + a_\lambda a_\varkappa &= 0 \\ a_\varkappa^\dagger a_\lambda^\dagger + a_\lambda^\dagger a_\varkappa^\dagger &= 0 \end{aligned}\right\} \tag{36}$$

und

$$a_\varkappa^\dagger a_\lambda + a_\lambda a_\varkappa^\dagger = \delta_{\varkappa\lambda}. \tag{40}$$

Wir wollen nun zeigen, daß diese ψ-Zahlrelationen die Operatoren a, $a^\dagger$ schon eindeutig bestimmen, wenn man sich auf irreduzible Matrizensysteme beschränkt und Matrizensysteme, die auseinander durch Ähnlichkeitstransformation hervorgehen, als nicht verschieden voneinander ansieht*.

Um dies einzusehen, bilden wir zunächst folgende Größen

$$\left.\begin{aligned} \alpha_\varkappa &= a_\varkappa + a_\varkappa^\dagger, \\ \alpha_{K+\varkappa} &= \frac{1}{i}(a_\varkappa - a_\varkappa^\dagger). \end{aligned}\right\} \tag{I}$$

Die $2K$ Matrizen α bestimmen umgekehrt die a eindeutig. Nun gilt für die $\alpha_\varkappa$ allgemein

$$\alpha_\varkappa \alpha_\lambda + \alpha_\lambda \alpha_\varkappa = 2\,\delta_{\varkappa\lambda}. \tag{II}$$

Man überzeugt sich z. B., wenn $\varkappa < K$, $\lambda < K$ ist, daß

$$\alpha_\varkappa \alpha_\lambda + \alpha_\lambda \alpha_\varkappa = (a_\varkappa + a_\varkappa^\dagger)\cdot(a_\lambda + a_\lambda^\dagger) + (a_\lambda + a_\lambda^\dagger)\cdot(a_\varkappa + a_\varkappa^\dagger) = 2\,\delta_{\varkappa\lambda}.$$

Man kann (II) auch schreiben

$$\left.\begin{aligned} \alpha_\varkappa^2 &= 1 \\ \alpha_\varkappa \alpha_\lambda &= -1\,\alpha_\lambda \alpha_\varkappa \quad \text{für} \quad \varkappa \neq \lambda, \end{aligned}\right\} \tag{II a}$$

* Dies ist die Transformation aller Matrizen a durch dieselbe Matrix S zu $S^{-1} a S$, also nach dem Sprachgebrauch der allgemeinen Quantenmechanik die kanonische Transformation der Matrizendarstellung.

was aber mit anderen Worten so viel bedeutet, daß die $2K$ Matrizen α zusammen mit der Matrix -1 eine Gruppe aufspannen. Wenn z. B. $K = 2$ ist, so hat diese Gruppe folgende Elemente

$$\left.\begin{array}{l} 1;\ \alpha_1,\ \alpha_2,\ \alpha_3,\ \alpha_4;\ \alpha_1\alpha_2,\ \alpha_1\alpha_3,\ \alpha_1\alpha_4,\ \alpha_2\alpha_3,\ \alpha_2\alpha_4,\ \alpha_3\alpha_4; \\ -1;\ -\alpha_1,\ -\alpha_2,\ -\alpha_3,\ -\alpha_4;\ -\alpha_1\alpha_2,\ -\alpha_1\alpha_3,\ -\alpha_1\alpha_4,\ -\alpha_2\alpha_3,\ -\alpha_2\alpha_4,\ -\alpha_3\alpha_4; \\ \quad \alpha_1\alpha_2\alpha_3,\ \alpha_1\alpha_3\alpha_4,\ \alpha_1\alpha_2\alpha_4,\ \alpha_2\alpha_3\alpha_4,\ \alpha_1\alpha_2\alpha_3\alpha_4. \\ \quad -\alpha_1\alpha_2\alpha_3,\ -\alpha_1\alpha_3\alpha_4,\ -\alpha_1\alpha_2\alpha_4,\ -\alpha_2\alpha_3\alpha_4,\ -\alpha_1\alpha_2\alpha_3\alpha_4. \end{array}\right\} \quad \text{(III)}$$

Das sind 32 Elemente, im allgemeinen 2^{2K+1} Elemente. Das irreduzible Matrizensystem, das (II) genügt, ist sicher eine irreduzible Darstellung dieser Gruppe (umgekehrt braucht es nicht der Fall zu sein, die Isomorphie kann ja mehrstufig sein). Wir werden nur die irreduziblen Darstellungen bestimmen.

Unsere Gruppe hat den Normalteiler $1, -1$ (das Zentrum), ihre Faktorgruppe vom Grade 2^{2K} ist abelsch. Sie hat also 2^{2K} dieser Abelschen Faktorgruppe entsprechende irreduzible Darstellungen vom Grade 1, die auch Darstellungen der ganzen Gruppe sind. Indessen kommen sie für uns nicht in Betracht, da sie den Gleichungen II nicht genügen (da sie ja kommutativ sind).

Wie viele Klassen hat unsere Gruppe? Die beiden Elemente 1 und -1 bilden je eine Klasse für sich, sonst ist aber jedes Element R mit $-1 \cdot R$ in einer Klasse. Besteht nämlich R in (III) aus einer ungeraden Anzahl von Faktoren, so ist $\alpha R \alpha^{-1} = -1 \cdot R$, wenn α in R nicht enthalten ist, wenn R aus einer geraden Anzahl von Faktoren besteht, ist $\alpha R \alpha^{-1} = -1 \cdot R$, wenn α in R enthalten ist. Die Anzahl der Klassen ist also 2^{2K+1}; dies ist auch die Anzahl der voneinander verschiedenen irreduziblen Darstellungen. Da wir 2^{2K} Darstellungen schon kennen und diese nicht für uns in Betracht kommen, kann es nur eine einzige, die letzte sein, die die Gleichungen (II) befriedigt, alle anderen Lösungen von (II) gehen daraus durch Ähnlichkeitstransformation hervor.

Wir bestimmen noch die Anzahl von Zeilen und Spalten, die Dimension dieser Darstellung. (Es muß dabei natürlich 2^K herauskommen.) In der Tat ist der Grad der Gruppe 2^{2K+1} gleich der Summe der Quadrate der Dimensionen ihrer Darstellungen. Es haben 2^{2K} die Dimension 1, die letzte muß die Dimension 2^K haben, damit $(2^K)^2 + 2^{2K} \cdot 1^2 = 2^{2K+1}$. Sie stimmt also tatsächlich mit unserem Matrizensystem (69) oder § 3 und 6 überein.

Göttingen, Institut für theoretische Physik.

SITZUNG VOM 23. JULI 1934.

Über die mit der Entstehung von Materie aus Strahlung verknüpften Ladungsschwankungen

Von

W. Heisenberg

Die Diracsche Theorie[1]) des Positrons hat gezeigt, daß Materie aus Strahlung entstehen kann, indem z. B. ein Lichtquant sich in ein negatives und ein positives Elektron verwandelt. Dieses auch experimentell bestätigte Ergebnis hat zur Folge, daß überall dort, wo zur Messung eines physikalischen Sachverhalts große elektromagnetische Felder benötigt werden, mit einer bisher nicht beachteten Störung des Beobachtungsobjektes durch das Beobachtungsmittel gerechnet werden muß, nämlich mit der Erzeugung von Materie durch den Meßapparat. So gering diese Störung für die üblichen Experimente auch sein mag, ihre Berücksichtigung ist für das Verständnis der Theorie des Positrons von prinzipieller Bedeutung.

Zu ihrer näheren Untersuchung in einem sehr einfachen Fall betrachten wir ein quantenmechanisches System von freien negativen Elektronen, die sich gegenseitig nicht merklich beeinflussen; ihre Anzahl pro ccm sei $\frac{N}{V}$. In einem Volumen $v (v \ll V)$ wird dann im Mittel die Ladung

$$\bar{e} = -\varepsilon N \frac{v}{V} \tag{1}$$

zu finden sein (ε ist der Absolutbetrag der Elementarladung). Für das mittlere Schwankungsquadrat der Ladung im Volumen v würde man nach der klassischen Statistik den Wert

$$\Delta e^2 = \varepsilon^2 N \frac{v}{V} \tag{2}$$

erwarten. Es soll nun gezeigt werden, daß sich nach der Diracschen Theorie im allgemeinen ein größeres Schwankungsquadrat ergibt. Der Überschuß gegenüber Gl. (2) ist auf die mögliche Entstehung von Materie bei der Messung der Ladung im Volumen v zurückzuführen.

1) P. A. Dirac, The principles of Quantum mechanics. p. 255. Oxford 1930. Proc. Cambr. Phil. Soc. 30, 150, 1934.

Die Eigenfunktionen der Elektronen hängen vom Ort $\mathfrak{r}$, der Zeit t und der Spinvariable σ ab, sie sollen $u_n(\mathfrak{r}, t, \sigma)$ heißen und in einem Volumen V ($V \gg v$) normiert sein. Die allgemeine Wellenfunktion der Materie wird dann

$$\psi(\mathfrak{r}, t, \sigma) = \sum_n a_n u_n(\mathfrak{r}, t, \sigma), \tag{3}$$

wobei wegen des Paulischen Ausschließungsprinzips die V.R.

$$a_n^* a_m + a_m a_n^* = \delta_{nm} \tag{4}$$

gelten. Daraus folgt

$$a_n^* a_n = N_n; \quad a_n a_n^* = 1 - N_n. \tag{5}$$

Für die Zustände negativer Energie ($E_n < 0$) soll noch eingeführt werden:

$$a_n' = a_n^*; \quad a_n'^* = a_n; \quad N_n' = 1 - N_n. \tag{6}$$

Es bedeutet dann N_n die Anzahl der Elektronen im Zustand n, N_n' die Anzahl der Positronen im Zustand n.

Für die Ladungsdichte ergibt sich nach der Diracschen Theorie der folgende Ausdruck[1]):

$$-\varepsilon \sum_\sigma \Big[\sum_{E_n > 0} N_n u_n^* u_n - \sum_{E_n < 0} N_n' u_n^* u_n + \sum_{n \neq m} a_n^* a_m u_n^* u_m \Big]. \tag{7}$$

Für die Ladung e im Volumen v erhält man daher

$$e = -\varepsilon \Big[\sum_{E_n > 0} N_n \cdot \frac{v}{V} - \sum_{E_n < 0} N_n' \frac{v}{V} + \sum_{n \neq m} a_n^* a_m \sum_\sigma \int_v d\mathfrak{r}\, u_n^*(\mathfrak{r}t\sigma) u_m(\mathfrak{r}t\sigma) \Big]. \tag{8}$$

Aus Gründen, die von Bohr und Rosenfeld[2]) ausführlich diskutiert wurden, soll der zeitliche Mittelwert von e über ein endliches Intervall T betrachtet werden, wobei auch die Grenzen des Intervalls eventuell noch unscharf gelassen werden. Wir führen daher eine Funktion $f(t)$ ein, die nur im Bereich $0 \lesssim t \lesssim T$ von Null verschieden ist und für die

$$\int f(t)\, dt = 1.$$

Auch die Grenzen des Volumens v sollen eventuell unscharf bleiben; es soll daher $g(\mathfrak{r})$ eine Funktion von $\mathfrak{r}$ bedeuten, die in v bis auf die Umgebung der Grenzen 1 ist, außen verschwindet und für die $\int d\mathfrak{r}\, g(\mathfrak{r}) = v$. Der entsprechende zeitliche und räumliche Mittelwert der Ladung e wird dann

1) Vgl. z. B. W. H. Furry u. J. R. Oppenheimer, Phys. Rev. 45, 245, 1934.
2) N. Bohr und L. Rosenfeld, Verh. der Kgl. Dän. Gesellsch. d. Wiss. XII, 8, 1933.

$$\bar{e} = -\varepsilon\left[\sum_{E_n>0} N_n \frac{v}{V} - \sum_{E_n<0} N_n' \frac{v}{V} + \sum_{n \neq m} a_n^* a_m \iint dt\, d\mathfrak{r} \sum f(t)\, g(\mathfrak{r})\, u_n^*(\mathfrak{r} t \sigma)\, u_m(\mathfrak{r} t \sigma)\right]. \quad (9)$$

Bildet man nun den Erwartungswert $\bar{\bar{e}}$ von $\bar{e}$ für denjenigen Zustand, bei dem die Anzahlen N_n bekannte c-Zahlen sind, so erhält man

$$\bar{\bar{e}} = -\varepsilon\left[\sum_{E_n>0} N_n - \sum_{E_n<0} N_n'\right]\frac{v}{V}. \quad (10)$$

in Übereinstimmung mit Gl. (1). Für den Erwartungswert des Schwankungsquadrats ergibt sich jedoch:

$$\begin{aligned}
\overline{(\Delta \bar{e})^2} &= \overline{\bar{e}^2} - \bar{\bar{e}}^2 = \varepsilon^2 \sum_{n \neq m} \sum_{n' \neq m'} \overline{a_n^* a_m a_{n'}^* a_{m'}} \sum_{\sigma\sigma'} \int dt \int dt' \int d\mathfrak{r} \\
&\qquad \int d\mathfrak{r}'\, f(t)\, f(t')\, g(\mathfrak{r})\, g(\mathfrak{r}')\, u_n^*(\mathfrak{r} t \sigma)\, u_m(\mathfrak{r} t \sigma)\, u_{n'}^*(\mathfrak{r}' t' \sigma')\, u_{m'}(\mathfrak{r}' t' \sigma') \\
&= \varepsilon^2 \sum_{n \neq m} N_n (1 - N_m) \sum_{\sigma\sigma'} \int dt \int dt' \int d\mathfrak{r} \qquad (11) \\
&\qquad \int d\mathfrak{r}'\, f(t)\, f(t')\, g(\mathfrak{r})\, g(\mathfrak{r}')\, u_n^*(\mathfrak{r} t \sigma)\, u_n(\mathfrak{r}' t' \sigma')\, u_m^*(\mathfrak{r}' t' \sigma')\, u_m(\mathfrak{r} t \sigma) \\
&= \varepsilon^2 \sum_{n \neq m} N_n (1 - N_m) J_{nm}\,.
\end{aligned}$$

Diesen Ausdruck kann man in drei Teile zerlegen. Der erste wäre bereits vorhanden, wenn die Erwartungswerte der N_n für $E_n > 0$ und N_n' für $E_n < 0$ alle verschwinden, d. h. im Vakuum (man beachte $J_{nm} = J_{mn}$):

$$\varepsilon^2 \sum_{E_n>0;\, E_m<0} J_{nm}\,. \quad (12)$$

Ein zweiter Teil wird, wenn nur negative Elektronen vorhanden sind ($N' = 0$):

$$\varepsilon^2 \sum_{E_n>0} N_n \left(\sum_{E_m>0} - \sum_{E_m<0}\right) J_{nm}\,. \quad (13)$$

Schließlich bleibt noch als dritter Teil

$$-\varepsilon^2 \sum_{E_n>0;\, E_m>0} N_n N_m J_{nm} \quad (14)$$

übrig. Dieser dritte Teil enthält jedoch, wie man aus (11) berechnet, den Faktor $\left(\frac{v}{V}\right)^2$; er kann daher gegenüber den beiden ersten vernachlässigt werden, wenn — wie angenommen wurde — $v \ll V$ ist.

Für die Berechnung des zweiten Teils (13) bemerken wir zunächst, daß

$$\left(\sum_{E_m>0} - \sum_{E_m<0}\right) u_m^*(\mathfrak{r}' t' \sigma')\, u_m(\mathfrak{r} t \sigma) \quad (15)$$

$$= \int \frac{d\mathfrak{p}}{h^3} \frac{1}{2} \left\{ \left(1 - \frac{\alpha_i p_i + \beta mc}{p_0}\right)_{\sigma\sigma'} e^{\frac{i}{\hbar}\left[\mathfrak{p}(\mathfrak{r}-\mathfrak{r}') - p_0 c (t-t')\right]} - \left(1 + \frac{\alpha_i p_i + \beta mc}{p_0}\right)_{\sigma\sigma'} e^{\frac{i}{\hbar}\left[\mathfrak{p}(\mathfrak{r}-\mathfrak{r}') + p_0 c (t-t')\right]} \right\}.$$

Es wird also

$$\left(\sum_{E_m>0}-\sum_{E_m<0}\right) J_{nm} = \sum_{\sigma\sigma'} \int dt \int dt' \int d\mathfrak{r} \int d\mathfrak{r}' \, f(t)\, f(t')\, g(\mathfrak{r})\, g(\mathfrak{r}') \cdot \int \frac{d\mathfrak{p}}{h^3} \frac{1}{2}\left\{\right\}$$

$$\cdot\, u_n^*(\mathfrak{r}\, t\, \sigma)\, u_n(\mathfrak{r}'\, t'\, \sigma') = \int \frac{d\mathfrak{p}}{h^3} \sum_{\sigma\sigma'} \frac{1}{2}\left\{\left(1 - \frac{\alpha_i p_i + \beta m c}{p_0}\right)_{\sigma\sigma'} |f(p_0 - p_0^n)|^2\right.$$

$$\left. - \left(1 + \frac{\alpha_i p_i + \beta m c}{p_0}\right)_{\sigma\sigma'} |f(p_0 + p_0^n)|^2\right\} |g(\mathfrak{p} - \mathfrak{p}^n)|^2\, b_n^*(\sigma)\, b_n(\sigma'),$$

wobei

$$f(p_0) = \int dt\, f(t)\, e^{\frac{i}{\hbar} p_0 c t}$$

$$g(\mathfrak{p}) = \int d\mathfrak{r}\, g(\mathfrak{r})\, e^{\frac{i}{\hbar} \mathfrak{p}\mathfrak{r}} \tag{16}$$

$$u_n(\mathfrak{r}\, t\, \sigma) = b_n(\sigma)\, e^{\frac{i}{\hbar}(\mathfrak{p}^n \mathfrak{r} - p_0^n t)}$$

gesetzt ist. Aus der Wellengleichung folgt dann weiter

$$\begin{aligned}\left(\sum_{E_m>0}-\sum_{E_m<0}\right) J_{nm} = \frac{1}{V}\int \frac{d\mathfrak{p}}{h^3} \frac{1}{2}\left\{\left(1 + \frac{p_i^n p_i + (mc)^2}{p_0 p_0^n}\right) |f(p_0 - p_0^n)|^2\right. \\ \left. - \left(1 - \frac{p_i^n p_i + (mc)^2}{p_0 p_0^n}\right) |f(p_0 + p_0^n)|^2\right\} |g(\mathfrak{p} - \mathfrak{p}^n)|^2 .\end{aligned} \tag{17}$$

Wenn die zum Zustand n gehörige Wellenlänge klein ist gegen die räumliche Ausdehnung des Gebiets v und wenn ferner die Zeit, über die gemittelt werden soll, so klein ist, daß die Elektronen in dieser Zeit nur Strecken durchlaufen, die ebenfalls klein sind im Verhältnis zur räumlichen Ausdehnung von v, so ist $g(\mathfrak{p}\text{-}\mathfrak{p}^n)$ als sehr schnell veränderlich gegenüber dem Bruch

$$\frac{p_i^n p_i + (mc)^2}{p_0 p_0^n}$$

anzusehen, ferner kann in dieser Annäherung $|f(p_0 - p_0^n)|^2 \sim 1$ gesetzt werden. Dann wird

$$\left(\sum_{E_m>0}-\sum_{E_m<0}\right) J_{nm} \approx \frac{1}{V}\int \frac{d\mathfrak{p}}{h^3} |g(\mathfrak{p} - \mathfrak{p}^n)|^2 = \frac{1}{V}\int d\mathfrak{r}\, g^2(\mathfrak{r}) \approx \frac{v}{V}, \tag{18}$$

und für den zweiten Teil des Schwankungsquadrats erhält man:

$$\varepsilon^2 \left(\sum_{E>0} N_n\right) \cdot \frac{v}{V}, \tag{19}$$

in Übereinstimmung mit Gl. (2).
Hierzu kommt nun noch der erste Teil, der auch im Vakuum auftritt:

$$\varepsilon^2 \sum_{E_n>0;\, E_m<0} J_{nm}.$$

Ähnlich wie in Gl. (15) findet man:

$$\left.\begin{aligned}\sum_{E_n>0;\,E_m<0} J_{nm} &= \int dt \int dt' \int d\mathfrak{r} \int d\mathfrak{r}' \sum_{\sigma\sigma'} f(t)\, f(t')\, g(\mathfrak{r})\, g(\mathfrak{r}') \int \frac{d\mathfrak{p}}{h^3} \int \frac{d\mathfrak{p}'}{h^3} \frac{1}{4}\left(1 - \frac{\alpha_i p_i + \beta m c}{p_0}\right) \\ &\qquad \left(1 + \frac{\alpha_i p_i' + \beta m c}{p_0'}\right) e^{\frac{i}{\hbar}[(\mathfrak{p}-\mathfrak{p}')(\mathfrak{r}-\mathfrak{r}') - (p_0+p_0')(t-t')]} \\ &= \int dt \int dt' \int d\mathfrak{r} \int d\mathfrak{r}'\, f(t)\, f(t')\, g(\mathfrak{r})\, g(\mathfrak{r}') \int \frac{d\mathfrak{p}}{h^3} \int \frac{d\mathfrak{p}'}{h^3} \frac{p_0 p_0' - \mathfrak{p}\mathfrak{p}' - m^2 c^2}{p_0 p_0'}\, e^{\frac{i}{\hbar}[(\mathfrak{p}-\mathfrak{p}')(\mathfrak{r}-\mathfrak{r}') - (p_0+p_0')(t-t')]} \\ &= \int \frac{d\mathfrak{p}}{h^3} \int \frac{d\mathfrak{p}}{h^3} |f(p_0+p_0')|^2 \cdot |g(\mathfrak{p}-\mathfrak{p}')|^2 \frac{p_0 p_0' - \mathfrak{p}\mathfrak{p}' - m^2 c^2}{p_0 p_0'}\,. \end{aligned}\right\} \quad (20)$$

Setzt man $g(\mathfrak{r}) = 1$ in einem rechteckigen, scharf umgrenzten Volumen mit den Seiten l_1, l_2, l_3 u. außerhalb Null, so wird

$$g(\mathfrak{p}) = \int d\mathfrak{r}\, g(\mathfrak{r})\, e^{\frac{i}{\hbar}\mathfrak{p}\mathfrak{r}} = \int_0^{l_1} dx \int_0^{l_2} dy \int_0^{l_3} dz\, e^{\frac{i}{\hbar}(p_x x + p_y y + p_z z)} = \left(\frac{e^{\frac{i}{\hbar}p_x l_1} - 1}{\frac{i}{\hbar}p_x}\right)\left(\frac{e^{\frac{i}{\hbar}p_y l_2} - 1}{\frac{i}{\hbar}p_y}\right)\left(\frac{e^{\frac{i}{\hbar}p_z l_3} - 1}{\frac{i}{\hbar}p_z}\right) \quad (21)$$

$$\text{u.} \quad |g^2(\mathfrak{p})| = \frac{\hbar^6}{p_x^2 p_y^2 p_z^2}\, 64 \sin^2\frac{p_x l_1}{2\hbar} \sin^2\frac{p_y l_2}{2\hbar} \sin^2\frac{p_z l_3}{2\hbar}\,.$$

Begrenzt man das Zeitintervall scharf, so wird

$$f(p_0) = \frac{1}{T}\int_0^T dt\, e^{\frac{i}{\hbar}p_0 c t} = \frac{e^{\frac{i}{\hbar}p_0 c T} - 1}{\frac{i}{\hbar}p_0 c T}\,; \quad |f(p_0)|^2 = 4 \cdot \frac{\sin^2\frac{p_0 c T}{2\hbar}}{\left(\frac{p_0 c T}{\hbar}\right)^2}\,. \quad (22)$$

Setzt man die Werte (21) und (22) in Gl. (20) ein, so divergiert das Integral auf der rechten Seite von Gl. (20). Man erkennt dies am einfachsten, indem man als neue Integrationsvariabeln $\mathfrak{p} - \mathfrak{p}' = \mathfrak{k}$ und $\frac{\mathfrak{p}+\mathfrak{p}'}{2} = \mathfrak{P}$ einführt; es wird dann:

$$\sum_{E_n>0;\,E_m<0} J_{nm} = \int \frac{d\mathfrak{k}}{h^3} \int \frac{d\mathfrak{P}}{h^3} |f(p_0+p_0')|^2 \cdot |g(\mathfrak{k})|^2 \frac{\sqrt{\left(\mathfrak{P}^2 + \frac{\mathfrak{k}^2}{4} + m^2c^2\right)^2 - (\mathfrak{P}\mathfrak{k})^2} - \mathfrak{P}^2 + \frac{\mathfrak{k}^2}{4} - m^2c^2}{\sqrt{\left(\mathfrak{P}^2 + \frac{\mathfrak{k}^2}{4} + m^2c^2\right)^2 - (\mathfrak{P}\mathfrak{k})^2}}\,. \quad (23)$$

Bei einem vorgegebenen großen Wert von $\mathfrak{k}$ ($k \gg mc$) ergibt die Integration über $\mathfrak{P}$, die konvergiert, einen Faktor der ungefähren Größe $\frac{k}{\hbar\,(cT)^2}$ zu $|\mathfrak{g}(\mathfrak{k})^2|$; die Integration über $\mathfrak{k}$ führt dann im wesentlichen auf das Integral

$$\int \frac{k}{k_x^2 \cdot k_y^2 \cdot k_z^2}\, dk_x\, dk_y\, dk_z \sin^2\frac{k_x l_1}{2\hbar} \sin^2\frac{k_y l_2}{2\hbar} \sin^2\frac{k_z l_3}{2\hbar}\,,$$

das divergiert. Das Schwankungsquadrat von $\bar{e}$ wird also unendlich groß, wenn man das Raum-Zeitgebiet, über das man die Mittelung der Ladung ausführt, scharf begrenzt. Dieses Ergebnis entspricht den Resultaten, die sich bei der Untersuchung der Energieschwankungen in einem Strahlungsfeld herausgestellt haben[1]). Auch dort erwies es sich als notwendig, das Raum-

1) Vgl. W. Heisenberg, Verh. d. Sächs. Ak. 83, 3, 1931.

gebiet, in dem die Energie aufgesucht werden soll, unscharf zu begrenzen. Für die Ladungsschwankungen der Gl. (20) genügt es, entweder die Grenzen des Zeitintervalls oder die des Raumintervalls in geeigneter Weise unscharf zu wählen. Nehmen wir z. B. an, daß die Größe $g(\mathfrak{r})$ in einem Gebiet der Breite b in der Umgebung des Randes von v etwa nach der Art einer Gaußschen Fehlerfunktion von 1 auf 0 abnimmt. Dann verschwindet $g(\mathfrak{p})$ für Werte von $p > \frac{\hbar}{b}$ ebenfalls wie die Fehlerfunktion; nimmt man weiter an, daß die Zeit T klein sei gegen $\frac{b}{c}$, so kann man in genügender Approximation den Bruch in (23) entwickeln für $k^2 \ll \mathfrak{P}^2 + m^2c^2$. Es ergibt sich dann

$$\sum_{E_n>0,\ E_m<0} J_{nm} = \int \frac{d\mathfrak{k}}{h^3} \int \frac{d\mathfrak{P}}{h^3} \, |f(2P)|^2 \, |g(\mathfrak{k})|^2 \, \frac{\mathfrak{k}^2 - \frac{(\mathfrak{P}\mathfrak{k})^2}{P^2}}{2(P^2 + m^2c^2)}. \tag{24}$$

Bis auf unwesentliche konstante Faktoren wird daraus

$$\frac{1}{\hbar^2 c T} \int \frac{d\mathfrak{k}}{h^3} \, |g(\mathfrak{k})|^2 \, \mathfrak{k}^2 \qquad \text{für } T \ll \frac{\hbar}{mc^2},$$

$$\frac{1}{\hbar (cT)^2 \cdot mc} \int \frac{d\mathfrak{k}}{h^3} \, |g(\mathfrak{k})|^2 \, \mathfrak{k}^2 \qquad \text{für } T \gg \frac{\hbar}{mc^2}.$$

Für die Größenordnung des ersten Teiles des Schwankungsquadrats erhält man daher $\left(\text{für } l_1 \sim l_2 \sim l_3 = \sqrt[3]{v}\right)$:

$$\begin{aligned} \overline{(\Delta \bar{e})^2} &\sim \frac{\varepsilon^2}{\hbar c T} \frac{\hbar}{b} v^{\frac{2}{3}} = \varepsilon^2 \frac{v^{\frac{2}{3}}}{cT \cdot b} && \text{für } T \ll \frac{\hbar}{mc^2} \\ \overline{(\Delta \bar{e})^2} &\sim \varepsilon^2 \frac{v^{\frac{2}{3}} \hbar}{(cT)^2 \cdot mc \cdot b} && \text{für } T \gg \frac{\hbar}{mc^2} \end{aligned} \tag{25}$$

Der Faktor $v^{\frac{2}{3}}$ zeigt hier deutlich, daß es sich bei diesen Schwankungen um einen Oberflächeneffekt handelt, der davon herrührt, daß an den Wänden, die das vorgegebene Volumen v abgrenzen, Materie entstehen kann. Er wird um so größer, je kleiner die Zeit gewählt wird, in der die Messung der Ladung vorgenommen werden soll und je schärfer die Begrenzung des Volumens ist. Diese Ergebnisse dürften eng zusammenhängen mit dem Umstand, daß die für die Erzeugung von Materie maßgebende Inhomogenität der Diracschen Gleichung die zweiten Ableitungen der elektromagnetischen Feldstärken enthält, daß also eine Unstetigkeit im ersten Differentialquotienten der Feldstärken bereits zur Entstehung von unendlich viel Materie Anlaß geben könnte.

Zusammenfassend sei festgestellt, daß bei der Messung der Ladung in einem vorgegebenen Raum-Zeitgebiet Schwankungen auftreten, die in der klassischen Theorie kein Analogon besitzen; sie werden verursacht durch die Materie, die an der Oberfläche des betrachteten Raumgebiets bei der Messung entsteht.

On the Self-Energy and the Electromagnetic Field of the Electron

V. F. WEISSKOPF
University of Rochester, Rochester, New York

(Received April 12, 1939)

The charge distribution, the electromagnetic field and the self-energy of an electron are investigated. It is found that, as a result of Dirac's positron theory, the charge and the magnetic dipole of the electron are extended over a finite region; the contributions of the spin and of the fluctuations of the radiation field to the self-energy are analyzed, and the reasons that the self-energy is only logarithmically infinite in positron theory are given. It is proved that the latter result holds to every approximation in an expansion of the self-energy in powers of e^2/hc. The self-energy of charged particles obeying Bose statistics is found to be quadratically divergent. Some evidence is given that the "critical length" of positron theory is as small as $h/(mc)\cdot\exp(-hc/e^2)$.

I. Introduction and Discussions of Results

THE self-energy of the electron is its total energy in free space when isolated from other particles or light quanta. It is given by the expression

$$W = T + (1/8\pi)\int (H^2+E^2)d\mathbf{r}. \quad (1)$$

Here T is the kinetic energy of the electron; H and E are the magnetic and electric field strengths. In classical electrodynamics the self-energy of an electron of radius a at rest and without spin is given by $W \sim mc^2 + e^2/a$ and consists solely of the energy of the rest mass and of its electrostatic field. This expression diverges linearly for an infinitely small radius. If the electron is in motion, other terms appear representing the energy produced by the magnetic field of the moving electron. These terms, of course, can be obtained by a Lorentz transformation of the former expression.

The quantum theory of the electron has put the problem of the self-energy in a critical state. There are three reasons for this:

(a) Quantum kinematics shows that the radius of the electron must be assumed to be zero. It is easily proved that the product of the charge densities at two different points, $\rho(\mathbf{r}-\boldsymbol{\xi}/2)\times\rho(\mathbf{r}+\boldsymbol{\xi}/2)$, is a delta-function $e^2\delta(\boldsymbol{\xi})$. In other words: if one electron alone is present, the probability of finding a charge density simultaneously at two different points is zero for every finite distance between the points. Thus the energy of the electrostatic field is infinite as

$$W_{st} = \lim_{(a\to 0)} e^2/a.$$

(b) The quantum theory of the relativistic electron attributes a magnetic moment to the electron, so that an electron at rest is surrounded by a magnetic field. The energy

$$U_{\text{mag}} = (1/8\pi)\int H^2 d\mathbf{r}$$

of this field is computed in Section III and the result is

$$U_{\text{mag}} = e^2h^2/(6\pi m^2c^2a^3).$$

This corresponds to the field energy of a magnetic dipole of the moment $eh/2mc$ which is spread over a volume of the dimensions a. The spin, however, does not only produce a magnetic field, it also gives rise to an alternating electric field. The closer analysis of the Dirac wave equation has shown[1] that the magnetic moment of the spin is produced by an irregular circular fluctuation movement (Zitterbewegung) of the electron which is superimposed to the translatory motion. The instantaneous value of the velocity is always found to be c. It must be expected that this motion will also create an alternating electric field. The existence of this field is demonstrated in Section III by the computation of the expression

$$U_{\text{el}} = (1/8\pi)\int E_s^2 d\mathbf{r}.$$

There E_s is the solenoidal part (div. $E_s=0$) of the electric field strength created by the electron. The fact that the above expression does not vanish for an electron at rest proves the existence

[1] E. Schroedinger, Berl. Ber. 1930, 418 (1930).

of a solenoidal field[2] apart from the irrotational electric field of the charge. The energies of the electric and magnetic fields of the spin are found to be equal. The spin movement does not, of course, give rise to a radiation. The time average of the Poynting vector is zero.

The electromagnetic field of the spin does not contribute to the self-energy of the electron. It is shown in Section IV, that the charge dependent part of the self-energy to a first approximation is given by

$$W'=\tfrac{1}{2}\int\left(\rho\phi-\frac{1}{c}\mathbf{i}\cdot\mathbf{A}\right)d\mathbf{r}=\frac{1}{8\pi}\int(E^2-H^2)d\mathbf{r}.$$

Here ρ and $\mathbf{i}$ are the charge and current densities, ϕ and $\mathbf{A}$ are the scalar and vector potential, respectively. If the self-energy is expressed in terms of the field energies, the electric and magnetic parts have opposite signs,[3] so that the contributions of the electric and of the magnetic fields of the spin cancel one another.

(c) The quantum theory of the electromagnetic field postulates the existence of field strength fluctuations in empty space. These give rise to an additional energy, which diverges more strongly than the electrostatic self-energy. The following crude calculation may demonstrate how this particular part of the self-energy arises: Let us consider an electron with radius a. The field fluctuations in a volume a^3 are of the order $E^2\sim hc/a^4$.[4] The mean frequency of the fluctuations is $\nu\sim c/a$. This field induces the electron to perform vibrations with an amplitude $x\sim eE/m\nu^2$ and an energy $W_{\text{fluct}}\sim e^2E^2/m\nu^2\sim e^2h/mca^2$. This energy diverges quadratically for infinitely small radius. The exact value is calculated in Section IV and is $W_{\text{fluct}}=\lim_{(a=0)}e^2h/\pi mca^2$.

A new situation is created by Dirac's theory of the positron: The self-energy diverges only logarithmically with infinitely small radius. This fact has been proved[5] only for the first approximation of the self-energy expanded in powers of e^2/hc. However it will be shown in Section VI that the divergence is logarithmic in every approximation. The main purpose of this paper is to show the physical significance of the logarithmic divergence and to demonstrate the reasons of its occurrence.

Let us consider the case of one electron embedded in the vacuum as described by the positron theory. The vacuum is represented by the state in which all negative energy states are filled with electrons. The charge density of these "vacuum electrons" is not observable in the unperturbed state of a field-free vacuum. However, the differences between the actual density and the unperturbed density are observable.

The presence of an electron in the vacuum causes a considerable change in the distribution of the vacuum electrons because of a peculiar effect of the Pauli exclusion principle. According to this principle it is impossible to find two or more electrons in a single cell of a volume h^3 in the phase space. If two electrons of equal spin are brought together to a small distance d, their momentum difference must be at least h/d. This effect is similar to a repulsive force which causes two particles with equal spin not to be found closer together than approximately one de Broglie wave-length.

As a consequence of this we find at the position of the electron a "hole" in the distribution of the vacuum electrons which completely compensates its charge. But we also find around the electron a cloud of higher charge density coming from the displaced electrons, which must be found one wave-length from the original electron. The total effect is a broadening of the charge of the electron over a region of the order h/mc as it is indicated schematically in Fig. 1. The product $\rho(\mathbf{r}-\xi/2)\times\rho(\mathbf{r}+\xi/2)$ is no longer zero for a finite distance ξ, and is given by the function

$$G(\xi)=e^2\frac{mc}{h}\frac{1}{\xi}\frac{\partial}{\partial\xi}\frac{i}{2\pi}H_0^{(1)}(imc\xi/h)$$

(Section II). Here $H_0^{(1)}(x)$ is the Hankel function of first kind. $G(\xi)$ has still a quadratic singularity

[2] A solenoidal electric field is necessarily an alternating field for its time average vanishes in a stationary state, whereas the time average of a magnetic field does not vanish if stationary currents are present.

[3] This at first sight unfamiliar result is connected with the well-known fact that a system of steady currents increases its magnetic field energy if it performs mechanical work, whereas a system of charges decreases its field energy by performing mechanical work.

[4] The fluctuations are of the order of magnitude of the field-strength of one light quantum with wave-length.

[5] V. Weisskopf, Zeits. f. Physik 89, 27 (1934); 90, 817 (1934).

for $\xi=0$. It is shown quantitatively in Section II, that this broadening of the charge distribution is just sufficient to reduce the electrostatic self-energy to a logarithmically divergent expression.

The broadening effect also changes the magnetic field distribution of the spin moment. In positron theory the magnetic field energy is given by

$$U_{\text{mag}}=\lim_{(a=0)}[e^2h/(2\pi mca^2)$$
$$-e^2mc/(4\pi h)\cdot \lg (h/mca)]. \quad (2)$$

This is equal to the field energy of a momentum distribution spread over a finite region, which is proportional to the spread of charge described above. The divergence, which is less strong than in the one-electron theory,[6] comes from the quadratic singularity of the distribution. The electric field energy of the spin, however, is not equal to the magnetic field energy because of the following effect, which is again based upon the exclusion principle. The vacuum electrons which are found in the neighborhood of the original electron, fluctuate with a phase opposite to the phase of the fluctuations of the original electron. This phase relation, applied to the circular fluctuation of the spin, decreases its total electric field by means of interference, but does not change the magnetic field of the spins since the latter is due to circular currents and is not dependent on the phase of the circular motion. Thus the total solenoidal electric field energy is reduced by interference if an electron is added to the vacuum. The electric field energy U_{el} of an electron in positron theory is therefore *negative* since it is the difference between the field energy of the vacuum plus one electron, and the energy of the vacuum alone. The exact calculations of Section III give $U_{\text{el}}=-U_{\text{mag}}$. Thus the contribution of the spin to the self-energy does not vanish in positron theory and is by Eq. (2)

$$W_{\text{sp}}=-2U_{\text{mag}}=-\lim_{(a=0)}[e^2h/(\pi mca^2)$$
$$-e^2mc/(2\pi h)\cdot \lg h/(mca)].$$

The broadening effect cannot, however, be applied to the energy W_{fluct}, which is the energy of

[6] We use the term "one-electron theory" for the description of the electron by means of the Dirac wave equation without filling up the negative energy states, in order to distinguish it from the "positron theory."

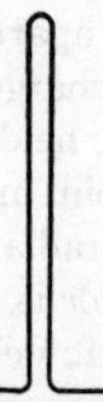

FIG. 1a. Schematic charge distribution of the electron.

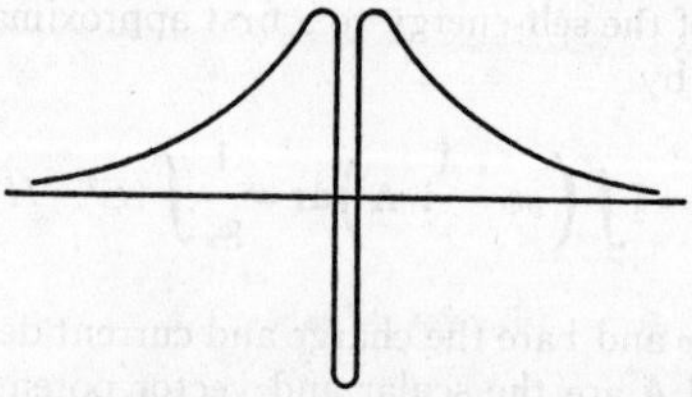

FIG. 1b. Schematic charge distribution of the vacuum electrons in the neighborhood of an electron.

the action of the electromagnetic field fluctuations upon the electron. The effect of an external field upon an electron in positron theory is to a first approximation the same as one expects for an electron with infinitely small radius, since the effect of the field upon the displaced vacuum electrons can be neglected. For instance, no destructive interference effect would occur in the interaction with a light wave whose wave-length is smaller than h/mc. The exclusion principle does not alter the interaction of an electron with the field as long as one considers that action to a first approximation to be the sum of independent actions at every point; it has only an effect on the probability of finding one particle in the neighborhood of another.

The energy W_{fluct} in positron theory is therefore not different from the same quantity in one electron theory as shown in Section IV. In the former theory, however, it is balanced by the spin energy W_{sp} the most strongly divergent terms of which are just oppositely equal to W_{fluct}. The sum of W_{sp} and W_{fluct} is only logarithmically divergent.

Thus according to positron theory the self-energy of an electron consists of three parts:

(a) The energy W_{st} of the Coulomb field, which diverges logarithmically because of the characteristic spread of charge.

(b) The energy W_{sp} of the oscillatory motion which produces the spin. This energy, although zero in the one-electron theory, is negative and quadratically divergent in the positron theory. This is because of the negative contribution of the magnetic field and the interference effect of the electric field of the vacuum electrons.

(c) The energy W_{fluct} of forced vibrations under the influence of the zero-point fluctuations of the radiation field. The energies (b) and (c) compensate each other to a logarithmic term.

It is interesting to apply similar considerations to the scalar theory of particles obeying the Bose statistics, as has been developed by Pauli and the author.[7] Here the probability of finding two equal particles closer than their wave-lengths is *larger* than at longer distances. The effect on the self-energy is therefore just the opposite. The influence of the particle on the vacuum causes a higher singularity in the charge distribution instead of the hole which balanced the original charge in the previous considerations. It is shown in Section V that this gives rise to a quadratically divergent energy of the Coulomb field of the particle. Thus the situation here is even worse than in the classical theory. The spin term obviously does not appear and the energy W_{fluct} is exactly equal to its value for a Fermi particle.

A few remarks might be added about the possible significance of the logarithmic divergence of the self-energy for the theory of the electron. It is proved in Section VI that every term in the expansion of the self-energy in powers of e^2/hc

$$W=\sum_n W^{(n)} \qquad (3)$$

diverges logarithmically with infinitely small electron radius and is approximately given by

$$W^{(n)}\sim z_n mc^2(e^2/hc)^n[\lg\,(h/mca)]^t, \quad t\leqq n.$$

Here the z_n are dimensionless constants which cannot easily be computed. It is therefore not sure, whether the series (3) converges even for finite a, but it is highly probable that it converges if $\delta=e^2/(hc)\cdot\lg\,(h/mca)<1$. One then would get $W=mc^2O(\delta)$ where $O(\delta)=1$ for a value of $\delta<1$. We then can define an electron radius in the same way as the classical radius e^2/mc^2 is defined, by putting the self-energy equal to mc^2. One obtains then roughly a value $a\sim h/(mc)\cdot\exp\,(-hc/e^2)$ which is about 10^{-58} times smaller than the classical electron radius. The "critical length" of the positron theory is thus infinitely smaller than usually assumed.

The situation is, however, entirely different for a particle with Bose statistics. Even the Coulombian part of the self-energy diverges to a first approximation as $W_{st}\sim e^2h/(mca^2)$ and requires a much larger critical length that is $a=(hc/e^2)^{-\frac{1}{2}}\cdot h/(mc)$, to keep it of the order of magnitude of mc^2. This may indicate that a theory of particles obeying Bose statistics must involve new features at this critical length, or at energies corresponding to this length; whereas a theory of particles obeying the exclusion principle is probably consistent down to much smaller lengths or up to much higher energies.

II. The Charge Distribution of the Electron

The charge distribution in the neighborhood of an electron can be determined from the expression

$$G(\xi)=\int\rho(\mathbf{r}-\xi/2)\rho(\mathbf{r}+\xi/2)d\mathbf{r}\,; \qquad (4)$$

here $\rho(\mathbf{r})$ is the charge density at the point $\mathbf{r}$. $G(\xi)$ is the probability of finding charge simultaneously at two points in a distance ξ. If applied to a situation in which one electron alone is present, direct information can be drawn from this expression concerning the charge distribution in the electron itself. The charge density is given by

$$\rho(\mathbf{r})=e\{\psi^*(\mathbf{r})\psi(\mathbf{r})\}-\sigma, \qquad (5)$$

where $\psi(\mathbf{r})$, the wave function, is a spinor with four components ψ_μ, $\mu=1, 2, 3, 4$. We write

$$\{\psi^*\psi\}=\sum_{\mu=1}^{4}\psi_\mu^*\psi_\mu$$

for the scalar product of two spinors. σ is the charge density of the unperturbed electrons in the negative energy states which is to be subtracted in the positron theory. In the one-electron theory σ is zero. The wave function ψ can be expanded in wave functions φ_q of the

[7] W. Pauli and V. Weisskopf, Helv. Phys. Acta 7, 709 (1934).

stationary states q of a free electron:

$$\psi(\mathbf{r})=\sum_q a_q\varphi_q(\mathbf{r}). \qquad (6)$$

The following relation holds for the φ_q

$$\{\varphi^*_q(\mathbf{r})\varphi_q(\mathbf{r})\}=1/V, \qquad (7)$$

where V is the total volume of the system. We denote functions with positive energy values by φ_{+q} and with negative energy values by φ_{-q}.

We apply the method of quantized waves and consider the ψ's as operators acting on eigenfunctions $c(\cdots N_q\cdots)$ whose variables are the numbers N_q of electrons in different states q. If the ψ's are written in the form (6), the a's are operators which fulfill the well-known relations:

$$a_q{}^*a_q=N_q, \quad a_qa_q{}^*=1-N_q. \qquad (8)$$

We now insert (6) into (5) and (5) into (4) and keep only terms which contain the products of two a's of the form (8) or the following combinations of four a's:

$$\begin{aligned}a_q{}^*a_qa_{q'}{}^*a_{q'}&=N_qN_{q'},\\ a_q{}^*a_{q'}a_{q'}{}^*a_q&=N_q(1-N_{q'}).\end{aligned} \qquad (9)$$

All other combinations do not contribute to the expectation value $\bar{G}(\xi)$ of $G(\xi)$ because they have no diagonal elements. We obtain then

$$\bar{G}(\xi)=e^2\sum_q\sum_{q'}N_qN_{q'}+e^2\sum_q\sum_{q'}N_q(1-N_{q'})$$
$$\times\int\{\varphi_q{}^*(\mathbf{r}_1)\varphi_{q'}(\mathbf{r}_1)\}\{\varphi_{q'}{}^*(\mathbf{r}_2)\varphi_q(\mathbf{r}_2)\}d\mathbf{r}$$
$$-2\sigma e\sum_q N_q+\sigma^2V. \qquad (10)$$

Here and in the following formulas we put $\mathbf{r}_1=\mathbf{r}-\xi/2$, $\mathbf{r}_2=\mathbf{r}+\xi/2$.

We first apply this expression to a single electron. We then put $\sigma=0$, and $N_{q_0}=1$ for $q=q_0$, $N_q=0$ for $q\neq q_0$:

$$\bar{G}(\xi)=e\sum_q\int\{\varphi_{q_0}{}^*(r_1)\varphi_q(r_1)\}\{\varphi_q{}^*(r_2)\varphi_{q_0}(r_2)\}d\mathbf{r}.$$

This expression can be evaluated by inserting the wave functions of a free electron. If q_0 is the state of an electron at rest, one obtains after replacing the sum over q by an integral over the momenta $\mathbf{p}$ of the states, the following result:

$$\bar{G}(\xi)=e^2\int d\mathbf{p}\frac{\exp i(\xi\cdot\mathbf{p})/h}{8\pi^3h^3}=e^2\delta(\xi). \qquad (11)$$

Thus in the one-electron theory, $\bar{G}(\xi)$ is equal to the δ-function.

We now apply (10) to the vacuum of the positron theory, that is, we set

$$N_{+q}=0, \quad N_{-q}=1, \quad \sigma=\sum_{-q}N_q.$$

It is easily seen, that the first, the third and the fourth term cancel each other. The terms remaining give

$$G_{\text{vac}}(\xi)=e^2\sum_{+q}\sum_{-q'}\int\{\varphi_{-q'}{}^*(r_1)\varphi_{+q}(r_1)\}$$
$$\times\{\varphi_{+q}{}^*(r_2)\varphi_{-q'}(r_2)\}d\mathbf{r}.$$

The fact that this expression is different from zero and even infinite in the vacuum is closely connected with the charge fluctuations of the empty space which have been investigated by Heisenberg and Oppenheimer.[8] Heisenberg has shown that the charge fluctuations are infinite if the region in which they are measured is sharply limited. This result is due to the electron pairs produced when the charge is measured in a sharply defined region.

We are at present interested in the expression $G(\xi)$ corresponding to the charge distribution of one electron. This can be obtained by calculating $G_{\text{vac}+1}(\xi)$ for the state in which one electron in the state $+q_0$ is present ($N_{q_0}=1$ all other $N_{+q}=0$, $N_{-q}=1$), and by subtracting the effect of the vacuum $G_{\text{vac}}(\xi)$:

$$\bar{G}(\xi)=G_{\text{vac}+1}(\xi)-G_{\text{vac}}(\xi)=$$
$$e^2\Big(\sum_{+q}-\sum_{-q}\Big)\int\{\varphi_{q_0}{}^*(r_1)\varphi_q(r_1)\}\{\varphi_q{}^*(r_2)\varphi_{q_0}(r_2)\}d\mathbf{r}.$$

If one inserts the actual solutions φ_q of Dirac's wave equation of the free electron, this expression can be readily evaluated. One obtains after replacing the sum by an integral as before:

$$\bar{G}(\xi)=e^2mc^2\int d\mathbf{p}\frac{\exp i(\xi\cdot\mathbf{p})/h}{8\pi^3h^3E(p)}; \qquad (12)$$

[8] W. Heisenberg, Verh. d. Sächs. Akad. **86**, 317 (1934); J. R. Oppenheimer, Phys. Rev. **47**, 144 (1934).

here $E(p)=c(p^2+m^2c^2)^{\frac{1}{2}}$. This integral can be evaluated and gives:

$$\tilde{G}(\xi)=e^2\frac{mc}{h}\frac{1}{\xi}\frac{\partial}{\partial\xi}\frac{i}{2\pi}H_0^{(1)}\left(\frac{imc}{h}\cdot\xi\right).$$

$H_0^{(1)}(x)$ is the Hankel function of first kind; this function has a logarithmic singularity for $x=0$ and falls off exponentially for $x\gg1$. We obtain thus

$$\tilde{G}(\xi)=\begin{cases}\dfrac{e^2}{4\pi^2}\dfrac{mc}{h}\dfrac{1}{\xi^2} & \text{for } \xi\ll h/mc\\[2ex] e^2\cdot\left(\dfrac{mc}{h}\right)^2(h/2\pi^3mc\xi^3)^{\frac{1}{2}}\cdot e^{-mc\xi/h} & \\ & \text{for } \xi\gg h/mc.\end{cases}$$

This expression replaces the delta-function of the one-electron theory and indicates a spread of charge over a finite region of the order of h/mc. It is of interest to construct a charge density $\bar{\rho}(r)$ for which

$$\int\bar{\rho}(r_1)\bar{\rho}(r_2)d\mathbf{r}=G(\xi).$$

This density is given by

$$\bar{\rho}=e\int d\mathbf{p}\left(\frac{mc^2}{E(p)}\right)^{\frac{1}{2}}\frac{\exp i(\boldsymbol{\xi}\cdot\mathbf{p})/h}{8\pi^3h^3},$$

and for

$$r\ll h/mc:\ \bar{\rho}\sim e\frac{mc}{h}2^{-5/2}\pi^{-3/2}r^{-5/2};$$

for $r\gg h/mc$, $\bar{\rho}$ falls off exponentially.

In order to show that this "spread of charge" does not reduce the effect of a periodical field with a short wave-length, let us consider the operator

$$\int\rho(\mathbf{r})\exp i\mathbf{k}\cdot\mathbf{r}\,d\mathbf{r},$$

which represents an interaction energy between the charge and a field of wave number k. By inserting (5), this operator can be written in the form

$$e\sum_q a_q{}^*a_{q'},\quad \mathbf{p}_{q'}=\mathbf{p}_q+\mathbf{k}$$

and gives rise to transitions from any occupied state q' to any unoccupied state q. These transitions take place quite independently of the ratio of h/k to the linear dimensions h/mc of the spread of charge.

The energy W_{st} of the electrostatic field can be calculated directly from $\tilde{G}(\xi)$:

$$W_{st}=\tfrac{1}{2}\int\frac{\tilde{G}(\xi)}{|\xi|}d\xi.$$

The quadratic singularity of $\tilde{G}(\xi)$ at $\xi=0$ gives rise to a logarithmic divergence of W_{st}. By substituting (12) and by performing the integration over ξ first, we obtain the result

$$W_{st}=\frac{e^2}{4\pi^2}\int d\mathbf{p}\frac{mc^2}{h^3E_{(p)}p^2}$$

$$=\lim_{(P=\infty)}\frac{e^2}{\pi hc}mc^2\lg\frac{P+(P^2+m^2c^2)^{\frac{1}{2}}}{mc}\qquad(13)$$

or by putting $P=h/a$, where a is a length giving the "dimensions" of the electron, we get

$$W_{st}\sim\lim_{(a=0)}\frac{e^2}{\pi hc}m^2c^2\lg\frac{h}{mca}.$$

III. The Electromagnetic Field of the Electron

We calculate in this section the solenoidal part, E_s and H, of the electromagnetic field produced by the electron. It is given by

$$\mathbf{E}_s=-\frac{1}{c}\frac{\partial\mathbf{A}_s'}{\partial t},\qquad \mathbf{H}=\text{curl }\mathbf{A}'.$$

Here $\mathbf{A}_s'$ is the solenoidal part of the vector potential $\mathbf{A}$ which is given by

$$\mathbf{A}'(\mathbf{r},t)=\frac{1}{c}\int\frac{\mathbf{i}(\mathbf{r}',t-|\mathbf{r}-\mathbf{r}'|/c)}{|\mathbf{r}-\mathbf{r}'|}d\mathbf{r}'.$$

A is primed to indicate that this field is produced by the electron. The current density $\mathbf{i}$ is defined by

$$i_x=ec\{\psi^*\alpha_x\psi\},\quad\text{etc.}$$

Here α_i are the well-known Dirac matrices. We consider in our approximation the wave functions ψ to be the solutions of the wave equations of the field-free electron. If we expand the wave functions according to (6) we obtain

$$A_x'(r,t)=e\int d\mathbf{r}'\frac{\sum_q\sum_{q'} a_q^*a_{q'}\{\varphi_q^*(r't')\alpha_x\varphi_{q'}(r't')\}}{|\mathbf{r}-\mathbf{r}'|},$$

$$t'=t-\frac{|r-r'|}{c}.$$

The wave functions φ_q of the free electron are

$$\varphi_q=\frac{1}{V^{\frac{1}{2}}}u_q\exp-\frac{1}{h}(i\mathbf{p}_q\cdot\mathbf{r}-iE_qt). \quad (14)$$

p_q and E_q are momentum and energy of the state q, V is the volume of the space considered, u_q is a normalized spinor. We obtain then for A_x'

$$A_x(r,t)=\frac{2\pi eh^2c^2}{V}\sum_q\sum_{q'}\frac{\{u_q^*\alpha_xu_{q'}\}a_q^*a_{q'}}{E_qE_{q'}-m^2c^4-c^2\mathbf{p}_q\cdot\mathbf{p}_{q'}}$$

$$\times\exp i[(\mathbf{p}_{q'}-\mathbf{p}_q)\cdot\mathbf{r}-(E_{q'}-E_q)t]/h. \quad (15)$$

The field strengths can immediately be computed from this expression. The time average of the magnetic field energy, U_{mag} is found to be

$$U_{\text{mag}}=\frac{1}{8\pi}\int(\text{curl }\mathbf{A}')^2d\mathbf{r}$$

$$=\frac{\pi}{2}e^2h^2c^4\sum_q\sum_{q'}(\mathbf{p}_q-\mathbf{p}_{q'})^2\cdot N_q(1-N_{q'})$$

$$\times\frac{\{u_q^*\alpha_su_{q'}\}\{u_{q'}^*\alpha_su_q\}}{[E_qE_{q'}-m^2c^4-c^2\mathbf{p}_q\cdot\mathbf{p}_{q'}]^2} \quad (16)$$

if one uses the relations (9). α_s is the component of $\boldsymbol{\alpha}$ which is perpendicular to $\mathbf{p}_q-\mathbf{p}_{q'}$. We apply this expression first to a single electron at rest ($N_q=1$, $q=q_0$; $N_q=0$, $q\neq q_0$):

$$U_{\text{mag}}=\frac{\pi}{2}e^2h^2\sum_q p_q^2\frac{\{u_{q_0}^*\alpha_su_q\}\{u_q^*\alpha_su_{q_0}\}}{m^2(E_q-m^2c^4)^2}.$$

One obtains after averaging over the spin directions and replacing the sum by an integral:

$$U_{\text{mag}}=\pi\frac{e^2h^2}{m^2c^2}\int\frac{d\mathbf{p}}{8\pi^3h^3}=\frac{1}{2\pi}\lim_{(a=0)}\frac{e^2h^2}{3m^2c^2a^3}. \quad (17)$$

This divergent expression corresponds to the field energy of a magnetic dipole density concentrated in a sphere of infinitely small radius a. If one now calculated expression (16) for the vacuum of the positron theory ($N_{+q}=0$, $N_{-q}=1$) one obtains a highly divergent expression which represents the magnetic field energy $U_{\text{mag}}(\text{Vac.})$ produced by the current fluctuations of the vacuum. We are interested in the field energy of one electron at rest which is obtained by calculating $U_{\text{mag}}(\text{Vac.}+1)$ for the state ($N_{q_0}=1$, $p_{q_0}=0$, $E_{q_0}=mc^2$ all other $N_{+q}=0$, $N_{-q}=1$) and by subtracting the effect of the vacuum:

$$\tilde{U}_{\text{mag}}=U_{\text{mag}}(\text{Vac.}+1)-U_{\text{mag}}(\text{Vac.})$$

$$=2\pi e^2h^2(\sum_{+q}-\sum_{-q})p_q^2\frac{\{u_{q_0}^*\alpha_su_q\}\{u_q^*\alpha_su_{q_0}\}}{m^2(E_q-m^2c^4)^2}$$

Averaging over the spin directions gives:

$$U_{\text{mag}}=\pi\frac{e^2h^2}{m}\int\frac{d\mathbf{p}}{8\pi^3h^3E(p)}$$

$$=\frac{1}{2\pi}\frac{e^2}{mch}\lim_{(P=\infty)}\left[PP_0-\frac{m^2c^2}{2}\lg\frac{P+P_0}{mc}\right]. \quad (18)$$

Here P_0 is defined by $P_0=(P^2+m^2c^2)^{\frac{1}{2}}$. This quadratically divergent expression is just what one would expect for the field energy of the magnetic dipole density of the electron if this density is equal to the charge distribution calculated in the previous section.

In order to show directly, that this magnetic field is equal to the "static" field of the magnetic moment of the spin, we consider the magnetic polarization (dipole density) $\mathbf{M}$,

$$M_x=\frac{eh}{2mc}\{\psi^*\beta\alpha_y\alpha_z\psi\},$$

and calculate the function

$$J(\xi)=\int\mathbf{M}(r-\xi/2)\cdot\mathbf{M}(r+\xi/2)dr,$$

which corresponds to $G(\xi)$ (see (4)) and which provides information about the "spin distribution" in the electron. If one evaluates this integral by the method which is used in Section II, one obtains

$$J(\xi)=\frac{3}{4}\frac{h^2}{m^2c^2}G(\xi),$$

which shows[9] that the spin is distributed in exactly the same way as the charge. The magnetic field energy of this distribution is given by

$$U_{\text{mag}}=\tfrac{1}{2}\int\frac{\text{div. }\mathbf{M}(r_1)\text{ div. }\mathbf{M}(r_2)}{|\xi|}d\mathbf{r}$$

and can be evaluated by the methods used for the other calculations in this section and leads to the expressions (17) for the one-electron theory and to (18) for the positron theory. Hence we are allowed to consider them as the field energy of the magnetic moment.

The energy of the solenoidal electric field strength is given by

$$U_{\text{el}}=\frac{1}{8\pi c^2}\int\left(\frac{\partial A_s}{\partial t}\right)^2 d\mathbf{r}$$

$$=\frac{\pi}{2}\frac{e^2h^2}{c^2}\sum_q\sum_{q'}(E_{q'}-E_q)^2$$

$$\frac{\{u_q{}^*\alpha_s u_{q'}\}\{u_{p'}{}^*\alpha_s u_q\}}{[E_qE_{q'}-m^2c^4-c^2\mathbf{p}_q\cdot\mathbf{p}_{q'}]^2}N_q(1-N_{q'}).$$

Applied to a single electron at rest, this expression gives

$$U_{\text{el}}=\pi\frac{e^2h^2}{m^2c^2}\int\frac{d\mathbf{p}}{8\pi^3h^3}=U_{\text{mag}}. \tag{19}$$

Applied to an electron at rest in positron theory, one obtains, however,

$$\tilde{U}_{\text{el}}=U_{\text{el}}(\text{Vac.}+1)-U_{\text{el}}(\text{Vac.})$$

$$=\frac{2\pi e^2h^2}{c^2}\left(\sum_{+q}-\sum_{-q}\right)(E_q-mc^2)^2$$

$$\times\frac{\{u_{q_0}{}^*\alpha_s u_q\}\{u_q{}^*\alpha_s u_{q_0}\}}{m^2(E_q-m^2c^4)^2}$$

and then

$$\tilde{U}_{\text{el}}=-\pi\frac{e^2h^2}{m}\int\frac{d\mathbf{p}}{8\pi^3h^3E(p)}=-\tilde{U}_{\text{mag}}. \tag{20}$$

The interpretation of this result is given in Section I.

[9] The factor $\frac{3}{4}$ is $s(s+1)$ for $s=\frac{1}{2}$.

IV. The Self-Energy of the Electron

The self-energy is calculated in this section by means of a method which is different from the usual perturbation method, in order to outline the physical significance of the different terms. It is similar to the method applied to this problem previously by the author.[5]

The Hamiltonian of a system of charged particles and their electromagnetic field can be written in the form

$$\mathfrak{H}=\tfrac{1}{2}\int\left[\frac{1}{c^2}\left(\frac{\partial\mathbf{A}}{\partial t}\right)^2+\sum_{i,\,k=1}^{3}\left(\frac{\partial A_i}{\partial x_k}\right)^2 -\frac{1}{c^2}\left(\frac{\partial\phi}{\partial t}\right)^2-(\text{grad. }\phi)^2\right]d\mathbf{r}$$

$$+\int\left(\rho\phi-\frac{1}{c}\mathbf{i}\cdot\mathbf{A}\right)d\mathbf{r}$$

$$+c\sum_i\int\{\psi_i{}^*(\alpha\cdot\mathbf{p}_i+\beta mc)\psi_i\}d\mathbf{r}.$$

The summation in the last term is performed over all particles. The solutions of this Hamiltonian are restricted by the condition

$$\partial\phi/\partial t+c\text{ div. }\mathbf{A}=0.$$

By introducing the field strengths instead of the potentials we obtain the expression

$$\mathfrak{H}=\frac{1}{8\pi}\int(E^2+H^2)d\mathbf{r}-\frac{1}{4\pi}\int\phi\cdot\text{div. }Ed\mathbf{r}$$

$$+\int\left(\rho\phi-\frac{1}{c}\mathbf{i}\cdot\mathbf{A}\right)d\mathbf{r}$$

$$+c\sum_i\int\{\psi_i{}^*(\alpha\cdot\mathbf{p}_i+\beta mc)\psi_i\}d\mathbf{r}.$$

This is equal to (1) if one uses the relations div. $E=4\pi\rho$ and

$$T=c\sum_i\int\{\psi_i{}^*[\alpha\cdot(\mathbf{p}-(e/c)\mathbf{A})+\beta mc]\psi\}d\mathbf{r}.$$

The interaction energy

$$\int(\rho\phi-(1/c)\mathbf{i}\cdot\mathbf{A})=eH'$$

between matter and field contains the electronic charge e explicitly as a linear factor, so that in the above notation H' is explicitly independent of e; $\partial H'/\partial e=0$. Let us consider the energy W_s of a stationary state s of this Hamiltonian. If the electronic charge e is increased by de the Hamiltonian gets the additional term $de\, H'$. According to perturbation theory the increase of W_s is $de\langle H'\rangle_{\text{Av}}$, where $\langle H'\rangle_{\text{Av}}$ is the time average of H' in the state s, assuming that the electronic charge has its original value e. We therefore get

$$W_s=W_s^{(0)}+\int_0^e \langle H'(e)\rangle_{\text{Av}}de, \qquad (21)$$

where $W_s^{(0)}$ is the value of the energy for $e=0$. We now expand $H'(e)$ in a power series of e:

$$H'(e)=H'^{(0)}+eH'^{(1)}+\cdots$$

and get from (21)

$$W_s=W_s^{(0)}+e\langle H'^{(0)}\rangle_{\text{Av}}+\frac{e^2}{2}\langle H'^{(1)}\rangle_{\text{Av}}+\cdots.$$

The second term is zero since W_s cannot depend upon the sign of e, and we obtain by neglecting all terms containing e in a higher power than the second

$$W'=W_s-W_s^{(0)}$$

$$=\frac{e^2}{2}\langle H'^{(1)}\rangle_{\text{Av}}=\frac{e}{2}\langle H'\rangle_{\text{Av}}=\tfrac{1}{2}\int\left(\langle\rho\phi\rangle_{\text{Av}}-\frac{1}{c}\langle \mathbf{i}\cdot\mathbf{A}\rangle_{\text{Av}}\right)d\mathbf{r}.$$

The potentials can be split into two parts

$$\mathbf{A}=\mathbf{A}_0+\mathbf{A}', \qquad \phi=\phi_0+\phi'.$$

$\mathbf{A}_0$ and ϕ_0 are the potentials of the field when no electron is present. In empty space, A_0 is the potential of the zero-point oscillations and $\phi_0=0$. A' and ϕ' is the field produced by the electron. We then get

$$W'=\tfrac{1}{2}\int\left(\langle\rho\phi'\rangle_{\text{Av}}-\frac{1}{c}\langle\mathbf{i}\cdot\mathbf{A}'\rangle_{\text{Av}}\right)d\mathbf{r}$$
$$-\frac{1}{2c}\int\langle\mathbf{i}\cdot\mathbf{A}_0\rangle_{\text{Av}}d\mathbf{r}.$$

The first term can be transformed by means of the relations

$$\Delta\phi'-\frac{1}{c^2}\ddot{\phi}'=-4\pi\rho, \quad \Delta\mathbf{A}'-\frac{1}{c^2}\ddot{\mathbf{A}}'=-4\pi\mathbf{i},$$

and we obtain

$$W'=\frac{1}{8\pi}\int(\langle E'^2\rangle_{\text{Av}}-\langle H'^2\rangle_{\text{Av}})d\mathbf{r}-\frac{1}{2c}\int\langle\mathbf{i}\cdot\mathbf{A}_0\rangle_{\text{Av}}d\mathbf{r}.$$

We consider now the state s to be the state of an electron at rest. The charge dependent part W' of the self-energy can then be written in the form

$$W'=W_{\text{st}}+W_{\text{sp}}+W_{\text{fluct}}. \qquad (22)$$

Here

$$W_{\text{st}}=\frac{1}{8\pi}\int E_{\text{st}}^2 d\mathbf{r}$$

is the static field energy of the irrotational field E_{st}; W_{sp} is defined by

$$W_{\text{sp}}=\frac{1}{8\pi}\int(\langle E_s'^2\rangle_{\text{Av}}-\langle H'^2\rangle_{\text{Av}})d\mathbf{r}=U_{\text{el}}-U_{\text{mag}}. \qquad (23)$$

It contains the contribution of the field produced by the spin and is calculated in the previous section; W_{fluct} is the energy produced by the fluctuations of the radiation field,

$$W_{\text{fluct}}=-\frac{1}{2c}\int\mathbf{i}\cdot\mathbf{A}_0 d\mathbf{r}.$$

In order to calculate W_{fluct}, we divide $\mathbf{i}$ into two parts $\mathbf{i}=\mathbf{i}_0+\mathbf{i}'$, where $\mathbf{i}_0$ is the current density for the field-free case. The term $\int\mathbf{i}_0\cdot\mathbf{A}_0 d\mathbf{r}$ vanishes when averaged over the time because of the absence of phase relations between $\mathbf{i}_0$ and $\mathbf{A}_0$. The remaining term

$$W_{\text{fluct}}=-\frac{1}{2c}\int\mathbf{i}'\cdot\mathbf{A}_0 d\mathbf{r}$$

can be evaluated as follows:

$\mathbf{i}'$ is to a first approximation given by

$$\mathbf{i}'=\{\psi_0{}^*\alpha\psi_1\}+\{\psi_1{}^*\alpha\psi_0\}. \qquad (24)$$

Here $\psi_0=\Sigma a_q\varphi_q$ is the wave function unperturbed by the field and ψ_1 is the first approximation of the perturbed wave function:

$$\psi_1 = \sum a_q \varphi_q'.$$

φ_q' can be calculated by means of the ordinary perturbation method. The interaction energy with an arbitrary field **A** is given by

$$-e\boldsymbol{\alpha}\mathbf{A} = -e\alpha\sum_k [A_k^{(+)} \exp i(\mathbf{k}\cdot\mathbf{r}+c|k|t) + A_k^{(-)} \exp -i(\mathbf{k}\cdot\mathbf{r}+c|k|t)].$$

The sum is taken over all wave numbers **k** of an orthogonal system of plane waves in the volume V. If we write the wave functions φ_q of the free electron in the form (14), we get for φ_q'

$$\varphi_q' = e\sum_{q'} \varphi_{q'} \left[\frac{\{u_{q'}^* \alpha u_q\} A_k^{(+)}}{E_q - E_{q'} + c|k|} e^{+ic|k|t} + \frac{\{u_{q'}^* \alpha u_q\} A_k^{(-)}}{E_q - E_{q'} - c|k|} e^{-ic|k|t} \right], \quad h\mathbf{k} = \mathbf{p}_{q'} - \mathbf{p}_q.$$

This expression is introduced into (24) and gives for the current

$$i_x' = \sum_q \sum_{q'} N_q \{u_q^* \alpha_x u_{q'}\}\{u_{q'}^* \boldsymbol{\alpha} u_q\} \times \left[\frac{\mathbf{A}^{(+)} \exp i(\mathbf{k}\cdot\mathbf{r}+c|k|t)}{E_q - E_{q'} + c|k|} + \frac{\mathbf{A}^{(-)} \exp -i(\mathbf{k}\cdot\mathbf{r}+c|k|t)}{E_q - E_{q'} - c(k)} \right] + \text{conj.}$$

We have retained only terms containing products (8) or (9). It is seen from this expression, that i' is the same in the one-electron theory and in the positron theory. In the latter case we have to consider $\bar{i} = i'(\text{Vac.}+1) - i'(\text{Vac.})$. The actual value can easily be evaluated for an electron at rest. One obtains

$$\mathbf{i}' = -\frac{1}{V}\frac{e^2}{mc}\mathbf{A},$$

which is immediately understood as the forced vibrations of a point charge under the action of an oscillating field **A**. W_{fluct} is directly obtained if one replaces **A** by the field fluctuations $\mathbf{A}_0$ of the vacuum:

$$W_{\text{fluct}} = -\frac{1}{2c}\int \mathbf{i}'\cdot\mathbf{A}_0 d\mathbf{r} = \frac{e^2}{2mc^2}\langle A_0^2\rangle_{\text{Av}}.$$

We have

$$\langle A_0^2\rangle_{\text{Av}} = \frac{1}{\pi^2}\int \frac{ch}{|k|} d\mathbf{k} \qquad (25)$$

and we get finally, on replacing $|k|$ by p/h,

$$W_{\text{fluct}} = \frac{e^2}{2\pi^2 mc^2}\int \frac{c d\mathbf{p}}{h|p|} = \frac{e^2}{\pi hc}\frac{1}{m}\lim_{(P=\infty)} P^2$$

We now collect the results obtained for the other parts of the self-energy of an electron at rest. The one-electron theory gives [Eqs. (11), (23), (19)]

$$W_{\text{st}} = \lim_{(a=0)} \frac{e^2}{a}, \quad W_{\text{sp}} = U_{\text{el}} - U_{\text{mag}} = 0.$$

The positron theory gives (Eqs. (13), (23), (20))

$$W_{\text{st}} = \lim_{(P=\infty)} \frac{e^2}{\pi hc} mc^2 \lg \frac{P+P_0}{mc},$$

$$W_{\text{sp}} = U_{\text{el}} - U_{\text{mag}} = -\frac{e^2}{\pi hc}\frac{1}{m}\lim_{(P=\infty)} \left[PP_0 - \frac{m^2c^2}{2}\lg\frac{P+P_0}{mc} \right].$$

W_{sp} is partly balanced by W_{fluct}. The total self-energy in positron theory is then given by

$$W' = \frac{3}{2\pi}\frac{e^2}{hc} mc^2 \lim_{(P=\infty)} \lg\frac{P+P_0}{mc} + \text{finite terms.} \qquad (26)$$

The self-energy of a free electron in motion can be obtained by a Lorentz transformation from (26). The direct calculation from the above methods is ambiguous because it leads to a difference of terms, each of which diverges quadratically. The factor of the logarithmically divergent difference of these terms depends essentially on the way in which the infinite terms are subtracted. The calculation of the self-energy of an electron at rest is not so much exposed to these ambiguities because of the spherical symmetry of the problem, which suggests only one natural way of subtracting two divergent integrals over the momentum space, namely, the subtraction of the contributions of concentric spherical shells around the center. It must be expected, that the value of the self-energy of a moving electron can only be covariant to the value (26) if one performs the subtraction appropriately. This is why the expressions for the self-energy obtained in reference 5 are apparently not relativistically covariant.

V. The Self-Energy of a Particle Obeying the Bose-Statistics

It has been shown that the quantization of the scalar wave equation of Klein and Gordon leads to a theory of elementary particles with Bose statistics and charges of both signs. The theory includes a description of pair creation and of all related phenomena. The quantitative results are not very different from the results of Dirac's positron theory. The formalism has been recently applied to particles with intrinsic angular momentum. It will be shown here that the calculation of the self-energy, however, gives results quite different from the positron theory. The energy of the electrostatic field of the electron is found to be more strongly divergent than in the classical theory; the energy of the radiation field diverges quadratically and is equal to the corresponding energy of a single electron in the one-electron theory. The qualitative arguments for this behavior are given in I. The following calculation is based on the formulas derived elsewhere.[7]

The operator of the charge density is given by

$$\rho = ie(\psi^*\pi^* - \psi\pi),$$

where ψ is the wave function and π its conjugate operator:

$$\pi = h\frac{\partial\psi^*}{\partial t},$$

$$\psi(r)\pi(r') - \pi(r')\psi(r) = \delta(r-r').$$

We introduce new variables by means of

$$\psi = \frac{1}{V^{\frac{1}{2}}}\sum_k q(\mathbf{k}) \exp i\mathbf{k}\cdot\mathbf{r},$$
$$\pi = \frac{1}{V^{\frac{1}{2}}}\sum_k p(\mathbf{k}) \exp -i\mathbf{k}\cdot\mathbf{r}. \tag{27}$$

Here $1/V^{\frac{1}{2}}\cdot\exp i\mathbf{kr}$ form a set of orthogonal functions in the volume V.

We further introduce

$$p(k) = \left(\frac{E(k)}{2}\right)^{\frac{1}{2}}(a^*(\mathbf{k}) + b(\mathbf{k})),$$
$$q(k) = -i\left(\frac{1}{2E(k)}\right)^{\frac{1}{2}}(a^*(\mathbf{k}) - b(\mathbf{k})), \tag{28}$$
$$E(k) = c(h^2k^2 + m^2c^2)^{\frac{1}{2}},$$

and obtain for

$$\rho(\mathbf{s}) = \frac{1}{V}\int \rho(\mathbf{r}) \exp(-i\mathbf{s}\cdot\mathbf{r})d\mathbf{r}$$

the following expression:

$$\rho(\mathbf{s}) = \frac{e}{2V}\sum_k \frac{E(k)+E(l)}{[E(l)E(k)]^{\frac{1}{2}}}[a^*(\mathbf{k})a(\mathbf{l}) - b^*(\mathbf{k})b(\mathbf{l})] + \frac{E(k)-E(l)}{[E(k)E(l)]^{\frac{1}{2}}}[a(\mathbf{k})b(\mathbf{l}) - a^*(\mathbf{k})b^*(\mathbf{l})], \tag{29}$$

where $\mathbf{l} = \mathbf{k} + \mathbf{s}$.

The $a(\mathbf{k})$ and $b(\mathbf{k})$ fulfill the relations

$$a^*(\mathbf{k})a(\mathbf{k}) = N(\mathbf{k}), \qquad b^*(\mathbf{k})b(\mathbf{k}) = M(\mathbf{k}),$$
$$a(\mathbf{k})a^*(\mathbf{k}) = 1 + N(\mathbf{k}), \qquad b(\mathbf{k})b^*(\mathbf{k}) = 1 + M(\mathbf{k}).$$

The $N(\mathbf{k})$'s are the numbers of positrons, the $M(\mathbf{k})$'s the number of negatrons in the states with the wave vector $\mathbf{k}$. The electrostatic self-energy is given by:

$$W_{st} = \tfrac{1}{2}\int \frac{\rho(\mathbf{r}-\xi/2)\rho(r+\xi/2)}{|\xi|}d\mathbf{r}d\xi = 2\pi V\sum_s \frac{\rho(-\mathbf{s})\rho(\mathbf{s})}{s^2}.$$

We obtain by introducing (29) and retaining only the diagonal terms:[10]

$$W_{st} = \frac{\pi e^2}{2V}\sum_k\sum_s \frac{1}{s^2}\left[\frac{(E(k)+E(l))^2}{E(k)E(l)} \times (N(\mathbf{k})[N(\mathbf{l})+1] + M(\mathbf{k})[M(\mathbf{l})+1]) + \frac{(E(\mathbf{k})-E(\mathbf{l}))^2}{E(k)E(l)} \times (N(\mathbf{k})M(\mathbf{l}) + [N(\mathbf{k})+1][M(\mathbf{l})+1])\right].$$

This expression does not vanish for the vacuum ($N(\mathbf{k}) = M(\mathbf{k}) = 0$ for every $\mathbf{k}$). We calculate the difference

$$\bar{W}_{st} = W_{st}(\text{Vac.}+1) - W_{st}(\text{Vac.})$$

---

[10] The term $s=0$ is omitted. It can easily be shown that this term does not contribute for $V \to \infty$.

and assume that the particle is at rest:

$$\tilde{W}_{\mathrm{st}}=\frac{e^2h}{16mc}\int ds\frac{1+2m^2c^2/h^2s^2}{(s^2+m^2c^2/h^2)^{\frac{1}{2}}}$$

$$=e^2\frac{h}{mc}\frac{\pi}{4}\lim_{(P=\infty)}\left[\frac{P^2}{h^2}+\frac{m^2c^2}{h^2}\lg\frac{P+P_0}{mc}\right].$$

This is an expression which diverges quadratically. By putting $P=h/a$ one obtains

$$\tilde{W}_{\mathrm{st}}\cong\frac{\pi}{4}mc^2\left(\frac{h}{mca}\right)^2.$$

We now show that the particle does not produce a solenoidal electric or magnetic field in the approximation considered here. The current density is given by Eq. (43) of reference 7:

$$\mathbf{i}=\mathbf{i}_1+\mathbf{i}_2,$$

$$\mathbf{i}_1=ihce(\psi\ \mathrm{grad.}\ \psi^*-\psi^*\ \mathrm{grad.}\ \psi),$$

$$\mathbf{i}_2=-2e^2\mathbf{A}\psi^*\psi.$$

We introduce the new variables (27) into i_1 and get

$$\mathbf{i}_1=\sum_s i(\mathbf{s})\exp i\mathbf{r}\cdot\mathbf{s},$$

$$i(\mathbf{s})=\tfrac{1}{2}hce\sum_k\frac{\mathbf{k}+\mathbf{l}}{[E(k)E(l)]^{\frac{1}{2}}}(a^*(\mathbf{k})a(\mathbf{l})+b(\mathbf{k})b^*(\mathbf{l})$$
$$-a(\mathbf{k})b(\mathbf{l})-b(\mathbf{k})a(\mathbf{l})),\quad \mathbf{l}=\mathbf{k}+\mathbf{s}.$$

$i_1(\mathbf{s})$ is proportional to $\mathbf{k}+\mathbf{l}$ and $\mathbf{i}_1$ is therefore irrotational for all transitions which start or end with a particle at rest. ($\mathbf{k}+\mathbf{l}$ is parallel to $\mathbf{s}$ if $\mathbf{k}=0$ or $\mathbf{l}=0$). Thus $\mathbf{i}_1$ does not produce a solenoidal field. $\mathbf{i}_2$ is proportional to e^2 so that its field does not come into consideration. The particle does not give rise to a solenoidal field as long as it is at rest since it has no magnetic spin moment.

The remaining term of the self-energy is

$$W_{\mathrm{fluct}}=-\frac{1}{2c}\int\mathbf{i}'\cdot\mathbf{A}_0d\mathbf{r}. \qquad (22)$$

$\mathbf{i}'$ is defined as the current density produced by the field $\mathbf{A}_0$. Here the first part $\mathbf{i}_1'$ can be shown to be again irrotational. The integral over the product of $\mathbf{i}_1'$ and the solenoidal vector $\mathbf{A}_0$ vanishes. The second part $\mathbf{i}_2'$ is in the required approximation directly given by

$$\mathbf{i}_2'=-2e^2\mathbf{A}_0\psi^*\psi.$$

We obtain then

$$W_{\mathrm{fluct}}=e^2\int\mathbf{A}_0{}^2\psi^*\psi d\mathbf{r}.$$

By introducing the new variables and retaining only diagonal elements we find

$$W_{\mathrm{fluct}}=\frac{e^2}{2}\langle A_0{}^2\rangle_{\mathrm{Av}}\sum_k\frac{N(\mathbf{k})+M(\mathbf{k})+1}{E(k)}$$

and finally

$$\tilde{W}_{\mathrm{fluct}}=W_{\mathrm{fluct}}(\mathrm{Vac.}+1)-W_{\mathrm{fluct}}(\mathrm{Vac.})$$

$$=\frac{e^2}{2mc^2}\langle A_0{}^2\rangle_{\mathrm{Av}}.$$

This is identical with the corresponding expression for the Dirac electron.

VI. The Higher Approximations of the Self-Energy in the Theory of the Positron

It will be proved in this section that the successive approximations of the self-energy of the electron vanish in the limit $m\rightarrow 0$. Furthermore it is shown that the divergence of the self-energy is logarithmic in every approximation.[11]

We consider the total system containing the electrons and the radiation field and calculate the energy $W(s)$ of the state s of this system. $W(s)$ can be expanded in a series of approximations $W(s)=\Sigma_n W^{(n)}(s)$ corresponding to an expansion in powers of the parameter e^2/hc. Since this procedure does not give zero for the vacuum in Dirac's positron theory, the self-energy of the electron must be defined as the difference between the energy $W(\mathrm{Vac.}+1)$ of the state in which one electron is present and the energy $W(\mathrm{Vac.})$ of the vacuum alone. We confine ourselves to the calculation of the electrodynamic

[11] Recently A. Mercier (Helv. Phys. Acta 12, 55 (1938)) has treated the same problem and has obtained a higher divergence. As he dces not compute the numerical factors of the divergent expressions, he cannot exclude a factor zero for the highest divergent terms. The following considerations, however, show that the highest nonvanishing terms diverge only logarithmically.

self-energy. The calculation of the electrostatic energy and the mixed terms in higher approximation can be made along the same lines.

We now consider the detailed form of the nth approximation $W_s^{(n)}$ of the energy of the state s. $W_s^{(n)}$ is a sum of terms containing a product of $2n$ matrix elements of the interaction energy which correspond to consecutive transitions of the total system from one state to another starting from the state s and returning to it. The terms have denominators that are products of energy differences between the original state s and intermediate states.

The self-energy of the electron $\tilde{W}^{(n)}$ is given by the difference

$$\tilde{W}^{(n)} = W^{(n)}(\text{Vac.}+1) - W^{(n)}(\text{Vac.}). \quad (30)$$

We now prove that $\tilde{W}^{(n)}=0$ for $m=0$ by comparing it with the self-energy of a positron $\tilde{W}'^{(n)}$ in the same state:

$$\tilde{W}'^{(n)} = W^{(n)}(\text{Vac.}-1) - W^{(n)}(\text{Vac.}) = \tilde{W}^{(n)}, \quad (31)$$

which is equal to the self-energy of the electron. There is no loss of generality if we confine our considerations to the self-energy of an electron at rest. The state (Vac.+1) is then specified by: every negative energy state and the lowest positive state s_{+0} occupied; (Vac.−1) means: every negative energy state except the highest one s_{-0} is occupied. We now show that

$$W^{(n)}(\text{Vac.}+1) + W^{(n)}(\text{Vac.}-1) = 2W^{(n)}(\text{Vac.}) \quad \text{for} \quad m=0. \quad (32)$$

Comparing $W^{(n)}(\text{Vac.}+1)$ and $W^{(n)}(\text{Vac.}-1)$ with $W^{(n)}(\text{Vac.})$ we notice:

(a) $W^{(n)}(\text{Vac.}-1)$ lacks all terms containing transitions of the electron in the state s_{-0}.

(b) $W^{(n)}(\text{Vac.}+1)$ contains additional terms from transitions of the additional electron in s_{+0}.

(c) $W^{(n)}(\text{Vac.}-1)$ contains additional terms from transitions of one of the vacuum electrons into the empty state s_{-0}.

(d) $W^{(n)}(\text{Vac.}+1)$ lacks terms containing transitions of the vacuum electrons into the state s_{+0} of the additional electron.

We now prove that the missing terms of (a) and (d) are in the limit $m\to 0$ identical with the additional terms of (b) and (c), respectively. The only difference between the pair (a), (b) and the pair (c), (d) consists in the fact that the specified transitions start from a positive or from a negative state, respectively. This fact does not affect the energy differences in the denominators in the limit $m\to 0$, as in this limit the energies of s_{+0} and s_{-0} are equal. It remains to show that the numerators are also unchanged. This can be seen in the following way: the transition matrix elements appearing in the denominator belong to a chain of consecutive transitions starting from and returning to the initial state of the system. Thus the transition elements from or to s_{+0} or s_{-0} belong to a chain of transitions of an electron starting from the state s_{+0} (or from s_{-0}, respectively) and returning to this state. The transition elements form the product:

$$P_{\pm} = \{\psi^*(\pm 0)H_1\psi(p_1)\}\{\psi^*(p_1)H_2\psi(p_2)\}\cdots \times\{\psi^*(p_{n-1})H_n\psi(\pm 0)\}. \quad (33)$$

Here H_i is the interaction energy with the light quantum which is emitted or absorbed with the ith transition and $\psi(p_i)$ is the wave function of the electron with the momentum p_i which performs the transition. $\psi(\pm 0)$ is the wave function of the state s_{+0} or s_{-0}, respectively. After averaging over the two spin states for every momentum p_i, this product can be written as trace of the following matrix:

$$P_{\pm} = \text{Trace}\left[H_1 \cdot \tfrac{1}{2}\left(1+\frac{\boldsymbol{\alpha}\cdot\mathbf{p}_1+\beta mc}{E_1}\right)\cdot \text{H}_2 \cdot \tfrac{1}{2}\left(1+\frac{\boldsymbol{\alpha}\cdot\mathbf{p}_2+\beta mc}{E_2}\right)\cdots H_n \tfrac{1}{2}(1\pm\beta)\right].$$

Here $E_i = \pm(p_i^2+m^2c^2)^{\frac{1}{2}}$ is the energy of the wave function $\psi(p_i)$. If we now go to the limit $m\to 0$, all terms with β disappear except those containing the β in the last parenthesis. (H_i does not contain the operator β.)[12] Since the trace of every expression containing α's but only one β is zero, we are allowed to omit the last also, and it becomes evident that $P_+ = P_-$. From (30), (31)

[12] This conclusion does not hold if one or more of the p_i's are $\sim mc$. As $\tilde{W}^{(n)}$ is an integral over all possible intermediate momenta p_i in the volume V, the terms with one or more $p_i \sim mc$ contain one or more factors $(mc)^3$ so that these contributions should be neglected.

and (32) it follows directly that

$$\tilde{W}^{(n)} \doteq 0 \quad \text{for} \quad m=0.$$

We now prove that this result infers a logarithmical divergence of $\tilde{W}^{(n)}$ for finite m. The terms of $\tilde{W}^{(n)}$ contain transitions in which n light quanta are emitted and absorbed. We write $W^{(n)}$ as an integral over the wave vectors $\mathbf{k}_1 \cdots \mathbf{k}_n$ of these light quanta:

$$\tilde{W}^{(n)} = \int_0^{P/h} d\mathbf{k}_1 \cdots \int_0^{P/h} d\mathbf{k}_n F(k_1 \cdots k_n) \quad (34)$$

and integrate to a finite limit $P/h \gg mc/h$ in order to make $\tilde{W}^{(n)}$ finite. The properties of $F(k_1 \cdots k_n)$ are very simple for all $p_i \gg mc$ where p_i are as above the momenta of the electrons which change their states in the transitions to intermediate states (except of course the momentum $p_0 = 0$ of the electron under consideration in its initial state). It can be shown that replacing every k_l $(l = 1 \cdots n)$ by $k_l' = xk_l$ gives

$$F(k_1' \cdots k_n') = x^N F(k_1 \cdots k_n) \quad \text{if} \quad p_i \gg mc,$$

where N is an integer. This result may be understood in the following way. By means of the substitution $k_l' = xk_l$ the momenta p_i of the excited electrons in the intermediate states are also multiplied by a since the p_i are sums or differences of the momenta of the absorbed or emitted light quanta. If now $k_l \gg mc$, all momenta p_i involved are large compared to mc, and the corresponding energies can be replaced by $c|p_i|$. This neglect of mc compared to p_i has as consequence that all energy differences $E_s - E_i$ between the initial state and the intermediate states are multiplied by x if one replaces k_l by k_l'. Since all terms of $\tilde{W}^{(n)}$ have $2n-1$ energy differences in the denominator, the latter is proportional to x^{2n-1}. The numerators consist of n matrix elements which form expressions like (33). From the fact that every H_i is proportional to $k_i^{-\frac{1}{2}}$ and from the structure of (33) it follows that the numerators are sums of terms proportional to $x^{-n}, x^{-n-1} \cdots x^{-2n}$. We get, therefore,

$$F(k_1' \cdots k_n') = \sum_{z=3n-1}^{z=5n-1} c_z x^{-z} \quad \text{for} \quad p_i \gg mc.$$

Not all of these c_z need to be different from zero. If c_ξ is the first coefficient different from zero, we can write

$$F(k_1' \cdots k_n') = c_\xi x^{-\xi} \cong x^{-\xi} F(k_1 \cdots k_n) \quad (35)$$

because we certainly can neglect the terms with $z > \xi$, which are smaller by the ratio $\sim (mc/p_i)^{z-\xi}$. We are now interested in the function $\tilde{W}^{(n)}(P)$. The relation (35) is not valid in the entire region of integration in (34). The regions in which the conditions $p_i \gg mc$ are not fulfilled are restricted to certain small areas in the $3n$-dimensional space of the wave vectors $\mathbf{k}_1 \cdots \mathbf{k}_n$. The contributions of these areas can be neglected for $P \gg mc$. We therefore get

$$\tilde{W}^{(n)}(xP) \cong \int_0^{P/h} d\mathbf{k}_1' \cdots \int_0^{P/h} d\mathbf{k}_n' F(k_1' \cdots k_n')$$
$$\cong x^{3n-\xi} \cdot W'^{(n)}(P) + \text{smaller terms}.$$

The additional smaller terms come from the regions of integration in which these considerations are not applicable and from the neglected terms. From this relation follows, that

$$\tilde{W}^{(n)} \cong c\left(\frac{e^2}{hc}\right)^n \cdot \frac{P^N}{(mc)^N} \cdot \left(\lg \frac{P}{mc}\right)^t \cdot mc^2 + \text{smaller terms}.$$

Here $N = 3n - \xi$, c is a numerical factor and $0 \leqq t \leqq n$ because any of the n integrations might give rise to a logarithm. The factors mc are applied in order to make the dimensions fit. It was proved that

$$\lim_{(m=0)} \tilde{W}^{(n)} = 0.$$

This is only possible if $N \leqq 0$. This is equivalent to the fact that $\tilde{W}^{(n)}$ does not diverge stronger than

$$\tilde{W}^{(n)} \sim \left(\frac{e^2}{hc}\right)^n mc^2 \left(\lg \frac{P}{mc}\right)^n.$$

THÉORIE DU POSITRON

PAR M. P.-A.-M. DIRAC.

La découverte récente de l'électron positif ou positron a ramené l'attention vers une théorie déjà ancienne sur les états d'énergie négative de l'électron, les résultats expérimentaux obtenus jusqu'ici se trouvant d'accord avec les prévisions de cette théorie.

La question des énergies négatives se pose dès que l'on étudie le mouvement d'une particule conformément au principe de relativité restreinte. Dans la mécanique non relativiste, l'énergie W d'une particule est donnée en fonction de sa vitesse v ou de sa quantité de mouvement p par :

$$W = \frac{1}{2} m v^2 = \frac{1}{2m} p^2,$$

ce qui correspond à un W toujours positif; mais en mécanique relativiste ces formules doivent être remplacées par :

$$W^2 = m^2 c^4 + c^2 p^2,$$

ou

$$W = c\sqrt{m^2 c^2 + p^2},$$

ce qui permet à W d'être positif ou négatif.

On fait d'ordinaire l'hypothèse supplémentaire que l'énergie W doit toujours être positive. Cela est admissible en théorie classique où les grandeurs varient toujours de manière continue et où W ne peut par conséquent jamais passer d'une de ses valeurs positives, qui doit être $\geqq mc^2$, à une de ses valeurs négatives qui doit être $\leqq -mc^2$. Dans la théorie quantique, au contraire, une variable peut subir des changements discontinus, de sorte que W peut passer d'une valeur positive à une valeur négative.

Il n'a pas été possible de développer une théorie quantique relativiste de l'électron dans laquelle les transitions d'une valeur positive à une valeur négative de l'énergie soient exclues. Il n'est donc plus possible d'admettre que l'énergie est toujours positive sans qu'il en résulte des inconséquences dans la théorie.

Dans ces conditions, deux possibilités nous restent ouvertes. Ou bien nous devons trouver une signification physique pour les états d'énergie négative, ou bien nous devons admettre que la théorie quantique relativiste est inexacte dans la mesure où elle prévoit des transitions entre les états d'énergie positive et ceux d'énergie négative. Or de semblables transitions sont en général prévues pour tous les processus mettant en jeu des échanges d'énergie de l'ordre de mc^2 et il ne semble y avoir aucune raison de principe contre l'applicabilité de la mécanique quantique actuelle à de semblables échanges d'énergie. Il est vrai que cette mécanique ne semble pas pouvoir s'appliquer aux phénomènes dans lesquels interviennent des distances de l'ordre du rayon classique de l'électron $\frac{e^2}{mc^2}$, puisque la théorie actuelle ne peut en aucune façon rendre compte de la structure de l'électron, mais de telles distances, considérées comme longueurs d'onde électroniques, correspondent à des énergies de l'ordre $\frac{hc}{e^2}\, mc^2$, beaucoup plus grandes que les changements en question. Il semble donc que la solution la plus raisonnable est de chercher un sens physique pour les états d'énergie négative.

Un électron dans un état d'énergie négative est un objet tout à fait étranger à notre expérience, mais que nous pouvons cependant étudier au point de vue théorique; nous pouvons, en particulier, prévoir son mouvement dans un champ électromagnétique quelconque donné. Le résultat du calcul, effectué soit en mécanique classique, soit en théorie quantique, est qu'un électron d'énergie négative est dévié par le champ exactement comme le serait un électron d'énergie positive s'il avait une charge électrique positive $+ e$ au lieu de la charge négative habituelle $- e$.

Ce résultat suggère immédiatement une assimilation entre l'électron d'énergie négative et le positron. On serait tenté d'admettre qu'un électron dans un état d'énergie négative constitue précisément un positron, mais cela n'est pas acceptable, parce que le positron observé n'a certainement pas une énergie cinétique négative.

Nous pouvons obtenir un meilleur résultat en utilisant le principe d'exclusion de Pauli, en vertu duquel un état quantique donné ne peut être occupé par plus d'un électron. Admettons que dans l'Univers tel que nous le connaissons, les états d'énergie négative soient presque tous occupés par des électrons, et que la distribution ainsi obtenue ne soit pas accessible à notre observation à cause de son uniformité dans toute l'étendue de l'espace. Dans ces conditions, tout état d'énergie négative non occupé représentant une rupture de cette uniformité, doit se révéler à l'observation comme une sorte de lacune. Il est possible d'admettre que ces lacunes constituent les positrons.

Cette hypothèse résout les difficultés principales de l'interprétation des états d'énergie négative. Une lacune dans la distribution des électrons d'énergie négative représente une énergie positive, puisqu'elle correspond à un défaut local d'énergie négative. De plus, le mouvement de cette lacune dans un champ électromagnétique quelconque est exactement le même que celui de l'électron nécessaire pour combler la lacune. Nous pouvons tirer de là deux conclusions : d'abord que le mouvement de la lacune peut être représenté par une fonction d'onde de Schrödinger analogue à celle qui représente le mouvement d'un électron, et ensuite, que la lacune se comporte dans un champ de la même manière qu'un électron positif d'énergie positive. Ainsi la lacune prend exactement l'aspect d'une particule ordinaire électrisée positivement et son identification avec le positron se présente comme tout à fait plausible.

Si notre hypothèse est correcte, nous devons pouvoir en déduire un certain nombre de conséquences expérimentalement vérifiables. Tout d'abord, la masse du positron doit être exactement égale à celle de l'électron et sa charge doit être exactement égale et opposée à celle de l'électron. De plus nous pouvons prévoir certains résultats concernant la création et la disparition des positrons.

Un électron ordinaire d'énergie positive ne peut pas sauter dans l'un des états occupés d'énergie négative, en raison du principe de Pauli; il peut, au contraire, sauter dans une lacune pour la combler. Ainsi un électron et un positron peuvent se détruire réciproquement. Leur énergie doit se retrouver sous forme de photons et il résulte des principes de conservation de l'énergie et de la quantité de mouvement que deux photons au moins doivent être produits. On peut calculer la probabilité pour qu'un tel processus ait lieu et obtenir ainsi la vie probable d'un positron se mouvant à travers une distribution donnée d'électrons. Le résultat est une vie moyenne de 3×10^{-7} sec pour un positron en mouvement lent dans l'air à la pression atmosphérique, cette durée moyenne augmentant avec la vitesse. La valeur ainsi obtenue est d'un ordre de grandeur compatible avec l'expérience, puisqu'elle est suffisante pour permettre à un positron rapide de traverser une chambre de condensation de Wilson sans y être en général détruit, et assez petite cependant pour que les positrons ne soient pas des objets communément présents au laboratoire.

Un électron et un positron peuvent s'annihiler réciproquement en donnant naissance à un seul photon si un noyau atomique est présent pour absorber la quantité de mouvement libérée. Le processus inverse consiste dans la production d'un positron et d'un électron par la rencontre d'un seul photon d'énergie suffisante avec un noyau atomique. On peut se le représenter comme un effet photo-électrique sur un des électrons d'énergie négative décrivant des orbites hyperboliques au voisinage du noyau; cet électron étant élevé vers un état d'énergie positive et apparaissant ainsi comme électron ordinaire, tandis qu'il laisse derrière lui une lacune se comportant comme un positron. La probabilité d'apparition d'un tel processus a été calculée approximativement par Oppenheimer et indépendamment par Peierls, et le résultat est d'un ordre de grandeur qui concorde avec les observations relatives à la production de positrons par des rayons γ durs tombant sur des noyaux lourds.

Pour que la conception que nous proposons des états d'énergie négative se développe en théorie complète, nous devons considérer non seulement le mouvement des électrons et des lacunes dans un champ, mais aussi la manière dont un champ électromagnétique est produit par les électrons et les lacunes. Pour cela, il est nécessaire d'introduire une hypothèse nouvelle puisque la conception ordinaire que chaque charge $-e$ sur un électron contribue à produire la densité électrique ρ qui détermine le champ électrique $\vec{E}$ conformément à l'équation de Maxwell :

$$\text{div}\,\vec{E} = 4\pi\rho \tag{1}$$

conduirait évidemment à un champ infini en tout point.

Faisons l'hypothèse que la distribution d'électrons dans laquelle aucun état d'énergie positive n'est occupé, tandis que tous les états d'énergie négative le sont ne produit aucun champ, et que ce sont les écarts à partir de cette distribution qui déterminent les champs conformément à l'équation (1). Dans cette hypothèse, un état d'énergie positive occupé produirait un champ correspondant à une charge négative $-e$ et un état d'énergie négative non occupé produirait un champ correspondant à une charge positive $+e$. Nous obtenons ainsi une nouvelle propriété des lacunes qui contribue à rendre vraisemblable notre assimilation de ces lacunes avec les positrons.

La nouvelle hypothèse est tout à fait satisfaisante lorsqu'il s'agit d'une région de l'espace où n'existe aucun champ, et où la distinction entre les états d'énergie positive et ceux d'énergie négative est nettement définie; mais elle doit être précisée lorsqu'il s'agit d'une région de l'espace où le champ électromagnétique n'est pas nul pour pouvoir conduire à des résultats libres de toute ambiguïté. Il faut spécifier mathématiquement quelle distribution d'électrons est supposée ne produire aucun champ et donner aussi une règle pour soustraire cette distribution de celle qui existe effectivement dans chaque problème particulier, de manière à obtenir une différence finie qui peut figurer dans (1), puisque, en général, l'opération mathématique de soustraction entre deux infinités est ambiguë.

Cette question n'a pas encore été examinée ni résolue pour le cas général d'un champ électromagnétique arbitraire. Il y a, cependant, un cas particulier dans lequel les hypothèses nécessaires semblent assez évidentes : celui d'un champ électrostatique permanent. Nous allons traiter ici ce cas, en supposant le champ suffisamment faible pour qu'une méthode de perturbation puisse être utilisée. Nous constaterons que la distribution qui ne produit aucun champ ne satisfait pas aux équations du mouvement. En retranchant cette distribution de celle qui satisfait à ces équations et qui correspond à un état où ne sont présents ni électrons ni positrons, nous obtiendrons une différence qui pourra être interprétée physiquement comme un effet de polarisation par le champ électrique de la distribution normale des électrons d'énergie négative.

Nous emploierons la méthode d'approximation de Hartree-Fock qui attribue à chaque électron sa propre fonction d'onde individuelle $\psi(q)$, et nous introduirons la matrice densité R définie par

$$(q' | \mathrm{R} | q'') = \Sigma_r \overline{\psi_r}(q') \psi_r(q''),$$

la sommation s'étendant à tous les électrons, c'est-à-dire à tous les états occupés. Toute distribution d'électrons peut être caractérisée par une semblable matrice, au degré de précision que comporte la méthode de Hartree-Fock. Cette représentation n'est pas relativiste puisque les valeurs q', q'' des variables associées à un élément de la matrice R correspondent à deux points différents de l'espace, mais à un même instant. Néanmoins elle convient pour notre problème actuel.

L'équation de mouvement pour R est [1] :

$$(2) \qquad i\hbar \dot{\mathrm{R}} = \mathrm{HR} - \mathrm{RH},$$

où H est l'hamiltonien pour un électron mobile dans le champ :

$$\mathrm{H} = c\rho_1(\boldsymbol{\sigma}, \mathbf{p}) + \rho_3 mc^2 - e\mathrm{V},$$

les ρ et σ étant les matrices de spin habituelles et V le potentiel électrostatique. V doit contenir une partie représentant la contribution au champ des autres électrons présents. La condition pour que la distribution satisfasse au principe d'exclusion est

$$(3) \qquad \mathrm{R}^2 = \mathrm{R}.$$

Représentons par R_0 la distribution qui est supposée ne produire

[1] *Cf.* Dirac, *Proc. Camb. Phil. Soc.*, t. **25**, 1929, p. 62, et **26**, 1930, p. 376.

aucun champ. L'hypothèse la plus immédiate pour R_0 est

$$R_0 = \frac{1}{2}\left(1 - \frac{W}{|W|}\right). \tag{4}$$

où W est l'énergie cinétique d'un électron :

$$W = c\rho_1(\boldsymbol{\sigma}, \mathbf{p}) + \rho_3 mc^2.$$

Ceci signifie que, dans une représentation matricielle où W est diagonale, R_0 sera également diagonale et aura pour éléments diagonaux 0 ou 1 suivant que l'énergie W est positive ou négative. C'est l'énergie cinétique W qui doit figurer dans (4) et non l'énergie totale H parce que, dans ce dernier cas, l'expression (4) serait modifiée seulement par l'addition d'une constante au potentiel électrique V et ne pourrait par conséquent avoir aucune signification physique.

Considérons un état permanent pour lequel l'équation de mouvement (2) se réduit à :

$$0 = HR - RH. \tag{5}$$

Cette équation n'est pas satisfaite par $R = R_0$ à moins que V ne soit une constante. Supposons que V soit une petite quantité du premier ordre et cherchons une solution de (3) et (5) de la forme $R = R_0 + R_1$ où R_1 soit une quantité du premier ordre. En négligeant les petites quantités du second ordre, l'équation (5) donne :

$$\begin{aligned} 0 &= (W - eV)(R_0 + R_1) - (R_0 + R_1)(W - eV) \\ &= WR_1 - R_1 W - e(VR_0 - R_0 V). \end{aligned} \tag{6}$$

L'opérateur $|W|$ peut être défini comme la racine carrée positive de W^2 ou $m^2 c^4 + c^2 \mathbf{p}^2$. Ainsi :

$$|W| = c(m^2 c^2 + \mathbf{p}^2)^{\frac{1}{2}}.$$

Si nous posons

$$\frac{W}{|W|} = \gamma,$$

nous avons

$$W = c\gamma(m^2 c^2 + \mathbf{p}^2)^{\frac{1}{2}}.$$

et aussi

$$R_0 = \frac{1}{2}(1 - \gamma).$$

L'équation (6) peut, par conséquent, s'écrire :

$$(7)\qquad \gamma(m^2c^2+\mathbf{p}^2)^{\frac{1}{2}}R_1 - R_1\gamma(m^2c^2+\mathbf{p}^2)^{\frac{1}{2}} = \frac{1}{2}\frac{e}{c}(\gamma V - V\gamma).$$

L'équation (3) donne :

$$(R_0+R_1)^2 = R_0+R_1,$$

$$R_0R_1+R_1R_0 = R_1,$$

qui se réduit à

$$\gamma R_1 + R_1\gamma = 0.$$

En utilisant cette relation et l'équation $\gamma^2 = 1$, nous déduisons de (7), après multiplication des deux membres à gauche par γ :

$$(m^2c^2+\mathbf{p}^2)^{\frac{1}{2}}R_1 + R_1(m^2c^2+\mathbf{p}^2)^{\frac{1}{2}} = \frac{1}{2}\frac{e}{c}(V - \gamma V\gamma).$$

La quantité qui nous intéresse est la densité électrique correspondant à la distribution R_1. Pour l'obtenir, nous devons former la somme diagonale de R_1, par rapport aux variables de spin et prendre ensuite l'élément diagonal général, multiplié par $-e$, de la matrice résultante par rapport aux variables de position x. Si D représente la somme diagonale par rapport aux variables de spin, nous avons, après un calcul simple :

$$\begin{aligned}(m^2c^2+\mathbf{p}^2)^{\frac{1}{2}}D(R_1) &+ D(R_1)(m^2c^2+\mathbf{p}^2)^{\frac{1}{2}}\\ &= \frac{1}{2}\frac{e}{c}D(V-\gamma V\gamma)\\ &= 2\frac{e}{c}\left\{V - \frac{1}{(m^2c^2+\mathbf{p}^2)^{\frac{1}{2}}}[(\mathbf{p}, V\mathbf{p}) + m^2c^2V]\frac{1}{(m^2c^2+\mathbf{p}^2)^{\frac{1}{2}}}\right\}.\end{aligned}$$

Si nous supposons maintenant l'emploi d'une représentation dans laquelle la matrice quantité de mouvement p est diagonale et si $(p' \mid D(R_1) \mid p'')$ désigne, dans ces conditions, l'élément général de la matrice $D(R_1)$, nous avons

$$\begin{aligned}(m^2c^2+p'^2)^{\frac{1}{2}}(p'\mid D(R_1)\mid p'') &+ (p'\mid D(R_1)\mid p'')(m^2c^2+p''^2)^{\frac{1}{2}}\\ &= 2\frac{e}{c}(p'\mid V\mid p'')\left\{1 - \frac{1}{(m^2c^2+p'^2)^{\frac{1}{2}}}[(\mathbf{p}', \mathbf{p}'')+m^2c^2]\frac{1}{(m^2c^2+p''^2)^{\frac{1}{2}}}\right\},\end{aligned}$$

ce qui donne

$$(8)\qquad (p'\mid D(R_1)\mid p'') = 2\frac{e}{c}(p'\mid V\mid p'')\,\frac{1-\dfrac{(\mathbf{p}', \mathbf{p}'')+m^2c^2}{(m^2c^2+p'^2)^{\frac{1}{2}}(m^2c^2+p''^2)^{\frac{1}{2}}}}{(m^2c^2+p'^2)^{\frac{1}{2}}+(m^2c^2+p''^2)^{\frac{1}{2}}}.$$

Nous pouvons maintenant transformer $D(R_1)$ dans une représentation pour laquelle les variables de position x sont diagonales et en calculer l'élément diagonal. En utilisant les lois habituelles de transformation, on obtient

$$(9) \qquad (x \mid D(R_1) \mid x) = \frac{1}{h^3} \iint e^{-i(\mathbf{x}, \mathbf{p}' - \mathbf{p}'')/h} (p' \mid D(R_1) \mid p'')\, dp'\, dp''.$$

Maintenant, puisque V n'est fonction que des variables de position x et pas des quantités de mouvement p, $(p' \mid V \mid p'')$ ne doit dépendre que de la différence $\mathbf{p}' - \mathbf{p}''$. Par suite, si nous substituons l'expression donnée par le second membre de (8) dans l'intégrale (9), et si nous prenons pour nouvelles variables d'intégration $\mathbf{p}' + \mathbf{p}''$ et $\mathbf{p}' - \mathbf{p}''$, nous pouvons effectuer l'intégration par rapport à $\mathbf{p}' + \mathbf{p}''$ en laissant V quelconque. Le résultat contient un infini logarithmique.

On pourrait croire, à première vue, que la présence de cet infini rend la théorie inacceptable. Cependant, nous ne pouvons pas supposer que la théorie s'applique lorsqu'il s'agit d'énergies supérieures à l'ordre de $137\, mc^2$, et la manière de procéder la plus raisonnable semble être de limiter arbitrairement le domaine d'intégration à une valeur de la quantité de mouvement $\frac{1}{2}(\mathbf{p}' + \mathbf{p}'')$ correspondant à des énergies électroniques de l'ordre indiqué. Cela revient, physiquement, à admettre que la distribution concernant les électrons d'énergie négative inférieure à un niveau d'environ $-137\, mc^2$ ne donne pas lieu à une polarisation par le champ électrique de la manière indiquée par notre théorie. La place exacte que nous attribuons à ce niveau d'énergie limite n'a pas grande importance puisque la valeur de ce niveau figure seulement dans un logarithme.

Si P est la grandeur du vecteur quantité de mouvement $\frac{1}{2}(\mathbf{p}' + \mathbf{p}'')$ à laquelle nous limitons le domaine d'intégration, le résultat final, obtenu après une intégration compliquée, est :

$$(10) \qquad -e(x \mid D(R_1) \mid x) = -\frac{e^2}{\hbar c} \frac{2}{3\pi} \left(\log \frac{2P}{mc} - \frac{5}{6} \right) \rho - \frac{4}{15\pi} \frac{e^2}{\hbar c} \left(\frac{\hbar}{mc} \right)^2 \nabla^2 \rho,$$

où ρ est la densité électrique produisant le potentiel V, de sorte que

$$\nabla^2 V = -4\pi\rho,$$

et où les termes contenant les dérivées de ρ d'ordre supérieur au second ont été négligés.

Le second membre de (10) donne la densité électrique provenant de la polarisation produite par l'action du champ sur la distribution des électrons d'énergie négative. Le terme important est le premier qui, pour $\frac{P}{mc} = 137$, est sensiblement $-\frac{e^2}{\hbar c}\rho$ ou $-\frac{1}{137}\rho$. Ceci signifie qu'il n'y a de densité produite par polarisation que dans les endroits où se trouve située la densité ρ productrice du champ et que la densité induite y neutralise une fraction d'environ $\frac{1}{137}$ de la densité productrice du champ. Le second terme dans le second membre de (10) représente une correction importante seulement lorsque la densité ρ varie rapidement avec la position et change de manière appréciable sur une distance de l'ordre de $\frac{\hbar}{mc}$.

Comme conséquence du calcul précédent, il semblerait que les charges électriques normalement observées sur les électrons, protons ou autres particules électrisées ne sont pas les charges véritables portées par ces particules et figurant dans les équations fondamentales, mais sont légèrement plus petites dans le rapport d'environ 136 à 137. Pour des processus comportant des échanges d'énergie de l'ordre de mc^2, il n'y aurait probablement pas le temps, cependant, pour que la polarisation des électrons d'énergie négative s'établisse de manière complète, de sorte qu'on doit s'attendre à ce que les charges observées soient plus voisines des charges réelles. Il en résulterait des déviations de l'ordre de 1 pour 100 dans des expressions telles que la formule de Klein-Nishina ou la formule de diffusion de Rutherford lorsque des énergies de l'ordre de mc^2 sont en jeu. Lorsque la vérification expérimentale de ces formules pourra être rendue suffisamment précise, on y trouvera un contrôle de l'exactitude de nos hypothèses sur le champ produit par la distribution des électrons d'énergie négative.

ÜBER DIE ELEKTRODYNAMIK DES VAKUUMS AUF GRUND DER QUANTENTHEORIE DES ELEKTRONS

VON

V. WEISSKOPF

Eines der wichtigsten Ergebnisse in der neueren Entwicklung der Elektronentheorie ist die Möglichkeit, elektromagnetische Feldenergie in Materie zu verwandeln. Ein Lichtquant z. B. kann bei Vorhandensein von andern elektromagnetischen Feldern im leeren Raum absorbiert und in Materie verwandelt werden, wobei ein Paar von Elektronen mit entgegengesetzter Ladung entsteht.

Die Erhaltung der Energie erfordert, falls das Feld, in welchem die Absorption vor sich geht, statisch ist, dass das absorbierte Lichtquant die ganze zur Erzeugung des Elektronenpaares notwendige Energie aufbringt. Die Frequenz derselben muss somit der Beziehung $h\nu = 2mc^2 + \varepsilon_1 + \varepsilon_2$ genügen, wobei mc^2 die Ruhenergie eines Elektrons und ε_1 und ε_2 die übrige Energie der beiden Elektronen ist. Diesen Fall haben wir z. B. bei der Erzeugung eines Elektronenpaares durch ein γ-Quant im Coulombfeld eines Atomkerns vor uns.

Die Absorption kann auch in Feldern stattfinden, die von andern Lichtquanten stammen, wobei die letzteren zur Energie des Elektronenpaares beitragen können, sodass in diesem Falle die Energie $2mc^2 + \varepsilon_1 + \varepsilon_2$ der beiden Elektronen gleich der Summe aller bei diesem Prozess absorbierten Lichtquanten sein muss.

Das Phänomen der Absorption von Licht im Vakuum

stellt eine wesentliche Abweichung von der Maxwell'schen Elektrodynamik dar. Das Vakuum sollte nämlich unabhängig von den dort herrschenden Feldern für eine Lichtwelle frei durchdringlich sein, da sich verschiedene Felder nach den Maxwell'schen Gleichungen infolge der Linearität derselben unabhängig überlagern können.

Es ist bereits ohne näheres Eingehen auf die spezielle Theorie verständlich, dass auch in solchen Feldern, die nicht die nötige Energie besitzen, um ein Elektronenpaar zu erzeugen, Abweichungen von der Maxwell'schen Elektrodynamik auftreten müssen: wenn hochfrequentes Licht in elektromagnetischen Feldern absorbiert werden kann, so wird man für Lichtstrahlen, deren Frequenz zur Paarerzeugung nicht ausreicht, eine Streuung oder Ablenkung erwarten, analog zur Streuung des Lichts an einem Atom, dessen kleinste Absorptionsfrequenz grösser als die des Lichts ist. Das Licht wird sich also beim Durchgang durch elektromagnetische Felder so verhalten, als ob das Vakuum unter der Einwirkung der Felder eine von der Einheit verschiedene Dielektrizitätskonstante erhalten würde.

Um diese Erscheinungen darstellen zu können, muss die Theorie dem leeren Raum gewisse Eigenschaften zuschreiben, die die erwähnten Abweichungen von der Maxwell'schen Elektrodynamik hervorrufen. Tatsächlich führt die relativistische Wellengleichung des Elektrons auch zu derartigen Folgerungen, wenn man die aus der Dirac'schen Wellengleichung folgenden Zustände mit negativer kinetischer Energie zur Beschreibung des Vakuums heranzieht.

Die Grundannahme der Dirac'schen Theorie des Positrons besteht darin, dass das physikalische Verhalten des Vakuums im gewissen Sinne beschrieben werden kann durch das Verhalten einer unendlichen Menge von Elektronen, — die

Vakuumelektronen — die sich in den Zuständen negativer kinetischer Energie befinden und sämtliche dieser Zustände besetzt halten. Die Übereinstimmung kann selbstverständlich nicht vollkommen sein, da die Vakuumelektronen eine unendliche Ladungs- und Stromdichte besitzen, die sicher keine physikalische Bedeutung haben darf. Es zeigt sich aber, dass z. B. die Paarerzeugung (und ihr Umkehrprozess) gut wiedergegeben wird als ein Sprung eines Vakuumelektrons in einen Zustand positiver Energie unter dem Einfluss elektromagnetischer Felder, wodurch es als ein reales Elektron in Ercheinung tritt, während das Vakuum um ein negatives Elektron ärmer geworden ist, was sich durch das Auftreten eines positiven Elektrons äussern muss. Die von diesem Bilde ausgehende Berechnung der Paarerzeugung und -Vernichtung zeigt eine gute Übereinstimmung mit der Erfahrung.

Die Berechnung der meisten andern Effekte, die aus der Positronentheorie folgen, stossen immer auf das Problem, in welchem Ausmass das Verhalten der Vakuumelektronen tatsächlich als das des Vakuums anzusehen ist. Dieses Problem wird noch durch den Umstand erschwert, dass Ladungs-, Strom- und Energiedichte der Vakuumelektronen unendlich sind, sodass es sich meistens darum handelt, von einer unendlichen Summe in eindeutiger Weise einen endlichen Teil abzutrennen und diesem Realität zuzuschreiben. Die Lösung dieses Problems wurde von Dirac und Heisenberg dadurch durchgeführt, dass sie eine wiederspruchsfreie Methode angaben, den physikalisch bedeutungsvollen Teil der Wirkungen der Vakuumelektronen zu bestimmen. Es wird im folgenden gezeigt, dass diese Bestimmung weitgehend frei von jeder Willkür ist, da sie in konsequenter Weise nur folgende Eigenschaften der Vakumelektronen als physikalisch bedeutungslos annimmt:

(I)
1) Die Energie der Vakuumelektronen im feldfreien Raum.
2) Die Ladungs- und Stromdichte der Vakuumelektronen im feldfreien Raum.
3) Eine räumlich und zeitlich konstante feldunabhängige elektrische und magnetische Polarisierbarkeit des Vakuums.

Diese Grössen[1] beziehen sich nur auf das feldfreie Vakuum, und es darf als selbstverständlich angesehen werden, dass diese keine physikalische Bedeutung haben können. Alle drei Grössen erweisen sich nach Summierung der Beiträge aller Vakuumselektronen als divergierende Summen. Es sei noch hinzugefügt, dass eine konstante Polarisierbarkeit in keiner Weise feststellbar wäre, sondern nur sämtliche Ladungs- und Feldstärkenwerte mit einem konstanten Faktor multiplizieren würde.

Wir werden im nächsten Abschnitt auf Grund dieser Annahmen die physikalischen Eigenschaften des Vakuums bei Anwesenheit von Feldern berechnen, die zeitlich und räumlich langsam veränderlich sind. Wir verstehen darunter solche Felder F, die sich auf Strecken der Länge $\frac{h}{mc}$ und in Zeiten der Länge $\frac{h}{mc^2}$ nur wenig verändern[2] und somit den Bedingungen

$$(1) \qquad \frac{h}{mc}|\operatorname{grad} F| << |F|, \qquad \frac{h}{mc^2}\left|\frac{\partial F}{\partial t}\right| << |F|$$

genügen. Bei Anwesenheit solcher Felder werden im allgemeinen keine Paare erzeugt, da die auftretenden Lichtquanten zu geringe Energie haben. Die Extremfälle, in denen die Strahlungsdichte so hoch ist, um das Zusammenwirken

[1] Die Annahmen, 1.) oder 2.) oder 3.) als bedeutungslos anzusehen, werden im folgenden mit I_1, I_2 bzw. I_3 zitiert.

[2] h ist die durch 2π geteilte PLANCK'sche Konstante.

von sehr vielen Quanten zu gestatten, oder in denen elektrostatische Felder mit Potentialdifferenzen von über $2mc^2$ vorhanden sind (in diesem Falle würden auf Grund des KLEIN'schen Paradoxons Paare entstehen) wollen wir von der Betrachtung ausschliessen. Unter diesen Umständen lassen sich die elektromagnetischen Eigenschaften des Vakuums durch eine feldabhängige elektrische und magnetische Polarisierbarkeit des leeren Raums darstellen, die z. B. zu einer Lichtbrechung in elektrischen Feldern oder zu einer Streuung von Licht an Licht führt. Der Dielektrizitäts- und Permeabilitätstensor des Vakuums hat dann für schwächere Feldstärken näherungsweise folgende Form: ($\vec{E}$, $\vec{H}$, $\vec{D}$, $\vec{B}$ sind die vier elektromagnetischen Feldgrössen[1].)

$$D_i = \sum_k \varepsilon_{ik} E_k, \quad H_i = \sum_k \mu_{ik} B_k$$

$$(2) \quad \varepsilon_{ik} = \delta_{ik} + \frac{e^4 h}{45\pi m^4 c^7}\left[2(E^2 - B^2)\,\delta_{ik} + 7 B_i B_k\right]$$

$$\mu_{ik} = \delta_{ik} + \frac{e^4 h}{45\pi m^4 c^7}\left[2(E^2 - B^2)\,\delta_{ik} - 7 E_i E_k\right]. \qquad \delta_{ik} = \begin{cases} 1, & i = k \\ 0, & i \neq k \end{cases}$$

Die Berechnung dieser Erscheinungen wurde von EULER u. KOCKEL[2] und von HEISENBERG u. EULER[3] bereits durchgeführt. Im nächsten Abschnitt sollen jedoch bedeutend einfachere Methoden angewendet werden. Ausserdem sollen die Eigenschaften des Vakuums auf Grund der skalaren relativistischen Wellengleichung des Elektrons von KLEIN u. GORDON berechnet werden. Diese Wellengleichung liefert nach PAULI u. WEISSKOPF[4] die Existenz positiver und negativer Partikel, und ihre Erzeugung und Vernichtung durch elektro-

[1] Es werden im folgenden nur dort Pfeile über Vektorgrössen gesetzt, wo Verwechslungen möglich sind.

[2] H. EULER u. B. KOCKEL, Naturwiss. **23**, 246, 1935; H. EULER, Ann. d. Phys. V. **26**, 398.

[3] W. HEISENBERG u. H. EULER, ZS. f. Phys. **38**, 714, 1936.

[4] W. PAULI u. V. WEISSKOPF, Helv. Phys. Acta. **7**, 710, 1934.

magnetische Felder ohne jede besondere Zusatzannahme. Jedoch besitzen diese Partikel keinen Spin und befolgen die Bosestatistik, weshalb diese Theorie nicht auf die realen Elektonen anwandbar ist. Es ist jedoch bemerkenswert, dass auch diese Theorie auf Eigenschaften des Vakuums führt, denen keine physikalische Bedeutung zukommen kann. So erhält man z. B. ebenfalls eine unendliche räumlich und zeitlich konstante feldunabhängige Polarisierbarkeit des Vakuums. Nach Weglassung der entsprechenden Glieder gelangt man zu ähnlichen Resultaten, wie die der Positronentheorie Dirac's. Die physikalischen Eigenschaften des Vakuums rühren in dieser Theorie von der »Nullpunktsenergie« der Materie her, die auch bei nichtvorhandenen Teilchen von den äusseren Feldstärken abhängt und somit ein Zusatzglied zu der reinen Maxwell'schen Feldenergie liefert.

Im 3. Abschnitt behandeln wir die Folgerungen aus der Dirac'schen Positronentheorie für den Fall allgemeiner äusserer Felder und wir zeigen, dass man auf Grund der genannten drei Annahmmen über die Wirkungen der Vakuumelektronen stets zu endlichen und eindeutigen Resultaten kommt. Die Heisenberg'schen Subtraktionsvorschriften erweisen sich als identisch mit diesen drei Annahmen und erscheinen somit bedeutend weniger willkürlich als es in der Literatur bisher angenommen wurde.

Alle folgende Rechnungen berücksichtigen nicht explizit die gegenseitigen Wechselwirkungen der Vakuumelektronen sondern betrachten ausschliesslich jedes einzelne Vakuumelektron allein unter der Einwirkung eines vorhandenen Feldes. Bei diesem Verfahren sind aber die gegenseitigen Einwirkungen nicht vollkommen vernachlässigt, da man das äussere Feld gar nicht von dem Feld trennen kann, das von den Vakuumselektronen selbst erzeugt ist, sodass das in die Rechnung eingehende Feld die Wirkungen der andern Vakuumelektronen zum Teil implizit enthält. Dieses Vorgehen ist analog zur HARTREE'schen Berechnung der Elektronenbahnen eines Atoms in dem Feld, das von den Elektronen selbst verändert wird. Zur expliziten Berechnung der Wechselwirkungen müsste man die Quantenelektrodynamik anwenden, d. h. die Quantelung der Wellenfelder vornehmen. Dies führt bekanntlich auch bereits ohne die Annahme unendlich vieler Vakuumselektronen zu Divergenzen und soll im folgenden nicht näher berührt werden.

II.

In diesem Abschnitt soll die Elektrodynamik des Vakuums für Felder behandelt werden, die den Bedingungen (1) genügen. Die Feldgleichungen sind durch die Angabe der Energiedichte U als Funktion der Feldstärken festgelegt. Wir bestimmen diese aus der Energiedichte $\tilde{U}$ der Vakuumelektronen, die für das Verhalten des Vakuums massgebend sein sollen.

Es ist vorteilhaft, auf die Lagrangefunktion L des elektromagnetischen Feldes zurückzugreifen, da diese durch die Forderung der relativistischen Invarianz schon weitgehend festgelegt ist. Zwischen der Lagrangefunktion L und der Energiedichte U bestehen folgende Beziehungen:

$$U = \sum_i E_i \frac{\partial L}{\partial E_i} - L. \tag{3}$$

In der MAXWELL'schen Elektrodynamik gilt:

$$L = \frac{1}{8\pi}(E^2 - B^2), \quad U = \frac{1}{8\pi}(E^2 + B^2).$$

Die Zusätze zu dieser Lagrangefunktion müssen ebenso wie diese selbst relativistische Invarianten sein. Solange wir uns nur auf langsam veränderliche Felder beschränken, (Bedingung (1)), werden diese Zusätze nur von den Werten der Feldstärken abhängen und nicht von deren Ableitungen. Sie können daher nur Funktionen der Invarianten $(E^2 - B^2)$ und $(EB)^2$ sein. Entwickeln wir die Zusätze nach Potenzen der Feldstärken bis zur 6. Ordnung, so erhalten wir:

$$L = \frac{1}{8\pi}(E^2 - B^2) + L'$$
$$L' = \alpha(E^2 - B^2)^2 + \beta(EB)^2 + \\ + \xi(E^2 - B^2)^3 + \zeta(E^2 - B^2)(EB)^2 + \cdots$$

und daher nach (3)

$$(4) \quad \begin{cases} U = \frac{1}{8\pi}(E^2 + B^2) + U' \\ U' = \alpha(E^2 - B^2)(3E^2 + B^2) + \beta(EB)^2 + \\ \quad + \xi(E^2 - B^2)^2(5E^2 + B^2) + \zeta(EB)^2(3E^2 - B^2) + \cdots \end{cases}$$

Der Zusatz zur Energiedichte ist somit durch die Invarianzeigenschaften weitgehend festgelegt; es wird also im folgenden nur notwendig sein die vorkommenden Konstanten $\alpha, \beta, \xi, \zeta, \ldots$ zu bestimmen. Diesen Ansätzen liegt schon die spezielle Annahme zugrunde, dass U' keine Glieder 2. Ordnung in den Feldstärken enthält, sondern nur höhere. Dies ist gleichbedeutend damit, dass das Vakuum keine von den Feldern unabhängige Polarisierbarkeit besitzt.

Die Rechnungen von Euler u. Kockel und von Heisenberg u. Euler liefern für die vier Konstanten die Werte:

$$\alpha = \frac{1}{360\pi^2}\frac{e^4 h}{m^4 c^7}, \quad \beta = 7\alpha, \quad \xi = \frac{1}{630\pi^2}\frac{e^6 h^3}{m^8 c^{13}}, \quad \zeta = \frac{13}{2}\xi.$$

Die in (2) angegebenen Dielektrizitäts- und Permeabilitätstensoren ergeben sich aus den Beziehungen:

$$D_i = 4\pi \frac{\partial L}{\partial E_i}, \quad H_i = -4\pi \frac{\partial L}{\partial B_i}.$$

Wir werden im folgenden diese Resultate auf eine wesentlich einfachere Weise herleiten.

Der Zusatz U' zur MAXWELL'schen Energiedichte des Vakuums soll durch den Zusatz $\tilde{U}'$ bestimmt sein, den die Vakuumelektronen beitragen. Die Energiedichte bei Anwesenheit von Elektronen in den Zuständen $\psi_1, \psi_2 \ldots \psi_i \ldots$ ist gegeben durch

$$U = \frac{1}{8\pi}(E^2 + B^2) + \tilde{U}'$$

$$\tilde{U}' = \sum_i \left\{\psi_i^*, \left[\left(\vec{\alpha}, \frac{hc}{i}\,\mathrm{grad} + e\vec{A}\right) + \beta mc^2\right]\psi_i\right\}$$

wobei $\vec{\alpha}, \beta$ die DIRAC'schen Matrizen und $\vec{A}$ das vektorielle Potential ist. Der Zusatz $\tilde{U}'$ zur MAXWELL'schen Dichte ist somit nicht gleich der ganzen materiellen Energiedichte U_{mat}

(5) $$U_{\mathrm{mat}} = ih \sum_i \left\{\psi_i^*, \frac{\partial}{\partial t}\psi_i\right\}$$

sondern

(6) $$\tilde{U}' = U_{\mathrm{mat}} - \sum_i \{\psi_i^*, eV\psi_i\}$$

wobei V das skalare Potential ist. Man kann $\tilde{U}'$ als die kinetische Energiedichte bezeichnen. Die gesamte materielle Energiedichte U_{mat} lässt sich, wie wir sehen werden, leicht berechnen; der zweite Term von (6) — die potentielle Energiedichte — ergibt sich aus U_{mat} in folgender Weise: Wenn

* Zwei Eigenfunktionen ψ und φ in geschwungenen Klammern: $\{\psi, \varphi\}$ bedeutet hier und im folgenden das innere Produkt der beiden Spinozen ψ u. φ: $\{\psi, \varphi\} = \sum_k \psi^k \varphi^k$, wobei k der Spinindex ist.

man sich das skalare Potential proportional zu dem konstanten Faktor λ denkt, so gilt:[1]

$$(7) \qquad \lambda \int \sum_i \{\psi_i^*, eV\psi_i\} d\tau = \lambda \frac{\partial}{\partial \lambda} \int U_{\text{mat}} d\tau$$

wobei die Integrationen sich über den ganzen Raum erstrecken. Im Grenzfall konstanter Felder, den wir hier wegen der Bedingungen (1) betrachten wollen, können wir die Feldstärke E selbst als den konstanten Faktor λ ansehen, und können ausserdem die Beziehung (7) auch auf die Energiedichten übertragen. Wir erhalten dann für die kinetische Energiedichte

$$(7\text{a}) \qquad \tilde{U}' = U_{\text{mat}} - E \frac{\partial U_{\text{mat}}}{\partial E}.$$

Vergleicht man dies mit (3) so sieht man, dass zwischen der materiellen und der kinetischen Energiedichte dieselbe Beziehung besteht, wie zwischen $-L$ und U. U_{mat} kann also hier dem durch die Vakuumelektronen hervorgerufenen Zusatz zur Lagrangefunktion gleichgesetzt werden:

$$(8) \qquad U_{\text{mat}} = -\tilde{L}'.$$

Da die Form von U' weitgehend durch die relativistischen Invarianzforderungen festgelegt ist, so genügt es, U' für ein

[1] Der Beweis läuft folgendermassen: Wenn der Energieoperator H von einem Parameter λ abhängig ist, so ändert sich das Diagonalelement H_{ii} des Energieoperators bei einer infinitisemalen adiabatischen Änderung $d\lambda$ von λ um:

$$dH_{ii} = \left(\frac{\partial H}{\partial \lambda}\right)_{ii} d\lambda.$$

Wenn wir nun setzen:

$$H = H_0 + \lambda eV$$

so gilt dann:

$$\lambda (eV)_{ii} = \lambda \frac{\partial H_{ii}}{\partial \lambda}.$$

spezielles Feld zu bestimmen. Wir wählen ein homogenes magnetisches Feld $B = (B_x, 0, 0)$ und ein dazu paralleles räumlich periodisches elektrostatisches Feld, dessen Potential durch

(9) $$V = V_0 e^{\frac{igx}{h}} + V_0{}^* e^{\frac{-igx}{h}}$$

gegeben ist. Wir vergleichen dann dieses Resultat mit dem allgemeinen Form (4) und werden daraus die Koeffizienten dieser Form bestimmen.

Heisenberg u. Euler wählen im Gegensatz hierzu ein konstantes elektrisches Feld, wodurch Schwierigkeiten infolge des Klein'schen Paradoxons entstehen: Jedes noch so schwache homogene elektrische Feld erzeugt Elektronenpaare, wenn es sich über den ganzen Raum erstreckt. Die Elektronenbesetzung der Energiezustände ist dann nicht exakt stationär. In der vorliegenden Rechnung kann durch die Periodizität vermieden werden, dass Potentialdifferenzen über $2\,mc^2$ vorkommen, sodass keine Paarerzeugungen stattfinden.

Die materielle Energiedichte ist bei voller Besetzung aller negativen Energiezustände gegeben durch

(10) $$U_{\text{mat}} = \sum_i W_i \{\psi_i{}^*, \psi_i\}$$

W_i ist die zur Eigenfunktion ψ_i gehörige Energie, summiert wird über alle negative Zustände. Die Summe ist selbstverständlich unendlich. Welcher endliche Teil dieser Summe von physikalischer Bedeutung ist, wird sich eindeutig aus dem expliziten Ausdruck für U_{mat} ergeben.

Die ψ_i befolgen die Wellengleichung:

(11) $$\left\{\frac{ih}{c}\frac{\partial}{\partial t} - \frac{eV}{c} + \alpha_x ih \frac{\partial}{\partial x} + K\right\}\psi = 0$$

$$(12) \qquad K = \alpha_y ih \frac{\partial}{\partial y} + \alpha_z \left[ih \frac{\partial}{\partial z} - \frac{e}{c} |B| y \right] - \beta mc.$$

Wir folgen vorläufig der Rechnung Heisenbergs u. Eulers l. c., wobei wir nur unwesentliche Änderungen in der Bedeutung der Variabeln anbringen.

Als Lösung setzen wir an:

$$(13) \qquad \psi_i = \frac{1}{\sqrt{2\pi h}} e^{\frac{i}{h} p_z z} \cdot u(y) X(x).$$

Der Operator K ergibt, zweimal auf ψ angewendet:

$$K^2 \psi = \left[-h^2 \frac{\partial^2}{\partial y^2} - i \alpha_y \alpha_z \frac{eh}{c} |B| + \left(p_z + \frac{e}{c} |B| y \right)^2 + m^2 c^2 \right] \psi.$$

Wir setzen nun

$$\eta = \left(y + \frac{2 p_z h}{b} \right) \sqrt{\frac{b}{2h^2}}, \quad b = \frac{2eh}{c} |B|.$$

b ist das Mass des Magnetfeldes. Durch Einführung von η erreichen wir, dass K^2 die Form einer Oszillator-Hamiltonfunktion erhält. Wir setzen daher

$$u(y) = \tilde{H}_n(\eta) \left(\frac{b}{2h^2} \right)^{1/4}$$

wobei $\tilde{H}_n(\eta)$ die n-te auf 1 normierte Oszillator-Eigenfunktion ist. Dann gilt $\int |u(y)|^2 dy = 1$ und

$$(14) \qquad K^2 \psi = \left\{ m^2 c^2 + b \left(n + \frac{1 - \sigma_x}{2} \right) \right\} \psi, \quad \sigma_x = i \alpha_y \alpha_z.$$

Es lässt sich nun eine Darstellung der 4-komponentigen ψ wählen, in der σ_x diagonal ist:

$$\sigma_x = \begin{pmatrix} 1 & 0 & 0 & 0 \\ 0 & 1 & 0 & 0 \\ 0 & 0 & -1 & 0 \\ 0 & 0 & 0 & -1 \end{pmatrix}.$$

Den ersten beiden Komponenten von ψ entspricht dann ein positiver, den beiden andern ein negativer Spin in der x-Richtung. Bei dieser Wahl zerfällt die Wellengleichung (11) in zwei getrennte Gleichungssysteme für die beiden Komponentenpaare mit gleichem Spin, sodass wir zwei Wellengleichungen mit zweireihigen Matrizen gewinnen. Der Operator K lässt sich dann in der Form $K = \gamma |K|$ schreiben, wobei γ eine zweireihige Matrix ist, die die Bedingung $\gamma^2 = 1$ erfüllt und $|K|$ die gewöhnliche Zahl

$$|K| = \sqrt{m^2 c^2 + b\left(n + \frac{1-\sigma_x}{2}\right)}$$

bedeutet, die vom Wert σ_x des Spins abhängt. Ebenso ist die in der Wellengleichung auftretende Matrix α_x zweireihig und ist mit γ antikommunativ: $\alpha_x \gamma + \gamma \alpha_x = 0$, da α_x nach (12) auch mit K antikommutativ ist. Die beiden Wellengleichungen lassen sich dann in der Form

$$\left\{\frac{ih}{c}\frac{\partial}{\partial t} + \alpha_x ih \frac{\partial}{\partial x} - \frac{e}{c} V + \gamma |K|\right\} \psi = 0 \qquad (16)$$

schreiben, wobei α_x und γ zweireihige Matrizen sind, die sich nur auf ein Komponentenpaar gleichen Spins beziehen. Der Unterschied in der Wellengleichung für die beiden Spinrichtungen liegt nur in dem verschiedenen Wert von $|K|$. Nachdem die Abhängigkeit von ψ von den Variabeln y and z durch (13) bereits festgelegt wurde, stellt (16) eine Wellengleichung für die Funktion $X(x)$ allein dar. Bisher ist der

Rechnungsgang im wesentlichen identisch mit dem Heisenbergs und Eulers.

Nun behandeln wir vorerst den Fall $V = 0$. Die Eigenwerte und die normierten Eigenfunktionen für (16) lauten

$$(17) \qquad X_n^{(\pm)}(p_x) = a^{(\pm)}(p_x) \frac{1}{\sqrt{2\pi h}} e^{\frac{ip_x x}{h}} \cdot e^{\frac{iW_n^{\pm}(p_x)}{h}t}$$

$$(18) \; W_n^{\pm}(p_x) = \pm c\sqrt{p_x^2 + |K|^2} = \pm c\sqrt{p_x^2 + m^2c^2 + b\left(n + \frac{1-\sigma_x}{2}\right)}.$$

Der obere Index (+) oder (—) unterscheidet die Zustände positiver und negativer Energie. $a^{\pm}(p)$ ist ein normierter 2-komponentiger »Spinor«. Die Gleichung (16) und ihre Lösungen (17), (18) stellen ein eindimensionales Analogon zu Dirac'gleichung dar, in dem $\gamma|K|\psi$ statt des Massengliedes $\beta mc \cdot \psi$ steht. Zu einem Impuls p_x gehören ein positiver und ein negativer Energiewert. (Die beiden andern Energiewerte liefert die Wellengleichung mit entgegengesetztem Spin).

Setzen wir nun diese Grössen in die Energiedichte (10) ein, so erhalten wir

$$U_{\text{mat}} = \sum_{\sigma=-1}^{+1} \sum_{n=0}^{\infty} \iint \frac{dp_x dp_z}{2\pi h} W_n^-(p_x) |\tilde{H}(\eta)|^2 \left(\frac{b}{2h^2}\right)^{\frac{1}{2}} |X_n^{(-)}(p_x)|^2.$$

Die Integration über p_z liefert infolge $dp_z = \sqrt{\frac{b}{2}}\, d\eta$ und $|X_n^{(-)}(p)|^2 = \frac{1}{2\pi h}$

$$(19) \qquad U_{\text{mat}} = \frac{b}{8\pi^2 h^3} \sum_{\sigma=-1}^{+1} \sum_{n=0}^{\infty} \int_{+\infty}^{-\infty} dp\, W_n^-(p).$$

Von hier ab schreiben wir p statt p_x.

Um die Summation durchzuführen, bilden wir

$$\sum_{\sigma=-1}^{+1} \sum_{n=0}^{\infty} W_n^- = W_0^- + 2 \sum_{n=1}^{\infty} W_n^- .$$

Wir verwenden nun die EULER'sche Summenformel für eine Funktion $F(x)$:

$$\frac{1}{2} F(a) + \sum_{r=1}^{N} F(a+rb) + \frac{1}{2} F(a+Nb) =$$

$$= \frac{1}{b}\left[\int_a^{a+Nb} F(x)\,dx - \sum_{m=1}^{\infty} (-)^m \frac{B_m}{(2m)!} b^{2m} \left\{F^{(2m-1)}(a+Nb) - F^{(2m-1)}(a)\right\}\right]$$

B_m ist die m-te BERNOULLI'sche Zahl. $F^{(m)}(x)$ ist die m-te Ableitung von $F(x)$. Wenn wir dies auf (19) anwenden, erhalten wir:

(20)
$$U_{\text{mat}} = \frac{1}{4\pi^2 h^3} \int dp \left[\int_0^{\infty} F(x)\,dx + \sum_{m=1}^{\infty} b^{2m} \frac{B_m}{(2m)!} (-)^m F^{(2m-1)}(0)\right]$$
$$F(x) = -c\sqrt{p^2 + m^2c^2 + x}.$$

In dem Spezialfall eines reinen Magnetfeldes, kann man nach (7a) U_{mat} und $\tilde{U}'$ gleich setzen. Dieser Ausdruck stellt bereits die Energiedichte dar, in einer Entwicklung nach Potenzen der magnetischen Feldstärke b. Nun ist es sehr leicht, jenen Teil des Beitrages $\tilde{U}'$ der Vakuumelektronen zu bestimmen, der für das wirkliche Vakuum massgebend sein soll: Das von b unabhängige Glied stellt die Energiedichte des feldfreien Vakuums dar und ist ein divergentes Integral; da die Energiedichte für das feldfreie Vakuum verschwinden muss, kann dieser Ausdruck keine reale Bedeutung haben. Weiter müssen die (übrigens auch divergierenden) Glieder mit b^2 weggelassen werden, da die Energiedichte keine

Glieder zweiter Ordnung in den Feldstärken besitzen soll. Das Weglassen dieser Glieder ist durch die Annahme begründet, dass die Polarisierbarkeit des Vakuums mit verschwindenden Feldern gegen Null strebt. Es sei hervorgehoben, dass die hier vorgenommenen Subtraktionen ausschliesslich auf triviale Annahmen über das feldlose Vakuum beruhen.

So erhalten wir für den Zusatz zur Maxwell'schen Energiedichte:

$$(21)\quad U' = -\frac{c}{4\pi^2 h^3}\sum_{m=2}^{\infty}\frac{B_m(-)^m}{(2m)!}b^{2m}\frac{1.3\ldots(4m-5)}{2^{2m-1}}\int_{-\infty}^{+\infty}\frac{dp}{(p^2+m^2c^2)^{\frac{4m-3}{2}}}.$$

Diese Potenzreihe lässt sich leicht durch die Potenzreihenentwicklung des hyperbolischen Ctg darstellen. Man erhält:

$$U' = \frac{1}{8\pi^2}\,mc^2\left(\frac{mc}{h}\right)^3\int_0^{\infty}\frac{d\eta}{\eta^3}e^{-\eta}\left\{\eta\,\mathfrak{B}\,\mathrm{Ctg}\,\eta\,\mathfrak{B}-1-\frac{\eta^2}{3}\mathfrak{B}^2\right\}$$

wobei $\mathfrak{B}$ die magnetische Feldstärke gemessen in Einheiten der kritischen Feldstärke $\frac{m^2c^3}{eh}$ ist:

$$\mathfrak{B} = \frac{eh}{m^2c^3}B.$$

Das erste und zweite Glied der Entwicklung liefert:

$$U' = -\frac{1}{360\pi^2}\frac{e^4h}{m^4c^7}B^4+\frac{1}{630\pi^2}\frac{e^6h^3}{m^8c^{13}}B^6+\cdots$$

Wenn wir dies mit jenen Gliedern von (4) vergleichen, die das Magnetfeld in 4. und 6. Potenz enthalten, bekommen wir:

$$\alpha = \frac{1}{360\pi^2}\frac{e^4h}{m^4c^7},\quad \xi = \frac{1}{630\pi^2}\frac{e^6h^3}{m^8c^{13}}.$$

Es werde nun das elektrische Feld mitberücksichtigt. Zu diesem Zweck lösen wir die Wellengleichung (16) für $X(x)$ mit der Born'schen Näherungsmethode. Wir erwarten, dass die vom Potential V abhängigen Teile von U_{mat} in der zweiten, zu V^2 proportionalen Näherung erscheinen. Wenn wir U_{mat} nach Potenzen von V entwickeln: $U_{\text{mat}} = U_{\text{mat}}^{(0)} + U_{\text{mat}}^{(1)} + \cdots$ so erhalten wir nach (10):

$$(22) \quad U_{\text{mat}}^{(2)} = \sum_i W_i^{-(2)} (|\psi_i|^2)^{(0)} + \sum_i W_i^{-(0)} (|\psi_i|^2)^{(2)}.$$

$W_i^{-(k)}$, $(|\psi_i|^2)^{(k)}$ sind die k-ten Näherungen in der entsprechenden Entwicklung von W_i^- und $|\psi_i|^2$. Man beachte, dass $W_i^{-(1)}$ in dem angegebenen elektrischen Feld verschwindet. Es lässt sich leicht zeigen, dass

$$\int (|\psi|^2)^{(2)}\, dx\, dy\, dz = 0$$

sodass der räumliche Mittelwert von U_{mat} nur durch das erste Glied in (22) gegeben ist:

$$\overline{U_{\text{mat}}^{(2)}} = \sum_i W_i^{-(2)} (|\psi_i|^2)^{(0)}.$$

$(|\psi_i|^2)^{(0)}$ wurde bereits im Falle des reinen Magnetfeldes berechnet und wir erhalten somit ganz analog zu (19)

$$\overline{U_{\text{mat}}^{(2)}} = \frac{b}{8\pi^2 h^3} \sum_{\sigma=-1}^{+1} \sum_{n=0}^{\infty} \int_{-\infty}^{+\infty} dp\, W_n^{-(2)}(p).$$

Der Wert von $W_n^{-(2)}$ lässt sich mit der Born'schen Näherungsmethode berechnen. Mit den Eigenfunktionen (17) ergibt sich:

$$(23)\left\{\begin{aligned} W_n^{-(2)}(p) = e^2 |V_0|^2 \Bigg[& \frac{|\{a^{(+)*}(p+g), a^{(-)}(p)\}|^2}{W_n^-(p) - W_n^+(p+g)} + \\ & + \frac{|\{a^{(-)*}(p+g), a^{(-)}(p)\}|^2}{W_n^-(p) - W_n^-(p+g)} \Bigg] + \\ & + (\text{dasselbe mit } -g). \end{aligned}\right.$$

Der zwischen den { }-Klammern stehende Ausdruck stellt ein skalares Produkt zweier zweikomponentiger Spinoren dar. Bei der Integration von (23) über p fallen die zweiten Glieder in den [ ]-Klammern weg, wenn man die Integration der Glieder mit $-g$ mit der Variabeln $p' = p - g$ ausführt:

$$(24)\left\{\begin{aligned} \int dp\, W_n^{-(2)} = e^2 |V_0|^2 \int dp \frac{|\{a^{(+)*}(p+g), a^{(-)}(p)\}|^2}{W_n^{(-)}(p) - W_n^{(+)}(p+g)} + \\ + (\text{dasselbe mit } -g). \end{aligned}\right.$$

Dieses Vorgehen ist im allgemeinen keineswegs eindeutig, da die Integration des zweiten Gliedes in den [ ]-Klammern von (23) zu einem divergenten Resultat führt, das aber nach Addition des entsprechenden Gliedes mit $-g$ endlich gemacht werden oder, wie in (24), zum Verschwinden gebracht werden kann, je nachdem in welcher Weise man die Integrationsvariabeln wählt. Diese Willkür berührt aber unsere Rechnung nicht, da wir nach Ausführung der Summation über n nur die zu b^2, b^4, etc. proportionalen Glieder verwenden, in denen auf Grund der EULER'schen Summenformel nur Ableitungen von $W_n^{-(2)}(p)$ nach n auftreten. Wie man sich leicht überzeugen kann, divergieren diese Ableitungen des zweiten Gliedes in den [ ]-Klammern nicht mehr bei der Integration über p, sodass das Resultat dieser Integration unabhängig von der Wahl der Integrationsvariabeln ist.

Es ergibt sich weiter aus (24):

$$\int dp\, W_n^{-(2)} = -e^2 |V_0|^2 \frac{g^2}{4} \int dp \frac{|K|^2}{c(p^2+|K|^2)^{\frac{5}{2}}}$$

wobei bereits eine Reihenentwicklung nach Potenzen von g durchgeführt wurde und die Glieder höherer als zweiter Ordnung weggelassen wurden. Dies bedeutet die Vernachlässigung der Ableitungen der Feldstärke auf Grund der Bedingungen (1). Ebenso, wie in der vorigen Rechnung, ist nun $\overline{U^{(2)}_{\text{mat}}}$ durch (20) gegeben, wenn man setzt:

$$F(x) = -e^2 |V_0|^2 \frac{g^2}{4} \frac{m^2c^2+x}{c(p^2+m^2c^2+x)^{\frac{5}{2}}}.$$

Man erhält dann, wenn man zuerst über p integriert:

$$\overline{U^{(2)}_{\text{mat}}} = -\frac{1}{4\pi^2 h^3 c} \frac{g^2}{3} e^2 |V_0|^2 \left[\int_0^\infty \frac{dx}{m^2c^2+x} + \right.$$

$$\left. + \sum_{m=1}^{\infty} b^{2m} \frac{B_m (-)^m}{(2m)!} \left(\frac{d^{2m-1}}{dx^{2m-1}} \frac{1}{m^2c^2+x} \right)_{x=0} \right].$$

Da dieser Ausdruck quadratisch in den elektrischen Feldstärken ist, erhalten wir für die kinetische Energiedichte nach (7a)

$$\overline{\tilde{U}'^{(2)}} = -\overline{U^{(2)}_{\text{mat}}}.$$

Aus den früher diskutierten Gründen können erst die Glieder 4. und höherer Ordnung in den Feldstärken für das Vakuum von physikalischer Bedeutung sein, sodass das divergierende Integral wegzulassen ist. Wir ersetzen nun V_0 durch die elektrische Feldstärke E:

$$\overline{E^2} = 2\frac{g^2}{h^2} |V_0|^2$$

wobei die Querstriche Raummittelungen bedeuten und erhalten für die ersten beiden Glieder:

$$(25)\quad U'^{(2)} = \frac{5}{360\pi^2}\frac{e^4 h}{m^4 c^7}E^2 B^2 - \frac{7}{2}\frac{1}{630\pi^2}\frac{e^6 h^3}{m^8 c^{13}}E^2 B^4 + \cdots$$

für den Grenzfall schwach veränderlichen Felder bei welchem die Raummittelungen weggelassen werden können.

Vergleicht man (25) mit den zu $E^2 B^2$ und $E^2 B^4$ proportionalen Gliedern in (4) so erhält man die Beziehungen

$$\beta - 2\alpha = \frac{5}{360}\frac{e^4 h}{m^4 c^7}, \quad 3\xi - \zeta = \frac{7}{2}\frac{1}{630\pi^2}\frac{e^6 h^3}{m^8 c^{13}},$$

und mit den bereits berechneten Werten von α und β

$$\beta = 7\alpha, \quad \zeta = \frac{13}{2}\xi.$$

Der in der magnetischen Feldstärke exakte Ausdruck für $U'^{(2)}$ ergibt sich zu

$$U'^{(2)} = \frac{1}{8\pi^2} mc^2 \left(\frac{mc}{h}\right)^3 \frac{1}{3}\mathfrak{E}^2 \int_0^\infty \frac{d\eta}{\eta} e^{-\eta}\{\eta\,\mathfrak{B}\operatorname{Ctg}\eta\,\mathfrak{B} - 1\},$$

wobei $\mathfrak{E} = \frac{eh}{m^2 c^3} E$ ist.

Die höheren Näherungen in E lassen sich leicht bis auf einen konstanten Faktor bestimmen. Denken wir die k-te Näherung $W_n^{-(k)}(p)$ der Energie des durch p und n gegebenen Zustandes bestimmt; sie wird auf Grund der Wellengleichung (16) die folgende Gestalt haben:

$$W_n^{-(k)}(p) = g^k e^k |V_0|^k \cdot G(c, h, |K|, p)$$

wobei G eine Funktion ist, in welcher nur die angegebenen Grössen vorkommen. $W^{(k)}$ muss infolge der Eichinvarianz mindestens k-ter Ordnung in g sein. Die höheren Potenzen von g sind vernachlässigt. Die Energiedichte in k-ter Ordnung wird dann:

$$(26)\quad U_{\mathrm{mat}}^{(k)} = \frac{-1}{4\pi^2 h^3} g^k e^k |V_0|^k \left[\int_0^{\infty} dx \int_{-\infty}^{+\infty} G dp + \sum_{m=1}^{\infty} b^{2m} \frac{B_m}{(2m)!} (-)^m \left(\frac{d^{2m-1}}{dx^{2m-1}} \int_{-\infty}^{+\infty} G dp \right)_{x=0} \right].$$

Das Integral über G muss die Dimension $(\text{Energie})^{-(k-1)}$ $(\text{Impuls})^{-(k-1)}$ haben, und darf nur mehr von den Grössen $c, h, |K|$ abhängen, was nur in der Form möglich ist:

$$\int_{-\infty}^{+\infty} G dp = f_k \frac{1}{c^{k-1} |K|^{2k-2}} = f_k \frac{1}{c^{k-1}} \frac{1}{(m^2 c^2 + x)^{k-1}}$$

wobei f_k ein Zahlenfaktor ist.

Wenn man dieses in (26) einsetzt, so lässt sich $U_{\mathrm{mat}}^{(k)}$ bis auf den Faktor f_k vollständig angeben.

Die Zahlenfaktoren f_k bestimmen sich aber leicht durch die Überlegung, dass U_{mat} nach (8) eine relalivistische Invariante sein muss. Da deswegen U_{mat} nur von $E^2 - B^2$ und $(EB)^2$ abhängen darf, muss z. B. der Koeffizient von E^k sich von dem Koeffizient von B^k nur durch den Faktor $(-)^{\frac{k}{2}}$ unterscheiden. Der letztere Koeffizient wurde bereits berechnet und ist durch (21) gegeben. Man erhält dann

$$f_{2m} = \frac{2^{3m-2} B^m}{m(2m-1)}$$

und kann damit die Darstellung berechnen, die HEISENBERG u. EULER für L' angegeben haben[1]:

[1] In der Frage der Konvergenz dieses Integrals verweisen wir auf die diesbezüglichen Bemerkungen in der Arbeit von HEISENBERG u. EULER S. 729.

$$L' = -\frac{1}{8\pi^2} mc^2 \left(\frac{mc}{h}\right)^3 \int_0^\infty \frac{d\eta}{\eta^3} e^{-\eta} \left\{ \eta \mathfrak{B} \operatorname{Ctg} \eta \mathfrak{B} \cdot \eta \mathfrak{E} \operatorname{ctg} \eta \mathfrak{E} - 1 + \frac{\eta^2}{3} (\mathfrak{E}^2 - \mathfrak{B}^2) \right\}$$

$$\mathfrak{E} = \frac{m^2 c^3}{eh} E, \quad \mathfrak{B} = \frac{m^2 c^3}{eh} B.$$

Dieser Ausdruck ist für parallele Felder berechnet. Um ihn auf beliebige Felder zu verallgemeinern, muss man ihn als Funktion der beiden Invarianten $E^2 - B^2$ und $(EB)^2$ schreiben. Dies ist nach Heisenberg und Euler in einfacher Weise mit der Beziehung

$$\operatorname{Ctg} \alpha \operatorname{ctg} \beta = -i \frac{\cos \sqrt{\beta^2 - \alpha^2 + 2 i \alpha \beta} + \text{conj}}{\cos \sqrt{\beta^2 - \alpha^2 + 2 i \alpha \beta} - \text{conj}}$$

möglich und man erhält

$$L' = \frac{1}{8\pi^2} \frac{e^2}{hc} \int_0^\infty e^{-\eta} \frac{d\eta}{\eta^3} \left\{ i \eta^2 (EB) \frac{\cos \eta \sqrt{\mathfrak{E}^2 - \mathfrak{B}^2 + 2 i (\mathfrak{E}\mathfrak{B})} + \text{conj}}{\cos \eta \sqrt{\mathfrak{E}^2 - \mathfrak{B}^2 + 2 i (\mathfrak{E}\mathfrak{B})} - \text{conj}} + \right.$$
$$\left. + \frac{m^4 c^6}{e^2 h^2} + \frac{\eta^2}{3} (B^2 - E^2) \right\}.$$

Wegen der Realität des Gesamtausdrucks ist dieser tatsächlich nur von $E^2 - B^2$ und $(EB)^2$ abhängig.

Die Berechnung der Energiedichte und Lagrangefunktion des Vakuums ist in der skalaren Theorie des Positrons mit den gleichen mathematischen Hilfsmitteln durchzuführen. Eine Energiedichte des Vakuums entsteht in dieser Theorie durch die Nullpunktsenergie der Materiewellen. Die Gesamtenergie ist nach Pauli und Weisskopf l. c. (Formel (29)) durch

$$E_{\text{mat}} = \sum_k W_k (N_k^+ + N_k^- + 1)$$

gegeben, wobei W_k die Energie des k-ten Zustandes ist und N_k^+ die Anzahl der Positronen, N_k^- die Anzahl der Elektronen ist, die diesem Zustand angehören. Im leeren Vakuum bleibt die Summe über alle Energien W_k übrig, wobei die Energie W_k des durch den Impuls p und der Quantenzahl n charakterisierten Zustandes in einem Magnetfeld B den Wert hat:

$$W_n^{\text{skal}}(p, B) = c\sqrt{p^2 + m^2 c^2 + b\left(n + \frac{1}{2}\right)}.$$

Die Summation über alle Zustände und Division durch das Gesamtvolumen führt zur Energiedichte, die sich leicht analog zu (19) ergibt:

$$U_{\text{mat}} = \frac{b}{8\pi^2 h^3} \sum_{n=0}^{\infty} \int_{-\infty}^{+\infty} dp\, W_n^{\text{skal}}(p, B).$$

Der einzige Unterschied gegen die frühere Rechnung besteht in dem Wegfallen der Summierung über die beiden Spinrichtungen. Nun verifiziert man leicht die folgende Beziehung zwischen der Energie $W_n^{\text{skal}}(p,B)$ in der skalaren und der Energie $W_n^-(p, B)$ in der DIRAC'schen Elektronentheorie:

$$2\sum_{n=0}^{N} W_n^{\text{skal}}(p, B) = \sum_{\sigma=-1}^{+1} \sum_{n=0}^{N} W_n^-(p, B) - \sum_{\sigma=-1}^{+1} \sum_{n=0}^{2N} W_n^-(p, B/2).$$

Wir können daher die Energiedichte in der skalaren Theorie $\tilde{U}'_{\text{skal}}$ durch die Energiedichte $\tilde{U}'$ der DIRAC'schen Positronentheorie in folgender Weise ausdrücken:

$$2\tilde{U}'_{\text{skal}}(B) = \tilde{U}'(B) - 2\tilde{U}'(B/2).$$

Man sieht daran, dass auch hier der von den Feldstärken unabhängige und der in ihnen quadratische Anteil unendlich ist. Der letztere liefert also eine unendliche von den Feldstärken unabhängige Polarisierbarkeit. Um für das feldlose Vakuum ein brauchbares Resultat zu erhalten, muss man wieder diese beiden Anteile streichen und erhält infolge der Beziehung

$$\mathrm{Ctg}\,\beta - 2\,\mathrm{Ctg}\frac{\beta}{2} = -\frac{1}{\mathrm{Sin}\,\beta}:$$

$$U'_{\text{skal}} = -\frac{1}{16\pi^2}mc^2\left(\frac{mc}{h}\right)^3\int_{(0)}^{\infty}\frac{d\eta}{\eta^3}e^{-\eta}\left\{\eta\mathfrak{B}\frac{1}{\mathrm{Sin}\,\eta\mathfrak{B}} - 1 + \frac{\eta^2}{6}\mathfrak{B}^2\right\}.$$

Die Durchführung einer analogen Störungsrechnung im elektrischen Feld führt in gleicher Weise zu einem Zusatz zur Lagrangefunktion des Feldes, der mit dem aus der Dirac'schen Positronentheorie gewonnenen sehr verwandt ist:

$$L'_{\text{skal}} = -\frac{1}{16\pi^2}\frac{e^2}{hc}\int_0^{\infty}\frac{d\eta}{\eta^3}e^{-\eta}\left\{\frac{2\,i\eta^2(EB)}{\cos\eta\sqrt{(\mathfrak{E}^2-\mathfrak{B}^2)+2i(\mathfrak{E}\mathfrak{B})} - \mathrm{conj}} + \right.$$

$$\left. + \frac{m^4c^6}{e^2h^2} - \frac{\eta^2}{6}(B^2 - E^2)\right\}.$$

Für die in (4) definierten Koeffizienten α, β erhält man daher:

$$\alpha = \frac{7}{16}\frac{1}{360\pi^2}\frac{e^4h}{m^4c^7}, \quad \beta = \frac{4}{7}\alpha.$$

Es sei hier noch auf folgende Eigenschaft der Lagrangefunktion des Vakuums hingewiesen. Für sehr grosse Feldstärken E oder B haben die höchsten Glieder des Zusatzes L' zur Maxwell'schen Lagrangefunktion in der Dirac'schen Theorie des Positrons die Form

$$L' \sim -\frac{e^2}{24\pi^2hc}E^2\lg\mathfrak{E} \quad \text{bzw.}\quad L' \sim \frac{e^2}{24\pi^2hc}B^2\lg\mathfrak{B}.$$

Das Verhältnis zwischen diesem Zusatz L' und der Maxwell'schen Lagrangefunktion $L_0 = \frac{1}{8\pi}(E^2 - B^2)$ ist somit logarithmisch in den Feldstärken für hohe Werte derselben und ist ausserdem mit dem Faktor $\frac{e^2}{hc}$ multipliziert:

$$\frac{L'}{L_0} \sim \frac{-e^2}{3\pi hc} \lg \mathfrak{E} \quad \text{bzw.} \quad \frac{L'}{L_0} \sim -\frac{e^2}{3\pi hc} \lg \mathfrak{B}.$$

Die Nichtlinearitäten der Feldgleichungen stellen somit auch bei Feldstärken, die wesentlich höher als die kritische Feldstärke $\frac{m^2 c^3}{eh}$ sind, nur kleine Korrektionen dar. Die in der Note von EULER und KOCKEL l. c. und in der Arbeit von EULER l. c. zitierte Verwandschaft der aus der Positronentheorie folgenden Nichtlinearität der Feldgleichungen mit der nichtlinearen Feldtheorie von BORN und INFELD[1] ist daher nur äusserlich. In der letzteren Theorie sind die MAXWELL'schen Gleichungen bei der kritischen Feldstärke $F_0 = \frac{m^2 c^4}{e^3}$ »am Rande des Elektrons« bereits vollkommen abgeändert, wodurch dann die endliche Selbstenergie einer Punktladung erreicht wird. Hier hingegen sind die Abweichungen von den MAXWELL'schen Feldgleichungen für Felder der Grösse F_0 noch sehr klein und wachsen viel zu langsam an, um eine ähnliche Rolle in den Selbstenergieproblem zu spielen. Die Extrapolation der vorliegenden Rechnungen auf die Felder am »Rande des Elektrons« ist allerdings nicht einwandfrei, da dort die Bedingungen (1) nicht erfüllt sind. Es ist jedoch nicht wahrscheinlich, dass eine exaktere Betrachtung in dieser Hinsicht ein wesentlich verschiedenes Resultat liefert.

III.

In diesem Abschnitt soll der Einfluss beliebiger Felder auf das Vakuum behandelt werden. Wir beschränken uns vorerst auf statische Felder. Die stationären Zustände des

[1] M. BORN u. L. INFELD, Proc. Roy. Soc. **143**, 410, 1933.

Elektrons auf Grund der Dirac'schen Wellengleichung und ihre Energieeigenwerte werden sich im allgemeinen in zwei Gruppen einteilen lassen, die bei einer adiabatischen Einschaltung des statischen Feldes aus den positiven bzw. aus den negativen Energieniveaus des freien Elektrons entstanden sind. Dies trifft z. B. im Coulombfeld eines Atomkernes zu und bei allen in der Natur vorkommenden statischen Feldern.

Es sind jedoch auch solche statische Felder angebbar, in denen eine derartige Einteilung versagt, da infolge der Felder Übergänge von negativen zu positiven Zuständen vorkommen. Ein bekanntes Beispiel hierfür ist eine Potentialstufe der Höhe $> 2mc^2$. Diese Ausnahmsfälle sind stationär nicht behandelbar und müssen als ein zeitabhängiges Feld betrachtet werden, das zu einer gegebenen Zeit eingeschaltet wird. Dies ist umsomehr schon deswegen notwendig, da derartige Felder infolge der fortwährende Paarentstehung gar nicht stationär aufrechterhalten werden könnten.

Im Falle dass aber das Eigenwertspektrum sich eindeutig in die beiden Gruppen einteilen lässt, kann man die Energiedichte U und die Strom-Ladungsdichte $\vec{i}$, ϱ der Vakuumelektronen nach den Formeln

$$
(29) \quad \left\{ \begin{aligned} U &= ih \sum_i \left\{\psi_i^*, \ \frac{\partial}{\partial t}\psi_i\right\} \\ \varrho &= e \sum_i \{\psi_i^*, \ \psi_i\} \\ \vec{i} &= e \sum_i \{\psi_i^*, \ c\vec{\alpha}\psi_i\} \end{aligned} \right.
$$

berechnen, wobei über die Zustände zu summieren ist, die den negativen Energiezuständen des freien Elektrons entsprechen. Die angeschriebenen Summen werden divergieren.

Wenn aber die physikalisch bedeutungslosen Teile abgetrennt werden, erhalten wir konvergente Ausdrücke.

Um diese Teile auf Grund der Annahmen (I) festzulegen, entwickeln wir die Summanden der Ausdrücke (29) nach Potenzen der äusseren Feldstärken in der Weise, dass wir uns die letzteren mit einem Faktor λ multipliziert denken und nach Potenzen dieses Faktors entwickeln. Dieses Verfahren ist identisch mit einer successiven Störungsrechnung, die von den freien Elektronen als nullte Näherung ausgeht.

Die Annahmen I_1, I_2 verlangen vor allem das Verschwinden der von λ unabhängigen Glieder, die aus den Beiträgen der vom Felde unabhängigen freien Vakuumelektronen bestehen. Wenn wir vorläufig nur jene freien Vakuumelektronen berücksichtigen, deren Impuls $|p| < P$ ist, so erhalten wir hierfür folgende Beiträge:[1]

$$(30)\quad \begin{cases} U_0 = -\dfrac{1}{4\pi^3 h^3}\displaystyle\int\limits_{|p|<P} d\vec{p}\, c\sqrt{p^2 + m^2 c^2}, \\ \varrho_0 = \dfrac{e}{4\pi^3 h^3}\displaystyle\int\limits_{|p|<P} d\vec{p}, \\ \vec{i}_0 = \dfrac{e}{4\pi^3 h^3}\displaystyle\int\limits_{|p|<P} d\vec{p}\, \dfrac{c\vec{p}}{\sqrt{p^2 + m^2 c^2}}. \end{cases}$$

Die Beiträge sämtlicher Elektronen — $P \to \infty$ — divergieren natürlich.

Durch das Abtrennen der von λ unabhängigen Glieder ist aber die Annahme I_2 noch nicht vollständig erfüllt. Die Ladungs- und Stromdichte ϱ_0, $\vec{i}_0$ der feldfreien Vakuum-

[1] Es ist nämlich der zum Impuls $\vec{p}$ gehörige Strom: $\dfrac{ec\vec{p}}{\sqrt{p^2+m^2c^2}}$ und die Anzahl der Zustände in Intervall $d\vec{p}$: $\dfrac{d\vec{p}}{4\pi^3 h^3}$.

elektronen äussert sich nämlich auch dadurch, dass sie bei Anwesenheit von Potentialen V, $\vec{A}$ einen Zusatz $\varrho_0 V$ und $(\vec{i}_0 \vec{A})$ zur Energiedichte liefert, der ebenfalls abgetrennt werden muss. Diese Zusätze treten auf, da auch die Energie und der Impuls der von den Feldern noch unbeeinflussten Vakuumelektronen bei Anwesenheit von Potentialen um den Betrag eV bzw. $\frac{e}{c}\vec{A}$ geändert wird.

Die Annahmen I_1 und I_2 sind daher erst vollkommen erfüllt, wenn man die wegzulassende Beiträge (30) folgendermassen modifiziert:

$$
(31)\left\{\begin{aligned}
U_0' &= -\frac{1}{4\pi^3 h^3}\int\limits_{|p|<P} d\vec{p}\left[c\sqrt{\left(p+\frac{e}{c}A\right)^2+m^2c^2}+eV\right]\\
\varrho_0' &= \frac{e}{4\pi^3 h^3}\int\limits_{|p|<P} d\vec{p}\\
\vec{i}_0' &= \frac{e}{4\pi^3 h^2}\int\limits_{|p|<P} d\vec{p}\,\frac{c\left(\vec{p}+\frac{e}{c}\vec{A}\right)}{\sqrt{\left(p+\frac{e}{c}A\right)^2+m^2c^2}},
\end{aligned}\right.
$$

wobei auch wieder nur der von freien Vakuumelektronen mit Impulsen $|p| < P$ herrührenden Teil angeschrieben ist. Nun ist noch die Bedingung I_3 zu erfüllen. Hierzu beachten wir, dass eine konstante feldunabhängige Polarisierbarkeit zu Gliedern in Energiedichte $U(x)$ führt, die proportional zu den Quadrat $E^2(x)$ und $B^2(x)$ der Feldstärken an der Stelle x sind. Ebenso führt sie zu einer Strom- und Ladungsdichte, die zu den ersten Ableitungen der Felder proportional ist, auf Grund der Beziehungen

$$
\begin{aligned}
i &= \operatorname{rot} M + \frac{dP}{dt}\\
\varrho &= \operatorname{div} P
\end{aligned}
$$

worin M und P die magnetischen und elektrischen Polarisationen sind, die im Falle einer konstanten Polarisierbarkeit zu den Feldern proportional sind. Um die Annahme I_3 zu erfüllen, müssen daher in der Energiedichte U der Vakuumelektronen die zu E^2 und zu B^2 proportionalen Glieder verschwinden und in der Strom-Ladungsdichte die zu den ersten Ableitungen proportionalen Glieder weggelassen werden. Es ist praktischer, die Form dieser Glieder nicht explizit anzugeben, sondern dieselben im Laufe der Rechnung an ihren Eigenschaften zu erkennen.

Als Erläuterung berechnen wir die Ladungs- und Stromdichte des Vakuums unter dem Einfluss eines elektrischen Potentials

$$V = V_0 e^{\frac{i(\vec{g}\,\vec{r})}{h}} + V_0{}^* e^{-\frac{i(\vec{g}\,\vec{r})}{h}} \tag{32}$$

und eines magnetischen Potentials

$$\vec{A} = \vec{A}_0 e^{\frac{i(\vec{g}\,\vec{r})}{h}} + \vec{A}_0{}^* e^{-\frac{i(\vec{g}\,\vec{r})}{h}}, \quad (\vec{A}_0, \vec{g}) = 0 \tag{33}$$

mit Hilfe der Störungstheorie. Diese Berechnungen sind bereits von HEISENBERG[1] und noch viel allgemeiner von SERBER[2] und PAULI u. ROSE[3] durchgeführt worden und sollen hier nur als Illustration zu unserer physikalischen Interpretation der Substraktionsterme dienen. Die Ladungsdichte ϱ ist bis zur ersten Ordnung gegeben durch $\varrho = \varrho^{(0)} + \varrho^{(1)}$,

$$\varrho^{(1)} = e \sum_i \sum_k \frac{H_{k\,i}\{\psi_i{}^*, \psi_k\}}{W_i - W_k} + \text{conj}$$

[1] HEISENBERG, Z. f. Phys. **90**, 209, 1934.
[2] R. SERBER, Phys. Rev. **48**, 49, 1935.
[3] W. PAULI u. M. ROSE, Phys. Rev. **49**, 462, 1936.

wobei die i über die besetzten, die k über die unbesetzten Zustände zu summieren sind, und H_{ik} das Matrixelement der Störungsenergie ist. Setzen wir das Potential (32) als Störung ein, so erhalten wir ($W(p) = c\sqrt{p^2 + m^2 c^2}$)

$$\varrho^{(1)} = \frac{-e^2 V_0}{8\pi^3 h^3} \cdot \int d\vec{p} \left\{ \frac{W(p)\, W(p+g) - c^2 (p, p+g) - m^2 c^4}{W(p)\, W(p+g)\, [W(p) + W(p+g)]} + \right.$$

$$\left. + \begin{pmatrix} \text{dasselbe} \\ \text{mit } -g \end{pmatrix} \right\} e^{\frac{i(\vec{g}\vec{r})}{h}} + \text{conj}.$$

Entwickelt man dies nach Potenzen von $\vec{g}$ so erhält man

$$\varrho^{(1)} = \frac{e^2 V_0}{8\pi^3 h^3} \int d\vec{p}\, \frac{c^2}{W^3(p)}\, e^{\frac{i(\vec{g}\vec{r})}{h}}$$

$$\left\{ \frac{g^2}{2} - \frac{c^2 (pg)^2}{W^2(p)} - \frac{c^2 g^4}{4 W^2(p)} + \frac{25}{16} \frac{c^4 (pg)^2 g^2}{W^4(p)} - \frac{21}{8} \frac{c^6 (pg)^4}{W^6(p)} + \cdots \right\} + \text{conj}.$$

Welcher Teil dieser Ladungsdichte hat nun physikalische Bedeutung? Wegen der Annahme I_2 muss ϱ_0 wegfallen; in $\varrho^{(1)}$ sind die Glieder mit g^2 proportional zur zweiten Ableitung von V und somit zur ersten Ableitung der Feldstärken und müssen wegen I_3 weggelassen werden. Man bemerke, dass auch nur diese Glieder zu Divergenzen führen. Der Rest liefert endliche Integrale und ist nach Heisenberg in der Form

$$(34) \quad \varrho^{(1)} = \frac{1}{60\pi^2} \frac{e^2}{hc} \left(\frac{h}{mc}\right)^2 \triangle \triangle V + \text{höhere Ableitungen von V}$$

zu schreiben. Die exakte Ausrechnung der ersten Näherung wurde von Serber und Pauli u. Rose l. c. geliefert.

Als weiteres Beispiel betrachten wir die Stromdichte $\vec{i}$ in der ersten Näherung des Feldes (33)

$$\vec{i}^{(1)} = e \sum_{ik} \frac{H_{ki} \{\psi_i^*, \vec{\alpha}\psi_k\}}{W_i - W_k} + \text{conj}$$

Es ergibt sich

$$\vec{i}^{(1)} = -e^2 \frac{c\vec{A}_0}{8\pi^3 h^3}\int d\vec{p}\, e^{\frac{i(\vec{g}\vec{r})}{h}}$$

$$\left\{\frac{W(p)\,W(p+g)+E^2(p)+c^2(pg)-2c^2(n,p+g)(n,p)}{W(p+g)\,W(p)\left[W(p+g)+W(p)\right]}+\right.$$

$$\left.+\begin{pmatrix}\text{dasselbe}\\ \text{mit } -g\end{pmatrix}\right\}+\text{conj}$$

wobei n der Einheitsvektor in der Richtung $\vec{A}$ ist. Dieses ergibt nach g entwickelt:

$$(35)\quad \begin{cases} \vec{i}^{(1)} = -\dfrac{e^2 c\vec{A}}{4\pi^3 h^3}\displaystyle\int d\vec{p}\,\frac{1}{W^3(p)} \\ \left\{W^2(p)-c^2(np)^2-\dfrac{c^2g^2}{2}+\dfrac{3}{4}\dfrac{c^4(pg)^2}{W^2(p)}-\dfrac{5}{2}\dfrac{c^6(np)^2(pg)^2}{W^4(p)}+\right. \\ \left.+\dfrac{3}{4}\dfrac{c^4(np)^2g^2}{W^2(p)}+\text{Glieder 4. u. höherer Ordnung in } g\right\}. \end{cases}$$

Hier treten auch von g unabhängige, also nicht eichinvariante Glieder auf. Diese sind aber mit den wegzulassenden Beiträgen $\vec{i}_0'$ aus (31) identisch: Entwickelt man nämlich $\vec{i}_0'$ nach $\vec{A}$, so erhält man:

$$\vec{i}_0' = \frac{e}{4\pi^3 h^3}\int d\vec{p}\left\{\frac{cp}{W(p)}+\frac{e\vec{A}}{W(p)}-\frac{c^2 p\,(e\vec{A},\vec{p})}{W^3(p)}+\cdots\right\}$$

$$= \vec{i}_0 + e^2\frac{c\vec{A}}{4\pi^3 h^3}\int d\vec{p}\,\frac{1}{W^3(p)}\left\{W^2(p)-c^2(np)^2+\cdots\right\}.$$

Die Glieder erster Ordnung in $\vec{A}$ stimmen mit den von g unabhängigen Gliedern in (35) überein. Die zu g^2 proportionalen Glieder von (35) werden ebenfalls weggelassen, der Rest ergibt das zu (34) entsprechende konvergierende Resultat:

$$\vec{i}^{(1)} = \frac{1}{60\pi^2} \frac{e^2}{hc} \left(\frac{h}{mc}\right)^2 \triangle \triangle \vec{A} + \text{höhere Ableitungen.}$$

Die beiden Beispiele sollen zeigen, dass bei einer Störungsrechnung die wegzulassenden Beiträge unmittelbar erkenntlich sind und dass die übrigen durch die Annahmen (I) nicht berührten Beiträge der Vakuumelektronen bei der Summierung zu keinen Divergenzen mehr führen. Die angeführten Beispiele beweisen dies zwar nur in erster Näherung. Die Überlegungen lassen sich aber ohne weiteres auf höhere Näherungen ausdehnen.

Die Behandlung zeitabhängiger Felder ist im wesentlichen nicht von dem obigen Verfahren verschieden. Es ist notwendig, die zeitabhängigen Felder von einem Zeitpunkt t_0 an wirken zu lassen, an welchem die Vakuumelektronen in feldfreien Zuständen waren, oder in solchen stationären Zuständen, die sich einwandfrei in besetzte und unbesetzte einteilen lassen. Die zeitliche Veränderung dieser Zustände von diesem Zeitpunkt t_0, an lässt sich dann mit Hilfe einer Störungsrechnung in Potenzen der äusseren Felder darstellen. Die Ausdrücke (31) und die aus der Bedingung (I_3) folgenden Glieder können dann abgetrennt werden, wobei der übrigbleibende Rest nicht mehr zu Divergenzen führt. Die Berechnung der Ladungs- und Stromdichte des Vakuums bei beliebigen zeithängigen Feldern in erster Näherung findet sich bei Serber l. c. und bei Pauli u. Rose l. c. Die abzuziehenden Teile werden dort aus der Heisenberg'schen Arbeit formal entnommen. Sie sind aber mit jenen, die aus den Annahmen I folgen, vollkommen identisch.

Wie äussert sich nun die Entstehung von Paaren durch zeitabhängige Felder? Die Paare kommen bei der Berechnung der Energie-, Strom- und Ladungsdichte nicht unmittelbar zum Ausdruck. Die Paarerzeugung zeigt sich nur in einer proportional zur Zeit wachsenden Gesamtenergie, die der Energie der entstehenden Elektronen entspricht. Die Ladungs- und Stromdichte ist nicht unmittelbar von der Paarentstehung beeinflusst, da stets positive und negative Elektronen zugleich erzeugt werden, die die Strom-Ladungsdichte erst dadurch beeinflussen, dass die äusseren Felder auf die entstandenen Elektronen je nach der Ladung verschieden einwirken.[1]

Es ist daher praktischer, die Paarentstehung durch äussere Felder direkt zu berechnen als Übergang eines Vakuumelektrons in einen Zustand positiver Energie. Die Entstehungswahrscheinlichkeit des Elektronenpaares ist dann identisch mit der Zunahme der Intensität der betreffenden Eigenfunktion positiver Energie bzw. mit der Abnahme der Intensität der entsprechenden Eigenfunktion negativer Energie infolge des Einwirkens der zeitabhängigen Felder auf die Zustände, die bis zur Zeit t_0 geherrscht haben. Die Berechnung wurde von Bethe u. Heitler[2], Hulme u. Jaeger[3] u. s. w. ausgeführt.

Die Paarvernichtung unter Lichtausstrahlung lässt sich wie jeder anderer spontane Ausstrahlungsprozess nur durch Quantelung der Wellenfelder behandeln, oder durch korrespondenzmässige Umkehrung des Lichtabsorptionsprozesses.

In der bisherigen Darstellung wurden die abzutrennenden Teile der Vakuumelektronen nicht explizit angegeben, sondern nur ihre Form und ihre Abhängigkeit von

[1] Die von Serber berechnete Strom- und Ladungsdichte bei paarerzeugenden Feldern ist daher dem Mitschwingen der Vakuumelektronen zuzuschreiben und ist nicht etwa die »erzeugte Strom-Ladungsdichte«. Die auftretenden Resonanznenner rühren daher, dass dieses Mitschwingen besonders stark ist, wenn die äussere Frequenz sich einer Absorptionsfrequenz des Vakuums nähert.

[2] H. Bethe u. W. Heitler, Proc. Roy. Soc. **146**, 84, 1934.

[3] H. R. Hulme u. J. C. Jaeger, Proc. Roy. Soc.

den äusseren Feldern bestimmt. Um sie explizit darzustellen muss man ein etwas anderes Verfahren wählen, da diese Teile ja divergente Ausdrücke enthalten. Hierzu eignet sich die von Dirac eingeführte Dichtematrix, die dann von Dirac und spezieller von Heisenberg l. c. auf dieses Problem angewendet wurde. Die Dichtematrix R ist durch folgenden Ausdruck gegeben:

$$(x', k' \,|\, R \,|\, x'' k'') = \sum_i \psi_i{}^* (x', k') \, \psi_i (x'' k'')$$

wobei x' und x'' zwei Raum-Zeitpunkte, k' und k'' zwei Spinindizes bedeuten. Die Summe soll über alle besetzten Zustände erstreckt werden. Aus dieser Matrix kann man dann leicht die Strom- und Ladungsdichte $\vec{i}$, ϱ und der Energie-Impuls-Tensor[1] U_ν^μ bilden auf Grund den Beziehungen

$$\vec{i} = \lim_{x' = x''} e \sum_{k' k''} (\vec{\alpha})_{k' k''} (x' k' \,|\, R \,|\, x'' k'')$$

$$\varrho = \lim_{x' = x''} e \sum_{k'} (x' k' \,|\, R \,|\, x'' k')$$

$$U_\nu^\mu = \lim_{x' = x''} \frac{1}{2} \left\{ ich \left[\frac{\partial}{\partial x'_\mu} - \frac{\partial}{\partial x''_\mu} \right] - e \left[A^\mu (x') + A^\mu (x'') \right] \right\}$$

$$\sum_{k' k''} (\alpha^\nu)_{k' k''} (x' k' \,|\, R \,|\, x'' k''). \qquad \alpha^4 = \text{Einheitsmatrix}$$

Die Dichte-Matrix hat den Vorteil, dass für $x' \neq x''$ die Summation über die Vakuumelektronen nicht divergieren, sondern einen Ausdruck ergeben der für $x' = x''$ singulär wird.

Man kann nun aus den Annahmen (I) eindeutig angeben, welche Teile der Dichtematrix der Vakuumelektronen für

[1] Der vollständige Energie-Impuls-Tensor besteht aus der Summe von U_ν^μ und dem Maxwell'schen Energie-Impuls-Tensor des Feldes. Die U_4^4-komponente ist daher nicht die gesamte materielle Energiedichte, sondern nur die kinetische.

$x' = x''$ in die wegzulassenden Teile übergehen und gewinnt auf diese Weise eine explizite Darstellung dieser Glieder.

Der physikalisch bedeutungslose Teil der Dichtematrix muss dann aus jenen Gliedern bestehen, die von den Feldstärken unabhängig sind, aus denen, die zu Gliedern in der Stromdichte führen, die zu den Ableitungen der Felder proportional sind und aus denen die zu Gliedern in der Energiedichte führen, die proportional zum Quadrat der Feldstärke sind. Ausserdem muss der abzuziehende Teil der Dichtematrix noch mit dem Faktor

$$u' = \exp\left[\frac{ie}{hc}\int_{x'}^{x''}\left(\sum_{i=1}^{3} A_i dx_i - V dt\right)\right]$$

multipliziert werden, wobei das Integral im Exponenten in gerader Linie vom Punkte x' zu dem Punkt x'' zu erstrecken ist. Dieser Faktor fügt zum abzuziehenden Energie-Impuls-Tensor gerade die Beitrage hinzu, die davon herrühren, dass die noch ungestörten Vakuumelektronen im Feld eine Zusatzenergie eV und einen Zusatzimpuls $\frac{e}{c}\vec{A}$ erhalten, und die infolge der Annahme I_2 mit abgezogen werden sollen.

Da der abzuziehende Teil der Dichtematrix bis auf den Faktor u' höchstens zweiter Ordnung in den Feldstärken ist, lässt er sich durch eine Störungsrechnung aus der Dichtematrix der freien Elektronen gewinnen. Diese im Prinzip einfache, in der Ausführung jedoch sehr komplizierte Rechnung liegt der Bestimmung dieser Matrix von Heisenbreg l. c. zu Grunde. Das Ergebnis lässt sich mathematisch einfacher formulieren, wenn man bei jeder Grösse stets das Mittel bildet aus der Berechnung mit Hilfe der vorliegenden Theorie und aus der Berechnung mit Hilfe einer Theorie in der die Elektronenladung positiv ist, und das negative

Elektron als »Loch« dargestellt wird. Das Resultat ist ja in beiden Fällen dasselbe. Die Dichtematrix R wird dann durch R′ ersetzt:

$$(x'k' \,|\, R' \,|\, x''k'') = \frac{1}{2}\Big\{\sum_i \psi_i^*(x'k')\,\psi_i(x''k'') - \\ - \sum_k \psi_k^*(x'k')\,\psi_k(x''k'')\Big\},$$

wobei die erste Summe über die besetzten, die zweite Summe über die unbesetzten Zustände zu erstrecken ist.

Der abzuziehende Teil $(x'k' \,|\, S \,|\, x''k'')$ hat dann die Form

$$(x'k' \,|\, S \,|\, x''k'') = u' S_0 + \frac{\bar{a}}{|x' - x''|^2} + \bar{b} \lg \frac{|x' - x''|^2}{C}.$$

Hierbei ist S_0 die Matrix R' für verschwindene Potentiale, $\bar{a}$ und $\bar{b}$ sind Funktionen der Feldstärken und ihrer Ableitungen, C ist eine Konstante. Diese Grössen sind bei Heisenberg l. c. und bei Heisenberg u. Euler l. c. explizit angegeben.

Für die Ausführungen spezieller Rechnungen ist es praktischer, nicht auf den expliziten Ausdruck Heisenberg's zurückzugreifen, sondern die wegzulassenden Glieder an ihrer Struktur zu erkennen. Dies ist vor allem deshalb einfacher, da die übrig bleibenden Ausdrücke nicht mehr bei $x' = x''$ singulär werden, sodass man für die Berechnung derselben gar nicht das formale Hilfsmittel der Dichtematrix benötigt. Die Summierungen über alle Vakuumelektronen führen hierbei nicht mehr zu divergenten Ausdrücken. Allerdings eignet sich die explizite Darstellung Heisenberg's gut dazu, die relativistische Invarianz und die Gültigkeit der Erhaltungssätze in dem Verfahren zu zeigen.

Es ist hieraus ersichtlich, dass die hier beschriebene Bestimmung der physikalischen Eigenschaften der Vakuumelektronen im wesentlichen keine Willkür enthält, da ausschliesslich nur jene Wirkungen derselben weggelassen werden, die infolge der Grundannahme der Positronentheorie wegfallen müssen: die Energie und die Ladung der von den Feldern ungestörten Vakuumelektronen, und die physikalisch sinnlose feldunabhängige konstante Polarisierbarkeit des Vakuums. Alle physikalisch sinnvollen Wirkungen der Vakuumelektronen werden mitberücksichtigt und führen zu konvergenten Ausdrücken. Man darf daraus wohl den Schluss ziehen, dass die Löchtertheorie des Positrons keine wesentlichen Schwierigkeiten für die Blektronentheorie mit sich geführt hat, solange man sich auf die Behandlung der ungequantelten Wellenfelder beschränkt.

Ich möchte an dieser Stelle den Herren Prof. Bohr, Heisenberg und Rosenfeld meinen herzlichsten Dank für viele Disskussionen aussprechen. Auch bin ich dem Rask-Ørsted-Fond Dank schuldig, der es mir ermöglicht hat, diese Arbeit am Institut for teoretisk Fysik in Kopenhagen auszuführen.

This paper deals with the modifications introduced into the electrodynamics of the vacuum by Dirac's theory of the positron. The behaviour of the vacuum can be described unambigously by assuming the existence of an infinite number of electrons occupying the negative energy states, provided that certain well defined effects of these electrons are omitted, but only those to which it is obvious that no physical meaning can be ascribed.

The results are identical with these of Heisenberg's and Dirac's mathematical method of obtaining finite expressions in positron theory. A simple method is given of calculating the polarisability of the vacuum for slowly varying fields.

Note on the Radiation Field of the Electron

F. Bloch and A. Nordsieck*
Stanford University, California

(Received May 14, 1937)

Previous methods of treating radiative corrections in nonstationary processes such as the scattering of an electron in an atomic field or the emission of a β-ray, by an expansion in powers of $e^2/\hbar c$, are defective in that they predict infinite low frequency corrections to the transition probabilities. This difficulty can be avoided by a method developed here which is based on the alternative assumption that $e^2\omega/mc^3$, $\hbar\omega/mc^2$ and $\hbar\omega/c\Delta p$ (ω=angular frequency of radiation, Δp=change in momentum of electron) are small compared to unity. In contrast to the expansion in powers of $e^2/\hbar c$, this permits the transition to the classical limit $\hbar=0$. External perturbations on the electron are treated in the Born approximation. It is shown that for frequencies such that the above three parameters are negligible the quantum mechanical calculation yields just the directly reinterpreted results of the classical formulae, namely that the total probability of a given change in the motion of the electron is unaffected by the interaction with radiation, and that the mean number of emitted quanta is infinite in such a way that the mean radiated energy is equal to the energy radiated classically in the corresponding trajectory.

I. Introduction

The quantum theory of radiation has been successfully applied to radiative emission and absorption processes. If the methods which lead to these results are used to obtain more general radiative corrections, a characteristic difficulty arises. This difficulty is clearly visible in the formulae given by Mott, Sommerfeld[1] and Bethe and Heitler[2] for the probability of scattering of an electron in a Coulomb field accompanied by the emission of a single light quantum. If the emitted quantum lies in a frequency range ω to $\omega+d\omega$, this probability is for small frequencies proportional to $d\omega/\omega$ independently of the angle of scattering. Taking these formulae literally and asking for the total probability of scattering with the emission of any light quantum, one therefore gets by integration over ω a result which diverges logarithmically in the low frequencies. The same difficulty appears in the radiative correction to the probability of β-decay,[3] and to that of other nonstationary processes.

This "infrared catastrophe" is obviously unrelated to the fundamental "ultraviolet" difficulties of quantum electrodynamics, exemplified by the divergent result for the self-energy of the electron. While the latter is already inherent in the classical theory, the former has no counterpart there. There is, however, a feature in the classical theory which indicates the cause of the difficulty: If for simplicity one considers only frequencies which are small compared to the reciprocal of the collision time, the mechanism of emission may be described as follows. The amplitude of each Fourier component of the proper field of the electron before the impact retains its value after the impact.[4] The difference between the new field and the field proper to the electron in its new motion is the emitted radiation. The significant point is now that I_ω, the radiated intensity per unit frequency interval, does not approach zero[5] as $\omega\rightarrow 0$. Hence $I_\omega/\hbar\omega$, which may be taken as an estimate of the mean number of light quanta emitted per unit frequency range, tends to infinity as $\omega\rightarrow 0$. Since the same result has to be expected in a rigorous quantum-theoretical treatment, one has to anticipate that only the probability for the simultaneous emission of infinitely many quanta can be finite; the probability of emission of any finite number of quanta must vanish.

* National Research Fellow.

[1] N. F. Mott., Proc. Camb. Phil. Soc. **27**, 255 (1931); A. Sommerfeld, Ann. d. Physik **11**, 257 (1931).

[2] H. Bethe and W. Heitler, Proc. Roy Soc. **A146**, 83 (1934).

[3] J. K. Knipp and G. E. Uhlenbeck, Physica **3**, 425 (1936); F. Bloch, Phys. Rev. **50**, 272 (1936); see pp. 276–7 in the latter.

[4] This is in strict analogy to the mechanical model of an oscillator initially held away from its equilibrium position by a constant force, the value of which suddenly changes. We shall see later that the same analogy holds in quantum mechanics, where it finds its mathematical expression in the expansion of the wave function of an oscillator with one equilibrium position in terms of the wave functions corresponding to a new equilibrium position.

[5] For a charged particle moving in a pure Coulomb field, $I_\omega \sim \log 1/\omega\tau$ as $\omega\rightarrow 0$, where τ is the collision time; in a field which falls off more rapidly than $1/r^2$ for large r, I_ω approaches a finite value.

In these circumstances the ordinary treatment of the interaction between electron and field by a consistent expansion in powers of e, the charge of the electron, is clearly inadequate, for such a procedure is based on the physical assumption that the probability of multiple light quantum emission decreases with increasing number of emitted quanta. In fact, however, if one considers only frequencies above a certain minimum frequency ω_0, the expansion parameter of such a treatment is $e^2/\hbar c\cdot(v/c)^2 \log E/\hbar\omega_0$ (E, v the kinetic energy resp. velocity of the electron),[6] which in the limit $\omega_0 \to 0$ violates the original assumption that it is small.

In order to avoid this expansion, a method is developed below in which the coupling between the electron and the electromagnetic waves of low frequency is not considered as a small perturbation. The method is analogous to the classical expansion in powers of $e^2\omega/mc^3$, the ratio of the electron radius to the wave-length considered, in which in first approximation the motion of the electron is treated as given and the reaction of the field on the electron is taken into account in higher orders. We shall show how this can be formulated in quantum mechanics as the solution in successive approximations of a system of two simultaneous differential equations; of these approximations only the one of lowest order is here needed and investigated. After thus having treated the system electron plus electromagnetic field, transitions of this system due to external forces on the electron can be treated by the ordinary method of small perturbations.[7]

II. Formulation of the Method

The Hamilton operator of the total system electron plus electromagnetic field, after elimination of the longitudinal parts of the electric field, is[8]

$$\mathcal{H}=c\{(\boldsymbol{\alpha},\ \mathbf{p}-(e/c)\mathbf{A})+\beta mc\} + (1/8\pi)\int((\mathbf{E}^{tr})^2+\mathbf{H}^2)dV.$$

We expand the vector potential $\mathbf{A}$ in the form

$$\mathbf{A}=2c(\pi\hbar/\Omega)^{\frac{1}{2}}\sum_{s\lambda}\omega_s^{-\frac{1}{2}}\boldsymbol{\varepsilon}_{s\lambda}(P_{s\lambda}\cos(\mathbf{k}_s,\mathbf{r}) + Q_{s\lambda}\sin(\mathbf{k}_s,\mathbf{r})),$$

where Ω is the volume within which the cyclical boundary conditions are applied, the summation index s characterizes the direction and circular frequency ω_s of the various waves with propagation vector $\mathbf{k}_s$, λ their state of polarization; and $\boldsymbol{\varepsilon}_{s\lambda}$ is a unit vector in the direction of polarization. The dynamical variables $P_{s\lambda}$ and $Q_{s\lambda}$ are related to the quantized amplitudes $a(\lambda, \mathbf{k})$ and $a^+(\lambda, \mathbf{k})$[9] according to

$$2^{-\frac{1}{2}}(P_{s\lambda}+iQ_{s\lambda})=a(\lambda,\mathbf{k}_s),$$

$$2^{-\frac{1}{2}}(P_{s\lambda}-iQ_{s\lambda})=a^+(\lambda,\mathbf{k}_s),$$

and obey the commutation laws

$$[P_{s\lambda}, Q_{s'\lambda'}]=-i\delta_{ss'}\delta_{\lambda\lambda'};$$

$$[P_{s\lambda}, P_{s'\lambda'}]=[Q_{s\lambda}, Q_{s'\lambda'}]=0.$$

We then obtain

$$\mathcal{H}=c\{(\boldsymbol{\alpha},\ \mathbf{p}-\sum_{s\lambda}\mathbf{a}_{s\lambda}[P_{s\lambda}\cos(\mathbf{k}_s,\mathbf{r}) + Q_{s\lambda}\sin(\mathbf{k}_s,\mathbf{r})])+\beta mc\} + \tfrac{1}{2}\sum_{s\lambda}(P_{s\lambda}^2+Q_{s\lambda}^2)\hbar\omega_s, \tag{1}$$

where

$$\mathbf{a}_{s\lambda}=2e(\pi\hbar/\Omega\omega_s)^{\frac{1}{2}}\boldsymbol{\varepsilon}_{s\lambda}. \tag{2}$$

The matrix vector $\boldsymbol{\alpha}$ of Dirac is to be interpreted physically as the velocity of the electron divided by c. If following the procedure in classical theory, we should in first approximation neglect the reaction of the electromagnetic field on the electron, we should be led to replace $\boldsymbol{\alpha}$ in (1) by the c number

$$\boldsymbol{\mu}=\mathbf{v}/c, \tag{3}$$

where $\mathbf{v}$ stands for the (constant) velocity of the electron in its undisturbed motion. The operator β has then at the same time to be replaced by $(1-\mu^2)^{\frac{1}{2}}$. But with c numbers substituted for $\boldsymbol{\alpha}$ and β in (1) the dynamical problem can immediately be solved. Indeed, the presence of the interaction energy in (1) then causes no mathe-

[6] This is the order of magnitude of the ratio of the probabilities of single quantum emission and radiationless process.

[7] A case in which the action of a scattering field on the electron is not considered as a small perturbation, is treated in the following paper by A. Nordsieck.

[8] W. Pauli, *Handbuch der Physik*, Vol. 24, p. 266.

[9] W. Pauli, *Handbuch der Physik* Vol. 24, p. 250 ff.; the notation used in the present paper is essentially the same as that of G. Wentzel, *Handbuch der Physik* Vol. 24, p. 740 ff.

matical difficulty, since being closely analogous to the potential of a given constant force on each field oscillator, it can be taken into account by merely adding functions of $\mathbf{r}$ to the $P_{s\lambda}$ and $Q_{s\lambda}$.

Of course, the above replacement is not rigorously justified. In order to enable one to take into account the error thus committed by successive approximations, we proceed as follows. The solution $\psi(\mathbf{r}, Q_{s\lambda})$ of the wave equation

$$\mathcal{H}\psi = E\psi \tag{4}$$

can be uniquely separated into two parts

$$\psi = \psi^+ + \psi^- \tag{5}$$

by the help of the operator

$$\Lambda = (\boldsymbol{\alpha}, \boldsymbol{\mu}) + \beta(1-\mu^2)^{\frac{1}{2}} \tag{6}$$

and the conditions

$$\Lambda\psi^+ = \psi^+; \quad \Lambda\psi^- = -\psi^-. \tag{7}$$

Using further the relations

$$\Lambda\boldsymbol{\alpha} = 2\boldsymbol{\mu} - \boldsymbol{\alpha}\Lambda; \quad \Lambda\beta = 2(1-\mu^2)^{\frac{1}{2}} - \beta\Lambda \tag{8}$$

and adding and subtracting Eqs. (4) and

$$\Lambda\mathcal{H}\psi = E\Lambda\psi \tag{4a}$$

we get

$$\begin{aligned}\{c(\boldsymbol{\mu}, \mathbf{p} - \sum_{s\lambda}\mathbf{a}_{s\lambda}[P_{s\lambda}\cos(\mathbf{k}_s, \mathbf{r}) + Q_{s\lambda}\sin(\mathbf{k}_s, \mathbf{r})]) \\ + mc^2(1-\mu^2)^{\frac{1}{2}} \pm \tfrac{1}{2}\sum_{s\lambda}(P_{s\lambda}^2 + Q_{s\lambda}^2)\hbar\omega_s \mp E\}\psi^{+,-} \\ = -c(\boldsymbol{\mu} \pm \boldsymbol{\alpha}, \mathbf{p} - mc(1-\mu^2)^{-\frac{1}{2}}\boldsymbol{\mu} \\ - \sum_{s\lambda}\mathbf{a}_{s\lambda}[P_{s\lambda}\cos(\mathbf{k}_s, \mathbf{r}) + Q_{s\lambda}\sin(\mathbf{k}_s, \mathbf{r})])\psi^{-,+}\end{aligned} \tag{9}$$

where the upper sign goes with the first superscript and the lower with the second.

The two functions ψ^+ and ψ^- in Eqs. (9) correspond in the limit as the interaction between the electron and the field tends to zero, to motions of the electron in states with momentum $mc(1-\mu^2)^{-\frac{1}{2}}\boldsymbol{\mu}$ and with energies $\pm mc^2(1-\mu^2)^{-\frac{1}{2}}$, respectively. The approximation mentioned before of first assuming the velocity of the electron to be a given c number $\boldsymbol{\mu}c$ is equivalent to an approximate solution of the equations (9) in which the right-hand sides are neglected and ψ^- is assumed to be zero. The latter assumption is made to obtain a state in which the energy of the electron is positive.

We accordingly solve the equation

$$\begin{aligned}\{c(\boldsymbol{\mu}, \mathbf{p} - \sum_{s\lambda}\mathbf{a}_{s\lambda}[P_{s\lambda}\cos(\mathbf{k}_s, \mathbf{r}) \\ + Q_{s\lambda}\sin(\mathbf{k}_s, \mathbf{r})]) + mc^2(1-\mu^2)^{\frac{1}{2}} \\ + \tfrac{1}{2}\sum_{s\lambda}(P_{s\lambda}^2 + Q_{s\lambda}^2)\hbar\omega_s - E\}u = 0.\end{aligned} \tag{10}$$

This can best be done by first making the canonical transformation

$$\begin{aligned}P_{s\lambda} &= P'_{s\lambda} + \sigma_{s\lambda}\cos(\mathbf{k}_s, \mathbf{r}); \\ Q_{s\lambda} &= Q'_{s\lambda} + \sigma_{s\lambda}\sin(\mathbf{k}_s, \mathbf{r}); \quad \mathbf{r} = \mathbf{r}'; \\ \mathbf{p} &= \mathbf{p}' - \sum_{s\lambda}\hbar\mathbf{k}_s\sigma_{s\lambda}[P'_{s\lambda}\cos(\mathbf{k}_s, \mathbf{r}) \\ &\quad + Q'_{s\lambda}\sin(\mathbf{k}_s, \mathbf{r}) + \tfrac{1}{2}\sigma_{s\lambda}].\end{aligned} \tag{11}$$

The corresponding transformation of the wave function is

$$\begin{aligned}u(\mathbf{r}, Q_{s\lambda}) = \exp\{i\sum_{s\lambda}\sigma_{s\lambda}\cos(\mathbf{k}_s, \mathbf{r}) \\ \times[Q'_{s\lambda} + \tfrac{1}{2}\sigma_{s\lambda}\sin(\mathbf{k}_s, \mathbf{r})]\}\cdot u'(\mathbf{r}, Q'_{s\lambda}).\end{aligned} \tag{12}$$

Choosing

$$\sigma_{s\lambda} = (\boldsymbol{\mu}, \mathbf{a}_{s\lambda})/\hbar(k_s - (\boldsymbol{\mu}, \mathbf{k}_s)) \tag{13}$$

one obtains for u' the simple equation

$$\begin{aligned}\{c(\boldsymbol{\mu}, \mathbf{p}') + mc^2(1-\mu^2)^{-\frac{1}{2}} + \tfrac{1}{2}\sum_{s\lambda}(P'^2_{s\lambda} + Q'^2_{s\lambda})\hbar\omega_s \\ - (c/2\hbar)\sum_{s\lambda}(\boldsymbol{\mu}, \mathbf{a}_{s\lambda})^2/(k_s - (\boldsymbol{\mu}, \mathbf{k}_s)) - E\}u' \doteq 0\end{aligned} \tag{14}$$

with the general solution

$$\begin{aligned}u' = \gamma(\boldsymbol{\mu})\Omega^{-\frac{1}{2}}\exp\{i/\hbar(mc(1-u^2)^{-\frac{1}{2}}\boldsymbol{\mu} + \mathbf{g}, \mathbf{r})\} \\ \times\prod_{s\lambda}h_{m_{s\lambda}}(Q'_{s\lambda}),\end{aligned} \tag{15}$$

$$\begin{aligned}E(\boldsymbol{\mu}, m_{s\lambda}, \mathbf{g}) = mc^2(1-\mu^2)^{-\frac{1}{2}} + c(\boldsymbol{\mu}, \mathbf{g}) \\ + \sum_{s\lambda}(m_{s\lambda} + \tfrac{1}{2})\hbar\omega_s - (c/2\hbar)\sum_{s\lambda}(\boldsymbol{\mu}, \mathbf{a}_{s\lambda})^2/(k_s - (\boldsymbol{\mu}, \mathbf{k}_s)).\end{aligned} \tag{16}$$

Here $\gamma(\boldsymbol{\mu})$ is a normalized four-component amplitude satisfying the relation $\Lambda\gamma = \gamma$.[10] $\mathbf{g}$ is an arbitrary vector introduced so that for a fixed value of $\boldsymbol{\mu}$ the functions (15) form a complete orthogonal set. We shall below use only such functions u' for which $\mathbf{g} = 0$, so that for vanishing interaction between electron and field we have to deal with an electron of momentum $mc(1-\mu^2)^{-\frac{1}{2}}\boldsymbol{\mu}$.

[10] γ may be taken to be proportional to any linear combination of the columns of $(1+\Lambda)$.

$h_m(x)$ is the normalized solution of the oscillator equation

$$h_m''-x^2h_m+(2m+1)h_m=0.$$

Going back to the original function u and the original variables $Q_{s\lambda}$, we have from (11), (12), and (15)

$$u(\mathbf{u},\, m_{s\lambda})=\gamma(\mathbf{u})\Omega^{-\frac{1}{2}}\exp\{i/\hbar(mc(1-\mu^2)^{-\frac{1}{2}}\mathbf{u},\,\mathbf{r})$$
$$+i\sum_{s\lambda}\sigma_{s\lambda}\cos(\mathbf{k}_s,\,\mathbf{r})[Q_{s\lambda}-\tfrac{1}{2}\sigma_{s\lambda}\sin(\mathbf{k}_s,\,\mathbf{r})]\}$$
$$\cdot\prod_{s\lambda}h_{m_{s\lambda}}(Q_{s\lambda}-\sigma_{s\lambda}\sin(\mathbf{k}_s,\,\mathbf{r})).\quad(17)$$

It is to be noted that the approximate validity of this solution is not conditioned by the smallness of the fine structure constant $e^2/\hbar c$, since we have neglected only the recoil of the electron and can therefore make the solution (17) the more accurate the larger a mass we assign to the electron. In order to obtain the number which must be supposed to be small in order to have (17) valid, one may calculate ψ^- in first approximation and compare its norm with the norm of ψ^+, which is unity. An estimate of norm (ψ^-) gives

$$\text{norm}\,(\psi^-)\sim e^2\omega_1/mc^3[f(\mu)(\hbar\omega_1/mc^2)$$
$$+g(\mu)(e^2\omega_1/mc^3)],$$

where ω_1 is the highest frequency taken into account, and f and g are functions of μ alone such that for small $\mu, f\sim 1/3\pi, g\sim(4/9\pi^2)\mu^2$ and for large $(1-\mu^2)^{-\frac{1}{2}}$, $f\sim(1/4\pi)(1-\mu^2)$, $g\sim(1/4\pi^2)(1-\mu^2)^2\log(1/(1-\mu^2))$. That the expansion parameter is not $e^2/\hbar c$ may also be seen by considering the classical limit, in which $\hbar$ is assumed to be zero; in this limit the shifts $\sigma_{s\lambda}\cos(\mathbf{k}_s,\,\mathbf{r})$ and $\sigma_{s\lambda}\sin(\mathbf{k}_s,\,\mathbf{r})$ of $P_{s\lambda}$ and $Q_{s\lambda}$ expressed in (17) correctly describe the superposition of the transverse part of the field produced by the uniformly moving electron, on the field of the external light waves.[11]

III. Application to Small Perturbations on the Electron; Transition Probabilities and Mean Radiated Energy

The solution (17) can be used to calculate transition probabilities due to small external perturbations acting on the electron. Consider a transition of the system between two states characterized by the velocities $\mathbf{v}=\mathbf{u}c$ and $\mathbf{w}=\nu c$ of the electron and the quantum numbers $m_{s\lambda}$ and $n_{s\lambda}$, respectively. The matrix element of a perturbation V operating on the coordinate and spin variables of the electron is given by

$$V(\mathbf{u},\, m_{s\lambda};\,\mathbf{v},\, n_{s\lambda})$$
$$=\frac{1}{\Omega}\int d\mathbf{r}\exp\{-i/\hbar(mc(1-\mu^2)^{-\frac{1}{2}}\mathbf{u},\,\mathbf{r})\}$$
$$\cdot\gamma^*(\mathbf{u})V\delta(\mathbf{v})\exp\{i/\hbar(mc(1-\nu^2)^{-\frac{1}{2}}\mathbf{v},\,\mathbf{r}\}$$
$$\cdot\prod_{s\lambda}I(\mathbf{k}_s;\,\sigma_{s\lambda},\,m_{s\lambda};\,\tau_{s\lambda},\,n_{s\lambda}),\quad(18)$$

where $\tau_{s\lambda}$ is the same function of $\mathbf{v}$ as $\sigma_{s\lambda}$ is of $\mathbf{u}$ (see (13)), $\delta(\mathbf{v})$ is a four-component amplitude for the state $\mathbf{v}$ corresponding to $\gamma(\mathbf{u})$ for the state $\mathbf{u}$, and

$$I(\mathbf{k};\,\sigma,\,m;\,\tau,\,n)=\int dQh_m(Q-\sigma\sin(\mathbf{k},\,\mathbf{r}))$$
$$\times\exp\{-i(\sigma-\tau)\cos(\mathbf{k},\,\mathbf{r})Q+\tfrac{1}{2}i(\sigma^2-\tau^2)$$
$$\times\sin(\mathbf{k},\,\mathbf{r})\cos(\mathbf{k},\,\mathbf{r})\}h_n(Q-\tau\sin(\mathbf{k},\,\mathbf{r}))\quad(19)$$
$$=\exp\{-i(m-n)(\mathbf{k},\,\mathbf{r})\}\cdot K$$

with

$$K(\sigma,\,m;\,\tau,\,n)=\exp\{-\tfrac{1}{4}(\sigma-\tau)^2\}$$
$$\times(m!n!)^{\frac{1}{2}}[-i2^{-\frac{1}{2}}(\sigma-\tau)]^{|m-n|}$$
$$\sum_{\zeta=0}^{l}\frac{[-\frac{1}{2}(\sigma-\tau)^2]^\zeta}{(l-\zeta)!\zeta!(|m-n|+\zeta)!},\quad(20)$$

where $l=m$ if $m\leq n$ and $l=n$ otherwise.[12] Putting the value (19) of I into (18), we can write the matrix element in the form

$$V(\mathbf{u},\, m_{s\lambda};\,\mathbf{v},\, n_{s\lambda})=\frac{1}{\Omega}F(\mathbf{q})\prod_{s\lambda}K(\sigma_{s\lambda},m_{s\lambda};\,\tau_{s\lambda},n_{s\lambda}),\quad(21)$$

where

$$F(\mathbf{q})=\int d\mathbf{r}\exp\{-i(\mathbf{q},\,\mathbf{r})\}\gamma^*(\mathbf{u})V\delta(\mathbf{v})\quad(22)$$

and $$\mathbf{q}=(mc/\hbar)[(1-\mu^2)^{-\frac{1}{2}}\mathbf{u}-(1-\nu^2)^{-\frac{1}{2}}\mathbf{v}]$$
$$+\sum_{s\lambda}(m_{s\lambda}-n_{s\lambda})\mathbf{k}_s.\quad(23)$$

[11] In fact, the quantities $\sigma_{s\lambda}$ can be obtained exactly by Fourier analysis of the transverse part of the classical vector potential of a uniformly moving electron.

[12] The integral over Q is best evaluated by using the generating function for the Hermitian polynomials.

From (21) follows the probability per unit time of the transition $\mathbf{u}, m_{s\lambda} \rightarrow \mathbf{v}, n_{s\lambda}$ in the form

$$T(\mathbf{u}, m_{s\lambda}; \mathbf{v}, n_{s\lambda}) = (2\pi/\hbar\Omega^2) \times D[mc^2(1-\mu^2)^{-\frac{1}{2}} - mc^2(1-\nu^2)^{-\frac{1}{2}} + \sum_{s\lambda}(m_{s\lambda}-n_{s\lambda})\hbar\omega_s] |F(\mathbf{q})|^2 \times \prod_{s\lambda} |K(\sigma_{s\lambda}, m_{s\lambda}; \tau_{s\lambda}, n_{s\lambda})|^2, \quad (24)$$

where D is Dirac's δ-function.[13]

To simplify the discussion we shall from now on neglect the last (electromagnetic) terms in (23) and in the argument of D in (24). We thus consider the changes in energy and momentum of the electromagnetic field to be small compared to the energy and the change in momentum of the electron, an assumption which is clearly justified for the treatment of the low frequencies.[14] Furthermore, we consider the state $\mathbf{u}, m_{s\lambda}$ for which all $m_{s\lambda}=0$; this means that initially no free quanta are present. The probability per unit time for a transition in which $n_{s\lambda}$ quanta of the kind s, λ are emitted and the electron after the transition moves in a direction within the element of solid angle $\sin \vartheta d\vartheta d\varphi$ is by (24) and (20)

$$U(\mathbf{u}; \mathbf{v}, n_{s\lambda}) \sin \vartheta d\vartheta d\varphi = (1/4\pi^2\hbar^4\Omega) \sin \vartheta d\vartheta d\varphi (m^2 vc/(1-\nu^2)) \times |F(\mathbf{q})|^2 \prod_{s\lambda} |K(\sigma_{s\lambda}, 0; \tau_{s\lambda}, n_{s\lambda})|^2 \quad (25)$$

$$= (1/4\pi^2\hbar^4\Omega) \sin \vartheta d\vartheta d\varphi (m^2 vc/(1-\nu^2)) |F(\mathbf{q})|^2 \times \prod_{s\lambda} \exp\{-\tfrac{1}{2}(\sigma_{s\lambda}-\tau_{s\lambda})^2\} \frac{[\frac{1}{2}(\sigma_{s\lambda}-\tau_{s\lambda})^2]^{n_{s\lambda}}}{n_{s\lambda}!}. \quad (26)$$

The result confirms what has been anticipated in the introduction, namely, that the probability of the emission of any finite number of quanta is zero. In order to show this, consider first the case in which no light quanta are emitted, i.e., where every $n_{s\lambda}=0$. We have then from (13) and (2)

$$\prod_{s\lambda} \exp\{-\tfrac{1}{2}(\sigma_{s\lambda}-\tau_{s\lambda})^2\} = \exp\left\{-(2\pi e^2/\hbar\Omega) \times \sum_{s\lambda} \omega_s^{-1}\left[\frac{(\mathbf{u}, \boldsymbol{\varepsilon}_{s\lambda})}{k_s-(\mathbf{u}, \mathbf{k}_s)} - \frac{(\mathbf{v}, \boldsymbol{\varepsilon}_{s\lambda})}{k_s-(\mathbf{v}, \mathbf{k}_s)}\right]^2\right\}. \quad (27)$$

The exponent becomes, after introducing the number of modes of vibration $\Omega/(2\pi)^3 \cdot d\mathbf{k}_s$ in the interval $d\mathbf{k}_s$ and carrying out the sum over λ,

$$-(e^2/4\pi^2\hbar c) \lim_{\omega_0 \to 0} \int_{\omega_0}^{\omega_1} \frac{d\omega_s}{\omega_s} \int_0^{\pi} \sin \vartheta_s d\vartheta_s \int_0^{2\pi} d\varphi_s \times \left[\left(\frac{\mathbf{u}}{1-\mu_s} - \frac{\mathbf{v}}{1-\nu_s}\right)^2 - \left(\frac{\mu_s}{1-\mu_s} - \frac{\nu_s}{1-\nu_s}\right)^2\right], \quad (28)$$

where ϑ_s, φ_s are the polar angles of $\mathbf{k}_s$ and μ_s, ν_s are the components of $\mathbf{u}, \mathbf{v}$ along $\mathbf{k}_s$ and where we have cut off the frequency spectrum at an upper limit ω_1. The integration over the angles gives a finite positive factor; the integration over ω_s, however, gives a result which is logarithmically infinite in the limit $\omega_0 \rightarrow 0$. Hence (27) vanishes. If a finite number of the $n_{s\lambda}$ in (26) are different from zero, (26) differs from (27) in only a finite number of factors, thus remaining zero.

However, the total probability of a transition $\mathbf{u} \rightarrow \mathbf{v}$ of the electron independently of the number of quanta emitted, which is given by the sum of (26) over all values of every $n_{s\lambda}$, is just

$$(1/4\pi^2\hbar^4\Omega) \sin \vartheta d\vartheta d\varphi (m^2 vc/(1-\nu^2)) |F(\mathbf{q})|^2, \quad (29)$$

which is the result which one would have obtained by neglecting entirely the interaction with the electromagnetic field. This result had physically to be expected since the change in momentum of the electromagnetic field has been neglected compared to that of the electron.

From the facts that the probability of the transition $\mathbf{u} \rightarrow \mathbf{v}$ accompanied by the emission of a finite number of light quanta is zero and that the total transition probability is finite, it follows that the mean total number of quanta emitted is infinite.

The mean radiated energy in the frequency interval $d\omega_s$ and in the angular range $d\vartheta_s$, $d\varphi_s$

[13] In the argument of the δ-function there should stand, rigorously speaking, $E(\mathbf{u}, m_{s\lambda}, 0) - E(\mathbf{v}, n_{s\lambda}, 0)$ where the values of E are given by (16). We have neglected the term $-(c/2h)\sum_{s\lambda}(\mathbf{u}, a_{s\lambda})^2/(k_s-(\mathbf{u}, \mathbf{k}_s))$ in (16), representing the kinetic electromagnetic self-energy of the electron, compared to the mass term $mc^2(1-\mu^2)^{-\frac{1}{2}}$. One may, in order to be consistent, imagine the frequency spectrum to be broken off at an upper limit ω_1 such that $e^2\omega_1/mc^3 \ll 1$. Since we are here concerned with difficulties arising from the low frequencies, this procedure, although connected with the fundamental difficulties of electrodynamics, does not affect our conclusions.

[14] For the treatment of higher frequencies one may use an expansion in powers of $e^2/\hbar c$.

when the electron is deflected into the element of solid angle $\sin\vartheta d\vartheta d\varphi$ is given by

$$(\Omega/(2\pi c)^3)\omega_s^2 d\omega_s \sin\vartheta_s d\vartheta_s d\varphi_s$$

$$\times \sum_{\lambda} \sum_{n_{s\lambda}=0}^{\infty} n_{s\lambda}\hbar\omega_s U \sin\vartheta d\vartheta d\varphi \quad (30)$$

$$= (1/4\pi^2\hbar^4\Omega)\sin\vartheta d\vartheta d\varphi(m^2\nu c/(1-\nu^2))\,|F(\mathbf{q})|^2$$

$$\cdot (e^2/4\pi^2 c)\left[\left(\frac{\mathbf{u}}{1-\mu_s}-\frac{\mathbf{v}}{1-\nu_s}\right)^2 - \left(\frac{\mu_s}{1-\mu_s}-\frac{\nu_s}{1-\nu_s}\right)^2\right] d\omega_s \sin\vartheta_s d\vartheta_s d\varphi_s. \quad (31)$$

This has been obtained by making use of (26), (13) and (2) and of the formula

$$\sum_{n=0}^{\infty} n e^{-x} x^n / n! = x.$$

On account of the extra factor ω_s in (30), the mean total energy, unlike the mean total number of quanta, is finite. Formula (31) is just the total probability that the electron be scattered into the element of solid angle $\sin\vartheta d\vartheta d\varphi$ multiplied by the amount of energy which it would radiate classically in such a deflection. The expansion in powers of $e^2/\hbar c$ in the limit of small frequencies, where alone Formula (31) may be supposed to be valid, leads to the same result for the mean energy radiated, though its results are entirely misleading so far as transition probabilities are concerned.[15]

The above considerations can be applied almost literally also to cases such as the theory of β-decay, in which the external perturbation does not act on the electron coordinates, but creates an electron. The only difference, so far as the electromagnetic field is concerned, consists in replacing $\mathbf{u}c$ by the velocity of the nucleus and $\mathbf{v}c$ by the velocity with which the electron is created. Here again the total probability of β-decay is unaltered by the interaction of the electron with the low frequency radiation and the mean radiated energy is in agreement with that calculated by expanding in powers of $e^2/\hbar c$.

[15] This remark does not affect the cross section for the emission of a high frequency quantum as calculated by Bethe and Heitler, and experimentally verified. However, the cross section so derived has to be interpreted, not as the probability that the high frequency quantum alone be emitted, but as the probability that this happens no matter how many other light quanta are emitted.

On the Intrinsic Moment of the Electron*

H. M. Foley and P. Kusch
Columbia University, New York, New York
December 26, 1947

IN a previous letter[1] we have reported the observation that the ratio of the g_J values of the $^2P_{3/2}$ and $^2P_{\frac{1}{2}}$ states of gallium has the value 2.00344; the value 2 for this ratio follows from Russell-Saunders coupling and the conventional spin and orbital gyromagnetic ratios. If each of these states is exactly described by Russell-Saunders coupling, this observation can only be explained by setting $(\delta_S - 2\delta_L) = 0.00229 \pm 0.00008$, where the electron spin g value is $g_S = 2 + \delta_S$, and the orbital momentum g value is $g_L = 1 + \delta_L$. Since each of these atomic states may be separately subject to configuration interaction perturbations, the interpretation of this result was not entirely clear.

A determination has now been made of the ratio of the g_J values of Na in the $^2S_{\frac{1}{2}}$ state and of Ga in the $^2P_{\frac{1}{2}}$ state. The experimental procedure was similar to that previously described.[1] The known hyperfine interaction constants of gallium[2] and sodium[3] were employed in the analysis of the data. We find for this ratio the value 3.00732 ± 0.00018 instead of the value 3. This result can be explained by making $(\delta_S - 2\delta_L) = 0.00244 \pm 0.00006$.

The agreement between the values of $(\delta_S - 2\delta_L)$ obtained by the two experiments makes it unlikely that one can account for the effect by perturbation of the states. The effect of configuration interaction on the g_J value of sodium is presumably negligible.[4] To explain our observed effect without modification of the conventional values of g_S or g_L introduces the rather unlikely requirement that both states of gallium be perturbed, and by amounts just great enough to give the agreement noted above.

From any experiment in which the ratio of the g_J values of atomic states is determined, it is possible to determine only the quantity $(\delta_S - 2\delta_L)$. If, on the basis of the correspondence principle we set δ_L equal to zero, we may state the result of our first experiment as

$$g_S = 2.00229 \pm 0.00008$$

and that of our recent experiment as

$$g_S = 2.00244 \pm 0.00006.$$

It is not possible, at the present time, to state whether the apparent discrepancy between these values is real. It is conceivable that some small perturbation of the states would give rise to a discrepancy of the indicated magnitude.

These results are not in agreement with the recent suggestion by Breit[6] as to the magnitude of the intrinsic moment of the electron.

* Publication assisted by the Ernest Kempton Adams Fund for Physical Research of Columbia University.

[1] P. Kusch and H. M. Foley, Phys. Rev. **72**, 1256 (1947).
[2] G. E. Becker and P. Kusch, to be published.
[3] S. Millman and P. Kusch, Phys. Rev. **58**, 438 (1940).
[4] M. Phillips, Phys. Rev. **60**, 100 (1941).
[5] Dr. J. Schwinger has very kindly informed us in advance of publication of his conclusion from theoretical studies that δ_L is zero whereas δ_S may not vanish.
[6] G. Breit, Phys. Rev. **72**, 984 (1947).

Fine Structure of the Hydrogen Atom by a Microwave Method* **

Willis E. Lamb, Jr. and Robert C. Retherford
Columbia Radiation Laboratory, Department of Physics, Columbia University, New York, New York
(Received June 18, 1947)

THE spectrum of the simplest atom, hydrogen, has a fine structure[1] which according to the Dirac wave equation for an electron moving in a Coulomb field is due to the combined effects of relativistic variation of mass with velocity and spin-orbit coupling. It has been considered one of the great triumphs of Dirac's theory that it gave the "right" fine structure of the energy levels. However, the experimental attempts to obtain a really detailed confirmation through a study of the Balmer lines have been frustrated by the large Doppler effect of the lines in comparison to the small splitting of the lower or $n=2$ states. The various spectroscopic workers have alternated between finding confirmation[2] of the theory and discrepancies[3] of as much as eight percent. More accurate information would clearly provide a delicate test of the form of the correct relativistic wave equation, as well as information on the possibility of line shifts due to coupling of the atom with the radiation field and clues to the nature of any non-Coulombic interaction between the elementary particles: electron and proton.

The calculated separation between the levels $2^2P_{\frac{1}{2}}$ and $2^2P_{3/2}$ is 0.365 cm^{-1} and corresponds to a wave-length of 2.74 cm. The great wartime advances in microwave techniques in the vicinity of three centimeters wave-length make possible the use of new physical tools for a study of the $n=2$ fine structure states of the hydrogen atom. A little consideration shows that it would be exceedingly difficult to detect the direct absorption of radiofrequency radiation by excited H atoms in a gas discharge because of their small population and the high background absorption due to electrons. Instead, we have found a method depending on a novel property of the $2^2S_{\frac{1}{2}}$ level. According to the Dirac theory, this state exactly coincides in energy with the $2^2P_{\frac{1}{2}}$ state which is the lower of the two P states. The S state in the absence of external electric fields is metastable. The radiative transition to the ground state $1^2S_{\frac{1}{2}}$ is forbidden by the selection rule $\Delta L = \pm 1$. Calculations of Breit and Teller[4] have shown that the most probable decay mechanism is double quantum emission with a lifetime of 1/7 second. This is to be contrasted with a lifetime of only 1.6×10^{-9} second for the non-metastable 2^2P states. The metastability is very much reduced in the presence of external electric fields[5] owing to Stark effect mixing of the S and P levels with resultant rapid decay of the combined state. If for any reason, the $2^2S_{\frac{1}{2}}$ level does not exactly coincide with the $2^2P_{\frac{1}{2}}$ level, the vulnerability of the state to external fields will be reduced. Such a removal of the accidental degeneracy may arise from any defect in the theory or may be brought about by the Zeeman splitting of the levels in an external magnetic field.

In brief, the experimental arrangement used is the following: Molecular hydrogen is thermally dissociated in a tungsten oven, and a jet of atoms emerges from a slit to be cross-bombarded by an electron stream. About one part in a hundred million of the atoms is thereby excited to the metastable $2^2S_{\frac{1}{2}}$ state. The metastable atoms (with a small recoil deflection) move on out of the bombardment region and are detected by the process of electron ejection from a metal target. The electron current is measured with an FP-54 electrometer tube and a sensitive galvanometer.

If the beam of metastable atoms is subjected to any perturbing fields which cause a transition to any of the 2^2P states, the atoms will decay while moving through a very small distance. As a result, the beam current will decrease, since the

* Publication assisted by the Ernest Kempton Adams Fund for Physical Research of Columbia University, New York.
** Work supported by the Signal Corps under contract number W 36-039 sc-32003.

[1] For a convenient account, see H. E. White, *Introduction to Atomic Spectra* (McGraw-Hill Book Company, New York, 1934), Chap. 8.

[2] J. W. Drinkwater, O. Richardson, and W. E. Williams, Proc. Roy. Soc. **174**, 164 (1940).

[3] W. V. Houston, Phys. Rev. **51**, 446 (1937); R. C. Williams, Phys. Rev. **54**, 558 (1938); S. Pasternack, Phys. Rev. **54**, 1113 (1938) has analyzed these results in terms of an upward shift of the S level by about 0.03 cm^{-1}.

[4] H. A. Bethe in *Handbuch der Physik*, Vol. 24/1, §43.

[5] G. Breit and E. Teller, Astrophys. J. **91**, 215 (1940).

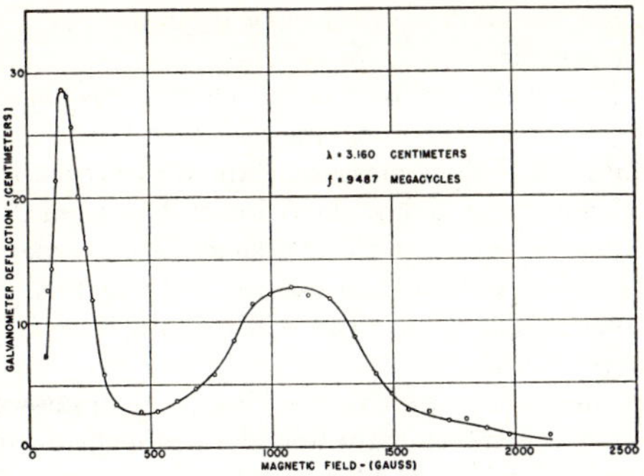

FIG. 1. A typical plot of galvanometer deflection due to interruption of the microwave radiation as a function of magnetic field. The magnetic field was calibrated with a flip coil and may be subject to some error which can be largely eliminated in a more refined apparatus. The width of the curves is probably due to the following causes: (1) the radiative line width of about 100 Mc/sec. of the ^2P states, (2) hyperfine splitting of the ^2S state which amounts to about 88 Mc/sec., (3) the use of an excessive intensity of radiation which gives increased absorption in the wings of the lines, and (4) inhomogeneity of the magnetic field. No transitions from the state $2^2S_{\frac{1}{2}}(m=-\frac{1}{2})$ have been observed, but atoms in this state may be quenched by stray electric fields because of the more nearly exact degeneracy with the Zeeman pattern of the ^2P states.

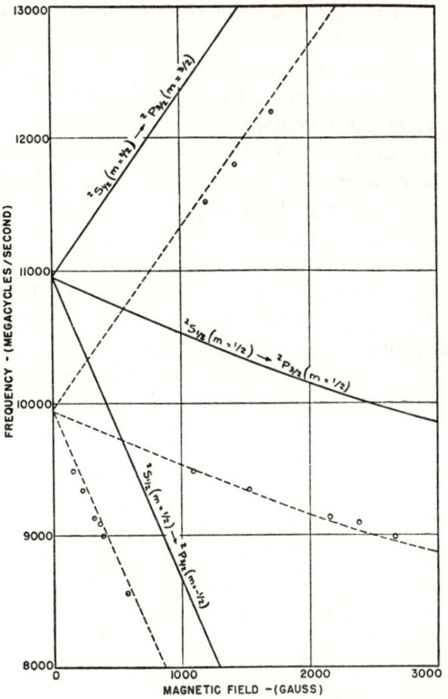

FIG. 2. Experimental values for resonance magnetic fields for various frequencies are shown by circles. The solid curves show three of the theoretically expected variations, and the broken curves are obtained by shifting these down by 1000 Mc/sec. This is done merely for the sake of comparison, and it is not implied that this would represent a "best fit." The plot covers only a small range of the frequency and magnetic field scale covered by our data, but a complete plot would not show up clearly on a small scale, and the shift indicated by the remainder of the data is quite compatible with a shift of 1000 Mc.

detector does not respond to atoms in the ground state. Such a transition may be induced by the application to the beam of a static electric field somewhere between source and detector. Transitions may also be induced by radiofrequency radiation for which *hν* corresponds to the energy difference between one of the Zeeman components of $2^2S_{\frac{1}{2}}$ and any component of either $2^2P_{\frac{1}{2}}$ or $2^2P_{3/2}$. Such measurements provide a precise method for the location of the $2^2S_{\frac{1}{2}}$ state relative to the *P* states, as well as the distance between the latter states.

We have observed an electrometer current of the order of 10^{-14} ampere which must be ascribed to metastable hydrogen atoms. The strong quenching effect of static electric fields has been observed, and the voltage gradient necessary for this has a reasonable dependence on magnetic field strength.

We have also observed the decrease in the beam of metastable atoms caused by microwaves in the wave-length range 2.4 to 18.5 cm in various magnetic fields. In the measurements, the frequency of the r-f is fixed, and the change in the galvanometer current due to interruption of the r-f is determined as a function of magnetic field strength. A typical curve of quenching *versus* magnetic field is shown in Fig. 1. We have plotted in Fig. 2 the resonance magnetic fields for various frequencies in the vicinity of 10,000 Mc/sec. The theoretically calculated curves for the Zeeman effect are drawn as solid curves, while for comparison with the observed points, the calculated curves have been shifted downward by 1000 Mc/sec. (broken curves). The results indicate clearly that, contrary to theory but in essential agreement with Pasternack's hypothesis,[3] the $2^2S_{\frac{1}{2}}$ state is higher than the $2^2P_{\frac{1}{2}}$ by about 1000 Mc/sec. (0.033 cm^{-1} or about 9 percent of the spin relativity doublet separation. The lower frequency transitions $^2S_{\frac{1}{2}}(m=\frac{1}{2})\rightarrow$ $^2P_{\frac{1}{2}}(m=\pm\frac{1}{2})$ have also been observed and agree well with such a shift of the $^2S_{\frac{1}{2}}$ level. With the present precision, we have not yet detected any discrepancy between the Dirac theory and the doublet separation of the P levels. (According to most of the imaginable theoretical explanations of the shift, the doublet separation would not be affected as much as the relative location of the S and P states.) With proposed refinements in sensitivity, magnetic field homogeneity, and calibration, it is hoped to locate the S level with respect to each P level to an accuracy of at least ten Mc/sec. By addition of these frequencies and assumption of the theoretical formula $\Delta\nu = \frac{1}{16}\alpha^2 R$ for the doublet separation, it should be possible to measure the square of the fine structure constant times the Rydberg frequency to an accuracy of 0.1 percent.

By a slight extension of the method, it is hoped to determine the hyperfine structure of the $2^2S_{\frac{1}{2}}$ state. All of these measurements will be repeated for deuterium and other hydrogen-like atoms.

A paper giving a fuller account of the experimental and theoretical details of the method is being prepared, and this will contain later and more accurate data.

The experiments described here were discussed at the Conference on the Foundations of Quantum Mechanics held at Shelter Island on June 1–3, 1947 which was sponsored by the National Academy of Sciences.

The Electromagnetic Shift of Energy Levels

H. A. Bethe
Cornell University, Ithaca, New York
(Received June 27, 1947)

By very beautiful experiments, Lamb and Retherford[1] have shown that the fine structure of the second quantum state of hydrogen does not agree with the prediction of the Dirac theory. The 2*s* level, which according to Dirac's theory should coincide with the $2p_{\frac{1}{2}}$ level, is actually higher than the latter by an amount of about 0.033 cm^{-1} or 1000 megacycles. This discrepancy had long been suspected from spectroscopic measurements.[2,3] However, so far no satisfactory theoretical explanation has been given. Kemble and Present, and Pasternack[4] have shown that the shift of the 2*s* level cannot be explained by a nuclear interaction of reasonable magnitude, and Uehling[5] has investigated the effect of the "polarization of the vacuum" in the Dirac hole theory, and has found that this effect also is much too small and has, in addition, the wrong sign.

Schwinger and Weisskopf, and Oppenheimer have suggested that a possible explanation might be the shift of energy levels by the interaction of the electron with the radiation field. This shift comes out infinite in all existing theories, and has therefore always been ignored. However, it is possible to identify the most strongly (linearly) divergent term in the level shift with an electromagnetic *mass* effect which must exist for a bound as well as for a free electron. This effect should

[1] Phys. Rev. **72**, 241 (1947).
[2] W. V. Houston, Phys. Rev. **51**, 446 (1937).
[3] R. C. Williams, Phys. Rev. **54**, 558 (1938).
[4] E. C. Kemble and R. D. Present, Phys. Rev. **44**, 1031 (1932); S. Pasternack, Phys. Rev. **54**, 1113 (1938).
[5] E. A. Uehling, Phys. Rev. **48**, 55 (1935).

properly be regarded as already included in the observed mass of the electron, and we must therefore subtract from the theoretical expression, the corresponding expression for a free electron of the same average kinetic energy. The result then diverges only logarithmically (instead of linearly) in non-relativistic theory: Accordingly, it may be expected that in the hole theory, in which the *main* term (self-energy of the electron) diverges only logarithmically, the result will be *convergent* after subtraction of the free electron expression.[6] This would set an effective upper limit of the order of mc^2 to the frequencies of light which effectively contribute to the shift of the level of a bound electron. I have not carried out the relativistic calculations, but I shall assume that such an effective relativistic limit exists.

The ordinary radiation theory gives the following result for the self-energy of an electron in a quantum state m, due to its interaction with transverse electromagnetic waves:

$$W=-(2e^2/3\pi\hbar c^3)$$
$$\times\int_0^K kdk\sum_n |\mathbf{v}_{mn}|^2/(E_n-E_m+k), \quad (1)$$

where $k=\hbar\omega$ is the energy of the quantum and $\mathbf{v}$ is the velocity of the electron which, in non-relativistic theory, is given by

$$\mathbf{v}=\mathbf{p}/m=(\hbar/im)\nabla. \quad (2)$$

Relativistically, $\mathbf{v}$ should be replaced by $c\boldsymbol{\alpha}$ where $\boldsymbol{\alpha}$ is the Dirac operator. Retardation has been neglected and can actually be shown to make no substantial difference. The sum in (1) goes over all atomic states n, the integral over all quantum energies k up to some maximum K to be discussed later.

For a free electron, $\mathbf{v}$ has only diagonal elements and (1) is replaced by

$$W_0=-(2e^2/3\pi\hbar c^3)\int kdk\mathbf{v}^2/k. \quad (3)$$

This expression represents the change of the kinetic energy of the electron for fixed momentum, due to the fact that electromagnetic mass is added to the mass of the electron. This electromagnetic mass is already contained in the experimental electron mass; the contribution (3) to the energy should therefore be disregarded. For a bound electron, $\mathbf{v}^2$ should be replaced by its expectation value, $(\mathbf{v}^2)_{mm}$. But the matrix elements of $\mathbf{v}$ satisfy the sum rule

$$\sum_n |\mathbf{v}_{mn}|^2=(\mathbf{v}^2)_{mm}. \quad (4)$$

Therefore the relevant part of the self-energy becomes

$$W'=W-W_0=+\frac{2e^2}{3\pi\hbar c^3}$$
$$\times\int_0^K dk\sum_n\frac{|\mathbf{v}_{mn}|^2(E_n-E_m)}{E_n-E_m+k}. \quad (5)$$

This we shall consider as a true shift of the levels due to radiation interaction.

It is convenient to integrate (5) first over k. Assuming K to be large compared with all energy differences E_n-E_m in the atom,

$$W'=\frac{2e^2}{3\pi\hbar c^3}\sum_n|\mathbf{v}_{mn}|^2(E_n-E_m)\ln\frac{K}{|E_n-E_m|}. \quad (6)$$

(If E_n-E_m is negative, it is easily seen that the principal value of the integral must be taken, as was done in (6).) Since we expect that relativity theory will provide a natural cut-off for the frequency k, we shall assume that in (6)

$$K\approx mc^2. \quad (7)$$

(This does not imply the same limit in Eqs. (2) and (3).) The argument in the logarithm in (6) is therefore very large; accordingly, it seems permissible to consider the logarithm as constant (independent of n) in first approximation.

We therefore should calculate

$$A=\sum_n A_{nm}=\sum_n|\mathbf{p}_{nm}|^2(E_n-E_m). \quad (8)$$

This sum is well known; it is

$$A=\sum|\mathbf{p}_{nm}|^2(E_n-E_m)$$
$$=-\hbar^2\int\psi_m{}^*\nabla V\cdot\nabla\psi_m d\tau$$
$$=\tfrac{1}{2}\hbar^2\int\nabla^2 V\psi_m{}^2 d\tau=2\pi\hbar^2e^2Z\psi_m{}^2(0), \quad (9)$$

[6] It was first suggested by Schwinger and Weisskopf that hole theory must be used to obtain convergence in this problem.

for a nuclear charge Z. For any electron with angular momentum $l \neq 0$, the wave function vanishes at the nucleus; therefore, the sum $A=0$. For example, for the $2p$ level the negative contribution $A_{1S,2P}$ balances the positive contributions from all other transitions. For a state with $l=0$, however,

$$\psi_m^2(0)=(Z/na)^3/\pi, \tag{10}$$

where n is the principal quantum number and a is the Bohr radius.

Inserting (10) and (9) into (6) and using relations between atomic constants, we get for an S state

$$W_{ns}'=\frac{8}{3\pi}\left(\frac{e^2}{\hbar c}\right)^3 \mathrm{Ry}\frac{Z^4}{n^3}\ln\frac{K}{\langle E_n-E_m\rangle_{\mathrm{AV}}}, \tag{11}$$

where Ry is the ionization energy of the ground state of hydrogen. The shift for the $2p$ state is negligible; the logarithm in (11) is replaced by a value of about -0.04. The average excitation energy $\langle E_n-E_m\rangle_{\mathrm{AV}}$ for the $2s$ state of hydrogen has been calculated numerically[7] and found to be 17.8 Ry, an amazingly high value. Using this figure and $K=mc^2$, the logarithm has the value 7.63, and we find

$$W_{ns}'=136\ln[K/(E_n-E_m)] =1040 \text{ megacycles.} \tag{12}$$

[7] I am indebted to Dr. Stehn and Miss Steward for the numerical calculations.

This is in excellent agreement with the observed value of 1000 megacycles.

A relativistic calculation to establish the limit K is in progress. Even without exact knowledge of K, however, the agreement is sufficiently good to give confidence in the basic theory. This shows

(1) that the level shift due to interaction with radiation is a real effect and is of finite magnitude,

(2) that the effect of the infinite electromagnetic mass of a point electron can be eliminated by proper identification of terms in the Dirac radiation theory,

(3) that an accurate experimental and theoretical investigation of the level shift may establish relativistic effects (e.g., Dirac hole theory). These effects will be of the order of unity in comparison with the logarithm in Eq. (11).

If the present theory is correct, the level shift should increase roughly as Z^4 but not quite so rapidly, because of the variation of $\langle E_n-E_m\rangle_{\mathrm{AV}}$ in the logarithm. For example, for He^+, the shift of the $2s$ level should be about 13 times its value for hydrogen, giving 0.43 cm^{-1}, and that of the $3s$ level about 0.13 cm^{-1}. For the x-ray levels LI and LII, this effect should be superposed upon the effect of screening which it partly compensates. An accurate theoretical calculation of the screening is being undertaken to establish this point.

This paper grew out of extensive discussions at the Theoretical Physics Conference on Shelter Island, June 2 to 4, 1947. The author wishes to express his appreciation to the National Academy of Science which sponsored this stimulating conference.

On Quantum-Electrodynamics and the Magnetic Moment of the Electron

JULIAN SCHWINGER
Harvard University, Cambridge, Massachusetts
December 30, 1947

ATTEMPTS to evaluate radiative corrections to electron phenomena have heretofore been beset by divergence difficulties, attributable to self-energy and vacuum polarization effects. Electrodynamics unquestionably requires revision at ultra-relativistic energies, but is presumably accurate at moderate relativistic energies. It would be desirable, therefore, to isolate those aspects of the current theory that essentially involve high energies, and are subject to modification by a more satisfactory theory, from aspects that involve only moderate energies and are thus relatively trustworthy. This goal has been achieved by transforming the Hamiltonian of current hole theory electrodynamics to exhibit explicitly the logarithmically divergent self-energy of a free electron, which arises from the virtual emission and absorption of light quanta. The electromagnetic self-energy of a free electron can be ascribed to an electromagnetic mass, which must be added to the mechanical mass of the electron. Indeed, the only meaningful statements of the theory involve this combination of masses, which is the experimental mass of a free electron. It might appear, from this point of view, that the divergence of the electromagnetic mass is unobjectionable, since the individual contributions to the experimental mass are unobservable. However, the transformation of the Hamiltonian is based on the assumption of a weak interaction between matter and radiation, which requires that the electromagnetic mass be a small correction ($\sim(e^2/\hbar c)m_0$) to the mechanical mass m_0.

The new Hamiltonian is superior to the original one in essentially three ways: it involves the experimental electron mass, rather than the unobservable mechanical mass; an electron now interacts with the radiation field only in the presence of an external field, that is, only an accelerated electron can emit or absorb a light quantum;* the interaction energy of an electron with an external field is now subject to a *finite* radiative correction. In connection with the last point, it is important to note that the inclusion of the electromagnetic mass with the mechanical mass does not avoid all divergences; the polarization of the vacuum produces a logarithmically divergent term proportional to the interaction energy of the electron in an external field. However, it has long been recognized that such a term is equivalent to altering the value of the electron charge by a constant factor, only the final value being properly identified with the experimental charge. Thus the interaction between matter and radiation produces a renormalization of the electron charge and mass, all divergences being contained in the renormalization factors.

The simplest example of a radiative correction is that for the energy of an electron in an external magnetic field. The detailed application of the theory shows that the radiative correction to the magnetic interaction energy corresponds to an additional magnetic moment associated with the electron spin, of magnitude $\delta\mu/\mu=(\frac{1}{2}\pi)e^2/\hbar c=0.001162$. It is indeed gratifying that recently acquired experimental data confirm this prediction. Measurements on the hyperfine splitting of the ground states of atomic hydrogen and deuterium[1] have yielded values that are definitely larger than those to be expected from the directly measured nuclear moments and an electron moment of one Bohr magneton. These discrepancies can be accounted for by a small additional electron spin magnetic moment.[2] Recalling that the nuclear moments have been calibrated in terms of the electron moment, we find the additional moment necessary to account for the measured hydrogen and deuterium hyperfine structures to be $\delta\mu/\mu=0.00126\pm0.00019$ and $\delta\mu/\mu=0.00131\pm0.00025$, respectively. These values are not in disagreement with the theoretical prediction. More precise conformation is provided by measurement of the g values for the $^2S_{\frac{1}{2}}$, $^2P_{\frac{1}{2}}$, and $^2P_{3/2}$ states of sodium and gallium.[3] To account for these results, it is necessary to ascribe the following additional spin magnetic moment to the electron, $\delta\mu/\mu=0.00118\pm0.00003$.

The radiative correction to the energy of an electron in a Coulomb field will produce a shift in the energy levels of hydrogen-like atoms, and modify the scattering of electrons in a Coulomb field. Such energy level displacements have recently been observed in the fine structures of hydrogen,[4] deuterium, and ionized helium.[5] The values yielded by our theory differ only slightly from those conjectured by Bethe[6] on the basis of a non-relativistic calculation, and are, thus, in good accord with experiment. Finally, the finite radiative correction to the elastic scattering of electrons by a Coulomb field provides a satisfactory termination to a subject that has been beset with much confusion.

A paper dealing with the details of this theory and its applications is in course of preparation.

* A classical non-relativistic theory of this type was discussed by H. A. Kramers at the Shelter Island Conference, held in June 1947 under the auspices of the National Academy of Sciences.

1 J. E. Nafe, E. B. Nelson, and I. I. Rabi, Phys. Rev. **71**, 914 (1947); D. E. Nagel, R. S. Julian, and J. R. Zacharias, Phys. Rev. **72**, 971 (1947).

2 G. Breit, Phys. Rev. **71**, 984 (1947). However, Breit has not correctly drawn the consequences of his empirical hypothesis. The effects of a nuclear magnetic field and a constant magnetic field do not involve different combinations of μ and $\delta\mu$.

3 P. Kusch and H. M. Foley, Phys. Rev. **72**, 1256 (1947), and further unpublished work.

4 W. E. Lamb, Jr. and R. C. Retherford, Phys. Rev. **72**, 241 (1947).

5 J. E. Mack and N. Austern, Phys. Rev. **72**, 972 (1947).

6 H. A. Bethe, Phys. Rev. **72**, 339 (1947).

On Radiative Corrections to Electron Scattering

JULIAN SCHWINGER
Harvard University, Cambridge, Massachusetts
January 21, 1949

RADIATIVE corrections to the electromagnetic properties of the electron produce energy level displacements and modify electron scattering cross sections. Although the high accuracy of radiofrequency spectroscopy facilitates the measurement of energy level displacements, as in the Lamb-Retherford experiment[1] and the evidence on the anomalous magnetic moment of the electron,[2] nevertheless the correction to the cross section for scattering of an electron in a Coulomb field is not without interest, since it permits a comparison between theory and experiment in the relativistic region, as compared to the non-relativistic domain to which the energy level measurements pertain.

The radiative correction to the cross section for essentially elastic scattering of an electron by a Coulomb field has been computed with the form of quantum electrodynamics developed in several recent papers.[3] In addition to the emission and absorption of virtual quanta, we include the real emission of quanta with maximum energy ΔE, which is small in comparison with $W=E-\mathrm{mc}^2$, the initial kinetic energy of the electron. In other words, we treat only those inelastic events in which a small fraction of the original energy is radiated. The contribution of the remainder of the inelastic processes can be derived from the well-known bremsstrahlung cross section, and is not of principal interest. The result, expressed as a fractional decrease in the differential cross section for scattering through an angle ϑ, is

$$\delta=2\alpha/\pi\left[\left(\log\frac{E}{\Delta E}-1\right)(K_0+K_1)+\tfrac{1}{2}K_0-K_1+\tfrac{1}{3}K_2-L+\tfrac{1}{2}\frac{(\mathrm{mc}^2/E)^2}{1-\beta^2\sin^2\vartheta/2}K_0\right], \quad (1)$$

where

$$K_0=[\lambda/(1+\lambda^2)^{1/2}]\log[(1+\lambda^2)^{1/2}+\lambda],\ \lambda=(p/\mathrm{mc})\sin(\vartheta/2),$$
$$K_1=[(1+\lambda^2)/\lambda^2]K_0-1, \quad (2)$$
$$K_2=[(1+\lambda^2)/\lambda^2]K_1-\tfrac{1}{3},$$

and

$$L=(\lambda^2+\tfrac{1}{2})\frac{(\mathrm{mc}^2/E)^2}{\beta}\int_0^1\left[\frac{\log\frac{1}{2}(1-\beta x)}{1+\beta x}-\frac{\log\frac{1}{2}(1+\beta x)}{1-\beta x}\right]\frac{dv}{x}-\tfrac{1}{2}\frac{(\mathrm{mc}^2/E)^2}{\beta}\left[\frac{\log\frac{1}{2}(1-\beta)}{1+\beta}-\frac{\log\frac{1}{2}(1+\beta)}{1-\beta}\right]. \quad (3)$$

Here

$$x=[1-\sin^2(\vartheta/2)(1-v^2)]^{1/2} \quad (4)$$

and $\beta=pc/E$. Note that δ diverges logarithmically in the limit $\Delta E\to 0$. It is well known that this difficulty stems from the neglect of processes involving more than one low frequency quantum.[4] Actually the essentially elastic scattering cross section approaches zero as $\Delta E\to 0$; that is, it never happens that a scattering event is unaccompanied by the emission of quanta. This is described by replacing the radiative correction factor $1-\delta$ with $e^{-\delta}$, which has the proper limiting behavior. The further terms in the series expansion of $e^{-\delta}$ express the effects of higher order processes involving the multiple emission of soft quanta. However, for practical purposes such a refinement is unnecessary. The accuracy with which the energy of a particle can be measured is such that the limit $\Delta E\to 0$ cannot be realized, and δ will be small in comparison with unity under presently accessible circumstances.

In the non-relativistic limit, $\beta\ll 1$,

$$K_n=[1/(2n+1)]\beta^2\sin^2(\vartheta/2),$$
$$L=[(4/3)(\log 2-1)-\tfrac{1}{3}]\beta^2\sin^2(\vartheta/2), \quad (5)$$

and

$$\delta=(8\alpha/3\pi)\beta^2\sin^2(\vartheta/2)[\log(\mathrm{mc}^2/2\Delta E)+(19/30)], \quad (6)$$

which increases linearly with the kinetic energy of the particle. For a slowly moving particle, it is an elementary matter to include the additional scattering produced by the real emission of quanta with energies in the interval from ΔE to W. One thereby obtains the following fractional decrease in the differential cross section for scattering through an angle ϑ, irrespective of the final energy:

$$\delta=(8\alpha/3\pi)\beta^2\sin^2(\vartheta/2)[\log(\mathrm{mc}^2/8W)+(19/30)+(\pi-\vartheta)\tan(\vartheta/2)+[\cos\vartheta/\cos^2(\vartheta/2)]\log\csc(\vartheta/2)]. \quad (7)$$

We may remark, parenthetically, that in the same non-relativistic approximation, the radiative correction to the energy of a particle moving in an external field with potential energy $V(r)$ is[5]

$$\begin{aligned}\delta E&=(\alpha/3\pi)[\log(\mathrm{mc}^2/2\Delta W)+(31/120)](\hbar/\mathrm{mc})^2\langle\nabla^2V\rangle+(\alpha/2\pi)(\hbar/2\mathrm{mc})\langle-i\beta\boldsymbol{\alpha}\cdot\nabla V\rangle\\&=(\alpha/3\pi)(\hbar/\mathrm{mc})^2[(\log(\mathrm{mc}^2/2\Delta W)+(19/30))\langle\nabla^2V\rangle+\tfrac{3}{4}\langle\boldsymbol{\sigma}\cdot\mathbf{L}(1/r)(dV/dr)\rangle],\end{aligned} \quad (8)$$

where $\mathbf{L}$ is the orbital angular momentum operator in in units of $\hbar$, and ΔW is an average excitation energy of the system.[6] Applied to the relative displacement of the $2^2S_{1/2}$ and $2^2P_{1/2}$ levels of hydrogen, this formula yields 1051 mc/sec., to be compared with the experimental value[7] of 1062 ± 5 mc/sec.

The extreme relativistic limit of (1) is

$$\delta = (4\alpha/\pi)[(\log(E/\Delta E) - (13/12)) \times (\log(2E/mc^2)\sin(\vartheta/2) - \tfrac{1}{2}) + (17/72) + \phi(\vartheta)], \quad (9)$$

where

$$\phi(\vartheta) = \tfrac{1}{2}\sin(\vartheta/2)\int_{\cos(\vartheta/2)}^{1}\left[\frac{\log\frac{1}{2}(1+x)}{1-x} - \frac{\log\frac{1}{2}(1-x)}{1+x}\right] \times \frac{dx}{(x^2-\cos^2(\vartheta/2))^{\frac{1}{2}}}. \quad (10)$$

The integral can be performed analytically for $\vartheta = \pi$, $\phi(\pi) = \pi^2/24$, but must be evaluated numerically for other angles. An approximation in excess, which has the correct asymptotic form at small angles, is provided by

$$\phi(\vartheta) \sim \frac{1-\cos(\vartheta/2)}{[2\cos(\vartheta/2)(1+\cos(\vartheta/2))]^{\frac{1}{2}}} \times \left[\log\frac{1}{2(1-\cos(\vartheta/2))} + \frac{1-\cos(\vartheta/2)}{2} + 1\right]. \quad (11)$$

This approximation is reasonably accurate even at $\vartheta = \frac{1}{2}\pi$, where the value yielded by (11) exceeds by only 8.6 percent the following result of a numerical calculation, $\phi(\pi/2) = 0.292$.

The asymptotic formula (9) is quite accurate for even moderate energies. Thus, with $\vartheta = \frac{1}{2}\pi$, $\Delta E = 10$ kev, and $W = 3.1$ Mev, which corresponds to $(E/mc^2)\sin(\vartheta/2) = 5$, the value of δ computed from (9) differs from the correct value, $\delta = 8.6\ 10^{-2}$, by only a few tenths of a percent. It is evident from this numerical result that radiative corrections to scattering cross sections can be quite appreciable. For the particular conditions chosen, ΔE can be materially increased (but still subject to $\Delta E \ll W$), without seriously impairing δ. Thus, with $\Delta E = 40$ kev, $\delta = 6.3\ 10^{-2}$, while $\Delta E = 80$ kev yields $\delta = 5.1\ 10^{-2}$. As to the energy dependence of δ, we remark that with a given accuracy in the determination of the energy, say $\Delta E/E = 0.04/3.6 = 1.1\ 10^{-2}$, an increase in the total energy by a factor of four produces an addition of $4.4\ 10^{-2}$ to δ. Thus, for a kinetic energy of 14 Mev, $\delta = 11\ 10^{-2}$.

The variation of δ with angle, at moderate energies, cannot be studied with the asymptotic formula (9) alone, for at small angles the condition $(p/mc)^2\sin^2(\vartheta/2) \gg 1$, which underlies this formula, will not be maintained. It is evident from (1) that δ is proportional to $\sin^2(\partial/2)$ at angles such that $(p/mc)^2\sin^2\frac{1}{2}\theta \ll 1$. However, for $W = 3.1$ Mev, $\Delta E = 40$ kev, and $\vartheta = \pi/4$, which corresponds to $(p/mc)\sin(\vartheta/2) = 2.7$, the correct value of δ, $4.2\ 10^{-2}$, exceeds that deduced from (9) by only 2 percent. For the same choice of W and ΔE, the value of δ associated with $\vartheta = 3\pi/4$ is $\delta = 7.2\ 10^{-2}$.

The verification of these predictions would provide valuable conformation for the relativistic aspects of the radiative corrections to the electromagnetic properties of the electron.

[1] W. E. Lamb and R. C. Retherford, Phys. Rev. **72**, 241 (1947).
[2] J. E. Nafe, E. B. Nelson, and I. I. Rabi, Phys. Rev. **71**, 914 (1947); D. E. Nagle, R. S. Julian, and J. R. Zacharias, Phys. Rev. **72**, 971 (1947); P. Kusch and H. M. Foley, Phys. Rev. **72**, 1256 (1947); **73**, 412 (1948).
[3] J. Schwinger, Phys. Rev. **74**, 1939 (1948); **75**, 651 (1949).
[4] F. Bloch and A. Nordsieck, Phys. Rev. **52**, 54 (1937).
[5] This result agrees with that obtained by an earlier method [J. Schwinger, Phys. Rev. **73**, 416 (1948)], and announced at the January 1948 meeting of the American Physical Society. However, in the previous method the contribution of the additional magnetic moment to the energy in an electric field had to be artificially corrected in order to obtain a Lorentz invariant result. This difficulty is attributable to the incorrect transformation properties of the electron self-energy obtained from the conventional Hamiltonian treatment, and is completely removed in the covariant formulation now employed. Independent calculations by J. B. French and V. F. Weisskopf [Phys. Rev. **75**, 338 (1949)], as well as N. M. Kroll and W. E. Lamb, Jr. [Phys. Rev. **75**, 388 (1949)], are also in agreement with Eq. (8).
[6] H. A. Bethe, Phys. Rev. **72**, 339 (1947).
[7] R. C. Retherford and W. E. Lamb, Jr., Bull. Am. Phys. Soc. **24**, No. 1, (1949).

ELECTRON THEORY

Report to the Solvay Conference for Physics at Brussels, Belgium September 27 to October 2, 1948

by J. R. OPPENHEIMER

In this report I shall try to give an account of the developments of the last year in electrodynamics. It will not be useful to give a complete presentation of the formalism; rather I shall try to pick out the essential logical points of the development, and raise at least some of the questions which may be open, and which bear on an evaluation of the scope of the recent developments, and their place in physical theory. I shall divide the report into three sections: (1) a brief summary of related past work in electrodynamics; (2) an account of the logical and procedural aspects of the recent developments; and (3) a series of remarks and questions on applications of these developments to nuclear problems and on the question of the closure of electrodynamics.

1. History

The problems with which we are concerned go back to the very beginnings of the quantum electrodynamics of Dirac, of Heisenberg and Pauli.(1) This theory, which strove to explore the consequences of complementarity for the electromagnetic field and its interactions with matter, led to great success in the understanding of emission, absorption and scattering processes, and led as well to a harmonious synthesis of the description of static fields and of light quantum phenomena. But it also led, as was almost at once recognized,(2) to paradoxical results, of which the infinite displacement of spectral terms and lines was an example. One recognized an analogy between these results and the infinite electromagnetic inertia of a point electron in classical theory, according to which electrons moving with different mean velocity should have energies infinitely displaced. Yet no attempt at a quantitative interpretation was made, nor was the question raised in a serious way of isolating from the infinite displacements new and typical finite parts clearly separable from the inertial effects. In fact such a program could hardly have been carried through before the discovery of pair production, and an understanding of the far-reaching differences in the actual problem of the singularities of quantum electrodynamics from the classical analogue of a point electron interacting with its field. In the former, the field and charge fluctuations of the vacuum—which clearly have no such classical counterpart—play a decisive part; whereas on the other hand the very phenomena of pair production, which so seriously limit the usefulness of a point model of the electron for distances small compared to its Compton wave length $\hbar/mc$, in some measure ameliorate, though they do not resolve, the problems of the infinite electromagnetic inertia and of the instability of the electron's charge distribution. These last points first were made clear by the self-energy calculations of Weisskopf,(3) and were still further emphasized by the finding, by Pais,(4) and by Sakata,(5) that to the order e^2 (and to this limitation we shall have repeatedly to return) the electron's self-energy could be made finite, and indeed small, and its stability insured, by introducing forces of small magnitude and essentially *arbitrarily* small range, corresponding to a new field, and quanta of arbitrarily high rest mass.(6)

On the other hand the decisive, if classically unfamiliar, role of vacuum fluctuations was perhaps first shown—albeit in a highly academic situation—by Rosenfeld's calculation(7) of the (infinite) gravitational energy of the light quantum, and came prominently into view with the discovery of the problem of the self-energy of the photon due to the current fluctuations of the electron-positron field, and the related problems of the (infinite) polarizability of that field. Here for the first time the notion of renormalization was introduced. The infinite polarization of vacuum refers in fact just to situations in which a classical definition of charge should be possible (weak, slowly varying fields); if the polarization were finite, the linear constant term could not be measured directly, nor measured in any classically interpretable experiment; only the sum of "true" and induced charge could be measured. Thus it seemed natural to ignore the infinite linear constant polarizability of vacuum, but to attach significance to the finite deviations from this polarization in rapidly varying and in strong fields.(8) Direct attempts to measure these deviations were not successful; they are in any case intimately related to those which do describe the Lamb-Retherford level shift,(9) but are too small and of wrong sign to account for the bulk of this observation.(10) But the renormalization procedure and philosophy here applied to charge was to prove, in its obvious extension to the electron's mass, the starting point for new developments.

In their application to level shifts, these developments, which could have been carried out at any time during the last fifteen years, required the impetus of experiment to stimulate and verify. Nevertheless, in other closely related problems, results were obtained essentially identical with those required to understand the Lamb-Retherford shift and the Schwinger corrections to the electron's gyromagnetic ratio.

Thus there is the problem—first studied by Bloch, Nordsieck,(11) Pauli and Fierz,(12) of the radiative corrections to the scattering of a slow electron (of velocity v) by a static potential V. The contribution of electromagnetic inertia is readily eliminated in non-relativistic calculations, and involves some subtlety in relativistic treatment only in the case of spin $^1/_2$ (rather than spin zero) charges.[13] It was even pointed out[14] that the new effects of radiation could be summarized by a small supplementary potential

$$\text{I.} \qquad \sim \left(\frac{2}{3}\right)\left(\frac{e^2}{\hbar c}\right)\left(\frac{\hbar}{mc}\right)^2 \Delta V \ln\left(\frac{c}{v}\right)$$

(where e, $\hbar$, m, c have their customary meaning). This of course gives the essential explanation of the Lamb shift.

On the other hand the anomalous g-value of the electron was foreshadowed by the remark,[15] that in meson theory, and even for neutral mesons, the coupling of nucleon spin and meson fluctuations would give to the sum of neutron and proton moments a value different from (and in non-relativistic estimates less than) the nuclear magneton.

Yet until the advent of reliable experiments on the electron's interaction, these points hardly attracted serious attention; and interest attached rather to exploring the possibilities of a consistent and reasonable modification of electrodynamics, which should preserve its agreement with experience, and yet, for high fields or short wave lengths, introduce such alterations as to make self-energies finite and the electron stable. In this it has proved decisive that it is *not* sufficient to develop a satisfactory classical analogue; rather one must cope directly with the specific quantum phenomena of fluctuation and pair production.[6] Within the framework of a continuum theory, with the point interactions of what Dirac[16] calls a "localizable" theory—no such satisfactory theory has been found; one may doubt whether, *within* this framework, such a theory can be formed that is expansible in powers of the electron's charge e. On the other hand, as mentioned earlier, many families of theories are possible which give satisfactory and consistent results to the order e^2.

A further general point which emerged from the study of electrodynamics is that—although the singularities occurring in solutions indicate that it is not a completed consistent theory, the structure of the theory itself gives no indication of a field strength, a maximum frequency of minimum length, beyond which it can no longer consistently be supposed to apply. This last remark holds in particular for the actual electron—for the theory of the Dirac electron-positron field coupled to the Maxwell field. For particles of lower and higher spin, some rough and necessarily ambiguous indications of limiting frequencies and fields do occur.

To these purely theoretical findings, there is a counterpart in experience. No credible evidence, despite much searching, indicates any departure, in the behaviour of electrons and gamma rays, from the expectations of theory. There are, it is true, the extremely weak couplings of β decay; there are the weak electromagnetic interactions of gamma rays, and electrons, with the mesons and nuclear matter. Yet none of these should give appreciable corrections to the present theory in its characteristic domains of application; they serve merely to suggest that for very small (nuclear) distances, and very high energies, electron theory and electrodynamics will no longer be so clearly separable from other atomic phenomena. In the theory of the electron and the electromagnetic field, we have to do with an almost closed, almost complete system, in which however we look precisely to the absence of complete closure to bring us away from the paradoxes that still inhere in it.

2. Procedures

The problem then is to see to what extent one can isolate, recognize and postpone the consideration of those quantities, like the electron's mass and charge, for which the present theory gives infinite results—results which, if finite, could hardly be compared with experience in a world in which arbitrary values of the ratio $e^2/\hbar c$ cannot occur. What one can hope to compare with experience is the totality of other consequences of the coupling of charge and field, consequences of which we need to ask: does theory give for them results which are finite, unambiguous and in agreement with experiment?

Judged by these criteria the earliest methods must be characterized as encouraging but inadequate. They rested, as have to date *all* treatments not severely limited throughout by the neglect of relativity, recoil, and pair formation, on an expansion in powers of e, going characteristically to the order e^2. One carried out the calculation of the problem in question; (for radiative scattering corrections, Lewis(17); for the Lamb shift, Lamb and Kroll,(18) Weisskopf and French,(19) Bethe(20); for the electron's g-value, Luttinger(21)); one also calculated to the same order the electron's electromagnetic mass, its charge, and the charge induced by external fields, and the light quantum mass; finally one asked for the effect of these changes in charge and mass on the problem in question, and sought to delete the corresponding terms from the direct calculation. Such a procedure would no doubt be satisfactory—if cumbersome—were all quantities involved finite and unambiguous. In fact, since mass and charge corrections are in general represented by logarithmically divergent integrals, the above outlined procedure serves to obtain finite, but not necessarily unique or correct, reactive corrections for the behaviour of an electron in an external field; and a special tact is necessary, such as that implicit in Luttinger's derivation of the electron's anomalous gyromagnetic ratio, if results are to be, not merely plausible, but unambiguous and sound. Since, in more complex problems, and in calculations carried to higher order in e, this straightforward procedure becomes more and more ambiguous, and the results more dependent on the choice of Lorentz frame and of gauge, more powerful methods are required. Their development has occurred in two steps, the first largely, the second almost wholly, due to Schwinger.(22)

ELECTRON THEORY

The first step is to introduce a change in representation, a contact transformation, which seeks, for a single electron not subject to external fields, and in the absence of light quanta, to describe the electron in terms of classically measurable charge e and mass m, and eliminate entirely all "virtual" interaction with the fluctuations of electromagnetic and pair fields. In the non-relativistic limit, as was discussed in connection with Kramer's report,[(23)] and as is more fully described in Bethe's,[(24)] this transformation can be carried out rigorously to all powers of e, without expansion; in fact, the unitary transformation is given by

II. $$U = \exp \frac{e}{mc}[\mathcal{Z} \cdot \nabla]$$

where $\mathcal{Z}$ is the (transverse) Hertz vector of the electromagnetic field minus the quasi-static field of the electron. When this formalism is applied to the problem of an electron in an external field, it yields reactive corrections which do not converge for frequencies $\nu > mc^2/\hbar$, thus indicating the need for a fuller consideration of typical relativistic effects.

This generalization is in fact straightforward; yet here it would appear essential that the power series expansion in e is no longer avoidable, not only because no such simple solution as II now exists, but because, owing to the possibilities of pair creation and annihilation, and of interactions of light quanta with each other, the very definition of states of single electrons or single photons depends essentially on the expansion in question.[(25)] However that may be, the work has so far been carried out only by treating $e^2/\hbar c$ as small, and essentially only to include corrections of the first order in that quantity.

In this form, the contact transformation clearly yields:

(a) an infinite term in the electron's electromagnetic inertia;
(b) an ambiguous light quantum self-energy;
(c) no other effects for a single electron or photon;
(d) interactions of order e^2 between electrons, positrons, and photons, which in this order, correspond to the familiar Møller interactions and Compton effect and pair production probabilities;
(e) an infinite vacuum polarizability;
(f) the familiar frequency-dependent finite polarizability for external electromagnetic fields;
(g) emission and absorption probabilities equivalent to those of the Dirac theory for an electron in an external e.m. field;
(h) new reactive corrections of order e^2 to the effective charge and current distribution of an electron, which correspond to vanishing total supplementary charge, and to currents of the order $e^3/\hbar c$ distributed over

dimensions of the order $\hbar/mc$, and which include the supplementary potential I, and the supplementary magnetic moment

$$\left(\frac{e^2}{2\pi\hbar c}\right)\left(\frac{e\hbar}{2mc}\right)\left(\vec{\sigma}\right)$$

as special (non-relativistic) limiting cases.

Were such calculations to be carried further, to higher order in e, they would lead to still further renormalizations of charge and mass, to the successive elimination of all "virtual" interactions, and to reactive corrections, in the form of an expansion in powers of $e^2/\hbar c$, to the probabilities of transitions: pair production, collisions, scattering, etc. Nevertheless, before such a program could be undertaken, or the physically interesting new terms (h) above be taken as correct, a new development is required. The reason for this is the following: the results (h) are not in general independent of gauge and Lorentz frame. Historically this was first discovered by comparison of the supplementary magnetic interaction energy in a uniform magnetostatic field H

$$\left(\frac{e^2}{2\pi\hbar c}\right)\left(\frac{e\hbar}{2mc}\right)\left(\vec{\sigma}\cdot\vec{H}\right)$$

with the supplementary (imaginary) electric dipole interaction which appeared with an electron in a homogeneous electric field E derived from a static scalar potential

$$\left(\frac{e^2}{6\pi\hbar c}\right)\left(\frac{e\hbar}{2mc}\right) i\rho_2\left(\vec{\sigma}\cdot\vec{E}\right)$$

a manifestly non-covariant result.

Now it is true that the fundamental equations of quantum-electrodynamics are gauge and Lorentz covariant. But they have in a strict sense no solutions expansible in powers of e. If one wishes to explore these solutions, bearing in mind that certain infinite terms will, in a later theory, no longer be infinite, one needs a covariant way of identifying these terms; and for that, not merely the field equations themselves, but the whole method of approximation and solution must at all stages preserve covariance. This means that the familiar Hamiltonian methods, which imply a fixed Lorentz frame t = constant, must be renounced; neither Lorentz frame nor gauge can be specified until after, in a given order in e, all terms have been identified, and those bearing on the definition of charge and mass recognized and relegated; then of course, in the actual calculation of transition probabilities and the reactive corrections to them, or in the determination of stationary states in fields which can be treated as static, and in the reactive corrections thereto, the introduction of a definite coordinate system and gauge for these no longer singular and completely well-defined terms can lead to no difficulty.

ELECTRON THEORY

It is probable that, at least to order e^2, more than one covariant formalism can be developed. Thus Stueckelberg's four-dimensional perturbation theory[26] would seem to offer a suitable starting point, as also do the related algorithms of Feynman.[27] But a method originally suggested by Tomonaga,[28] and independently developed and applied by Schwinger,[22] would seem, apart from its practicality, to have the advantage of very great generality and a complete conceptual consistency. It has been shown by Dyson[29] how Feynman's algorithms can be derived from the Tomonaga equations.

The easiest way to come to this is to start with the equations of motion of the coupled Dirac and Maxwell field. These are gauge and Lorentz covariants. The commutation laws, through which the typical quantum features are introduced, can readily be rewritten in covariant form to show: (1) at points outside the light cone from each other, all field quantities commute; and (2) the integral over an *arbitrary space-like* hypersurface yields a simple finite value for the commutator of a field variable at a variable point on the hypersurface, and that of another field variable at a fixed point on the hypersurface.

In this Heisenberg representation, the state vector is of course constant; commutators of field quantities separated by time-like intervals, depending on the solution of the coupled equation of motion, can not be known *a priori*; and no direct progress at either a rigorous or an approximate solution in powers of e has been made.* But a simple change to a mixed representation, that introduced by Tomonaga and called by Schwinger the "interaction representation," makes it possible to carry out the covariant analogue of the power series contact transformation of the Hamiltonian theory.

The change of representation involved is a contact transformation to a system in which the state vector is no longer constant, but in which it would be constant if there were no coupling between the fields, i.e., if the elementary charge $e=0$. The basis of this representation is the solution of the uncoupled field equations, which, together with their commutators at all relative positions, are of course well known. This transformation leads directly to the Tomonaga equation for the variation of the state vector Ψ:

III. $$i\hbar \frac{\delta \Psi}{\delta \sigma} = -\frac{1}{c} j^{\mu(P)} A_{\mu}^{(P)} \Psi$$

Here σ is an arbitrary space-like surface through the point P. $\delta\Psi$ is the variation in Ψ when a small variation is made in σ, localized near the point P; $\delta\sigma$ is the four-volume between varied and unvaried surfaces; $A_{\mu}^{(P)}$ is the

* Author's note, 1956. Approximate solutions of the Heisenberg equations of motion were obtained by Yang and Feldman, *Phys. Rev.*, **79**, 972, 1950; and Källén, *Arkiv För Fysik*, **2**, 371, 1950.

operator of the four-vector electromagnetic potential at P; $j\mu(P)$ is the (charge-symmetrized) operator of electron-positron four-vector current density at the same point.

It may be of interest, in judging the range of applicability of these methods, to note that in the theory of the charged particle of zero spin (the scalar and not Dirac pair field), the Tomonaga equation does not have the simple form III; the operator on Ψ on the right involves explicitly an arbitrary time-like unit vector.(30)

Schwinger's program is then to eliminate the terms of order e, e^2, and so, in so far as possible, from the right-hand side of III. As before, only the "virtual" transitions can be eliminated by contact transformation; the real transitions of course remain, but with transition amplitudes eventually themselves modified by reactive corrections.

Apart from the obvious resulting covariance of mass and charge corrections, a new point appears for the light quantum self-energy, which now appears in the form of a product of a factor which must be zero on invariance grounds, and an infinite factor. As long as this term is identifiable, it must of course be zero in any gauge and Lorentz invariant formulation; in these calculations for the first time it is possible to make it zero. Yet even here, if one attempts to evaluate directly the product of zero factor and infinite integral, indeterminate, infinite, or even finite(31) values may result. A somewhat similar situation obtains in the problem, so much studied by Pais, of the direct evaluation of the stress in the electron's rest system, where a direct calculation yields the value $(-e^2/2\pi\hbar c)mc^2$, instead of the value zero which follows at once as the limit of the zero value holding uniformly, in this order e^2, for the theory rendered convergent by the f-quantum hypothesis, even for arbitrarily high f-quantum mass. These examples, far from casting doubt on the usefulness of the formalism, may just serve to emphasize the importance of identifying and evaluating such terms without any specialization of coordinate system, and utilizing throughout the covariance of the theory.

To order e^2, one again finds the terms (a) to (h) listed above; the covariance of the new reactive terms is now apparent; and they exhibit themselves again but more clearly as supplementary currents, corresponding to charge distribution of order $e^3/\hbar c$ (but vanishing total charge) extended throughout the interior of the light cones about the electron's position, and of spatial dimensions $\sim \hbar/mc$; inversely, they may also be interpreted as corrections of relative order $e^2/\hbar c$ and static range $\hbar/mc$ to the external fields. The supplementary currents immediately make possible simple treatments of the electron in external fields (where neither the electron's velocity, nor the derivatives of the fields need be treated small), and so give corrections for emission, absorption and scattering processes to the extent at least in which the fields may be classically described(32); the reactive corrections to the Møller interaction and to pair production can probably not be derived without carrying the contact transformation to order e^4, since for these typical exchange effects, not included in the classical description of fields, must be expected to appear.

At the moment, to my best present knowledge, the reactive corrections agree with the S level displacements of H to about 1%, the present limit of experimental accuracy. For ionized helium, and for the correction to the electron's g-value, the agreement is again within experimental precision, which in this case, however, is not yet so high.

3. Questions

Even this brief summary of developments will lead us to ask a number of questions:

(1) Can the development be carried further, to higher powers of e, (a) with finite results, (b) with unique results, (c) with results in agreement with experiment?

(2) Can the procedure be freed of the expansion in e, and carried out rigorously?

(3) How general is the circumstance that the only quantities which are not, in this theory, finite, are those like the electromagnetic inertia of electrons, and the polarization effects of charge, which cannot directly be measured within the framework of the theory? Will this hold for charged particles of other spin?

(4) Can these methods be applied to the Yukawa-meson fields of nucleons? Does the resulting power series in the coupling constant converge at all? Do the corrections improve agreement with experience? Can one expect that when the coupling is large there is any valid content to the Maxwell-Yukawa analogy?

(5) In what sense, or to what extent, is electrodynamics—the theory of Dirac pairs and the e.m. field—"closed"?

There is very little experience to draw on for answering this battery of questions. So far there has not yet been a complete treatment of the electron problem in order higher than e^2, although preliminary study(33) indicates that here too the physically interesting corrections will be finite.

The experience in the meson fields is still very limited. With the pseudoscalar theory, Case(34) has indeed shown that the magnetic moment of the neutron is finite (this has nothing to do with the present technical developments), and that the sum of neutron and proton moments, minus the nuclear magneton (which is the analogue of the electron's anomalous g-value) is of the same order as the neutron moment, finite, and in disagreement with experience. The proton-neutron mass difference is infinite and of the wrong sign; the reactive corrections to nuclear forces, formally analogous to the corrections to the Moller interaction, have not been evaluated. Despite these discouragements, it would seem premature to evaluate the prospects without further evidence.

Yet it is tempting to suppose that these new successes of electrodynamics, which extend its range very considerably beyond what had earlier been believed possible, can themselves be traced to a rather simple general feature. As we have noted, both from the formal and from the physical side, electrodynamics is an almost closed subject; changes limited to very small distances, and having little effect even in the typical relativistic domain $E \sim mc^2$, could suffice to make a consistent theory; in fact, only weak and remote interactions appear to carry us out of the domain of electrodynamics, into that of the mesons, the nuclei, and the other elementary particles. Similar successes could perhaps be expected for those mesons (which may well also be described by Dirac-fields), which also show only weak non-electromagnetic interactions. But for mesons and nucleons generally, we are in a quite new world, where the special features of almost complete closure that characterizes electrodynamics are quite absent. That electrodynamics is also not quite closed is indicated, not alone by the fact that for finite $e^2/\hbar c$ the present theory is not after all self-consistent, but equally by the existence of those small interactions with other forms of matter to which we must in the end look for a clue, both for consistency, and for the actual value of the electron's charge.

I hope that even these speculations may suffice as a stimulus and an introduction to further discussion.

ELECTRON THEORY

References

1. Heisenberg and Pauli, *Zeits. f. Physik.*, **56**, 1, 1929.
2. J. R. Oppenheimer, *Phys. Rev.*, **35**, 461, 1930.
3. V. Weisskopf, *Zeits. f. Physik.*, **90**, 817, 1934.
4. A. Pais, *Verhandelingen Roy. Ac., Amsterdam*, **19**, 1, 1946.
5. Sakata and Hara, *Progr. Theor. Phys.*, **2**, 30, 1947
6. For a recent summary of the state of theory, see A. Pais, *Developments in the Theory of the Electron*, Princeton University Press, 1948.
7. L. Rosenfeld, *Zeits. f. Physik.*, **65**, 589, 1930.
8. General treatments: R. Serber, *Phys. Rev.*, **48**, 49, 1938, and V. Weisskopf, *Kgl. Dansk. Vidensk. Selskab. Math.-fys. Medd.*, **14**, 6, 1936.
9. Lamb and Retherford, *Phys. Rev.*, **72**, 241, 1947.
10. E. Uehling, *Phys. Rev.*, **48**, 55, 1935.
11. Bloch and Nordsieck, *Phys. Rev.*, **52**, 54, 1937.
12. Pauli and Fierz, *Il Nuovo Cimento*, **15**, 167, 1938.
13. S. Dancoff, *Phys. Rev.*, **55**, 959, 1939; H. Lewis, *Phys. Rev.*, **73**, 173, 1948.
14. Shelter Island Conference, June, 1947.
15. Fröhlich, Heitler and Kemmer, *Proc. Roy. Soc.*, A **166**, 154, 1938
16. P. Dirac, *Phys. Rev.*, **73**, 1092, 1948.
17. H. Lewis, *Phys. Rev.*, **73**, 173, 1948.
18. Lamb and Kroll, *Phys. Rev.*, in press.
19. Weisskopf and French, *Phys. Rev.*, in press.
20. H. Bethe, *Phys. Rev.*, **72**, 339, 1947.
21. P. Luttinger, *Phys. Rev.*, **74**, 893, 1948.
22. J. Schwinger, *Phys. Rev.*, **74**, 1439, 1948, and in press.
23. Report to the 8th Solvay Conference.
24. Report to the 8th Solvay Conference.
25. This may be seen very strikingly in writing down an explicit solution for the Tomonaga equation III below. Formally it is:

$$\Psi(\sigma) = \text{``}\exp\text{''}\left[\frac{i}{hc}\int_{\sigma_0}^{\sigma} j_\mu A^\mu \, d_4x\right]\Psi(\sigma_0)$$

In order to define the "exp", we have at present no other resort than to approximate by a power series, where the ordering of the non-commuting factors for $j_\mu A^\mu$ at different points of space-time can be simply prescribed (e.g., the later factor to the left). Cf. especially F. J. Dyson, *Phys. Rev.*, in press.
26. Stueckelberg, *Ann. der Phys.*, **21**, 367, 1934.
27. R. Feynman, *Phys. Rev.*, **74**, 1430, 1948.
28. S. Tomonaga, *Progr. Theor. Phys.*, **1**, 27, 109, 1946.
29. F. Dyson, *Phys. Rev.*, in press.
30. Kanesawa and Tomonaga, *Progr. Theor. Phys.*, **3**, 1, 107, 1948.
31. G. Wentzel, *Phys. Rev.*, **74**, 1070, 1948.
32. See for instance results reported to this conference by Pauli on corrections to the Compton effect for long wave lengths.
33. F. Dyson, *Phys. Rev.*, in press.
* Author's note, 1956. Questions 1(a) and 1(b) were indeed answered by Dyson, *Phys. Rev.*, **75**, 1736, 1949.
34. K. Case, *Phys. Rev.*, **74**, 1884, 1948.

PROGRESS OF THEORETICAL PHYSICS
Vol. I, No. 2, Aug.–Sept., 1946

On a Relativistically Invariant Formulation of the Quantum Theory of Wave Fields *

by S. TOMONAGA

1. The formalism of the ordinary quantum theory of wave fields

Recently Yukawa(1) has made a comprehensive consideration about the basis of the quantum theory of wave fields. In his article he has pointed out the fact that the existing formalism of the quantum field theory is not yet perfectly relativistic.

Let $v(xyz)$ be the quantity specifying the field, and $\lambda(xyz)$ denote its canonical conjugate. Then the quantum theory requires commutation relations of the form:

$$\begin{cases}[v(xyzt),\ v(x'y'z't)]=[\lambda(xyzt),\ \lambda(x'y'z't)]=0\\ [v(xyzt),\ \lambda(x'y'z't)]=i\hbar\delta(x-x')\delta(y-y')\delta(z-z'),\end{cases} \tag{1}$$ †

but these have quite non-relativistic forms.

The equations (1) give the commutation relations between the quantities at different points (xyz) and $(x'y'z')$ at the same instant of time t. The concept "same instant of time at different points" has, however, a definite meaning only if one specifies some definite Lorentz frame of reference. Thus this is not a relativistically invariant concept.

Further, the Schrödinger equation for the ψ-vector representing the state of the system has the form:

$$\left(\bar{H}+\frac{\hbar}{i}\frac{\partial}{\partial t}\right)\psi=0, \tag{2}$$

where $\bar{H}$ is the operator representing the total energy of the field which is given by the space integral of a function of v and λ. As we adopt here the Schrödinger picture, v and λ are operators independent of time. The vector representing the state is in this picture a function of the time, and its dependence on t is determined by (2).

* Translated from the paper, *Bull. I. P. C. R.* (*Riken-iho*), **22** (1943), 545, appeared originally in Japanese.

† $[A, B]=AB-BA$. We assume that the field obeys the Bose statistics. Our considerations apply also to the case of Fermi statistics.

Also the differential equation (2) is no less non-relativistic. In this equation the time variable t plays a role quite distinct from the space co-ordinates x, y and z. This situation is closely connected with the fact that the notion of probability amplitude does not fit with the relativity theory.

As is well known, the vector ψ has, as the probability amplitude, the following physical meaning: Suppose the representation which makes the field quantity $v(xyz)$ diagonal. Let $\psi[v'(xyz)]$ denote the representative of ψ in this representation.* Then the representative $\psi[v'(xyz)]$ is called probability amplitude, and its absolute square

$$W[v'(xyz)] = |\psi[v'(xyz)]|^2 \tag{3}$$

gives the relative probability of $v(xyz)$ having the specified functional form $v'(xyz)$ at the instant of time t. In other words: Suppose a plane† which is parallel to the xyz-plane and intercepts the time axis at t. Then the probability that the field has the specified functional form $v'(xyz)$ on this plane is given by (3).

As one sees, a plane parallel to the xyz-plane plays here a significant role. But such a plane is defined only by referring to a certain frame of reference. Thus the probability amplitude is not a relativistically invariant concept in the space-time world.

2. Four-dimensional form of the commutation relations

As stated above, the laws of the quantum theory of wave fields are usually expressed as mathematical relations between quantities having their meanings only in some specified Lorentz frame of reference. But since it is proved that the whole contents of the theory are of course relativistically invariant, it must be certainly possible to build up the theory on the basis of concepts having relativistic space-time meanings. Thus, in his considera-tion, Yukawa has required with Dirac(2) a generalization of the notion of probability amplitude to fit with the relativity theory. We shall now show below that the generalization of the theory on these lines is in fact possible to the relativistically necessary and sufficient extent. Our results are,

* We use the square brackets to indicate a functional. Thus $\psi[v'(xyz)]$ means that ψ is a functional of the variable function $v'(xyz)$. When we use ordinary parentheses ( ), as $\psi(v'(xyz))$, we consider ψ as an ordinary function of the function $v'(xyz)$. For example: the energy density is written as $H(v(xyz), \lambda(xyz))$ and this is also a function of x, y and z, whereas the total energy $H = \int H(v(xyz), \lambda(xyz))dv$ is a functional of $v(xyz)$ and $\lambda(xyz)$ and is written as $\bar{H}[v(xyz), \lambda(xyz)]$.

† We call a three-dimensional manifold in the four-dimensional space-time world simply a "surface."

however, not so general as expected by Dirac and by Yukawa, but are already sufficiently general in so far as it is required by the relativity theory.

Let us suppose for simplicity that there are only two fields interacting with each other. The case of a greater number of fields can also be treated in the same way. Let v_1 and v_2 denote the quantities specifying the fields. The canonically conjugate quantities are λ_1 and λ_2 respectively. Then between these quantities the commutation relations

$$\begin{cases} [v_r(xyzt),\ v_s(x'y'z't)]=0 \\ [\lambda_r(xyzt),\ \lambda_s(x'y'z't)=0 \qquad\qquad r,\ s=1,\ 2 \\ [v_r(xyzt),\ \lambda_s(x'y'z't)]=i\hbar\delta(x-x')\delta(y-y')\delta(z-z')\delta_{rs} \end{cases} \tag{4}$$

must hold. The ψ-vector satisfies the Schrödinger equation

$$\left(\bar{H}_1+\bar{H}_2+\bar{H}_{12}+\frac{\hbar}{i}\frac{\partial}{\partial t}\right)\psi=0. \tag{5}$$

In this equation $\bar{H}_1$ and $\bar{H}_2$ mean respectively the energy of the first and the second field. $\bar{H}_1$ is given by the space integral of a function of v_1 and λ_1, $\bar{H}_2$ by the space integral of a function of v_2 and λ_2. Further, $\bar{H}_{12}$ is the interaction energy of the fields and is given by the space integral of a function of both v_1, λ_1 and v_2, λ_2. We assume (i) that the integrand of $\bar{H}_{12}$, i.e. the interaction-energy density, is a scalar quantity, and (ii) that the energy densities at two different points (but at the same instant of time) commute with each other. In general, these two facts follow from the single assumption: the interaction term in the Lagrangian does not contain the time derivatives of v_1 and v_2.

If this energy density is denoted by H_{12}, then we have

$$\bar{H}_{12}=\int H_{12}\,dx\,dy\,dz. \tag{6}$$

As we adopt here the Schrödinger picture, the quantities v and λ in H_1, H_2 and H_{12} are all operators independent of time.

Thus far we have merely summarized the well-known facts. Now, as the first stage of making the theory relativistic, we suppose the unitary operator

$$U=\exp\left\{\frac{i}{\hbar}(\bar{H}_1+\bar{H}_2)t\right\} \tag{7}$$

and introduce the following unitary transformations of v and λ, and the corresponding transformation of ψ:

$$\begin{cases} V_r=Uv_rU^{-1},\quad \Lambda_r=U\lambda_rU^{-1} \\ \qquad \Psi=U\psi. \end{cases} \qquad r=1,\ 2 \tag{8}$$

As stated above, v and λ in (5) are quantities independent of time. But V and Λ obtained from them by means of (8) contain t through U. Thus they depend on t by

$$\begin{cases} i\hbar\dot{V}_r=V_r\bar{H}_r-\bar{H}_rV_r \\ i\hbar\dot{\Lambda}_r=\Lambda_r\bar{H}_r-\bar{H}_r\Lambda_r. \end{cases} \qquad r=1,\ 2 \tag{9}$$

PROGRESS OF THEORETICAL PHYSICS

These equations must necessarily have covariant forms against Lorentz transformations, because they are just the field equations for the fields when they are left alone without interacting with each other.

Now, the solutions of these "vacuum equations," the equations which the fields must satisfy when they are left alone, together with the commutation relations (4), give rise to the relations of the following forms:

$$\begin{cases} [V_r(xyzt),\ V_s(x'y'z't')] = A_{rs}(x-x', y-y', z-z', t-t') \\ [\Lambda_r(xyzt),\ \Lambda_s(x'y'z't')] = B_{rs}(x-x', y-y', z-z', t-t') \\ [V_r(xyzt),\ \Lambda_s(x'y'z't')] = C_{rs}(x-x', y-y', z-z', t-t') \end{cases} \tag{10}$$

where A_{rs}, B_{rs} and C_{rs} are functions which are combinations of the so-called four-dimensional δ-functions and their derivatives.(3) One denotes usually these four-dimensional δ-functions by $D_r(xyzt)$, $r=1, 2$. They are defined by

$$D_r(xyzt) = \frac{1}{16\pi^3}\iiint\left\{\frac{e^{i(k_x x+k_y y+k_z z+ck_r t)}}{ik_r} - \frac{e^{i(k_x x+k_y y+k_z z-ck_r t)}}{ik_r}\right\}dk_x\, dk_y\, dk_z \tag{11}$$

with

$$k_r = \sqrt{k_x{}^2+k_y{}^2+k_z{}^2+\varkappa_r{}^2}, \tag{12}$$

$\varkappa_r$ being the constant characteristic to the field r. It can be easily proved that these functions are relativistically invariant.*

Since (10) gives, in contrast with (4), the commutation relations between the fields at two different world points $(xyzt)$ and $(x'y'z't')$, it contains no more the notion of same instant of time. Therefore, (10) is sufficiently relativistic presupposing no special frame of reference. We call (10) four-dimensional form of the commutation relations.

One property of $D(xyzt)$ will be mentioned here: When the world point $(xyzt)$ lies outside the light cone whose vertex is at the origin, then $D(xyzt)$ vanishes identically:

$$D(xyzt) = 0 \quad \text{for} \quad x^2+y^2+z^2-c^2t^2 > 0. \tag{13}$$

It follows directly from (13) that, if the world point $(x'y'z't')$ lies outside the light cone whose vertex is at the world point $(xyzt)$, the right-hand

* Suppose that a surface in the $k_x k_y k_z k$-space is defined by means of the equation $k^2 = k_x{}^2+k_y{}^2+k_z{}^2+\varkappa^2$. Then this surface has the invariant meaning in this space, since $k_x{}^2+k_y{}^2+k_z{}^2-k^2$ is invariant against Lorentz transformations. The area of the surface element of this surface is given by

$$dS = \sqrt{\left(\frac{\partial k}{\partial k_x}\right)^2+\left(\frac{\partial k}{\partial k_y}\right)^2+\left(\frac{\partial k}{\partial k_z}\right)^2-1}\; dk_x\, dk_y\, dk_z = \varkappa\frac{dk_x\, dk_y\, dk_z}{k}.$$

Now, since dS has the invariant meaning, we can thus conclude that $\dfrac{dk_x\, dk_y\, dk_z}{k}$ is an invariant, and this implies that the function defined by (11) is invariant.

sides of (10) always vanish. In words: Suppose two world points P and P'. When these points lie outside each other's light cones, the field quantities at P and field quantities at P' commute with each other.

3. Generalization of the Schrödinger equation

Next we observe the vector Ψ obtained from ψ by means of the unitary transformation U. We see from (5), (7) and (8) that this Ψ, considered as a function of t, satisfies

$$\left\{\int H_{12}(V_1(xyzt), \Lambda_1(xyzt), V_2(xyzt), \Lambda_2(xyzt))\, dx\, dy\, dz + \frac{\hbar}{i}\frac{\partial}{\partial t}\right\}\Psi = 0. \quad (14)$$

One sees that t plays also here a role distinct from x, y and z: also here a plane parallel to the xyz-plane has a special significance. So we must in some way remove this unsatisfactory feature of the theory.

This improvement can be attained in the way similar to that in which Dirac(4) has built up the so-called many-time formalism of the quantum mechanics. We will now recall this theory.

The Schrödinger equation for the system containing N charged particles interacting with the electromagnetic field is given by

$$\left\{\bar{H}_{el} + \sum_{n=1}^{N} H_n(q_n, p_n, \mathfrak{a}(q_n)) + \frac{\hbar}{i}\frac{\partial}{\partial t}\right\}\psi = 0. \quad (15)$$

Here $\bar{H}_{el}$ means the energy of the electromagnetic field, H_n the energy of the nth particle. H_n contains, besides the kinetic energy of the nth particle, the interaction energy between this particle and the field through $\mathfrak{a}(q_n)$, q_n being the coordinates of the particle and $\mathfrak{a}$ the potential of the field. p_n in (15) means as usual the momentum of the nth particle.

We consider now the unitary operator

$$u = \exp\left\{\frac{i}{\hbar}\bar{H}_{el}t\right\} \quad (16)$$

and introduce the unitary transformation of $\mathfrak{a}$:

$$\mathfrak{A} = u\mathfrak{a}u^{-1} \quad (17)$$

and the corresponding transformation of ψ:

$$\Phi = u\psi. \quad (18)$$

Then we see that Φ satisfies the equation

$$\left\{\sum_n H_n(q_n, p_n, \mathfrak{A}(q_n, t)) + \frac{\hbar}{i}\frac{\partial}{\partial t}\right\}\Phi = 0. \quad (19)$$

In contrast with $\mathfrak{a}$, which was independent of time (Schrödinger picture), $\mathfrak{A}$ contains t through u. To emphasize this, we have written t explicitly as the argument of $\mathfrak{A}$. We can prove that $\mathfrak{A}$ satisfies the Maxwell equations in vacuo (accurately speaking, we need special considerations for the equation div $\mathfrak{E}=0$).

The equation (19) is the starting point of the many-time theory. In this theory one introduces then the function $\Phi(q_1t_1, q_2t_2, \cdots, q_N, t_N)$ containing as many time variables $t_1, t_2, \cdots t_N$ as the number of the particles in place of the function $\Phi(q_1, q_2, \cdots, q_N, t)$ containing only one time variable,* and suppose that this $\Phi(q_1t_1, q_2t_2, \cdots, q_Nt_N)$ satisfies simultaneously the following N equations:

$$\left\{H_n(q_n, p_n, \mathfrak{A}(q_n, t_n)) + \frac{\hbar}{i}\frac{\partial}{\partial t_n}\right\}\Phi(q_1t_1, q_2t_2, \cdots, q_Nt_N) = 0$$
$$n = 1, 2, \cdots, N. \tag{20}$$

This $\Phi(t_1, t_2, \cdots, t_N)$, which is a fundamental quantity in the many-time theory, is related to the ordinary probability amplitude $\Phi(t)$ by

$$\Phi(t) = \Phi(t, t, \cdots, t). \tag{21}$$

Now, the simultaneous equations (20) can be solved when and only when the N^2 conditions

$$(H_nH_{n'} - H_{n'}H_n)\Phi(q_1t_1, q_2t_2, \cdots, q_Nt_N) = 0 \tag{22}$$

are satisfied for all pairs of n and n'. If the world point (q_nt_n) lies outside the light cone whose vertex is at the point $(q_{n'}t_{n'})$, we can prove that $H_nH_{n'} - H_{n'}H_n = 0$. As the result, the function satisfying (20) can exist in the region where

$$(q_n - q_{n'})^2 - c^2(t_n - t_{n'})^2 \geqq 0 \tag{23}$$

is satisfied simultaneously for all values of n and n'.

According to Bloch(5) we can give $\Phi(q_1t_1, q_2t_2, \cdots, q_Nt_N)$ a physical meaning when its arguments lie in the region given by (23). Namely

$$W(q_1t_1, q_2t_2, \cdots, q_Nt_N) = |\Phi(q_1t_1, q_2t_2, \cdots, q_Nt_N)|^2 \tag{24}$$

gives the relative probability that one finds the value q_1 in the measurement of the position of the first particle at the instant of time t_1, the value q_2 in the measurement of the position of the second particle at the instant of time $t_2, \cdots$ and the value q_N in the measurement of the position of the Nth particle at the instant of time t_N.

This is the outline of the many-time formalism of the quantum mechanics.

We will now return to our main subject. If we compare our equation (14) with the equation (19) of the many-time theory, we notice a marked similarity between these two equations. In (19) stands the suffix n, which designates the particle, while in (14) stand the variables x, y and z, which

* Here we suppose the representation which makes the coordinates $q_1, q_2, \cdots, q_N$ diagonal. Thus the vector Φ is represented by a function of these coordinates.

designate the position in space. Further, Φ is a function of the N independent variables $q_1, q_2, \cdots, q_N$, q_n giving the position of the nth particle, while Ψ is a functional of the infinitely many "independent variables" $v_1(xyz)$ and $v_2(xyz)$, $v_1(xyz)$ and $v_2(xyz)$ giving the fields at the position (xyz). Corresponding to the sum $\sum_n H_n$ in (19) the integral $\int H_{12}\, dx\, dy\, dz$ stands in (14). In this way, to the suffix n in (19) which takes the values 1, 2, 3, $\cdots$, N correspond the variables x, y and z which take continuously all values from $-\infty$ to $+\infty$.

Such a similarity suggests that we introduce infinitely many time variables t_{xyz}, which we may call local time,* each for one position (xyz) in the space, just as we have introduced N time variables, particle times, $t_1, t_2, \cdots, t_N$, each for one particle. The only difference is that we use in our case infinitely many time variables whereas we have used N time variables in the ordinary many-time theory.

Corresponding to the transition from the use of the function with one time variable to the use of the function of N time variables, we must now consider the transition from the use of $\Psi(t)$ to the use of a functional $\Psi[t_{xyz}]$ of infinitely many time variables t_{xyz}.

We now regard t_{xyz} as a function of (xyz) and consider its variation ε_{xyz} which differs from zero only in a small domain V_0 in the neighbourhood of the point $(x_0y_0z_0)$. We will define the partial differential coefficient of the functional $\Psi[t_{xyz}]$ with respect to the variable $t_{x_0y_0z_0}$ in the following manner:

$$\frac{\delta \Psi}{\delta t_{x_0y_0z_0}} = \lim_{\substack{\varepsilon \to 0 \\ V_0 \to 0}} \frac{\Psi[t_{xyz}+\varepsilon_{xyz}] - \Psi[t_{xyz}]}{\iiint \varepsilon_{xyz}\, dx\, dy\, dz} \tag{25}$$

We then generalize (14), and regard

$$\left\{ H_{12}(x, y, z, t) + \frac{\hbar}{i} \frac{\delta}{\delta t_{xyz}} \right\} \Psi = 0, \tag{26}$$

the infinitely many simultaneous equations corresponding to the N equations (20), as the fundamental equations of our theory. In (26) we have written, for simplicity, $H_{12}(x, y, z, t)$ in place of $H_{12}(V_1(xyz, t), V_2(xyz, t), \cdots)$. In general, when we have a function $F(V, \Lambda)$ of V and Λ, we will write simply $F(x, y, z, t)$ for $F(V(xyz, t_{xyz}), \Lambda\,(xyz, t_{xyz}))$, or still simpler $F(P)$, P denoting the world point with the coordinates (xyz, t_{xyz}). Thus $F(P')$ means $F(x', y', z', t')$ or, more precisely, $F(V(x'y'z', t_{x'y'z'}), (x'y'z', t_{x'y'z'}))$.

* The notion of local time of this kind has been occasionally introduced by Stueckelberg.[6]

We will now adopt the equation (26) as the basis of our theory. For $V_1(P)$, $V_2(P)$, $\Lambda_1(P)$ and $\Lambda_2(P)$ in H_{12} the commutation relations (10) hold, where $D(xyzt)$, has the property (13). As the consequence, we have

$$H_{12}(P)H_{12}(P')-H_{12}(P')H_{12}(P)=0 \tag{27}$$

when the point P lies a finite distance apart from P' and outside the light cone whose vertex is at P. Further, from our assumption (ii) the relation (27) holds also when P and P' are two adjacent points approaching in a space-like direction. Thus our system of equations (26) is integrable when the surface defined by the equations $t=t_{xyz}$, considering t_{xyz} as a function of x, y and z, is space-like.

In this way, a functional of the variable surface in the space-time world is determined by the functional partial differential equations (26). Corresponding to the relation (21) in case of many-time theory, $\Psi[t_{xyz}]$ reduces to the ordinary $\Psi(t)$ when the surface reduces to a plane parallel to the xyz-plane.

The dependent variable surface $t=t_{xyz}$ can be of any (space-like) form in the space-time world, and we need not presuppose any Lorentz frame of reference to define such a surface. Therefore, this $\Psi[t_{xyz}]$ is a relativistically invariant concept. The restriction that the surface must be space-like makes no trouble, since the property that a surface is space-like or time-like does not depend on a special choice of the reference system. It is not necessary, from the standpoint of the relativity theory, to admit also time-like surfaces for the variable surface, as was required by Dirac and by Yukawa. Thus we consider that $\Psi[t_{xyz}]$ introduced above is already the sufficient generalization of the ordinary ψ-vector, and assume that the quantum-theoretical state* of the fields is represented by this functional vector.

Let C denote the surface defined by the equation $t=t_{xyz}$. Then Ψ is a functional of the surface C. We write this as $\Psi[C]$. On C we take a point P, whose coordinates are $(xyz,\ t_{xyz})$, and suppose a surface C' which overlaps C except in a small domain about P. We denote the volume of the small world lying between C and C' by $d\omega_P$. Then we may write (25) also in the form:

$$\frac{\delta\Psi[C]}{\delta C_P}=\lim_{C'\to C}\frac{\Psi[C']-\Psi[C]}{d\omega_P}. \tag{28}$$

Then (26) can be written in the form:

$$\left\{H_{12}(P)+\frac{\hbar}{i}\frac{\delta}{\delta C_P}\right\}\Psi[C]=0. \tag{29}$$

* The word state is here used in the relativistic space-time meaning. Cf. Dirac's book (second edition), § 6.

This equation (29) has now a perfect space-time form. In the first place, H_{12} is a scalar according to our assumption (i); in the second place, the commutation relations between $V(P)$ and $\Lambda(P)$ contained in H_{12} has the four-dimensional forms as (10), and finally the differentiation $\frac{\delta}{\delta C_P}$ is defined by (28) quite independently of any frame of reference.

A direct conclusion drawn from (29) is that $\Psi[C']$ is obtained from $\Psi[C]$ by the following infinitesimal transformation:

$$\Psi[C'] = \left\{1 - \frac{i}{\hbar} H_{12}(P)\, d\omega_P\right\} \Psi[C]. \tag{30}$$

When there exist in the space-time world two surfaces C_1 and C_2 a finite distance apart, we need only to repeat the infinitesimal transformations in order to obtain $\Psi[C_2]$ from $\Psi[C_1]$. Thus

$$\Psi[C_2] = \prod_{C_1}^{C_2} \left\{1 - \frac{i}{\hbar} H_{12}(P)\, d\omega_P\right\} \Psi[C_1]. \tag{31}$$

The meaning of this equation is as follows: We divide the world region lying between C_1 and C_2 into small elements $d\omega_P$ (it is necessary that each world element be surrounded by two space-like surfaces). We consider for each world element the infinitesimal transformation $1 - \frac{i}{\hbar} H_{12}(P)\, d\omega_P$. Then we take the product of these transformations, the order of the factor being taken from C_1 to C_2. This product transforms then $\Psi[C_1]$ into $\Psi[C_2]$.

The surfaces C_1 and C_2 must here be both space-like, but otherwise they may have any form and any configuration. Thus C_2 does not necessarily lie afterward against C_1; C_1 and C_2 may even cross with each other.

The relation of the form (31) has been already introduced by Heisenberg.(7) It can be regarded as the integral form of our generalized Schrödinger equation (29).

4. Generalized probability amplitude

We must now find the physical meaning of the functional $\Psi[C]$. As regards this we can follow a method similar to that of Bloch in the case of ordinary many-time theory. Besides the fact that in our case there appear infinitely many time variables, one point differs from Bloch's case: in (16) the unitary operator u is commutable with the coordinates $q_1, q_2, \cdots, q_N$, while our U is not commutable with the field quantities $v_1(xyz)$ and $v_2(xyz)$. Noting this difference and treating the continuum infinity as the limit of a denumerable infinity by some artifice, for instance, by the procedure of Heisenberg and Pauli,(8) Bloch's consideration can be applied also here almost without any alteration. We shall give here only the results.

Let us suppose that the fields are in the state represented by a vector $\Psi[C]$. We suppose that we make measurements of a function $f(v_1, v_2, \lambda_1, \lambda_2)$ at every point on a surface C_1 in the space-time world. Let P_1 denote the variable point on C_1, then, if $f(P_1)$ at any two "values" of P_1 commute with each other, the measurements of f at each of these two points do not interfere with each other. Our first conclusion says that in this case the expectation value of $f(P_1)$ is given by

$$\overline{f(P_1)}=((\Psi[C_1],f(P_1)\Psi[C_1])) \tag{32}$$

where $f(P_1)$ means $f(V_1(P_1), \cdots)$ according to our convention on page `8, and the symbol $((A, B))$ with double parentheses is the scalar product of two vectors A and B. It is impossible in cases of continuously many degrees of freedom to represent this scalar product by an integral of the product of two functions. For this purpose we must replace the continuum infinity by an at least denumerable infinity.

More generally, we suppose a functional $F[f(P_1)]$ of the independent variable function $f(P_1)$, regarding $f(P_1)$ as a function of P_1. Then the expectation value of this F is given by

$$\overline{F[f(P_1)]}=((\Psi[C_1],F[f(P_1)]\Psi[C_1])). \tag{33}$$

A physically interesting F is the projective operator $M[v_1'(P_1), v_2'(P_1); V_1(P_1), V_2(P_1)]$ belonging to the "eigen-value" $v_1'(P_1), v_2'(P_1)$ of $V_1(P_1)$, $V_2(P_1)$. Then its expectation value

$$\begin{aligned}&\overline{M[v_1'(P_1), v_2'(P_1); V_1(P_1), V_2(P_1)]}\\&\quad=((\Psi[C_1], M[v_1'(P_1), v_2'(P_1); V_1(P_1), V_2(P_1)]\Psi[C_1]))\end{aligned} \tag{34}$$

gives the probability that the field 1 and the field 2 have respectively the functional form $v_1'(P_1)$ and $v_2'(P_1)$ on the surface C_1. As C_1 is assumed to be space-like, the measurement of the functional M is possible (the measurements of $V_1(P_1)$ and $V_2(P_1)$ at all points on C_1 mean just the measurement of M).

Thus far we have made no mention of the representation of $\Psi[C]$. We use now the special representation in which $V_1(P_1)$ at all points on C_1 are simultaneously diagonal. It is always possible to make all $V_1(P_1)$ and $V_2(P_1)$ diagonal when the surface C_1 is space-like. In this representation $\Psi[C_1]$ is represented by a functional $\Psi[v_1'(P_1), v_2'(P_1); C_1]$ of the eigen-values $v_1'(P_1)$ and $v_2'(P_1)$ of $V_1(P_1)$ and $V_2(P_1)$. The projection operator M has in this representation such diagonal form that (34) is simplified as follows

$$\begin{aligned}W[v_1'(P_1), v_2'(P_1)]&=\overline{M[v_1'(P_1), v_2'(P_1); V_1(P_1), V_2(P_1)]}\\&=|\Psi[v_1'(P_1), v_2'(P_1); C_1]|^2.\end{aligned} \tag{35}$$

In this sense we can call $\Psi[v_1'(P_1), v_2'(P_1); C_1]$ the "generalized probability amplitude."

5. Generalized transformation functional

We have stated above that between $\Psi[C_1]$ and $\Psi[C_2]$ the relation (31) holds, where C_1 and C_2 are two space-like surfaces in the space-time world. We see thus that the transformation operator

$$T[C_2; C_1] = \prod_{C_1}^{C_2}\left(1 - \frac{i}{\hbar}H_{12}\, d\omega\right) \tag{36}$$

plays an important role. It is evident that this operator also has a space-time meaning.

Just as the special representative of the ψ-vector, the probability amplitude, has a distinct physical meaning, so there is a special representation in which the representative of the transformation operator $T[C_2; C_1]$ has a distinct physical meaning.

We now introduce the mixed representative of $T[C_2; C_1]$ whose rows refer to the representation in which $V_1(P_1)$ and $V_2(P_1)$ at all points on C_1 become diagonal and whose column refer to the representation in which $V_1(P_2)$ and $V_2(P_2)$ at all points on C_2 become diagonal. We denote this representation by

$$[v_1''(P_2), v_2''(P_2) | T[C_2; C_1] | v_1'(P_1), v_2'(P_1)], \tag{37}^*$$

or simpler:

$$[v_1''(P_2), v_2''(P_2) | v_1'(P_1), v_2'(P_1)]. \tag{38}^*$$

If we note here the relation (35), we see that we can give the matrix elements of this representation the following meaning: One measures the field quantities V_1 and V_2 at all points on C_2 when the fields are prepared in such a way that they have certainly the values $v_1'(P_1)$ and $v_2'(P_1)$ at all points on C_1. Then

$$W[v_1''(P_2), v_2''(P_2); v_1'(P_1), v_2'(P_1)] = |[v_1''(P_2), v_2''(P_2) | v_1'(P_1), v_2'(P_1)]|^2 \tag{39}$$

gives the probability that one obtains the result $v_1''(P_2)$ and $v_2''(P_2)$ in this measurement. In this proposition we have assumed that C_2 lies afterward against C_1.

From this physical interpretation we may regard the matrix element (37), or (38), considered as a functional of $v_1''(P_2)$, $v_2''(P_2)$ and $v_1'(P_1)$, $v_2'(P_1)$, as the generalization of the ordinary transformation function $(q_{t_2}''|q_{t_1}')$.

As a special case it may happen that C_2 lies apart from C_1 only in a portion S_2 and a portion S_1 of C_2 and C_1 respectively, the other parts of C_1 and C_2 overlapping with each other.

In this case the matrix elements of $T[C_2; C_1]$ depend only on the values of the fields on the portions S_1 and S_2 of the surfaces C_1 and C_2. In this

* As the matrix elements are functionals of $v(P)$, we use here the square brackets.

case we need for calculating $T[C_2; C_1]$ to take the product in (36) only in the closed domain surrounded by S_1 and S_2, thus

$$T[S_2; S_1] = \prod_{S_1}^{S_2}\left(1 - \frac{i}{\hbar}H_{12}\, d\omega\right). \tag{40}$$

The matrix elements of the mixed representation of this T is a functional of $v_1'(p_1)$, $v_2'(p_1)$ and $v_1''(p_2)$, $v_2''(p_2)$, where p_1 denotes the moving point of the portion S_1, and p_2 the moving point on the portion S_2. This matrix is independent on the field quantities on the other portions of the surfaces C_1 and C_2.

The matrix element of $T[S_2; S_1]$ regarded as a functional of $v_1'(p_1)$, $v_2'(p_1)$ and $v_1''(p_2)$, $v_2''(p_2)$ has the properties of g.t.f. (generalized transformation functional) of Dirac. But in defining our g.t.f. we had to restrict the surfaces S_1 and S_2 to be space-like, while Dirac has required his g.t.f. to be defined also referring to the time-like surfaces. As mentioned above, however, such a generalization as required by Dirac is superfluous so far as concerns the relativity theory.

It is to be noted that for the physical interpretation of $[v_1''(P_2), v_2''(P_2)| v_1'(P_1), v_2'(P_1)]$ it is not necessary to assume C_2 to lie afterward against C_1. Also when the inverse is the case, we can as well give the physical meaning for W of (39): One measures the field quantities V_1 and V_2 at all points on C_2 when the fields are prepared in such a way that they would have certainly the values $v_1'(P_1)$ and $v_2'(P_1)$ at all points on C_1 if the fields were left alone until C_1 without being measured before on C_2. Then W gives the probability that one finds the results $v_1''(P_2)$ and $v_2''(P_2)$ in this measurement on C_2.

6. Concluding remarks

We have thus shown that the quantum theory of wave fields can be really brought into a form which reveals directly the invariance of the theory against Lorentz transformations. The reason why the ordinary formalism of the quantum field theory is so unsatisfactory is that it has been built up in a way much too analogous to the ordinary non-relativistic mechanics. In this ordinary formalism of the quantum theory of fields the theory is divided into two distinct sections: the section giving the kinematical relations between various quantities at the same instant of time, and the section determining the causal relations between quantities at different instants of time. Thus the commutation relations (1) belong to the first section and the Schrödinger equation (2) to the second.

As stated before, this way of separating the theory into two sections is very unrelativistic, since here the concept "same instant of time" plays a distinct role.

Also in our formalism the theory is divided into two sections, but now the separation is introduced in another place. One section gives the laws of behaviour of the fields when they are left alone, and the other gives the laws determining the deviation from this behaviour due to interactions. This way of separating the theory can be carried out relativistically.

Although in this way the theory can be brought into more satisfactory form, no new contents are added thereby. So, the well-known divergence difficulties of the theory are inherited also by our theory. Indeed, our fundamental equations (29) admit only catastrophic solutions, as can be seen directly in the fact that the unavoidable infinity due to non-vanishing zero-point amplitudes of the fields inheres in the operator $H_{12}(P)$. Thus, a more profound modification of the theory is required in order to remove this fundamental difficulty.

It is expected that such a modification of the theory could possibly be introduced by some revision of the concept of interaction, because we meet no such difficulty when we deal with the non-interacting fields. This revision would then have the result that in the separation of the theory into two sections, one for free fields and one for interactions, some uncertainty would be introduced. This seems to be implied by the very fact that, when we formulate the quantum field theory in a relativistically satisfactory manner, this way of separation has revealed itself as the fundamental element of the theory.

Physics Department,
Tokyo Bunrika University.

References

1. H. Yukawa, *Kagaku*, **12**, 251, 282 and 322, 1942.
2. P. A. M. Dirac, *Phys. Z. USSR.*, **3**, 64, 1933.
3. W. Pauli, *Solvay Berichte*, 1939.
4. P. A. M. Dirac, *Proc. Roy. Soc. London*, **136**, 453, 1932.
5. F. Bloch, *Phys. Z. USSR.*, **5**, 301, 1943.
6. E. Stueckelberg, *Helv. Phys. Acta*, **11**, 225, § 5, 1938.
7. W. Heisenberg, *Z. Phys.*, **110**, 251, 1938.
8. W. Heisenberg and W. Pauli, *Z. Phys.*, **56**, 1, 1929.

Quantum Electrodynamics. III. The Electromagnetic Properties of the Electron —Radiative Corrections to Scattering

JULIAN SCHWINGER
Harvard University, Cambridge, Massachusetts
(Received May 26, 1949)

The discussion of vacuum polarization in the previous paper of this series was confined to that produced by the field of a prescribed current distribution. We now consider the induction of current in the vacuum by an electron, which is a dynamical system and an entity indistinguishable from the particles associated with vacuum fluctuations. The additional current thus attributed to an electron implies an alteration in its electromagnetic properties which will be revealed by scattering in a Coulomb field and by energy level displacements. This paper is concerned with the computation of the second-order corrections to the current operator and the application to electron scattering. Radiative corrections to energy levels will be treated in the next paper of the series. Following a canonical transformation which effectively renormalizes the electron mass, the correction to the current operator produced by the coupling with the electromagnetic field is developed in a power series, of which first- and second-order terms are retained. One thus obtains second-order modifications in the current operator which are of the same general nature as the previously treated vacuum polarization current, save for a contribution that has the form of a dipole current. The latter implies a fractional increase of $\alpha/2\pi$ in the spin magnetic moment of the electron. The only flaw in the second-order current correction is a logarithmic divergence attributable to an infra-red catastrophe. It is remarked that, in the presence of an external field, the first-order current correction will introduce a compensating divergence. Thus, the second-order corrections to particle electromagnetic properties cannot be completely stated without regard for the manner of exhibiting them by an external field. Accordingly, we consider in the second section the interaction of three systems, the matter field, the electromagnetic field, and a given current distribution. It is shown that this situation can be described in terms of an external potential coupled to the current operator, as modified by the interaction with the vacuum electromagnetic field. Application is made to the scattering of an electron by an external field, in which the latter is regarded as a small perturbation. It is found convenient to calculate the total rate at which collisions occur and then identify the cross sections for individual events. The correction to the cross section for radiationless scattering is determined by the second-order correction to the current operator, while scattering that is accompanied by single quantum emission is a consequence of the first-order current correction. The final object of calculation is the differential cross section for scattering through a given angle with a prescribed maximum energy loss, which is completely free of divergences. Detailed evaluations are given in two situations, the essentially elastic scattering of an electron, in which only a small fraction of the kinetic energy is radiated, and the scattering of a slowly moving electron with unrestricted energy loss. The Appendix is devoted to an alternative treatment of the polarization of the vacuum by an external field. The conditions imposed on the induced current by the charge conservation and gauge invariance requirements are examined. It is found that the fulfillment of these formal properties requires the vanishing of an integral that is not absolutely convergent, but naturally vanishes for reasons of symmetry. This null integral is then used to simplify the expression for the induced current in such a manner that direct calculation yields a gauge invariant result. The induced current contains a logarithmically divergent multiple of the external current, which implies that a non-vanishing total charge, proportional to the external charge, is induced in the vacuum. The apparent contradiction with charge conservation is resolved by showing that a compensating charge escapes to infinity. Finally, the expression for the electromagnetic mass of the electron is treated with the methods developed in this paper.

A COVARIANT form of quantum electrodynamics has been developed, and applied to two elementary vacuum fluctuation phenomena in the previous articles of this series.[1] These applications were the polarization of the vacuum, expressing the modifications in the properties of an electromagnetic field arising from its interaction with the matter field vacuum fluctuations, and the electromagnetic mass of the electron, embodying the corrections to the mechanical properties of the matter field, in its single particle aspect, that are produced by the vacuum fluctuations of the electromagnetic field. In these problems, the divergences that mar the theory are found to be concealed in unobservable charge and mass renormalization factors.

The previous discussion of the polarization of the vacuum was concerned with a given current distribution, one that is not affected by the dynamical reactions of the electron-positron matter field. We shall now consider the more complicated situation in which the original current is that ascribed to an electron or positron—a dynamical system, and an entity indistinguishable from the particles associated with the matter field vacuum fluctuations. The changed electromagnetic properties of the particle will be exhibited in an external field, and may be compared with the experimental indications of deviations from the Dirac theory that were briefly discussed in I. To avoid a work of excessive length, this discussion will be given in two papers. In this paper we shall construct the current operator as modified, to the second order, by the coupling with the vacuum electromagnetic field. This will be applied to compute the radiative correction to the scattering of an electron by a Coulomb field.[2] The second paper will deal with the effects of radiative corrections on energy levels.

1. SECOND-ORDER CORRECTIONS TO THE CURRENT OPERATOR

We shall evaluate the second-order modifications of the current operator produced by the coupling between

[1] Julian Schwinger, "Quantum Electrodynamics. I," Phys. Rev. **74**, 1439 (1948); "Quantum Electrodynamics. II," Phys. Rev. **75**, 651 (1949).

[2] A short account of the results has already been published, Julian Schwinger, Phys. Rev. **75**, 898 (1949).

the matter and electromagnetic fields. The latter is described by

$$i\hbar c\frac{\delta\Psi[\sigma]}{\delta\sigma(x)}=\mathcal{H}(x)\Psi[\sigma],$$

$$\mathcal{H}(x)=-\frac{1}{c}j_\mu(x)A_\mu(x). \tag{1.1}$$

Among the effects produced by this coupling is the electromagnetic mass of the electron, as contained in the self-energy operator $\mathcal{H}_{1,0}(x)$. In order to describe the electron in terms of the experimental mass, we write (1.1) as

$$i\hbar c\frac{\delta\Psi[\sigma]}{\delta\sigma(x)}=(\mathcal{H}_{1,0}(x)+\mathcal{K}(x))\Psi[\sigma], \tag{1.2}$$

where

$$\mathcal{K}(x)=\mathcal{H}(x)-\mathcal{H}_{1,0}(x). \tag{1.3}$$

The canonical transformation

$$\Psi[\sigma]\rightarrow W[\sigma]\Psi[\sigma],$$

$$i\hbar c\frac{\delta W[\sigma]}{\delta\sigma(x)}=\mathcal{H}_{1,0}(x)W[\sigma], \tag{1.4}$$

then replaces (1.2) with

$$i\hbar c\frac{\delta\Psi[\sigma]}{\delta\sigma(x)}=W^{-1}[\sigma]\mathcal{K}(x)W[\sigma]\Psi[\sigma], \tag{1.5}$$

while the operator representing the current becomes $W^{-1}[\sigma]j_\mu(x)W[\sigma]$. Now, as we have shown in II, the spinor $W^{-1}[\sigma]\psi(x)W[\sigma]$ obeys the Dirac equation for a particle of mass $m=m_0+\delta m$, the experimental mass of the electron. Accordingly, the expectation value of the current operator can be computed as

$$\langle j_\mu(x)\rangle=(\Psi[\sigma],j_\mu(x)\Psi[\sigma]), \tag{1.6}$$

where

$$i\hbar c\frac{\delta\Psi[\sigma]}{\delta\sigma(x)}=\mathcal{K}(x)\Psi[\sigma], \tag{1.7}$$

with the understanding that the experimental electron mass is to be employed.

If a solution of the latter equation is constructed in the form

$$\Psi[\sigma]=U[\sigma]\Psi_0, \tag{1.8}$$

the expectation value of the current operator becomes

$$\langle j_\mu(x)\rangle=(\Psi_0,U^{-1}[\sigma]j_\mu(x)U[\sigma]\Psi_0)=(\Psi_0,\mathbf{j}_\mu(x)\Psi_0), \tag{1.9}$$

in which the latter version describes the effect of the coupling between the fields by changing the current operator into

$$\mathbf{j}_\mu(x)=U^{-1}[\sigma]j_\mu(x)U[\sigma]. \tag{1.10}$$

The unitary operator $U[\sigma]$ obeys the equation of motion

$$i\hbar c\frac{\delta U[\sigma]}{\delta\sigma(x)}=\mathcal{K}(x)U[\sigma], \tag{1.11}$$

which may be supplemented by the boundary condition

$$U[-\infty]=1, \tag{1.12}$$

in accordance with the supposition that coupling between the two fields is adiabatically established in the remote past.

The operator $\mathbf{j}_\mu(x)$ can now be evaluated by remarking that

$$\begin{aligned}\mathbf{j}_\mu(x)&=j_\mu(x)+\int_{-\infty}^{\sigma}d\omega'\frac{\delta}{\delta\sigma'(x')}(U^{-1}[\sigma']j_\mu(x)U[\sigma'])\\&=j_\mu(x)-\frac{i}{\hbar c}\int_{-\infty}^{\sigma}d\omega'U^{-1}[\sigma'][j_\mu(x),\mathcal{K}(x')]U[\sigma'].\end{aligned} \tag{1.13}$$

This process can be continued according to

$$\int_{-\infty}^{\sigma}d\omega'U^{-1}[\sigma'][j_\mu(x),\mathcal{K}(x')]U[\sigma']=\int_{-\infty}^{\sigma}d\omega'[j_\mu(x),\mathcal{K}(x')]+\int_{-\infty}^{\sigma}d\omega'\int_{-\infty}^{\sigma'}d\omega''\frac{\delta}{\delta\sigma''(x'')}(U^{-1}[\sigma''][j_\mu(x),\mathcal{K}(x')]U[\sigma'']), \tag{1.14}$$

and yields $\mathbf{j}_\mu(x)$ in the form of an infinite series,

$$\mathbf{j}_\mu(x)=j_\mu(x)+\left(-\frac{i}{\hbar c}\right)\int_{-\infty}^{\sigma}d\omega'[j_\mu(x),\mathcal{K}(x')]+\left(-\frac{i}{\hbar c}\right)^2\int_{-\infty}^{\sigma}d\omega'\int_{-\infty}^{\sigma'}d\omega''[[j_\mu(x),\mathcal{K}(x')],\mathcal{K}(x'')]+\cdots. \tag{1.15}$$

An equivalent procedure, which exhibits $\mathbf{j}_\mu(x)$ in a form that is more symmetrical between past and future, is based on the following observation,

$$\int_{-\infty}^{\infty} d\omega' \epsilon[\sigma,\sigma'] \frac{\delta}{\delta\sigma'(x')}(U^{-1}[\sigma'] j_\mu(x) U[\sigma']) = \int_{-\infty}^{\sigma} d\omega' \frac{\delta}{\delta\sigma'(x')}(U^{-1}[\sigma'] j_\mu(x) U[\sigma'])$$

$$+\int_{\infty}^{\sigma} d\omega' \frac{\delta}{\delta\sigma'(x')}(U^{-1}[\sigma'] j_\mu(x) U[\sigma']) = (\mathbf{j}_\mu(x) - j_\mu(x)) + (\mathbf{j}_\mu(x) - U^{-1}[\infty] j_\mu(x) U[\infty]), \quad (1.16)$$

or

$$\mathbf{j}_\mu(x) = \tfrac{1}{2}(j_\mu(x) + S^{-1} j_\mu(x) S) + \left(-\frac{i}{2\hbar c}\right) \int_{-\infty}^{\infty} d\omega' \epsilon[\sigma,\sigma'] U^{-1}[\sigma'] [j_\mu(x), \mathfrak{K}(x')] U[\sigma'], \quad (1.17)$$

where

$$S = U[\infty], \quad U[-\infty] = 1 \quad (1.18)$$

is the collision operator which describes the real transitions that permanently alter the state of the system. This process can be continued and finally yields

$$\mathbf{j}_\mu(x) = \tfrac{1}{2}(k_\mu(x) + S^{-1} k_\mu(x) S), \quad (1.19)$$

in which

$$k_\mu(x) = j_\mu(x) + \left(-\frac{i}{2\hbar c}\right) \int_{-\infty}^{\infty} d\omega' \epsilon[\sigma,\sigma'] [j_\mu(x), \mathfrak{K}(x')]$$

$$+\left(-\frac{i}{2\hbar c}\right)^2 \int_{-\infty}^{\infty} d\omega' d\omega'' \epsilon[\sigma,\sigma'] \epsilon[\sigma',\sigma''] [[j_\mu(x), \mathfrak{K}(x')], \mathfrak{K}(x'')] + \cdots. \quad (1.20)$$

The further terms in the series are not required to compute the second-order correction of the current operator.

The collision operator S can be constructed in a similar manner. Thus,

$$S - 1 = \int_{-\infty}^{\infty} d\omega \frac{\delta}{\delta\sigma(x)} U[\sigma] = -\frac{i}{\hbar c} \int_{-\infty}^{\infty} d\omega \mathfrak{K}(x) U[\sigma], \quad (1.21)$$

and

$$U[\sigma] - \tfrac{1}{2}(S+1) = \frac{1}{2}\int_{-\infty}^{\infty} d\omega' \epsilon[\sigma,\sigma'] \frac{\delta}{\delta\sigma'(x')} U[\sigma'] = -\frac{i}{2\hbar c} \int_{-\infty}^{\infty} d\omega' \epsilon[\sigma,\sigma'] \mathfrak{K}(x') U[\sigma'], \quad (1.22)$$

whence,

$$S - 1 = \left(-\frac{i}{2\hbar c}\right) \int_{-\infty}^{\infty} d\omega \mathfrak{K}(x)(S+1) + 2\left(-\frac{i}{2\hbar c}\right)^2 \int_{-\infty}^{\infty} d\omega d\omega' \epsilon[\sigma,\sigma'] \mathfrak{K}(x) \mathfrak{K}(x') U[\sigma']. \quad (1.23)$$

Continuing in this manner, we obtain

$$\frac{S-1}{S+1} = \left(-\frac{i}{2\hbar c}\right) \int_{-\infty}^{\infty} d\omega \mathfrak{K}(x) + \left(-\frac{i}{2\hbar c}\right)^2 \int_{-\infty}^{\infty} d\omega d\omega' \epsilon[\sigma,\sigma'] \mathfrak{K}(x) \mathfrak{K}(x') + \cdots. \quad (1.24)$$

Only the indicated terms need be retained for the desired degree of approximation. In view of the absence of real first-order effects, as expressed by

$$\int_{-\infty}^{\infty} \mathfrak{H}(x) d\omega = 0, \quad (1.25)$$

the leading terms in (1.24) are of the second order:

$$\frac{S-1}{S+1} = -\frac{i}{2\hbar c} \int_{-\infty}^{\infty} d\omega \left[-\frac{i}{4\hbar c} \int_{-\infty}^{\infty} \epsilon(x-x') [\mathfrak{H}(x), \mathfrak{H}(x')] d\omega' - \mathfrak{H}_{1,0}(x) \right]. \quad (1.26)$$

According to (II 3.14) and (II 3.71), (the vacuum term $\mathfrak{H}_{0,0}$ is of no consequence),

$$-\frac{i}{4\hbar c} \int_{-\infty}^{\infty} \epsilon(x-x') [\mathfrak{H}(x), \mathfrak{H}(x')] d\omega' = \mathfrak{H}_{1,0}(x) + \mathfrak{H}_{2,0}(x) + \mathfrak{H}_{1,1}(x), \quad (1.27)$$

whence,

$$\frac{S-1}{S+1} = -\frac{i}{2\hbar c} \int_{-\infty}^{\infty} [\mathfrak{H}_{2,0}(x) + \mathfrak{H}_{1,1}(x)] d\omega, \quad (1.28)$$

which describes the real effects involving either two particles or one particle and a light quantum. Since we shall be concerned only with second-order effects referring to a single particle and the absence of light quanta, such real processes do not come into play and S is effectively unity. Consequently, the current operator is modified only by virtual processes, and is completely symmetrical between past and future. Thus, to the desired order of approximation,

$$\mathbf{j}_\mu(x)=j_\mu(x)-\frac{i}{2\hbar c}\int_{-\infty}^{\infty}d\omega'\epsilon[\sigma,\sigma'][j_\mu(x),\mathcal{H}(x')]+\frac{i}{2\hbar c}\int_{-\infty}^{\infty}d\omega'\epsilon[\sigma,\sigma'][j_\mu(x),\mathcal{H}_{1,0}(x')]$$
$$+\left(-\frac{i}{2\hbar c}\right)^2\int_{-\infty}^{\infty}d\omega'd\omega''\epsilon[\sigma,\sigma']\epsilon[\sigma',\sigma''][[j_\mu(x),\mathcal{H}(x')],\mathcal{H}(x'')]. \quad (1.29)$$

The correction to the current operator may now be written

$$\mathbf{j}_\mu(x)-j_\mu(x)=\delta j_\mu^{(1)}(x)+\delta j_\mu^{(2)}(x), \quad (1.30)$$

where

$$\delta j_\mu^{(1)}(x)=\frac{i}{2\hbar c^2}\int_{-\infty}^{\infty}d\omega'\epsilon(x-x')[j_\mu(x),j_\nu(x')]A_\nu(x'), \quad (1.31)$$

and

$$(\delta j_\mu^{(2)}(x))_{1,0}=-\frac{1}{4\hbar^2c^4}\int_{-\infty}^{\infty}d\omega'd\omega''\epsilon[\sigma,\sigma']\epsilon[\sigma',\sigma''][[j_\mu(x),j_\nu(x')]A_\nu(x'),j_\lambda(x'')A_\lambda(x'')]_{1,0}$$
$$+\frac{i}{2\hbar c}\int_{-\infty}^{\infty}d\omega'\epsilon(x-x')[j_\mu(x),\mathcal{H}_{1,0}(x')]_1 \quad (1.32)$$

are the first- and second-order corrections, respectively. In the latter, the subscripts emphasize that we are only concerned with second-order effects involving a single particle and no light quanta. To simplify (1.32), note that

$$[[j_\mu(x),j_\nu(x')]A_\nu(x'),j_\lambda(x'')A_\lambda(x'')]_{1,0}=\tfrac{1}{2}[A_\nu(x'),A_\lambda(x'')]\{[j_\mu(x),j_\nu(x')],j_\lambda(x'')\}_1$$
$$+\tfrac{1}{2}\{A_\nu(x'),A_\lambda(x'')\}_0[[j_\mu(x),j_\nu(x')],j_\lambda(x'')]_1, \quad (1.33)$$

whence

$$(\delta j_\mu^{(2)}(x))_{1,0}=\frac{i}{4\hbar c^3}\int_{-\infty}^{\infty}d\omega'd\omega''\epsilon(x-x')\bar{D}(x'-x'')\{[j_\mu(x),j_\nu(x')],j_\nu(x'')\}_1$$
$$-\frac{1}{8\hbar c^3}\int_{-\infty}^{\infty}d\omega'd\omega''\epsilon[\sigma,\sigma']\epsilon[\sigma',\sigma'']D^{(1)}(x'-x'')[[j_\mu(x),j_\nu(x')],j_\nu(x'')]_1+\frac{i}{2\hbar c}\int_{-\infty}^{\infty}d\omega'\epsilon(x-x')[j_\mu(x),\mathcal{H}_{1,0}(x')]_1, \quad (1.34)$$

in consequence of

$$-\tfrac{1}{2}\epsilon(x'-x'')[A_\nu(x'),A_\lambda(x'')]=i\hbar c\delta_{\nu\lambda}\bar{D}(x'-x'') \quad (1.35)$$

and

$$\{A_\nu(x'),A_\lambda(x'')\}_0=\hbar c\delta_{\nu\lambda}D^{(1)}(x'-x''). \quad (1.36)$$

The double commutator in (1.34) is easily evaluated,

$$[[j_\mu(x),j_\nu(x')],j_\nu(x'')]=-e^3c^3[\bar{\psi}(x)\gamma_\mu S(x-x')\gamma_\nu\psi(x')-\bar{\psi}(x')\gamma_\nu S(x'-x)\gamma_\mu\psi(x),\bar{\psi}(x'')\gamma_\nu\psi(x'')]$$
$$=ie^3c^3(\bar{\psi}(x)\gamma_\mu S(x-x')\gamma_\nu S(x'-x'')\gamma_\nu\psi(x'')+\bar{\psi}(x'')\gamma_\nu S(x''-x')\gamma_\nu S(x'-x)\gamma_\mu\psi(x)$$
$$-\bar{\psi}(x')\gamma_\nu S(x'-x)\gamma_\mu S(x-x'')\gamma_\nu\psi(x'')-\bar{\psi}(x'')\gamma_\nu S(x''-x)\gamma_\mu S(x-x')\gamma_\nu\psi(x')). \quad (1.37)$$

The one-particle part of $\{[j_\mu(x),j_\nu(x')],j_\nu(x'')\}$ can be constructed in the manner employed in II. We have only to notice that $[j_\mu(x),j_\nu(x')]$ has a non-vanishing vacuum expectation value. Thus,

$$\{[j_\mu(x),j_\nu(x')],j_\nu(x'')\}_1-2[j_\mu(x),j_\nu(x')]_0 j_\nu(x'')$$
$$=-e^3c^3\{(\bar{\psi}(x)\gamma_\mu S(x-x')\gamma_\nu\psi(x')-\bar{\psi}(x')\gamma_\nu S(x'-x)\gamma_\mu\psi(x))_1,(\bar{\psi}(x'')\gamma_\nu\psi(x''))_1\}_1$$
$$=-e^3c^3(\bar{\psi}(x)\gamma_\mu S(x-x')\gamma_\nu\{\psi(x'),\bar{\psi}(x'')\}_0\gamma_\nu\psi(x'')-\bar{\psi}(x'')\gamma_\nu\{\psi(x''),\bar{\psi}(x')\}_0\gamma_\nu S(x'-x)\gamma_\mu\psi(x)$$
$$-\bar{\psi}(x')\gamma_\nu S(x'-x)\gamma_\mu\{\psi(x),\bar{\psi}(x'')\}_0\gamma_\nu\psi(x'')+\bar{\psi}(x'')\gamma_\nu\{\psi(x''),\bar{\psi}(x)\}_0\gamma_\mu S(x-x')\gamma_\nu\psi(x'))_1, \quad (1.38)$$

and

$$\{[j_\mu(x),j_\nu(x')],j_\nu(x'')\}_1=2[j_\mu(x),j_\nu(x')]_0 j_\nu(x'')$$
$$+e^3c^3(\bar\psi(x)\gamma_\mu S(x-x')\gamma_\nu S^{(1)}(x'-x'')\gamma_\nu\psi(x'')-\bar\psi(x'')\gamma_\nu S^{(1)}(x''-x')\gamma_\nu S(x'-x)\gamma_\mu\psi(x)$$
$$-\bar\psi(x')\gamma_\nu S(x'-x)\gamma_\mu S^{(1)}(x-x'')\gamma_\nu\psi(x'')+\bar\psi(x'')\gamma_\nu S^{(1)}(x''-x)\gamma_\mu S(x-x')\gamma_\nu\psi(x'))_1. \quad (1.39)$$

On inserting (1.37) and (1.39) into (1.34), we obtain

$$(\delta j_\mu{}^{(2)}(x))_{1,0}=\frac{i}{2\hbar c^2}\int_{-\infty}^{\infty}d\omega'd\omega''\epsilon(x-x')[j_\mu(x),j_\nu(x')]_0\bar D(x'-x'')j_\nu(x'')$$
$$-\frac{ie^3}{2\hbar}\int_{-\infty}^{\infty}d\omega'd\omega''(\bar\psi(x')\gamma_\nu\bar S(x'-x)\gamma_\mu S^{(1)}(x-x'')\gamma_\nu\psi(x'')+\bar\psi(x')\gamma_\nu S^{(1)}(x'-x)\gamma_\mu\bar S(x-x'')\gamma_\nu\psi(x''))_1\bar D(x'-x'')$$
$$-\frac{ie^3}{2\hbar}\int_{-\infty}^{\infty}d\omega'd\omega''(\bar\psi(x')\gamma_\nu\bar S(x'-x)\gamma_\mu\bar S(x-x'')\gamma_\nu\psi(x''))_1D^{(1)}(x'-x'')$$
$$+\frac{ie}{\hbar}\int_{-\infty}^{\infty}d\omega'(\bar\psi(x)\gamma_\mu\bar S(x-x')\phi(x')+\bar\phi(x')\bar S(x'-x)\gamma_\mu\psi(x))_1$$
$$-\frac{e}{2\hbar}\int_{-\infty}^{\infty}d\omega'\epsilon(x-x')(\bar\psi(x)\gamma_\mu[\psi(x),\mathcal{H}_{1,0}(x')]+[\bar\psi(x),\mathcal{H}_{1,0}(x')]\gamma_\mu\psi(x))_1, \quad (1.40)$$

where (see II (3.78))

$$\phi(x)=-\frac{e^2}{2}\int_{-\infty}^{\infty}d\omega'\gamma_\nu(\bar D(x-x')S^{(1)}(x-x')+D^{(1)}(x-x')\bar S(x-x'))\gamma_\nu\psi(x')$$
$$=\delta mc^2\psi(x). \quad (1.41)$$

The third term of (1.40) is derived from

$$\frac{ie^3}{8\hbar}\int_{-\infty}^{\infty}d\omega'd\omega''(\epsilon[\sigma,\sigma']-\epsilon[\sigma,\sigma''])\epsilon[\sigma',\sigma''](\bar\psi(x')\gamma_\nu S(x'-x)\gamma_\mu S(x-x'')\gamma_\nu\psi(x''))_1D^{(1)}(x'-x'') \quad (1.42)$$

with the aid of the identity

$$(\epsilon[\sigma,\sigma']-\epsilon[\sigma,\sigma''])\epsilon[\sigma',\sigma'']=\epsilon[\sigma,\sigma']\epsilon[\sigma,\sigma'']-1, \quad (1.43)$$

and the null value of

$$\int_{-\infty}^{\infty}d\omega'd\omega''(\bar\psi(x')\gamma_\nu S(x'-x)\gamma_\mu S(x-x'')\gamma_\nu\psi(x''))_1D^{(1)}(x'-x'')$$
$$=-\frac{1}{\hbar c}\int_{-\infty}^{\infty}d\omega'd\omega''(\bar\psi(x')\gamma_\nu\{\psi(x'),\bar\psi(x)\}\gamma_\mu\{\psi(x),\bar\psi(x'')\}\gamma_\lambda\psi(x''))_1\{A_\nu(x'),A_\lambda(x'')\}_0. \quad (1.44)$$

The latter is an immediate consequence of (1.25), expressing the absence of real first-order transitions.

The insertion of the expression for $\mathcal{H}_{1,0}(x)$, (II (3.77)),

$$\mathcal{H}_{1,0}(x)=\tfrac{1}{2}(\bar\psi(x)\phi(x)+\bar\phi(x)\psi(x))_1, \quad (1.45)$$

enables the last two terms of (1.40) to be combined into

$$\frac{ie}{2\hbar}(\bar\psi(x)\gamma_\mu\chi(x)+\bar\chi(x)\gamma_\mu\psi(x))_1, \quad (1.46)$$

where

$$\chi(x)=\int_{-\infty}^{\infty}d\omega'\left[\bar S(x-x')\phi(x')+\frac{i}{2}\epsilon(x-x')\{\psi(x),\bar\phi(x')\}\psi(x')\right]. \quad (1.47)$$

It will first be observed that the integrand of (1.47) vanishes, in virtue of the relation $\phi(x)=\delta mc^2\psi(x)$, since

$$\frac{i}{2}\epsilon(x-x')\{\psi(x),\bar\phi(x')\}\psi(x')=-\bar S(x-x')\delta mc^2\psi(x'). \quad (1.48)$$

On the other hand, integrals of the form

$$\int_{-\infty}^{\infty} d\omega' \bar{S}(x-x')\psi(x') \tag{1.49}$$

are divergent, since $\psi(x')$ and $\bar{S}(x-x')$, respectively, obey homogeneous and inhomogeneous equations associated with the same differential operator. It is convenient to express the latter integral as the limit of the finite quantity obtained by an alteration of the differential equation satisfied by $\psi(x)$, in which the mass parameter κ is replaced by $\kappa+\delta\kappa$ and the limit $\delta\kappa\to 0$ is taken. The differential equations

$$\left(\gamma_\mu \frac{\partial}{\partial x_\mu'}+\kappa+\delta\kappa\right)\psi(x')=0, \quad \frac{\partial}{\partial x_\mu'}\bar{S}(x-x')\gamma_\mu-\kappa\bar{S}(x-x')=\delta(x-x'), \tag{1.50}$$

imply the relation

$$\frac{\partial}{\partial x_\mu'}[\bar{S}(x-x')\gamma_\mu\psi(x')]+\delta\kappa\bar{S}(x-x')\psi(x')=\delta(x-x')\psi(x), \tag{1.51}$$

whence

$$\int_{-\infty}^{\infty} d\omega' \bar{S}(x-x')\psi(x')=\operatorname*{Lim}_{\delta\kappa\to 0}\frac{1}{\delta\kappa}\psi(x). \tag{1.52}$$

It may be inferred that a non-vanishing value will be obtained for $\chi(x)$ if the spinor, of which (1.47) is a linear function, is subject to the Dirac equation with the mass parameter $\kappa+\delta\kappa$, and $\delta\kappa$ is allowed to approach zero. Indeed, according to (1.48) and (1.52),

$$\begin{aligned}\chi(x)&=\operatorname*{Lim}_{\delta\kappa\to 0}\int_{-\infty}^{\infty} d\omega' \bar{S}(x-x')(\phi(x')-\delta mc^2\psi(x'))\\ &=\operatorname*{Lim}_{\delta\kappa\to 0}\frac{1}{\delta\kappa}(\phi(x)-\delta mc^2\psi(x)),\end{aligned} \tag{1.53}$$

or

$$\chi(x)=\operatorname*{Lim}_{\delta\kappa\to 0}\frac{1}{\delta\kappa}\left[-\frac{e^2}{2}\int_{-\infty}^{\infty} d\omega'\gamma_\nu(\bar{D}(x-x')S^{(1)}(x-x')+D^{(1)}(x-x')\bar{S}(x-x'))\gamma_\nu\psi(x')-\delta mc^2\psi(x)\right]. \tag{1.54}$$

A suitable representation of the solution of the Dirac equation with an altered mass parameter, $\psi_{\kappa+\delta\kappa}(x')$, in terms of the actual spinor, $\psi_\kappa(x')$, is provided by

$$\psi_{\kappa+\delta\kappa}(x')=\psi_\kappa(x')+\frac{\delta\kappa}{\kappa}(x_\lambda'-x_\lambda)\frac{\partial}{\partial x_\lambda'}\psi_\kappa(x') \tag{1.55}$$

since

$$\begin{aligned}\left(\gamma_\mu\frac{\partial}{\partial x_\mu'}+\kappa\right)\psi_{\kappa+\delta\kappa}(x')&=\frac{\delta\kappa}{\kappa}\gamma_\mu\frac{\partial}{\partial x_\mu'}\psi_\kappa(x')\\ &=-\delta\kappa\psi_\kappa(x'),\end{aligned} \tag{1.56}$$

which establishes the validity of (1.55) to the first order in $\delta\kappa$. The latter is so constructed that $\psi_{\kappa+\delta\kappa}(x)=\psi_\kappa(x)$, whence,

$$\chi(x)=\frac{e^2}{2\kappa}\int_{-\infty}^{\infty} d\omega'\gamma_\nu(\bar{D}(x-x')S^{(1)}(x-x')+D^{(1)}(x-x')\bar{S}(x-x'))\gamma_\nu(x_\lambda-x_\lambda')\frac{\partial}{\partial x_\lambda'}\psi(x'). \tag{1.57}$$

The resulting expression for the second-order correction to the current operator is

$$(\delta j_\mu^{(2)}(x))_{1,0}=\frac{i}{2\hbar c^2}\int_{-\infty}^{\infty} d\omega'\epsilon(x-x')[j_\mu(x),j_\nu(x')]_0\delta A_\nu(x')-\frac{ie^3}{2\hbar}\int_{-\infty}^{\infty} d\omega' d\omega''(\bar{\psi}(x')K_\mu(x'-x,x-x'')\psi(x''))_1, \tag{1.58}$$

where

$$\delta A_\mu(x)=\frac{1}{c}\int_{-\infty}^{\infty} d\omega'\,\bar{D}(x-x')j_\mu(x'), \tag{1.59}$$

and

$$K_\mu(x'-x,x-x'')=K_\mu^{(1)}(x'-x,x-x'')+K_\mu^{(2)}(x'-x,x-x''). \tag{1.60}$$

Here

$$K_\mu^{(1)}(\xi,\eta)=\gamma_\nu(\bar{S}(\xi)\gamma_\mu S^{(1)}(\eta)\,\bar{D}(\xi+\eta)+S^{(1)}(\xi)\gamma_\mu\bar{S}(\eta)\,\bar{D}(\xi+\eta)+\bar{S}(\xi)\gamma_\mu\bar{S}(\eta)D^{(1)}(\xi+\eta))\gamma_\nu \tag{1.61}$$

and

$$K_\mu^{(2)}(\xi,\eta)=-\gamma_\mu\delta(\xi)\frac{1}{2\kappa}\frac{\partial}{\partial\eta_\lambda}\eta_\lambda\gamma_\nu(\bar{D}(\eta)S^{(1)}(\eta)+D^{(1)}(\eta)\bar{S}(\eta))\gamma_\nu-\frac{1}{2\kappa}\frac{\partial}{\partial\xi_\lambda}\xi_\lambda\gamma_\nu(\bar{D}(\xi)S^{(1)}(\xi)+D^{(1)}(\xi)\bar{S}(\xi))\gamma_\nu\delta(\eta)\gamma_\mu. \quad (1.62)$$

An equivalent form can be given in terms of the functions

$$S_\pm(x)=\bar{S}(x)\pm\frac{i}{2}S^{(1)}(x),$$

$$D_\pm(x)=\bar{D}(x)\pm\frac{i}{2}D^{(1)}(x). \quad (1.63)$$

Thus,

$$K_\mu^{(1)}(\xi,\eta)=\frac{1}{i}\gamma_\nu(S_+(\xi)\gamma_\mu S_+(\eta)D_+(\xi+\eta)-S_-(\xi)\gamma_\mu S_-(\eta)D_-(\xi+\eta))\gamma_\nu, \quad (1.64)$$

and

$$K_\mu^{(2)}(\xi,\eta)=-\gamma_\mu\delta(\xi)\frac{1}{2\kappa}\frac{\partial}{\partial\eta_\lambda}\eta_\lambda\frac{1}{i}\gamma_\nu(D_+(\eta)S_+(\eta)-D_-(\eta)S_-(\eta))\gamma_\nu$$

$$-\frac{1}{2\kappa}\frac{\partial}{\partial\xi_\lambda}\xi_\lambda\frac{1}{i}\gamma_\nu(D_+(\xi)S_+(\xi)-D_-(\xi)S_-(\xi))\gamma_\nu\delta(\eta)\gamma_\mu. \quad (1.65)$$

The first term of (1.58) is the current induced by the electromagnetic field that accompanies a given current distribution, as discussed in II. It is the second part of (1.58), expressing the additional effects involved when the current is associated with the matter field, rather than an external system, that merits our attention.

In order to evaluate $K_\mu(x'-x,x-x'')$, we shall substitute Fourier integral representations for the various functions involved, (II (A.10), (A.31)),

$$\bar{S}(x)=\frac{1}{(2\pi)^4}\int(dk)e^{ikx}(i\gamma k-\kappa)\frac{1}{k^2+\kappa^2},$$

$$S^{(1)}(x)=\frac{1}{(2\pi)^3}\int(dk)e^{ikx}(i\gamma k-\kappa)\delta(k^2+\kappa^2),$$

$$\bar{D}(x)=\frac{1}{(2\pi)^4}\int(dk)e^{ikx}\frac{1}{k^2},$$

$$D^{(1)}(x)=\frac{1}{(2\pi)^3}\int(dk)e^{ikx}\delta(k^2), \quad (1.66)$$

in which the principal part of $1/(k^2+\kappa^2)$ and $1/k^2$ is understood. We have employed the simplified notation ab to denote $a_\mu b_\mu$, the scalar product of two four-vectors. The functions (1.63) have the following Fourier integral representations,

$$S_\pm(x)=\frac{1}{(2\pi)^4}\int(dk)e^{ikx}(i\gamma k-\kappa)\left(\frac{1}{k^2+\kappa^2}\pm\pi i\delta(k^2+\kappa^2)\right),$$

$$D_\pm(x)=\frac{1}{(2\pi)^4}\int(dk)e^{ikx}\left(\frac{1}{k^2}\pm\pi i\delta(k^2)\right). \quad (1.67)$$

These expressions can be written more compactly by observing that

$$\lim_{\epsilon\to+0}\frac{1}{\xi\mp i\epsilon}=\lim_{\epsilon\to+0}\left(\frac{\xi}{\xi^2+\epsilon^2}\pm i\frac{\epsilon}{\xi^2+\epsilon^2}\right)$$

$$=P\frac{1}{\xi}\pm\pi i\delta(\xi), \quad (1.68)$$

whence,

$$S_{\pm}(x)=\frac{1}{(2\pi)^4}\int(dk)e^{ikx}(i\gamma k-\kappa)\frac{1}{k^2+\kappa^2\mp i\epsilon},$$

$$D_{\pm}(x)=\frac{1}{(2\pi)^4}\int(dk)e^{ikx}\frac{1}{k^2\mp i\epsilon},\tag{1.69}$$

in which the limit $\epsilon\to+0$ is understood.

The form obtained for $K_\mu{}^{(1)}$ from (1.61), is

$$K_\mu{}^{(1)}(x'-x,x-x'')=\frac{1}{(2\pi)^{11}}\int(dk)(dk')(dk'')e^{i(k+k')(x'-x)}e^{i(k+k'')(x-x'')}$$
$$\times\gamma_\nu(i\gamma k'-\kappa)\gamma_\mu(i\gamma k''-\kappa)\gamma_\nu\left[\frac{\delta(k'^2+\kappa^2)}{(k''^2+\kappa^2)k^2}+\frac{\delta(k''^2+\kappa^2)}{(k'^2+\kappa^2)k^2}+\frac{\delta(k^2)}{(k'^2+\kappa^2)(k''^2+\kappa^2)}\right].\tag{1.70}$$

It is convenient to replace k_μ' and k_μ'' by

$$p_\mu'=k_\mu+k_\mu',\quad p_\mu''=k_\mu+k_\mu'',\tag{1.71}$$

which enter directly in the coordinate dependence of the Fourier integral. Since $K_\mu{}^{(1)}(x'-x,x-x'')$ is to be multiplied by $\bar\psi(x')$, $\psi(x'')$ and integrated with respect to x' and x'', only such values of p_μ' and p_μ'' occur for which

$$p'^2+\kappa^2=p''^2+\kappa^2=0.\tag{1.72}$$

As a result of this transformation,

$$K_\mu{}^{(1)}(x'-x,x-x'')=\frac{1}{(2\pi)^{11}}\int(dk)(dp')(dp'')e^{ip'(x'-x)}e^{ip''(x-x'')}\gamma_\nu(i\gamma(p'-k)-\kappa)\gamma_\mu(i\gamma(p''-k)-\kappa)\gamma_\nu$$
$$\times\left[\frac{\delta(k^2-2kp')}{(2kp'-2kp'')(2kp')}+\frac{\delta(k^2-2kp'')}{(2kp''-2kp')(2kp'')}+\frac{\delta(k^2)}{(2kp')(2kp'')}\right].\tag{1.73}$$

The last factor in (1.73) can be simplified by writing it as

$$\frac{1}{2k(p'-p'')}\left[\frac{1}{2kp'}(\delta(k^2-2kp')-\delta(k^2))-\frac{1}{2kp''}(\delta(k^2-2kp'')-\delta(k^2))\right],\tag{1.74}$$

and observing that

$$\frac{1}{2kp}(\delta(k^2-2kp)-\delta(k^2))=-\int_0^1du\delta'(k^2-2kpu),\tag{1.75}$$

whence (1.74) becomes

$$-\frac{1}{2k(p'-p'')}\int_0^1du[\delta'(k^2-2kp'u)-\delta'(k^2-2kp''u)].\tag{1.76}$$

This, in turn, can be represented more compactly as

$$\frac{1}{2}\int_{-1}^1dv\int_0^1udu\delta''(k^2-k(p'+p''+(p'-p'')v)u).\tag{1.77}$$

Therefore,

$$K_\mu{}^{(1)}(x'-x,x-x'')=\frac{1}{2(2\pi)^{11}}\int_{-1}^1dv\int_0^1udu\int(dk)(dp')(dp'')e^{ip'(x'-x)}e^{ip''(x-x'')}$$
$$\times\gamma_\nu(i\gamma(p'-k)-\kappa)\gamma_\mu(i\gamma(p''-k)-\kappa)\gamma_\nu\delta''(k^2-k(p'+p''+(p'-p'')v)u).\tag{1.78}$$

If the expression (1.64) is employed for $K_\mu{}^{(1)}$, the bracketed factors in (1.70) and (1.73) are replaced by

$$\frac{1}{\pi}Im\frac{1}{k'^2+\kappa^2-i\epsilon}\,\frac{1}{k''^2+\kappa^2-i\epsilon}\,\frac{1}{k^2-i\epsilon}=\frac{1}{\pi}Im\frac{1}{k^2-2kp'-i\epsilon}\,\frac{1}{k^2-2kp''-i\epsilon}\,\frac{1}{k^2-i\epsilon}.\tag{1.79}$$

However,

$$\frac{1}{k^2-2kp'-i\epsilon}\,\frac{1}{k^2-2kp''-i\epsilon}\,\frac{1}{k^2-i\epsilon}=\frac{1}{2k(p'-p'')}\left[\frac{1}{2kp'}\left(\frac{1}{k^2-2kp'-i\epsilon}-\frac{1}{k^2-i\epsilon}\right)\right.$$
$$\left.-\frac{1}{2kp''}\left(\frac{1}{k^2-2kp''-i\epsilon}-\frac{1}{k^2-i\epsilon}\right)\right], \quad (1.80)$$

and, on extracting the imaginary part divided by π, we again encounter (1.74).

The second part of K_μ, (1.62), can also be readily expressed in Fourier integral form

$$K_\mu^{(2)}(x'-x,x-x'')=\frac{1}{(2\pi)^{11}}\int(dk)(dp')(dp'')e^{ip'(x'-x)}e^{ip''(x-x'')}\left[\frac{1}{2\kappa}p_\lambda'\frac{\partial}{\partial p_\lambda'}\gamma_\nu(i\gamma(p'-k)-\kappa)\gamma_\nu\left(\frac{\delta((k-p')^2+\kappa^2)}{k^2}\right.\right.$$
$$\left.\left.+\frac{\delta(k^2)}{(k-p')^2+\kappa^2}\right)\gamma_\mu+\gamma_\mu\frac{1}{2\kappa}p_\lambda''\frac{\partial}{\partial p_\lambda''}\gamma_\nu(i\gamma(p''-k)-\kappa)\gamma_\nu\left(\frac{\delta((k-p'')^2+\kappa^2)}{k^2}+\frac{\delta(k^2)}{(k-p'')^2+\kappa^2}\right)\right]. \quad (1.81)$$

To evaluate the derivatives with respect to p_λ' and p_λ'', we observe that

$$p_\lambda\frac{\partial}{\partial p_\lambda}(i\gamma(p-k)+\kappa)(i\gamma(p-k)-\kappa)f((p-k)^2+\kappa^2)=0, \quad (1.82)$$

where $f(x)$ is $\delta(x)$ or $1/x$. On differentiating and multiplying to the left by $i\gamma(p-k)-\kappa$, we obtain

$$p_\lambda\frac{\partial}{\partial p_\lambda}(i\gamma(p-k)-\kappa)f((p-k)^2+\kappa^2)=(i\gamma(p-k)-\kappa)i\gamma p(i\gamma(p-k)-\kappa)\frac{f(k^2-2kp)}{k^2-2kp}. \quad (1.83)$$

Consequently,

$$p_\lambda\frac{\partial}{\partial p_\lambda}\gamma_\nu(i\gamma(p-k)-\kappa)\gamma_\nu\left(\frac{\delta((p-k)^2+\kappa^2)}{k^2}+\frac{\delta(k^2)}{(p-k)^2+\kappa^2}\right)$$
$$=-\gamma_\nu(i\gamma(p-k)-\kappa)i\gamma p(i\gamma(p-k)-\kappa)\gamma_\nu\left(\frac{\delta'(k^2-2kp)}{k^2}-\frac{(\delta k^2)}{(2kp)^2}\right), \quad (1.84)$$

in virtue of the delta-function property

$$\delta'(x)=-\frac{\delta(x)}{x}. \quad (1.85)$$

Furthermore,

$$\frac{\delta'(k^2-2kp)}{k^2}-\frac{\delta(k^2)}{(2kp)^2}=-\frac{\partial}{\partial(2kp)}\left(\frac{\delta(k^2-2kp)}{k^2}-\frac{\delta(k^2)}{2kp}\right)=-\int_0^1 udu\delta''(k^2-2kpu), \quad (1.86)$$

according to (1.75). Therefore, (1.81) becomes

$$K_\mu^{(2)}(x'-x,x-x'')=\frac{1}{(2\pi)^{11}}\int_0^1 udu\int(dk)(dp')(dp'')e^{ip'(x'-x)}e^{ip''(x-x'')}$$
$$\times\left[\delta''(k^2-2kp'u)\frac{1}{2\kappa}\gamma_\nu(i\gamma(p'-k)-\kappa)i\gamma p'(i\gamma(p'-k)-\kappa)\gamma_\nu\gamma_\mu\right.$$
$$\left.+\gamma_\mu\frac{1}{2\kappa}\gamma_\nu(i\gamma(p'-k)-\kappa)i\gamma p''(i\gamma(p''-k)-\kappa)\gamma_\nu\delta''(k^2-2kp''u)\right]. \quad (1.87)$$

The transformation

$$k_\mu\rightarrow k_\mu+(p_\mu'+p_\mu''+(p_\mu'-p_\mu'')v)\frac{u}{2} \quad (1.88)$$

now brings the delta-function of (1.78) into the form

$$\delta''(k^2+\lambda^2u^2), \quad (1.89)$$

where

$$\lambda^2=\kappa^2\left(1+\frac{(p'-p'')^2}{4\kappa^2}(1-v^2)\right), \tag{1.90}$$

in virtue of the relations

$$\left(\frac{p'+p''}{2}\right)^2+\left(\frac{p'-p''}{2}\right)^2+\kappa^2=0, \quad (p'+p'')(p'-p'')=0. \tag{1.91}$$

As a consequence of this transformation, the factor involving the Dirac matrices in (1.78) becomes

$$\gamma_\nu\left(i\gamma\left(p'-\frac{p'+p''}{2}u-\frac{p'-p''}{2}uv\right)-\kappa\right)\gamma_\mu\left(i\gamma\left(p''-\frac{p'+p''}{2}u-\frac{p'-p''}{2}uv\right)-\kappa\right)\gamma_\nu-\gamma_\mu k^2. \tag{1.92}$$

In writing this result we have exploited the symmetry of the delta-function (1.89), in connection with the k integration, and discarded terms linear in k_λ, while replacing $k_\lambda k_\sigma$ by $\frac{1}{4}\delta_{\lambda\sigma}k^2$. The following property of the Dirac matrices has also been used,

$$\gamma_\lambda\gamma_\mu\gamma_\lambda=-2\gamma_\mu. \tag{1.93}$$

The factor (1.92) can be further simplified by omitting the terms linear in v, which will vanish on integration, and rearranging the remaining terms to obtain

$$\begin{aligned}&4\kappa^2\gamma_\mu(1-u-\tfrac{1}{2}u^2)-\gamma_\mu k^2+2\kappa(u-u^2)\sigma_{\mu\nu}(p_\nu'-p_\nu'')+2(p'-p'')^2\gamma_\mu\left(1-u+\frac{1-v^2}{4}u^2\right)\\&+i(1-u^2v^2)(p_\mu'-p_\mu'')((i\gamma p'+\kappa)-(i\gamma p''+\kappa))-2(1-u)\left[(i\gamma p'+\kappa)\left(\kappa(1+u)\gamma_\mu+ip_\mu''+i\frac{1-u}{2}(p_\mu'+p_\mu'')\right)\right.\\&\left.-(i\gamma p'+\kappa)\gamma_\mu(i\gamma p''+\kappa)+\left(\kappa(1+u)\gamma_\mu+ip_\mu'+i\frac{1-u}{2}(p_\mu'+p_\mu'')\right)(i\gamma p''+\kappa)\right].\end{aligned} \tag{1.94}$$

Now, a right-hand factor $i\gamma p''+\kappa$ is equivalent to $-\gamma_\lambda(\partial/\partial x_\lambda'')+\kappa$ operating on $K_\mu(x'-x,x-x'')$, which annihilates $\psi(x'')$ on integration by parts. Similarly, a left-hand factor $i\gamma p'+\kappa$ annihilates $\bar\psi(x')$. As a consequence of the Dirac equation, therefore,

$$\begin{aligned}K_\mu^{(1)}(x'-x,x-x'')=&\frac{1}{(2\pi)^{11}}\int_{-1}^{1}dv\int_0^1 udu\int(dk)(dp')(dp'')e^{ip'(x'-x)}e^{ip''(x-x'')}\delta''(k^2+\lambda^2u^2)\\&\times\left[2\kappa^2\gamma_\mu(1-u-\tfrac{1}{2}u^2)-\tfrac{1}{2}\gamma_\mu k^2+\kappa(u-u^2)\sigma_{\mu\nu}(p_\nu'-p_\nu'')+(p'-p'')^2\gamma_\mu\left(1-u+\frac{1-v^2}{4}u^2\right)\right].\end{aligned} \tag{1.95}$$

Transformations analogous to (1.88) can be introduced in the two terms of (1.87), namely $k_\mu\to k_\mu+p_\mu'u$, and $k_\mu\to k_\mu+p_\mu''u$. Both delta functions then become $\delta''(k^2+\kappa^2u^2)$, while the factors involving the Dirac matrices simplify according to

$$\frac{1}{2\kappa}\gamma_\nu(i\gamma(p-k)-\kappa)i\gamma p(i\gamma(p-k)-\kappa)\gamma_\nu\to-2\kappa^2(1-u-\tfrac{1}{2}u^2)+\tfrac{1}{2}k^2-(2\kappa^2(1-u+\tfrac{1}{2}u^2)+\tfrac{1}{2}k^2)\frac{1}{\kappa}(i\gamma p+\kappa), \tag{1.96}$$

where p is p' or p'' for the two terms of (1.87). In consequence of the Dirac equation, therefore,

$$\begin{aligned}K_\mu^{(2)}(x'-x,x-x'')=-&\frac{2}{(2\pi)^{11}}\int_0^1 udu\int(dk)(dp')(dp'')e^{ip'(x'-x)}e^{p''(x-x'')}\delta''(k^2+\kappa^2u^2)\\&\times[2\kappa^2\gamma_\mu(1-u-\tfrac{1}{2}u^2)-\tfrac{1}{2}\gamma_\mu k^2].\end{aligned} \tag{1.97}$$

To combine $K_\mu^{(1)}$ and $K_\mu^{(2)}$, it is sufficient to perform an integration by parts with respect to v for the first two terms of (1.95), as indicated by

$$\int_{-1}^{1}dv\delta''(k^2+\lambda^2u^2)=2\delta''(k^2+\kappa^2u^2)-\int_{1}^{1}dvv\frac{\partial}{\partial v}\delta''(k^2+\lambda^2u^2). \tag{1.98}$$

The integrated terms precisely cancel $K_\mu^{(2)}$. If the v differentiation is explicitly performed for the second term of (1.95), the k integral thus encountered is

$$\int (dk)k^2\delta'''(k^2+\lambda^2u^2)=\frac{1}{2}\int (dk)k_\nu\frac{\partial}{\partial k_\nu}\delta''(k^2+\lambda^2u^2)$$
$$=-2\int (dk)\delta''(k^2+\lambda^2u^2). \tag{1.99}$$

Hence,

$$K_\mu(x'-x,x-x'')=\frac{1}{(2\pi)^{11}}\int_{-1}^{1} dv\int_0^1 udu\int (dk)(dp')(dp'')e^{ip'(x'-x)}e^{ip''(x-x'')}\bigg[(p'-p'')^2\gamma_\mu\bigg(1-u+\frac{1+v^2}{4}u^2\bigg)$$
$$+\kappa(u-u^2)\sigma_{\mu\nu}(p_\nu'-p_\nu'')-2\kappa^2\gamma_\mu(1-u-\tfrac{1}{2}u^2)v\frac{\partial}{\partial v}\bigg]\delta''(k^2+\lambda^2u^2). \tag{1.100}$$

The integration with respect to k may now be effected. According to the integral representation,

$$\delta(k^2+\lambda^2u^2)=\frac{1}{2\pi}\int_{-\infty}^{\infty} dwe^{iw(k^2+\lambda^2u^2)}, \tag{1.101}$$

we have

$$\int (dk)\delta''(k^2+\lambda^2u^2)=-\frac{1}{2\pi}\int_{-\infty}^{\infty} w^2dwe^{iw\lambda^2u^2}\int (dk)e^{iwk^2}$$
$$=-\frac{\pi i}{2}\int_{-\infty}^{\infty} dw\frac{w}{|w|}e^{iw\lambda^2u^2}$$
$$=\frac{\pi}{\lambda^2u^2}. \tag{1.102}$$

However, it should be noticed that we are then required to evaluate integrals with respect to u of the form

$$\int_0^1 u^{n+1}du\int (dk)\delta''(k^2+\lambda^2u^2)=\frac{\pi}{\lambda^2}\int_0^1 u^{n-1}du, \tag{1.103}$$

in which n may be 0, 1 or 2. For $n=0$, the integral is logarithmically divergent.

In order to ascertain the significance of this divergence, we shall interchange the operations used in obtaining (1.103), thus producing a more easily interpreted divergent k integral. For $n=0$, (1.103) reads

$$\frac{1}{2\lambda^2}\int (dk)\int_0^1 du\frac{\partial}{\partial u}\delta'(k^2+\lambda^2u^2)=\frac{1}{2\lambda^2}\int (dk)[\delta'(k^2+\lambda^2)-\delta'(k^2)]. \tag{1.104}$$

One may express this invariant integral, in three-dimensional notation, as

$$\frac{1}{2\lambda^2}\int (\mathbf{dk})dk_0\frac{1}{2k_0}\frac{\partial}{\partial k_0}[\delta(k_0^2-\mathbf{k}^2)-\delta(k_0^2-\mathbf{k}^2-\lambda^2)]=\frac{1}{4\lambda^2}\int (\mathbf{dk})dk_0\frac{1}{k_0^2}[\delta(k_0^2-\mathbf{k}^2)-\delta(k_0^2-\mathbf{k}^2-\lambda^2)], \tag{1.105}$$

in which the delta-functions describe the energy-momentum relations of a light quantum, and of a particle with mass $\hbar\lambda/c$. On performing the k_0 integration, (1.105) becomes

$$\frac{1}{4\lambda^2}\int (\mathbf{dk})\bigg[\frac{1}{|\mathbf{k}|^3}-\frac{1}{(\mathbf{k}^2+\lambda^2)^{\frac{3}{2}}}\bigg]=\frac{\pi}{\lambda^2}\bigg[\int_0^\infty dk\bigg(\frac{1}{k}-\frac{1}{(k^2+\lambda^2)^{\frac{1}{2}}}\bigg)+1\bigg], \tag{1.106}$$

in which form it is evident that the divergence is associated with zero frequency light quanta—an "infra-red catastrophe." As we shall later demonstrate, this divergence is entirely spurious, and is removed on properly including the effects of $\delta j_\mu^{(1)}(x)$, the first-order correction to the current operator. The divergent integral (1.106) can be expressed in terms of an invariant minimum light quantum wave number, $k_{\min}$, as

$$\frac{\pi}{\lambda^2}\left(\log\frac{\lambda}{2k_{\min}}+1\right). \tag{1.107}$$

With the k and u integrations thus performed, $K_\mu(x'-x,x-x'')$ becomes

$$K_\mu(x'-x,x-x'')=\frac{1}{(2\pi)^{10}}\int_0^1 dv\int(dp')(dp'')\exp\left[i\frac{p'+p''}{2}(x'-x'')\right]\exp\left[i(p''-p')\left(x-\frac{x'+x''}{2}\right)\right]$$
$$\times\left[\frac{(p'-p'')^2}{\kappa^2}\gamma_\mu\frac{1+v^2}{2}\left(\log\frac{\kappa}{2k_{\min}}+1+\tfrac{1}{2}\log\left(1+\frac{(p'-p'')^2}{4\kappa^2}(1-v^2)\right)\right)-\frac{(p'-p'')^2}{4\kappa^2}\gamma_\mu\right.$$
$$\left.+\frac{1}{2\kappa}\sigma_{\mu\nu}(p_\nu'-p_\nu'')\right]\frac{1}{1+((p'-p'')^2/4\kappa^2)(1-v^2)}, \tag{1.108}$$

in which we have evaluated the third term of (1.100) by writing

$$\frac{1}{1+((p'-p'')^2/4\kappa^2)(1-v^2)}=1-\frac{((p'-p'')^2/4\kappa^2)(1-v^2)}{1+((p'-p'')^2/4\kappa^2)(1-v^2)}, \tag{1.109}$$

and performing the v differentiation for the term obtained from the first part of (1.109), while reversing the integration by parts for the term produced by the second part of (1.109). It will now be observed that the integrand of (1.108) involves only $p_\lambda'-p_\lambda''$. It is then useful to introduce the new variables

$$P_\lambda=\frac{p_\lambda'+p_\lambda''}{2},\quad p_\lambda=p_\lambda''-p_\lambda', \tag{1.110}$$

since the P integration can be immediately performed, yielding $\delta(x'-x'')$. In this way, we obtain

$$K_\mu(x'-x,x-x'')=-\frac{1}{8\pi^2}\gamma_\mu\delta(x'-x'')\frac{1}{\kappa^2}\Box^2\left[\log\frac{\kappa}{2k_{\min}}(F_0(x-x')+F_1(x-x'))\right.$$
$$\left.+\tfrac{1}{2}F_0(x-x')+F_1(x-x')+\tfrac{1}{2}G(x-x')\right]+\frac{i}{8\pi^2}\delta(x'-x'')\sigma_{\mu\nu}\frac{\partial}{\partial x_\nu}F_0(x-x'), \tag{1.111}$$

where

$$F_n(x)=\int_0^1 dv v^{2n}\frac{1}{(2\pi)^4}\int(dp)\frac{e^{ipx}}{1+(p^2/4k^2)(1-v^2)}$$
$$=16\kappa^2\int_0^1 dv\frac{v^{2n}}{(1-v^2)^2}\bar{\Delta}\left(\frac{2}{(1-v^2)^{\frac{1}{2}}}x\right) \tag{1.112}$$

and

$$G(x)=\int_0^1 dv(1+v^2)\frac{1}{(2\pi)^4}\int(dp)e^{ipx}\frac{\log(1+(p^2/4\kappa^2)(1-v^2))}{1+(p^2/4\kappa^2)(1-v^2)}$$
$$=8\Box^2\int_0^1 dv\frac{1+v^2}{1-v^2}\int_0^1\frac{udu}{1-u^2}\left[\frac{1}{u^2}\bar{\Delta}\left(\frac{2}{(1-v^2)^{\frac{1}{2}}u}x\right)-\bar{\Delta}\left(\frac{2}{(1-v^2)^{\frac{1}{2}}}x\right)\right]. \tag{1.113}$$

Finally, then,

$$-\frac{ie^3}{2h}\int d\omega'd\omega''(\bar{\psi}(x')K_\mu(x'-x,x-x'')\psi(x''))_1=\frac{\alpha}{4\pi}\log\frac{\kappa}{2k_{\min}}\frac{1}{\kappa^2}\Box^2\int[F_0(x-x')+F_1(x-x')]j_\mu(x')d\omega'$$
$$+\frac{\alpha}{4\pi}\frac{1}{\kappa^2}\Box^2\int[\tfrac{1}{2}F_0(x-x')+F_1(x-x')+\tfrac{1}{2}G(x-x')]j_\mu(x')d\omega'+\frac{\alpha}{2\pi}c\frac{\partial}{\partial x_\nu}\int F_0(x-x')m_{\mu\nu}(x')d\omega', \tag{1.114}$$

in which

$$m_{\mu\nu}(x)=\frac{e}{2\kappa}(\bar{\psi}(x)\sigma_{\mu\nu}\psi(x))_1$$

$$=\frac{e}{2\kappa}\tfrac{1}{2}[\bar{\psi}(x)\sigma_{\mu\nu}\psi(x)-\bar{\psi}'(x)\sigma_{\mu\nu}\psi'(x)]. \qquad (1.115)$$

Expressed in the same notation, the first term of (1.58) is (see II (2.44)),

$$\frac{i}{2\hbar c^2}\int d\omega'\epsilon(x-x')[j_\mu(x),j_\nu(x')]_0\delta A_\nu(x')=-\frac{\alpha}{4\pi}\frac{1}{\kappa^2}\Box^2\int[F_1(x-x')-\tfrac{1}{3}F_2(x-x')]j_\mu(x')d\omega', \qquad (1.116)$$

from which we have omitted the charge renormalization term, with the understanding that the value of e is to be correspondingly altered. A rederivation of this result, employing methods akin to those presented in this paper, is given in the Appendix. Evidently the new contributions to the one particle current operator, as given in (1.114), are of the same general nature as the previously considered effect, (1.116), with the exception of the last term in (1.114). This is an addition to the current vector of the form

$$c(\partial/\partial x)_\nu\delta m_{\mu\nu}(x), \qquad (1.117)$$

where

$$\delta m_{\mu\nu}(x)=\alpha/2\pi\int F_0(x-x')m_{\mu\nu}(x')d\omega'. \qquad (1.118)$$

A current vector of this type can be interpreted as a dipole current, derived from an antisymmetrical dipole tensor $\delta m_{\mu\nu}$ which combines electric and magnetic dipole moment densities. The tensor $m_{\mu\nu}$ is that characteristic of the Dirac theory, in which intrinsic dipole moments are related to the antisymmetrical spin tensor $\sigma_{\mu\nu}$, the factor of proportionality being

$$\mu_0=e/2\kappa=e\hbar/2mc, \qquad (1.119)$$

the Bohr magneton. According to (1.118), the correction to the dipole tensor at a point involves an average of $m_{\mu\nu}$ over the vicinity of that point. If all quantities are slowly varying, relative to $\hbar/mc$ and $\hbar/mc^2$ as units of length and time, an expansion in ascending powers of $\Box^2$ can be constructed, as in II (2.47). For this purpose, it is sufficient to expand the denominator in the first form of (1.112), thus obtaining

$$F_n(x)=\frac{1}{2n+1}\delta(x)+\frac{1}{(2n+1)(2n+3)}\frac{1}{2\kappa^2}\Box^2\delta(x)+\cdots. \qquad (1.120)$$

Hence,

$$\delta m_{\mu\nu}(x)=\frac{\alpha}{2\pi}\left[m_{\mu\nu}(x)+\frac{1}{6\kappa^2}\Box^2 m_{\mu\nu}(x)+\cdots\right], \qquad (1.121)$$

and, under conditions that permit the neglect of all but the first term in this series, an electron will act as though it possessed an additional spin magnetic moment[3]

$$\delta\mu=(\alpha/2\pi)\mu_0. \qquad (1.122)$$

The comparison of this prediction with experiment will be discussed in the sequel to this paper.

The final result for the second-order correction to the one particle current operator is

$$(\delta j_\mu^{(2)}(x))_{1,0}=\frac{\alpha}{4\pi}\log\frac{\kappa}{2k_{\min}}\frac{1}{\kappa^2}\Box^2\int[F_0(x-x')+F_1(x-x')]j_\mu(x')d\omega'$$
$$+\frac{\alpha}{4\pi}\frac{1}{\kappa^2}\Box^2\int[\tfrac{1}{2}F_0(x-x')+\tfrac{1}{3}F_2(x-x')+\tfrac{1}{2}G(x-x')]j_\mu(x')d\omega'+c\frac{\partial}{\partial x_\nu}\delta m_{\mu\nu}(x). \qquad (1.123)$$

Under conditions of slow variation $((1/\kappa^2)\Box^2 j_\mu,\ m_{\mu\nu}\ll j_\mu, m_{\mu\nu})$, this reduces to

$$(\delta j_\mu^{(2)}(x))_{1,0}=\frac{\alpha}{3\pi}\left(\log\frac{\kappa}{2k_{\min}}+\frac{17}{40}\right)\frac{1}{\kappa^2}\Box^2 j_\mu(x)+\frac{\alpha}{2\pi}c\frac{\partial}{\partial x_\nu}m_{\mu\nu}(x), \qquad (1.124)$$

in virtue of (1.120), and the analogous expansion of $G(x)$,

$$G(x)=-\frac{1}{5\kappa^2}\Box^2\delta(x)+\cdots. \qquad (1.125)$$

It will be noted that the total charge computed from $(\delta j_\mu^{(2)}(x))_{1,0}$ is zero, in agreement with evident charge conservation requirements, and the formal property that the operator of total charge commutes with all

[3] This result was announced at the January, 1948 meeting of the American Physical Society. The formula is misprinted in a published note, J. Schwinger, Phys. Rev. 73, 416 (1948). The misprint has unfortunately been copied by L. Rosenfeld in his book, *Nuclear Forces* (Interscience Publishers, Inc., New York, 1949), p. 438.

one-particle operators. The apparent contradiction between these statements and the existence of the charge renormalization term is discussed in the Appendix, where it is shown that a compensating charge is created at infinity.

Our result, (1.123), is marred only by the appearance of the logarithmic divergence associated with zero frequency quanta. It should be remarked, however, that $(\delta j_\mu^{(2)}(x))_{1,0}$ is not a complete description of the radiative corrections under discussion. In order to measure the correction to the current, it is necessary to impose an external field. This will induce the emission of quanta, as described by $\delta j_\mu^{(1)}(x)$, among the effects of which is a compensating low frequency divergence. It will be apparent that, as a consequence of the "infra-red catastrophe," the second-order corrections to particle electromagnetic properties cannot be completely stated without regard for the manner of exhibiting them by an external field. We therefore turn to a discussion of the behavior of a single particle in an external field, as modified by the vacuum fluctuations of the electromagnetic field.

2. RADIATIVE CORRECTIONS TO ELECTRON SCATTERING

We shall now be concerned with the interaction of three systems—the matter field, the electromagnetic field, and a given current distribution. The latter may be associated with a nucleus or a macroscopic apparatus, two situations in which the reaction on the current distribution may have a negligible effect. A description of this state of affairs, in the interaction representation, is given by

$$ihc\frac{\delta\Psi[\sigma]}{\delta\sigma(x)}=\left[-\frac{1}{c}(j_\mu(x)+J_\mu(x))A_\mu(x)\right]\Psi[\sigma], \tag{2.1a}$$

$$\left[\frac{\partial A_\mu(x')}{\partial x_\mu'}-\frac{1}{c}\int_\sigma D(x'-x)(j_\mu(x)+J_\mu(x))d\sigma_\mu\right]\Psi[\sigma]=0, \tag{2.1b}$$

where $j_\mu(x)$ and $J_\mu(x)$ are the current vectors associated with the matter field and the external system, respectively. Both current distributions are coupled to the electromagnetic field, as characterized by $A_\mu(x)$. An equally valid way of stating matters is in terms of an external electromagnetic field acting on the matter field current distribution:

$$ihc\frac{\delta\Psi[\sigma]}{\delta\sigma(x)}=\left[-\frac{1}{c}j_\mu(x)(A_\mu(x)+A_\mu^{(e)}(x))\right]\Psi[\sigma], \tag{2.2a}$$

$$\left[\frac{\partial A_\mu(x')}{\partial x_\mu'}-\frac{1}{c}\int_\sigma D(x'-x)j_\mu(x)d\sigma_\mu\right]\Psi[\sigma]=0, \tag{2.2b}$$

where

$$\Box^2 A_\mu^{(e)}(x)=-\frac{1}{c}J_\mu(x),\quad \frac{\partial A_\mu^{(e)}(x)}{\partial x_\mu}=0. \tag{2.3}$$

The equivalence of the two descriptions is established by showing that (2.2) is obtained from (2.1) by a canonical transformation, namely,

$$\Psi[\sigma]\rightarrow e^{-iJ[\sigma]}\Psi[\sigma], \tag{2.4}$$

with $J[\sigma]$ determined by

$$hc\frac{\delta J[\sigma]}{\delta\sigma(x)}=-\frac{1}{c}J_\mu(x)A_\mu(x). \tag{2.5}$$

The functional $J[\sigma]$ is explicitly exhibited as

$$J[\sigma]=-\frac{1}{hc^2}\int_{-\infty}^{\sigma}J_\mu(x')A_\mu(x')d\omega', \tag{2.6}$$

in which the choice of lower limit corresponds to selecting the retarded potentials for the electromagnetic field generated by the given current distribution. The equation of motion satisfied by the new state vector is

$$ihc\frac{\delta\Psi[\sigma]}{\delta\sigma(x)}+ihce^{iJ[\sigma]}\frac{\delta e^{-iJ[\sigma]}}{\delta\sigma(x)}\Psi[\sigma]=\left[-\frac{1}{c}(j_\mu(x)+J_\mu(x))e^{iJ[\sigma]}A_\mu(x)e^{-iJ[\sigma]}\right]\Psi[\sigma]. \tag{2.7}$$

Now

$$e^{iJ[\sigma]}A_\mu(x)e^{-iJ[\sigma]}=A_\mu(x)+i[J[\sigma],A_\mu(x)]-\tfrac{1}{2}[J[\sigma],[J[\sigma],A_\mu(x)]]+\cdots$$
$$=A_\mu(x)-\frac{1}{c}\int_{-\infty}^{\sigma}D(x-x')J_\mu(x')d\omega'$$
$$=A_\mu(x)+A_\mu^{(e)}(x),\qquad(2.8)$$

in which the series ends after two terms since the components of $J_\mu(x)$ are mutually commutative, in view of the prescribed nature of this current distribution. It is easily seen that

$$A_\mu^{(e)}(x)=-\frac{1}{c}\int_{-\infty}^{\sigma}D(x-x')J_\mu(x')d\omega'\qquad(2.9)$$

obeys (2.3). Indeed,

$$\Box^2 A_\mu^{(e)}(x)=-\frac{1}{c}\frac{\partial}{\partial x_\nu}\int_{-\infty}^{\sigma}\frac{\partial D(x-x')}{\partial x_\nu'}J_\mu(x')d\omega'$$
$$=-\frac{1}{c}\int_\sigma d\sigma_\nu'\frac{\partial D(x-x')}{\partial x_\nu'}J_\mu(x')$$
$$=-\frac{1}{c}J_\mu(x),\qquad(2.10)$$

and

$$\frac{\partial A_\mu^{(e)}(x)}{\partial x_\mu}=\frac{1}{c}\int_{-\infty}^{\sigma}d\omega'\frac{\partial}{\partial x_\mu'}(D(x-x')J_\mu(x'))$$
$$=0.\qquad(2.11)$$

Furthermore,

$$i\hbar c e^{iJ[\sigma]}\frac{\delta e^{-iJ[\sigma]}}{\delta\sigma(x)}=\hbar c\frac{\delta J[\sigma]}{\delta\sigma(x)}+\frac{i\hbar c}{2}\left[J[\sigma],\frac{\delta J[\sigma]}{\delta\sigma(x)}\right]+\cdots$$
$$=-\frac{1}{c}J_\mu(x)A_\mu(x)-\frac{1}{2c}J_\mu(x)A_\mu^{(e)}(x),\qquad(2.12)$$

and the transformed equation of motion therefore reads

$$i\hbar c\frac{\delta\Psi[\sigma]}{\delta\sigma(x)}=\left[-\frac{1}{c}j_\mu(x)(A_\mu(x)+A_\mu^{(e)}(x))-\frac{1}{2c}J_\mu(x)A_\mu^{(e)}(x)\right]\Psi[\sigma],\qquad(2.13)$$

which is equivalent to (2.2a), since the term $-(1/2c)J_\mu(x)A_\mu^{(e)}(x)$, describing the self-action of the given current distribution, has no dynamical consequences and can be omitted.

In a similar way, the supplementary condition (2.1b) is transformed into

$$\left[e^{iJ[\sigma]}\frac{\partial A_\mu(x')}{\partial x_\mu'}e^{-iJ[\sigma]}-\frac{1}{c}\int_\sigma D(x'-x)(j_\mu(x)+J_\mu(x))d\sigma_\mu\right]\Psi[\sigma]=0,\qquad(2.14)$$

wherein

$$e^{iJ[\sigma]}\frac{\partial A_\mu(x')}{\partial x_\mu'}e^{-iJ[\sigma]}=\frac{\partial A_\mu(x')}{\partial x_\mu'}+i\left[J[\sigma],\frac{\partial A_\mu(x')}{\partial x_\mu'}\right]=\frac{\partial A_\mu(x')}{\partial x_\mu'}-\frac{1}{c}\int_{-\infty}^{\sigma}\frac{\partial D(x'-x'')}{\partial x_\mu'}J_\mu(x'')d\omega''.\qquad(2.15)$$

However,

$$-\frac{1}{c}\int_{-\infty}^{\sigma}\frac{\partial D(x'-x'')}{\partial x_\mu'}J_\mu(x'')d\omega''=\frac{1}{c}\int_{-\infty}^{\sigma}\frac{\partial}{\partial x_\mu''}(D(x'-x'')J_\mu(x''))d\omega''=\frac{1}{c}\int_\sigma D(x'-x)J_\mu(x)d\sigma_\mu,\qquad(2.16)$$

which verifies (2.2b).

One can bring (2.2) into a form which enables the results of the previous section to be utilized. The mass renormalization transformation

$$\Psi[\sigma]\to W[\sigma]\Psi[\sigma],\quad i\hbar c\frac{[\delta W\sigma]}{\delta\sigma(x)}=\mathcal{H}_{1,0}(x)W[\sigma],\qquad(2.17)$$

replaces (2.2a) with

$$i\hbar c\frac{\delta\Psi[\sigma]}{\delta\sigma(x)}=[\mathcal{K}(x)+\mathcal{H}^{(e)}(x)]\Psi[\sigma], \tag{2.18}$$

where (see (1.3))

$$\mathcal{K}(x)=\mathcal{H}(x)-\mathcal{H}_{1,0}(x), \tag{2.19}$$

$$\mathcal{H}^{(e)}(x)=-\frac{1}{c}j_\mu(x)A_\mu^{(e)}(x), \tag{2.20}$$

and the Dirac equation for $\psi(x)$ now involves the experimental mass. The further transformation

$$\Psi[\sigma]=U[\sigma]\Phi[\sigma], \tag{2.21}$$

where

$$i\hbar c\frac{\delta U[\sigma]}{\delta\sigma(x)}=\mathcal{K}(x)U[\sigma],\quad U[-\infty]=1, \tag{2.22}$$

is the analog of (1.8), save that $\Phi[\sigma]$ varies in the presence of an external field,

$$\begin{aligned} i\hbar c\frac{\delta\Phi[\sigma]}{\delta\sigma(x)}&=U^{-1}[\sigma]\mathcal{H}^{(e)}(x)U[\sigma]\Phi[\sigma]\\ &=-\frac{1}{c}\mathbf{j}_\mu(x)A_\mu^{(e)}(x)\Phi[\sigma], \end{aligned}\tag{2.23}$$

in response to the coupling with the current operator $\mathbf{j}_\mu(x)$. The latter contains the modifications produced by the vacuum electromagnetic field. The supplementary condition (2.2b) appears as

$$\left[U^{-1}[\sigma]\frac{\partial A_\mu(x')}{\partial x_\mu'}U[\sigma]-\frac{1}{c}\int_\sigma D(x'-x)\mathbf{j}_\mu(x)d\sigma_\mu\right]\Phi[\sigma]=0, \tag{2.24}$$

in consequence of these transformations. However,

$$\begin{aligned} U^{-1}[\sigma]\frac{\partial A_\mu(x')}{\partial x_\mu'}U[\sigma]-\frac{\partial A_\mu(x')}{\partial x_\mu'}&=\int_{-\infty}^{\sigma}d\omega''\frac{\delta}{\delta\sigma''(x'')}\left(U^{-1}[\sigma'']\frac{\partial A_\mu(x')}{\partial x_\mu'}U[\sigma'']\right)\\ &=\frac{i}{\hbar c^2}\int_{-\infty}^{\sigma}d\omega''U^{-1}[\sigma'']\left[\frac{\partial A_\mu(x')}{\partial x_\mu'},A_\nu(x'')\right]j_\nu(x'')U[\sigma'']=\frac{1}{c}\int_{-\infty}^{\sigma}d\omega''\frac{\partial}{\partial x_\mu''}(D(x'-x'')\mathbf{j}_\mu(x''))\\ &=\frac{1}{c}\int_\sigma D(x'-x)\mathbf{j}_\mu(x)d\sigma_\mu, \end{aligned}\tag{2.25}$$

so that the supplementary condition associated with (2.23) is simply

$$\frac{\partial A_\mu(x')}{\partial x_\mu'}\Phi[\sigma]=0. \tag{2.26}$$

As the first application of (2.23), we shall consider the scattering of an electron produced by its interaction with an external field, in which the latter is regarded as a small perturbation.[4] We shall restrict the external potential to be that of a time independent field, which will eventually be specialized to the Coulomb field of a stationary nucleus.

A solution of (2.23) can be constructed in the form

$$\Phi[\sigma]=R[\sigma]\Phi_1, \tag{2.27}$$

where

$$i\hbar c\frac{\delta R[\sigma]}{\delta\sigma(x)}=H(x)R[\sigma], \tag{2.28}$$

and

$$R[\sigma]\to 1,\quad \sigma\to-\infty. \tag{2.29}$$

[4] Radiative corrections to scattering have been discussed by many authors. That a finite correction is obtained after a renormalization of charge and mass was independently observed by Z. Koba and S. Tomonaga, Prog. Theor. Phys. **3**, 290 (1948); H. W. Lewis, Phys. Rev. **73**, 173 (1948); and J. Schwinger, Phys. Rev. **73**, 416 (1948). See also R. P. Feynman, Phys. Rev. **74**, 1430 (1948).

The state vector Φ_1 characterizes the initial state of the system, composed of one electron with definite energy and momentum, and no light quanta. The total probability, per unit time, that a scattering process occurs, can be obtained by evaluating the time rate of decrease of the probability that the system remain in the initial state,

$$w=-c\int dv\frac{\delta}{\delta\sigma(x)}|(\Phi_1,\Phi[\sigma])|^2=-c\int dv\frac{\delta}{\delta\sigma(x)}|(\Phi_1,R[\sigma]\Phi_1)|^2. \tag{2.30}$$

The integration is extended over the surface $t=\text{const.}$, with dv the three-dimensional volume element. Now

$$i\hbar c\frac{\delta}{\delta\sigma(x)}|(\Phi_1,R[\sigma]\Phi_1)|^2=(\Phi_1,R^{-1}[\sigma]\Phi_1)(\Phi_1,H(x)R[\sigma]\Phi_1)-(\Phi_1,R[\sigma]\Phi_1)(\Phi_1,R^{-1}[\sigma]H(x)\Phi_1). \tag{2.31}$$

In view of the treatment of $H(x)$ as a small perturbation, it is sufficient to write

$$R[\sigma]=1-\frac{i}{\hbar c}\int_{-\infty}^{\sigma}H(x')d\omega',\quad R^{-1}[\sigma]=1+\frac{i}{\hbar c}\int_{-\infty}^{\sigma}H(x')d\omega'. \tag{2.32}$$

It will also be useful to introduce

$$H'(x)=H(x)-(\Phi_1,H(x)\Phi_1), \tag{2.33}$$

which possesses a vanishing diagonal matrix element for the initial state, and obtain

$$R[\sigma]=\exp\left[-\frac{i}{\hbar c}\int_{-\infty}^{\sigma}(1|H(x')|1)d\omega'\right]\left(1-\frac{i}{\hbar c}\int_{-\infty}^{\sigma}H'(x')d\omega'\right). \tag{2.34}$$

The phase factor evidently has no effect in (2.31), and can be omitted. The latter is also unaffected if $H(x)$ is replaced by $H'(x)$. Hence to the accuracy of first-order perturbation theory, we have

$$-\frac{\delta}{\delta\sigma(x)}|(\Phi_1,R[\sigma]\Phi_1)|^2=\frac{1}{\hbar^2c^2}\left(1\left|H'(x)\int_{-\infty}^{\sigma}H'(x')d\omega'+\int_{-\infty}^{\sigma}H'(x')d\omega'H'(x)\right|1\right), \tag{2.35}$$

and

$$w=\frac{1}{\hbar^2c}\int dvdv'\left(1\left|H'(x)\int_{-\infty}^{x_0}H'(x')dx_0'+\int_{-\infty}^{x_0}H'(x')dx_0'H'(x)\right|1\right). \tag{2.36}$$

We may now remark that a diagonal matrix element for a state of definite energy must be invariant with respect to time displacements, whence

$$\int dvdv'\left(1\left|\int_{\infty}^{x_0}H'(x')dx_0'H'(x)\right|1\right)=\int dvdv'\left(1\left|H'(x)\int_{x_0}^{\infty}H'(x')dx_0'\right|1\right), \tag{2.37}$$

and

$$w=\frac{1}{\hbar^2c}\int dvdv'\left(1\left|H'(x)\int_{-\infty}^{\infty}H'(x')dx_0'\right|1\right). \tag{2.38}$$

This result is perfectly equivalent to the more conventional perturbation formula in which the rate of transition from the initial state is expressed as a sum of transition rates to all possible final states of equal energy. The energy conservation law is here expressed by the time integration, and the summation over all states other than the original is provided for by the removal from $H(x)$ of the diagonal matrix element. Our basic formula for calculating the transition rate for scattering of a particle by a time independent potential is thus

$$w=\frac{1}{\hbar^2c^3}\int dvdv'A_\mu^{(e)}(\mathbf{r})A_\nu^{(e)}(\mathbf{r}')\left(1\left|\mathbf{j}_\mu(x)\int_{-\infty}^{\infty}\mathbf{j}_\nu(x')dx_0'\right|1\right). \tag{2.39}$$

We have not indicated that the diagonal matrix element is to be subtracted from $\mathbf{j}_\mu(x)$, since it is sufficient to remove, in the final result, those transitions in which no change of state occurs.

We have shown in the first section that, to the second order in e,

$$\mathbf{j}_\mu(x)=j_\mu(x)+\delta j_\mu^{(1)}(x)+\delta j_\mu^{(2)}(x), \tag{2.40}$$

where

$$\delta j_\mu^{(1)}(x)=\frac{i}{2\hbar c^2}\int_{-\infty}^{\infty}\epsilon(x-x')[j_\mu(x),j_\nu(x')]_1 A_\nu(x')d\omega'$$

$$=\frac{ie^2}{\hbar}\int_{-\infty}^{\infty}[\bar\psi(x)\gamma_\mu\bar S(x-x')\gamma_\nu\psi(x')+\bar\psi(x')\gamma_\nu\bar S(x'-x)\gamma_\mu\psi(x)]_1 A_\nu(x')d\omega', \tag{2.41}$$

and

$$(\delta j_\mu^{(2)}(x))_{1,0}=iec\int_{-\infty}^{\infty}[\bar\psi(x')\Gamma_\mu(x-x')\psi(x')]_1 d\omega'. \tag{2.42}$$

Here

$$\Gamma_\mu(x)=\frac{\alpha}{4\pi}\gamma_\mu\log\frac{\kappa}{2k_{\min}}\frac{1}{\kappa^2}\Box^2(F_0(x)+F_1(x))+\frac{\alpha}{4\pi}\gamma_\mu\frac{1}{\kappa^2}\Box^2(\tfrac{1}{2}F_0(x)+\tfrac{1}{3}F_2(x)+\tfrac{1}{2}G(x))-i\frac{\alpha}{4\pi}\sigma_{\mu\nu}\frac{1}{\kappa}\frac{\partial}{\partial x_\nu}F_0(x). \tag{2.43}$$

It is only the indicated portion of $\delta j_\mu^{(2)}(x)$, referring to one particle and no light quanta, that need be retained to compute the second-order correction to the scattering cross section for an external field, since only this part of $\delta j_\mu^{(2)}(x)$ is coherent with $j_\mu(x)$.

The total rate of transition from the initial state can now be written as

$$w=w_0+w_1, \tag{2.44}$$

where

$$w_0=\frac{1}{\hbar^2c^3}\int dvdv'A_\mu^{(e)}(\mathbf{r})A_\nu^{(e)}(\mathbf{r}')\left(1\left|(j_\mu(x)+(\delta j_\mu^{(2)}(x))_{1,0})\int_{-\infty}^{\infty}(j_\nu(x')+(\delta j_\nu^{(2)}(x'))_{1,0})dx_0'\right|1\right) \tag{2.45}$$

describes the rate of radiationless scattering, while

$$w_1=\frac{1}{\hbar^2c^3}\int dvdv'A_\mu^{(e)}(\mathbf{r})A_\nu^{(e)}(\mathbf{r}')\left(1\left|\delta j_\mu^{(1)}(x)\int_{-\infty}^{\infty}\delta j_\nu^{(1)}(x')dx_0'\right|1\right) \tag{2.46}$$

accounts for scattering that is accompanied by single quantum emission.

To indicate the manner in which the perturbation formulas are to be used, we consider the evaluation of

$$\int_{-\infty}^{\infty}(1|j_\mu(x)j_\nu(x')|1)dx_0'. \tag{2.47}$$

This can be written as

$$-e^2c^2\int_{-\infty}^{\infty}(1|\bar\psi(x)\gamma_\mu\psi(x)\bar\psi(x')\gamma_\nu\psi(x')|1)dx_0', \tag{2.48}$$

in which it is understood that one omits the processes in which $\bar\psi(x)\psi(x)$, or $\bar\psi(x')\psi(x')$ induces no change in state. Now $\psi(x')$ can either annul the original electron, in which case $\bar\psi(x')\psi(x')$ causes an electron transition to some final state, or $\psi(x')$ generates a positron, in which event $\bar\psi(x')\psi(x')$ induces the creation of a pair. However, the latter process is incompatible with the energy conservation that is enforced by the time integration, and can therefore be omitted. Hence, it only occurs that $\psi(x')$ annihilates the original electron, whence $\psi(x')\Phi_1$ is a multiple of the vacuum state vector. The same comment applies to $\bar\psi^\dagger(x)\Phi_1=\gamma_4\psi(x)\Phi_1$. Therefore, only the vacuum expectation value of the operator $\psi(x)\bar\psi(x')$ is required in (2.48). Furthermore, since only one state of the matter field is initially excited, as described by the wave function ue^{ipx}, we arrive at the result

$$\int_{-\infty}^{\infty}(1|j_\mu(x)j_\nu(x')|1)dx_0'=\frac{e^2c^2}{(2\pi)^2}\int(dq)\delta(q_0-p_0)\delta(q^2+\kappa^2)\bar u\gamma_\mu(i\gamma q-\kappa)\gamma_\nu u e^{i(\mathbf{p}-\mathbf{q})\cdot(\mathbf{r}'-\mathbf{r})}, \tag{2.49}$$

on employing the relation

$$\langle\psi_\alpha(x)\bar\psi_\beta(x')\rangle_0=-iS_{\alpha\beta}^{(+)}(x-x')=-\frac{1}{(2\pi)^3}\int_{q_0>0}(dq)\delta(q^2+\kappa^2)(i\gamma q-\kappa)_{\alpha\beta}e^{iq(x-x')}. \tag{2.50}$$

Before further simplifying this expression, we shall consider the analogous evaluations of

$$\int_{-\infty}^{\infty}(1|j_\mu(x)(\delta j_\nu^{(2)}(x'))_{1,0}|1)dx_0' \tag{2.51a}$$

and

$$\int_{-\infty}^{\infty}(1|(\delta j_\mu^{(2)}(x))_{1,0}j_\nu(x')|1)dx_0' \tag{2.51b}$$

which describe the radiationless corrections to the scattering process. Now (2.51a) can be written

$$-e^2c^2\int_{-\infty}^{\infty}dx_0'\int d\omega''(1|\bar\psi(x)\gamma_\mu\psi(x)\bar\psi(x'')\Gamma_\nu(x'-x'')\psi(x'')|1), \tag{2.52}$$

which, according to the arguments presented in connection with (2.47), becomes

$$-e^2c^2\int_{-\infty}^{\infty}dx_0'\int d\omega''\bar u\gamma_\mu\langle\psi(x)\bar\psi(x'')\rangle_0\Gamma_\nu(x'-x'')ue^{ip(x''-x)}$$
$$=\frac{e^2c^2}{(2\pi)^3}\int_{-\infty}^{\infty}dx_0'\int d\omega''\int_{q_0>0}(dq)\delta(q^2+\kappa^2)\bar u\gamma_\mu(i\gamma q-\kappa)\Gamma_\nu(x'-x'')ue^{i(p-q)(x''-x)}. \tag{2.53}$$

On introducing the Fourier transform of $\Gamma_\nu(x)$:

$$\Gamma_\nu(p-q)=\int e^{-i(p-q)x}\Gamma_\nu(x)d\omega, \tag{2.54}$$

we obtain

$$\int_{-\infty}^{\infty}(1|j_\mu(x)(\delta j_\nu^{(2)}(x'))_{1,0}|1)dx_0'=\frac{e^2c^2}{(2\pi)^2}\int(dq)\delta(q_0-p_0)\delta(q^2+\kappa^2)\bar u\gamma_\mu(i\gamma q-\kappa)\Gamma_\nu(p-q)ue^{i(\mathbf{p}-\mathbf{q})\cdot(\mathbf{r}'-\mathbf{r})}. \tag{2.55}$$

The result of combining (2.49) and (2.55) with the analogous evaluation of (2.51b) is expressed by

$$w_0=\frac{1}{(2\pi)^2}\frac{e^2}{\hbar^2c}\int dq_0(d\mathbf{q})\delta(q_0-p_0)\delta(q^2+\kappa^2)\int e^{-i(\mathbf{p}-\mathbf{q})\cdot\mathbf{r}}A_\mu^{(e)}(\mathbf{r})dv\int e^{i(\mathbf{p}-\mathbf{q})\cdot\mathbf{r}'}A_\nu^{(e)}(\mathbf{r}')dv'$$
$$\times\bar u(\gamma_\mu+\Gamma_\mu(q-p))(i\gamma q-\kappa)(\gamma_\nu+\Gamma_\nu(p-q))u. \tag{2.56}$$

On performing the integration with respect to q_0 and $|\mathbf{q}|$, we obtain w_0 in the form of an integral extended over all directions of the vector $\mathbf{q}$, other than the incident direction:

$$w_0=\frac{1}{8\pi^2}\frac{e^2}{\hbar^2c}\int d\Omega|\mathbf{p}|\int e^{-i(\mathbf{p}-\mathbf{q})\cdot\mathbf{r}}A_\mu^{(e)}(\mathbf{r})dv\int e^{i(\mathbf{p}-\mathbf{q})\cdot\mathbf{r}'}A_\nu^{(e)}(\mathbf{r}')dv'\bar u(\gamma_\mu+\Gamma_\mu(q-p))(i\gamma q-\kappa)(\gamma_\nu+\Gamma_\nu(p-q))u. \tag{2.57}$$

This must be interpreted as the rate of transition from the initial state, expressed as the probability per unit time for a deflection into an arbitrary element of solid angle. A further simplification can be introduced by averaging (2.57) with respect to the two spin states in which the incident electron may occur. For this purpose, we require the average of $u_\alpha\bar u_\beta$ for the two polarization states associated with a given energy and momentum. It can be inferred from the anticommutator

$$\{\psi_\alpha(x),\bar\psi_\beta(x')\}=\frac{1}{i}S_{\alpha\beta}(x-x')=-\frac{1}{(2\pi)^3}\int(dp)\delta(p^2+\kappa)\epsilon(p)(i\gamma p-\kappa)_{\alpha\beta}e^{ip(x-x')}, \tag{2.58}$$

which exhibits, with equal weight, the contributions of all states of a particle, that

$$\langle u_\alpha\bar u_\beta\rangle=A(i\gamma p-\kappa)_{\alpha\beta}, \tag{2.59}$$

for a state with wave number four-vector p_μ. The constant A is conveniently evaluated for our purpose in terms of the expectation value of the particle flux vector in the initial state,

$$\mathbf{S}^{(inc)}=(1|ic(\bar\psi(x)\gamma\psi(x))_1|1)$$
$$=ic\bar u\gamma u. \tag{2.60}$$

Thus,

$$\mathbf{S}^{(inc)}=ic\gamma_{\beta\alpha}u_\alpha\bar u_\beta\rightarrow icA\,Tr\gamma(i\gamma p-\kappa)$$
$$=-4cA\mathbf{p}, \tag{2.61}$$

so that

$$\langle u_\alpha \bar{u}_\beta\rangle = -\frac{1}{4c}\frac{|\mathbf{S}^{(inc)}|}{|\mathbf{p}|}(i\gamma p-\kappa)_{\alpha\beta}. \tag{2.62}$$

This leads to the following expression for the total rate of transition from the initial state,

$$w_0=\frac{1}{8\pi^2}\frac{e^2}{\hbar^2c^2}|\mathbf{S}^{(inc)}|\int d\Omega\left|\int e^{i(\mathbf{p}-\mathbf{q})\cdot\mathbf{r}}\frac{Ze}{4\pi r}dv\right|^2 \tfrac{1}{4}Tr[(i\gamma p-\kappa)(\gamma_4+\Gamma_4(q-p))(i\gamma q-\kappa)(\gamma_4+\Gamma_4(p-q))], \tag{2.63}$$

in which we have also specialized to the Coulomb potential of a stationary nucleus. We may now infer that the differential cross section for radiationless scattering through the angle ϑ into a unit solid angle is

$$\frac{d\sigma_0(\vartheta)}{d\Omega}=2\left[\frac{Z\alpha}{(\mathbf{p}-\mathbf{q})^2}\right]^2\tfrac{1}{4}Tr[(i\gamma p-\kappa)(\gamma_4+\Gamma_4(q-p))(i\gamma q-\kappa)(\gamma_4+\Gamma_4(p-q))]. \tag{2.64}$$

The Fourier transform of Γ_4 is conveniently written in the form

$$\Gamma_4(p-q)=-\frac{\alpha}{4\pi}\gamma_4\left[4\lambda^2A(\lambda)+\frac{i}{\kappa}\gamma\cdot(\mathbf{p}-\mathbf{q})F_0(\lambda)\right], \tag{2.65}$$

where

$$A(\lambda)=\log\frac{\kappa}{2k_{\min}}(F_0(\lambda)+F_1(\lambda))+\tfrac{1}{2}(F_0(\lambda)+\tfrac{2}{3}F_2(\lambda)+G(\lambda)). \tag{2.66}$$

Here

$$\lambda=\frac{|\mathbf{p}-\mathbf{q}|}{2\kappa}=\frac{|\mathbf{p}|}{\kappa}\sin\frac{\vartheta}{2}, \tag{2.67}$$

and

$$F_0(\lambda)=\int_0^1\frac{dv}{1+\lambda^2(1-v^2)}=\frac{\log((1+\lambda^2)^{\frac{1}{2}}+\lambda)}{(1+\lambda^2)^{\frac{1}{2}}\lambda} \tag{2.68a}$$

$$F_1(\lambda)=\int_0^1\frac{v^2dv}{1+\lambda^2(1-v^2)}=\left(1+\frac{1}{\lambda^2}\right)F_0(\lambda)-\frac{1}{\lambda^2} \tag{2.68b}$$

$$F_2(\lambda)=\int_0^1\frac{v^4dv}{1+\lambda^2(1-v^2)}=\left(1+\frac{1}{\lambda^2}\right)F_1(\lambda)-\frac{1}{3\lambda^2}. \tag{2.68c}$$

The more complicated transform, $G(\lambda)$ is not required in the following development. The trace of the Dirac matrices contained in (2.64) is easily computed:

$$\tfrac{1}{4}Tr[(i\gamma p-\kappa)(\gamma_4+\Gamma_4(q-p))(i\gamma q-\kappa)(\gamma_4+\Gamma_4(p-q))]=2(p_0^2-\kappa^2\lambda^2)\left(1-\frac{2\alpha}{\pi}\lambda^2A(\lambda)\right)-\frac{2\alpha}{\pi}\kappa^2\lambda^2F_0(\lambda), \tag{2.69}$$

whence

$$\frac{d\sigma_0(\vartheta)}{d\Omega}=\left(\frac{Z\alpha}{2|\mathbf{p}|\beta}\csc^2\frac{\vartheta}{2}\right)^2\left(1-\beta^2\sin^2\frac{\vartheta}{2}\right)\left[1-\frac{2\alpha}{\pi}\lambda^2A(\lambda)-\frac{\alpha}{\pi}\frac{\kappa^2}{p_0^2-\kappa^2\lambda^2}\lambda^2F_0(\lambda)\right] \tag{2.70}$$

in which $\beta=|\mathbf{p}|/p_0$ is the speed of the particle relative to c.

To evaluate the rate at which transitions occur accompanied by radiation, we consider

$$\left(1\left|\delta j_\mu^{(1)}(x)\int_{-\infty}^{\infty}\delta j_\nu^{(1)}(x')dx_0'\right|1\right)=-\frac{e^4}{\hbar^2}\int d\omega''d\omega'''\left(1\left|[\bar{\psi}(x)\gamma_\mu\bar{S}(x-x'')\gamma_\lambda\psi(x'')+\bar{\psi}(x'')\gamma_\lambda\bar{S}(x''-x)\gamma_\mu\psi(x)]_1\right.\right.$$

$$\left.\left.\times\int_{-\infty}^{\infty}dx_0'[\bar{\psi}(x')\gamma_\nu\bar{S}(x'-x''')\gamma_\sigma\psi(x''')+\bar{\psi}(x''')\gamma_\sigma\bar{S}(x'''-x')\gamma_\nu\psi(x')]_1A_\lambda(x'')A_\sigma(x''')\right|1\right). \tag{2.71}$$

Since the state vector Φ_1 is characterized by an absence of quanta, only the following vacuum expectation value is required for the electromagnetic field,

$$\langle A_\lambda(x'')A_\sigma(x''')\rangle_0=i\hbar c\delta_{\lambda\sigma}D^{(+)}(x''-x''')=\frac{\hbar c}{(2\pi)^3}\delta_{\lambda\sigma}\int_{k_0>0}(dk)\delta(k^2)e^{ik(x''-x''')}. \tag{2.72}$$

The matter field operators are treated as before, with the result

$$\left(1\left|\delta j_\mu^{(1)}(x)\int_{-\infty}^{\infty}\delta j_\mu^{(1)}(x')dx_0'\right|1\right)=\frac{e^4}{\hbar^2}\frac{\hbar c}{(2\pi)^6}\int_{q_0,k_0>0}(dq)\delta(q^2+\kappa^2)(dk)\delta(k^2)\int_{-\infty}^{\infty}dx_0'e^{i(p-q-k)(x'-x)}$$
$$\times\bar{u}(\gamma_\mu\bar{S}(q+k)\gamma_\lambda+\gamma_\lambda\bar{S}(p-k)\gamma_\mu)(i\gamma q-\kappa)(\gamma_\nu\bar{S}(p-k)\gamma_\lambda+\gamma_\lambda\bar{S}(q+k)\gamma_\nu)u. \quad (2.73)$$

Here

$$\bar{S}(q+k)=\int e^{-i(q+k)x}\bar{S}(x)d\omega=\frac{i\gamma(q+k)-\kappa}{2qk} \quad (2.74)$$

and

$$\bar{S}(p-k)=-\frac{i\gamma(p-k)-\kappa}{2pk} \quad (2.75)$$

are Fourier transforms of $\bar{S}(x)$. The integration with respect to x_0' imposes the energy conservation law

$$p_0=q_0+k_0, \quad (2.76)$$

which is evidently that of a light quantum emission process. The integration with respect to q_0 and the magnitude of $\mathbf{q}$ can now be performed, leaving one with an expression for w_1 in the form of an integral extended over all directions of the scattered electron, and all light quanta, as restricted by energy conservation. On averaging with respect to the polarization of the incident electron, and specializing to the Coulomb field of a nucleus, one obtains

$$w_1=\frac{\alpha}{\pi^2}|\mathbf{S}^{(inc)}|\int_{k_0>0}d\Omega(dk)\delta(k^2)\frac{|\mathbf{q}|}{|\mathbf{p}|}\left[\frac{Z\alpha}{(\mathbf{p}-\mathbf{q}-\mathbf{k})^2}\right]^2\tfrac{1}{4}Tr\left[(i\gamma p-\kappa)\right.$$
$$\left.\times\left(\gamma_4\frac{i\gamma(q+k)-\kappa}{2qk}\gamma_\lambda-\gamma_\lambda\frac{i\gamma(p-k)-\kappa}{2pk}\gamma_4\right)(i\gamma q-\kappa)\left(\gamma_4\frac{i\gamma(p-k)-\kappa}{2pk}\gamma_\lambda-\gamma_\lambda\frac{i\gamma(q+k)-\kappa}{2qk}\gamma_4\right)\right]. \quad (2.77)$$

It may then be inferred that the differential cross section for radiative scattering through the angle ϑ, in which the energy loss does not exceed ΔE, is

$$\frac{d\sigma_1(\vartheta,\Delta E)}{d\Omega}=\frac{\alpha}{\pi^2}\int_{k_0=0}^{k_0=K}(dk)\delta(k^2)\frac{|\mathbf{q}|}{|\mathbf{p}|}\left[\frac{Z\alpha}{(\mathbf{p}-\mathbf{q}-\mathbf{k})^2}\right]^2\tfrac{1}{4}Tr\left[(i\gamma p-\kappa)\right.$$
$$\left.\times\left(\gamma_4\left(\frac{q_\lambda}{qk}-\frac{p_\lambda}{pk}\right)+\gamma_4\frac{\gamma k}{2qk}\gamma_\lambda+\gamma_\lambda\frac{\gamma k}{2pk}\gamma_4\right)(i\gamma q-\kappa)\left(\gamma_4\left(\frac{q_\lambda}{qk}-\frac{p_\lambda}{pk}\right)+\gamma_\lambda\frac{\gamma k}{2qk}\gamma_4+\gamma_4\frac{\gamma k}{2pk}\gamma_\lambda\right)\right], \quad (2.78)$$

where

$$K=\Delta E/\hbar c. \quad (2.79)$$

We shall first consider the simple situation in which the emitted radiation exerts a negligible reaction on the electron. That is to say, we shall treat the essentially elastic scattering of an electron, in which only a small fraction of the electron kinetic energy is radiated. Under these circumstances, which are expressed by $\Delta E\ll W=E-mc^2$, (2.79) simplifies to

$$\frac{d\sigma_1(\vartheta,\Delta E)}{d\Omega}=\left(\frac{Z\alpha}{2|\mathbf{p}|\beta}\csc^2\frac{\vartheta}{2}\right)^2\left(1-\beta^2\sin^2\frac{\vartheta}{2}\right)\frac{\alpha}{2\pi^2}\int_{k_0=0}^{k_0=K}(dk)\delta(k^2)\left(\frac{p}{pk}-\frac{q}{qk}\right)^2. \quad (2.80)$$

Now

$$\left(\frac{p}{pk}-\frac{q}{qk}\right)^2=\frac{(p-q)^2}{(pk)(qk)}+\kappa^2\left(\frac{2}{(pk)(qk)}-\frac{1}{(pk)^2}-\frac{1}{(qk)^2}\right), \quad (2.81)$$

and

$$\frac{1}{(pk)(qk)}=\frac{1}{(q-p)k}\left(\frac{1}{pk}-\frac{1}{qk}\right)=\frac{1}{2}\int_{-1}^{1}\frac{dv}{\left[\left(\frac{p+q}{2}+\frac{p-q}{2}v\right)k\right]^2}, \quad (2.82)$$

from which one deduces, on integration by parts, that

$$\frac{2}{(pk)(qk)}-\frac{1}{(pk)^2}-\frac{1}{(qk)^2}=-\int_{-1}^{1}dvv\frac{\partial}{\partial v}\frac{1}{\left[\left(\frac{p+q}{2}+\frac{p-q}{2}v\right)k\right]^2}. \quad (2.83)$$

Therefore,

$$\int (dk)\delta(k^2)\left(\frac{p}{pk}-\frac{q}{qk}\right)^2=\kappa^2\int_{-1}^{1} dv\left(\frac{(p-q)^2}{2\kappa^2}-v\frac{\partial}{\partial v}\right)\int (dk)\frac{\delta(k^2)}{\left[\left(\frac{p+q}{2}+\frac{p-q}{2}v\right)k\right]^2}. \tag{2.84}$$

The k integration in the latter equation can be written as

$$-\int (dk)\frac{\delta(k^2)}{\left(\frac{p+q}{2}+\frac{p-q}{2}v\right)^2}\left(\frac{p+q}{2}+\frac{p-q}{2}v\right)_\lambda\frac{\partial}{\partial k_\lambda}\frac{1}{\left(\frac{p+q}{2}+\frac{p-q}{2}v\right)k}$$

$$=\frac{1}{\kappa^2}\frac{1}{1+[(p-q)^2/4\kappa^2](1-v^2)}\left[\int (dk)\frac{\partial}{\partial k_\lambda}\left[\delta(k^2)\frac{\left(\frac{p+q}{2}+\frac{p-q}{2}v\right)_\lambda}{\left(\frac{p+q}{2}+\frac{p-q}{2}v\right)k}\right]-2\int (dk)\delta'(k^2)\right]. \tag{2.85}$$

However,

$$-2\int (dk)\delta'(k^2)=\int\frac{(dk)}{k_0}\frac{\partial}{\partial k_0}\delta(k^2)=\int (dk)\frac{\partial}{\partial k_0}\left(\frac{\delta(k^2)}{k_0}\right)+\int\frac{(dk)}{k_0{}^2}\delta(k^2), \tag{2.86}$$

so that

$$\int (dk)\delta(k^2)\left(\frac{p}{pk}-\frac{q}{qk}\right)^2=\int_{-1}^{1} dv\left(\frac{(p-q)^2}{2\kappa^2}-v\frac{\partial}{\partial v}\right)\frac{1}{1+[(p-q)^2/4\kappa^2](1-v^2)}\left[\int\frac{(dk)}{k_0{}^2}\delta(k^2)\right.$$

$$\left.+\int (dk)\frac{\partial}{\partial k_0}\left\{\delta(k^2)\left[\frac{\left(\frac{p+q}{2}+\frac{p-q}{2}v\right)_0}{\left(\frac{p+q}{2}+\frac{p-q}{2}v\right)k}+\frac{1}{k_0}\right]\right\}\right], \tag{2.87}$$

in which we have discarded terms that obviously vanish on integration over the domain $0<k_0<K$.

The first bracketed integral in (2.87), when expressed in three-dimensional notation, becomes

$$\int\frac{(d\mathbf{k})dk_0}{k_0{}^2}\delta(\mathbf{k}^2-k_0{}^2)=2\pi\int_{k_{\min}}^{k}\frac{dk_0}{k_0}=2\pi\log\frac{K}{k_{\min}}, \tag{2.88}$$

in which we have again introduced an invariant minimum light quantum wave number to characterize a logarithmic divergence associated with the "infra-red catastrophe." A similar expression of the second bracketed integral in (2.87) yields.

$$\int (d\mathbf{k})\delta(\mathbf{k}^2-K^2)\left[\frac{1}{K}-\frac{p_0}{p_0K-\left(\frac{\mathbf{p}+\mathbf{q}}{2}+\frac{\mathbf{p}-\mathbf{q}}{2}v\right)\cdot\mathbf{k}}\right]$$

$$=2\pi\left[1-\frac{1}{2}\frac{p_0}{(\mathbf{p}^2-[(\mathbf{p}-\mathbf{q})^2/4](1-v^2))^{\frac{1}{2}}}\log\frac{p_0+(\mathbf{p}^2-[(\mathbf{p}-\mathbf{q})^2/4](1-v^2))^{\frac{1}{2}}}{p_0-(\mathbf{p}^2-[(\mathbf{p}-\mathbf{q})^2/4](1-v^2))^{\frac{1}{2}}}\right]. \tag{2.89}$$

Therefore,

$$\int_{k_0=0}^{k_0=K}(dk)\delta(k^2)(p/pk-q/qk)^2$$

$$=4\pi\int_0^1 dv\left(\frac{(\mathbf{p}-\mathbf{q})^2}{2\kappa^2}-v\frac{\partial}{\partial v}\right)\frac{1}{1+[(\mathbf{p}-\mathbf{q})^2/4\kappa^2](1-v^2)}\left[\log\frac{K}{k_{\min}}+1-\frac{1}{2\beta\xi}\log\frac{1+\beta\xi}{1-\beta\xi}\right], \tag{2.90}$$

where

$$\xi=\left(1-\sin^2\frac{\vartheta}{2}(1-v^2)\right)^{\frac{1}{2}}. \tag{2.91}$$

We may now employ the identity

$$\frac{1-1/2\beta\xi\log(1+\beta\xi/1-\beta\xi)}{1+[(\mathbf{p}-\mathbf{q})^2/4\kappa^2](1-v^2)}=\frac{1}{2}\frac{\log(1+[(\mathbf{p}-\mathbf{q})^2/4\kappa^2](1-v^2))}{1+[(\mathbf{p}-\mathbf{q})^2/4\kappa^2](1-v^2)}-\frac{\log(2p_0/\kappa)-1}{1+[(\mathbf{p}-\mathbf{q})^2/4\kappa^2](1-v^2)}$$
$$+\frac{\kappa^2}{p_0^2}\frac{1}{2\beta\xi}\left[\frac{\log\frac{1-\beta\xi}{2}}{1+\beta\xi}-\frac{\log\frac{1+\beta\xi}{2}}{1-\beta\xi}\right] \tag{2.92}$$

to cast (2.90) into the form

$$\int_{k_0=0}^{k_0=K}(dk)\delta(k^2)\left(\frac{p}{pk}-\frac{q}{qk}\right)^2=4\pi\frac{\mathbf{p}^2}{\kappa^2}\sin^2\frac{\vartheta}{2}\left[\left(\log\frac{K}{k_{\min}}-\log\frac{2p_0}{\kappa}+1\right)(F_0+F_1)+F_1+\tfrac{1}{2}G+H\right], \tag{2.93}$$

where

$$H=\left(1+\frac{1}{2\lambda^2}\right)\frac{\kappa^2}{p_0^2}\int_0^1\frac{dv}{\beta\xi}\left[\frac{\log\frac{1-\beta\xi}{2}}{1+\beta\xi}-\frac{\log\frac{1+\beta\xi}{2}}{1-\beta\xi}\right]-\frac{1}{2\lambda^2}\frac{\kappa^2}{p_0^2}\frac{1}{\beta}\left[\frac{\log\frac{1-\beta}{2}}{1+\beta}-\frac{\log\frac{1+\beta}{2}}{1-\beta}\right]. \tag{2.94}$$

The function H approaches a constant in the limit of small velocities,

$$\beta\ll1:\quad H=\frac{4}{3}(\log2-1)-\frac{1}{9}. \tag{2.95}$$

At high energies, H is approximated by

$$p_0/\kappa\gg1:\quad H=-\frac{\kappa^2}{p_0^2}f(\vartheta), \tag{2.96}$$

with

$$f(\vartheta)=\frac{1}{\sin\vartheta/2}\int_{\cos\vartheta/2}^1\left[\frac{\log\frac{1+\xi}{2}}{1-\xi}-\frac{\log\frac{1-\xi}{2}}{1+\xi}\right]\frac{d\xi}{(\xi^2-\cos^2(\vartheta/2))^{\frac{1}{2}}}. \tag{2.97}$$

This integral can be performed analytically for $\vartheta=\pi$,

$$f(\pi)=\pi^2/12, \tag{2.98}$$

but must be evaluated numerically for other angles. An approximation in excess, which has the correct asymptotic form at small angles, is provided by

$$f(\vartheta)\sim\left(\frac{2}{\cos(\vartheta/2)(1+\cos(\vartheta/2))^3}\right)^{\frac{1}{2}}\left[\log\frac{1}{2(1-\cos(\vartheta/2))}+\frac{1-\cos(\vartheta/2)}{2}+1\right] \tag{2.99}$$

This formula is reasonably accurate even for $\vartheta=\pi/2$, where the value yielded by (2.99) exceeds by only 8.6 percent the following result of a numerical calculation:

$$f(\pi/2)=1.167. \tag{2.100}$$

The total differential cross section for scattering through the angle ϑ, in which the energy loss does not exceed ΔE, is

$$\frac{d\sigma(\vartheta,\Delta E)}{d\Omega}=\frac{d\sigma_0(\vartheta)}{d\Omega}+\frac{d\sigma_1(\vartheta,\Delta E)}{d\Omega}=\left(\frac{Z\alpha}{2|\mathbf{p}|\beta}\csc^2\frac{\vartheta}{2}\right)^2\left(1-\beta^2\sin^2\frac{\vartheta}{2}\right)(1-\delta(\vartheta,\Delta E)), \tag{2.101}$$

where $\delta(\vartheta,\Delta E)$ is the desired fractional decrease in the cross section produced by radiative effects. For essentially elastic scattering, we obtain

$$\delta(\vartheta,\Delta E\ll W)=\frac{2\alpha}{\pi}\beta^2\frac{p_0^2}{\kappa^2}\sin^2\frac{\vartheta}{2}\left[\left(\log\frac{E}{\Delta E}-1\right)(F_0+F_1)+\frac{1}{2}F_0-F_1+\frac{1}{3}F_2-H+\frac{1}{2}\frac{\kappa^2/p_0^2}{1-\beta^2\sin^2(\vartheta/2)}F_0\right], \tag{2.102}$$

on combining (2.70) with (2.80). It will be noted that the infra-red catastrophe, as characterized by $k_{\min}$ has disappeared. However, it is possible, in principle, to consider the limit $\Delta E\to 0$, which would make δ diverge logarithmically. It is well known that this difficulty stems from the neglect of processes involving more than one low frequency quantum.[5] Actually, the essentially elastic scattering cross section must approach zero as $\Delta E\to 0$; that is, it never happens that a scattering event is unaccompanied by the emission of quanta. This is described by replacing the radiative correction factor $1-\delta$ with $e^{-\delta}$, which has the proper limiting behavior as $\Delta E\to 0$. The further terms in the series expansion of $e^{-\delta}$ express the effects of higher order processes involving the multiple emission of soft quanta. However, for practical purposes, such a refinement is unnecessary. The accuracy with which the energy of a particle can be measured is such that the limit $\Delta E\to 0$ cannot be realized, and δ will be small in comparison with unity under presently accessible circumstances.

For a slowly moving particle,

$$\beta\ll 1: \delta(\vartheta,\Delta E\ll W)=\frac{8\alpha}{3\pi}\beta^2\sin^2\frac{\vartheta}{2}\left[\log\frac{mc^2}{2\Delta E}+\frac{19}{30}\right], \tag{2.103}$$

according to (2.95), the limiting form of H and the corresponding limiting form of F_n:

$$\lambda\ll 1:\quad F_n=\frac{1}{2n+1}. \tag{2.104}$$

The radiative correction thus increases linearly with the kinetic energy of the particle. In the extreme relativistic region, on the other hand,

$$\frac{p_0}{\kappa}\sin\frac{\vartheta}{2}\gg 1:$$

$$\delta(\vartheta,\Delta E\ll W)=\frac{4\alpha}{\pi}\left[\left(\log\frac{E}{\Delta E}-\frac{13}{12}\right)\left(\log\frac{2p_0}{K}\sin\frac{\vartheta}{2}-\frac{1}{2}\right)+\frac{17}{72}+\frac{1}{2}\sin^2\frac{\vartheta}{2}f(\vartheta)\right], \tag{2.105}$$

which has a logarithmic dependence on the particle energy. The asymptotic form (2.105) is quite accurate for even moderate energies. Thus, with $\vartheta=\pi/2$, $\Delta E=10$ kev and $W=3.1$ Mev, which corresponds to $(p_0/\kappa)\times\sin\vartheta/2=5$, the value of δ computed from (2.105) differs from the correct value,

$$\delta=8.6\ 10^{-2}, \tag{2.106}$$

by only a fraction of a percent. It is evident from this numerical result that radiative corrections to scattering cross sections can be quite appreciable. For the particular conditions chosen, ΔE can be materially increased (but still subject to $\Delta E\ll W$), without seriously impairing δ. Thus, with $\Delta E=40$ kev, $\delta=6.3\ 10^{-2}$, while $\Delta E=80$ kev yields $\delta=5.1\ 10^{-2}$. As to the energy dependence of δ, we remark that with a given accuracy in the determination of the energy, $\Delta E/E$, δ varies linearly with the logarithm of the energy. Thus, with $\Delta E/E=0.04/3.6=1.1\ 10^{-2}$, an increase in the total energy by a factor of four produces an addition of $4.4\ 10^{-2}$ to δ, whence $\delta=11\ 10^{-2}$ for $W=14$ Mev, and $\delta=15\ 10^{-2}$ for $W=57$ Mev.

The angular dependence of δ at relativistic energies is not fully described by the asymptotic formula (2.105), since the underlying condition, $(p_0/\kappa)\sin\vartheta/2\ll 1$, cannot be maintained with diminishing ϑ. Indeed, δ is proportional to $\sin^2\vartheta/2$ at angles such that $(p_0/\kappa)\sin\vartheta/2\ll 1$. However (2.105) can be used over a wide angular range, even at moderate energies. Thus, with $W=3.1$ Mev, $\Delta E=40$ kev, and $\vartheta=\pi/4$, which corresponds to $(p_0/\kappa)\sin\vartheta/2=2.7$, the value of δ deduced from (2.105) exceeds by only 2 percent the correct value, $\delta=4.2\ 10^{-2}$. We may note that under the same energy conditions, but with $\vartheta=3\pi/4$, $\delta=7.2\ 10^{-2}$. The angular dependence of δ may be particularly suitable for an experimental test of these predictions, which involve the relativistic aspects of the radiative corrections to the electromagnetic properties of the electron.

We have thus far considered only the essentially elastic scattering of an electron, in which radiative corrections arise primarily from virtual processes. If we wish to compute the differential cross section for scattering with an arbitrary maximum energy loss ΔE, it is only necessary to augment the essentially elastic cross section, in which the maximum energy loss is $\Delta E'\ll W$, by the cross section for scattering with the emission of a light quantum in the energy range from $\Delta E'$ to ΔE. The latter process involves the well-known bremsstrahlung cross section which, of course, is the content of (2.79). This will be illustrated by the calculation of the differential cross section for the scattering of a slowly moving electron, irrespective of the final energy. The differential cross section per unit solid angle for scattering of an electron through the angle ϑ, in which a light quantum is emitted in the energy range from $\Delta E'$ to W, is

$$\frac{\alpha}{\pi^2}\int_{\Delta E'/\hbar c}^{W/\hbar c}\frac{dk_0}{k_0}\int d\omega\frac{|\mathbf{q}|}{|\mathbf{p}|}\left[\frac{Z\alpha}{(\mathbf{p}-\mathbf{q})^2}\right]^2[(\mathbf{p}-\mathbf{q})^2-(\mathbf{n}\cdot(\mathbf{p}-\mathbf{q}))^2], \tag{2.107}$$

according to the non-relativistic limit of (2.78). Here $d\omega$ is an element of solid angle associated with the direction of the unit vector $\mathbf{n}=\mathbf{k}/k_0$, and

$$|\mathbf{q}|=(\mathbf{p}^2-2\kappa k_0)^{\frac{1}{2}}. \tag{2.108}$$

[5] F. Bloch and A. Nordsieck, Phys. Rev. **52**, 54 (1937).

On performing the integration over all emission directions of the light quantum, and introducing the new variable of integration, $x=|\mathbf{q}|/|\mathbf{p}|$, (2.107) becomes

$$\frac{8\alpha}{3\pi}\left(\frac{Z\alpha}{|\mathbf{p}|}\right)^2\int_0^{(1-\Delta E'/W)^{\frac{1}{2}}}\frac{x}{1+x^2-2x\cos\vartheta}\frac{2xdx}{1-x^2}$$
$$=\left(\frac{Z\alpha}{2|\mathbf{p}|\beta}\csc^2\frac{\vartheta}{2}\right)^2\frac{8\alpha}{3\pi}\beta^2\sin^2\frac{\vartheta}{2}\left[\log\frac{W}{\Delta E'}-\int_0^1\frac{1-x}{1+x^2-2x\cos\vartheta}\frac{2xdx}{1+x}\right]. \tag{2.109}$$

Thus the contribution to δ produced by emission of quanta with energies in the range from $\Delta E'$ to W is

$$-\frac{8\alpha}{3\pi}\beta^2\sin^2\frac{\vartheta}{2}\left[\log\frac{4W}{\Delta E'}-(\pi-\vartheta)\tan\frac{\vartheta}{2}-\frac{\cos\vartheta}{\cos^2\vartheta/2}\log\csc\frac{\vartheta}{2}\right]. \tag{2.110}$$

On adding this to $\delta(\vartheta,\Delta E')$, as given by (2.103), we obtain the desired result:

$$\beta\ll 1:\quad \delta(\vartheta,W)=\frac{8\alpha}{3\pi}\beta^2\sin^2\frac{\vartheta}{2}\left[\log\frac{1}{4\beta^2}+\frac{19}{30}+(\pi-\vartheta)\tan\frac{\vartheta}{2}+\frac{\cos\vartheta}{\cos^2\vartheta/2}\log\csc\frac{\vartheta}{2}\right]. \tag{2.111}$$

It may be remarked, finally, that the analogous meso-nuclear phenomenon, the radiative correction to nucleon-nucleon scattering associated with virtual meson emission, will be a relatively more significant effect in view of the stronger couplings involved. This may well be the explanation of the discrepancy between the observed neutron-proton scattering cross section for high energy neutrons and the larger theoretical values computed from various assumed interaction potentials.[6]

APPENDIX

In this section, we shall first give an alternative treatment of the polarization of the vacuum by an external field, employing the methods developed in the preceding pages. It is desired to compute the expectation value of $j_\mu(x)$,

$$\langle j_\mu(x)\rangle=(\Psi[\sigma],j_\mu(x)\Psi[\sigma]), \tag{A.1}$$

where $\Psi[\sigma]$ obeys

$$i\hbar c\frac{\delta\Psi[\sigma]}{\delta\sigma(x)}=-\frac{1}{c}j_\mu(x)A_\mu(x)\Psi[\sigma], \tag{A.2}$$

and $A_\mu(x)$ is the potential of a prescribed current distribution. The physical situation can be described as follows. In the remote past, the matter field is uncoupled from the external electromagnetic field, and the state vector is that of the vacuum,

$$\Psi[-\infty]=\Psi_0. \tag{A.3}$$

It is supposed that the coupling is adiabatically switched on, and that the external field does not induce real pair creation. The latter restriction implies that the final state of the matter field, after the coupling is adiabatically switched off, is simply Ψ_0, whence

$$\Psi[\infty]-\Psi[-\infty]=(S-1)\Psi_0=0. \tag{A.4}$$

A solution of (A.2), in the form

$$\Psi[\sigma]=U[\sigma]\Psi_0, \tag{A.5}$$

may be constructed, where

$$i\hbar c\frac{\delta U[\sigma]}{\delta\sigma(x)}=-\frac{1}{c}j_\mu(x)A_\mu(x)U[\sigma], \tag{A.6}$$

and

$$U[\infty]=S,\quad U[-\infty]=1. \tag{A.7}$$

The current induced in the vacuum is then written as

$$\langle j_\mu(x)\rangle=\langle U^{-1}[\sigma]j_\mu(x)U[\sigma]\rangle_0. \tag{A.8}$$

Now

$$\frac{1}{2}\int_{-\infty}^{\infty}d\omega'\epsilon[\sigma,\sigma']\frac{\delta}{\delta\sigma'(x')}U^{-1}[\sigma']j_\mu(x)U[\sigma']=U^{-1}[\sigma]j_\mu(x)U[\sigma]-\frac{1}{2}(j_\mu(x)+S^{-1}j_\mu(x)S), \tag{A.9}$$

whence

$$U^{-1}[\sigma]j_\mu(x)U[\sigma]=\frac{1}{2}(j_\mu(x)+S^{-1}j_\mu(x)S)+\frac{i}{2\hbar c^2}\int_{-\infty}^{\infty}d\omega'\epsilon[\sigma,\sigma']U^{-1}[\sigma'][j_\mu(x),j_\nu(x')]U[\sigma']A_\nu(x'). \tag{A.10}$$

On placing $U[\sigma']=1$ on the right side of (A.10), one obtains the first approximation in a treatment that regards the disturbance of the vacuum as small. Hence,

$$\langle j_\mu(x)\rangle=\frac{i}{2\hbar c^2}\int_{-\infty}^{\infty}d\omega'\epsilon(x-x')[j_\mu(x),j_\nu(x')]_0A_\nu(x'), \tag{A.11}$$

in view of (A.4) and the absence of a current in the unperturbed vacuum. We shall, for convenience, write this formula as

$$\langle j_\mu(x)\rangle=\frac{4e^2}{\hbar}\int G_{\mu\nu}(x-x')A_\nu(x')d\omega', \tag{A.12}$$

where, according to II (2.10),

$$G_{\mu\nu}(x-x')=\frac{1}{8}Tr[S^{(1)}(x'-x)\gamma_\mu\bar{S}(x-x')\gamma_\nu+\bar{S}(x'-x)\gamma_\mu S^{(1)}(x-x')\gamma_\nu]. \tag{A.13}$$

The introduction of the Fourier integral representations for the functions $S^{(1)}$ and $\bar{S}$, combined with the trace evaluation (see II (2.10))

$$\frac{1}{8}Tr[(-i\gamma k'+\kappa)\gamma_\mu(i\gamma k''+\kappa)\gamma_\nu+(-i\gamma k''+\kappa)\gamma_\mu(i\gamma k'+\kappa)\gamma_\nu]=k_\mu'k_\nu''+k_\nu'k_\mu''-\delta_{\mu\nu}(k'k''-\kappa^2), \tag{A.14}$$

[6] A summary is given by L. Rosenfeld, Nuclear Forces (Interscience Publishers, Inc., New York, 1949), pp. 450, 454.

yields the following expression for $G_{\mu\nu}(x)$,

$$G_{\mu\nu}(x)=\frac{1}{(2\pi)^7}\int (dk')(dk'')e^{i(k'+k'')x}\frac{\delta(k''^2+\kappa^2)}{k'^2+\kappa^2}[k_\mu'k_\nu''+k_\nu'k_\mu''-\delta_{\mu\nu}(k'k''-\kappa^2)]. \tag{A.15}$$

It is instructive to examine the conditions imposed on $G_{\mu\nu}(x)$ by the related requirements of charge conservation and gauge invariance. The former evidently demands that

$$\frac{\partial}{\partial x_\mu}G_{\mu\nu}(x)=0. \tag{A.16}$$

The requirement of gauge invariance is that the induced current be unaffected by the gauge transformation

$$A_\mu(x)\rightarrow A_\mu(x)-\frac{\partial\Lambda(x)}{\partial x_\mu}, \tag{A.17}$$

or that

$$\int G_{\mu\nu}(x-x')\frac{\partial\Lambda(x')}{\partial x_\nu'}d\omega'=\int\frac{\partial}{\partial x_\nu}G_{\mu\nu}(x-x')\Lambda(x')d\omega'=0, \tag{A.18}$$

in which the absence of an integrated term is a consequence of the adiabatic removal of the coupling in the remote past and future. Evidently (A.18) is satisfied in virtue of (A.16), since

$$G_{\mu\nu}(x)=G_{\nu\mu}(x). \tag{A.19}$$

On computing $\partial G_{\mu\nu}(x)/\partial x_\mu$ from (A.15), we obtain

$$\frac{\partial}{\partial x_\mu}G_{\mu\nu}(x)=\frac{i}{(2\pi)^7}\int (dk')(dk'')e^{i(k'+k'')x}k_\nu''\delta(k''^2+\kappa^2)$$
$$=\frac{i}{(2\pi)^7}\int (dk)e^{ikx}\int (dk'')k_\nu''\delta(k''^2+\kappa^2), \tag{A.20}$$

where

$$k_\mu=k_\mu'+k_\mu'', \tag{A.21}$$

which is indeed zero if

$$\int (dk'')k_\nu''\delta(k''^2+\kappa^2)=0. \tag{A.22}$$

Although the latter integral is strictly divergent, the value of zero is unambiguously obtained from any limiting process in which the delta-function is replaced by a suitable non-singular function. In this sense, the requirements of charge conservation and gauge invariance are satisfied. It may be noted that the same integral is encountered in evaluating the current in the unperturbed vacuum, II (1.73),

$$\langle j_\mu(x)\rangle_0=\frac{iec}{2}Tr\gamma_\mu S^{(1)}(0)$$
$$=\frac{2ec}{(2\pi)^3}\int (dk'')k_\mu''\delta(k''^2+\kappa^2), \tag{A.23}$$

which must also be zero.

We return to the evaluation of $G_{\mu\nu}(x)$, and utilize the identity

$$k'k''-\kappa^2=2\frac{(kk')(kk'')}{k^2}-(k'^2+\kappa^2)\frac{kk''}{k^2}-(k''^2+\kappa^2)\frac{kk'}{k^2} \tag{A.24}$$

to simplify (A.15). The third term in the latter expression makes no contribution in view of the null value of $(k''^2+\kappa^2)\delta(k''^2+\kappa^2)$, while the second term produces a contribution of $G_{\mu\nu}(x)$ of the form.

$$\int (dk)(dk'')e^{ikx}\frac{kk''}{k^2}\delta(k''^2+\kappa^2), \tag{A.25}$$

which must also be zero, in consequence of (A.22). Hence,

$$G_{\mu\nu}(x)=\frac{1}{(2\pi)^7}\int (dk')(dk'')e^{ikx}\left(\frac{\delta(k'^2+\kappa^2)}{k''^2+\kappa^2}+\frac{\delta(k''^2+\kappa^2)}{k'^2+\kappa^2}\right)\left(\frac{k_\mu'k_\nu''+k_\nu'k_\mu''}{2}-\delta_{\mu\nu}\frac{(kk')(kk'')}{k^2}\right). \tag{A.26}$$

The delta-function factor can be simplified in the manner introduced in the text:

$$\frac{\delta(k'^2+\kappa^2)}{k''^2+\kappa^2}+\frac{\delta(k''^2+\kappa^2)}{k'^2+\kappa^2}=\frac{1}{k''^2-k'^2}(\delta(k'^2+\kappa^2)-\delta(k''^2+\kappa^2))=-\tfrac{1}{2}\int_{-1}^{1}dv\delta'\left(\frac{k'^2+k''^2}{2}+\frac{k'^2-k''^2}{2}v+\kappa^2\right). \tag{A.27}$$

The introduction of the new variables k_μ and p_μ, as defined by

$$k_\mu'=\tfrac{1}{2}k_\mu+\left(p_\mu-\frac{v}{2}k_\mu\right)$$
$$k_\mu''=\tfrac{1}{2}k_\mu-\left(p_\mu-\frac{v}{2}k_\mu\right), \tag{A.28}$$

then brings $G_{\mu\nu}(x)$ into the form

$$G_{\mu\nu}(x)=-\frac{1}{8(2\pi)^7}\int_{-1}^{1}dv(1-v^2)\int (dk)(dp)e^{ikx}(k_\mu k_\nu-\delta_{\mu\nu}k^2)\delta'\left(p^2+\kappa^2+\frac{k^2}{4}(1-v^2)\right), \tag{A.29}$$

where, in virtue of the dependence of the delta-function on p^2 alone, terms linear in p_μ have been discarded, and $p_\mu p_\nu$ has been replaced by $\frac{1}{4}\delta_{\mu\nu}p^2$. It is thereby shown that

$$G_{\mu\nu}(x)=\left(\frac{\partial}{\partial x_\mu}\frac{\partial}{\partial x_\nu}-\delta_{\mu\nu}\Box^2\right)G(x), \tag{A.30}$$

with

$$G(x)=\frac{1}{8(2\pi)^7}\int_{-1}^{1}dv(1-v^2)\int (dk)(dp)e^{ikx}\delta'\left(p^2+\kappa^2+\frac{k^2}{4}(1-v^2)\right). \tag{A.31}$$

The divergent and convergent parts of $G(x)$ can be separated by a partial integration with respect to v,

$$G(x)=\frac{1}{6(2\pi)^3}\int (dp)\delta'(p^2+\kappa^2)\delta(x)-\Box^2\frac{1}{8(2\pi)^7}\int_0^1 v^2\left(1-\frac{v^2}{3}\right)dv\int (dk)e^{ikx}\int (dp)\delta''\left(p^2+\kappa^2+\frac{k^2}{4}(1-v^2)\right). \tag{A.32}$$

The invariant, logarithmically divergent integral that occurs in the first term of (A.32) can be expressed in three-dimensional notation as

$$\int(dp)\delta'(p^2+\kappa^2)=-\int(d\mathbf{p})dp_0\frac{1}{2p_0}\frac{\partial}{\partial p_0}\delta(p_0{}^2-\mathbf{p}^2-\kappa^2)=-\tfrac{1}{4}\int\frac{(d\mathbf{p})}{(\mathbf{p}^2+\kappa^2)^{\frac{3}{2}}}=-2\pi\lim_{P\to\infty}\left(\log\frac{P_0+P}{\kappa}-1\right), \tag{A.33}$$

where

$$P_0=(P^2+\kappa^2)^{\frac{1}{2}}. \tag{A.34}$$

The convergent second integral of (A.32) is then obtained by differentiating (A.33) with respect to κ^2 and replacing the latter by $\kappa^2+(k^2/4)(1-v^2)$,

$$\int(dp)\delta''\left(p^2+\kappa^2+\frac{k^2}{4}(1-v^2)\right)=\frac{\pi}{\kappa^2+(k^2/4)(1-v^2)}. \tag{A.35}$$

With these evaluations, $G(x)$ becomes

$$G(x)=-\frac{1}{24\pi^2}\lim_{P\to\infty}\left(\log\frac{P_0+P}{\kappa}-1\right)\delta(x)-\frac{1}{64\pi^2}\frac{1}{\kappa^2}\Box^2(F_1(x)-\tfrac{1}{3}F_2(x)), \tag{A.36}$$

where

$$F_n(x)=\int_0^1 v^{2n}dv\frac{1}{(2\pi)^4}\int(dk)\frac{e^{ikx}}{1+(k^2/4\kappa^2)(1-v^2)}. \tag{A.37}$$

Finally, we may insert (A.30) into (A.12) and integrate by parts to obtain

$$\langle j_\mu(x)\rangle=16\pi\alpha\int G(x-x')J_\nu(x')d\omega', \tag{A.38}$$

where $J_\mu(x)$ is the current vector that generates the external electromagnetic field. The expression of $G(x)$ contained in (A.36) then yields

$$\langle j_\mu(x)\rangle=-\frac{2\alpha}{3\pi}\lim_{P\to\infty}\left(\log\frac{P_0+P}{\kappa}-1\right)J_\mu(x)-\frac{\alpha}{4\pi}\frac{1}{\kappa^2}\Box^2\int(F_1(x-x')-\tfrac{1}{3}F_2(x-x'))J_\mu(x')d\omega', \tag{A.39}$$

in which the first term represents the logarithmically divergent renormalization of charge.

It should be remarked that the existence of a charge renormalization term would appear to contradict the conservation of charge, since it implies that a non-vanishing total charge is induced in the vacuum. Indeed, a formal evaluation of the total induced charge would yield zero,

$$\begin{aligned}\frac{1}{c}\int\langle j_\mu(x)\rangle d\sigma_\mu&=\frac{i}{2\hbar c^2}\int\epsilon(x-x')\left[\frac{1}{c}\int j_\mu(x)d\sigma_\mu,j_\nu(x')\right]_0 A_\nu(x')d\omega'\\&=0,\end{aligned} \tag{A.40}$$

since the operator of the total charge commutes with the current vector at an arbitrary point. The expression of $\langle j_\mu(x)\rangle$ as

$$\langle j_\mu(x)\rangle=\frac{\partial}{\partial x_\nu}\frac{4e^2}{\hbar}\int G(x-x')F_{\mu\nu}(x')d\omega', \tag{A.41}$$

where

$$F_{\mu\nu}=\frac{\partial}{\partial x_\mu}A_\nu-\frac{\partial}{\partial x_\nu}A_\mu, \tag{A.42}$$

is formally consistent with the result since

$$\begin{aligned}\frac{1}{c}\int\langle j_\mu(x)\rangle d\sigma_\mu&=\frac{2e^2}{\hbar}\int\left(d\sigma_\mu\frac{\partial}{\partial x_\nu}-d\sigma_\nu\frac{\partial}{\partial x_\mu}\right)\int G(x-x')F_{\mu\nu}(x')d\omega'\\&=0,\end{aligned} \tag{A.43}$$

in view of the theorem

$$\int\left(d\sigma_\mu\frac{\partial}{\partial x_\nu}F(x)-d\sigma_\nu\frac{\partial}{\partial x_\mu}F(x)\right)=0. \tag{A.44}$$

However, it is evident that these formal manipulations are only justified if the integrand in (A.44) decreases sufficiently rapidly in space-like directions, which is not fulfilled for the field strengths generated by a charge distribution of non-vanishing total charge.

This difficulty can be avoided by treating the actual electromagnetic field as the limit of a spatially confined field, for which the total induced charge is zero. A convenient way to accomplish this is to introduce a finite light quantum mass, which is eventually allowed to vanish. We thus write the potentials generated by the given charge distribution as

$$A_\mu(x)=\frac{1}{c}\int\bar{D}(x-x')J_\mu(x')d\omega',\quad \frac{\partial A_\mu(x)}{\partial x_\mu}=0, \tag{A.45}$$

where

$$\bar{D}(x)=\lim_{\epsilon\to 0}\frac{1}{(2\pi)^4}\int\frac{e^{ikx}}{k^2+\epsilon^2}(dk). \tag{A.46}$$

The induced current can then be exhibited in the form

$$\begin{aligned}\langle j_\mu(x)\rangle&=-\frac{4e^2}{\hbar c}\int G(x-x')\Box'^2\bar{D}(x'-x'')J_\mu(x'')d\omega'd\omega''\\&=\frac{4e^2}{\hbar c}\frac{1}{(2\pi)^4}\int e^{ik\xi}G(k)k^2\bar{D}(k)J_\mu(x-\xi)(dk)(d\xi),\end{aligned} \tag{A.47}$$

in which the second version involves the Fourier transforms of the functions, $G(x)$ and $\bar{D}(x)$,

$$\begin{aligned}G(k)&=\frac{1}{(4\pi)^3}\int_{-1}^1 dv(1-v^2)\int(dp)\delta'\left(p^2+\kappa^2+\frac{k^2}{4}(1-v^2)\right)\\\bar{D}(k)&=\lim_{\epsilon\to 0}\frac{1}{k^2+\epsilon^2}.\end{aligned} \tag{A.48}$$

The total induced charge can then be calculated in terms of the total external charge,

$$Q=\frac{1}{c}\int J_\mu(x-\xi)d\sigma_\mu, \tag{A.49}$$

which expression is independent of ξ. Therefore,

$$\frac{1}{c}\int\langle j_\mu(x)\rangle d\sigma_\mu=\frac{4e^2}{\hbar c}Q\int G(k)k^2\bar{D}(k)\delta(k)(dk)$$
$$=\frac{4e^2}{\hbar c}Q\lim_{\epsilon\to 0}\left(\frac{k^2}{k^2+\epsilon^2}G(k)\right)_{k_\mu=0}. \tag{A.50}$$

If ϵ is placed equal to zero before evaluating the Fourier transforms at $k_\mu=0$, we obtain the previously computed non-vanishing induced charge

$$\delta Q=\frac{4e^2}{\hbar c}Q[G(k)]_{k_\mu=0}=\frac{\alpha}{3\pi^2}Q\int(dp)\delta'(p^2+\kappa^2). \tag{A.51}$$

On the other hand, if the limiting process $\epsilon\to 0$ is reserved to the end of the calculation, we evidently find $\delta Q=0$.

The implications of this limiting process may be further indicated by noting that, in the first term of (A.39), $J_\mu(x)$ will be replaced by

$$J_\mu(x)-\epsilon^2 cA_\mu(x), \tag{A.52}$$

in virtue of the differential equation

$$(\Box^2-\epsilon^2)A_\mu(x)=-\frac{1}{c}J_\mu(x) \tag{A.53}$$

obeyed by the potential (A.45). Now (A.52) reduces to $J_\mu(x)$ at any point as $\epsilon\to 0$. Yet the total charge computed from (A.52) is zero. This is illustrated by the charge density associated, according to (A.52), with a point charge at the origin:

$$\lim_{\epsilon\to 0} e\left(\delta(\mathbf{r})-\epsilon^2\frac{e^{-\epsilon r}}{4\pi r}\right). \tag{A.54}$$

We may conclude that in the process of vacuum polarization, a non-vanishing, and indeed divergent charge is attached to the original charge distribution, and a compensating charge is created at infinity.

We shall finally apply the computational methods of this paper to evaluate the invariant expression for the electromagnetic mass,

$$\delta mc^2\psi(x)=-\frac{e^2}{2}\int\gamma_\mu[\bar{D}(x-x')S^{(1)}(x-x')+D^{(1)}(x-x')\bar{S}(x-x')]\gamma_\mu\psi(x')d\omega'. \tag{A.55}$$

The insertion of the Fourier integral representations yields

$$\delta mc^2\psi(x)=-\frac{e^2}{2}\frac{1}{(2\pi)^7}\int(dk)(dk')e^{i(k+k')(x-x')}\gamma_\mu(i\gamma k'-\kappa)\gamma_\mu\left(\frac{\delta(k'^2+\kappa^2)}{k^2}+\frac{\delta(k^2)}{k'^2+\kappa^2}\right)\psi(x')d\omega'. \tag{A.56}$$

This becomes

$$-\frac{e^2}{2}\frac{1}{(2\pi)^7}\int(dk)(dp)e^{ip(x-x')}\gamma_\mu(i\gamma(p-k)-\kappa)\gamma_\mu\left(\frac{\delta(k^2-2pk)}{2pk}-\frac{\delta(k^2)}{2pk}\right)\psi(x')d\omega' \tag{A.57}$$

on introducing

$$p_\mu=k_\mu+k_\mu', \tag{A.58}$$

which is effectively subject to the restriction

$$p^2+\kappa^2=0, \tag{A.59}$$

in view of the wave equation satisfied by $\psi(x')$. Now

$$\frac{1}{2pk}[\delta(k^2-2pk)-\delta(k^2)]=-\int_0^1\delta'(k^2-2pku)du, \tag{A.60}$$

and

$$\gamma_\mu(i\gamma(p-k)-\kappa)\gamma_\mu=-2(i\gamma(p-k)+2\kappa), \tag{A.61}$$

whence

$$\delta mc^2\psi(x)=\frac{e^2}{(2\pi)^7}\int(dk)(dp)\int_0^1 du e^{ip(x-x')}(i\gamma k-\kappa)\delta'(k^2-2pku)\psi(x')d\omega', \tag{A.62}$$

in which we have employed the fact that $i\gamma p+\kappa$ is equivalent to $\gamma_\mu(\partial/\partial x_\mu')+\kappa$ applied to $\psi(x')$ and is thus effectively equal to zero. The transformation

$$k_\mu\to k_\mu+p_\mu u \tag{A.63}$$

then yields

$$\delta mc^2\psi(x)=\frac{e^2}{(2\pi)^7}\int(dk)(dp)\int_0^1 du e^{ip(x-x')}(i\gamma pu-\kappa)\delta'(k^2+\kappa^2u^2)\psi(x')d\omega'$$
$$=-\frac{e^2}{(2\pi)^3}\kappa\int_0^1 du\int(dk)(1+u)\delta'(k^2+\kappa^2u^2)\psi(x) \tag{A.64}$$

Hence,

$$\delta m/m=-\frac{\alpha}{2\pi^2}\int_0^1(1+u)du\int(dk)\delta'(k^2+\kappa^2u^2)=-\frac{3\alpha}{4\pi^2}\int(dk)\delta'(k^2+\kappa^2)+\frac{\alpha}{2\pi^2}\int_0^1(2u^2+u^3)du\kappa^2\int(dk)\delta''(k^2+\kappa^2u^2), \tag{A.65}$$

and

$$\delta m/m=\frac{3\alpha}{2\pi}\lim_{K\to\infty}\left(\log\frac{K_0+K}{\kappa}-\tfrac{1}{6}\right). \tag{A.66}$$

according to the integrals (A.33) and (A.35).

On Infinite Field Reactions in Quantum Field Theory

SIN-ITIRO TOMONAGA
Physics Institute, Tokyo Bunrika Daigaku, Tokyo, Japan
June 1, 1948

IN interpreting the level-shift of the hydrogen atom in terms of the radiative reaction, Bethe[1] proposed a method of dealing with this problem without touching the inherent divergency of the current quantum field theory. By his theory it has become possible to treat the problem involving field reactions for the first time in close connection with the reliable experimental data. On the other hand, Lewis[2] and Epstein[3] analyzed the infinities occurring in the radiative correction to the scattering cross-section of an electron in an external field of force and found that also these infinities could be got over by means of a procedure similar to Bethe's. In a recent issue of the PHYSICAL REVIEW, Schwinger[4] pointed out that in these problems the method of canonical transformation was useful, by which the separation from the fields of the parts belonging to free photons and non-radiating electrons was performed, a method which can be regarded as a relativistic generalization of the procedure used by Bloch and Nordsieck[5] as well as by Pauli and Fierz[6] in the discussions about the self-field of an electron.

Almost the same line of attack was taken independently of these American authors also by our Tokyo group and some results are now in course of publication in our English journal, *Progress of Theoretical Physics*. Under the unfavorable conditions after wartime, however, it will be a long time before these papers will appear in print. So I should like to give here a brief summary of the state and views of our investigations.

We first treated the effect of field reactions in the collision problem on the f-field theory of Pais[7]—or the C-meson theory of Sakata,[8] which had been developed by Sakata independently of Pais—intending to examine whether this theory which had been put forward aiming at the elimination of the infinity occurring in the self-energy of charged particles, was also capable of cancelling out the infinity occurring in the scattering cross section of an electron. We found that this was really the case: the f-field was proved capable of compensating one part of the infinite reactions of the radiation field in this phenomenon. The other part of infinities, that is, of the positron theoretical origin, however, could not be eliminated by introducing this new field.[9]

Then the above mentioned work of Bethe appeared. We tried[10] to formulate Bethe's procedure in a mathematically more closed form by using a relativistic generalization of the canonical transformation of Pauli and Fierz. By this transformation an infinite term could be separated in the Hamiltonian density, and this term, we found, had the same form as the mass term in the Dirac equation, being bilinear in ψ^* and ψ. Because of its structure this infinite term can be amalgamated into the mass term of the Dirac equation and one can reinterpret the electron mass in such a way that the compound mass is just what we observe, corresponding to the idea of Bethe. This procedure violates neither the invariance of the theory nor the integrability condition of the generalized Schrödinger equation.[11] In order to investigate the effectiveness of this amalgamation of infinity into the electron mass, we applied the method to the problem lying close at our hand, i.e., the radiative correction to the scattering cross section of an electron.[12] We found this amalgamation was in fact effective in eliminating the non-positron theoretical infinity in this problem, a fact that was pointed out independently by Lewis. It was also shown that the infinity of the positron theoretical origin could be eliminated by reinterpreting the scattering external potential or by reinterpreting the electron charge.

The result of the canonical transformation shows on the other hand that there occurs an infinity of another type, a term containing electromagnetic potential bilinearly. Because of this structure this infinity is to be attributed to the vacuum polarization effect. In order to see the role to be played by this effect in collision phenomena we analyzed the infinities occurring in the e^2-correction to the Klein-Nishina formula.[13] In this problem, we found, besides the infinities of the types mentioned above, an infinity which is closely related to the above mentioned vacuum effect. Infinity of this kind can be, in fact, driven away from the cross section when we substract beforehand the infinite term of the vacuum type from the Hamiltonian. But for this subtraction we cannot find a reasoning so natural and plausible as that used in the case of mass-type and charge-type infinities, where the subtraction was considered as an amalgamation. This is because it would necessarily result in a drastic change of the Maxwell equation for the radiation.

A way out of this difficulty was suggested:[14] it might be possible to introduce some fields which would give rise to the vacuum effect with the opposite sign so that a compensation method similar to the f-field theory might be used here. In fact, one finds, applying the same method, that a Pauli-Weisskopf field has this property.[15] An alternative possibility is to consider, in the style of Dirac's theory of the classical electron, that the "original equation" for the radiation contained, in the same way as the "original mass" of the electron, in itself an infinity with the opposite sign so that, supplemented with the infinity appearing as the result of the interaction, the equation for the observable field becomes just of the Maxwellian form.

The calculation of the level-shift of a bound electron was also undertaken.[16] This work is not yet completed but it was confirmed that the result converges by virtue of our subtraction prescription. We found further, in agreement with Schwinger, that a part of the radiative correction to the energy can be interpreted as caused by an anomalous moment of the electron the existence of which had been expected by Breit.[17]

We hope that various postwar difficulties will soon be settled and that our results will appear in print in the near future.

1 H. A. Bethe, Phys. Rev. **72**, 339 (1947).
2 H. W. Lewis, Phys. Rev. **73**, 173 (1948).
3 S. T. Epstein, Phys. Rev. **73**, 177 (1948).
4 J. Schwinger, Phys. Rev. **73**, 415 (1948).
5 F. Bloch and A. Nordsieck, Phys. Rev. **52**, 54 (1937).
6 W. Pauli and M. Fierz, Nuovo Cimiento, **15**, No. 3, 1 (1938).
7 A. Pais, Phys. Rev. **68**, 227 (1946).
8 S. Sakata, Prog. Theor. Phys. **2**, 30 (1947).
9 D. Ito, Z. Koba and S. Tomonaga, Prog. Theor. Phys. **2**, 216, 217 (L) (1947).
10 T. Tati and S. Tomonaga, lecture at the symposium on the theory of elementary particles, Nov. 1947. Full account of this lecture will be published in Progress of Theoretical Physics.
11 S. Tomonaga, Prog. Theor. Phys. **1**, 27 (1946); Z. Koba, T. Tati and S. Tomonaga, Prog. Theor. Phys. **2**, 101, 198 (1947).
12 Z. Koba and S. Tomonaga, Prog. Theor. Phys. **2**, 218 (L) (1947).
13 Z. Koba and G. Takeda, appearing in Prog. Theor. Phys.
14 M. Taketani, private conversation.
15 K. Baba, M. Sasaki and R. Suzuki, to be published in Prog. Theor. Phys. It was first pointed out by Sakata and Umesawa in the Nagoya University that the Pauli-Weisskopf field gives rise to a positive self-energy of a photon in contrast to a negative one due to the electron field and this would result in the compensation of the vacuum effect mentioned above.
16 Y. Nambu, to be published in Prog. Theor. Phys.
17 G. Breit, Phys. Rev. **72**, 984 (1947).

Note on the Above Letter: In transmitting to the Physical Review the accompaning review by Tomonaga of the remarkable work carried out in Japan in recent years, there is one technical note that may be helpful.

Tomonaga remarks in the fifth paragraph from the end that in addition to the infinite terms which may be recognized as contributions to mass and charge, there are other infinities which appear, particularly in the corrections to the Klein-Nishina formula. These have to do with the familiar problem of the light quantum self-energy. As long experience and the recent discussions of Schwinger and others have shown, the very greatest care must be taken in evaluating such self-energies lest, instead of the zero value which they should have, they give non-gauge covariant, non-covariant, in general infinite results. From manuscripts kindly sent by Tomonaga, I would conclude that the difficulties referred to in this note result from an insufficiently cautious treatment, and therefore inadequate identification, of light quantum self-energies.

J. R. OPPENHEIMER
Institute for Advanced Study
Princeton, New Jersey

On the Invariant Regularization in Relativistic Quantum Theory

W. PAULI AND F. VILLARS
Swiss Federal Institute of Technology, Zurich, Switzerland

(Received May 10, 1949)

The formal method of regularization of mathematical expressions of sums of products of different types of δ-functions is first applied to the example of vacuum polarization. It is emphasized that only a regularization of the whole expression without factorization leads to gauge invariant results. It is further shown, that for the regularization of the expression for the magnetic moment of the electron, a single auxiliary mass is sufficient, provided that different functions of the same particle (e.g., the photon functions $\bar{D}$ and $D^{(1)}$) are regularized in the same way and that the regularization of products of two electron functions is never factorized. The result is then the same as that of using Schwinger's method of introducing suitable parameters as new integration variables in the argument of δ-functions, without using any auxiliary masses.

§1. INTRODUCTION

In spite of many successes of the new relativistically invariant formalism of quantum electrodynamics,[1] which is based on the idea of "renormalization" of mass and charge, there are still some problems of uniqueness left, which need further clarification. The most important one seems to us to be the problem of the self-energy of the photon, which was raised by Wentzel's[2] remark that the formal application of Schwinger's original technique of integration to the resulting integral gives a finite result different from zero for this self-energy. This problem is formally contained in the more general problem of the gauge invariance for the resulting current due to vacuum-polarization by an arbitrary external field (not necessarily by a light wave). Schwinger[3] has shown that this current is given by

$$\langle j_\mu(x)\rangle = -\frac{i}{2}\int d^4x' \langle [j_\mu(x), j_\nu(x')]\rangle_0 \epsilon(x-x') A_\nu^{\text{ext}}(x')$$

when $\epsilon(x)=\pm 1$ for $t \gtrless 0$; $A_\nu^{\text{ext}}(x)$ is the vector potential of the external field and $\langle [j_\mu(x), j_\nu(x')]\rangle_0$ is the vacuum expectation value of the commulator of $j_\mu(x)$ with $j_\nu(x')$. The condition for the gauge invariance of this expression for $\langle j_\mu(x)\rangle$ (which includes the vanishing of the photon self-energy as a special case) is:

$$\partial/\partial x_\mu \{\langle [j_\mu(x), j_\nu(x')]\rangle_0 \epsilon(x-x')\} = 0.$$

Schwinger tried to prove the validity of this condition, after reducing it to the form

$$\langle [j_\mu(x), j_\nu(x')]\rangle_0 [\partial\epsilon(x-x')/\partial x_\mu] = 0,$$

by the argument that a time-like component of the current commutes with j_ν at all points of a space-like surface.[4] The specialization of the general invariant form of the commutator to this case, however, gives a result proportional to

$$\delta^{(3)}(x-x')(\partial\Delta^{(1)}/\partial x_\nu),$$

which is indeterminate due to the singularity of $\partial\Delta^{(1)}/\partial x_\nu$ on the light cone, which has the form $\sim x_\nu/(x_\sigma x_\sigma)$. The whole expression may therefore be written as

$$\delta^{(4)}(x-x')(\partial\Delta^{(1)}/\partial x_\nu),$$

in agreement with the straightforward computation (see §2 below).

The occurrence of products of functions with a δ-type singularity and with a pole is typical of the new formalism and seems to be the main source of the remaining uniqueness problems.

In order to overcome these ambiguities we apply in the following the method of regularization of Δ-functions (or products of them) with the help of an introduction of auxiliary masses. This method has already a long history. Much work has been done to compensate the infinities in the self-energy of the electron with the help of auxiliary fields corresponding to other neutral particles with finite rest-masses interacting with the electrons.[5] Some authors assumed formally a negative energy of the free auxiliary particles, while others did not need these artificial assumptions and could obtain the necessary compensations by using the different sign of the self-energy of the electron due to its interaction with different kinds of fields (for instance scalar fields *vs.* vector fields). We shall denote these theories, in which the auxiliary particles with finite masses and positive energy are assumed to be observable in principle and are described by observables entering the Hamil-

[1] S. Tomonaga, Prog. Theor. Phys. **1**, 27 (1946). J. Schwinger, Phys. Rev. **74**, 1439 (1948); Phys. Rev. **75**, 651 (1949); Phys. Rev. **75**, 1912 (1949). These papers are quoted in the following as SI, SII, SIII. Our notations follow as closely as possible those of these papers. For the definitions and the properties of the functions Δ, **Δ**: (**Δ** is identical with $\bar{\Delta}$ in SII), and $\Delta^{(1)}$ see particularly the appendix of SII. In this paper natural units $\hbar=c=1$ are used throughout. F. J. Dyson, Phys. Rev. **75**, 486 (1949), and Phys. Rev. **75**, 1736 (1949). In the following quoted as DI, and DII.

[2] G. Wentzel, Phys. Rev. **74**, 1070 (1948).

[3] SII, Eq. (2.19).

[4] SII, Eq. (2.29).

[5] Compare for older literature (including his own contributions), A. Pais, *The Development of the Theory of the Electron* (Princeton University Press, Princeton, New Jersey, 1948).

tonian explicitly as "realistic," in contrast to "formalistic" theories, in which the auxiliary masses are used merely as mathematical parameters which may finally tend to infinity. Recently the "realistic" standpoint was extended to the problem of the cancellation of the singularities in the vacuum polarization, due to virtual electron-positron pairs generated by external fields, by introducing auxiliary pairs of particles with opposite electric charges and masses different from that of the electron.[6] It was shown that the signs of the polarization effect allow compensation of the singularities only if the auxiliary particles are assumed to obey Bose-Statistics. Until now it was not possible to carry through the "realistic" standpoint to include all possible effects in higher order approximations in the fine-structure constant, nor is it proven that this problem is not overdetermined. Presumably a consistent "realistic" theory will only be possible if, from the very beginning, all observables entering the theory have commutation rules and vacuum expectation values free from singularities, i.e., different from the Δ and $\Delta^{(1)}$ functions which obey a wave equation corresponding to a given mass value. Until now, however, it has not been possible to carry through such a program.

At the present stage of our knowledge it is therefore of interest to investigate further the "formalistic" use of auxiliary masses in relativistic quantum theory. This was done independently by Feynman[7] and by Stueckelberg and Rivier.[8] The latter authors use (more generally) an arbitrary number of auxiliary masses, while the former introduces only a single large auxiliary mass, which was sufficient for his particular problem, the regularization of the self-energy of the electron. From the well-known expansions of the Δ- and $\Delta^{(1)}$-functions near the light cone it can easily be seen (see §2) that in the linear combinations:

$$\Delta_R(x)=\sum_i c_i\Delta(x;M_i)$$

and

$$\Delta_R^{(1)}(x)=\sum_i c_i\Delta^{(1)}(x;M_i),$$

the strongest singularities cancel, if

$$\sum_i c_i=0,$$

and the remaining singularities (finite jumps and logarithmic singularities) also cancel if in addition the condition

$$\sum_i c_iM_i^2=0$$

holds. If the first condition alone is sufficient to guarantee the regularity of a certain result, it is obvious that a single auxiliary mass M_1 (besides the original electron mass $M_0=m$) with $c_1=-c_0=-1$ is all that is necessary. It should not be forgotten, however, that Feynman's success in using a single auxiliary mass in the problem of the self-energy of the electron implies the assumption that in the corresponding expression[9] resulting from the invariant form of perturbation theory the photon functions $\bar{D}$ and $D^{(1)}$ have both to be regularized with the *same* auxiliary mass. (A formal alternative would be to leave the photon-functions unchanged, but to regularize the electron-functions Δ and $\Delta^{(1)}$ with the same auxiliary mass, or to regularize the whole expression without factorization and with one auxiliary mass.)

The application of the formal method of mass-regularization to the problem of vacuum-polarization[10] (§4) shows that not only the use of a single auxiliary mass is here insufficient, but that any regularization of Δ- or $\Delta^{(1)}$-functions as separate factors leads to results that are not gauge invariant. As was shown by Rayski[11] only the regularization of the whole expression for the resulting current (without factorization) gives satisfactory results in this case. The formal use of continuous mass distributions is here particularly suited to illustrate the connection between the different results of Wentzel and Schwinger for the photon self-energy.

In §5 the example of the correction to the magnetic moment of the electron, which is one of the main results of Schwinger, is treated from the "formalistic" standpoint of mass regularization. We agree with Schwinger that the use of auxiliary masses is not necessary in this case if the computations in momentum space are made with sufficient care (see additional remark A). In any case (different from the situation in the problem of vacuum-polarization), the use of a single auxiliary mass is here sufficient to avoid any ambiguity, provided that the same mass is applied both to the $\bar{D}$ and the $D^{(1)}$-functions of the photon, analogous to Feynman's method for the self-energy of the electron, or that the regularization is applied to the products of two Δ- and $\Delta^{(1)}$-functions without factorization[12] (see reference 20).

[6] G. Rayski, Acta Phys. Polonica **9**, 129 (1948). (Only light waves as external electromagnetic field are considered in this paper.) Umesawa, Yukawa, and Yamada, Prog. Theor. Phys. **3**, No. 3, 317 (1948).

[7] R. P. Feynman, Pocono Conference 1948; Phys. Rev. **74**, 1439 (1948). Applications by V. F. Weisskopf and J. B. French, Phys. Rev. **75**, 1240 (1949).

[8] E. C. G. Stueckelberg and D. Rivier, Phys. Rev. **74**, 218 and 986 (1948). D. Rivier, Helv. Phys. Acta **XXII**, 265 (1949).

[9] SII, Eqs. (3.77) and (3.82).

[10] Dyson (see DII) applies to this problem a method of regularization without use of auxiliary masses, which is more similar to the methods used in the earlier stages of positron theory.

[11] Rayski made this proposal in the summer of 1948, during his investigations on the photon self-energy of *Bosons* (see reference 6). With his friendly consent we later resumed his work and generalized the method for arbitrary external fields (not necessarily light waves).

[12] The problem of the magnetic moment of nucleons due to a mesonic interaction, which shows a close analogy to the problem of the magnetic moment of the electron due to electromagnetic interaction, is not treated in this paper. Stueckelberg-Rivier give (see reference 8) a formula for the magnetic moment of the neutron which they characterize as not leading to a definite numerical value. A justification of this may, in principle, be seen in the fact that the most general form of regularization with auxiliary masses must always lead to an arbitrary value for integrals of this type. On the other hand the mentioned general analogy between the two cases makes it plausible that the same mathematical methods which lead to an unambiguous definition of the magnetic moment of the electron will also lead to an unique definition of the value of the theoretical results for the magnetic moments of the nucleons (at least for scalar and pseudo-

Both groups of authors (Stueckelberg-Rivier[8] and Feynman-Dyson[13]) seem to ascribe to a particular combination of Δ-functions, which describes outgoing waves for the future and incoming waves for the past, an important or even fundamental significance. As this question can be left open for the purpose of this paper, we discuss Dyson's expression for the magnetic moment of the electron, in which the function Δ_c for the electron and D_c for the photon alone occur,[14] only in a brief additional remark (B; §5). We believe that in order to investigate the range of applicability of the particular function Δ_c, the discussion of more complicated examples will be necessary.

Summarizing, one must admit that the additional rules which the "formalistic" standpoint has to use (e.g., to apply the same mass values for **Δ**- and $\Delta^{(1)}$-functions, and not to factorize the regularization of products of **Δ**- and $\Delta^{(1)}$-functions corresponding to pairs of charged particles) could be immediately understood from the "realistic" standpoint and appears as if borrowed from the latter.[15] It seems very likely that the "formalistic" standpoint used in this paper and by other workers can only be a transitional stage of the theory, and that the auxiliary masses will eventually either be entirely eliminated, or the "realistic" standpoint will be so much improved that the theory will not contain any further accidental compensations.

§2. THE BASIC CONCEPTS OF REGULARIZATION

In an invariant perturbation theory, such as the one introduced by Schwinger into quantum electrodynamics, the two invariant functions, **Δ** and $\Delta^{(1)}$, play an essential role. Vacuum expectation values of properly symmetrized products of field operators are expressed in terms of $\Delta^{(1)}$, while **Δ** appears in connection with the covariant formulation of commutation rules.

The handling of expressions involving **Δ**- and $\Delta^{(1)}$-functions exhibit some characteristic difficulties, which may be summarized as follows:

(a) The occurrence of *indeterminate* expressions as a consequence of the coincidence of the δ-type singularity of $\mathbf{\Delta}(x)$ with the pole of $\Delta^{(1)}(x)$ on the light cone. Only a properly defined limiting process may give them a definite meaning.

(b) The necessity of taking into account, in the course of the calculation, the "covariance" of some diverging (however formally covariant) expression in order to split off a finite part. This too may be done in a proper way only after these expressions have been made finite by a regularization process.

Since both difficulties are connected with the singular features of the **Δ**- and $\Delta^{(1)}$-functions on the light cone, an invariant elimination of these singularities may be helpful in an attempt to escape the above-mentioned complications. Looking for such a device, one is guided by the dependence of **Δ** and $\Delta^{(1)}$ on the rest mass of the corresponding field. This dependence is exhibited in the integral representations:

$$\Delta^{(1)}(x) = -(m^2/2\pi^2)\int_0^\infty d\alpha \sin[\lambda m^2\cdot\alpha + (1/4\alpha)], \quad \text{(1a)}$$

$$\mathbf{\Delta}(x) = (m^2/4\pi^2)\int_0^\infty d\alpha \cos[\lambda m^2\cdot\alpha + (1/4\alpha)], \quad \text{(1b)}$$

where

$$\lambda = -x_\mu x_\mu,^* \quad (2)$$

which show that both $\Delta^{(1)}$ and **Δ** are of the form:

$$m^2\cdot fu(\lambda m^2).$$

From this it follows that δ-type singularities ($\delta(\lambda)$) and first order poles ($1/\lambda$) are independent of m, whereas finite jumps and logarithmic singularities are proportional to m^2. Since these are the only types of singularities occurring in **Δ** and $\Delta^{(1)}$, they may be avoided by introducing the regularized invariant functions $\mathbf{\Delta}_R$ and $\Delta_R^{(1)}$ is:

$$\Delta_R^{(1)} = \sum_i c_i\Delta^{(1)}(M_i), \quad \mathbf{\Delta}_R = \sum_i c_i\mathbf{\Delta}(M_i), \quad (3)$$

where c_i satisfies the conditions:

$$\sum_i c_i = 0 \quad \text{(I)}$$

$$\sum_i c_i M_i^2 = 0. \quad \text{(Ia)}$$

In order to exhibit more clearly the efficacy of these conditions we give the development of $\Delta^{(1)}$ and **Δ** for small λ (omitting all terms vanishing for $\lambda=0$):

$$\Delta^{(1)}(x) = \frac{1}{4\pi}\left\{\frac{-2}{\lambda} + m^2 \log\frac{\gamma}{2}(|\lambda|m^2)^{\frac{1}{2}} - \frac{m^2}{2} + \cdots\right\} \quad (4)$$

$$\mathbf{\Delta}(x) = \frac{1}{4\pi}\left\{\delta(\lambda) + \left(\frac{m^2}{4} + \cdots\right)\theta^+(\lambda)\right\} \quad (5)$$

where

$$\theta^+(x) = \begin{Bmatrix}1\\0\end{Bmatrix} \quad \text{for} \quad x \gtrless 0.$$

scalar mesons). Meanwhile K. M. Case, Phys. Rev. **76**, 1 (1949), obtains an unambiguous result (from invariant perturbation theory) for the magnetic nucleon moments which agrees completely with those of Luttinger (Helv. Phys. Acta **XXI**, 483 (1948)). He does not give the details of his evaluation of the integrals, for which no auxiliary masses are needed.

[13] Compare DI and DII. The function in question is denoted with D_c by Stueckelberg-Rivier, with D_F by Dyson and with Δ_F by Case. We use the notation Δ_c and D_c for the corresponding electron and photon functions respectively.

[14] DI, Section X, formula for L.

[15] The interesting problem of the "self-stress" of the electron (see A. Pais, reference 5; in an unpublished letter of last year Pais gave the result that in the theory of holes the value of this self-stress is finite, namely $\alpha/2\pi\cdot m$ (α=fine-structure constant), but not zero, as special relativity requires for the total stress of a closed system) may throw more light on the relations between the two standpoints. Detailed calculations by one of us (F.V.) gave the result that a formal regularization with auxiliary masses does not change the finite value of Pais for the self-stress; one has therefore either to consider the localization of energy in space and time as a non-physical concept in quantum theory and to admit only the energy-momentum vector (which is already integrated over space-time), or one has to ascribe to the compensating auxiliary masses a physical reality such that their contribution to the stress in the intermediate states compensates the other part of the self-stress of the electron.

* In the following $x_4 = ix_0 = it$.

It is easily seen that $\mathbf{\Delta}_R$ vanishes for $\lambda=0$, such that

$$\Delta(x)=-2\epsilon(x)\mathbf{\Delta}(x)$$

is regularized too. $\Delta_R^{(1)}$, however, takes the value

$$\frac{1}{4\pi}\sum_i c_i M_i^2 \log M_i \quad \text{for} \quad \lambda=0.$$

It is the meaning of the regularization prescriptions that the first term in the series (3) represents the non-regularized function itself, i.e., that

$$c_0=1, \quad M_0=m$$

and that all M_i ($i>0$) should finally tend to ∞ (according to the "formalistic" standpoint, adopted in the following). The coefficients c_i need hereby not remain finite. We shall, however, impose the condition

$$\sum_i{}' (|c_i|/M_i^2)\to 0 \tag{6}$$

which ensures that

$$\sum_i{}' c_i F(M_i^2)\to 0 \quad \text{if only}$$
$$|M_i^2 F(M_i^2)|<A \quad \text{for all} \quad i>0.$$

For the purposes of a general discussion it may sometimes be advantageous to replace the discrete spectrum of auxiliary masses by a continuous one (including, or course, the discrete as a special case):

$$\mathbf{\Delta}_R(x)=\int_{-\infty}^{+\infty} d\kappa\rho(\kappa)\mathbf{\Delta}(x;\kappa), \quad \text{etc.} \tag{7}$$

where κ has the signification of the square of a mass.

The conditions (I, Ia) now read:

$$\int d\kappa\rho(\kappa)=0 \tag{I$'$}$$

$$\int d\kappa\kappa\rho(\kappa)=0. \tag{I$'$a}$$

On writing $\rho(\kappa)$ as $\delta(\kappa-m^2)+\rho_1(\kappa)$ condition (6) takes the form

$$\int_a^\infty d\kappa(|\rho_1(\kappa)|/\kappa)\to 0. \quad a>0 \tag{8}$$

An alternative possibility of regularization is contained in the prescription

$$F_R=\int d\kappa\rho(\kappa)F(\Delta^{(1)}(\kappa), \mathbf{\Delta}(\kappa)) \tag{9}$$

where F represents some (bilinear or higher order) form in $\mathbf{\Delta}$, $\Delta^{(1)}$, $d\mathbf{\Delta}/d\lambda$, $d\Delta^{(1)}/d\lambda$.

The use of the prescription (9) needs further explanation: The form F may contain both Δ-(electron) and D-(photon) functions. We may then either regularize the expression F as a whole (both Δ- and D-functions without factorization):

$$F_R=\int d\kappa\rho(\kappa)F(\Delta^{(1)}(\kappa'), D^{(1)}(\kappa); \mathbf{\Delta}(\kappa'), \bar{D}(\kappa))$$
$$\kappa'=(m^2+\kappa)^{\frac{1}{2}},$$

or we may also regularize F only with respect to the D-function referring to one type of field, e.g.:

$$F_R=\int d\kappa\rho(\kappa)F(\Delta^{(1)}(m), D^{(1)}(\kappa); \mathbf{\Delta}(m), \bar{D}(\kappa)).$$

Whenever in this latter case F is linear with respect to the D-function to be regularized, (9) reduces to (3) (i.e., the introduction of individually regularized D-functions), but implies the important additional rule that the same regulator $\rho(\kappa)$ has to be applied to both $\bar{D}$ and $D^{(1)}$. If, on the contrary, F is bilinear with respect to the field in question, all these bilinear terms have to be regularized without factorization and with the same regulator. In this latter case the conditions (I$'$), (I$'$a) are then, in general, not sufficient to remove all singularities from (9). They eliminate however the strongest ones, especially those of the type $\delta(\lambda)/\lambda$.

The rule (9), interpreted in the above-explained sense, will be adopted in the following throughout. It is this rule that assures the gauge invariance of the polarization current in the problem of vacuum polarization—in contrast to the results of (3).

One may object that this prescription suffers from a lack of uniqueness, but this apparent deficiency affects only the mass and charge renormalization terms. Hereby we mean, more precisely, that after mass and charge terms have been removed, all additional corrective terms *shall be independent of the way they are regularized, and shall, of course, be independent of the parameters* c_i *and* M_i *in the limit* $M_i\to\infty$ (*or* $\rho_1(\kappa)\to 0$ *for any finite* κ). This is not the case if in the form F, (9), the individual summands are regularized independently and differently:

$$F=F_1+F_2; \quad F_R=\int d\kappa\rho_a(\kappa)F_1(\kappa)+\int d\kappa\rho_b(\kappa)F_2(\kappa);$$

a quite arbitrary result may then be obtained, as will be shown later on (see §5, additional remark C). The charge and mass terms themselves, however, depend on the way they are regularized, since they depend on

$$\sum_i c_i \log M_i \quad \text{or} \quad \int d\kappa\rho(\kappa)\log\kappa. \tag{10}$$

In connection with the use of the Fourier-integral representation of $\mathbf{\Delta}$ and $\Delta^{(1)}$ (as in (1a, b)) for computational purposes, it is convenient to have conditions (I$'$, I$'$a) expressed in terms of the Fourier transformed of $\rho(\kappa)$:

$$R(a)=\int d\kappa\rho(\kappa)e^{-ia\kappa}. \tag{11}$$

The conditions (I′, I′a) then read

$$R(0)=0 \qquad (\mathrm{I}'')$$
$$R'(0)=0 \qquad (\mathrm{I}''\mathrm{a})$$

while the integral (10) is transformed into

$$-\frac{1}{2}\int_{-\infty}^{+\infty}\frac{da}{a}\epsilon(a)R(a). \qquad (10')$$

$$(\epsilon(a)=\pm 1 \quad \text{for} \quad a\gtrless 0).$$

§3. THE INVARIANT PERTURBATION THEORY

Let $\psi(x)$, $\bar{\psi}(x)=\psi^*(x)\beta$ be the quantized operators of the electron field and $A_\mu(x)$ the four potential representing the radiation field. ψ, $\bar{\psi}$ and A_μ are supposed to satisfy the equations of the uncoupled fields:

$$[\gamma_\nu(\partial/\partial x_\nu)+m]\psi=0, \qquad (12)$$

$$(\partial\bar{\psi}/\partial x_\nu)\gamma^\nu-m\bar{\psi}=0, \qquad (12\mathrm{a})$$

$$\Box^2 A_\mu=0, \qquad (13)$$

$$(\partial A_\mu/\partial x_\mu)\Psi(t)=0, \qquad (14)$$

and accordingly, obey the commutation relations:

$$[A_\mu(x),A_\nu(x')]=i\delta_{\mu\nu}D(x-x') \qquad (15)$$

$$\{\psi_\alpha(x),\bar{\psi}_\beta(x')\}=\frac{1}{i}S_{\alpha\beta}(x-x')$$

$$=\frac{1}{i}\left(\gamma^\nu\frac{\partial}{\partial x_\nu}-m\right)_{\alpha\beta}\Delta(x-x'). \qquad (16)$$

$$[A,B]=AB-BA, \quad \{A,B\}=AB+BA.$$

The auxiliary condition (14) involves the state vector Ψ of the system, whose equation of motion is given by

$$i(\partial\Psi/\partial t)=H\Psi. \qquad (17)$$

Herein H represents the interaction energy:

$$H=-\int d^3x j_\mu(x)A_\mu(x)$$

$$=-ie\int d^3x\bar{\psi}(x)\gamma^\mu\psi(x)\cdot A_\mu(x). \qquad (18)$$

The solution of (17) may be achieved in successive approximations by a set of unitary transformations:

$$\Psi=e^{-iS_1(t)}\Psi_1=e^{-iS_1(t)}e^{-iS_2(t)}\Psi_2=\cdots=U(t)\Psi_K, \quad (19)$$

where Ψ_K is time-independent in the desired approximation and thus represents some "free-particle"-state.

The change in properties of an operator Ω referring say to the field a, due to the interaction of a with the vacuum of the field b, is then expressed in the vacuum expectation value (with respect to b) of the transformed operator $\mathbf{\Omega}$:

$$\langle\mathbf{\Omega}\rangle_{\mathrm{vac}(b)}=\langle U^{-1}\Omega U\rangle_{\mathrm{vac}(b)}. \qquad (20)$$

Sometimes it may be convenient to separate the electromagnetic field (and its sources) from the system under consideration and treat it as a given (c-number) field: $A_\mu^{\mathrm{ext}}(x)$, satisfying:

$$\Box^2 A_\mu^{\mathrm{ext}}(x)=-J_\mu^{\mathrm{ext}}(x). \qquad (21)$$

As before, "states" are represented by a time-independent state vector Ψ_K and a transformation formally analogous to (19), introducing Ψ_K, exhibits the change, induced by the presence of A_μ^{ext}, in the expectation value of an operator Ω.

The transformation u, (19), shall be written, more precisely, as

$$u(t)=e^{-iS_1(t)}N_1e^{-iS_2(t)}N_2\cdots, \qquad (22)$$

where $S_1(t)$ is thought to remove the first-order coupling, $S_2(t)$ the remaining second-order interaction, etc. Between these steps may take place renormalization transformations N, defining new matter field operators which obey equations with adjusted mass.

According to this program, S_1 is defined by:

$$\dot{S}_1=H$$

leaving for Ψ_1 the equation of motion

$$i(\partial\Psi_1/\partial t)=(i/2[S_1,H]+\cdots)\Psi_1. \qquad (23)$$

Then

$$\dot{S}_2=i/2[S_1,H]-H_{\mathrm{self}} \qquad (24)$$

leading thus to a state vector Ψ_2 which is constant in time up to and including terms in e^2, provided a mass renormalization has taken place, removing H_{self} from $i/2[S_1,H]$.

Restricting ourselves to this order of approximation, a transformed operator $\mathbf{\Omega}$, according to (20), may be written as follows:

$$\mathbf{\Omega}=\Omega+i[S_1,\Omega]-\tfrac{1}{2}[S_1[S_1,\Omega]]+i[S_2,\Omega]$$

$$\mathbf{\Omega}(t)=\Omega(t)+i\int^t dt'[H(t'),\Omega(t)]$$
$$-\tfrac{1}{2}\int^t dt'\int^t dt''[H(t')[H(t''),\Omega(t)]]$$
$$+i\int^t dt'\left[\frac{i}{2}[S_1(t'),H(t')]-H_s(t'),\Omega(t)\right]$$

or

$$\mathbf{\Omega}(t)=\Omega(t)+i\int^t dt'[H(t'),\Omega(t)]$$
$$-\int^t dt'\int^{t'} dt''[H(t'')[H(t'),\Omega(t)]]$$
$$-i\int^t dt'[H_s(t'),\Omega(t)]. \quad (25)^{**}$$

** To obtain this latter form a simplification due to Schwinger has been used, which employs Jacobi's identity.

With the help of the commutators (15), (16) and the derived relation:

$$[\bar{\psi}(x)\gamma^\mu\psi(x), \bar{\psi}(x')\gamma^\nu\psi(x')] = 1/i\{\bar{\psi}(x)\gamma^\mu S(x-x')\gamma^\nu\psi(x') - \bar{\psi}(x')\gamma^\nu S(x'-x)\gamma^\mu\psi(x)\}, \quad (26)$$

the transformed operator $\mathbf{\Omega}$ is easily evaluated. To have its expectation value for a definite state (for instance the vacuum with respect to one of the fields in question) we need still know the vacuum expectation values of the properly symmetrized products:

$$\langle\{A_\mu(x), A_\nu(x')\}\rangle_0 = \delta_{\mu\nu}D^{(1)}(x-x') \quad (27)$$

$$\langle[\psi_\alpha(x), \bar{\psi}_\beta(x')]\rangle_0 = -S_{\alpha\beta}^{(1)}(x-x'). \quad (28)$$

As examples which will be discussed later on, we give:

(a) The current induced in the (matter-) vacuum by an external electromagnetic field A_μ^{ext}, or its source J_μ^{ext}. The desired approximation is achieved with S_1 and yields immediately:

$$\langle j_\mu(x)\rangle = i\int^t dt'\langle[H^{\text{ext}}(t'), j_\mu(x)]\rangle_0$$

$$= -4e^2\int d^4x' K_{\mu\nu}(x-x')A_\nu^{\text{ext}}(x') \quad (29)$$

where

$$K_{\mu\nu}(x-x') = \frac{1}{8ie^2}\langle[j_\mu(x), j_\nu(x')]\rangle_0\epsilon(x-x'). \quad (30)^{***}$$

(b) The e^2-radiative correction to the expression for the current associated with the matter field. According to (25) this correction is contained in

$$\Delta j_\mu(x) = -\int^t dt'\int^{t'} dt''\langle[H(t'')[H(t'), j_\mu(x)]]\rangle_{\text{phot vac}}$$

$$-i\int^t dt'[H_{\text{self}}(t'), j_\mu(x)]. \quad (31)$$

The one particle part (for the definition of the concept of the one particle part of an operator, see SII, page 671) of $\Delta j_\mu(x)$ includes a term of the form

$$i_\mu^{(s)}(x) = \text{const.}\frac{\partial}{\partial x_\nu}(\bar{\psi}(x)\sigma^{\mu\nu}\psi(x)), \quad (32)$$

describing an anomalous g-factor of the electron.

§4. VACUUM POLARIZATION AND PHOTON SELF-ENERGY

In this section the tensor $K_{\mu\nu}(x-x')$ (30) shall be investigated. Invariance of the induced current with respect to the gauge transformation $A_\mu \rightarrow A_\mu + (\partial\Lambda/\partial x_\mu)$ gives

$$0 = \int d^4x' K_{\mu\nu}(x-x')(\partial\Lambda/\partial x_\nu'),$$

which requires:

$$\frac{\partial K_{\mu\nu}(x)}{\partial x_\nu} = 0. \quad (33)$$

In momentum space, where (29) reads:

$$\langle j_\mu(p)\rangle = -4e^2K_{\mu\nu}(p)A_\nu(p), \quad (34)$$

the condition (33) is equivalent to

$$K_{\mu\nu}(p)\cdot p_\nu = 0. \quad (33')$$

On writing:

$$K_{\mu\nu}(p) = K_1 p_\mu p_\nu + K_2\delta_{\mu\nu}p_\lambda p_\lambda. \quad (35)$$

Equation (33′) is equivalent to the condition

$$K_1 = -K_2 \quad (36)$$

(both K_1 and K_2 are still functions of the invariant $p_\lambda p_\lambda$). Since A_μ^{ext} and the current J_μ^{ext} generating this external field are connected through

$$J_\mu^{\text{ext}}(p) = (\delta_{\mu\nu}p_\lambda p_\lambda - p_\mu p_\nu)A_\mu^{\text{ext}}(p),$$

condition (36) assures that:

$$\langle j_\mu(p)\rangle = 4e^2 K_1 J_\mu^{\text{ext}}(p). \quad (37)$$

An evaluation of $K_{\mu\nu}$ with the help of relations (26) and (28) gives:

$$K_{\mu\nu}(x) = \frac{\partial\Delta}{\partial x_\mu}\cdot\frac{\partial\Delta^{(1)}}{\partial x_\nu} + \frac{\partial\Delta}{\partial x_\nu}\cdot\frac{\partial\Delta^{(1)}}{\partial x_\mu} - \delta_{\mu\nu}\left(\frac{\partial\Delta}{\partial x_\lambda}\cdot\frac{\partial\Delta^{(1)}}{\partial x_\lambda} + m^2\Delta^{(1)}\Delta\right). \quad (38a)$$

From the expansions (4, 5) it follows that this expression is indeterminate on the light cone, due to terms of the type $\delta(\lambda)/\lambda$, and the relation (33) yields, on account of

$$(\Box^2 - m^2)\Delta^{(1)} = 0, \quad (\Box^2 - m^2)\Delta = -\delta(x);$$

$$\frac{\partial K_{\mu\nu}(x)}{\partial x_\nu} = -\frac{\partial\Delta^{(1)}}{\partial x_\mu}\delta(x),$$

which is indeterminate, too.

In momentum space, where $K_{\mu\nu}$ is given by

$$K_{\mu\nu}(p) = \frac{1}{(2\pi)^3}\int d^4k\frac{\delta(k_\lambda k_\lambda + m^2)}{(k_\lambda - p_\lambda)^2 + m^2}\{2k_\mu k_\nu - k_\mu p_\nu - k_\nu p_\nu - \delta_{\mu\nu}(-k_\lambda p_\lambda + k_\lambda k_\lambda + m^2)\} \quad (38b)$$

this ambiguity is less manifest, since

$$K_{\mu\nu}(p)\cdot p_\nu = \frac{1}{(2\pi)^3}\int d^4k k_\mu\delta(k_\lambda k_\lambda + m^2) \quad (39)$$

*** The factor $\epsilon(x-x')$ is introduced by writing:

$$\int_{-\infty}^t dt' = \tfrac{1}{2}\int_{-\infty}^{+\infty} dt'\epsilon(t-t') + \tfrac{1}{2}\int_{-\infty}^{+\infty} dt';$$

the second integral vanishes, if no real transition is induced by the external field.

(hereby terms of the form $\delta(k_\lambda k_\lambda+m^2)\cdot(k_\lambda k_\lambda+m^2)$ have been omitted), an expression which may well be put equal to zero for symmetry reasons. Since, however, $K_{\mu\nu}(\rho)$ is represented by a divergent Fourier-integral, this property may be lost in the course of a direct computation of $K_{\mu\nu}$. Note that conditions (I,I a) are just those necessary to make the integral (39) convergent!

An evaluation of $K_{\mu\nu}$ is most conveniently done with the help of the Fourier-integral representations:[16]

$$\Delta^{(1)}(k)=2\pi\delta(k_\lambda k_\lambda+m^2)=\int_{-\infty}^{+\infty}d\alpha\,\exp[i\alpha(k_\lambda k_\lambda+m^2)] \tag{40}$$

$$\mathbf{\Delta}(k)=P\frac{1}{k_\lambda k_\lambda+m^2}=\frac{1}{2i}\int_{-\infty}^{+\infty}d\beta\epsilon(\beta)\,\exp[i\beta(k_\lambda k_\lambda+m^2)] \tag{41}$$

(P=principal value). After introducing a new variable

$$k'=k-(\beta/\alpha+\beta)\cdot p$$

and with

$$\int d^4k\,\exp(iak_\lambda k_\lambda)=(i\pi^2/a^2)\epsilon(a)$$

$$\int d^4kk_\mu k_\nu\,\exp(iak_\lambda k_\lambda)=-\delta_{\mu\nu}(\pi^2/2a^3)\epsilon(a),$$

we are left (after symmetrizing the integrand with respect to α and β) with

$$K_{\mu\nu}(p)=\frac{\pi^2}{4(2\pi)^4}\int\int d\alpha d\beta(\epsilon(\alpha)+\epsilon(\beta))\frac{\epsilon(\alpha+\beta)}{(\alpha+\beta)^2}\times\exp\left[i\frac{\alpha\beta}{\alpha+\beta}p_\lambda p_\lambda+i(\alpha+\beta)m^2\right]\left\{\frac{-2\alpha\beta}{(\alpha+\beta)^2}p_\mu p_\nu+\delta_{\mu\nu}\left(\frac{\alpha\beta}{(\alpha+\beta)^2}p_\lambda p_\lambda-m^2-\frac{i}{\alpha+\beta}\right)\right\}. \tag{42}$$

It is at this step that our regularization device comes into play, replacing $\exp[i(\alpha+\beta)m^2]$ by $R(\alpha+\beta)$, according to (9) and (11), and $m^2\exp[i(\alpha+\beta)m^2]$ by $(1/i)R'(\alpha+\beta)$.[17] The regularized Eq. (42) reads then:

$$K_{R(\mu\nu)}=\frac{\pi^2}{4(2\pi)^4}\int\int d\alpha d\beta(\epsilon(\alpha)+\epsilon(\beta))\frac{\epsilon(\alpha+\beta)}{(\alpha+\beta)^2}\times\exp\left(i\frac{\alpha\beta}{\alpha+\beta}p_\lambda p_\lambda\right)\left\{\frac{2\alpha\beta}{(\alpha+\beta)^2}R(\alpha+\beta)\times(-p_\mu p_\nu+\delta_{\mu\nu}p_\lambda p_\lambda)-\delta_{\mu\nu}\left[\frac{\alpha\beta}{(\alpha+\beta)^2}p_\lambda p_\lambda R(\alpha+\beta)+i\frac{R(\alpha+\beta)}{\alpha+\beta}-iR'(\alpha+\beta)\right]\right\}.$$

Expressed in terms of the variables

$$z=\alpha+\beta$$
$$y=1/z(\alpha-\beta),$$

which give

$$\epsilon(\alpha)+\epsilon(\beta)=2\epsilon(z)\quad\text{for}\quad|y|\leq1,$$

and 0 elsewhere; and

$$d\alpha d\beta=\tfrac{1}{2}|z|dzdy$$

$K_{R(\mu\nu)}$ reads:

$$K_{R(\mu\nu)}=\frac{\pi^2}{4(2\pi)^4}\int_{-1}^{+1}dy\int_{-\infty}^{+\infty}\frac{dz}{z}\epsilon(z)\exp\left[i\frac{z}{4}(1-y^2)p_\lambda p_\lambda\right]\times\left\{R(z)\frac{1-y^2}{2}(-p_\mu p_\nu+\delta_{\mu\nu}p_\lambda p_\lambda)-\delta_{\mu\nu}\left[R(z)\frac{1-y^2}{4}p_\lambda p_\lambda+\frac{i}{z}R(z)-iR'(z)\right]\right\}. \tag{43}$$

---

[16] S. T. Ma (Phys. Rev. **75**, 1264 (1949)) has evaluated $K_{\mu\nu}(p)$ by means of elementary momentum-space integrations, using the method due to Pauli and Rose (Phys. Rev. **49**, 462 (1936)). His results are neither gauge nor Lorentz-invariant, due to the presence of an additional constant term Γ in his expression for K_{ii} ($i=1, 2, 3$). As a consequence, this term appears also in the trace of $K_{\mu\nu}$:

$$K_{\mu\mu}=-3p_\mu p_\mu K_1+3\Gamma,$$

whereas, according to (35) and (36) $K_{\mu\mu}$ should be proportional to $(\rho_\mu\rho_\mu)$. But it is easily shown that the introduction of a regulator in his calculations makes the additional term Γ vanish. Indeed, we have

$$(K_{\lambda\lambda})_R=(2\pi)^{-3}\Sigma_i\,c_i\int d^4k\delta(k_\lambda k_\lambda+M_i^2)\left(\frac{p_\lambda p_\lambda-2M_i^2}{p_\lambda p_\lambda-2\kappa_\lambda p_\lambda}-1\right).$$

It follows for a time like vector $(p_\lambda=o,ip_0)$:

$$(K_{\lambda\lambda})_R=p_0^2(2\pi)^{-3}\Sigma_i\,c_i\int_0^\infty\frac{dkk^2}{\Omega_i(k)}\frac{2\Omega_i^2-M_i^2}{\Omega_i^2(p_0^2-4\Omega_i^2)}-2(2\pi)^{-3}\Sigma_i\,c_i\int_0^\infty\frac{k^2dk}{\Omega_i}\qquad[\Omega_i=(k^2+M_i^2)^{\frac{1}{2}}].$$

It is this second term that destroys covariance and gauge invariance; but since it can be written as

$$-(1/(2\pi)^3)\Sigma_i\,c_i(K^2-\tfrac{1}{2}M_i^2)_{K\to\infty}$$

we see that it vanishes on account of the two conditions (I, Ia).

[17] It may perhaps be helpful to show how a factorized regulator destroys gauge invariance. Taking a discrete spectrum of auxiliary masses, we have:

$$\frac{\partial K_{R(\mu\nu)}}{\partial x_\nu}=\sum_{ij}c_ic_j\left\{\frac{\partial\mathbf{\Delta}_i}{\partial x_\mu}\cdot M_j^2\Delta_j^{(1)}+M_i^2\mathbf{\Delta}_i\frac{\partial\Delta_j^{(1)}}{\partial x_\mu}-\frac{\partial}{\partial x_\mu}(M_iM_j\mathbf{\Delta}_i\Delta_j^{(1)})-\delta(x)\frac{\partial\Delta_j^{(1)}}{\partial x_\mu}\right\}$$
$$=\sum_{ij}c_ic_j(M_j-M_i)\left\{M_j\frac{\partial\mathbf{\Delta}_i}{\partial x_\mu}\cdot\Delta_j^{(1)}-M_i\mathbf{\Delta}_i\frac{\partial\Delta_j^{(1)}}{\partial x_\mu}\right\}$$

which never vanishes identically.

The contribution of the last term in the bracket may be written as

$$\frac{\pi^2}{4(2\pi)^4}\delta_{\mu\nu}\int_{-1}^{+1}dy\int_{-\infty}^{+\infty}dz\epsilon(z)$$
$$\times\frac{d}{dz}\left\{\frac{R(z)}{z}i\exp\left[i\frac{z}{4}(1-y^2)p_\lambda p_\lambda\right]\right\}$$
$$=\frac{-2i\pi^2}{4(2\pi)^4}\delta_{\mu\nu}\int_{-1}^{+1}dy\left(\frac{R(z)}{z}\right)_{z=0}=\frac{-i\pi^2}{(2\pi)^4}R'(0). \quad (44)$$

According to our regularization condition (I″a) this term is equal to zero and we are therefore left with an expression that has the required form (compare (35) and (36)):

$$K_{R(\mu\nu)}(p)=K_1(p_\lambda p_\lambda)\{p_\mu p_\nu-\delta_{\mu\nu}p_\lambda p_\lambda\}$$

where K_1 is given by:

$$K_1(p_\lambda p_\lambda)=\frac{-\pi^2}{4(2\pi)^4}\int_{-\infty}^{+\infty}\frac{dz}{z}\epsilon(z)R(z)$$
$$\times\int_{-1}^{+1}dy\frac{1-y^2}{2}\exp\left[i\frac{z}{4}(1-y^2)p_\lambda p_\lambda\right]. \quad (45)$$

The first term in the expansion of (45) in powers of $p_\lambda p_\lambda$,

$$K_1(p_\lambda p_\lambda)=K_1^{(0)}+p_\lambda p_\lambda K_1^{(1)}+\cdots$$

gives the charge renormalization (compare (37)):

$$\delta e=4e^2K_1^{(0)}=\frac{-1}{3}\frac{\alpha}{2\pi}\int_{-\infty}^{+\infty}\frac{dz}{z}\epsilon(z)R(z). \quad (46)$$

This may again be expressed, according to (10) and (10′), as

$$\delta e=\frac{\alpha}{3\pi}\int_{-\infty}^{+\infty}d\kappa\rho(\kappa)\log|\kappa|. \quad (46')$$

The connection with Wentzel's result for the photon self-energy is now most easily established. Since the field of a light wave is not connected with any current, $\langle j_\mu(x)\rangle_{\text{ind}}$ vanishes according to (37), unless $K_{\mu\nu}$ has not the required form (35), (36). From (43) and (44) it is made clear that in this case the induced current is given by

$$\langle j_\mu(x)\rangle_{\text{ind}}=-4e^2\cdot\frac{-i\pi^2}{(2\pi)^4}R'(0)A_\mu{}^{\text{rad}}(x).$$

The photon self-energy, defined as

$$H_{\text{self}}^{(\text{ph})}=-\tfrac{1}{2}\int d^3x[\langle j_\mu(x)\rangle_{\text{ind}}A_\mu{}^{\text{rad}}(x)]_{1\text{ph}-0\text{ph}}$$

becomes thus:

$$H_{\text{self}}^{(\text{ph})}=\frac{e^2}{8\pi^2}(-iR'(0))\int d^3x[A_\mu{}^{\text{rad}}(x)\cdot A_\mu{}^{\text{rad}}(x)]_{1\text{ph}-0\text{ph}}.$$

Since $R'(0)$ is zero, the regularized photon self-energy vanishes, as it should; without regularization $R'(0)=im^2$, which gives, together with

$$\int d^3x[A_\mu(x)\cdot A_\mu(x)]_{1\text{ph}-0\text{ph}}=1/|k|$$

for a photon of momentum k

$$H_{\text{self}}{}^{(p)}=\alpha/2\pi\cdot m^2/|k|$$

which is exactly Wentzel's result.

ADDITIONAL REMARKS

(A) In (38′) we may be tempted to omit the term

$$I(p)=\int d^4k\frac{\delta(k_\lambda k_\lambda+m^2)}{(k_\lambda-p_\lambda)^2+m^2}(k_\lambda k_\lambda+m^2)$$

which means putting $\Delta(x)(-\Box^2\Delta^{(1)}(x)+m^2\Delta^{(1)}(x))$ equal to zero, or its regularized counterpart:

$$\int d\kappa\rho(\kappa)\Delta(x;\kappa)(-\Box^2\Delta^{(1)}(x;\kappa)+\kappa\Delta^{(1)}(x;\kappa))=0. \quad (47)$$

An evaluation of $I_R(p)$ along the lines of the above calculations yields:

$$I_R(p)=\sim\int_{-1}^{+1}dy\int_{-\infty}^{+\infty}\frac{dz}{z}\epsilon(z)\exp\left[i\frac{z}{4}(1-y^2)p_\lambda p_\lambda\right]$$
$$\times\left[\left(\frac{-2}{z}+i\frac{1+y^2}{4}p_\lambda p_\lambda\right)R(z)+R'(z)\right]$$
$$+ip_\lambda p_\lambda\int_1^\infty dyy\int_{-\infty}^{+\infty}\frac{dz}{z}\epsilon(z)R(z)$$
$$\times\exp\left[i\frac{z}{4}(1-y^2)p_\lambda p_\lambda\right] \quad (48)$$
$$=\int_{-1}^{+1}dy\int_{-\infty}^{+\infty}dz\epsilon(z)\frac{d}{dz}\left\{\frac{R(z)}{z}\exp\left[i\frac{z}{4}(1-y^2)p_\lambda p_\lambda\right]\right\}$$
$$+\int_{-\infty}^{+\infty}\frac{dz}{z^2}\epsilon(z)R(z)\int_{-1}^{+1}dy$$
$$\times\frac{d}{dy}\left\{y\left(1-\exp\left[-\frac{iz}{4}(1-y^2)p_\lambda p_\lambda\right]\right)\right\}$$
$$=2R'(0)+0=0 \text{ on account of (II″a).}$$

Thus it may be seen that the identity (47) holds only because of our *stronger regularization condition.* It is of special interest to mention that it holds for a zero vector ($p_\lambda p_\lambda=0$) only if the limit $p_\lambda p_\lambda\to 0$ is carried through after the integration. Putting $p_\lambda p_\lambda=0$ in $I(p)$, (48), from the very beginning gives rise to an additional term

$$\sim\int_{-\infty}^{+\infty}(dz/z^2)\epsilon(z)R(z).$$

In consequence of this the non-regularized expression $I(0)$ becomes then *infinite,* whereas $I_R(0)$ is zero only if we put:

$$\int (dz/z^2)\epsilon(z)R(z)=0 \qquad (49)$$

or

$$\int d\kappa\rho(\kappa)\kappa\log|\kappa|=0. \qquad (49')$$

Thus Wentzel's result may as well have been infinite. From a physical point of view it appears, however, more natural to consider the case $\rho_\lambda\rho_\lambda=0$ as a limiting case of a non-zero vector, than to introduce the new condition (49) or (49').[18]

(B) At this point we shall briefly consider the case of the electron self-energy. In terms of $\boldsymbol{\Delta}(k)$, $\Delta^{(1)}(k)$, $\bar{D}(k)$, $D^{(1)}(k)$ the operator Δm reads:

$$\Delta m=\frac{\alpha}{2\pi^2}\int d^4k(i\gamma k+m)$$

$$\times\left(\frac{\delta(k_\lambda k_\lambda)}{(k_\lambda+q_\lambda)^2+m^2}+\frac{\delta((k_\lambda+q_\lambda)^2+m^2)}{k_\lambda k_\lambda}\right)(q_\nu q_\lambda=-m^2).$$

This expression is readily transformed into

$$\Delta m=\frac{m\alpha}{8\pi}\int_{-\infty}^{+\infty}\frac{dz}{z}\epsilon(z)\int_{-1}^{+1}dy\frac{3-y}{2}\exp\left[iz\left(\frac{1-y}{2}\right)^2\right].$$

Regularization of both D- and Δ-functions without factorization corresponds to the introduction of a regulator $R(z)$. $R(z)$ here reduces to 1 in the case of non-regularization, in contrast to the definition of $R(z)$ previously used; in the case considered here, the expression to be regularized contains both D- and Δ- functions, and κ takes the signification of an additional contribution to the square of the mass; the mass attributed to D is therefore $\kappa^{\frac{1}{2}}$, to Δ however $(m^2+\kappa)^{\frac{1}{2}}$). Δm shall now be written as

$$\Delta m=\frac{m\alpha}{16\pi}\int_{-1}^{+1}dy(3-y)\int\frac{dz}{z}\epsilon(z)\left(\exp\left[iz\left(\frac{1-y}{2}\right)^2\right]-e^{iz}\right)$$

$$+\frac{m\alpha}{16\pi}\int\frac{dz}{z}\epsilon(z)e^{iz}R(z)\int_{-1}^{+1}dy(3-y).$$

Since the first integral converges, no regulator is introduced. With the help of the formula

$$\log|a|=-\tfrac{1}{2}\int_{-\infty}^{+\infty}(dz/z)\epsilon(z)(e^{iaz}-e^{iz})$$

we have at once:

$$-(m\alpha/8\pi)\int_{-1}^{+1}dy(3-y)\log\left(\frac{1-y}{2}\right)^2=5\cdot(m\alpha/4\pi).$$

[18] The situation is exactly the same in the case of the evaluation of the trace $K_{\lambda\lambda}$ by elementary K-space integrations (compare footnote (16)). The constant Γ vanishes only if the additional assumption (which corresponds to (49'))

$$\Sigma_i\, c_iM_i^2\log M_i=0$$

is made, unless this particular case is again considered as the limiting case of a non-zero vector.

If, in the second, divergent integral we introduce the *special regulator*

$$R(z)=1-\exp\left[iz\left(\frac{1}{\gamma\omega_0}-1\right)\right]$$

(which satisfies (I''): $R(0)=0$, but not $R'(0)=0$), we obtain

$$-3\cdot\frac{m\alpha}{8\pi}\int\frac{dz}{z}\epsilon(z)\left(\exp\left(iz\frac{1}{\gamma\omega_0}\right)-e^{iz}\right)=\frac{6m\alpha}{8\pi}\log\left|\frac{1}{\gamma\omega_0}\right|$$

and thus

$$\Delta m=3m\alpha/2\pi(\tfrac{1}{2}\log|1/\gamma\omega_0|+\tfrac{5}{6}),$$

which is Schwinger's result (SII, Eq. (3.97)).

An alternative possibility is to regularize the photon-D-functions alone, as was done by Feynman.[7] In this case it is more convenient to introduce a regulator $\rho(\kappa)$; $(\Delta m)_R$ may then be written as

$$(\Delta m)_R=\int d\kappa\rho(\kappa)\Delta m(\kappa),$$

and

$$\Delta m(\kappa)=\frac{m\alpha}{16\pi}\int_{-1}^{+1}dy(3-y)\int\frac{dz}{z}\epsilon(z)$$

$$\times\exp\left[iz\left(\frac{\kappa}{2}(1+y)+\left(\frac{1-y}{2}\right)^2\right)\right].$$

(κ is a dimensionless mass parameter, i.e., $m(\kappa)^{\frac{1}{2}}$ is the auxiliary mass.) Taking into account condition (I'), we can omit all terms independent of κ; it follows therefore that

$$\Delta m(\kappa)=\frac{-m\alpha}{8\pi}\int_{-1}^{+1}dy(3-y)\log\left(\frac{\kappa}{2}(1+y)+\left(\frac{1-y}{2}\right)^2\right)$$

$$=\frac{-m\alpha}{8\pi}\int_{-1}^{+1}dy(y-5)(y-1)\frac{\kappa+(y-1)}{2\kappa(1+y)+(1-y)^2}$$

$$=\frac{-m\alpha}{8\pi}F(\kappa)$$

$$F(\kappa)=\begin{cases}-10+4\pi(\kappa)^{\frac{1}{2}}+0(\kappa) & \kappa\ll 1\\[2ex] 6\log\kappa-7+0\left(\dfrac{1}{\kappa}\log\kappa\right) & \kappa\gg 1.\end{cases}$$

With the help of one auxiliary mass:

$$\rho(\kappa)=\delta(\kappa)-\delta\left(\kappa-\frac{M^2}{m^2}\right)$$

which is sufficient to satisfy (I'), we obtain then

$$\Delta m=\frac{3m\alpha}{2\pi}\left(\log\frac{M}{m}+\frac{1}{4}\right)$$

which is Feynman's result.

§5. THE MAGNETIC MOMENT OF THE ELECTRON

The e^2 radiative correction Δj_μ to the current, as given by (31) may be written as

$$\Delta j_\mu(p)=F(p_\lambda p_\lambda)\cdot\frac{1}{(2\pi)^4}\int d^4q\bar{u}(p+q)\gamma^\mu u(q)$$

$$+G(p_\lambda p_\lambda)p_\nu\frac{1}{(2\pi)^4}\int d^4q\bar{u}(p+q)\gamma^{[\mu\nu]}u(q), \quad (50)$$

here $u(q)$, $\bar{u}(q')$ are the Fourier-amplitudes of ψ and $\bar{\psi}$ respectively, and $\gamma^{[\mu\nu]}=\frac{1}{2}(\gamma^\mu\gamma^\nu-\gamma^\nu\gamma^\mu)$.

For small values of p, F and G may be developed in powers of $p_\lambda p_\lambda$ (note that p is the difference in momentum ascribed to $\bar{\psi}$ and ψ, respectively). Whereas $F(0)$ describes again a charge renormalization, the term corresponding to $G(0)$ exhibits an extra current, which, in x-space, has the more familiar form (32), and describes the radiative correction to the electron's magnetic moment in a homogeneous external field.

From the well-known decomposition of the current due to Gordon[19]

$$ie\bar{u}(p+q)\gamma^\mu u(q)=e/2m\{(p_\mu+2q_\mu)\bar{u}(p+q)u(q) -p_\lambda\bar{u}(p+q)\gamma^{[\mu\lambda]}u(q)\} \quad (51)$$

it is seen that the radiative correction to the magnetic moment may be expressed in terms of an anomalous g-factor

$$\Delta g=-(4m/e)G(0), \quad (52)$$

since the unperturbed electron is characterized by $g=2$. The relevant terms of (31) (containing $G(p_\lambda p_\lambda)$) can be written as follows:

$$-(ie^3/2)\int\int d^4\xi d^4\eta\bar{\psi}(x+\xi)\gamma^\alpha[\bar{D}(\xi-\eta)\bar{S}(\xi)\gamma^\mu S^{(1)}(-\eta) +\bar{D}(\xi-\eta)S^{(1)}(\xi)\gamma^\mu\bar{S}(-\eta) +D^{(1)}(\xi-\eta)\bar{S}(\xi)\gamma^\mu\bar{S}(-\eta)]\gamma^\alpha\psi(x+\eta)$$

or, in momentum space (writing q' for $p+q$),

$$\frac{ie^3}{(2\pi)^3}\bar{u}(q')\int d^4k\Omega_\mu(k)[\bar{D}(k)\mathbf{\Delta}(k-q)\Delta^{(1)}(k-q') +\bar{D}(k)\Delta^{(1)}(k-q)\mathbf{\Delta}(k-q') +D^{(1)}(k)\mathbf{\Delta}(k-q)\mathbf{\Delta}(k-q')]u(q) \quad (53)$$

where

$$\Omega_\mu(k)=2ik_\mu(i\gamma\cdot k-m)-2i(q_\mu'+q_\mu)(i\gamma\cdot k)+\sim\gamma^\mu.$$

Let us introduce

$$T_{\sigma\tau}=\int d^4kk_\sigma k_\tau[D] \quad (54)$$

$$V_\sigma=\int d^4kk_\sigma[D]. \quad (54a)$$

([D] is the bracket in (53).) Since [D] depends on the two parameters q and q', or on

$$p=q'-q \quad\text{and}\quad Q=q'+q,$$

$T_{\sigma\tau}$ and V_σ may be written as:

$$T_{\sigma\tau}=\delta_{\sigma\tau}I_0+p_\sigma p_\tau I_1+Q_\sigma Q_\tau I_2 \quad (55)$$

$$V_\sigma=Q_\sigma J_2. \quad (55a)$$

All invariants thus introduced are still to be considered as functions of $p_\lambda p_\lambda$ (the invariant $Q_\lambda Q_\lambda$ reduces to the former by $Q_\lambda Q_\lambda=-(4m^2+p_\lambda p_\lambda)$). With the help of (51) and the relations:

$$\bar{u}(q')\gamma\cdot Qu(q)=2im\bar{u}(q')u(q), \quad \bar{u}(q')\gamma\cdot pu(q)=0$$

we obtain from (54) and (55):

$$\frac{ie^3}{(2\pi)^3}(-2im)(2I_2(p)-J_2(p))p_\lambda\bar{u}(p+q)\gamma^{[\mu\lambda]}u(q)+\sim\gamma\mu$$

thus yielding

$$G(p_\lambda p_\lambda)=\frac{2me^3}{(2\pi)^3}(2I_2(p)-J_2(p))$$

and, according to (52):

$$(\alpha=e^2/4\pi=1/137)$$

$$\Delta g=-4m^2/\pi(2I_2(0)-J_2(0))(\alpha/\pi). \quad (56)$$

The computation of the integrals involved in I_2 and J_2 may be carried through in different ways. A regularization device which guarantees the convergence of the expressions (54) for $T_{\sigma\tau}$ and V_σ allows the introduction of a special coordinate system, characterized by

$$p=0, \quad Q=(o, 2im)$$

in which

$$I_2(0)=1/4m^2(T_{11}-T_{44}), \quad J_2(0)=(1/2m)V_4. \quad (57)$$

The most convenient regularization method is that which regularizes [D] as a whole:

$$([D])_R=\textstyle\sum_i c_i[D]_i$$

(according to (9)).[20] The evaluation of Eq. (57) in k-space is elementary. With the notations

$$x=k/m, \quad \mu_i=M_i/m, \quad \Omega_i=(x^2+\mu_i^2)^{\frac{1}{2}}, \quad \mathbf{\Omega}_i=(1+x^2+\mu_i^2)^{\frac{1}{2}}$$

we are led to:

$$(I_2(0))_R=\frac{\pi}{4m^2}\int_0^\infty x^2dx\textstyle\sum_i c_i\left\{\frac{\Omega_i^2+\frac{1}{3}x^2}{\Omega_i^3}-\frac{2+\Omega_i^2+\frac{1}{3}x^2}{\mathbf{\Omega}_i^3}\right\} \quad (58)$$

$$(J_2(0))_R=\frac{\pi}{2m^2}\int_0^\infty x^2dx\textstyle\sum_i c_i\frac{1}{\Omega_i^2\mathbf{\Omega}_i^3}. \quad (58a)$$

Equation (58a) converges without regularization and yields

$$J_2(0)=\pi/2m^2$$

a value which remains unaltered under regularization,

[19] Compare Handbuch der Physik XXIV/1. 238.

[20] A control calculation, carried through with a regulator affecting only the electron Δ-functions in [D] gave exactly the same results.

since all terms $i>0$ vanish at least as c_i/M_i^2 (compare (6)). This is not the case, however, for (58), which becomes

$$(I_2(0,)_R=\frac{\pi}{12m^2}\left(2\sum_i c_i-\tfrac{3}{2}\sum_{(i>0)}{}' c_i\right). \quad (59)$$

The expression (58) proves thus to be convergent without regularization; we need only put $c_0=1$, $c,(i>0)=0$, to obtain

$$I_2(0)=\pi/6m^2$$

whereas the properly regularized expression yields

$$(I_2(0))_R=\pi/8m^2.$$

Therefore

$$2I_2(0)-J_2(0)=-\pi/6m^2,$$

whereas

$$2(I_2(0))_R-J_2(0)=-\pi/4m^2.$$

The regularized value, introduced in (56), gives

$$\Delta g=\alpha/\pi$$

in agreement with Schwinger's result.[21]

From this example it should be made clear how cautious one should be in handling divergent "covariant" expressions. Indeed, any decomposition, as done in (55), is by no means "covariant," as long as the expressions under discussion has not been made convergent by some invariant regularization. Once this has been done, one may quietly enjoy of all the facilities presented by a properly chosen coordinate system, as was done in (57).

ADDITIONAL REMARKS

(A) The various possibilities of evaluating the integrals I_2 and J_2 arise from the different possibilities of rewriting the [D]-bracket in (54), (54a). It is easily seen from the definitions (40), (41) that for $q=q_1'$, [D] may be written as

$$[D]=\int_0^1 duu\delta''(k_\lambda k_\lambda-2k_\lambda q_\lambda u).^{22} \quad (60)$$

The regularization of [D], as a whole, as was done in the above calculations, consists then in replacing (60) by

$$([D])_R=\int_0^1 duu\sum_i c_i\delta''(k_\lambda k_\lambda-2k_\lambda q_\lambda u+M_i^2). \quad (61)$$

Equation (61) may now be introduced into (54):

$$(T_{\sigma\tau})_R=\int_0^1 duu\int d^4kk_\sigma k_\tau\sum_i c_i\delta''(k_\lambda k_\lambda-2k_\lambda q_\lambda u+M_i^2)$$

$$=\int_0^1 duu\int d^4k(k_\sigma k_\tau+u^2q_\sigma q_\tau)$$

$$\times\sum_i c_i\delta''(k_\lambda k_\lambda+m^2u^2+M_i^2). \quad (62)$$

[21] J. Schwinger, Phys. Rev. 73, 416 (1948).
[22] This method is due to Schwinger; (SIII).

Thus $I_2(0)$ is immediately isolated:

$$I_2(0)=\tfrac{1}{4}\int_0^1 duu^3\sum_i c_i\int d^4k\delta''(k_\lambda k_\lambda+m^2u^2+M_i^2)$$

$$=\frac{\pi}{4}\sum_i c_i\int_0^1 duu^3\frac{1}{m^2u^2+M_i^2}=\frac{\pi}{8m^2}+0\left(\frac{c_i}{M_i^2}\right)_{i>0}.$$

The use of a regulator appears to be superfluous in this case, but from a general point of view it is doubtful whether a transformation of variables as needed to obtain (62) can be justified *a priori*.

(B) [D] may, according to Dyson, also be written as

$$-\tfrac{1}{4}Re(D_c(k)\Delta_c(k-q)\Delta_c(k-q'))$$

where

$$\Delta_c(k)=\Delta_1-2i\boldsymbol{\Delta}=2\pi\delta_+(k^2+m^2),\ (\text{and } D_c=D^{(1)}-2i\bar{D}).$$

With the elementary (complex) k-space integration, however, one immediately falls back on the formula (58), (58a).

(C) Finally we illustrate the possibility of obtaining a quite arbitrary result by regularizing in a different way different terms of a form $F(\boldsymbol{\Delta},\Delta^{(1)})$. Let us just take, as an example, the bracket [D], which for $q'=q$ may be written as follows:

$$[D]=\frac{\delta(k_\lambda k_\lambda)}{(2k_\lambda q_\lambda)^2}-\frac{1}{k_\lambda k_\lambda}\delta'(k_\lambda k_\lambda-2k_\lambda q_\lambda)=[D_1]+[D_2].$$

We need now only note that if R_1 and R_2 are two regulators, the operation $R=\frac{1}{2}(R_1+R_2)$ is a regulator too, whereas $\bar{R}=\frac{1}{2}(R_1-R_2)$ is an operator, corresponding to a mass spectrum which satisfies (I, Ia), but contains only auxiliary masses M_i; we may now write

$$R_1[D_1]+R_2[D_2]=R[D]+\bar{R}([D_1]-[D_2]).$$

The additional terms depending on $\bar{R}$ are by no means zero in general, but depend on the structure of $\bar{R}$ which shall be characterized by a spectrum M_i and coefficients γ_i. In evaluating $I_2(0)$ (54), (55), as an example, the additional terms due to $\bar{R}$ are:

$$\pi/4m^2(\textstyle\sum_i\gamma_i\mu_i^2\log\mu_i-\tfrac{2}{3}),$$

an expression which is completely indeterminate.

ACKNOWLEDGMENTS

We have to acknowledge Professor J. Schwinger, Dr. F. J. Dyson, Professor V. F. Weisskopf, Dr. S. T. Ma, and Dr. K. M. Case for making their publications accessible to us prior to publication; Dr. G. Rayski for valuable help, and Professor E. C. G. Stueckelberg and Dr. G. Rivier for interesting discussions.

On Gauge Invariance and Vacuum Polarization

JULIAN SCHWINGER
Harvard University, Cambridge, Massachusetts
(Received December 22, 1950)

This paper is based on the elementary remark that the extraction of gauge invariant results from a formally gauge invariant theory is ensured if one employs methods of solution that involve only gauge covariant quantities. We illustrate this statement in connection with the problem of vacuum polarization by a prescribed electromagnetic field. The vacuum current of a charged Dirac field, which can be expressed in terms of the Green's function of that field, implies an addition to the action integral of the electromagnetic field. Now these quantities can be related to the dynamical properties of a "particle" with space-time coordinates that depend upon a proper-time parameter. The proper-time equations of motion involve only electromagnetic field strengths, and provide a suitable gauge invariant basis for treating problems. Rigorous solutions of the equations of motion can be obtained for a constant field, and for a plane wave field. A renormalization of field strength and charge, applied to the modified lagrange function for constant fields, yields a finite, gauge invariant result which implies nonlinear properties for the electromagnetic field in the vacuum. The contribution of a zero spin charged field is also stated. After the same field strength renormalization, the modified physical quantities describing a plane wave in the vacuum reduce to just those of the maxwell field; there are no nonlinear phenomena for a single plane wave, of arbitrary strength and spectral composition. The results obtained for constant (that is, slowly varying fields), are then applied to treat the two-photon disintegration of a spin zero neutral meson arising from the polarization of the proton vacuum. We obtain approximate, gauge invariant expressions for the effective interaction between the meson and the electromagnetic field, in which the nuclear coupling may be scalar, pseudoscalar, or pseudovector in nature. The direct verification of equivalence between the pseudoscalar and pseudovector interactions only requires a proper statement of the limiting processes involved. For arbitrarily varying fields, perturbation methods can be applied to the equations of motion, as discussed in Appendix A, or one can employ an expansion in powers of the potential vector. The latter automatically yields gauge invariant results, provided only that the proper-time integration is reserved to the last. This indicates that the significant aspect of the proper-time method is its isolation of divergences in integrals with respect to the proper-time parameter, which is independent of the coordinate system and of the gauge. The connection between the proper-time method and the technique of "invariant regularization" is discussed. Incidentally, the probability of actual pair creation is obtained from the imaginary part of the electromagnetic field action integral. Finally, as an application of the Green's function for a constant field, we construct the mass operator of an electron in a weak, homogeneous external field, and derive the additional spin magnetic moment of $\alpha/2\pi$ magnetons by means of a perturbation calculation in which proper-mass plays the customary role of energy.

I. INTRODUCTION

QUANTUM electrodynamics is characterized by several formal invariance properties, notably relativistic and gauge invariance. Yet specific calculations by conventional methods may yield results that violate these requirements, in consequence of the divergences inherent in present field theories. Such difficulties concerning relativistic invariance have been avoided by employing formulations of the theory that are explicitly invariant under coordinate transformations, and by maintaining this generality through the course of calculations. The preservation of gauge invariance has apparently been considered to be a more formidable task. It should be evident, however, that the two problems are quite analogous, and that gauge invariance difficulties naturally disappear when methods of solution are adopted that involve only gauge invariant quantities.

We shall illustrate this assertion by applying such a gauge invariant method to treat several aspects of the problem of vacuum polarization by a prescribed electromagnetic field. The calculation of the current associated with the vacuum of a charged particle field involves the construction of the Green's function for the particle field in the prescribed electromagnetic field. This vacuum current can be exhibited as the variation of an action integral with respect to the potential vector, which action effectively adds to that of the maxwell field in describing the behavior of electromagnetic fields in the vacuum. We shall relate these problems to the solution of particle equations of motion with a proper-time parameter. The equations of motion, which involve only electromagnetic field strengths, provide the desired gauge invariant basis for our discussion.

Explicit solutions can be obtained in the two situations of constant fields, and fields propagated with the speed of light in the form of a plane wave.[1] For constant (that is, slowly varying) fields, a renormalization of field strength and charge yields a modified lagrange function differing from that of the maxwell field by terms that imply a nonlinear behavior for the electromagnetic field. The result agrees precisely with one obtained some time ago by other methods and a somewhat different viewpoint.[2] The modified physical quantities characterizing the plane wave in the vacuum revert to those of the maxwell field after the same field strength renormalization. For weak arbitrarily varying fields, perturbation methods can be applied to the equations of motion. This will be discussed in Appendix A.

The consequences thus obtained are useful in connection with a class of problems in which gauge invari-

[1] That the Dirac equation can be solved exactly, in the field of a plane wave, was recognized by D. M. Volkow, Z. Physik **94**, 25 (1935).

[2] W. Heisenberg and H. Euler, Z. Physik **98**, 714 (1936). V. Weisskopf, Kgl. Danske Videnskab. Selskabs. Mat.-fys. Medd. **14**, No. 6 (1936).

ance difficulties have been encountered[3]—the multiple photon disintegration of a neutral meson. Without further extensive calculation, we shall obtain approximate gauge invariant expressions for the interaction of a zero-spin, neutral meson with two photons, where the intermediate nuclear interaction may be scalar, or the equivalent pseudoscalar and pseudovector couplings.

The utility of the proper-time technique to be exploited in this paper, apart from its value in obtaining rigorous solutions in a few special cases, lies in its isolation of the divergent aspects of a calculation in integrals with respect to the proper-time, a parameter that makes no reference to the coordinate system or the gauge. Indeed, we shall show that the customary perturbation procedure of expansion in powers of the potential vector does yield gauge invariant results, provided only that the proper-time integration is reserved to the last. The technique of "invariant regularization"[4] represents a partial realization of this proper-time method through the use of specially weighted integrals over the conjugate quantity, the square of the proper mass.

Finally, in Appendix B we shall employ the Green's function of an electron in a weak, homogeneous, external field to calculate the second-order electromagnetic mass, thereby providing a simple derivation of the second-order correction to the electron magnetic moment.

II. GENERAL THEORY

The field equations, commutation relations, and current vector of the Dirac field are given by[5]

$$\gamma_\mu(-i\partial_\mu - eA_\mu(x))\psi(x) + m\psi(x) = 0, \quad (i\partial_\mu - eA_\mu(x))\bar{\psi}(x)\gamma_\mu + m\bar{\psi}(x) = 0, \tag{2.1}$$

$$\{\psi(\mathbf{x}, x_0), \bar{\psi}(\mathbf{x}', x_0)\} = \gamma_0\delta(\mathbf{x}-\mathbf{x}'), \tag{2.2}$$

$$j_\mu(x) = \tfrac{1}{2}e[\bar{\psi}(x), \gamma_\mu\psi(x)], \tag{2.3}$$

where

$$\tfrac{1}{2}\{\gamma_\mu, \gamma_\nu\} = -\delta_{\mu\nu} \tag{2.4}$$

and

$$\gamma_0 = -i\gamma_4, \quad \gamma_0^2 = 1. \tag{2.5}$$

The structure of the current operator,

$$j_\mu(x) = -e(\gamma_\mu)_{\beta\alpha}\tfrac{1}{2}[\psi_\alpha(x), \bar{\psi}_\beta(x)], \tag{2.6}$$

which arises from an explicit charge symmetrization, can be related to a time symmetrization by introducing chronologically ordered operators. Thus, with the notation

$$(A(x_0)B(x_0'))_+ = \begin{cases} A(x_0)B(x_0'), & x_0 > x_0' \\ B(x_0')A(x_0), & x_0 < x_0', \end{cases} \tag{2.7}$$

and

$$\epsilon(x-x') = \begin{cases} 1, & x_0 > x_0' \\ -1, & x_0 < x_0', \end{cases} \tag{2.8}$$

we have

$$(\psi_\alpha(x)\bar{\psi}_\beta(x'))_+\epsilon(x-x') = \begin{cases} \psi_\alpha(x)\bar{\psi}_\beta(x'), & x_0 > x_0' \\ -\bar{\psi}_\beta(x')\psi_\alpha(x), & x_0 < x_0'. \end{cases} \tag{2.9}$$

Therefore

$$\tfrac{1}{2}[\psi_\alpha(x), \bar{\psi}_\beta(x)] = (\psi_\alpha(x)\bar{\psi}_\beta(x'))_+\epsilon(x-x')]_{x'\to x}, \tag{2.10}$$

provided one takes the average of the forms obtained by letting x' approach x from the future, and from the past. The quantity of actual interest here is the expectation value of $j_\mu(x)$ in the vacuum of the Dirac field,

$$\langle j_\mu(x)\rangle = ie\, \mathrm{tr}\gamma_\mu G(x, x')]_{x'\to x}, \tag{2.11}$$

where

$$G(x, x') = i\langle(\psi(x)\bar{\psi}(x'))_+\rangle\epsilon(x-x'), \tag{2.12}$$

and tr indicates the diagonal sum with respect to the spinor indices.

The function $G(x, x')$ satisfies an inhomogeneous differential equation which is obtained by noting that

$$[\gamma(-i\partial - eA(x)) + m]G(x\, x') = \langle\gamma_0\{\psi(x), \bar{\psi}(x')\}\rangle\delta(x_0 - x_0'), \tag{2.13}$$

where the right side expresses the discontinuous change in form of $G(x, x')$ as x_0 is altered from $x_0'-0$ to $x_0'+0$. According to Eq. (2.2), therefore, we have

$$[\gamma(-i\partial - eA(x)) + m]G(x, x') = \delta(x-x'); \tag{2.14}$$

that is, $G(x, x')$ is a Green's function for the Dirac field. We shall not discuss which particular Green's function this is, as specified by the associated boundary conditions, since no ambiguity enters if actual pair creation in the vacuum does not occur, which we shall expressly assume.

It is useful to regard $G(x, x')$ as the matrix element of an operator G, in which states are labeled by space-time coordinates as well as by the suppressed spinor indices:

$$G(x, x') = (x|G|x'). \tag{2.15}$$

The defining differential equations for the Green's function is then considered to be a matrix element of the operator equation

$$(\gamma\Pi + m)G = 1, \tag{2.16}$$

where

$$\Pi_\mu = p_\mu - eA_\mu \tag{2.17}$$

is characterized by the operator properties

$$[x_\mu, \Pi_\nu] = i\delta_{\mu\nu}, \quad [\Pi_\mu, \Pi_\nu] = ieF_{\mu\nu}, \tag{2.18}$$

and

$$F_{\mu\nu} = \partial_\mu A_\nu - \partial_\nu A_\mu \tag{2.19}$$

is the antisymmetrical field strength tensor.

With this symbolism, it is easy to show that the

[3] H. Fukuda and Y. Miyamoto, Prog. Theor. Phys. 4, 347 (1949).
[4] W. Pauli and F. Villars, Revs. Modern Phys. 21, 434 (1949).
[5] We employ units in which $\hbar = c = 1$. Note that also $\bar{\psi} = \psi^\dagger\gamma_0$, since $\gamma_0\gamma_\mu$, $\mu = 0, 1, 2, 3$ form hermitian matrices.

vacuum current vector,

$$\langle j_\mu(x)\rangle = ie\,\mathrm{tr}\gamma_\mu(x|G|x), \tag{2.20}$$

is obtained from an action integral by variation of $A_\mu(x)$. This is accomplished by exhibiting

$$\delta W^{(1)} = \int (dx)\delta A_\mu(x)\langle j_\mu(x)\rangle = ie\,\mathrm{Tr}\gamma\delta AG \tag{2.21}$$

as a total differential, subject to $\delta A_\mu(x)$ vanishing at infinity. In the second version of $\delta W^{(1)}$, δA_μ denotes the operator with the matrix elements

$$(x|\delta A_\mu|x') = \delta(x-x')\delta A_\mu(x), \tag{2.22}$$

and Tr indicates the complete diagonal symmation, including spinor indices and the continuous space-time coordinates. Now

$$-e\gamma\delta A = \delta(\gamma\Pi + m), \tag{2.23}$$

and

$$G = \frac{1}{\gamma\Pi + m} = i\int_0^\infty ds\,\exp\{-i(\gamma\Pi+m)s\}, \tag{2.24}$$

so that

$$ie\,\mathrm{Tr}\gamma\delta AG = \delta\left[i\int_0^\infty dss^{-1}\,\mathrm{Tr}\exp\{-i(\gamma\Pi+m)s\}\right], \tag{2.25}$$

in virtue of the fundamental property of the trace,

$$\mathrm{Tr}AB = \mathrm{Tr}BA. \tag{2.26}$$

Thus, to within an additive constant,

$$W^{(1)} = i\int_0^\infty dss^{-1}e^{-ims}\,\mathrm{Tr}\exp\{-i\gamma\Pi s\} = \int (dx)\mathcal{L}^{(1)}(x), \tag{2.27}$$

where the lagrange function $\mathcal{L}^{(1)}(x)$ is given by

$$\mathcal{L}^{(1)}(x) = i\int_0^\infty dss^{-1}e^{-ims}\,\mathrm{tr}(x|\exp\{-i\gamma\Pi s\}|x). \tag{2.28}$$

An alternative representation, and the one we shall actually employ for calculations, is obtained by writing

$$G = (-\gamma\Pi+m)[m^2-(\gamma\Pi)^2]^{-1} = [m^2-(\gamma\Pi)^2]^{-1}(-\gamma\Pi+m) \tag{2.29}$$

or

$$G = (-\gamma\Pi+m)i\int_0^\infty ds\,\exp[-i(m^2-(\gamma\Pi)^2)s] = i\int_0^\infty ds\,\exp[-i(m^2-(\gamma\Pi)^2)s](-\gamma\Pi+m). \tag{2.30}$$

In virtue of the vanishing trace of an odd number of γ-factors, we have

$$ie\,\mathrm{Tr}\gamma\delta AG = -\mathrm{Tr}\delta(\gamma\Pi)\gamma\Pi\int_0^\infty ds\,\exp[-i(m^2-(\gamma\Pi)^2)s] = \delta\left[\tfrac{1}{2}i\int_0^\infty dss^{-1}\exp[-i(m^2-(\gamma\Pi)^2)s]\right], \tag{2.31}$$

which again involves the fundamental trace property (2.26). Thus,

$$\mathcal{L}^{(1)}(x) = \tfrac{1}{2}i\int_0^\infty dss^{-1}\exp(-im^2s)\,\mathrm{tr}(x|U(s)|x),\quad U(s) = \exp(-i\mathcal{H}s), \tag{2.32}$$

where

$$\mathcal{H} = -(\gamma\Pi)^2 = \Pi_\mu{}^2 - \tfrac{1}{2}e\sigma_{\mu\nu}F_{\mu\nu}, \tag{2.33}$$

and

$$\sigma_{\mu\nu} = \tfrac{1}{2}i[\gamma_\mu, \gamma_\nu]. \tag{2.34}$$

We now see that the construction of $G(x, x')$ and $\mathcal{L}^{(1)}(x)$ devolves upon the evaluation of

$$(x'|U(s)|x'') = (x(s)'|x(0)''). \tag{2.35}$$

The latter notation emphasizes that $U(s)$ may be regarded as the operator describing the development of a system governed by the "hamiltonian," $\mathcal{H}$, in the "time" s, the matrix element of $U(s)$ being the transformation function from a state in which $x_\mu(s=0)$ has the value x_μ'' to a state in which $x_\mu(s)$ has the value x_μ'. Thus, we are led to an associated dynamical problem in which the space-time coordinates of a "particle" depend upon a proper time parameter, in a manner determined by the equations of motion

$$\begin{aligned} dx_\mu/ds &= -i[x_\mu, \mathcal{H}] = 2\Pi_\mu, \\ d\Pi_\mu/ds &= -i[\Pi_\mu, \mathcal{H}] = e(F_{\mu\nu}\Pi_\nu + \Pi_\nu F_{\mu\nu}) + \tfrac{1}{2}e\sigma_{\lambda\nu}(\partial F_{\lambda\nu}/\partial x_\mu) = 2eF_{\mu\nu}\Pi_\nu \\ &\quad - ie(\partial F_{\mu\nu}/\partial x_\nu) + \tfrac{1}{2}e\sigma_{\lambda\nu}(\partial F_{\lambda\nu}/\partial x_\mu). \end{aligned} \tag{2.36}$$

The transformation function is characterized by the differential equations,[6]

$$i\partial_s(x(s)'|x(0)'') = (x(s)'|\mathcal{H}|x(0)''), \tag{2.37}$$

$$(-i\partial_\mu' - eA_\mu(x'))(x(s')|x(0)'') = (x(s)'|\Pi_\mu(s)|x(0)''), \tag{2.38}$$

$$(i\partial_\mu'' - eA_\mu(x''))(x(s)'|x(0)'') = (x(s)'|\Pi_\mu(0)|x(0)''), \tag{2.39}$$

and the boundary condition

$$(x(s)'|x(0)'')]_{s\to 0} = \delta(x'-x''). \tag{2.40}$$

[6] A proper time wave equation, in conjunction with the second-order Dirac operator, has been discussed by V. Fock, Physik. Z. Sowjetunion **12**, 404 (1937). See also Y. Nambu, Prog. Theor. Phys. **5**, 82 (1950).

We shall now illustrate, for the elementary situation $F_{\mu\nu}=0$, the procedure which will be employed in the following sections for constructing the transformation function.

The equations of motion read

$$d\Pi_\mu/ds=0,\quad dx_\mu/ds=2\Pi_\mu, \tag{2.41}$$

whence

$$\Pi_\mu(s)=\Pi_\mu(0), \tag{2.42}$$

and

$$(x_\mu(s)-x_\mu(0))/s=2\Pi_\mu(0). \tag{2.43}$$

Therefore

$$\begin{aligned}\mathcal{H}=\Pi^2=\tfrac{1}{4}s^{-2}(x(s)-x(0))^2=\tfrac{1}{4}s^{-2}[x^2(s)\\ -2x(s)x(0)+x^2(0)]+\tfrac{1}{4}s^{-2}[x(s),x(0)]\\ =\tfrac{1}{4}s^{-2}[x^2(s)-2x(s)x(0)+x^2(0)]-2is^{-1},\end{aligned} \tag{2.44}$$

since

$$[x_\mu(s),x_\nu(0)]=[x_\mu(0)+2s\Pi_\mu(0),x_\nu(0)]=-2is\delta_{\mu\nu}, \tag{2.45}$$

Having ordered the coordinate operators so that $x(s)$ everywhere stands to the left of $x(0)$, we can immediately evaluate the matrix element of $\mathcal{H}$ in Eq. (2.37), thus obtaining

$$\begin{aligned}i\partial_s(x(s)'|x(0)'')\\ =[\tfrac{1}{4}s^{-2}(x'-x'')^2-2is^{-1}](x(s)'|x(0)''),\end{aligned} \tag{2.46}$$

the solution of which is

$$(x(s)'|x(0)'')=C(x',x'')s^{-2}\exp[i\tfrac{1}{4}(x'-x'')^2/s]. \tag{2.47}$$

To determine the function $C(x',x'')$, we note that

$$\begin{aligned}(x(s)'|\Pi_\mu(s)|x(0)'')=(x(s)'|\Pi_\mu(0)|x(0)'')\\ =((x_\mu'-x_\mu'')/2s)(x(s)'|x(0)''),\end{aligned} \tag{2.48}$$

which, in conjunction with Eqs. (2.38) and (2.39), implies that

$$\begin{aligned}(-i\partial_\mu'-eA_\mu(x'))C(x',x'')\\ =(i\partial_\mu''-eA_\mu(x''))C(x',x'')=0,\end{aligned} \tag{2.49}$$

or

$$\begin{aligned}C(x',x'')=C\Phi(x',x''),\\ \Phi(x',x'')=\exp\left[ie\int_{x''}^{x'}dx_\mu A_\mu(x)\right].\end{aligned} \tag{2.50}$$

The line integral in Eq. (2.50) is independent of the integration path, since $F_{\mu\nu}=0$. Finally, the constant C is fixed by the boundary condition (2.40). It is evident that Eq. (2.47) does have the character of a delta-function as s approaches zero, provided

$$Cs^{-2}\int(dx)\exp(i\tfrac{1}{4}x^2/s)=1; \tag{2.51}$$

that is,

$$C=-i(4\pi)^{-2}. \tag{2.52}$$

Therefore,

$$\begin{aligned}(x(s)'|x(0)'')=-i(4\pi)^{-2}\Phi(x',x'')s^{-2}\\ \times\exp[i\tfrac{1}{4}(x'-x'')^2/s],\end{aligned} \tag{2.53}$$

and the Green's function is obtained as

$$\begin{aligned}G(x',x'')=i\int_0^\infty ds\exp(-im^2s)\\ \times(x(s)'|(-\gamma\Pi+m)|x(0)'')\\ =(4\pi)^{-2}\Phi(x',x'')\int_0^\infty dss^{-2}\exp(-im^2s)\\ \times\left(-\gamma\frac{(x'-x'')}{2s}+m\right)\exp\left[i\tfrac{1}{4}\frac{(x'-x'')^2}{s}\right].\end{aligned} \tag{2.54}$$

An equivalent, and more familiar procedure, is to employ the representation labeled by the eigenvalues of Π_μ. Now

$$\begin{aligned}(\Pi(s)'|\Pi(0)'')=(\Pi(0)'|U(s)|\Pi(0)'')\\ =\delta(\Pi'-\Pi'')\exp(-i\Pi'^2s),\end{aligned} \tag{2.55}$$

while $(x(s)'|\Pi(s)')$ is determined by

$$\begin{aligned}(-i\partial_\mu'-eA_\mu(x'))(x(s)'|\Pi(s)')\\ =\Pi_\mu'(x(s)'|\Pi(s)'),\end{aligned} \tag{2.56}$$

and the normalization condition

$$\int(\Pi(s)'|x(s)')(dx')(x(s)'|\Pi(s)'')=\delta(\Pi'-\Pi''), \tag{2.57}$$

to be

$$\begin{aligned}(x(s)'|\Pi(s)')\\ =(2\pi)^{-2}\exp\left[ie\int_0^{x'}dxA\right]\exp(ix'\Pi').\end{aligned} \tag{2.58}$$

Therefore,

$$\begin{aligned}(x(s)'|x(0)'')\\ =\int(x(s)'|\Pi(s)')(d\Pi')(\Pi(s)'|\Pi(0)'')\\ \times(d\Pi'')(\Pi(0)''|x(0)'')\\ =(2\pi)^{-4}\Phi(x',x'')\int(d\Pi')\\ \times\exp[i(x'-x'')\Pi'-i\Pi'^2s],\end{aligned} \tag{2.59}$$

and

$$\begin{aligned}G(x',x'')=i(2\pi)^{-4}\Phi(x',x'')\\ \times\int_0^\infty ds\int(d\Pi')\exp[i(x'-x'')\Pi']\\ \times(-\gamma\Pi'+m)\exp[-i(\Pi'^2+m^2)s],\end{aligned} \tag{2.60}$$

which reduce to Eqs. (2.53) and (2.54) on performance of the Π' integration.

III. CONSTANT FIELDS

The equations of motion (2.36) here simplify to

$$dx_\mu/ds=2\Pi_\mu,\quad d\Pi_\mu/ds=2eF_{\mu\nu}\Pi_\nu, \tag{3.1}$$

or, in matrix notation,

$$dx/ds=2\Pi,\quad d\Pi/ds=2eF\Pi. \tag{3.2}$$

The symbolic solution of these equations is

$$\Pi(s)=e^{2eFs}\Pi(0),$$
$$x(s)-x(0)=[(e^{2eFs}-1)/eF]\Pi(0), \tag{3.3}$$

whence

$$\Pi(0)=eF(e^{2eFs}-1)^{-1}(x(s)-x(0))$$
$$=\tfrac{1}{2}eFe^{-eFs}\sinh^{-1}(eFs)(x(s)-x(0)), \tag{3.4}$$

and

$$\Pi(s)=\tfrac{1}{2}eFe^{eFs}\sinh^{-1}(eFs)(x(s)-x(0))$$
$$=(x(s)-x(0))\tfrac{1}{2}eFe^{-eFs}\sinh^{-1}(eFs). \tag{3.5}$$

The latter form involves the fact that

$$\tilde{F}=-F,\quad (F_{\mu\nu}=-F_{\nu\mu}). \tag{3.6}$$

We now consider

$$\mathcal{H}+\tfrac{1}{2}e\sigma F=\Pi^2(s)=(x(s)-x(0))K(x(s)-x(0)),$$
$$K=\tfrac{1}{4}e^2F^2\sinh^{-2}(eFs). \tag{3.7}$$

In rearranging the order of these operators, the following commutator is required:

$$[x(s),x(0)]=[x(0)+(eF)^{-1}(e^{2eFs}-1)\Pi(0),x(0)]$$
$$=i(eF)^{-1}(e^{2eFs}-1). \tag{3.8}$$

Thus

$$\mathcal{H}+\tfrac{1}{2}e\sigma F=x(s)Kx(s)-2x(s)Kx(0)+x(0)Kx(0)$$
$$-\tfrac{1}{2}i\,\mathrm{tr}eF\coth(eFs), \tag{3.9}$$

where tr again denotes a diagonal summation, and we have employed the fact that

$$\mathrm{tr}(F)=0, \tag{3.10}$$

which follows from Eq. (3.6). The resulting differential equation (2.37)

$$i\partial_s(x(s)'|x(0)'')=[-\tfrac{1}{2}e\sigma F+(x'-x'')K(x'-x'')$$
$$-\tfrac{1}{2}i\,\mathrm{tr}eF\coth(eFs)](x(s)'|x(0)''), \tag{3.11}$$

has the solution

$$(x(s)'|x(0)'')=C(x',x'')e^{-L(s)}s^{-2}$$
$$\times\exp[\tfrac{1}{4}i(x'-x'')eF\coth(eFs)(x'-x'')]$$
$$\cdot\exp(i\tfrac{1}{2}e\sigma Fs), \tag{3.12}$$

$$L(s)=\tfrac{1}{2}\,\mathrm{tr}\ln[(eFs)^{-1}\sinh(eFs)].$$

To determine $C(x',x'')$, we employ

$$(x(s)'|\Pi(s)|x(0)'')=\tfrac{1}{2}[eF\coth(eFs)+eF]$$
$$\times(x'-x'')(x(s)'|x(0)''), \tag{3.13}$$

and

$$(x(s)'|\Pi(0)|x(0)'')=\tfrac{1}{2}[eF\coth(eFs)-eF]$$
$$\times(x'-x'')(x(s)'|x(0)''), \tag{3.14}$$

in conjunction with Eq. (3.12), to obtain the differential equations

$$[-i\partial_\mu'-eA_\mu(x')-\tfrac{1}{2}eF_{\mu\nu}(x'-x'')_\nu]C(x',x'')=0, \tag{3.15}$$

$$[i\partial_\mu''-eA_\mu(x'')-\tfrac{1}{2}eF_{\mu\nu}(x'-x'')_\nu]C(x',x'')=0. \tag{3.16}$$

The solution of Eq. (3.15) has the form

$$C(x',x'')$$
$$=C(x'')\exp\left[ie\int_{x''}^{x'}dx(A(x)+\tfrac{1}{2}F(x-x''))\right], \tag{3.17}$$

in which the integral is independent of the integration path, since $A_\mu(x)+\frac{1}{2}F_{\mu\nu}(x-x'')_\nu$ has a vanishing curl. However, by restricting the integration path to be a straight line connecting x' and x'', we may, in virtue of Eq. (3.6), simply write

$$C(x',x'')=C\Phi(x',x''),$$
$$\Phi(x',x'')=\exp\left[ie\int_{x''}^{x'}dxA(x)\right], \tag{3.18}$$

and, with C a constant, attain the solution of (3.15) and (3.16). The constant C has the value

$$C=-i(4\pi)^{-2} \tag{3.19}$$

since the limiting form of $(x(s)'|x(0)'')$ as $s\to 0$ is independent of the external field.

Finally, then

$$(x(s)'|x(0)'')=-i(4\pi)^{-2}\Phi(x',x'')e^{-L(s)}s^{-2}$$
$$\times\exp[i\tfrac{1}{4}(x'-x'')eF\coth(eFs)(x'-x'')]$$
$$\cdot\exp[i\tfrac{1}{2}e\sigma Fs], \tag{3.20}$$

and the Green's function $G(x',x'')$ is obtained from (2.30) in the two equivalent forms,

$$G(x',x'')=i\int_0^\infty ds\exp(-im^2s)$$
$$\times[-\gamma_\mu(x(s)'|\Pi_\mu(s)|x(0)'')$$
$$+m(x(s)'|x(0)'')]$$
$$=i\int_0^\infty ds\exp(-im^2s)$$
$$\times[-(x(s)'|\Pi_\mu(0)|x(0)'')\gamma_\mu$$
$$+m(x(s)'|x(0)'')], \tag{3.21}$$

which will be given explicitly on substituting Eqs. (3.13), (3.14), and (3.20).

The lagrange function $\mathcal{L}^{(1)}(x)$ is now computed as

$$\mathcal{L}^{(1)}(x)=\tfrac{1}{2}i\int_0^\infty ds s^{-1}\exp(-im^2s)$$
$$\times\mathrm{tr}(x(s)'|x(0)'')]_{x',x''\to x}$$
$$=(1/32\pi^2)\int_0^\infty ds s^{-3}\exp(-im^2s)$$
$$\times e^{-L(s)}\,\mathrm{tr}\exp(i\tfrac{1}{2}e\sigma Fs). \tag{3.22}$$

We may exhibit this more explicitly as a real quantity by a deformation of the integration path, which is effectively the substitution $s\rightarrow -is$:

$$\mathfrak{L}^{(1)}(x)=-(1/32\pi^2)\int_0^\infty dss^{-3}\exp(-m^2s)$$
$$\times e^{-l(s)}\operatorname{tr}\exp(\tfrac{1}{2}e\sigma Fs),$$
$$l(s)=\tfrac{1}{2}\operatorname{tr}\ln[(eFs)^{-1}\sin(eFs)]. \tag{3.23}$$

Indeed, we could have initially employed the integral representation

$$[m^2-(\gamma\Pi)^2]^{-1}=\int_0^\infty ds\exp[-(m^2-(\gamma\Pi)^2)s], \tag{3.24}$$

which exists in consequence of the restriction on real pair creation. This, however, would have obscured the proper time interpretation.

To evaluate the Dirac matrix trace, we employ the following spin matrix property:

$$\tfrac{1}{2}\{\sigma_{\mu\nu},\sigma_{\lambda\kappa}\}=\delta_{\mu\lambda}\delta_{\nu\kappa}-\delta_{\mu\kappa}\delta_{\nu\lambda}+i\epsilon_{\mu\nu\lambda\kappa}\gamma_5, \tag{3.25}$$

where

$$\gamma_5=i\gamma_1\gamma_2\gamma_3\gamma_4,\quad \gamma_5^2=-1, \tag{3.26}$$

and $\epsilon_{\mu\nu\lambda\kappa}$ is 1, or -1, if $(\mu\nu\lambda\kappa)$ forms an even, or odd permutation of (1234), and is zero otherwise. In terms of the dual field strength tensor,

$$F_{\mu\nu}^*=\tfrac{1}{2}i\epsilon_{\mu\nu\lambda\kappa}F_{\lambda\kappa}, \tag{3.27}$$

we have

$$(\tfrac{1}{2}\sigma_{\mu\nu}F_{\mu\nu})^2=\tfrac{1}{2}F_{\mu\nu}^2+\tfrac{1}{2}\gamma_5F_{\mu\nu}F_{\mu\nu}^*, \tag{3.28}$$

thus introducing the fundamental scalar

$$\mathfrak{F}=\tfrac{1}{4}F_{\mu\nu}^2=\tfrac{1}{2}(\mathbf{H}^2-\mathbf{E}^2), \tag{3.29}$$

and pseudoscalar

$$\mathfrak{G}=\tfrac{1}{4}F_{\mu\nu}F_{\mu\nu}^*=\mathbf{E}\cdot\mathbf{H} \tag{3.30}$$

constructed from the field strengths. Since

$$(\tfrac{1}{2}\sigma F)^2=2(\mathfrak{F}+\gamma_5\mathfrak{G}), \tag{3.31}$$

and $\gamma_5^2=-1$, it follows that $\frac{1}{2}\sigma F$ has the four eigenvalues

$$(\tfrac{1}{2}\sigma F)'=\pm(2(\mathfrak{F}\pm i\mathfrak{G}))^{\frac{1}{2}}. \tag{3.32}$$

Therefore,

$$\operatorname{tr}\exp(\tfrac{1}{2}e\sigma Fs)=4\operatorname{Re}\cosh es(2(\mathfrak{F}+i\mathfrak{G}))^{\frac{1}{2}}$$
$$\equiv 4\operatorname{Re}\cosh es\mathbf{X}, \tag{3.33}$$

where Re denotes the real part of the subsequent expression. Note, incidentally, that

$$\mathbf{X}^2=(\mathbf{H}+i\mathbf{E})^2. \tag{3.34}$$

The eigenvalues of the matrix $F=(F_{\mu\nu})$ are required for the construction of $\exp(-l(s))$. They can be obtained with the aid of the easily verifiable relations,

$$F_{\mu\lambda}F_{\lambda\nu}^*=-\delta_{\mu\nu}\mathfrak{G}, \tag{3.35}$$

and

$$F_{\mu\lambda}^*F_{\lambda\nu}^*-F_{\mu\lambda}F_{\lambda\nu}=2\delta_{\mu\nu}\mathfrak{F}. \tag{3.36}$$

From the eigenvalue equation

$$F_{\mu\nu}\psi_\nu=F'\psi_\mu, \tag{3.37}$$

and its equivalent according to Eq. (3.35),

$$F_{\mu\nu}^*\psi_\nu=-(1/F')\mathfrak{G}\psi_\mu, \tag{3.38}$$

we obtain by iteration:

$$F_{\mu\lambda}F_{\lambda\nu}\psi_\nu=(F')^2\psi_\mu,\quad F_{\mu\lambda}^*F_{\lambda\nu}^*\psi_\nu=(1/(F')^2)\mathfrak{G}^2\psi_\mu. \tag{3.39}$$

The identity (3.36) then yields the eigenvalue equation

$$(F')^4+2\mathfrak{F}(F')^2-\mathfrak{G}^2=0, \tag{3.40}$$

which has the solutions $\pm F^{(1)}$, $\pm F^{(2)}$, with

$$\begin{aligned}F^{(1)}&=(i/\sqrt{2})[(\mathfrak{F}+i\mathfrak{G})^{\frac{1}{2}}+(\mathfrak{F}-i\mathfrak{G})^{\frac{1}{2}}],\\ F^{(2)}&=(i/\sqrt{2})[(\mathfrak{F}+i\mathfrak{G})^{\frac{1}{2}}-(\mathfrak{F}-i\mathfrak{G})^{\frac{1}{2}}].\end{aligned} \tag{3.41}$$

Expressed in terms of these eigenvalues,

$$e^{-l(s)}=(es)^2F^{(1)}F^{(2)}/\sin(eF^{(1)}s)\sin(eF^{(2)}s)$$
$$=\frac{2(es)^2F^{(1)}F^{(2)}}{\cos es(F^{(1)}-F^{(2)})-\cos es(F^{(1)}+F^{(2)})}, \tag{3.42}$$

or

$$e^{-l(s)}=(es)^2\mathfrak{G}/\operatorname{Im}\cosh es\mathbf{X}, \tag{3.43}$$

where Im designates the imaginary part of the following expression.

The final result for $\mathfrak{L}^{(1)}$ is

$$\mathfrak{L}^{(1)}=-\frac{1}{8\pi^2}\int_0^\infty dss^{-3}\exp(-m^2s)$$
$$\times\left[(es)^2\mathfrak{G}\frac{\operatorname{Re}\cosh es\mathbf{X}}{\operatorname{Im}\cosh es\mathbf{X}}-1\right], \tag{3.44}$$

in which we have supplied the additive constant necessary to make $\mathfrak{L}^{(1)}$ vanish in the absence of a field. The first term in the expansion of $\mathfrak{L}^{(1)}$ for weak fields is

$$\mathfrak{L}^{(1)}\simeq-\frac{e^2}{12\pi^2}\int_0^\infty dss^{-1}\exp(-m^2s)\mathfrak{F}. \tag{3.45}$$

On separating this explicitly, and adding the lagrange function of the maxwell field,

$$\mathfrak{L}^{(0)}=-\mathfrak{F}=\tfrac{1}{2}(\mathbf{E}^2-\mathbf{H}^2), \tag{3.46}$$

we obtain the total lagrange function

$$\mathfrak{L}=-\left[1+\frac{e^2}{12\pi^2}\int_0^\infty dss^{-1}\exp(-m^2s)\right]\mathfrak{F}$$
$$-\frac{1}{8\pi^2}\int_0^\infty dss^{-3}\exp(-m^2s)$$
$$\times\left[(es)^2\mathfrak{G}\frac{\operatorname{Re}\cosh es\mathbf{X}}{\operatorname{Im}\cosh es\mathbf{X}}-1-\tfrac{2}{3}(es)^2\mathfrak{F}\right]. \tag{3.47}$$

The logarithmically divergent factor that multiplies the maxwell lagrange function may be absorbed by a change of scale for all fields, and a corresponding scale change, or renormalization, of charge. If we identify the quantities thus far employed by a zero subscript, and introduce new units of field strength and charge according to

$$\mathfrak{F}+i\mathfrak{G}=(1+Ce_0^2)(\mathfrak{F}_0+i\mathfrak{G}_0),\qquad e^2=e_0^2/(1+Ce_0^2), \tag{3.48}$$

$$C=\frac{1}{12\pi^2}\int_0^\infty ds\, s^{-1}\exp(-m^2s),$$

we obtain the finite, gauge invariant result

$$\mathfrak{L}=-\mathfrak{F}-\frac{1}{8\pi^2}\int_0^\infty dss^{-3}\exp(-m^2s)\times\left[(es)^2\mathfrak{G}\frac{\text{Re}\cosh esX}{\text{Im}\cosh esX}-1-\tfrac{2}{3}(es)^2\mathfrak{F}\right]$$

$$=\tfrac{1}{2}(\mathbf{E}^2-\mathbf{H}^2)+\frac{2\alpha^2}{45}\frac{(\hbar/mc)^3}{mc^2}\times[(\mathbf{E}^2-\mathbf{H}^2)^2+7(\mathbf{E}\cdot\mathbf{H})^2]+\cdots. \tag{3.49}$$

In the latter expansion, the conventional rationalized units have been reinstated, and $\alpha=e^2/4\pi\hbar c$.

Incidentally, the addition to the lagrange function produced by a spin zero charged field is obtained from Eq. (3.23) by omitting the Dirac trace, and multiplying by (-2). Thus,

$$\mathfrak{L}_{\text{spin }0}{}^{(1)}=\frac{1}{16\pi^2}\int_0^\infty dss^{-3}\exp(-\mu^2s)\times\left[\frac{(es)^2\mathfrak{G}}{\text{Im}\cosh esX}-1\right], \tag{3.50}$$

in which μ designates the mass of the spinless particle, and an additive constant has been supplied as in Eq. (3.44). The first term in the expansion for weak fields is separated explicitly by writing

$$\mathfrak{L}_0{}^{(1)}=-\frac{e^2}{48\pi^2}\int_0^\infty dss^{-1}\exp(-\mu^2s)\mathfrak{F}+\frac{1}{16\pi^2}\int_0^\infty dss^{-3}\exp(-\mu^2s)\times\left[\frac{(es)^2\mathfrak{G}}{\text{Im}\cosh esX}-1+\tfrac{1}{3}(es)^2\mathfrak{F}\right]. \tag{3.51}$$

If we take into account the existence of both spin 0 and spin $\frac{1}{2}$ charged fields,

$$\mathfrak{L}=\mathfrak{L}^{(0)}+\mathfrak{L}_0{}^{(1)}+\mathfrak{L}_{\frac{1}{2}}{}^{(1)}, \tag{3.52}$$

and a renormalization of the form (3.48), with

$$C=C_0+C_{\frac{1}{2}}=\frac{1}{48\pi^2}\int_0^\infty dss^{-1}\exp(-\mu^2s)+\frac{1}{12\pi^2}\int_0^\infty dss^{-1}\exp(-m^2s), \tag{3.53}$$

yields

$$\mathfrak{L}=-\mathfrak{F}-\frac{1}{8\pi^2}\int_0^\infty dss^{-3}\exp(-m^2s)\times[(es)^2\mathfrak{G}(\text{Re}/\text{Im})-1-\tfrac{2}{3}(es)^2\mathfrak{F}]+\frac{1}{16\pi^2}\int_0^\infty dss^{-3}\exp(-\mu^2s)\times[(es)^2\mathfrak{G}(1/\text{Im})-1+\tfrac{1}{3}(es)^2\mathfrak{F}]$$

$$=\tfrac{1}{2}(\mathbf{E}^2-\mathbf{H}^2)+\frac{2\alpha^2}{45}\frac{(\hbar/mc)^3}{mc^2}[(\mathbf{E}^2-\mathbf{H}^2)^2+7(\mathbf{E}\cdot\mathbf{H})^2]+\frac{\alpha^2}{90}\frac{(\hbar/\mu c)^3}{\mu c^2}\left[\frac{7}{4}(\mathbf{E}^2-\mathbf{H}^2)^2+(\mathbf{E}\cdot\mathbf{H})^2\right]+\cdots. \tag{3.54}$$

The physical quantities characterizing the field are comprised in the energy momentum tensor

$$\begin{aligned}T_{\mu\nu}&=\delta_{\mu\nu}\mathfrak{L}-(\partial\mathfrak{L}/\partial F_{\mu\lambda})F_{\nu\lambda}\\&=-(F_{\mu\lambda}F_{\nu\lambda}-\delta_{\mu\nu}\tfrac{1}{4}F_{\lambda\kappa}{}^2)(\partial\mathfrak{L}/\partial\mathfrak{F})\\&\quad+\delta_{\mu\nu}(\mathfrak{L}-\mathfrak{F}(\partial\mathfrak{L}/\partial\mathfrak{F})-\mathfrak{G}\partial\mathfrak{L}/\partial\mathfrak{G}).\end{aligned} \tag{3.55}$$

The maxwell tensor

$$T_{\mu\nu}{}^{(M)}=F_{\mu\nu}F_{\nu\lambda}-\delta_{\mu\nu}\tfrac{1}{4}F_{\lambda\kappa}{}^2 \tag{3.56}$$

is obtained from $\mathfrak{L}=-\mathfrak{F}$, the weak field approximation of Eq. (3.49). The next terms in the expansion of $\mathfrak{L}$ yield

$$T_{\mu\nu}=T_{\mu\nu}{}^M\left(1-\frac{16}{45}\alpha^2\frac{(\hbar/mc)^3}{mc^2}\mathfrak{F}\right)-\delta_{\mu\nu}\frac{2}{45}\alpha^2\frac{(\hbar/mc)^3}{mc^2}(4\mathfrak{F}^2+7\mathfrak{G}^2)+\cdots. \tag{3.57}$$

IV. PLANE WAVE FIELDS

A plane wave, traveling with the speed of light, is characterized by the field strength tensor

$$F_{\mu\nu}=f_{\mu\nu}F(\xi),\quad \xi=n_\mu x_\mu, \tag{4.1}$$

where n_μ is a null vector,

$$n_\mu{}^2=0, \tag{4.2}$$

and $F(\xi)$ is an arbitrary function. The constant tensor $f_{\mu\nu}$, and its dual $f_{\mu\nu}{}^*$, are restricted by the conditions

$$n_\mu f_{\mu\nu}=0,\quad n_\mu f_{\mu\nu}{}^*=0, \tag{4.3}$$

from which are derived

$$f_{\mu\lambda}f_{\lambda\nu}{}^*=0, \quad f_{\mu\lambda}f_{\lambda\nu}=f_{\mu\lambda}{}^*f_{\lambda\nu}{}^*=-n_\mu n_\nu. \quad (4.4)$$

The latter statement also includes a convention concerning the scale of $f_{\mu\nu}$.

The proper time equations of motion in this external field,

$$dx_\mu/ds=2\Pi_\mu,$$
$$d\Pi_\mu/ds=2eF(\xi)f_{\mu\nu}\Pi_\nu+n_\mu eF'(\xi)\tfrac{1}{2}\sigma_{\lambda\nu}f_{\mu\nu}, \quad (4.5)$$

admit several first integrals. Thus

$$d(n_\mu\Pi_\mu(s))/ds=0, \quad (4.6)$$

and

$$d(f_{\mu\nu}{}^*\Pi_\nu(s))/ds=0. \quad (4.7)$$

In addition,

$$d(f_{\mu\nu}\Pi_\nu(s))/ds=-n_\mu eF(\xi)d\xi/ds, \quad (4.8)$$

since

$$d\xi/ds=2n\Pi; \quad (4.9)$$

and therefore

$$d(f_{\mu\nu}\Pi_\nu+n_\mu eA(\xi))/ds=0, \quad (4.10)$$

where

$$dA(\xi)/d\xi=F(\xi). \quad (4.11)$$

In arriving at Eq. (4.10), it is necessary to recognize that $d\xi/ds$ commutes with ξ, in virtue of

$$[\xi, n\Pi]=[n_\mu x_\mu, n_\nu\Pi_\nu]=in_\mu{}^2=0. \quad (4.12)$$

Since $n\Pi$ is a constant of the motion, Eq. (4.9) can be integrated to yield

$$(\xi(s)-\xi(0))/s=2n\Pi, \quad (4.13)$$

from which we infer that

$$[\xi(s), \xi(0)]=2s[n\Pi, \xi(0)]=0. \quad (4.14)$$

The constant vector encountered on integration of Eq. (4.10),

$$f_{\mu\nu}\Pi_\nu+n_\mu eA(\xi)=C_\mu, \quad (4.15)$$

has the following evident properties:

$$n_\mu C_\mu=0, \quad f_{\mu\nu}{}^*C_\nu=0, \quad f_{\mu\nu}C_\nu=-n_\mu n\Pi,$$
$$C_\mu{}^2=(n\Pi)^2. \quad (4.16)$$

The elimination of $f_{\mu\nu}\Pi_\nu$ from the equation of motion, with the aid of Eq. (4.15), gives

$$d\Pi_\mu/ds=(d/d\xi)[2C_\mu eA(\xi)-n_\mu e^2A^2(\xi)+n_\mu eF(\xi)\tfrac{1}{2}\sigma f], \quad (4.17)$$

whence

$$\Pi_\mu=\tfrac{1}{2}dx_\mu/ds=(1/2n\Pi)[2C_\mu eA(\xi)-n_\mu e^2A^2(\xi)+n_\mu eF(\xi)\tfrac{1}{2}\sigma f]+D_\mu, \quad (4.18)$$

where D_μ is an integration constant. Note, incidentally, that

$$f_{\mu\nu}{}^*\Pi_\nu=f_{\mu\nu}{}^*D_\nu, \quad (4.19)$$

which is independent of s, in agreement with Eq. (4.7).

On integrating Eq. (4.18) with respect to s, we find that

$$x_\mu(s)-x_\mu(0)=\frac{1}{2(n\Pi)^2}\int_{\xi(0)}^{\xi(s)}d\xi[2C_\mu eA(\xi)-n_\mu e^2A^2(\xi)+n_\mu eF(\xi)\tfrac{1}{2}\sigma f]+2D_\mu s. \quad (4.20)$$

With the constant D_μ determined by Eq. (4.20), Eq. (4.18) states that

$$\Pi_\mu(s)=\frac{(x_\mu(s)-x_\mu(0))}{2s}+\frac{s}{(\xi(s)-\xi(0))}\times[2C_\mu eA(\xi(s))-n_\mu e^2A^2(\xi(s))+n_\mu eF(\xi(s))\tfrac{1}{2}\sigma f]$$
$$-\frac{s}{(\xi(s)-\xi(0))^2}\int_{\xi(0)}^{\xi(s)}d\xi[2C_\mu eA(\xi)-n_\mu e^2A^2(\xi)+n_\mu eF(\xi)\tfrac{1}{2}\sigma f]. \quad (4.21)$$

We can finally evaluate C_μ as

$$C_\mu=f_{\mu\nu}\Pi_\nu+n_\mu eA$$
$$=\frac{f_{\mu\nu}(x_\nu(s)-x_\nu(0))}{2s}+\frac{n_\mu}{\xi(s)-\xi(0)}\int_{\xi(0)}^{\xi(s)}d\xi eA(\xi). \quad (4.22)$$

The commutation properties of these operators are involved in the construction of the transformation function. As is already indicated in the commutativity of $\xi(s)$ and $\xi(0)$, these commutation relations are greatly simplified by the special nature of the external field. Thus to evaluate $[x_\mu(0), x_\mu(s)]$, we employ Eq. (4.21) to express $x_\mu(0)$ in terms of $x_\mu(s)$, $\Pi_\mu(s)$, $\xi(s)$, and $\xi(0)$. Now

$$[\xi(0), x_\mu(s)]=[\xi(s)-2sn\Pi, x_\mu(s)]=2isn_\mu, \quad (4.23)$$

and, in virtue of $n_\mu C_\mu=n_\mu{}^2=0$, we have simply

$$[x_\mu(s), x_\mu(0)]=[-2s\Pi_\mu(s), x_\mu(s)]=8is. \quad (4.24)$$

No other nonvanishing commutator intervenes in bringing $\mathcal{H}$ to the form

$$\mathcal{H}=\tfrac{1}{4}s^{-2}(x_\mu{}^2(s)-2x_\mu(s)x_\mu(0)+x_\mu{}^2(0))-2is^{-1}+\frac{1}{\xi(s)-\xi(0)}\int_{\xi(0)}^{\xi(s)}d\xi[e^2A^2(\xi)-eF(\xi)\tfrac{1}{2}\sigma f]$$
$$-\frac{1}{(\xi(s)-\xi(0))^2}\left[\int_{\xi(0)}^{\xi(s)}d\xi eA(\xi)\right]^2, \quad (4.25)$$

in which a constant added to $A(\xi)$ is without effect, as required by the corresponding ambiguity of Eq. (4.11).

The solution of the differential equation (2.37) is

$$(x(s)'|x(0)'')=C(x', x'')s^{-2}\exp[i\tfrac{1}{4}(x'-x'')^2/s]$$
$$\times\exp\left[-is/(\xi'-\xi'')\int_{\xi''}^{\xi'}d\xi[e^2A^2-eF\tfrac{1}{2}\sigma f]\right]$$
$$\times\exp\left[is\left(1/(\xi'-\xi'')\int_{\xi''}^{\xi'}d\xi eA\right)^2\right], \quad (4.26)$$

where

$$\xi'=n_\mu x_\mu', \quad \xi''=n_\mu x_\mu''. \tag{4.27}$$

The function $C(x', x'')$ is determined by the differential equations (2.38) and (2.39), in conjunction with Eq. (4.26). Thus,

$$\left[-i\partial_\mu'-eA_\mu(x')-f_{\mu\nu}(x'-x'')_\nu\frac{1}{\xi'-\xi''}\left(eA(\xi')-\frac{1}{\xi'-\xi''}\int_{\xi''}^{\xi'}d\xi eA(\xi)\right)\right]C(x', x'')=0. \tag{4.28}$$

The solution of this equation will be obtained in terms of a line integral which is independent of the integration path. Again choosing the path to be a straight line, we find simply

$$C(x', x'')=C\exp\left[ie\int_{x''}^{x'}dx_\mu A_\mu(x)\right]=C\Phi(x', x''), \tag{4.29}$$

in view of the antisymmetry of $f_{\mu\nu}$. It is evident that

$$C=-i(4\pi)^{-2}. \tag{4.30}$$

Only the behavior of the transformation function for $x_\mu'\simeq x_\mu''$ is of actual interest in applications to vacuum polarization phenomena. Now for $\xi'\simeq\xi''$,

$$\frac{1}{\xi'-\xi''}\int_{\xi''}^{\xi'}d\xi A^2(\xi)-\left[\frac{1}{\xi'-\xi''}\int_{\xi''}^{\xi'}d\xi A(\xi)\right]^2 \simeq\tfrac{1}{12}(\xi'-\xi'')^2F^2[\tfrac{1}{2}(\xi'+\xi'')], \tag{4.31}$$

and

$$\begin{aligned}(\xi'-\xi'')^2F^2&=(x'-x'')_\mu n_\mu F^2 n_\nu(x'-x'')_\nu\\ &=-(x'-x'')_\mu F_{\mu\lambda}F_{\lambda\nu}(x'-x'')_\nu,\end{aligned} \tag{4.32}$$

according to Eqs. (4.1) and (4.4). Therefore, for $x_\mu'\simeq x_\mu''$

$$\begin{aligned}(x(s)'|x(0)'')\simeq&-i(4\pi)^{-2}\Phi(x', x'')s^{-2}\\ &\times\exp[i\tfrac{1}{4}s^{-1}(x'-x'')_\mu(\delta_{\mu\nu}+\tfrac{1}{3}(es)^2F_{\mu\lambda}F_{\lambda\nu})(x'-x'')_\nu]\\ &\times\exp(\tfrac{1}{2}ie\sigma_{\mu\nu}F_{\mu\nu}s),\end{aligned} \tag{4.33}$$

which is identical with the transformation function for a constant field, as simplified by the special characteristics of the field now under consideration; namely,

$$\mathfrak{F}=0, \quad \mathfrak{G}=0. \tag{4.34}$$

We can conclude, without further calculation, that the physical quantities characterizing the plane wave field, the components of the energy-momentum tensor $T_{\mu\nu}$, will be identical in form with those of a constant field that obeys Eq. (4.34). On referring to Eq. (3.57), we see that $T_{\mu\nu}$ for a plane wave is just that of the maxwell field, which may be simplified further to

$$T_{\mu\nu}=F_{\mu\lambda}F_{\nu\lambda}=n_\mu n_\nu F^2(\xi). \tag{4.35}$$

Thus, there are no nonlinear vacuum phenomena for a single plane wave, of arbitrary strength and spectral composition.

V. γ-DECAY OF NEUTRAL MESONS

In this section we shall apply the results of our proper-time method to compute the effective coupling between a zero spin neutral meson and the electromagnetic field, as produced by the polarization of the proton vacuum. This interaction manifests itself in a spontaneous decay of the neutral meson into two photons.

The lagrange function for a spinless neutral meson field, in scalar interaction with the proton-antiproton field, is given by

$$\mathfrak{L}=-\tfrac{1}{2}[(\partial_\nu\phi)^2+\mu^2\phi^2]-g\phi\tfrac{1}{2}[\bar\psi, \psi]. \tag{5.1}$$

To find an approximate expression for the resultant coupling between the neutral meson field and the electromagnetic field, we replace $\frac{1}{2}[\bar\psi, \psi]$ by its vacuum expectation value, calculated in the presence of a homogeneous electromagnetic field. The use of the latter to represent the photons emitted in the spontaneous neutral meson decay introduces a small error, which is measured by the square of the meson-proton mass ratio, $(\mu/M)^2\simeq1/40$. On the other hand, by ignoring the effect of the meson field on the proton vacuum, we obtain only the initial approximation of a perturbation treatment. Now

$$\begin{aligned}\langle\tfrac{1}{2}[\bar\psi(x), \psi(x)]\rangle&=i\,\mathrm{tr}G(x, x)\\ &=-M\int_0^\infty ds\exp(-iM^2s)\,\mathrm{tr}(x|U(s)|x)\\ &=-\partial\mathfrak{L}^{(1)}(x)/\partial M,\end{aligned} \tag{5.2}$$

according to Eqs. (2.30) and (2.31). Thus, the effective lagrange function coupling term between the neutral meson and the electromagnetic field is given by

$$\mathfrak{L}'(x)=g\phi(x)\partial\mathfrak{L}^{(1)}(x)/\partial M, \tag{5.3}$$

which clearly also follows directly from the proton field equation of motion,

$$[\gamma(-i\partial-eA)+M+g\phi]\psi=0, \tag{5.4}$$

in the approximation which treats $\phi(x)$ as a weak, slowly varying, prescribed field. If we retain only the leading term in the expansion of $\mathfrak{L}^{(1)}$ for weak fields, Eq. (3.45), we have

$$\begin{aligned}\partial\mathfrak{L}^{(1)}/\partial M&\simeq(e^2/6\pi^2)M\int_0^\infty ds\exp(-M^2s)\mathfrak{F}\\ &=(2\alpha/3\pi)(1/M)\mathfrak{F}.\end{aligned} \tag{5.5}$$

Therefore the effective coupling term is

$$\mathfrak{L}'=(\alpha/3\pi)(g/M)\phi(\mathbf{H}^2-\mathbf{E}^2), \tag{5.6}$$

which describes the decay of a stationary meson, into two parallel polarized photons, at the rate

$$1/\tau=(\alpha^2/144\pi^3)(g^2/\hbar c)(\mu/M)^2(\mu c^2/\hbar). \tag{5.7}$$

A pseudoscalar interaction between the spinless neutral meson field and the proton field is described by the term

$$g\phi(x)\tfrac{1}{2}[\bar{\psi}(x), \gamma_5\psi(x)] \tag{5.8}$$

in the lagrange function. For our purposes, this is replaced by

$$\begin{aligned}\mathfrak{L}'(x) &= g\phi(x)\langle\tfrac{1}{2}[\bar{\psi}(x), \gamma_5\psi(x)]\rangle \\ &= ig\phi(x)\ \mathrm{tr}\gamma_5 G(x, x) \\ &= -g\phi(x)M\int_0^\infty ds\, \exp(-iM^2 s) \\ &\qquad \times \mathrm{tr}\gamma_5(x|U(s)|x).\end{aligned} \tag{5.9}$$

The transformation function (3.20), with $-is$ substituted for s, yields

$$\mathfrak{L}' = -g\phi M(4\pi)^{-2}\int_0^\infty ds s^{-2}\exp(-M^2 s)e^{-l(s)} \times \mathrm{tr}\gamma_5 \exp(\tfrac{1}{2}e\sigma F s). \tag{5.10}$$

Now, the eigenvalues of $\frac{1}{2}\sigma F$, as related to those of γ_5 by Eq. (3.31), give

$$\mathrm{tr}\gamma_5 \exp(\tfrac{1}{2}e\sigma F s) = -4\ \mathrm{Im}\cosh esX. \tag{5.11}$$

In view of Eq. (3.43), we obtain, without further approximation, simply

$$\begin{aligned}\mathfrak{L}' &= g\phi(e^2/4\pi^2)M\int_0^\infty ds\, \exp(-M^2 s)\mathcal{G} \\ &= (\alpha/\pi)(g/M)\phi \mathbf{E}\cdot\mathbf{H}.\end{aligned} \tag{5.12}$$

This effective coupling term implies the decay of a stationary neutral meson, into two perpendicularly polarized photons, at the rate

$$1/\tau\text{-}(\alpha^2/64\pi^3)(g^2/\hbar c)(\mu/M)^2(\mu c^2/\hbar). \tag{5.13}$$

The pseudovector interaction term,

$$(g/2M)\partial_\mu\phi(x)(1/2i)[\bar{\psi}(x), \gamma_5\gamma_\mu\psi(x)], \tag{5.14}$$

is formally equivalent to (5.8) for the problem under discussion, in the approximation to which it is being treated. This is demonstrated by a partial integration, combined with the use of the Dirac equation (2.1). Yet it has been found difficult[3,7] to verify the equivalence in the actual results of calculation. Such discrepancies between formal and explicit calculations may be produced by insufficient attention to the limiting processes implicit in the formalism. We shall demonstrate that, with appropriate care, the proper equivalence between the pseudoscalar and pseudovector couplings is indeed exhibited.

The effective pseudovector interaction between the meson and electromagnetic field is given by

$$\begin{aligned}\mathfrak{L}'(x) &= (g/2M)\partial_\mu\phi(x)\langle(1/2i)[\bar{\psi}(x), \gamma_5\gamma_\mu\psi(x)]\rangle \\ &= (g/2M)\partial_\mu\phi(x)\ \mathrm{tr}\gamma_5\gamma_\mu G(x, x) \\ &\to -(g/2M)\phi(x)\partial_\mu[\mathrm{tr}\gamma_5\gamma_\mu G(x, x)],\end{aligned} \tag{5.15}$$

where the last version represents the results of integrating by parts. We now remark that this derivative has the following meaning:

$$\begin{aligned}\partial_\mu[\mathrm{tr}\gamma_5\gamma_\mu G(x, x)] = \lim_{x', x''\to x} &[(\partial_\mu' - ieA_\mu(x')) \\ &+ (\partial_\mu'' + ieA_\mu(x''))]\ \mathrm{tr}\gamma_5\gamma_\mu G(x', x''),\end{aligned} \tag{5.16}$$

in which the structure of the right side is dictated by the requirement that only gauge covariant quantities be employed. We shall verify that the straightforward evaluation of Eq. (5.16) yields the pseudoscalar coupling (5.12), without further difficulty.

According to Eq. (3.21)

$$\begin{aligned}\mathrm{tr}\gamma_5\gamma_\mu G(x', x'') &= -i\ \mathrm{tr}\gamma_5\gamma_\mu\gamma_\nu\int_0^\infty ds\, \exp(-iM^2 s) \\ &\qquad \times (x(s)'|\Pi_\nu(s)|x(0)'') \\ &= -i\ \mathrm{tr}\gamma_\nu\gamma_5\gamma_\mu\int_0^\infty ds\, \exp(-iM^2 s) \\ &\qquad \times (x(s)'|\Pi_\nu(0)|x(0)'').\end{aligned} \tag{5.17}$$

The result of averaging these two equivalent expressions is

$$\begin{aligned}\mathrm{tr}\gamma_5\gamma_\mu G(x', x'') &= i\ \mathrm{tr}\gamma_5\int_0^\infty ds\, \exp(-iM^2 s) \\ &\qquad \times (x(s)'|\tfrac{1}{2}(\Pi_\nu(s) - \Pi_\nu(0))|x(0)'') \\ &\quad - \mathrm{tr}\gamma_5\sigma_{\mu\nu}\int_0^\infty ds\, \exp(-iM^2 s) \\ &\qquad \times (x(s)'|\tfrac{1}{2}(\Pi_\nu(s) + \Pi_\nu(0))|x(0)'').\end{aligned} \tag{5.18}$$

We shall be content to evaluate Eq. (5.18) in the approximation of weak fields. On referring to Eqs. (3.4), (3.5), and (3.20), it is apparent that the leading term in this approximation is

$$\begin{aligned}\mathrm{tr}\gamma_5\gamma_\mu G(x', x'') &= -(e/64\pi^2)\ \mathrm{tr}\gamma_5\sigma_{\mu\nu}\sigma_{\lambda\kappa}(x' - x'')_\nu F_{\lambda\kappa}\Phi(x', x'') \\ &\quad \times\int_0^\infty ds s^{-2}\exp(-iM^2 s)\exp[i\tfrac{1}{4}(x' - x'')^2/s] \\ &= (e/8\pi^2)F_{\mu\nu}{}^*(x' - x'')_\nu\Phi(x', x'') \\ &\quad \times\int_0^\infty ds s^{-2}\exp(-iM^2 s)\exp[i\tfrac{1}{4}(x' - x'')^2/s],\end{aligned} \tag{5.19}$$

---

[7] J. Steinberger, Phys. Rev. **76**, 1180 (1949).

with the aid of Eqs. (3.25) and (3.27). Since we are concerned with the behavior of this quantity only for $x'\simeq x''$, we may evaluate the proper time integral by an appropriate simplification. For $x'\simeq x''$,

$$\int_0^\infty dss^{-2}\exp(-iM^2s)\exp[i\tfrac{1}{4}(x'-x'')^2/s]$$

$$\simeq\int_0^\infty dss^{-2}\exp[i\tfrac{1}{4}(x'-x'')^2/s]$$

$$=\int_0^\infty d(s^{-1})\exp[i\tfrac{1}{4}(x'-x'')^2s^{-1}]$$

$$=4i/(x'-x'')^2. \tag{5.20}$$

Therefore,

$$\mathrm{tr}\gamma_5\gamma_\mu G(x',x'')\simeq(ie/2\pi^2)\Phi(x',x'')F_{\mu\nu}{}^*(x'-x'')_\nu(x'-x'')^{-2}. \tag{5.21}$$

To obtain the quantity of actual interest, Eq. (5.16), we observe that

$$[(\partial_\mu'-ieA_\mu(x'))+(\partial_\mu''+ieA_\mu(x''))]\Phi(x',x'')\times F_{\mu\nu}{}^*(x'-x'')_\nu(x'-x'')^{-2}=ie\Phi(x',x'')F_{\mu\nu}{}^*(x'-x'')_\nu F_{\mu\lambda}(x'-x'')_\lambda(x'-x'')^{-2}, \tag{5.22}$$

according to Eqs. (3.15) and (3.16). But, in view of Eq. (3.35),

$$F_{\mu\nu}{}^*(x'-x'')_\nu F_{\mu\lambda}(x'-x'')_\lambda=\mathcal{G}(x'-x'')^2, \tag{5.23}$$

and

$$\partial_\mu[\mathrm{tr}\gamma_5\gamma_\mu G(x,x)]=-(e^2/2\pi^2)\mathcal{G}\lim_{x'\to x''}\Phi(x',x'')=-(2\alpha/\pi)\mathcal{G}. \tag{5.24}$$

Thus, Eq. (5.15) yields

$$\mathcal{L}'=(\alpha/\pi)(g/M)\phi\mathbf{E}\cdot\mathbf{H}, \tag{5.25}$$

in complete agreement with Eq. (5.12).

VI. PERTURBATION THEORY

We shall now discuss the approximate evaluation of

$$W^{(1)}=i\tfrac{1}{2}\int_0^\infty dss^{-1}\exp(-im^2s)\,\mathrm{Tr}U(s), \tag{6.1}$$

by an expansion in powers of eA_μ and $eF_{\mu\nu}$. For this purpose, we write

$$\mathcal{H}=\mathcal{H}_0+\mathcal{H}_1, \tag{6.2}$$

where

$$\mathcal{H}_0=p^2 \tag{6.3}$$

and

$$\mathcal{H}_1=-e(pA+Ap)-\tfrac{1}{2}e\sigma F+e^2A^2. \tag{6.4}$$

To obtain the expansion of $\mathrm{Tr}U(s)$ in powers of $\mathcal{H}_1$, we observe that $U(s)$ obeys the differential equation

$$i\partial_sU(s)=(\mathcal{H}_0+\mathcal{H}_1)U(s). \tag{6.5}$$

The related operator

$$V(s)=U_0^{-1}(s)U(s), \tag{6.6}$$

where

$$U_0(s)=\exp(-i\mathcal{H}_0s), \tag{6.7}$$

is determined by

$$i\partial_sV(s)=U_0^{-1}(s)\mathcal{H}_1U_0(s)V(s) \tag{6.8}$$

and

$$V(0)=1. \tag{6.9}$$

One can combine Eqs. (6.8) and (6.9) in the integral equation

$$V(s)=1-i\int_0^s ds'U_0^{-1}(s')\mathcal{H}_1U_0(s')V(s'), \tag{6.10}$$

and construct the solution by iteration:

$$V(s)=1-i\int_0^s ds'U_0^{-1}(s')\mathcal{H}_1U_0(s')+(-i)^2\int_0^s ds'U_0^{-1}(s')\mathcal{H}_1U_0(s')\times\int_0^{s'}ds''U_0^{-1}(s'')\mathcal{H}_1U_0(s'')+\cdots. \tag{6.11}$$

On introducing new variables of integration, $u_1, u_2, \cdots$, according to

$$s'=su_1,\quad s''=s'u_2,\ \cdots, \tag{6.12}$$

we obtain the expansion

$$U(s)=\exp(-i\mathcal{H}s)$$

$$=U_0(s)+(-is)\int_0^1 du_1U_0((1-u_1)s)\mathcal{H}_1U_0(u_1s)+\cdots+(-is)^n\int_0^1 u_1^{n-1}du_1\cdots\int_0^1 du_n\times U_0((1-u_1)s)\mathcal{H}_1U_0(u_1(1-u_1)s)\cdots\times U_0(u_1\cdots u_{n-1}(1-u_n)s)\times\mathcal{H}_1U_0(u_1\cdots u_ns)+\cdots. \tag{6.13}$$

Instead of taking the trace of this expression directly, which would involve further simplification, we remark that

$$\mathrm{Tr}U(s)-\mathrm{Tr}U_0(s)=-is\int_0^1 d\lambda\,\mathrm{Tr}[\mathcal{H}_1\exp(-i(\mathcal{H}_0+\lambda\mathcal{H}_1)s)] \tag{6.14}$$

and insert the expansion (6.13) for $\exp[-i(\mathcal{H}_0+\lambda\mathcal{H}_1)s]$.

Thus,

$$\mathrm{Tr}U(s)=\mathrm{Tr}U_0(s)+(-is)\,\mathrm{Tr}[\mathcal{H}_1U_0(s)]$$

$$+\tfrac{1}{2}(-is)^2\int_0^1 du_1\,\mathrm{Tr}[\mathcal{H}_1U_0((1-u_1)s)\mathcal{H}_1U_0(u_1s)]+\cdots$$

$$+\frac{(-is)^{n+1}}{n+1}\int_0^1 u_1^{\,n-1}du_1\cdots\int_0^1 du_n$$

$$\times\mathrm{Tr}[\mathcal{H}_1U_0((1-u_1)s)\mathcal{H}_1\cdots$$

$$\times\mathcal{H}_1U_0(u_1\cdots u_ns)]+\cdots. \quad (6.15)$$

We shall retain only the first nonvanishing field dependent terms in this expansion:

$$W^{(1)}=\tfrac{1}{2}ie^2\int_0^\infty dss^{-1}\exp(-im^2s)$$

$$\times\Big\{-is\,\mathrm{Tr}[A^2\exp(-ip^2s)]$$

$$+\tfrac{1}{2}(-is)^2\int_{-1}^1\tfrac{1}{2}dv\,\mathrm{Tr}[(pA+Ap)\exp(-ip^2\tfrac{1}{2}(1-v)s)$$

$$\times(pA+Ap)\exp(-ip^2\tfrac{1}{2}(1+v)s)]$$

$$+\tfrac{1}{2}(-is)^2\int_{-1}^1\tfrac{1}{2}dv\,\mathrm{Tr}[\tfrac{1}{2}\sigma F\exp(-ip^2\tfrac{1}{2}(1-v)s)$$

$$\times\tfrac{1}{2}\sigma F\exp(-ip^2\tfrac{1}{2}(1+v)s)]\Big\}. \quad (6.16)$$

For convenience, the variable u_1 has been replaced by $\frac{1}{2}(1+v)$. The evaluation of these traces is naturally performed in a momentum representation. The matrix elements of the coordinate dependent field quantities depend only on momentum differences,

$$(p+\tfrac{1}{2}k|A_\mu|p-\tfrac{1}{2}k)=(2\pi)^{-4}\int(dx)e^{-ikx}A_\mu(x)$$

$$\equiv(2\pi)^{-2}A_\mu(k), \quad (6.17)$$

and

$$(p|A_\mu^2|p)=(2\pi)^{-4}\int(dx)A_\mu^2(x)$$

$$=(2\pi)^{-4}\int(dk)A_\mu(-k)A_\mu(k). \quad (6.18)$$

Therefore

$$W^{(1)}=\frac{2ie^2}{(2\pi)^4}\int_0^\infty dss^{-1}\exp(-im^2s)\Big\{-is\int(dk)$$

$$\times A_\mu(-k)A_\mu(k)\int(dp)\exp(-ip^2s)$$

$$+\tfrac{1}{2}(-is)^2\int_{-1}^1\tfrac{1}{2}dv\int(dk)\int(dp)2p_\mu A_\mu(-k)$$

$$\times\exp(-i(p+\tfrac{1}{2}k)^2\tfrac{1}{2}(1-v)s]$$

$$\times 2p_\nu A_\nu(k)\exp[-i(p-\tfrac{1}{2}k)^2\tfrac{1}{2}(1+v)s]$$

$$+\tfrac{1}{2}(-is)^2\int_{-1}^1\tfrac{1}{2}dv\int(dk)\int(dp)\,\tfrac{1}{4}\,\mathrm{tr}\tfrac{1}{2}\sigma F(-k)$$

$$\times\exp[-i(p+\tfrac{1}{2}k)^2\tfrac{1}{2}(1-v)s]\tfrac{1}{2}\sigma F(k)$$

$$\times\exp[-i(p-\tfrac{1}{2}k)^2\tfrac{1}{2}(1+v)s]\Big\}. \quad (6.19)$$

We thus encounter the elementary integrals

$$\int(dp\exp(-ip^2s)=-i\pi^2s^{-2}, \quad (6.20)$$

$$\int(dp)\exp[-i(p^2+(k^2/4))s+ipkvs]$$

$$=-i\pi^2s^{-2}\exp[-i(k^2/4)(1-v^2)s], \quad (6.21)$$

and

$$\int(dp)p_\mu p_\nu\exp[-i(p^2+(k^2/4))s+ipkvs]$$

$$=-\exp(-i\tfrac{1}{4}k^2s)(vs)^{-2}(\partial/\partial k_\mu)(\partial/\partial k_\nu)$$

$$\times\int(dp)\exp(-ip^2s+ipkvs)$$

$$=-i\pi^2s^{-2}(-i\tfrac{1}{2}s^{-1}\delta_{\mu\nu}+\tfrac{1}{4}v^2k_\mu k_\nu)$$

$$\times\exp[-i\tfrac{1}{4}k^2(1-v^2)s]. \quad (6.22)$$

It is convenient to replace the $\delta_{\mu\nu}$ term of the last integral by an expression which is equivalent to it in virtue of the integration with respect to v. Now

$$\int_{-1}^1\tfrac{1}{2}dv\exp[-i\tfrac{1}{4}k^2(1-v^2)s]$$

$$=1-is\tfrac{1}{2}k^2\int_{-1}^1\tfrac{1}{2}dvv^2\exp[-i\tfrac{1}{4}k^2(1-v^2)s], \quad (6.23)$$

so that, effectively

$$\int (dp)p_\mu p_\nu \exp[-i(p^2+\tfrac{1}{4}k^2)s+ipkvs]$$

$$=-\tfrac{1}{2}\pi^2 s^{-3}\delta_{\mu\nu}-i\pi^2 s^{-2}\tfrac{1}{4}v^2(k_\mu k_\nu-\delta_{\mu\nu}k^2)$$
$$\times\exp[-i\tfrac{1}{4}k^2(1-v^2)s]. \quad (6.24)$$

On inserting the values of the various integrals, and noticing that

$$(k_\mu k_\nu-\delta_{\mu\nu}k^2)A_\mu(-k)A_\nu(k)=-\tfrac{1}{2}F_{\mu\nu}(-k)F_{\mu\nu}(k), \quad (6.25)$$

we obtain immediately the gauge invariant form (with $s\to -is$)

$$W^{(1)}=-\frac{e^2}{(4\pi)^2}\int (dk)\tfrac{1}{2}F_{\mu\nu}(-k)F_{\mu\nu}(k)\int_0^1 dv(1-v^2)$$
$$\times\int_0^\infty dss^{-1}\exp\{-[m^2+\tfrac{1}{4}k^2(1-v^2)]s\}. \quad (6.26)$$

This has been achieved without any special device, other than that of reserving the proper-time integration to the last.

A significant separation of terms is produced by a partial integration with respect to v, according to

$$\int_0^1 dv(1-v^2)\int_0^\infty dss^{-1}\exp\{-[m^2+\tfrac{1}{4}k^2(1-v^2)]s\}$$
$$=\tfrac{2}{3}\int_0^\infty dss^{-1}\exp(-m^2 s)-\tfrac{1}{2}k^2\int_0^1 dv(v^2-\tfrac{1}{3}v^4)$$
$$\times\int_0^\infty ds\exp\{-[m^2+\tfrac{1}{4}k^2(1-v^2)]s\}. \quad (6.27)$$

Adding the action integral of the maxwell field, which is expressed in momentum space by

$$W^{(0)}=-\int (dk)\tfrac{1}{4}F_{\mu\nu}(-k)F_{\mu\nu}(k), \quad (6.28)$$

we obtain the modified action integral,

$$W=-\left[1+\frac{e^2}{12\pi^2}\int_0^\infty dss^{-1}\exp(-m^2 s)\right]$$
$$\times\int (dk)\tfrac{1}{4}F_{\mu\nu}(-k)F_{\mu\nu}(k)$$
$$+\frac{e^2}{(4\pi)^2}\int (dk)\tfrac{1}{4}F_{\mu\nu}(-k)F_{\mu\nu}(k)k^2$$
$$\times\int_0^1 dv\frac{v^2(1-\tfrac{1}{3}v^2)}{m^2+\tfrac{1}{4}k^2(1-v^2)}. \quad (6.29)$$

The field strength and charge renormalization contained in Eq. (3.48) then produces the finite gauge invariant result,[8]

$$W=-\int (dk)\tfrac{1}{4}F_{\mu\nu}(-k)F_{\mu\nu}(k)$$
$$\times\left[1-\frac{\alpha}{4\pi}\frac{k^2}{m^2}\int_0^1 dv\frac{v^2(1-\tfrac{1}{3}v^2)}{1+(k^2/4m^2)(1-v^2)}\right]. \quad (6.30)$$

The restriction which we have thus far imposed, that no actual pair creation occurs, corresponds to the requirement that $1+(k^2/4m^2)(1-v^2)$ never vanishes. This will be true if $-k^2<4m^2$, for all k_μ contained in the fourier representation of the field. Indeed, it is evident from energy and momentum considerations that to produce a pair by the absorption of a single quantum the momentum vector of the latter must be time-like and must have a magnitude exceeding $2m$. We shall now simply remark that, to extend our results to pair-producing fields, it is merely necessary to add an infinitesimal negative imaginary constant to the denominator of Eq. (6.30) and interpret the positive imaginary contribution to W thus obtained with the statement that

$$|e^{iW}|^2=e^{-2\,\mathrm{Im}W} \quad (6.31)$$

represents the probability that no actual pair creation occurs during the history of the field. The infinitesimal imaginary constant, as employed in

$$\lim_{\epsilon\to+0}\frac{1}{x-i\epsilon}=P\frac{1}{x}+\pi i\delta(x), \quad (6.32)$$

represents a familiar device for dealing with real processes. We obtain from Eq. (6.30) that

$$2\,\mathrm{Im}W=\tfrac{1}{2}\alpha\int (dk)\tfrac{1}{4}F_{\mu\nu}(-k)F_{\mu\nu}(k)\frac{k^2}{m^2}\int_0^1 dvv^2$$
$$\times\left(1-\frac{v^2}{3}\right)\delta\left[1+\frac{k^2}{4m^2}(1-v^2)\right]$$
$$=\alpha\int_{-k^2>4m^2}(dk)(-\tfrac{1}{4})F_{\mu\nu}(-k)F_{\mu\nu}(k)$$
$$\times\left(1-\frac{4m^2}{(-k^2)}\right)^{\frac{1}{2}}\frac{1}{3}\left(2+\frac{4m^2}{(-k^2)}\right). \quad (6.33)$$

For the weak fields that are being considered, Eq. (6.33) is just the probability that a pair is created by the field. It should be noticed, incidentally, that

$$-\tfrac{1}{4}F_{\mu\nu}(-k)F_{\mu\nu}(k)=\tfrac{1}{2}[|\mathbf{E}(k)|^2-|\mathbf{H}(k)|^2] \quad (6.34)$$

[8] The corresponding result for a spin zero charged field is obtained by omitting the spin term of Eq. (6.19), and multiplying the remainder with $(-\tfrac{1}{2})$. This effectively substitutes $\tfrac{1}{6}v^4$ for $(v^2-\tfrac{1}{3}v^4)$, in Eq. (6.30).

is actually positive for a pair-generating field. This follows, for example, from the vanishing of the magnetic field in the special coordinate system where k_μ has only a temporal component.

An alternative version of Eq. (6.33) is obtained by replacing the field with the current required to generate this field, according to the maxwell equations

$$ik_\mu F_{\mu\nu}(k)=-J_\nu(k),$$
$$k_\mu F_{\nu\lambda}(k)+k_\nu F_{\lambda\mu}(k)+k_\lambda F_{\mu\nu}(k)=0. \quad (6.35)$$

Now

$$k_\lambda^2 F_{\mu\nu}(-k)F_{\mu\nu}(k)=2k_\nu F_{\nu\mu}(-k)k_\lambda F_{\lambda\mu}(k)$$
$$=2J_\mu(-k)J_\mu(k), \quad (6.36)$$

so that[9]

$$2\,\mathrm{Im}W=(\alpha/8m^2)\int_{-k^2>4m^2}(dk)J_\mu(-k)J_\mu(k)$$
$$\times(1-\gamma)^{\frac{1}{2}}\gamma\tfrac{1}{3}(2+\gamma), \quad (6.37)$$

where

$$\gamma=4m^2/(-k^2). \quad (6.38)$$

It is now appropriate to notice that the integral (3.49), representing the lagrange function for a uniform field, has singularities, unless $\mathcal{G}=0$, $\mathfrak{F}>0$, corresponding to a pure magnetic field in an appropriate coordinate system. This is the analytic expression of the fact that pairs are created by a uniform electric field. In particular, for $\mathcal{G}=0$, $-2\mathfrak{F}=\mathcal{E}^2>0$, which invariantly characterizes a pure electric field, the lagrange function proper time integral,

$$\mathfrak{L}=\tfrac{1}{2}\mathcal{E}^2-(1/8\pi^2)\int_0^\infty ds s^{-3}\exp(-m^2s)$$
$$\times[e\mathcal{E}s\cot(e\mathcal{E}s)-1+\tfrac{1}{3}(e\mathcal{E}s)^2], \quad (6.39)$$

has singularities at

$$s=s_n=n\pi/e\mathcal{E},\quad n=1,2,\cdots. \quad (6.40)$$

If the integration path is considered to lie above the real axis, which is an alternative version of the device embodied in Eq. (6.32), we obtain a positive imaginary contribution to $\mathfrak{L}$,

$$2\,\mathrm{Im}\mathfrak{L}=\frac{1}{4\pi}\sum_{n=1}^\infty s_n^{-2}\exp(-m^2s_n)$$
$$=\frac{\alpha^2}{\pi^2}\mathcal{E}^2\sum_{n=1}^\infty n^{-2}\exp\left(\frac{-n\pi m^2}{e\mathcal{E}}\right). \quad (6.41)$$

This is the probability, per unit time and per unit volume, that a pair is created by the constant electric field.

We must now consider, in the framework of this special problem, the connection between the proper time method and that of "invariant regularization." The vacuum polarization addition to the action integral has the general structure

$$W^{(1)}=\int(dk)A_\mu(-k)K_{\mu\nu}(k,m^2)A_\nu(k). \quad (6.42)$$

The proper-time technique yields the coefficient $K_{\mu\nu}(k,m^2)$ in the form

$$K_{\mu\nu}(k,m^2)]_P=\int_0^\infty ds\exp(-im^2s)K_{\mu\nu}(k,s), \quad (6.43)$$

where $K_{\mu\nu}(k,s)$ is a finite, gauge invariant quantity; infinities appear only in the final stage of integrating s to the origin. In effect, this method substitutes a lower limit, s_0, in the proper time integral and reserves the limit, $s_0\to 0$, to the end of the calculation. If, on the contrary, the proper-time technique is not explicitly introduced, $K_{\mu\nu}(k,m^2)$ will be represented by divergent integrals which lead, in general, to non-gauge invariant results. The regulator technique avoids the difficulty by introducing a suitable weighted integration with respect to the square of the proper mass, thus substituting for $K_{\mu\nu}(k,m^2)$, the quantity

$$K_{\mu\nu}(k,m^2)]_R=\int_{-\infty}^\infty d\kappa\rho(\kappa)K_{\mu\nu}(k,\kappa). \quad (6.44)$$

The "regulator" $\rho(\kappa)$ must reduce to $\delta(\kappa-m^2)$, in an appropriate limit, and will produce gauge invariant results in this problem if the following integral conditions are satisfied:

$$\int_{-\infty}^\infty d\kappa\rho(\kappa)=0,\qquad \int_{-\infty}^\infty d\kappa\kappa\rho(\kappa)=0. \quad (6.45)$$

Expressed in terms of the fourier transformed quantities,

$$R(s)=\int_{-\infty}^\infty d\kappa e^{-i\kappa s}\rho(\kappa),$$
$$K_{\mu\nu}(k,s)=(1/2\pi)\int_{-\infty}^\infty d\kappa e^{i\kappa s}K_{\mu\nu}(k,\kappa), \quad (6.46)$$

we have

$$K_{\mu\nu}(k,m^2)]_R=\int_{-\infty}^\infty dsR(s)K_{\mu\nu}(k,s), \quad (6.47)$$

while the conditions on $\rho(\kappa)$ appear as

$$R(0)=0,\quad R'(0)=0,\quad R(s)\to\exp(-im^2s). \quad (6.48)$$

Now observe that the proper time method yields $K_{\mu\nu}(k,m^2)$ in the form (6.47), with

$$K_{\mu\nu}(k,s)=0,\quad s<0, \quad (6.49)$$

and

$$R(s)=\exp(-im^2s),\quad s>s_0$$
$$=0,\quad s<s_0. \quad (6.50)$$

[9] A simple example, to which this formula may be applied, is the creation of a pair in a nuclear $j=0\to 0$ transition. J. R. Oppenheimer and J. Schwinger, Phys. Rev. 56, 1066 (1939).

This $R(s)$, and all its derivatives, vanishes at the origin, thus satisfying the regulator conditions as $s_0 \to 0$. It appears, then, that regularization is a procedure for inserting, into a calculation that does not employ it, enough of the structure provided by the proper time representation to ensure gauge invariant results.

APPENDIX A

It is our purpose here to use the proper time equations of motion (2.36) for the computation of the current induced in the vacuum by a weak, arbitrarily varying field:

$$F_{\mu\nu}(x) = [1/(2\pi)^2]\int (dk) e^{ikx} F_{\mu\nu}(k). \tag{A.1}$$

In the absence of a field, the equations of motions are solved by

$$\Pi_\mu(s) = \Pi_\mu(0), \quad x_\mu(s) = x_\mu(0) + 2\Pi_\mu(0)s. \tag{A.2}$$

As a first approximation for weak fields, we accordingly write

$$d\Pi_\mu(s)/ds = [e/(2\pi)^2]\int (dk) F_{\mu\nu}(k)\{e^{ik(x(0)+2\Pi(0)s)}, \Pi_\nu(0)\}$$
$$+ e/(2\pi)^2 \int (dk) ik_\mu \tfrac{1}{2}\sigma_{\lambda\nu} F_{\lambda\nu}(k) e^{ik(x(0)+2\Pi(0)s)}. \tag{A.3}$$

On integrating with respect to s, one obtains

$$\Pi_\mu(s) - \Pi_\mu(0) = [e/(2\pi)^2]\int (dk) F_{\mu\nu}(k) \times \int_0^s ds' \{e^{ik(x(0)+2\Pi(0)s')}, \Pi_\nu(0)\}$$
$$+ e/(2\pi)^2 \int (dk) ik_\mu \tfrac{1}{2}\sigma_{\lambda\nu} F_{\lambda\nu}(k) \times \int_0^s ds' e^{ik(x(0)+2\Pi(0)s')}. \tag{A.4}$$

A second integration yields

$$\frac{x_\mu(s) - x_\mu(0)}{2s} = \Pi_\mu(0) + e/(2\pi)^2 \int (dk) F_{\mu\nu}(k) \times \int_0^s ds'(1 - s'/s)\{e^{ik(x(0)+2\Pi(0)s')}, \Pi_\nu(0)\}$$
$$+ \frac{e}{(2\pi)^2}\int (dk) ik_\mu \tfrac{1}{2}\sigma F(k) \times \int_0^s (ds'\left(1 - \frac{s'}{s}\right) e^{ik(x(0)+2\Pi(0)s')}, \tag{A.5}$$

and therefore

$$\tfrac{1}{2}(\Pi_\mu(s) + \Pi_\mu(0)) = \frac{x_\mu(s) - x_\mu(0)}{2s} + \frac{e}{(2\pi)^2}\int (dk) F_{\mu\nu}(k) \times \int_0^s ds'\left(\frac{s'}{s} - \frac{1}{2}\right)\{e^{ik(x(0)+2\Pi(0)s')}, \Pi_\nu(0)\}$$
$$+ \frac{e}{(2\pi)^2}\int (dk) ik_\mu \tfrac{1}{2}\sigma F(k) \times \int_0^s ds'\left(\frac{s'}{s} - \frac{1}{2}\right) e^{ik(x(0)+2\Pi(0)s')}. \tag{A.6}$$

The induced current is equivalently expressed by

$$\langle j_\mu(x)\rangle = e \operatorname{tr}\gamma_\mu\left(x \left| (\gamma\Pi - m)\int_0^\infty ds \exp(-im^2 s) U(s) \right| x\right)$$
$$= e\int_0^\infty ds \exp(-im^2 s) \operatorname{tr}\gamma_\mu\gamma_\nu (x(s)'|\Pi_\nu(s)|x(0)'')]_{x',x''\to x}, \tag{A.7}$$

and

$$\langle j_\mu(x)\rangle = e \operatorname{tr}\gamma_\mu\left(x \left| \int_0^\infty ds \exp(-im^2 s) U(s)(\gamma\Pi - m) \right| x\right)$$
$$= e\int_0^\infty ds \exp(-im^2 s) \operatorname{tr}\gamma_\nu\gamma_\mu (x(s)'|\Pi_\nu(0)|x(0)'')]_{x',x''\to x}. \tag{A.8}$$

On averaging the two forms, we find that

$$\langle j_\mu(x)\rangle = -e\int_0^\infty ds \exp(-im^2 s) \times \operatorname{tr}(x(s)'|\tfrac{1}{2}(\Pi_\mu(s) + \Pi_\mu(0))|x(0)'')]_{x',x''\to x}$$
$$- ie\int_0^\infty ds \exp(-im^2 s) \times \operatorname{tr}\sigma_{\mu\nu}(x(s)'|\tfrac{1}{2}(\Pi_\nu(s) - \Pi_\nu(0))|x(0)'')]_{x',x''\to x}. \tag{A.9}$$

It may be noted here that no current exists in the absence of a field, since

$$\lim_{x'-x''\to\pm 0} (x(s)'|x_\mu(s) - x_\mu(0)|x(0)'') = 0, \tag{A.10}$$

and, therefore, only the transformation function in the absence of a field is required for the first-order evaluation of Eq. (A.9). Now

$$\operatorname{tr}\sigma_{\mu\nu}(x(s)'|\tfrac{1}{2}(\Pi_\nu(s) - \Pi_\nu(0))|x(0)'')$$
$$= [2e/(2\pi)^2]\int (dk)(\partial F_{\mu\nu}/\partial x_\nu)(k) s\int_{-1}^1 \tfrac{1}{2}dv \times (x(s)'|\exp[i(kx(s)\tfrac{1}{2}(1+v) + kx(0)\tfrac{1}{2}(1-v))]|x(0)''), \tag{A.11}$$

in which the variable s' has been replaced by v, according to

$$s' = s(1+v)/2. \tag{A.12}$$

The operators $kx(s)$ and $kx(0)$ do not commute:

$$[kx(s), kx(0)] = 2s[k\Pi(0), kx(0)] = -2isk^2. \tag{A.13}$$

We may, however, employ the easily established theorem,

$$e^{A+B} = e^A e^B e^{-\frac{1}{2}[A,B]}, \tag{A.14}$$

for operators A and B that commute with their commutator $[A, B]$. Thus,

$$\exp[i(kx(s)\tfrac{1}{2}(1+v) + kx(0)\tfrac{1}{2}(1-v))]$$
$$= \exp[ikx(s)\tfrac{1}{2}(1+v)]\exp[ikx(0)\tfrac{1}{2}(1-v)] \times \exp[-ik^2\tfrac{1}{4}(1-v^2)s], \tag{A.15}$$

and

$$\operatorname{tr}\sigma_{\mu\nu}(x(s)'|\tfrac{1}{2}(\Pi_\nu(s) - \Pi_\nu(0))|x(0)'')]_{x',x''\to x}$$
$$= [2e/(2\pi)^2]\int (dk) e^{ikx}(\partial F_{\mu\nu}/\partial x_\nu)(k) s\int_{-1}^1 \tfrac{1}{2}dv \times \exp[-ik^2\tfrac{1}{4}(1-v^2)s](-i)/(4\pi)^2 s^2. \tag{A.16}$$

A similar treatment applies to

$$\operatorname{tr}(x(s)'|\tfrac{1}{2}(\Pi_\nu(s) + \Pi_\nu(0))|x(0)'')]_{x',x''\to x}$$
$$= \frac{2e}{(2\pi)^2}\int (dk) F_{\mu\nu}(k) s\int_{-1}^1 \tfrac{1}{2}dv v (x(s')'|\{\exp[i(kx(s)\tfrac{1}{2}(1+v) + kx(0)\tfrac{1}{2}(1-v))], (x_\nu(s) - x_\nu(0))/2s\}|x'(0))]_{x',x''\to x}. \tag{A.17}$$

With the aid of the commutation relations,

$$\begin{aligned} [e^{ikx(0)\frac{1}{2}(1-v)}, x_\nu(s)] &= -k_\nu(1-v)s e^{ikx(0)\frac{1}{2}(1-v)}, \\ [e^{ikx(s)\frac{1}{2}(1+v)}, x_\nu(0)] &= k_\nu(1+v)s e^{ikx(s)\frac{1}{2}(1+v)}, \end{aligned} \tag{A.18}$$

this reduces to

$$\operatorname{tr}(x(s)'|\tfrac{1}{2}(\Pi_\mu(s) + \Pi_\mu(0))|x(0)'')]_{x',x''\to x}$$
$$= -[2ie/(2\pi)^2]\int (dk) e^{ikx}(\partial F_{\mu\nu}/\partial x_\nu)(k) s\int_{-1}^1 \tfrac{1}{2}dv v^2 \times \exp[-ik^2\tfrac{1}{4}(1-v^2)s](-i)/(4\pi)^2 s^2. \tag{A.19}$$

We have thus obtained

$$\langle j_\mu(x)\rangle = -(\alpha/2\pi)(2\pi)^{-2}\int (dk) e^{ikx}(\partial F_{\mu\nu}/\partial x_\nu)(k)\int_0^1 dv(1-v^2) \times \int_0^\infty ds s^{-1}\exp\{-[m^2 + \tfrac{1}{4}k^2(1-v^2)]s\}, \tag{A.20}$$

in which the substitution $s \to -is$ has again been introduced. This is precisely the current derived from the action integral $W^{(1)}$ of Eq. (6.26), and further discussion proceeds as in Sec. VI.

APPENDIX B

An electron in interaction with its proper radiation field, and an external field, is described by the modified Dirac equation,[10]

$$\gamma_\mu(-i\partial_\mu-eA_\mu(x))\psi(x)+\int(dx')M(x,x')\psi(x')=0. \quad (B.1)$$

To the second order in e, the mass operator, $M(x, x')$, is given by

$$M(x,x')=m_0\delta(x-x')+ie^2\gamma_\mu G(x,x')\gamma_\mu D_+(x-x'). \quad (B.2)$$

Here $G(x, x')$ is the Green's function of the Dirac equation in the external field, and $D_+(x-x')$ is a photon Green's function, expressed by

$$D_+(x-x')=(4\pi)^{-2}\int_0^\infty dtt^{-2}\exp[i\tfrac{1}{4}(x-x')^2/t]. \quad (B.3)$$

We shall suppose the external field to be weak and uniform. Under these conditions, the transformation function $(x(s)|x(0)')$, involved in the construction of $G(x, x')$, may be approximated by

$$(x(s)|x(0)')\simeq -i(4\pi)^{-2}\Phi(x,x')s^{-2} \times\exp[i\tfrac{1}{4}(x-x')^2/s]\exp(i\tfrac{1}{2}e\sigma F); \quad (B.4)$$

that is, terms linear in the field strengths enter only through the Dirac spin magnetic moment. The corresponding simplification of the Green's function, obtain by averaging the two equivalent forms in Eq. (3.21), is

$$G(x,x')\simeq(4\pi)^{-2}\Phi(x,x')\int_0^\infty dss^{-2}\exp(-im^2s) \times\exp[i\tfrac{1}{4}(x-x')^2/s]\tfrac{1}{2}\left\{\frac{-\gamma(x-x')}{2s}+m,\exp(i\tfrac{1}{2}e\sigma F)\right\}. \quad (B.5)$$

The mass operator is thus approximately represented by

$$M(x,x')=m_0\delta(x-x')+[ie^2/(4\pi)^4]\Phi(x,x')\int_0^\infty dss^{-2}\int_0^\infty dtt^{-2} \times\exp(-im^2s)\exp\left[i\tfrac{1}{4}(x-x')^2\left(\frac{1}{s}+\frac{1}{t}\right)\right] \times\gamma_\lambda\tfrac{1}{2}\left\{\frac{-\gamma(x-x')}{2s}+m,\exp(i\tfrac{1}{2}e\sigma F)\right\}\gamma_\lambda, \quad (B.6)$$

or

$$M(x,x')=m_0\delta(x-x')+[ie^2/(4\pi)^4]\Phi(x,x') \times\int_0^\infty dss^{-2}\exp(-im^2s)\int_0^s dww^{-2}\exp[i\tfrac{1}{4}(x-x')^2/w] \times[-4m-s^{-1}\gamma(x-x')+\tfrac{1}{2}i\{\gamma(x-x'),\tfrac{1}{2}e\sigma F\}], \quad (B.7)$$

in which we have replaced t by the variable w,

$$w^{-1}=s^{-1}+t^{-1}, \quad (B.8)$$

and employed properties of the Dirac matrices, notably

$$\gamma_\lambda\sigma_{\mu\nu}\gamma_\lambda=0. \quad (B.9)$$

We shall also write

$$(x-x')_\mu\Phi(x,x')\exp[i\tfrac{1}{4}(x-x')^2/w] =2w(-i\partial_\mu-eA_\mu(x)-\tfrac{1}{2}eF_{\mu\nu}(x-x')_\nu)\Phi(x,x')\exp[i\tfrac{1}{4}(x-x')^2/w] \simeq[2w(-i\partial_\mu-eA_\mu(x))-2w^2eF_{\mu\nu}(-i\partial_\nu-eA_\nu(x))] \times\Phi(x,x')\exp[i\tfrac{1}{4}(x-x')^2/w], \quad (B.10)$$

which gives

$$M(x,x')=m_0\delta(x-x')+[e^2/(4\pi)^2]\int_0^\infty dss^{-2}\exp(-im^2s) \times\int_0^s dw[2m(2-w/s)+(2w/s)(\gamma(-i\partial-eA)+m) -2mw(1-w/s)i\tfrac{1}{2}e\sigma F-iw(1+w/s) \times\{\gamma(-i\partial-eA)+m,\tfrac{1}{2}e\sigma F\}](x(w)|x(0)'), \quad (B.11)$$

in virtue of the relation

$$[\gamma(-i\partial-eA),\tfrac{1}{2}\sigma F]=2i\gamma F(-i\partial-eA). \quad (B.12)$$

We now introduce a perturbation procedure in which the mass operator assumes the role customarily played by the energy. To evaluate $\int(dx')M(x,x')\psi(x')$, we replace $\psi(x')$ by the unperturbed wave function, a solution of the Dirac equation associated with the mass m (we need not distinguish, to this approximation, between the actual mass m and the mechanical mass m_0). The x' integration can be effected immediately,

$$\int(x(w)|x(0)')(dx')\psi(x')=\int(x|U(w)|x')(dx')\psi(x') =\exp(im^2w)\psi(x), \quad (B.13)$$

since $\psi(x)$ is an eigenfunction of $\mathcal{H}$, with the eigenvalue $-m^2$. Therefore, on discarding all terms containing the operator of the Dirac equation, which will not contribute to

$$\int(dx)(dx')\psi(x)M(x,x')\psi(x'),$$

we obtain

$$[\gamma(-i\partial-eA)+m-\mu'\tfrac{1}{2}\sigma F]\psi=0, \quad (B.14)$$

where

$$m=m_0+(\alpha/2\pi)m\int_0^\infty dss^{-1}\int_0^s dws^{-1}(2-w/s) \times\exp[-im^2(s-w)] \quad (B.15)$$

represents the mass of a free electron, and

$$\mu'=(\alpha/2\pi)emi\int_0^\infty ds\int_0^s(dw/s)(w/s)(1-w/s) \times\exp[-im^2(s-w)] \quad (B.16)$$

describes an additional spin magnetic moment. Both integrals are conveniently evaluated by introducing

$$u=1-w/s, \quad (B.17)$$

and making the replacement $s\to -is$, which yields

$$m=m_0+(\alpha/2\pi)m\int_0^\infty dss^{-1}\int_0^1 du(1+u)\exp(-m^2us) =m_0+(3\alpha/4\pi)m\left[\int_0^\infty dss^{-1}\exp(-m^2s)+\tfrac{5}{6}\right], \quad (B.18)$$

and

$$\mu'=(\alpha/2\pi)em\int_0^\infty ds\int_0^1 duu(1-u)\exp(-m^2us) =(\alpha/2\pi)(e/m)\int_0^1 du(1-u)=(\alpha/2\pi)(e\hbar/2mc). \quad (B.19)$$

We thus derive the spin magnetic moment of $\alpha/2\pi$ magnetons produced by second-order electromagnetic mass effects.

[10] The concepts employed here will be discussed at length in later publications.

The Theory of Positrons

R. P. FEYNMAN
Department of Physics, Cornell University, Ithaca, New York

(Received April 8, 1949)

The problem of the behavior of positrons and electrons in given external potentials, neglecting their mutual interaction, is analyzed by replacing the theory of holes by a reinterpretation of the solutions of the Dirac equation. It is possible to write down a complete solution of the problem in terms of boundary conditions on the wave function, and this solution contains automatically all the possibilities of virtual (and real) pair formation and annihilation together with the ordinary scattering processes, including the correct relative signs of the various terms.

In this solution, the "negative energy states" appear in a form which may be pictured (as by Stückelberg) in space-time as waves traveling away from the external potential backwards in time. Experimentally, such a wave corresponds to a positron approaching the potential and annihilating the electron. A particle moving forward in time (electron) in a potential may be scattered forward in time (ordinary scattering) or backward (pair annihilation). When moving backward (positron) it may be scattered backward in time (positron scattering) or forward (pair production). For such a particle the amplitude for transition from an initial to a final state is analyzed to any order in the potential by considering it to undergo a sequence of such scatterings.

The amplitude for a process involving many such particles is the product of the transition amplitudes for each particle. The exclusion principle requires that antisymmetric combinations of amplitudes be chosen for those complete processes which differ only by exchange of particles. It seems that a consistent interpretation is only possible if the exclusion principle is adopted. The exclusion principle need not be taken into account in intermediate states. Vacuum problems do not arise for charges which do not interact with one another, but these are analyzed nevertheless in anticipation of application to quantum electrodynamics.

The results are also expressed in momentum-energy variables. Equivalence to the second quantization theory of holes is proved in an appendix.

1. INTRODUCTION

THIS is the first of a set of papers dealing with the solution of problems in quantum electrodynamics. The main principle is to deal directly with the solutions to the Hamiltonian differential equations rather than with these equations themselves. Here we treat simply the motion of electrons and positrons in given external potentials. In a second paper we consider the interactions of these particles, that is, quantum electrodynamics.

The problem of charges in a fixed potential is usually treated by the method of second quantization of the electron field, using the ideas of the theory of holes. Instead we show that by a suitable choice and interpretation of the solutions of Dirac's equation the problem may be equally well treated in a manner which is fundamentally no more complicated than Schrödinger's method of dealing with one or more particles. The various creation and annihilation operators in the conventional electron field view are required because the number of particles is not conserved, i.e., pairs may be created or destroyed. On the other hand charge is conserved which suggests that if we follow the charge, not the particle, the results can be simplified.

In the approximation of classical relativistic theory the creation of an electron pair (electron A, positron B) might be represented by the start of two world lines from the point of creation, 1. The world lines of the positron will then continue until it annihilates another electron, C, at a world point 2. Between the times t_1 and t_2 there are then three world lines, before and after only one. However, the world lines of C, B, and A together form one continuous line albeit the "positron part" B of this continuous line is directed backwards in time. Following the charge rather than the particles corresponds to considering this continuous world line as a whole rather than breaking it up into its pieces. It is as though a bombardier flying low over a road suddenly sees three roads and it is only when two of them come together and disappear again that he realizes that he has simply passed over a long switchback in a single road.

This over-all space-time point of view leads to considerable simplification in many problems. One can take into account at the same time processes which ordinarily would have to be considered separately. For example, when considering the scattering of an electron by a potential one automatically takes into account the effects of virtual pair productions. The same equation, Dirac's, which describes the deflection of the world line of an electron in a field, can also describe the deflection (and in just as simple a manner) when it is large enough to reverse the time-sense of the world line, and thereby correspond to pair annihilation. Quantum mechanically the direction of the world lines is replaced by the direction of propagation of waves.

This view is quite different from that of the Hamiltonian method which considers the future as developing continuously from out of the past. Here we imagine the entire space-time history laid out, and that we just become aware of increasing portions of it successively. In a scattering problem this over-all view of the complete scattering process is similar to the S-matrix viewpoint of Heisenberg. The temporal order of events during the scattering, which is analyzed in such detail by the Hamiltonian differential equation, is irrelevant. The relation of these viewpoints will be discussed much more fully in the introduction to the second paper, in which the more complicated interactions are analyzed.

The development stemmed from the idea that in nonrelativistic quantum mechanics the amplitude for a given process can be considered as the sum of an amplitude for each space-time path available.[1] In view of the fact that in classical physics positrons could be viewed as electrons proceeding along world lines toward the past (reference 7) the attempt was made to remove, in the relativistic case, the restriction that the paths must proceed always in one direction in time. It was discovered that the results could be even more easily understood from a more familiar physical viewpoint, that of scattered waves. This viewpoint is the one used in this paper. After the equations were worked out physically the proof of the equivalence to the second quantization theory was found.[2]

First we discuss the relation of the Hamiltonian differential equation to its solution, using for an example the Schrödinger equation. Next we deal in an analogous way with the Dirac equation and show how the solutions may be interpreted to apply to positrons. The interpretation seems not to be consistent unless the electrons obey the exclusion principle. (Charges obeying the Klein-Gordon equations can be described in an analogous manner, but here consistency apparently requires Bose statistics.)[3] A representation in momentum and energy variables which is useful for the calculation of matrix elements is described. A proof of the equivalence of the method to the theory of holes in second quantization is given in the Appendix.

2. GREEN'S FUNCTION TREATMENT OF SCHRÖDINGER'S EQUATION

We begin by a brief discussion of the relation of the non-relativistic wave equation to its solution. The ideas will then be extended to relativistic particles, satisfying Dirac's equation, and finally in the succeeding paper to interacting relativistic particles, that is, quantum electrodynamics.

The Schrödinger equation

$$i\partial\psi/\partial t = H\psi, \tag{1}$$

describes the change in the wave function ψ in an infinitesimal time Δt as due to the operation of an operator $\exp(-iH\Delta t)$. One can ask also, if $\psi(\mathbf{x}_1, t_1)$ is the wave function at $\mathbf{x}_1$ at time t_1, what is the wave function at time $t_2 > t_1$? It can always be written as

$$\psi(\mathbf{x}_2, t_2) = \int K(\mathbf{x}_2, t_2; \mathbf{x}_1, t_1)\psi(\mathbf{x}_1, t_1)d^3\mathbf{x}_1, \tag{2}$$

where K is a Green's function for the linear Eq. (1). (We have limited ourselves to a single particle of coordinate $\mathbf{x}$, but the equations are obviously of greater generality.) If H is a constant operator having eigenvalues E_n, eigenfunctions ϕ_n so that $\psi(\mathbf{x}, t_1)$ can be expanded as $\sum_n C_n\phi_n(\mathbf{x})$, then $\psi(\mathbf{x}, t_2) = \exp(-iE_n(t_2-t_1)) \times C_n\phi_n(\mathbf{x})$. Since $C_n = \int \phi_n{}^*(\mathbf{x}_1)\psi(\mathbf{x}_1, t_1)d^3\mathbf{x}_1$, one finds (where we write 1 for $\mathbf{x}_1$, t_1 and 2 for $\mathbf{x}_2$, t_2) in this case

$$K(2, 1) = \sum_n \phi_n(\mathbf{x}_2)\phi_n{}^*(\mathbf{x}_1)\exp(-iE_n(t_2-t_1)), \tag{3}$$

for $t_2 > t_1$. We shall find it convenient for $t_2 < t_1$ to define $K(2, 1) = 0$ (Eq. (2) is then not valid for $t_2 < t_1$). It is then readily shown that in general K can be defined by that solution of

$$(i\partial/\partial t_2 - H_2)K(2, 1) = i\delta(2, 1), \tag{4}$$

which is zero for $t_2 < t_1$, where $\delta(2, 1) = \delta(t_2-t_1)\delta(x_2-x_1) \times\delta(y_2-y_1)\delta(z_2-z_1)$ and the subscript 2 on H_2 means that the operator acts on the variables of 2 of $K(2, 1)$. When H is not constant, (2) and (4) are valid but K is less easy to evaluate than (3).[4]

We can call $K(2, 1)$ the total amplitude for arrival at $\mathbf{x}_2$, t_2 starting from $\mathbf{x}_1$, t_1. (It results from adding an amplitude, $\exp iS$, for each space time path between these points, where S is the action along the path.[1]) The transition amplitude for finding a particle in state $\chi(\mathbf{x}_2, t_2)$ at time t_2, if at t_1 it was in $\psi(\mathbf{x}_1, t_1)$, is

$$\int \chi^*(2)K(2, 1)\psi(1)d^3\mathbf{x}_1 d^3\mathbf{x}_2. \tag{5}$$

A quantum mechanical system is described equally well by specifying the function K, or by specifying the Hamiltonian H from which it results. For some purposes the specification in terms of K is easier to use and visualize. We desire eventually to discuss quantum electrodynamics from this point of view.

To gain a greater familiarity with the K function and the point of view it suggests, we consider a simple perturbation problem. Imagine we have a particle in a weak potential $U(\mathbf{x}, t)$, a function of position and time. We wish to calculate $K(2, 1)$ if U differs from zero only for t between t_1 and t_2. We shall expand K in increasing powers of U:

$$K(2, 1) = K_0(2, 1) + K^{(1)}(2, 1) + K^{(2)}(2, 1) + \cdots. \tag{6}$$

To zero order in U, K is that for a free particle, $K_0(2, 1)$.[4] To study the first order correction $K^{(1)}(2, 1)$, first consider the case that U differs from zero only for the infinitesimal time interval Δt_3 between some time t_3 and $t_3 + \Delta t_3 (t_1 < t_3 < t_2)$. Then if $\psi(1)$ is the wave function at $\mathbf{x}_1$, t_1, the wave function at $\mathbf{x}_3$, t_3 is

$$\psi(3) = \int K_0(3, 1)\psi(1)d^3\mathbf{x}_1, \tag{7}$$

since from t_1 to t_3 the particle is free. For the short interval Δt_3 we solve (1) as

$$\begin{aligned}\psi(\mathbf{x}, t_3+\Delta t_3) &= \exp(-iH\Delta t_3)\psi(\mathbf{x}, t_3)\\ &= (1 - iH_0\Delta t_3 - iU\Delta t_3)\psi(\mathbf{x}, t_3),\end{aligned}$$

[1] R. P. Feynman, Rev. Mod. Phys. **20**, 367 (1948).

[2] The equivalence of the entire procedure (including photon interactions) with the work of Schwinger and Tomonaga has been demonstrated by F. J. Dyson, Phys. Rev. **75**, 486 (1949).

[3] These are special examples of the general relation of spin and statistics deduced by W. Pauli, Phys. Rev. **58**, 716 (1940).

[4] For a non-relativistic free particle, where $\phi_n = \exp(i\mathbf{p}\cdot\mathbf{x})$, $E_n = \mathbf{p}^2/2m$, (3) gives, as is well known

$$\begin{aligned}K_0(2, 1) &= \int \exp[-(i\mathbf{p}\cdot\mathbf{x}_1 - i\mathbf{p}\cdot\mathbf{x}_2) - i\mathbf{p}^2(t_2-t_1)/2m]d^3\mathbf{p}(2\pi)^{-3}\\ &= (2\pi i m^{-1}(t_2-t_1))^{-\frac{3}{2}}\exp(\tfrac{1}{2}im(\mathbf{x}_2-\mathbf{x}_1)^2(t_2-t_1)^{-1})\end{aligned}$$

for $t_2 > t_1$, and $K_0 = 0$ for $t_2 < t_1$.

where we put $H=H_0+U$, H_0 being the Hamiltonian of a free particle. Thus $\psi(\mathbf{x}, t_3+\Delta t_3)$ differs from what it would be if the potential were zero (namely $(1-iH_0\Delta t_3)\psi(\mathbf{x}, t_3)$) by the extra piece

$$\Delta\psi=-iU(\mathbf{x}_3, t_3)\cdot\psi(\mathbf{x}_3, t_3)\Delta t_3, \tag{8}$$

which we shall call the amplitude scattered by the potential. The wave function at 2 is given by

$$\psi(\mathbf{x}_2, t_2)=\int K_0(\mathbf{x}_2, t_2; \mathbf{x}_3, t_3+\Delta t_3)\psi(\mathbf{x}_3, t_3+\Delta t_3)d^3\mathbf{x}_3,$$

since after $t_3+\Delta t_3$ the particle is again free. Therefore the change in the wave function at 2 brought about by the potential is (substitute (7) into (8) and (8) into the equation for $\psi(\mathbf{x}_2, t_2)$):

$$\Delta\psi(2)=-i\int K_0(2, 3)U(3)K_0(3, 1)\psi(1)d^3\mathbf{x}_1d^3\mathbf{x}_3\Delta t_3.$$

In the case that the potential exists for an extended time, it may be looked upon as a sum of effects from each interval Δt_3 so that the total effect is obtained by integrating over t_3 as well as $\mathbf{x}_3$. From the definition (2) of K then, we find

$$K^{(1)}(2, 1)=-i\int K_0(2, 3)U(3)K_0(3, 1)d\tau_3, \tag{9}$$

where the integral can now be extended over all space and time, $d\tau_3=d^3\mathbf{x}_3dt_3$. Automatically there will be no contribution if t_3 is outside the range t_1 to t_2 because of our definition, $K_0(2, 1)=0$ for $t_2<t_1$.

We can understand the result (6), (9) this way. We can imagine that a particle travels as a free particle from point to point, but is scattered by the potential U. Thus the total amplitude for arrival at 2 from 1 can be considered as the sum of the amplitudes for various alternative routes. It may go directly from 1 to 2 (amplitude $K_0(2, 1)$, giving the zero order term in (6)). Or (see Fig. 1(a)) it may go from 1 to 3 (amplitude $K_0(3, 1)$), get scattered there by the potential (scattering amplitude $-iU(3)$ per unit volume and time) and then go from 3 to 2 (amplitude $K_0(2, 3)$). This may occur for any point 3 so that summing over these alternatives gives (9).

Again, it may be scattered twice by the potential (Fig. 1(b)). It goes from 1 to 3 ($K_0(3, 1)$), gets scattered there ($-iU(3)$) then proceeds to some other point, 4, in space time (amplitude $K_0(4, 3)$) is scattered again ($-iU(4)$) and then proceeds to 2 ($K_0(2, 4)$). Summing over all possible places and times for 3, 4 find that the second order contribution to the total amplitude $K^{(2)}(2, 1)$ is

$$(-i)^2\int\int K_0(2, 4)U(4)K_0(4, 3)$$
$$\times U(3)K_0(3, 1)d\tau_3d\tau_4. \tag{10}$$

This can be readily verified directly from (1) just as (9)

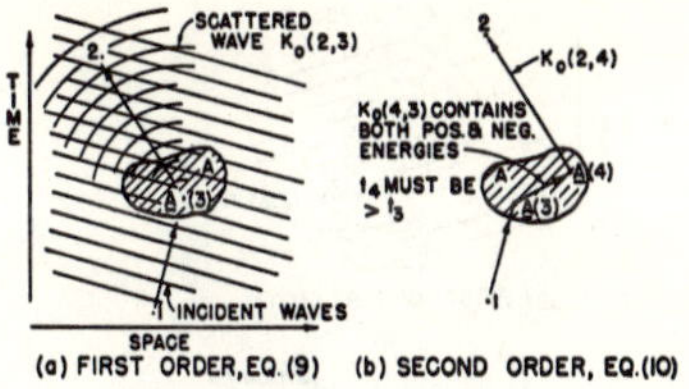

FIG. 1. The Schrödinger (and Dirac) equation can be visualized as describing the fact that plane waves are scattered successively by a potential. Figure 1 (a) illustrates the situation in first order. $K_0(2, 3)$ is the amplitude for a free particle starting at point 3 to arrive at 2. The shaded region indicates the presence of the potential A which scatters at 3 with amplitude $-iA(3)$ per cm³sec. (Eq. (9)). In (b) is illustrated the second order process (Eq. (10)), the waves scattered at 3 are scattered again at 4. However, in Dirac one-electron theory $K_0(4, 3)$ would represent electrons both of positive and of negative energies proceeding from 3 to 4. This is remedied by choosing a different scattering kernel $K_+(4, 3)$, Fig. 2.

was. One can in this way obviously write down any of the terms of the expansion (6).[5]

3. TREATMENT OF THE DIRAC EQUATION

We shall now extend the method of the last section to apply to the Dirac equation. All that would seem to be necessary in the previous equations is to consider H as the Dirac Hamiltonian, ψ as a symbol with four indices (for each particle). Then K_0 can still be defined by (3) or (4) and is now a 4–4 matrix which operating on the initial wave function, gives the final wave function. In (10), $U(3)$ can be generalized to $A_4(3)-\boldsymbol{\alpha}\cdot\mathbf{A}(3)$ where A_4, $\mathbf{A}$ are the scalar and vector potential (times e, the electron charge) and $\boldsymbol{\alpha}$ are Dirac matrices.

To discuss this we shall define a convenient relativistic notation. We represent four-vectors like $\mathbf{x}$, t by a symbol x_μ, where $\mu=1, 2, 3, 4$ and $x_4=t$ is real. Thus the vector and scalar potential (times e) $\mathbf{A}$, A_4 is A_μ. The four matrices $\beta\boldsymbol{\alpha}$, β can be considered as transforming as a four vector γ_μ (our γ_μ differs from Pauli's by a factor i for $\mu=1, 2, 3$). We use the summation convention $a_\mu b_\mu=a_4b_4-a_1b_1-a_2b_2-a_3b_3=a\cdot b$. In particular if a_μ is any four vector (but not a matrix) we write $\boldsymbol{a}=a_\mu\gamma_\mu$ so that $\boldsymbol{a}$ is a matrix associated with a vector ($\boldsymbol{a}$ will often be used in place of a_μ as a symbol for the vector). The γ_μ satisfy $\gamma_\mu\gamma_\nu+\gamma_\nu\gamma_\mu=2\delta_{\mu\nu}$ where $\delta_{44}=+1$, $\delta_{11}=\delta_{22}=\delta_{33}=-1$, and the other $\delta_{\mu\nu}$ are zero. As a consequence of our summation convention $\delta_{\mu\nu}a_\nu=a_\mu$ and $\delta_{\mu\mu}=4$. Note that $\boldsymbol{ab}+\boldsymbol{ba}=2a\cdot b$ and that $\boldsymbol{a}^2=a_\mu a_\mu=a\cdot a$ is a pure number. The symbol $\partial/\partial x_\mu$ will mean $\partial/\partial t$ for $\mu=4$, and $-\partial/\partial x$, $-\partial/\partial y$, $-\partial/\partial z$ for $\mu=1, 2, 3$. Call $\nabla=\gamma_\mu\partial/\partial x_\mu=\beta\partial/\partial t+\beta\boldsymbol{\alpha}\cdot\boldsymbol{\nabla}$. We shall imagine

[5] We are simply solving by successive approximations an integral equation (deducible directly from (1) with $H=H_0+U$ and (4) with $H=H_0$),

$$\psi(2)=-i\int K_0(2, 3)U(3)\psi(3)d\tau_3+\int K_0(2, 1)\psi(1)d^3\mathbf{x}_1,$$

where the first integral extends over all space and all times t_3 greater than the t_1 appearing in the second term, and $t_2>t_1$.

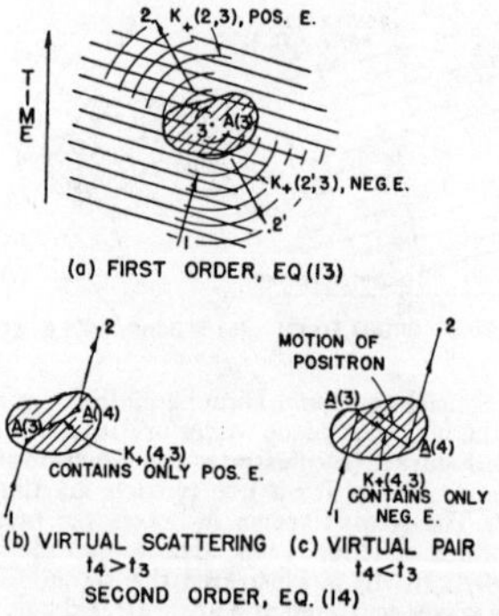

FIG. 2. The Dirac equation permits another solution $K_+(2, 1)$ if one considers that waves scattered by the potential can proceed backwards in time as in Fig. 2 (a). This is interpreted in the second order processes (b), (c), by noting that there is now the possibility (c) of virtual pair production at 4, the positron going to 3 to be annihilated. This can be pictured as similar to ordinary scattering (b) except that the electron is scattered backwards in time from 3 to 4. The waves scattered from 3 to 2′ in (a) represent the possibility of a positron arriving at 3 from 2′ and annihilating the electron from 1. This view is proved equivalent to hole theory: electrons traveling backwards in time are recognized as positrons.

hereafter, purely for relativistic convenience, that ϕ_n^* in (3) is replaced by its adjoint $\bar{\phi}_n=\phi_n^*\beta$.

Thus the Dirac equation for a particle, mass m, in an external field $A=A_\mu\gamma_\mu$ is

$$(i\nabla-m)\psi=A\psi, \qquad (11)$$

and Eq. (4) determining the propagation of a free particle becomes

$$(i\nabla_2-m)K_+(2,1)=i\delta(2,1), \qquad (12)$$

the index 2 on ∇_2 indicating differentiation with respect to the coordinates $x_{2\mu}$ which are represented as 2 in $K_+(2, 1)$ and $\delta(2, 1)$.

The function $K_+(2, 1)$ is defined in the absence of a field. If a potential A is acting a similar function, say $K_+^{(A)}(2, 1)$ can be defined. It differs from $K_+(2, 1)$ by a first order correction given by the analogue of (9) namely

$$K_+^{(1)}(2,1)=-i\int K_+(2,3)A(3)K_+(3,1)d\tau_3, \qquad (13)$$

representing the amplitude to go from 1 to 3 as a free particle, get scattered there by the potential (now the matrix $A(3)$ instead of $U(3)$) and continue to 2 as free. The second order correction, analogous to (10) is

$$K_+^{(2)}(2,1)=-\int\int K_+(2,4)A(4) \times K_+(4,3)A(3)K_+(3,1)d\tau_4d\tau_3, \qquad (14)$$

and so on. In general $K_+^{(A)}$ satisfies

$$(i\nabla_2-A(2)-m)K_+^{(A)}(2,1)=i\delta(2,1), \qquad (15)$$

and the successive terms (13), (14) are the power series expansion of the integral equation

$$K_+^{(A)}(2,1)=K_+(2,1) -i\int K_+(2,3)A(3)K_+^{(A)}(3,1)d\tau_3, \qquad (16)$$

which it also satisfies.

We would now expect to choose, for the special solution of (12), $K_+=K_0$ where $K_0(2, 1)$ vanishes for $t_2<t_1$ and for $t_2>t_1$ is given by (3) where ϕ_n and E_n are the eigenfunctions and energy values of a particle satisfying Dirac's equation, and ϕ_n^* is replaced by $\bar{\phi}_n$.

The formulas arising from this choice, however, suffer from the drawback that they apply to the one electron theory of Dirac rather than to the hole theory of the positron. For example, consider as in Fig. 1(a) an electron after being scattered by a potential in a small region 3 of space time. The one electron theory says (as does (3) with $K_+=K_0$) that the scattered amplitude at another point 2 will proceed toward positive times with both positive and negative energies, that is with both positive and negative rates of change of phase. No wave is scattered to times previous to the time of scattering. These are just the properties of $K_0(2, 3)$.

On the other hand, according to the positron theory negative energy states are not available to the electron after the scattering. Therefore the choice $K_+=K_0$ is unsatisfactory. But there are other solutions of (12). We shall choose the solution defining $K_+(2, 1)$ so that *$K_+(2, 1)$ for $t_2>t_1$ is the sum of (3) over positive energy states only.* Now this new solution must satisfy (12) for all times in order that the representation be complete. It must therefore differ from the old solution K_0 by a solution of the homogeneous Dirac equation. It is clear from the definition that the difference K_0-K_+ is the sum of (3) over all negative energy states, as long as $t_2>t_1$. But this difference must be a solution of the homogeneous Dirac equation for all times and must therefore be represented by the same sum over negative energy states also for $t_2<t_1$. Since $K_0=0$ in this case, it follows that our new kernel, *$K_+(2, 1)$, for $t_2<t_1$ is the negative of the sum (3) over negative energy states.* That is,

$$\begin{aligned}K_+(2,1)&=\textstyle\sum_{POS\ E_n}\phi_n(2)\bar{\phi}_n(1)\\&\quad\times\exp(-iE_n(t_2-t_1)) \quad\text{for}\quad t_2>t_1\\&=-\textstyle\sum_{NEG\ E_n}\phi_n(2)\bar{\phi}_n(1)\\&\quad\times\exp(-iE_n(t_2-t_1)) \quad\text{for}\quad t_2<t_1.\end{aligned} \qquad (17)$$

With this choice of K_+ our equations such as (13) and (14) will now give results equivalent to those of the positron hole theory.

That (14), for example, is the correct second order expression for finding at 2 an electron originally at 1 according to the positron theory may be seen as follows (Fig. 2). Assume as a special example that $t_2>t_1$ and that the potential vanishes except in interval t_2-t_1 so that t_4 and t_3 both lie between t_1 and t_2.

First suppose $t_4>t_3$ (Fig. 2(b)). Then (since $t_3>t_1$)

the electron assumed originally in a positive energy state propagates in that state (by $K_+(3, 1)$) to position 3 where it gets scattered ($A(3)$). It then proceeds to 4, which it must do as a positive energy electron. This is correctly described by (14) for $K_+(4, 3)$ contains only positive energy components in its expansion, as $t_4 > t_3$. After being scattered at 4 it then proceeds on to 2, again necessarily in a positive energy state, as $t_2 > t_4$.

In positron theory there is an additional contribution due to the possibility of virtual pair production (Fig. 2(c)). A pair could be created by the potential $A(4)$ at 4, the electron of which is that found later at 2. The positron (or rather, the hole) proceeds to 3 where it annihilates the electron which has arrived there from 1.

This alternative is already included in (14) as contributions for which $t_4 < t_3$, and its study will lead us to an interpretation of $K_+(4, 3)$ for $t_4 < t_3$. The factor $K_+(2, 4)$ describes the electron (after the pair production at 4) proceeding from 4 to 2. Likewise $K_+(3, 1)$ represents the electron proceeding from 1 to 3. $K_+(4, 3)$ must therefore represent the propagation of the positron or hole from 4 to 3. That it does so is clear. The fact that in hole theory the hole proceeds in the manner of and electron of negative energy is reflected in the fact that $K_+(4, 3)$ for $t_4 < t_3$ is (minus) the sum of only negative energy components. In hole theory the real energy of these intermediate states is, of course, positive. This is true here too, since in the phases $\exp(-iE_n(t_4 - t_3))$ defining $K_+(4, 3)$ in (17), E_n is negative but so is $t_4 - t_3$. That is, the contributions vary with t_3 as $\exp(-i|E_n|(t_3 - t_4))$ as they would if the energy of the intermediate state were $|E_n|$. The fact that the entire sum is taken as negative in computing $K_+(4, 3)$ is reflected in the fact that in hole theory the amplitude has its sign reversed in accordance with the Pauli principle and the fact that the electron arriving at 2 has been exchanged with one in the sea.[6] To this, and to higher orders, all processes involving virtual pairs are correctly described in this way.

The expressions such as (14) can still be described as a passage of the electron from 1 to 3 ($K_+(3, 1)$), scattering at 3 by $A(3)$, proceeding to 4 ($K_+(4, 3)$), scattering again, $A(4)$, arriving finally at 2. The scatterings may, however, be toward both future and past times, an electron propagating backwards in time being recognized as a positron.

This therefore suggests that negative energy components created by scattering in a potential be considered as waves propagating from the scattering point toward the past, and that such waves represent the propagation of a positron annihilating the electron in the potential.[7]

With this interpretation real pair production is also described correctly (see Fig. 3). For example in (13) if $t_1 < t_3 < t_2$ the equation gives the amplitude that if at time t_1 one electron is present at 1, then at time t_2 just one electron will be present (having been scattered at 3) and it will be at 2. On the other hand if t_2 is less than t_3, for example, if $t_2 = t_1 < t_3$, the same expression gives the amplitude that a pair, electron at 1, positron at 2 will annihilate at 3, and subsequently no particles will be present. Likewise if t_2 and t_1 exceed t_3 we have (minus) the amplitude for finding a single pair, electron at 2, positron at 1 created by $A(3)$ from a vacuum. If $t_1 > t_3 > t_2$, (13) describes the scattering of a positron. All these amplitudes are relative to the amplitude that a vacuum will remain a vacuum, which is taken as unity. (This will be discussed more fully later.)

The analogue of (2) can be easily worked out.[8] It is,

$$\psi(2) = \int K_+(2, 1) N(1) \psi(1) d^3V_1, \tag{18}$$

where d^3V_1 is the volume element of the closed 3-dimensional surface of a region of space time containing

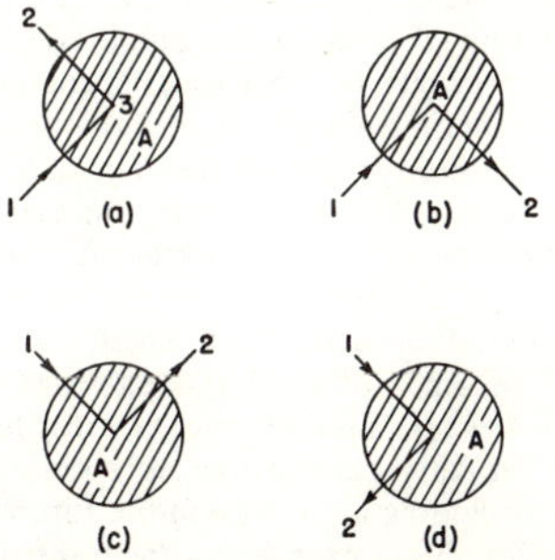

FIG. 3. Several different processes can be described by the same formula depending on the time relations of the variables t_2, t_1. Thus $P_v|K_+^{(A)}(2, 1)|^2$ is the probability that: (a) An electron at 1 will be scattered at 2 (and no other pairs form in vacuum). (b) Electron at 1 and positron at 2 annihilate leaving nothing. (c) A single pair at 1 and 2 is created from vacuum. (d) A positron at 2 is scattered to 1. ($K_+^{(A)}(2, 1)$ is the sum of the effects of scattering in the potential to all orders. P_v is a normalizing constant.)

---

[6] It has often been noted that the one-electron theory apparently gives the same matrix elements for this process as does hole theory. The problem is one of interpretation, especially in a way that will also give correct results for other processes, e.g., self-energy.

[7] The idea that positrons can be represented as electrons with proper time reversed relative to true time has been discussed by the author and others, particularly by Stückelberg. E. C. C. Stückelberg, Helv. Phys. Acta **15**, 23 (1942); R. P. Feynman, Phys. Rev. **74**, 939 (1948). The fact that classically the action (proper time) increases continuously as one follows a trajectory is reflected in quantum mechanics in the fact that the phase, which is $|E_n||t_2 - t_1|$, always increases as the particle proceeds from one scattering point to the next.

[8] By multiplying (12) on the right by $(-i\nabla_1 - m)$ and noting that $\nabla_1\delta(2, 1) = -\nabla_2\delta(2, 1)$ show that $K_+(2, 1)$ also satisfies $K_+(2, 1)(-i\nabla_1 - m) = i\delta(2, 1)$, where the ∇_1 operates on variable 1 in $K_+(2, 1)$ but is written after that function to keep the correct order of the γ matrices. Multiply this equation by $\psi(1)$ and Eq. (11) (with $A = 0$, calling the variables 1) by $K_+(2, 1)$, subtract and integrate over a region of space-time. The integral on the left-hand side can be transformed to an integral over the surface of the region. The right-hand side is $\psi(2)$ if the point 2 lies within the region, and is zero otherwise. (What happens when the 3-surface contains a light line and hence has no unique normal need not concern us as these points can be made to occur so far away from 2 that their contribution vanishes.)

point 2, and $N(1)$ is $N_\mu(1)\gamma_\mu$ where $N_\mu(1)$ is the *inward* drawn unit normal to the surface at the point 1. That is, the wave function $\psi(2)$ (in this case for a free particle) is determined at any point inside a four-dimensional region if its values on the surface of that region are specified.

To interpret this, consider the case that the 3-surface consists essentially of all space at some time say $t=0$ previous to t_2, and of all space at the time $T>t_2$. The cylinder connecting these to complete the closure of the surface may be very distant from $\mathbf{x}_2$ so that it gives no appreciable contribution (as $K_+(2,1)$ decreases exponentially in space-like directions). Hence, if $\gamma_4=\beta$, since the inward drawn normals N will be β and $-\beta$,

$$\psi(2)=\int K_+(2,1)\beta\psi(1)d^3\mathbf{x}_1 - \int K_+(2,1')\beta\psi(1')d^3\mathbf{x}_{1'}, \quad (19)$$

where $t_1=0$, $t_{1'}=T$. Only positive energy (electron) components in $\psi(1)$ contribute to the first integral and only negative energy (positron) components of $\psi(1')$ to the second. That is, the amplitude for finding a charge at 2 is determined both by the amplitude for finding an electron previous to the measurement and by the amplitude for finding a positron after the measurement. This might be interpreted as meaning that even in a problem involving but one charge the amplitude for finding the charge at 2 is not determined when the only thing known in the amplitude for finding an electron (or a positron) at an earlier time. There may have been no electron present initially but a pair was created in the measurement (or also by other external fields). The amplitude for this contingency is specified by the amplitude for finding a positron in the future.

We can also obtain expressions for transition amplitudes, like (5). For example if at $t=0$ we have an electron present in a state with (positive energy) wave function $f(\mathbf{x})$, what is the amplitude for finding it at $t=T$ with the (positive energy) wave function $g(\mathbf{x})$? The amplitude for finding the electron anywhere after $t=0$ is given by (19) with $\psi(1)$ replaced by $f(\mathbf{x})$, the second integral vanishing. Hence, the transition element to find it in state $g(\mathbf{x})$ is, in analogy to (5), just ($t_2=T$, $t_1=0$)

$$\int \bar{g}(\mathbf{x}_2)\beta K_+(2,1)\beta f(\mathbf{x}_1)d^3\mathbf{x}_1d^3\mathbf{x}_2, \quad (20)$$

since $g^*=\bar{g}\beta$.

If a potential acts somewhere in the interval between 0 and T, K_+ is replaced by $K_+^{(A)}$. Thus the first order effect on the transition amplitude is, from (13),

$$-i\int \bar{g}(\mathbf{x}_2)\beta K_+(2,3)A(3)K_+(3,1)\beta f(\mathbf{x}_1)d^3\mathbf{x}_1d^3\mathbf{x}_2. \quad (21)$$

Expressions such as this can be simplified and the 3-surface integrals, which are inconvenient for relativistic calculations, can be removed as follows. Instead of defining a state by the wave function $f(\mathbf{x})$, which it has at a given time $t_1=0$, we define the state by the function $f(1)$ of four variables $\mathbf{x}_1$, t_1 which is a solution of the free particle equation for all t_1 and is $f(\mathbf{x}_1)$ for $t_1=0$. The final state is likewise defined by a function $g(2)$ over-all space-time. Then our surface integrals can be performed since $\int K_+(3,1)\beta f(\mathbf{x}_1)d^3\mathbf{x}_1=f(3)$ and $\int \bar{g}(\mathbf{x}_2)\beta d^3\mathbf{x}_2K_+(2,3)=\bar{g}(3)$. There results

$$-i\int \bar{g}(3)A(3)f(3)d\tau_3, \quad (22)$$

the integral now being over-all space-time. The transition amplitude to second order (from (14)) is

$$-\int\int \bar{g}(2)A(2)K_+(2,1)A(1)f(1)d\tau_1d\tau_2, \quad (23)$$

for the particle arriving at 1 with amplitude $f(1)$ is scattered ($A(1)$), progresses to 2, ($K_+(2,1)$), and is scattered again ($A(2)$), and we then ask for the amplitude that it is in state $g(2)$. If $g(2)$ is a negative energy state we are solving a problem of annihilation of electron in $f(1)$, positron in $g(2)$, etc.

We have been emphasizing scattering problems, but obviously the motion in a fixed potential V, say in a hydrogen atom, can also be dealt with. If it is first viewed as a scattering problem we can ask for the amplitude, $\phi_k(1)$, that an electron with original free wave function was scattered k times in the potential V either forward or backward in time to arrive at 1. Then the amplitude after one more scattering is

$$\phi_{k+1}(2)=-i\int K_+(2,1)V(1)\phi_k(1)d\tau_1. \quad (24)$$

An equation for the total amplitude

$$\psi(1)=\sum_{k=0}^{\infty}\phi_k(1)$$

for arriving at 1 either directly or after any number of scatterings is obtained by summing (24) over all k from 0 to ∞;

$$\psi(2)=\phi_0(2)-i\int K_+(2,1)V(1)\psi(1)d\tau_1. \quad (25)$$

Viewed as a steady state problem we may wish, for example, to find that initial condition ϕ_0 (or better just the ψ) which leads to a periodic motion of ψ. This is most practically done, of course, by solving the Dirac equation,

$$(i\nabla-m)\psi(1)=V(1)\psi(1), \quad (26)$$

deduced from (25) by operating on both sides by $i\nabla_2-m$, thereby eliminating the ϕ_0, and using (12). This illustrates the relation between the points of view.

For many problems the total potential $A+V$ may be split conveniently into a fixed one, V, and another, A, considered as a perturbation. If $K_+^{(V)}$ is defined as in

(16) with V for A, expressions such as (23) are valid and useful with K_+ replaced by $K_+^{(V)}$ and the functions $f(1)$, $g(2)$ replaced by solutions for all space and time of the Dirac Eq. (26) in the potential V (rather than free particle wave functions).

4. PROBLEMS INVOLVING SEVERAL CHARGES

We wish next to consider the case that there are two (or more) distinct charges (in addition to pairs they may produce in virtual states). In a succeeding paper we discuss the interaction between such charges. Here we assume that they do not interact. In this case each particle behaves independently of the other. We can expect that if we have two particles a and b, the amplitude that particle a goes from $\mathbf{x}_1$ at t_1, to $\mathbf{x}_3$ at t_3 while b goes from $\mathbf{x}_2$ at t_2 to $\mathbf{x}_4$ at t_4 is the product

$$K(3,4;1,2)=K_{+a}(3,1)K_{+b}(4,2).$$

The symbols a, b simply indicate that the matrices appearing in the K_+ apply to the Dirac four component spinors corresponding to particle a or b respectively (the wave function now having 16 indices). In a potential K_{+a} and K_{+b} become $K_{+a}^{(A)}$ and $K_{+b}^{(A)}$ where $K_{+a}^{(A)}$ is defined and calculated as for a single particle. They commute. Hereafter the a, b can be omitted; the space time variable appearing in the kernels suffice to define on what they operate.

The particles are identical however and satisfy the exclusion principle. The principle requires only that one calculate $K(3,4;1,2)-K(4,3;1,2)$ to get the net amplitude for arrival of charges at 3, 4. (It is normalized assuming that when an integral is performed over points 3 and 4, for example, since the electrons represented are identical, one divides by 2.) This expression is correct for positrons also (Fig. 4). For example the amplitude that an electron and a positron found initially at $\mathbf{x}_1$ and $\mathbf{x}_4$ (say $t_1=t_4$) are later found at $\mathbf{x}_3$ and $\mathbf{x}_2$ (with $t_2=t_3>t_1$) is given by the same expression

$$K_+^{(A)}(3,1)K_+^{(A)}(4,2)-K_+^{(A)}(4,1)K_+^{(A)}(3,2). \quad (27)$$

The first term represents the amplitude that the electron proceeds from 1 to 3 and the positron from 4 to 2 (Fig. 4(c)), while the second term represents the interfering amplitude that the pair at 1, 4 annihilate and what is found at 3, 2 is a pair newly created in the potential. The generalization to several particles is clear. There is an additional factor $K_+^{(A)}$ for each particle, and antisymmetric combinations are always taken.

No account need be taken of the exclusion principle in intermediate states. As an example consider again expression (14) for $t_2>t_1$ and suppose $t_4<t_3$ so that the situation represented (Fig. 2(c)) is that a pair is made at 4 with the electron proceeding to 2, and the positron to 3 where it annihilates the electron arriving from 1. It may be objected that if it happens that the electron created at 4 is in the same state as the one coming from 1, then the process cannot occur because of the exclusion principle and we should not have included it in our term (14). We shall see, however, that considering the exclusion principle also requires another change which reinstates the quantity.

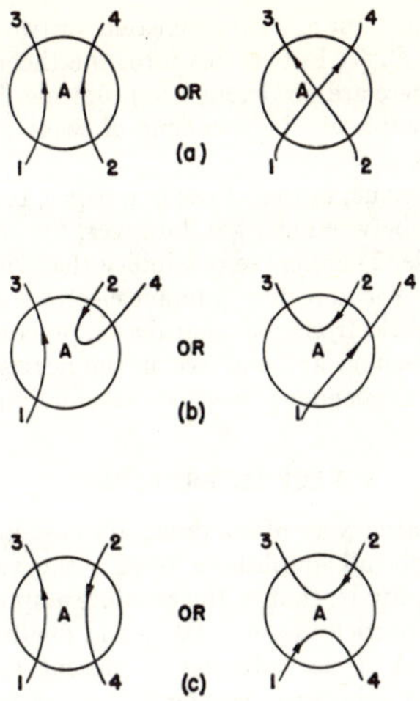

FIG. 4. Some problems involving two distinct charges (in addition to virtual pairs they may produce): $P_v|K_+^{(A)}(3,1)K_+^{(A)}(4,2)-K_+^{(A)}(4,1)K_+^{(A)}(3,2)|^2$ is the probability that: (a) Electrons at 1 and 2 are scattered to 3, 4 (and no pairs are formed). (b) Starting with an electron at 1 a single pair is formed, positron at 2, electrons at 3, 4. (c) A pair at 1, 4 is found at 3, 2, etc. The exclusion principle requires that the amplitudes for processes involving exchange of two electrons be subtracted.

For we are computing amplitudes relative to the amplitude that a vacuum at t_1 will still be a vacuum at t_2. We are interested in the alteration in this amplitude due to the presence of an electron at 1. Now one process that can be visualized as occurring in the vacuum is the creation of a pair at 4 followed by a re-annihilation of the *same* pair at 3 (a process which we shall call a closed loop path). But if a real electron is present in a certain state 1, those pairs for which the electron was created in state 1 in the vacuum must now be excluded. We must therefore subtract from our relative amplitude the term corresponding to this process. But this just reinstates the quantity which it was argued should not have been included in (14), the necessary minus sign coming automatically from the definition of K_+. It is obviously simpler to disregard the exclusion principle completely in the intermediate states.

All the amplitudes are relative and their squares give the relative probabilities of the various phenomena. Absolute probabilities result if one multiplies each of the probabilities by P_v, the true probability that if one has no particles present initially there will be none finally. This quantity P_v can be calculated by normalizing the relative probabilities such that the sum of the probabilities of all mutually exclusive alternatives is unity. (For example if one starts with a vacuum one can calculate the relative probability that there remains a vacuum (unity), or one pair is created, or two pairs, etc. The sum is P_v^{-1}.) Put in this form the theory is complete and there are no divergence problems. Real processes are completely independent of what goes on in the vacuum.

When we come, in the succeeding paper, to deal with interactions between charges, however, the situation is not so simple. There is the possibility that virtual electrons in the vacuum may interact electromagnetically with the real electrons. For that reason processes occurring in the vacuum are analyzed in the next section, in which an independent method of obtaining P_v is discussed.

5. VACUUM PROBLEMS

An alternative way of obtaining absolute amplitudes is to multiply all amplitudes by C_v, the vacuum to vacuum amplitude, that is, the absolute amplitude that there be no particles both initially and finally. We can assume $C_v=1$ if no potential is present during the interval, and otherwise we compute it as follows. It differs from unity because, for example, a pair could be created which eventually annihilates itself again. Such a path would appear as a closed loop on a space-time diagram. The sum of the amplitudes resulting from all such single closed loops we call L. To a first approximation L is

$$L^{(1)}=-\frac{1}{2}\int\int Sp[K_+(2,1)\mathbf{A}(1)\times K_+(1,2)\mathbf{A}(2)]d\tau_1 d\tau_2. \quad (28)$$

For a pair could be created say at 1, the electron and positron could both go on to 2 and there annihilate. The spur, Sp, is taken since one has to sum over all possible spins for the pair. The factor $\frac{1}{2}$ arises from the fact that the same loop could be considered as starting at either potential, and the minus sign results since the interactors are each $-i\mathbf{A}$. The next order term would be[9]

$$L^{(2)}=+(i/3)\int\int\int Sp[K_+(2,1)\mathbf{A}(1)\times K_+(1,3)\mathbf{A}(3)K_+(3,2)\mathbf{A}(2)]d\tau_1 d\tau_2 d\tau_3,$$

etc. The sum of all such terms gives L.[10]

In addition to these single loops we have the possibility that two independent pairs may be created and each pair may annihilate itself again. That is, there may be formed in the vacuum two closed loops, and the contribution in amplitude from this alternative is just the product of the contribution from each of the loops considered singly. The total contribution from all such pairs of loops (it is still consistent to disregard the exclusion principle for these virtual states) is $L^2/2$ for in L^2 we count every pair of loops twice. The total vacuum-vacuum amplitude is then

$$C_v=1-L+L^2/2-L^3/6+\cdots=\exp(-L), \quad (30)$$

the successive terms representing the amplitude from zero, one, two, etc., loops. The fact that the contribution to C_v of single loops is $-L$ is a consequence of the Pauli principle. For example, consider a situation in which two pairs of particles are created. Then these pairs later destroy themselves so that we have two loops. The electrons could, at a given time, be interchanged forming a kind of figure eight which is a single loop. The fact that the interchange must change the sign of the contribution requires that the terms in C_v appear with alternate signs. (The exclusion principle is also responsible in a similar way for the fact that the amplitude for a pair creation is $-K_+$ rather than $+K_+$.) Symmetrical statistics would lead to

$$C_v=1+L+L^2/2=\exp(+L).$$

The quantity L has an infinite imaginary part (from $L^{(1)}$, higher orders are finite). We will discuss this in connection with vacuum polarization in the succeeding paper. This has no effect on the normalization constant for the probability that a vacuum remain vacuum is given by

$$P_v=|C_v|^2=\exp(-2\cdot\text{real part of } L),$$

from (30). This value agrees with the one calculated directly by renormalizing probabilities. The real part of L appears to be positive as a consequence of the Dirac equation and properties of K_+ so that P_v is less than one. Bose statistics gives $C_v=\exp(+L)$ and consequently a value of P_v greater than unity which appears meaningless if the quantities are interpreted as we have done here. Our choice of K_+ apparently requires the exclusion principle.

Charges obeying the Klein-Gordon equation can be equally well treated by the methods which are discussed here for the Dirac electrons. How this is done is discussed in more detail in the succeeding paper. The real part of L comes out negative for this equation so that in this case Bose statistics appear to be required for consistency.[3]

---

[9] This term actually vanishes as can be seen as follows. In any spur the sign of all γ matrices may be reversed. Reversing the sign of γ in $K_+(2,1)$ changes it to the transpose of $K_+(1,2)$ so that the order of all factors and variables is reversed. Since the integral is taken over all τ_1, τ_2, and τ_3 this has no effect and we are left with $(-1)^3$ from changing the sign of $\mathbf{A}$. Thus the spur equals its negative. Loops with an odd number of potential interactors give zero. Physically this is because for each loop the electron can go around one way or in the opposite direction and we must add these amplitudes. But reversing the motion of an electron makes it behave like a positive charge thus changing the sign of each potential interaction, so that the sum is zero if the number of interactions is odd. This theorem is due to W. H. Furry, Phys. Rev. **51**, 125 (1937).

[10] A closed expression for L in terms of $K_+^{(A)}$ is hard to obtain because of the factor $(1/n)$ in the nth term. However, the perturbation in L, ΔL due to a small change in potential $\Delta\mathbf{A}$, is easy to express. The $(1/n)$ is canceled by the fact that $\Delta\mathbf{A}$ can appear in any of the n potentials. The result after summing over n by (13), (14) and using (16) is

$$\Delta L=-i\int Sp[(K_+^{(A)}(1,1)-K_+(1,1))\Delta\mathbf{A}(1)]d\tau_1. \quad (29)$$

The term $K_+(1,1)$ actually integrates to zero.

6. ENERGY-MOMENTUM REPRESENTATION

The practical evaluation of the matrix elements in some problems is often simplified by working with momentum and energy variables rather than space and time. This is because the function $K_+(2, 1)$ is fairly complicated but we shall find that its Fourier transform is very simple, namely $(i/4\pi^2)(p-m)^{-1}$ that is

$$K_+(2, 1)=(i/4\pi^2)\int (p-m)^{-1}\exp(-ip\cdot x_{21})d^4p, \quad (31)$$

where $p\cdot x_{21}=p\cdot x_2-p\cdot x_1=p_\mu x_{2\mu}-p_\mu x_{1\mu}$, $p=p_\mu\gamma_\mu$, and d^4p means $(2\pi)^{-2}dp_1dp_2dp_3dp_4$, the integral over all p. That this is true can be seen immediately from (12), for the representation of the operator $i\nabla-m$ in energy (p_4) and momentum ($p_{1,2,3}$) space is $p-m$ and the transform of $\delta(2, 1)$ is a constant. The reciprocal matrix $(p-m)^{-1}$ can be interpreted as $(p+m)(p^2-m^2)^{-1}$ for $p^2-m^2=(p-m)(p+m)$ is a pure number not involving γ matrices. Hence if one wishes one can write

$$K_+(2, 1)=i(i\nabla_2+m)I_+(2, 1),$$

where

$$I_+(2, 1)=(2\pi)^{-2}\int (p^2-m^2)^{-1}\exp(-ip\cdot x_{21})d^4p, \quad (32)$$

is not a matrix operator but a function satisfying

$$\Box_2^2 I_+(2, 1)-m^2I_+(2, 1)=\delta(2, 1), \quad (33)$$

where $-\Box_2^2=(\nabla_2)^2=(\partial/\partial x_{2\mu})(\partial/\partial x_{2\mu})$.

The integrals (31) and (32) are not yet completely defined for there are poles in the integrand when $p^2-m^2=0$. We can define how these poles are to be evaluated by the rule that *m is considered to have an infinitesimal negative imaginary part.* That is m, is replaced by $m-i\delta$ and the limit taken as $\delta\to 0$ from above. This can be seen by imagining that we calculate K_+ by integrating on p_4 first. If we call $E=+(m^2+p_1^2+p_2^2+p_3^2)^{\frac{1}{2}}$ then the integrals involve p_4 essentially as $\int \exp(-ip_4(t_2-t_1))dp_4(p_4^2-E^2)^{-1}$ which has poles at $p_4=+E$ and $p_4=-E$. The replacement of m by $m-i\delta$ means that E has a small negative imaginary part; the first pole is below, the second above the real axis. Now if $t_2-t_1>0$ the contour can be completed around the semicircle below the real axis thus giving a residue from the $p_4=+E$ pole, or $-(2E)^{-1}\exp(-iE(t_2-t_1))$. If $t_2-t_1<0$ the upper semicircle must be used, and $p_4=-E$ at the pole, so that the function varies in each case as required by the other definition (17).

Other solutions of (12) result from other prescriptions. For example if p_4 in the factor $(p^2-m^2)^{-1}$ is considered to have a positive imaginary part K_+ becomes replaced by K_0, the Dirac one-electron kernel, zero for $t_2<t_1$. Explicitly the function is[11] ($\mathbf{x}$, $t=x_{21\mu}$)

$$I_+(\mathbf{x}, t)=-(4\pi)^{-1}\delta(s^2)+(m/8\pi s)H_1^{(2)}(ms), \quad (34)$$

where $s=+(t^2-\mathbf{x}^2)^{\frac{1}{2}}$ for $t^2>\mathbf{x}^2$ and $s=-i(x^2-t^2)^{\frac{1}{2}}$ for $t^2<x^2$, $H_1^{(2)}$ is the Hankel function and $\delta(s^2)$ is the Dirac delta function of s^2. It behaves asymptotically as $\exp(-ims)$, decaying exponentially in space-like directions.[12]

By means of such transforms the matrix elements like (22), (23) are easily worked out. A free particle wave function for an electron of momentum p_1 is $u_1\exp(-ip_1\cdot x)$ where u_1 is a constant spinor satisfying the Dirac equation $p_1u_1=mu_1$ so that $p_1^2=m^2$. The matrix element (22) for going from a state p_1, u_1 to a state of momentum p_2, spinor u_2, is $-4\pi^2i(\tilde{u}_2a(q)u_1)$ where we have imagined A expanded in a Fourier integral

$$A(1)=\int a(q)\exp(-iq\cdot x_1)d^4q,$$

and we select the component of momentum $q=p_2-p_1$.

The second order term (23) is the matrix element between u_1 and u_2 of

$$-4\pi^2i\int (a(p_2-p_1-q))(p_1+q-m)^{-1}a(q)d^4q, \quad (35)$$

since the electron of momentum p_1 may pick up q from the potential $a(q)$, propagate with momentum p_1+q (factor $(p_1+q-m)^{-1}$) until it is scattered again by the potential, $a(p_2-p_1-q)$, picking up the remaining momentum, p_2-p_1-q, to bring the total to p_2. Since all values of q are possible, one integrates over q.

These same matrices apply directly to positron problems, for if the time component of, say, p_1 is negative the state represents a positron of four-momentum $-p_1$, and we are describing pair production if p_2 is an electron, i.e., has positive time component, etc.

The probability of an event whose matrix element is $(\tilde{u}_2Mu_1)$ is proportional to the absolute square. This may also be written $(\tilde{u}_1\bar{M}u_2)(\tilde{u}_2Mu_1)$, where $\bar{M}$ is M with the operators written in opposite order and explicit appearance of i changed to $-i$($\bar{M}$ is β times the complex conjugate transpose of βM). For many problems we are not concerned about the spin of the final state. Then we can sum the probability over the two u_2 corresponding to the two spin directions. This is not a complete set because p_2 has another eigenvalue, $-m$. To permit summing over all states we can insert the projection operator $(2m)^{-1}(p_2+m)$ and so obtain $(2m)^{-1}(\tilde{u}_1\bar{M}(p_2+m)Mu_1)$ for the probability of transition from p_1, u_1, to p_2 with arbitrary spin. If the incident state is unpolarized we can sum on its spins too, and obtain

$$(2m)^{-2}Sp[(p_1+m)\bar{M}(p_2+m)M] \quad (36)$$

for (twice) the probability that an electron of arbitrary spin with momentum p_1 will make transition to p_2. The expressions are all valid for positrons when p's with

[11] $I_+(\mathbf{x}, t)$ is $(2i)^{-1}(D_1(\mathbf{x}, t)-iD(\mathbf{x}, t))$ where D_1 and D are the functions defined by W. Pauli, Rev. Mod. Phys. **13**, 203 (1941).

[12] If the $-i\delta$ is kept with m here too the function I_+ approaches zero for infinite positive and negative times. This may be useful in general analyses in avoiding complications from infinitely remote surfaces.

negative energies are inserted, and the situation interpreted in accordance with the timing relations discussed above. (We have used functions normalized to $(\bar{u}u)=1$ instead of the conventional $(\bar{u}\beta u)=(u^*u)=1$. On our scale $(\bar{u}\beta u)=\text{energy}/m$ so the probabilities must be corrected by the appropriate factors.)

The author has many people to thank for fruitful conversations about this subject, particularly H. A. Bethe and F. J. Dyson.

APPENDIX

a. Deduction from Second Quantization

In this section we shall show the equivalence of this theory with the hole theory of the positron.[2] According to the theory of second quantization of the electron field in a given potential,[13] the state of this field at any time is represented by a wave function χ satisfying

$$i\partial\chi/\partial t=H\chi,$$

where $H=\int\Psi^*(\mathbf{x})(\boldsymbol{\alpha}\cdot(-i\boldsymbol{\nabla}-\mathbf{A})+A_4+m\beta)\Psi(\mathbf{x})d^3\mathbf{x}$ and $\Psi(\mathbf{x})$ is an operator annihilating an electron at position $\mathbf{x}$, while $\Psi^*(\mathbf{x})$ is the corresponding creation operator. We contemplate a situation in which at $t=0$ we have present some electrons in states represented by ordinary spinor functions $f_1(\mathbf{x}), f_2(\mathbf{x}), \cdots$ assumed orthogonal, and some positrons. These are described as holes in the negative energy sea, the electrons which would normally fill the holes having wave functions $p_1(\mathbf{x}), p_2(\mathbf{x}), \cdots$. We ask, at time T what is the amplitude that we find electrons in states $g_1(\mathbf{x}), g_2(\mathbf{x}), \cdots$ and holes at $q_1(\mathbf{x}), q_2(\mathbf{x}), \cdots$. If the initial and final state vectors representing this situation are χ_i and χ_f respectively, we wish to calculate the matrix element

$$R=\left(\chi_f^*\exp\left(-i\int_0^T Hdt\right)\chi_i\right)=(\chi_f^*S\chi_i). \tag{37}$$

We assume that the potential A differs from zero only for times between 0 and T so that a vacuum can be defined at these times. If χ_0 represents the vacuum state (that is, all negative energy states filled, all positive energies empty), the amplitude for having a vacuum at time T, if we had one at $t=0$, is

$$C_v=(\chi_0^*S\chi_0), \tag{38}$$

writing S for $\exp(-i\int_0^T Hdt)$. Our problem is to evaluate R and show that it is a simple factor times C_v, and that the factor involves the $K_+^{(A)}$ functions in the way discussed in the previous sections.

To do this we first express χ_i in terms of χ_0. The operator

$$\Phi^*=\int\Psi^*(\mathbf{x})\phi(\mathbf{x})d^3\mathbf{x}, \tag{39}$$

creates an electron with wave function $\phi(\mathbf{x})$. Likewise $\Phi=\int\phi^*(\mathbf{x})\times\Psi(\mathbf{x})d^3\mathbf{x}$ annihilates one with wave function $\phi(\mathbf{x})$. Hence state χ_i is $\chi_i=F_1^*F_2^*\cdots P_1P_2\cdots\chi_0$ while the final state is $G_1^*G_2^*\cdots\times Q_1Q_2\cdots\chi_0$ where F_i, G_i, P_i, Q_i are operators defined like Φ, in (39), but with f_i, g_i, p_i, q_i replacing ϕ; for the initial state would result from the vacuum if we created the electrons in $f_1, f_2, \cdots$ and annihilated those in $p_1, p_2, \cdots$. Hence we must find

$$R=(\chi_0^*\cdots Q_2^*Q_1^*\cdots G_2G_1SF_1^*F_2^*\cdots P_1P_2\cdots\chi_0). \tag{40}$$

To simplify this we shall have to use commutation relations between a Φ^* operator and S. To this end consider $\exp(-i\int_0^t Hdt')\Phi^*\times\exp(+i\int_0^t Hdt')$ and expand this quantity in terms of $\Psi^*(\mathbf{x})$, giving $\int\Psi^*(\mathbf{x})\phi(\mathbf{x},t)d^3\mathbf{x}$, (which defines $\phi(\mathbf{x},t)$). Now multiply this equation by $\exp(+i\int_0^t Hdt')\cdots\exp(-i\int_0^t Hdt')$ and find

$$\int\Psi^*(\mathbf{x})\phi(\mathbf{x})d^3\mathbf{x}=\int\Psi^*(\mathbf{x},t)\phi(\mathbf{x},t)d^3\mathbf{x}, \tag{41}$$

where we have defined $\Psi(\mathbf{x},t)$ by $\Psi(\mathbf{x},t)=\exp(+i\int_0^t Hdt')\Psi(\mathbf{x})\times\exp(-i\int_0^t Hdt')$. As is well known $\Psi(\mathbf{x},t)$ satisfies the Dirac equation, (differentiate $\Psi(\mathbf{x},t)$ with respect to t and use commutation relations of H and Ψ)

$$i\partial\Psi(\mathbf{x},t)/\partial t=(\boldsymbol{\alpha}\cdot(-i\boldsymbol{\nabla}-\mathbf{A})+A_4+m\beta)\Psi(\mathbf{x},t). \tag{42}$$

Consequently $\phi(\mathbf{x},t)$ must also satisfy the Dirac equation (differentiate (41) with respect to t, use (42) and integrate by parts).

That is, if $\phi(\mathbf{x},T)$ is that solution of the Dirac equation at time T which is $\phi(\mathbf{x})$ at $t=0$, and if we define $\Phi^*=\int\Psi^*(\mathbf{x})\phi(\mathbf{x})d^3\mathbf{x}$ and $\Phi'^*=\int\Psi^*(\mathbf{x})\phi(\mathbf{x},T)d^3\mathbf{x}$ then $\Phi'^*=S\Phi^*S^{-1}$, or

$$S\Phi^*=\Phi'^*S. \tag{43}$$

The principle on which the proof will be based can now be illustrated by a simple example. Suppose we have just one electron initially and finally and ask for

$$r=(\chi_0^*GSF^*\chi_0). \tag{44}$$

We might try putting F^* through the operator S using (43), $SF^*=F'^*S$, where f' in $F'^*=\int\Psi^*(\mathbf{x})f'(\mathbf{x})d^3\mathbf{x}$ is the wave function at T arising from $f(\mathbf{x})$ at 0. Then

$$r=(\chi_0^*GF'^*S\chi_0)=\int g^*(\mathbf{x})f'(\mathbf{x})d^3\mathbf{x}\cdot C_v-(\chi_0^*F'^*GS\chi_0), \tag{45}$$

where the second expression has been obtained by use of the definition (38) of C_v and the general commutation relation

$$GF^*+F^*G=\int g^*(\mathbf{x})f(\mathbf{x})d^3\mathbf{x},$$

which is a consequence of the properties of $\Psi(\mathbf{x})$ (the others are $FG=-GF$ and $F^*G^*=-G^*F^*$). Now $\chi_0^*F'^*$ in the last term in (45) is the complex conjugate of $F'\chi_0$. Thus if f' contained only positive energy components, $F'\chi_0$ would vanish and we would have reduced r to a factor times C_v. But F', as worked out here, does contain negative energy components created in the potential A and the method must be slightly modified.

Before putting F^* through the operator we shall add to it another operator F''^* arising from a function $f''(x)$ containing *only negative* energy components and so chosen that the resulting f' has *only positive* ones. That is we want

$$S(F_{\text{pos}}^*+F_{\text{neg}}''^*)=F_{\text{pos}}'^*S, \tag{46}$$

where the "pos" and "neg" serve as reminders of the sign of the energy components contained in the operators. This we can now use in the form

$$SF_{\text{pos}}^*=F_{\text{pos}}'^*S-SF_{\text{neg}}''^*. \tag{47}$$

In our one electron problem this substitution replaces r by two terms

$$r=(\chi_0^*GF_{\text{pos}}'^*S\chi_0)-(\chi_0^*GSF_{\text{neg}}''^*\chi_0).$$

The first of these reduces to

$$r=\int g^*(\mathbf{x})f_{\text{pos}}'(\mathbf{x})d^3\mathbf{x}\cdot C_v,$$

as above, for $F_{\text{pos}}'\chi_0$ is now zero, while the second is zero since the creation operator $F_{\text{neg}}''^*$ gives zero when acting on the vacuum state as all negative energies are full. This is the central idea of the demonstration.

The problem presented by (46) is this: Given a function $f_{\text{pos}}(\mathbf{x})$ at time 0, to find the amount, f_{neg}'', of negative energy component which must be added in order that the solution of Dirac's equation at time T will have only positive energy components, f_{pos}'. This is a boundary value problem for which the kernel $K_+^{(A)}$ is designed. We know the positive energy components initially, f_{pos}, and the negative ones finally (zero). The positive ones finally are therefore (using (19))

$$f_{\text{pos}}'(\mathbf{x}_2)=\int K_+^{(A)}(2,1)\beta f_{\text{pos}}(\mathbf{x}_1)d^3\mathbf{x}_1, \tag{48}$$

where $t_2=T$, $t_1=0$. Similarly, the negative ones initially are

$$f_{\text{neg}}''(\mathbf{x}_2)=\int K_+^{(A)}(2,1)\beta f_{\text{pos}}(\mathbf{x}_1)d^3\mathbf{x}_1-f_{\text{pos}}(\mathbf{x}_2), \tag{49}$$

where t_2 approaches zero from above, and $t_1=0$. The $f_{\text{pos}}(\mathbf{x}_2)$ is

[13] See, for example, G. Wentzel, *Einfuhrung in die Quantentheorie der Wellenfelder* (Franz Deuticke, Leipzig, 1943), Chapter V.

subtracted to keep in $f_{\text{neg}}''(\mathbf{x}_2)$ only those waves which return from the potential and not those arriving directly at t_2 from the $K_+(2,1)$ part of $K_+^{(A)}(2,1)$, as $t_2\to 0$. We could also have written

$$f_{\text{neg}}''(\mathbf{x}_2)=\int[K_+^{(A)}(2,1)-K_+(2,1)]\beta f_{\text{pos}}(\mathbf{x}_1)d^3\mathbf{x}_1. \quad (50)$$

Therefore the one-electron problem, $r=\int g^*(\mathbf{x})f_{\text{pos}}'(\mathbf{x})d^3\mathbf{x}\cdot C_v$, gives by (48)

$$r=C_v\int g^*(\mathbf{x}_2)K_+^{(A)}(2,1)\beta f(\mathbf{x}_1)d^3\mathbf{x}_1d^3\mathbf{x}_2,$$

as expected in accordance with the reasoning of the previous sections (i.e., (20) with $K_+^{(A)}$ replacing K_+).

The proof is readily extended to the more general expression R, (40), which can be analyzed by induction. First one replaces F_1^* by a relation such as (47) obtaining two terms

$$R=(\chi_0^*\cdots Q_2^*Q_1^*\cdots G_2G_1F_{1\text{pos}}'^*SF_2^*\cdots P_1P_2\cdots\chi_0)$$
$$-(\chi_0^*\cdots Q_2^*Q_1^*\cdots G_2G_1SF_{1\text{neg}}''^*F_2^*\cdots P_1P_2\cdots\chi_0).$$

In the first term the order of $F_{1\text{pos}}'^*$ and G_1 is then interchanged, producing an additional term $\int g_1^*(\mathbf{x})f_{1\text{pos}}'(\mathbf{x})d^3\mathbf{x}$ times an expression with one less electron in initial and final state. Next it is exchanged with G_2 producing an addition $-\int g_2^*(\mathbf{x})f_{1\text{pos}}'(\mathbf{x})d^3\mathbf{x}$ times a similar term, etc. Finally on reaching the Q_1^* with which it anticommutes it can be simply moved over to juxtaposition with χ_0^* where it gives zero. The second term is similarly handled by moving $F_{1\text{neg}}''^*$ through anti commuting F_2^*, etc., until it reaches P_1. Then it is exchanged with P_1 to produce an additional simpler term with a factor $\mp\int p_1^*(\mathbf{x})f_{1\text{neg}}''(\mathbf{x})d^3\mathbf{x}$ or $\mp\int p_1^*(\mathbf{x}_2)K_+^{(A)}(2,1)\beta f_1(\mathbf{x}_1)d^3\mathbf{x}_1d^3\mathbf{x}_2$ from (49), with $t_2=t_1=0$ (the extra $f_1(\mathbf{x}_2)$ in (49) gives zero as it is orthogonal to $p_1(\mathbf{x}_2)$). This describes in the expected manner the annihilation of the pair, electron f_1, positron p_1. The $F_{\text{neg}}''^*$ is moved in this way successively through the P's until it gives zero when acting on χ_0. Thus R is reduced, with the expected factors (and with alternating signs as required by the exclusion principle), to simpler terms containing two less operators which may in turn be further reduced by using F_2^* in a similar manner, etc. After all the F^* are used the Q^*'s can be reduced in a similar manner. They are moved through the S in the opposite direction in such a manner as to produce a purely negative energy operator at time 0, using relations analogous to (46) to (49). After all this is done we are left simply with the expected factor times C_v (assuming the net charge is the same in initial and final state.)

In this way we have written the solution to the general problem of the motion of electrons in given potentials. The factor C_v is obtained by normalization. However for photon fields it is desirable to have an explicit form for C_v in terms of the potentials. This is given by (30) and (29) and it is readily demonstrated that this also is correct according to second quantization.

b. Analysis of the Vacuum Problem

We shall calculate C_v from second quantization by induction considering a series of problems each containing a potential distribution more nearly like the one we wish. Suppose we know C_v for a problem like the one we want and having the same potentials for time t between some t_0 and T, but having potential zero for times from 0 to t_0. Call this $C_v(t_0)$, the corresponding Hamiltonian H_{t_0} and the sum of contributions for all single loops, $L(t_0)$. Then for $t_0=T$ we have zero potential at all times, no pairs can be produced, $L(T)=0$ and $C_v(T)=1$. For $t_0=0$ we have the complete problem, so that $C_v(0)$ is what is defined as C_v in (38). Generally we have,

$$C_v(t_0)=\left(\chi_0^*\exp\left(-i\int_0^T H_{t_0}dt\right)\chi_0\right)$$
$$=\left(\chi_0^*\exp\left(-i\int_{t_0}^T H_{t_0}dt\right)\chi_0\right),$$

since H_{t_0} is identical to the constant vacuum Hamiltonian H_T for $t<t_0$ and χ_0 is an eigenfunction of H_T with an eigenvalue (energy of vacuum) which we can take as zero.

The value of $C_v(t_0-\Delta t_0)$ arises from the Hamiltonian $H_{t_0-\Delta t_0}$ which differs from H_{t_0} just by having an extra potential during the short interval Δt_0. Hence, to first order in Δt_0, we have

$$C_v(t_0-\Delta t_0)=\left(\chi_0^*\exp\left(-i\int_{t_0-\Delta t_0}^T H_{t_0-\Delta t_0}dt\right)\chi_0\right)$$
$$=\left(\chi_0^*\exp\left(-i\int_{t_0}^T H_{t_0}dt\right)\left[1-i\Delta t_0\int\Psi^*(\mathbf{x})\right.\right.$$
$$\left.\left.\times(-\boldsymbol{\alpha}\cdot\mathbf{A}(\mathbf{x},t_0)+A_4(\mathbf{x},t_0))\Psi(\mathbf{x})d^3\mathbf{x}\right]\chi_0\right);$$

we therefore obtain for the derivative of C_v the expression

$$-dC_v(t_0)/dt_0=-i\left(\chi_0^*\exp\left(-i\int_{t_0}^T H_{t_0}dt\right)\right.$$
$$\left.\times\int\Psi^*(\mathbf{x})\beta A(\mathbf{x},t_0)\Psi(\mathbf{x})d^3\mathbf{x}\chi_0\right), \quad (51)$$

which will be reduced to a simple factor times $C_v(t_0)$ by methods analogous to those used in reducing R. The operator Ψ can be imagined to be split into two pieces Ψ_{pos} and Ψ_{neg} operating on positive and negative energy states respectively. The Ψ_{pos} on χ_0 gives zero so we are left with two terms in the current density, $\Psi_{\text{pos}}^*\beta A\Psi_{\text{neg}}$ and $\Psi_{\text{neg}}^*\beta A\Psi_{\text{neg}}$. The latter $\Psi_{\text{neg}}^*\beta A\Psi_{\text{neg}}$ is just the expectation value of βA taken over all negative energy states (minus $\Psi_{\text{neg}}\beta A\Psi_{\text{neg}}^*$ which gives zero acting on χ_0). This is the effect of the vacuum expectation current of the electrons in the sea which we should have subtracted from our original Hamiltonian in the customary way.

The remaining term $\Psi_{\text{pos}}^*\beta A\Psi_{\text{neg}}$, or its equivalent $\Psi_{\text{pos}}^*\beta A\Psi$ can be considered as $\Psi^*(\mathbf{x})\mathbf{f}_{\text{pos}}(\mathbf{x})$ where $\mathbf{f}_{\text{pos}}(\mathbf{x})$ is written for the positive energy component of the operator $\beta A\Psi(\mathbf{x})$. Now this operator, $\Psi^*(\mathbf{x})\mathbf{f}_{\text{pos}}(\mathbf{x})$, or more precisely just the $\Psi^*(\mathbf{x})$ part of it, can be pushed through the $\exp(-i\int_{t_0}^T Hdt)$ in a manner exactly analogous to (47) when f is a function. (An alternative derivation results from the consideration that the operator $\Psi(\mathbf{x},t)$ which satisfies the Dirac equation also satisfies the linear integral equations which are equivalent to it.) That is, (51) can be written by (48), (50),

$$-dC_v(t_0)/dt_0=-i\left(\chi_0^*\int\int\Psi^*(\mathbf{x}_2)K_+^{(A)}(2,1)\right.$$
$$\left.\times\exp\left(-i\int_{t_0}^T Hdt\right)A(1)\Psi(\mathbf{x}_1)d^3\mathbf{x}_1d^3\mathbf{x}_2\chi_0\right)$$
$$+i\left(\chi_0^*\exp\left(-i\int_{t_0}^T Hdt\right)\int\int\Psi^*(\mathbf{x}_2)[K_+^{(A)}(2,1)\right.$$
$$\left.-K_+(2,1)]A(1)\Psi(\mathbf{x}_1)d^3\mathbf{x}_1d^3\mathbf{x}_2\chi_0\right),$$

where in the first term $t_2=T$, and in the second $t_2\to t_0=t_1$. The (A) in $K_+^{(A)}$ refers to that part of the potential A after t_0. The first term vanishes for it involves (from the $K_+^{(A)}(2,1)$) only positive energy components of Ψ^*, which give zero operating into χ_0^*. In the second term only negative components of $\Psi^*(\mathbf{x}_2)$ appear. If, then $\Psi^*(\mathbf{x}_2)$ is interchanged in order with $\Psi(\mathbf{x}_1)$ it will give zero operating on χ_0, and only the term,

$$-dC_v(t_0)/dt_0=+i\int Sp[(K_+^{(A)}(1,1)$$
$$-K_+(1,1))A(1)]d^3\mathbf{x}_1\cdot C_v(t_0), \quad (52)$$

will remain, from the usual commutation relation of Ψ^* and Ψ.

The factor of $C_v(t_0)$ in (52) times $-\Delta t_0$ is, according to (29) (reference 10), just $L(t_0-\Delta t_0)-L(t_0)$ since this difference arises from the extra potential $\Delta A=A$ during the short time interval Δt_0. Hence $-dC_v(t_0)/dt_0=+(dL(t_0)/dt_0)C_v(t_0)$ so that integration from $t_0=T$ to $t_0=0$ establishes (30).

Starting from the theory of the electromagnetic field in second quantization, a deduction of the equations for quantum electrodynamics which appear in the succeeding paper may be worked out using very similar principles. The Pauli-Weisskopf theory of the Klein-Gordon equation can apparently be analyzed in essentially the same way as that used here for Dirac electrons.

Space-Time Approach to Quantum Electrodynamics

R. P. Feynman
Department of Physics, Cornell University, Ithaca, New York
(Received May 9, 1949)

In this paper two things are done. (1) It is shown that a considerable simplification can be attained in writing down matrix elements for complex processes in electrodynamics. Further, a physical point of view is available which permits them to be written down directly for any specific problem. Being simply a restatement of conventional electrodynamics, however, the matrix elements diverge for complex processes. (2) Electrodynamics is modified by altering the interaction of electrons at short distances. All matrix elements are now finite, with the exception of those relating to problems of vacuum polarization. The latter are evaluated in a manner suggested by Pauli and Bethe, which gives finite results for these matrices also. The only effects sensitive to the modification are changes in mass and charge of the electrons. Such changes could not be directly observed. Phenomena directly observable, are insensitive to the details of the modification used (except at extreme energies). For such phenomena, a limit can be taken as the range of the modification goes to zero. The results then agree with those of Schwinger. A complete, unambiguous, and presumably consistent, method is therefore available for the calculation of all processes involving electrons and photons.

The simplification in writing the expressions results from an emphasis on the over-all space-time view resulting from a study of the solution of the equations of electrodynamics. The relation of this to the more conventional Hamiltonian point of view is discussed. It would be very difficult to make the modification which is proposed if one insisted on having the equations in Hamiltonian form.

The methods apply as well to charges obeying the Klein-Gordon equation, and to the various meson theories of nuclear forces. Illustrative examples are given. Although a modification like that used in electrodynamics can make all matrices finite for all of the meson theories, for some of the theories it is no longer true that all directly observable phenomena are insensitive to the details of the modification used.

The actual evaluation of integrals appearing in the matrix elements may be facilitated, in the simpler cases, by methods described in the appendix.

THIS paper should be considered as a direct continuation of a preceding one[1] (I) in which the motion of electrons, neglecting interaction, was analyzed, by dealing directly with the *solution* of the Hamiltonian differential equations. Here the same technique is applied to include interactions and in that way to express in simple terms the solution of problems in quantum electrodynamics.

For most practical calculations in quantum electrodynamics the solution is ordinarily expressed in terms of a matrix element. The matrix is worked out as an expansion in powers of $e^2/\hbar c$, the successive terms corresponding to the inclusion of an increasing number of virtual quanta. It appears that a considerable simplification can be achieved in writing down these matrix elements for complex processes. Furthermore, each term in the expansion can be written down and understood directly from a physical point of view, similar to the space-time view in I. It is the purpose of this paper to describe how this may be done. We shall also discuss methods of handling the divergent integrals which appear in these matrix elements.

The simplification in the formulae results mainly from the fact that previous methods unnecessarily separated into individual terms processes that were closely related physically. For example, in the exchange of a quantum between two electrons there were two terms depending on which electron emitted and which absorbed the quantum. Yet, in the virtual states considered, timing relations are not significant. Olny the order of operators in the matrix must be maintained. We have seen (I), that in addition, processes in which virtual pairs are produced can be combined with others in which only positive energy electrons are involved. Further, the effects of longitudinal and transverse waves can be combined together. The separations previously made were on an unrelativistic basis (reflected in the circumstance that apparently momentum but not energy is conserved in intermediate states). When the terms are combined and simplified, the relativistic invariance of the result is self-evident.

We begin by discussing the solution in space and time of the Schrödinger equation for particles interacting instantaneously. The results are immediately generalizable to delayed interactions of relativistic electrons and we represent in that way the laws of quantum electrodynamics. We can then see how the matrix element for any process can be written down directly. In particular, the self-energy expression is written down.

So far, nothing has been done other than a restatement of conventional electrodynamics in other terms. Therefore, the self-energy diverges. A modification[2] in interaction between charges is next made, and it is shown that the self-energy is made convergent and corresponds to a correction to the electron mass. After the mass correction is made, other real processes are finite and insensitive to the "width" of the cut-off in the interaction.[3]

Unfortunately, the modification proposed is not completely satisfactory theoretically (it leads to some difficulties of conservation of energy). It does, however, seem consistent and satisfactory to define the matrix

[1] R. P. Feynman, Phys. Rev. **76**, 749 (1949), hereafter called I.

[2] For a discussion of this modification in classical physics see R. P. Feynman, Phys. Rev. **74** 939 (1948), hereafter referred to as A.

[3] A brief summary of the methods and results will be found in R. P. Feynman, Phys. Rev. **74**, 1430 (1948), hereafter referred to as B.

element for all real processes as the limit of that computed here as the cut-off width goes to zero. A similar technique suggested by Pauli and by Bethe can be applied to problems of vacuum polarization (resulting in a renormalization of charge) but again a strict physical basis for the rules of convergence is not known.

After mass and charge renormalization, the limit of zero cut-off width can be taken for all real processes. The results are then equivalent to those of Schwinger[4] who does not make explicit use of the convergence factors. The method of Schwinger is to identify the terms corresponding to corrections in mass and charge and, previous to their evaluation, to remove them from the expressions for real processes. This has the advantage of showing that the results can be strictly independent of particular cut-off methods. On the other hand, many of the properties of the integrals are analyzed using formal properties of invariant propagation functions. But one of the properties is that the integrals are infinite and it is not clear to what extent this invalidates the demonstrations. A practical advantage of the present method is that ambiguities can be more easily resolved; simply by direct calculation of the otherwise divergent integrals. Nevertheless, it is not at all clear that the convergence factors do not upset the physical consistency of the theory. Although in the limit the two methods agree, neither method appears to be thoroughly satisfactory theoretically. Nevertheless, it does appear that we now have available a complete and definite method for the calculation of physical processes to any order in quantum electrodynamics.

Since we can write down the solution to any physical problem, we have a complete theory which could stand by itself. It will be theoretically incomplete, however, in two respects. First, although each term of increasing order in $e^2/\hbar c$ can be written down it would be desirable to see some way of expressing things in finite form to all orders in $e^2/\hbar c$ at once. Second, although it will be physically evident that the results obtained are equivalent to those obtained by conventional electrodynamics the mathematical proof of this is not included. Both of these limitations will be removed in a subsequent paper (see also Dyson[4]).

Briefly the genesis of this theory was this. The conventional electrodynamics was expressed in the Lagrangian form of quantum mechanics described in the Reviews of Modern Physics.[5] The motion of the field oscillators could be integrated out (as described in Section 13 of that paper), the result being an expression of the delayed interaction of the particles. Next the modification of the delta-function interaction could be made directly from the analogy to the classical case.[2] This was still not complete because the Lagrangian method had been worked out in detail only for particles obeying the non-relativistic Schrödinger equation. It was then modified in accordance with the requirements of the Dirac equation and the phenomenon of pair creation. This was made easier by the reinterpretation of the theory of holes (I). Finally for practical calculations the expressions were developed in a power series in $e^2/\hbar c$. It was apparent that each term in the series had a simple physical interpretation. Since the result was easier to understand than the derivation, it was thought best to publish the results first in this paper. Considerable time has been spent to make these first two papers as complete and as physically plausible as possible without relying on the Lagrangian method, because it is not generally familiar. It is realized that such a description cannot carry the conviction of truth which would accompany the derivation. On the other hand, in the interest of keeping simple things simple the derivation will appear in a separate paper.

The possible application of these methods to the various meson theories is discussed briefly. The formulas corresponding to a charge particle of zero spin moving in accordance with the Klein Gordon equation are also given. In an Appendix a method is given for calculating the integrals appearing in the matrix elements for the simpler processes.

The point of view which is taken here of the interaction of charges differs from the more usual point of view of field theory. Furthermore, the familiar Hamiltonian form of quantum mechanics must be compared to the over-all space-time view used here. The first section is, therefore, devoted to a discussion of the relations of these viewpoints.

1. COMPARISON WITH THE HAMILTONIAN METHOD

Electrodynamics can be looked upon in two equivalent and complementary ways. One is as the description of the behavior of a field (Maxwell's equations). The other is as a description of a direct interaction at a distance (albeit delayed in time) between charges (the solutions of Lienard and Wiechert). From the latter point of view light is considered as an interaction of the charges in the source with those in the absorber. This is an impractical point of view because many kinds of sources produce the same kind of effects. The field point of view separates these aspects into two simpler problems, production of light, and absorption of light. On the other hand, the field point of view is less practical when dealing with close collisions of particles (or their action on themselves). For here the source and absorber are not readily distinguishable, there is an intimate exchange of quanta. The fields are so closely determined by the motions of the particles that it is just as well not to separate the question into two problems but to consider the process as a direct interaction. Roughly, the field point of view is most practical for problems involv-

[4] J. Schwinger, Phys. Rev. **74**, 1439 (1948), Phys. Rev. **75**, 651 (1949). A proof of this equivalence is given by F. J. Dyson, Phys. Rev. **75**, 486 (1949).

[5] R. P. Feynman, Rev. Mod. Phys. **20**, 367 (1948). The application to electrodynamics is described in detail by H. J. Groenewold, Koninklijke Nederlandsche Akademia van Weteschappen. Proceedings Vol. LII, 3 (226) 1949.

ing real quanta, while the interaction view is best for the discussion of the virtual quanta involved. We shall emphasize the interaction viewpoint in this paper, first because it is less familiar and therefore requires more discussion, and second because the important aspect in the problems with which we shall deal is the effect of virtual quanta.

The Hamiltonian method is not well adapted to represent the direct action at a distance between charges because that action is delayed. The Hamiltonian method represents the future as developing out of the present. If the values of a complete set of quantities are known now, their values can be computed at the next instant in time. If particles interact through a delayed interaction, however, one cannot predict the future by simply knowing the present motion of the particles. One would also have to know what the motions of the particles were in the past in view of the interaction this may have on the future motions. This is done in the Hamiltonian electrodynamics, of course, by requiring that one specify besides the present motion of the particles, the values of a host of new variables (the coordinates of the field oscillators) to keep track of that aspect of the past motions of the particles which determines their future behavior. The use of the Hamiltonian forces one to choose the field viewpoint rather than the interaction viewpoint.

In many problems, for example, the close collisions of particles, we are not interested in the precise temporal sequence of events. It is not of interest to be able to say how the situation would look at each instant of time during a collision and how it progresses from instant to instant. Such ideas are only useful for events taking a long time and for which we can readily obtain information during the intervening period. For collisions it is much easier to treat the process as a whole.[6] The Møller interaction matrix for the the collision of two electrons is not essentially more complicated than the non-relativistic Rutherford formula, yet the mathematical machinery used to obtain the former from quantum electrodynamics is vastly more complicated than Schrödinger's equation with the e^2/r_{12} interaction needed to obtain the latter. The difference is only that in the latter the action is instantaneous so that the Hamiltonian method requires no extra variables, while in the former relativistic case it is delayed and the Hamiltonian method is very cumbersome.

We shall be discussing the solutions of equations rather than the time differential equations from which they come. We shall discover that the solutions, because of the over-all space-time view that they permit, are as easy to understand when interactions are delayed as when they are instantaneous.

As a further point, relativistic invariance will be self-evident. The Hamiltonian form of the equations develops the future from the instantaneous present. But for different observers in relative motion the instantaneous present is different, and corresponds to a different 3-dimensional cut of space-time. Thus the temporal analyses of different observers is different and their Hamiltonian equations are developing the process in different ways. These differences are irrelevant, however, for the solution is the same in any space time frame. By forsaking the Hamiltonian method, the wedding of relativity and quantum mechanics can be accomplished most naturally.

We illustrate these points in the next section by studying the solution of Schrödinger's equation for non-relativistic particles interacting by an instantaneous Coulomb potential (Eq. 2). When the solution is modified to include the effects of delay in the interaction and the relativistic properties of the electrons we obtain an expression of the laws of quantum electrodynamics (Eq. 4).

2. THE INTERACTION BETWEEN CHARGES

We study by the same methods as in I, the interaction of two particles using the same notation as I. We start by considering the non-relativistic case described by the Schrödinger equation (I, Eq. 1). The wave function at a given time is a function $\psi(\mathbf{x}_a, \mathbf{x}_b, t)$ of the coordinates $\mathbf{x}_a$ and $\mathbf{x}_b$ of each particle. Thus call $K(\mathbf{x}_a, \mathbf{x}_b, t; \mathbf{x}_a', \mathbf{x}_b', t')$ the amplitude that particle a at $\mathbf{x}_a'$ at time t' will get to $\mathbf{x}_a$ at t while particle b at $\mathbf{x}_b'$ at t' gets to $\mathbf{x}_b$ at t. If the particles are free and do not interact this is

$$K(\mathbf{x}_a, \mathbf{x}_b, t; \mathbf{x}_a', \mathbf{x}_b', t') = K_{0a}(\mathbf{x}_a, t; \mathbf{x}_a', t')K_{0b}(\mathbf{x}_b, t; \mathbf{x}_b', t')$$

where K_{0a} is the K_0 function for particle a considered as free. In *this* case we can obviously define a quantity like K, but for which the time t need not be the same for particles a and b (likewise for t'); e.g.,

$$K_0(3, 4; 1, 2) = K_{0a}(3, 1)K_{0b}(4, 2) \tag{1}$$

can be thought of as the amplitude that particle a goes from $\mathbf{x}_1$ at t_1 to $\mathbf{x}_3$ at t_3 and that particle b goes from $\mathbf{x}_2$ at t_2 to $\mathbf{x}_4$ at t_4.

When the particles do interact, one can only define the quantity $K(3, 4; 1, 2)$ precisely if the interaction vanishes between t_1 and t_2 and also between t_3 and t_4. In a real physical system such is not the case. There is such an enormous advantage, however, to the concept that we shall continue to use it, imagining that we can neglect the effect of interactions between t_1 and t_2 and between t_3 and t_4. For practical problems this means choosing such long time intervals t_3-t_1 and t_4-t_2 that the extra interactions near the end points have small relative effects. As an example, in a scattering problem it may well be that the particles are so well separated initially and finally that the interaction at these times is negligible. Again energy values can be defined by the average rate of change of phase over such long time intervals that errors initially and finally can be neglected. Inasmuch as any physical problem can be defined in terms of scattering processes we do not lose much in

[6] This is the viewpoint of the theory of the S matrix of Heisenberg.

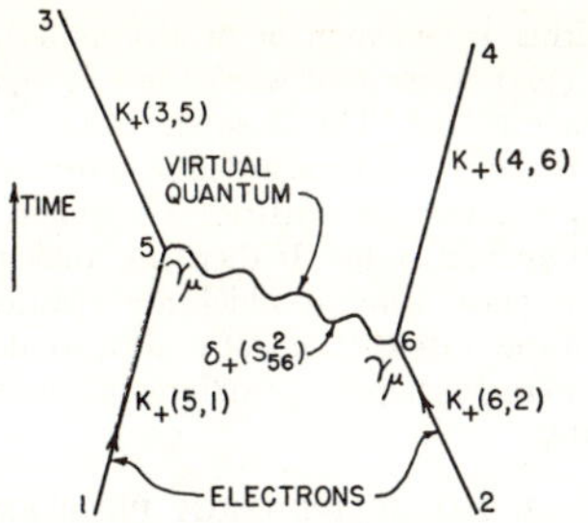

FIG. 1. The fundamental interaction Eq. (4). Exchange of one quantum between two electrons.

a general theoretical sense by this approximation. If it is not made it is not easy to study interacting particles relativistically, for there is nothing significant in choosing $t_1=t_3$ if $\mathbf{x}_1\neq\mathbf{x}_3$, as absolute simultaneity of events at a distance cannot be defined invariantly. It is essentially to avoid this approximation that the complicated structure of the older quantum electrodynamics has been built up. We wish to describe electrodynamics as a delayed interaction between particles. If we can make the approximation of assuming a meaning to $K(3,4;1,2)$ the results of this interaction can be expressed very simply.

To see how this may be done, imagine first that the interaction is simply that given by a Coulomb potential e^2/r where r is the distance between the particles. If this be turned on only for a very short time Δt_0 at time t_0, the first order correction to $K(3, 4; 1, 2)$ can be worked out exactly as was Eq. (9) of I by an obvious generalization to two particles:

$$K^{(1)}(3,4;1,2)=-ie^2\int\int K_{0a}(3,5)K_{0b}(4,6)r_{56}^{-1}\times K_{0a}(5,1)K_{0b}(6,2)d^3\mathbf{x}_5d^3\mathbf{x}_6\Delta t_0,$$

where $t_5=t_6=t_0$. If now the potential were on at all times (so that strictly K is not defined unless $t_4=t_3$ and $t_1=t_2$), the first-order effect is obtained by integrating on t_0, which we can write as an integral over both t_5 and t_6 if we include a delta-function $\delta(t_5-t_6)$ to insure contribution only when $t_5=t_6$. Hence, the first-order effect of interaction is (calling $t_5-t_6=t_{56}$):

$$K^{(1)}(3,4;1,2)=-ie^2\int\int K_{0a}(3,5)K_{0b}(4,6)r_{56}^{-1}\times\delta(t_{56})K_{0a}(5,1)K_{0b}(6,2)d\tau_5d\tau_6, \quad (2)$$

where $d\tau=d^3\mathbf{x}dt$.

We know, however, in classical electrodynamics, that the Coulomb potential does not act instantaneously, but is delayed by a time r_{56}, taking the speed of light as unity. This suggests simply replacing $r_{56}^{-1}\delta(t_{56})$ in (2) by something like $r_{56}^{-1}\delta(t_{56}-r_{56})$ to represent the delay in the effect of b on a.

This turns out to be not quite right,[7] for when this interaction is represented by photons they must be of only positive energy, while the Fourier transform of $\delta(t_{56}-r_{56})$ contains frequencies of both signs. It should instead be replaced by $\delta_+(t_{56}-r_{56})$ where

$$\delta_+(x)=\int_0^\infty e^{-i\omega x}d\omega/\pi=\lim_{\epsilon\to 0}\frac{(\pi i)^{-1}}{x-i\epsilon}=\delta(x)+(\pi ix)^{-1}. \quad (3)$$

This is to be averaged with $r_{56}^{-1}\delta_+(-t_{56}-r_{56})$ which arises when $t_5<t_6$ and corresponds to a emitting the quantum which b receives. Since

$$(2r)^{-1}(\delta_+(t-r)+\delta_+(-t-r))=\delta_+(t^2-r^2),$$

this means $r_{56}^{-1}\delta(t_{56})$ is replaced by $\delta_+(s_{56}^2)$ where $s_{56}^2=t_{56}^2-r_{56}^2$ is the square of the relativistically invariant interval between points 5 and 6. Since in classical electrodynamics there is also an interaction through the vector potential, the complete interaction (see A, Eq. (1)) should be $(1-(\mathbf{v}_5\cdot\mathbf{v}_6)\delta_+(s_{56}^2)$, or in the relativistic case,

$$(1-\boldsymbol{\alpha}_a\cdot\boldsymbol{\alpha}_b)\delta_+(s_{56}^2)=\beta_a\beta_b\gamma_{a\mu}\gamma_{b\mu}\delta_+(s_{56}^2).$$

Hence we have for electrons obeying the Dirac equation,

$$K^{(1)}(3,4;1,2)=-ie^2\int\int K_{+a}(3,5)K_{+b}(4,6)\gamma_{a\mu}\gamma_{b\mu}\times\delta_+(s_{56}^2)K_{+a}(5,1)K_{+b}(6,2)d\tau_5d\tau_6, \quad (4)$$

where $\gamma_{a\mu}$ and $\gamma_{b\mu}$ are the Dirac matrices applying to the spinor corresponding to particles a and b, respectively (the factor $\beta_a\beta_b$ being absorbed in the definition, I Eq. (17), of K_+).

This is our fundamental equation for electrodynamics. It describes the effect of exchange of one quantum (therefore first order in e^2) between two electrons. It will serve as a prototype enabling us to write down the corresponding quantities involving the exchange of two or more quanta between two electrons or the interaction of an electron with itself. It is a consequence of conventional electrodynamics. Relativistic invariance is clear. Since one sums over μ it contains the effects of both longitudinal and transverse waves in a relativistically symmetrical way.

We shall now interpret Eq. (4) in a manner which will permit us to write down the higher order terms. It can be understood (see Fig. 1) as saying that the amplitude for "a" to go from 1 to 3 and "b" to go from 2 to 4 is altered to first order because they can exchange a quantum. Thus, "a" can go to 5 (amplitude $K_+(5, 1)$)

[7] It, and a like term for the effect of a on b, leads to a theory which, in the classical limit, exhibits interaction through half-advanced and half-retarded potentials. Classically, this is equivalent to purely retarded effects within a closed box from which no light escapes (e.g., see A, or J. A. Wheeler and R. P. Feynman, Rev. Mod. Phys. **17**, 157 (1945)). Analogous theorems exist in quantum mechanics but it would lead us too far astray to discuss them now.

emit a quantum (longitudinal, transverse, or scalar $\gamma_{a\mu}$) and then proceed to 3 ($K_+(3, 5)$). Meantime "b" goes to 6 ($K_+(6, 2)$), absorbs the quantum ($\gamma_{b\mu}$) and proceeds to 4 ($K_+(4, 6)$). The quantum meanwhile proceeds from 5 to 6, which it does with amplitude $\delta_+(s_{56}{}^2)$. We must sum over all the possible quantum polarizations μ and positions and times of emission 5, and of absorption 6. Actually if $t_5 > t_6$ it would be better to say that "a" absorbs and "b" emits but no attention need be paid to these matters, as all such alternatives are automatically contained in (4).

The correct terms of higher order in e^2 or involving larger numbers of electrons (interacting with themselves or in pairs) can be written down by the same kind of reasoning. They will be illustrated by examples as we proceed. In a succeeding paper they will all be deduced from conventional quantum electrodynamics.

Calculation, from (4), of the transition element between positive energy free electron states gives the Möller scattering of two electrons, when account is taken of the Pauli principle.

The exclusion principle for interacting charges is handled in exactly the same way as for non-interacting charges (I). For example, for two charges it requires only that one calculate $K(3, 4; 1, 2) - K(4, 3; 1, 2)$ to get the net amplitude for arrival of charges at 3 and 4. It is disregarded in intermediate states. The interference effects for scattering of electrons by positrons discussed by Bhabha will be seen to result directly in this formulation. The formulas are interpreted to apply to positrons in the manner discussed in I.

As our primary concern will be for processes in which the quanta are virtual we shall not include here the detailed analysis of processes involving real quanta in initial or final state, and shall content ourselves by only stating the rules applying to them.[8] The result of the analysis is, as expected, that they can be included by the same line of reasoning as is used in discussing the virtual processes, provided the quantities are normalized in the usual manner to represent single quanta. For example, the amplitude that an electron in going from 1 to 2 absorbs a quantum whose vector potential, suitably normalized, is $c_\mu \exp(-ik\cdot x) = C_\mu(x)$ is just the expression (I, Eq. (13)) for scattering in a potential with $\mathbf{A}$ (3) replaced by $\mathbf{C}$ (3). Each quantum interacts only once (either in emission or in absorption), terms like (I, Eq. (14)) occur only when there is more than one quantum involved. The Bose statistics of the quanta can, in all cases, be disregarded in intermediate states. The only effect of the statistics is to change the weight of initial or final states. If there are among quanta, in the initial state, some n which are identical then the weight of the state is $(1/n!)$ of what it would be if these quanta were considered as different (similarly for the final state).

3. THE SELF-ENERGY PROBLEM

Having a term representing the mutual interaction of a pair of charges, we must include similar terms to represent the interaction of a charge with itself. For under some circumstances what appears to be two distinct electrons may, according to I, be viewed also as a single electron (namely in case one electron was created in a pair with a positron destined to annihilate the other electron). Thus to the interaction between such electrons must correspond the possibility of the action of an electron on itself.[9]

This interaction is the heart of the self energy problem. Consider to first order in e^2 the action of an electron on itself in an otherwise force free region. The amplitude $K(2, 1)$ for a single particle to get from 1 to 2 differs from $K_+(2, 1)$ to first order in e^2 by a term

$$K^{(1)}(2, 1) = -ie^2 \int\int K_+(2, 4)\gamma_\mu K_+(4, 3)\gamma_\mu \times K_+(3, 1)d\tau_3 d\tau_4 \delta_+(s_{43}{}^2). \quad (6)$$

It arises because the electron instead of going from 1 directly to 2, may go (Fig. 2) first to 3, ($K_+(3, 1)$), emit a quantum (γ_μ), proceed to 4, ($K_+(4, 3)$), absorb it (γ_μ), and finally arrive at 2 ($K_+(2, 4)$). The quantum must go from 3 to 4 ($\delta_+(s_{43}{}^2)$).

This is related to the self-energy of a free electron in the following manner. Suppose initially, time t_1, we have an electron in state $f(1)$ which we imagine to be a positive energy solution of Dirac's equation for a free particle. After a long time $t_2 - t_1$ the perturbation will alter

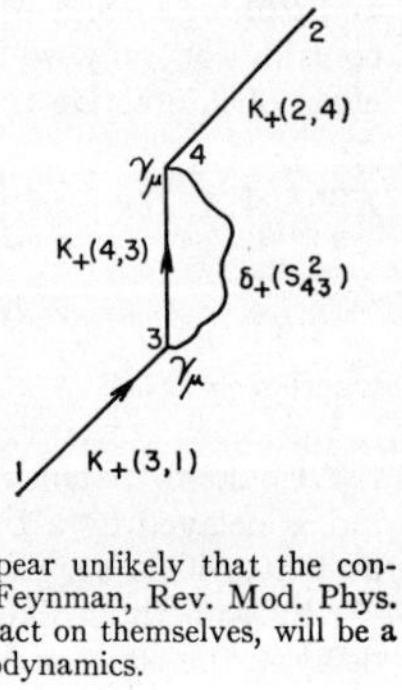

FIG. 2. Interaction of an electron with itself, Eq. (6).

[8] Although in the expressions stemming from (4) the quanta are virtual, this is not actually a theoretical limitation. One way to deduce the correct rules for real quanta from (4) is to note that in a closed system all quanta can be considered as virtual (i.e., they have a known source and are eventually absorbed) so that in such a system the present description is complete and equivalent to the conventional one. In particular, the relation of the Einstein A and B coefficients can be deduced. A more practical direct deduction of the expressions for real quanta will be given in the subsequent paper. It might be noted that (4) can be rewritten as describing the action on a, $K^{(1)}(3, 1) = i\int K_+(3, 5) \times \mathbf{A}(5)K_+(5, 1)d\tau_5$ of the potential $A_\mu(5) = e^2\int K_+(4, 6)\delta_+(s_{56}{}^2)\gamma_\mu \times K_+(6, 2)d\tau_6$ arising from Maxwell's equations $-\Box^2 A_\mu = 4\pi j_\mu$ from a "current" $j_\mu(6) = e^2 K_+(4, 6)\gamma_\mu K_+(6, 2)$ produced by particle b in going from 2 to 4. This is virtue of the fact that δ_+ satisfies

$$-\Box_2{}^2\delta_+(s_{21}{}^2) = 4\pi\delta(2, 1). \quad (5)$$

[9] These considerations make it appear unlikely that the contention of J. A. Wheeler and R. P. Feynman, Rev. Mod. Phys. **17**, 157 (1945), that electrons do not act on themselves, will be a successful concept in quantum electrodynamics.

the wave function, which can then be looked upon as a superposition of free particle solutions (actually it only contains f). The amplitude that $g(2)$ is contained is calculated as in (I, Eq. (21)). The diagonal element ($g=f$) is therefore

$$\int\int \bar{f}(2)\beta K^{(1)}(2,1)\beta f(1)d^3\mathbf{x}_1 d^3\mathbf{x}_2. \quad (7)$$

The time interval $T=t_2-t_1$ (and the spatial volume V over which one integrates) must be taken very large, for the expressions are only approximate (analogous to the situation for two interacting charges).[10] This is because, for example, we are dealing incorrectly with quanta emitted just before t_2 which would normally be reabsorbed at times after t_2.

If $K^{(1)}(2,1)$ from (6) is actually substituted into (7) the surface integrals can be performed as was done in obtaining I, Eq. (22) resulting in

$$-ie^2\int\int \bar{f}(4)\gamma_\mu K_+(4,3)\gamma_\mu f(3)\delta_+(s_{43}^2)d\tau_3 d\tau_4. \quad (8)$$

Putting for $f(1)$ the plane wave $u\exp(-ip\cdot x_1)$ where p_μ is the energy (p_4) and momentum of the electron ($p^2=m^2$), and u is a constant 4-index symbol, (8) becomes

$$-ie^2\int\int(\bar{u}\gamma_\mu K_+(4,3)\gamma_\mu u)$$
$$\times\exp(ip\cdot(x_4-x_3))\delta_+(s_{43}^2)d\tau_3 d\tau_4,$$

the integrals extending over the volume V and time interval T. Since $K_+(4,3)$ depends only on the difference of the coordinates of 4 and 3, $x_{43\mu}$, the integral on 4 gives a result (except near the surfaces of the region) independent of 3. When integrated on 3, therefore, the result is of order VT. The effect is proportional to V, for the wave functions have been normalized to unit volume. If normalized to volume V, the result would simply be proportional to T. This is expected, for if the effect were equivalent to a change in energy ΔE, the amplitude for arrival in f at t_2 is altered by a factor $\exp(-i\Delta E(t_2-t_1))$, or to first order by the difference $-i(\Delta E)T$. Hence, we have

$$\Delta E=e^2\int(\bar{u}\gamma_\mu K_+(4,3)\gamma_\mu u)\exp(ip\cdot x_{43})\delta_+(s_{43}^2)d\tau_4, \quad (9)$$

integrated over all space-time $d\tau_4$. This expression will be simplified presently. In interpreting (9) we have tacitly assumed that the wave functions are normalized so that $(u^*u)=(\bar{u}\gamma_4 u)=1$. The equation may therefore be made independent of the normalization by writing the left side as $(\Delta E)(\bar{u}\gamma_4 u)$, or since $(\bar{u}\gamma_4 u)=(E/m)(\bar{u}u)$ and $m\Delta m=E\Delta E$, as $\Delta m(\bar{u}u)$ where Δm is an equivalent change in mass of the electron. In this form invariance is obvious.

One can likewise obtain an expression for the energy shift for an electron in a hydrogen atom. Simply replace K_+ in (8), by $K_+^{(V)}$, the exact kernel for an electron in the potential, $V=\beta e^2/r$, of the atom, and f by a wave function (of space and time) for an atomic state. In general the ΔE which results is not real. The imaginary part is negative and in $\exp(-i\Delta ET)$ produces an exponentially decreasing amplitude with time. This is because we are asking for the amplitude that an atom initially with no photon in the field, will still appear after time T with no photon. If the atom is in a state which can radiate, this amplitude must decay with time. The imaginary part of ΔE when calculated does indeed give the correct rate of radiation from atomic states. It is zero for the ground state and for a free electron.

In the non-relativistic region the expression for ΔE can be worked out as has been done by Bethe.[11] In the relativistic region (points 4 and 3 as close together as a Compton wave-length) the $K_+^{(V)}$ which should appear in (8) can be replaced to first order in V by K_+ plus $K_+^{(1)}(2,1)$ given in I, Eq. (13). The problem is then very similar to the radiationless scattering problem discussed below.

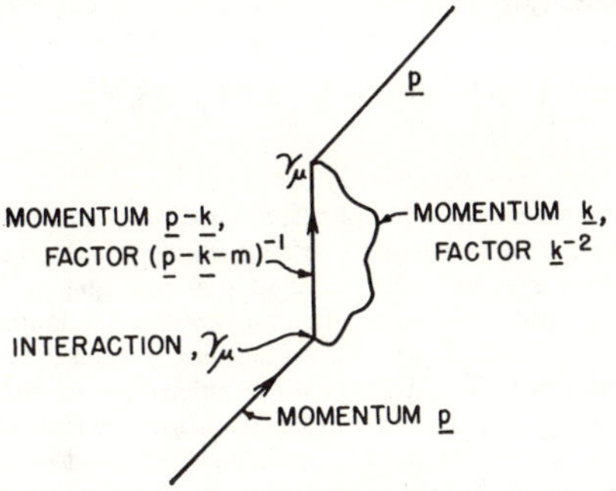

FIG. 3. Interaction of an electron with itself. Momentum space, Eq. (11).

4. EXPRESSION IN MOMENTUM AND ENERGY SPACE

The evaluation of (9), as well as all the other more complicated expressions arising in these problems, is very much simplified by working in the momentum and energy variables, rather than space and time. For this we shall need the Fourier Transform of $\delta_+(s_{21}^2)$ which is

$$-\delta_+(s_{21}^2)=\pi^{-1}\int\exp(-ik\cdot x_{21})k^{-2}d^4k, \quad (10)$$

which can be obtained from (3) and (5) or from I, Eq. (32) noting that $I_+(2,1)$ for $m^2=0$ is $\delta_+(s_{21}^2)$ from

[10] This is discussed in reference 5 in which it is pointed out that the concept of a wave function loses accuracy if there are delayed self-actions.

[11] H. A. Bethe, Phys. Rev. 72, 339 (1947).

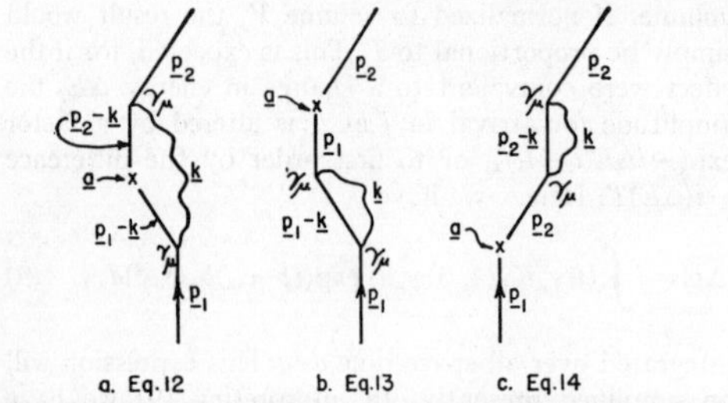

FIG. 4. Radiative correction to scattering, momentum space.

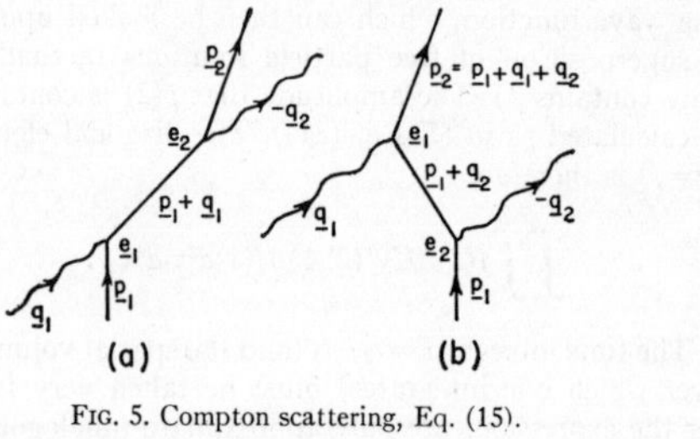

FIG. 5. Compton scattering, Eq. (15).

I, Eq. (34). The k^{-2} means $(k\cdot k)^{-1}$ or more precisely the limit as $\delta\rightarrow 0$ of $(k\cdot k+i\delta)^{-1}$. Further d^4k means $(2\pi)^{-2}dk_1dk_2dk_3dk_4$. If we imagine that quanta are particles of zero mass, then we can make the general rule that all poles are to be resolved by considering the masses of the particles and quanta to have infinitesimal negative imaginary parts.

Using these results we see that the self-energy (9) is the matrix element between $\bar{u}$ and u of the matrix

$$(e^2/\pi i)\int \gamma_\mu(p-k-m)^{-1}\gamma_\mu k^{-2}d^4k, \tag{11}$$

where we have used the expression (I, Eq. (31)) for the Fourier transform of K_+. This form for the self-energy is easier to work with than is (9).

The equation can be understood by imagining (Fig. 3) that the electron of momentum p emits (γ_μ) a quantum of momentum k, and makes its way now with momentum $p-k$ to the next event (factor $(p-k-m)^{-1}$) which is to absorb the quantum (another γ_μ). The amplitude of propagation of quanta is k^{-2}. (There is a factor $e^2/\pi i$ for each virtual quantum). One integrates over all quanta. The reason an electron of momentum p propagates as $1/(p-m)$ is that this operator is the reciprocal of the Dirac equation operator, and we are simply solving this equation. Likewise light goes as $1/k^2$, for this is the reciprocal D'Alembertian operator of the wave equation of light. The first γ_μ represents the current which generates the vector potential, while the second is the velocity operator by which this potential is multiplied in the Dirac equation when an external field acts on an electron.

Using the same line of reasoning, other problems may be set up directly in momentum space. For example, consider the scattering in a potential $A=A_\mu\gamma_\mu$ varying in space and time as $a\exp(-iq\cdot x)$. An electron initially in state of momentum $p_1=p_{1\mu}\gamma_\mu$ will be deflected to state p_2 where $p_2=p_1+q$. The zero-order answer is simply the matrix element of a between states 1 and 2. We next ask for the first order (in e^2) radiative correction due to virtual radiation of one quantum. There are several ways this can happen. First for the case illustrated in Fig. 4(a), find the matrix:

$$(e^2/\pi i)\int \gamma_\mu(p_2-k-m)^{-1}a(p_1-k-m)^{-1}\gamma_\mu k^{-2}d^4k. \tag{12}$$

For in this case, first[12] a quantum of momentum k is emitted (γ_μ), the electron then having momentum p_1-k and hence propagating with factor $(p_1-k-m)^{-1}$. Next it is scattered by the potential (matrix a) receiving additional momentum q, propagating on then (factor $(p_2-k-m)^{-1}$) with the new momentum until the quantum is reabsorbed (γ_μ). The quantum propagates from emission to absorption (k^{-2}) and we integrate over all quanta (d^4k), and sum on polarization μ. When this is integrated on k_4, the result can be shown to be exactly equal to the expressions (16) and (17) given in B for the same process, the various terms coming from residues of the poles of the integrand (12).

Or again if the quantum is both emitted and reabsorbed before the scattering takes place one finds (Fig. 4(b))

$$(e^2/\pi i)\int a(p_1-m)^{-1}\gamma_\mu(p_1-k-m)^{-1}\gamma_\mu k^{-2}d^4k, \tag{13}$$

or if both emission and absorption occur after the scattering, (Fig. 4(c))

$$(e^2/\pi i)\int \gamma_\mu(p_2-k-m)^{-1}\gamma_\mu(p_2-m)^{-1}ak^{-2}d^4k. \tag{14}$$

These terms are discussed in detail below.

We have now achieved our simplification of the form of writing matrix elements arising from virtual processes. Processes in which a number of real quanta is given initially and finally offer no problem (assuming correct normalization). For example, consider the Compton effect (Fig. 5(a)) in which an electron in state p_1 absorbs a quantum of momentum q_1, polarization vector $e_{1\mu}$ so that its interaction is $e_{1\mu}\gamma_\mu=e_1$, and emits a second quantum of momentum $-q_2$, polarization e_2 to arrive in final state of momentum p_2. The matrix for

[12] First, next, etc., here refer not to the order in true time but to the succession of events along the trajectory of the electron. That is, more precisely, to the order of appearance of the matrices in the expressions.

this process is $e_2(p_1+q_1-m)^{-1}e_1$. The total matrix for the Compton effect is, then,

$$e_2(p_1+q_1-m)^{-1}e_1+e_1(p_1+q_2-m)^{-1}e_2, \qquad (15)$$

the second term arising because the emission of e_2 may also precede the absorption of e_1 (Fig. 5(b)). One takes matrix elements of this between initial and final electron states $(p_1+q_1=p_2-q_2)$, to obtain the Klein Nishina formula. Pair annihilation with emission of two quanta, etc., are given by the same matrix, positron states being those with negative time component of p. Whether quanta are absorbed or emitted depends on whether the time component of q is positive or negative.

5. THE CONVERGENCE OF PROCESSES WITH VIRTUAL QUANTA

These expressions are, as has been indicated, no more than a re-expression of conventional quantum electrodynamics. As a consequence, many of them are meaningless. For example, the self-energy expression (9) or (11) gives an infinite result when evaluated. The infinity arises, apparently, from the coincidence of the δ-function singularities in $K_+(4, 3)$ and $\delta_+(s_{43}^2)$. Only at this point is it necessary to make a real departure from conventional electrodynamics, a departure other than simply rewriting expressions in a simpler form.

We desire to make a modification of quantum electrodynamics analogous to the modification of classical electrodynamics described in a previous article, A. There the $\delta(s_{12}^2)$ appearing in the action of interaction was replaced by $f(s_{12}^2)$ where $f(x)$ is a function of small width and great height.

The obvious corresponding modification in the quantum theory is to replace the $\delta_+(s^2)$ appearing the quantum mechanical interaction by a new function $f_+(s^2)$. We can postulate that if the Fourier transform of the classical $f(s_{12}^2)$ is the integral over all k of $F(k^2)\exp(-ik\cdot x_{12})d^4k$, then the Fourier transform of $f_+(s^2)$ is the same integral taken over only positive frequencies k_4 for $t_2>t_1$ and over only negative ones for $t_2<t_1$ in analogy to the relation of $\delta_+(s^2)$ to $\delta(s^2)$. The function $f(s^2)=f(x\cdot x)$ can be written* as

$$f(x\cdot x)=(2\pi)^{-2}\int_{k_4=0}^{\infty}\int \sin(k_4|x_4|)\times\cos(\mathbf{K}\cdot\mathbf{x})dk_4d^3\mathbf{K}g(k\cdot k),$$

where $g(k\cdot k)$ is k_4^{-1} times the density of oscillators and may be expressed for positive k_4 as (A, Eq. (16))

$$g(k^2)=\int_0^\infty (\delta(k^2)-\delta(k^2-\lambda^2))G(\lambda)d\lambda,$$

where $\int_0^\infty G(\lambda)d\lambda=1$ and G involves values of λ large compared to m. This simply means that the amplitude for propagation of quanta of momentum k is

$$-F_+(k^2)=\pi^{-1}\int_0^\infty (k^{-2}-(k^2-\lambda^2)^{-1})G(\lambda)d\lambda,$$

rather than k^{-2}. That is, writing $F_+(k^2)=-\pi^{-1}k^{-2}C(k^2)$,

$$-f_+(s_{12}^2)=\pi^{-1}\int \exp(-ik\cdot x_{12})k^{-2}C(k^2)d^4k. \qquad (16)$$

Every integral over an intermediate quantum which previously involved a factor d^4k/k^2 is now supplied with a convergence factor $C(k^2)$ where

$$C(k^2)=\int_0^\infty -\lambda^2(k^2-\lambda^2)^{-1}G(\lambda)d\lambda. \qquad (17)$$

The poles are defined by replacing k^2 by $k^2+i\delta$ in the limit $\delta\rightarrow 0$. That is λ^2 may be assumed to have an infinitesimal negative imaginary part.

The function $f_+(s_{12}^2)$ may still have a discontinuity in value on the light cone. This is of no influence for the Dirac electron. For a particle satisfying the Klein Gordon equation, however, the interaction involves gradients of the potential which reinstates the δ function if f has discontinuities. The condition that f is to have no discontinuity in value on the light cone implies $k^2C(k^2)$ approaches zero as k^2 approaches infinity. In terms of $G(\lambda)$ the condition is

$$\int_0^\infty \lambda^2G(\lambda)d\lambda=0. \qquad (18)$$

This condition will also be used in discussing the convergence of vacuum polarization integrals.

The expression for the self-energy matrix is now

$$(e^2/\pi i)\int \gamma_\mu(p-k-m)^{-1}\gamma_\mu k^{-2}d^4kC(k^2), \qquad (19)$$

which, since $C(k^2)$ falls off at least as rapidly as $1/k^2$, converges. For practical purposes we shall suppose hereafter that $C(k^2)$ is simply $-\lambda^2/(k^2-\lambda^2)$ implying that some average (with weight $G(\lambda)d\lambda$) over values of λ may be taken afterwards. Since in all processes the quantum momentum will be contained in at least one extra factor of the form $(p-k-m)^{-1}$ representing propagation of an electron while that quantum is in the field, we can expect all such integrals with their convergence factors to converge and that the result of all such processes will now be finite and definite (excepting the processes with closed loops, discussed below, in which the diverging integrals are over the momenta of the electrons rather than the quanta).

The integral of (19) with $C(k^2)=-\lambda^2(k^2-\lambda^2)^{-1}$ noting that $p^2=m^2$, $\lambda\gg m$ and dropping terms of order m/λ, is (see Appendix A)

$$(e^2/2\pi)[4m(\ln(\lambda/m)+\tfrac{1}{2})-p(\ln(\lambda/m)+5/4)]. \qquad (20)$$

* This relation is given incorrectly in A, equation just preceding 16.

When applied to a state of an electron of momentum p satisfying $pu=mu$, it gives for the change in mass (as in B, Eq. (9))

$$\Delta m = m(e^2/2\pi)(3\ln(\lambda/m)+\tfrac{3}{4}). \tag{21}$$

6. RADIATIVE CORRECTIONS TO SCATTERING

We can now complete the discussion of the radiative corrections to scattering. In the integrals we include the convergence factor $C(k^2)$, so that they converge for large k. Integral (12) is also not convergent because of the well-known infra-red catastrophy. For this reason we calculate (as discussed in B) the value of the integral assuming the photons to have a small mass $\lambda_{\min}\ll m\ll\lambda$. The integral (12) becomes

$$(e^2/\pi i)\int \gamma_\mu(p_2-k-m)^{-1}a(p_1-k-m)^{-1} \times\gamma_\mu(k^2-\lambda_{\min}^2)^{-1}d^4kC(k^2-\lambda_{\min}^2),$$

which when integrated (see Appendix B) gives $(e^2/2\pi)$ times

$$\left[2\left(\ln\frac{m}{\lambda_{\min}}-1\right)\left(1-\frac{2\theta}{\tan 2\theta}\right)+\theta\tan\theta + \frac{4}{\tan 2\theta}\int_0^\theta \alpha\tan\alpha d\alpha\right]a + \frac{1}{4m}(qa-aq)\frac{2\theta}{\sin 2\theta}+ra, \tag{22}$$

where $(q^2)^{\frac{1}{2}}=2m\sin\theta$ and we have assumed the matrix to operate between states of momentum p_1 and $p_2=p_1+q$ and have neglected terms of order $\lambda_{\min}/m$, m/λ, and q^2/λ^2. Here the only dependence on the convergence factor is in the term ra, where

$$r=\ln(\lambda/m)+9/4-2\ln(m/\lambda_{\min}). \tag{23}$$

As we shall see in a moment, the other terms (13), (14) give contributions which just cancel the ra term. The remaining terms give for small q,

$$(e^2/4\pi)\left(\frac{1}{2m}(qa-aq)+\frac{4q^2}{3m^2}a\left(\ln\frac{m}{\lambda_{\min}}-\frac{3}{8}\right)\right), \tag{24}$$

which shows the change in magnetic moment and the Lamb shift as interpreted in more detail in B.[13]

We must now study the remaining terms (13) and (14). The integral on k in (13) can be performed (after multiplication by $C(k^2)$) since it involves nothing but the integral (19) for the self-energy and the result is allowed to operate on the initial state u_1, (so that $p_1u_1=mu_1$). Hence the factor following $a(p_1-m)^{-1}$ will be just Δm. But, if one now tries to expand $1/(p_1-m) = (p_1+m)/(p_1^2-m^2)$ one obtains an infinite result, since $p_1^2=m^2$. This is, however, just what is expected physically. For the quantum can be emitted and absorbed at any time previous to the scattering. Such a process has the effect of a change in mass of the electron in the state 1. It therefore changes the energy by ΔE and the amplitude to first order in ΔE by $-i\Delta E\cdot t$ where t is the time it is acting, which is infinite. That is, the major effect of this term would be canceled by the effect of change of mass Δm.

The situation can be analyzed in the following manner. We suppose that the electron approaching the scattering potential a has not been free for an infinite time, but at some time far past suffered a scattering by a potential b. If we limit our discussion to the effects of Δm and of the virtual radiation of one quantum between two such scatterings each of the effects will be finite, though large, and their difference is determinate. The propagation from b to a is represented by a matrix

$$a(p'-m)^{-1}b, \tag{25}$$

in which one is to integrate possibly over p' (depending on details of the situation). (If the time is long between b and a, the energy is very nearly determined so that p'^2 is very nearly m^2.)

We shall compare the effect on the matrix (25) of the virtual quanta and of the change of mass Δm. The effect of a virtual quantum is

$$(e^2/\pi i)\int a(p'-m)^{-1}\gamma_\mu(p'-k-m)^{-1} \times\gamma_\mu(p'-m)^{-1}bk^{-2}d^4kC(k^2), \tag{26}$$

while that of a change of mass can be written

$$a(p'-m)^{-1}\Delta m(p'-m)^{-1}b, \tag{27}$$

and we are interested in the difference (26)-(27). A simple and direct method of making this comparison is just to evaluate the integral on k in (26) and subtract from the result the expression (27) where Δm is given in (21). The remainder can be expressed as a multiple $-r(p'^2)$ of the unperturbed amplitude (25);

$$-r(p'^2)a(p'-m)^{-1}b. \tag{28}$$

This has the same result (to this order) as replacing the potentials a and b in (25) by $(1-\frac{1}{2}r(p'^2))a$ and

---

[13] That the result given in B in Eq. (19) was in error was repeatedly pointed out to the author, in private communication, by V. F. Weisskopf and J. B. French, as their calculation, completed simultaneously with the author's early in 1948, gave a different result. French has finally shown that although the expression for the radiationless scattering B, Eq. (18) or (24) above is correct, it was incorrectly joined onto Bethe's non-relativistic result. He shows that the relation $\ln 2k_{\max}-1=\ln\lambda_{\min}$ used by the author should have been $\ln 2k_{\max}-5/6=\ln\lambda_{\min}$. This results in adding a term $-(1/6)$ to the logarithm in B, Eq. (19) so that the result now agrees with that of J. B. French and V. F. Weisskopf, Phys. Rev. **75**, 1240 (1949) and N. H. Kroll and W. E. Lamb, Phys. Rev. **75**, 388 (1949). The author feels unhappily responsible for the very considerable delay in the publication of French's result occasioned by this error. This footnote is appropriately numbered.

$(1-\frac{1}{2}r(p'^2))b$. In the limit, then, as $p'^2 \to m^2$ the net effect on the scattering is $-\frac{1}{2}ra$ where r, the limit of $r(p'^2)$ as $p'^2 \to m^2$ (assuming the integrals have an infra-red cut-off), turns out to be just equal to that given in (23). An equal term $-\frac{1}{2}ra$ arises from virtual transitions after the scattering (14) so that the entire ra term in (22) is canceled.

The reason that r is just the value of (12) when $q^2=0$ can also be seen without a direct calculation as follows: Let us call p the vector of length m in the direction of p' so that if $p'^2=m(1+\epsilon)^2$ we have $p'=(1+\epsilon)p$ and we take ϵ as very small, being of order T^{-1} where T is the time between the scatterings b and a. Since $(p'-m)^{-1} = (p'+m)/(p'^2-m^2) \approx (p+m)/2m^2\epsilon$, the quantity (25) is of order ϵ^{-1} or T. We shall compute corrections to it only to its own order (ϵ^{-1}) in the limit $\epsilon \to 0$. The term (27) can be written approximately[14] as

$$(e^2/\pi i)\int a(p'-m)^{-1}\gamma_\mu(p-k-m)^{-1} \times\gamma_\mu(p'-m)^{-1}bk^{-2}d^4kC(k^2),$$

using the expression (19) for Δm. The net of the two effects is therefore approximately[15]

$$-(e^2/\pi i)\int a(p'-m)^{-1}\gamma_\mu(p-k-m)^{-1}\epsilon p(p-k-m)^{-1} \times\gamma_\mu(p'-m)^{-1}bk^{-2}d^4kC(k^2),$$

a term now of order $1/\epsilon$ (since $(p'-m)^{-1} \approx (p+m) \times(2m^2\epsilon)^{-1}$) and therefore the one desired in the limit. Comparison to (28) gives for r the expression

$$(p_1+m/2m)\int \gamma_\mu(p_1-k-m)^{-1}(p_1m^{-1})(p_1-k-m)^{-1} \times\gamma_\mu k^{-2}d^4kC(k^2). \quad (29)$$

The integral can be immediately evaluated, since it is the same as the integral (12), but with $q=0$, for a replaced by p_1/m. The result is therefore $r\cdot(p_1/m)$ which when acting on the state u_1 is just r, as $p_1u_1=mu_1$. For the same reason the term $(p_1+m)/2m$ in (29) is effectively 1 and we are left with $-r$ of (23).[16]

In more complex problems starting with a free electron the same type of term arises from the effects of a virtual emission and absorption both previous to the other processes. They, therefore, simply lead to the same factor r so that the expression (23) may be used directly and these renormalization integrals need not be computed afresh for each problem.

In this problem of the radiative corrections to scattering the net result is insensitive to the cut-off. This means, of course, that by a simple rearrangement of terms previous to the integration we could have avoided the use of the convergence factors completely (see for example Lewis[17]). The problem was solved in the manner here in order to illustrate how the use of such convergence factors, even when they are actually unnecessary, may facilitate analysis somewhat by removing the effort and ambiguities that may be involved in trying to rearrange the otherwise divergent terms.

The replacement of δ_+ by f_+ given in (16), (17) is not determined by the analogy with the classical problem. In the classical limit only the real part of δ_+ (i.e., just δ) is easy to interpret. But by what should the imaginary part, $1/(\pi i s^2)$, of δ_+ be replaced? The choice we have made here (in defining, as we have, the location of the poles of (17)) is arbitrary and almost certainly incorrect. If the radiation resistance is calculated for an atom, as the imaginary part of (8), the result depends slightly on the function f_+. On the other hand the light radiated at very large distances from a source is independent of f_+. The total energy absorbed by distant absorbers will not check with the energy loss of the source. We are in a situation analogous to that in the classical theory if the entire f function is made to contain only retarded contributions (see A, Appendix). One desires instead the analogue of $\langle F\rangle_{\text{ret}}$ of A. This problem is being studied.

One can say therefore, that this attempt to find a consistent modification of quantum electrodynamics is incomplete (see also the question of closed loops, below). For it could turn out that any correct form of f_+ which will guarantee energy conservation may at the same time not be able to make the self-energy integral finite. The desire to make the methods of simplifying the calculation of quantum electrodynamic processes more widely available has prompted this publication before an analysis of the correct form for f_+ is complete. One might try to take the position that, since the energy discrepancies discussed vanish in the limit $\lambda \to \infty$, the correct physics might be considered to be that obtained by letting $\lambda \to \infty$ after mass renormalization. I have no proof of the mathematical consistency of this procedure, but the presumption is very strong that it is satisfactory. (It is also strong that a satisfactory form for f_+ can be found.)

7. THE PROBLEM OF VACUUM POLARIZATION

In the analysis of the radiative corrections to scattering one type of term was not considered. The potential

---

[14] The expression is not exact because the substitution of Δm by the integral in (19) is valid only if p operates on a state such that p can be replaced by m. The error, however, is of order $a(p'-m)^{-1}(p-m)(p'-m)^{-1}b$ which is $a((1+\epsilon)p+m)(p-m) \times((1+\epsilon)p+m)p(2\epsilon+\epsilon^2)^{-2}m^{-4}$. But since $p^2=m^2$, we have $p(p-m) = -m(p-m)=(p-m)p$ so the net result is approximately $a(p-m)b/4m^2$ and is not of order $1/\epsilon$ but smaller, so that its effect drops out in the limit.

[15] We have used, to first order, the general expansion (valid for any operators A, B)

$$(A+B)^{-1}=A^{-1}-A^{-1}BA^{-1}+A^{-1}BA^{-1}BA^{-1}-\cdots$$

with $A=p-k-m$ and $B=p'-p=\epsilon p$ to expand the difference of $(p'-k-m)^{-1}$ and $(p-k-m)^{-1}$.

[16] The renormalization terms appearing B, Eqs. (14), (15) when translated directly into the present notation do not give twice (29) but give this expression with the central p_1m^{-1} factor replaced by $m\gamma_4/E_1$ where $E_1=p_{1\mu}$ for $\mu=4$. When integrated it therefore gives $ra((p_1+m)/2m)(m\gamma_4/E_1)$ or $ra-ra(m\gamma_4/E_1)(p_1-m)/2m$. (Since $p_1\gamma_4+\gamma_4p_1=2E_1$) which gives just ra, since $p_1u_1=mu_1$.

[17] H. W. Lewis, Phys. Rev. 73, 173 (1948).

which we can assume to vary as $a_\mu \exp(-iq\cdot x)$ creates a pair of electrons (see Fig. 6), momenta p_a, $-p_b$. This pair then reannihilates, emitting a quantum $q=p_b-p_a$, which quantum scatters the original electron from state 1 to state 2. The matrix element for this process (and the others which can be obtained by rearranging the order in time of the various events) is

$$-(e^2/\pi i)(\bar{u}_2\gamma_\mu u_1)\int Sp[(p_a+q-m)^{-1} \times\gamma_\nu(p_a-m)^{-1}\gamma_\mu]d^4p_a q^{-2}C(q^2)a_\nu. \quad (30)$$

This is because the potential produces the pair with amplitude proportional to $a_\nu\gamma_\nu$, the electrons of momenta p_a and $-(p_a+q)$ proceed from there to annihilate, producing a quantum (factor γ_μ) which propagates (factor $q^{-2}C(q^2)$) over to the other electron, by which it is absorbed (matrix element of γ_μ between states 1 and 2 of the original electron $(\bar{u}_2\gamma_\mu u_1)$). All momenta p_a and spin states of the virtual electron are admitted, which means the spur and the integral on d^4p_a are calculated.

One can imagine that the closed loop path of the positron-electron produces a current

$$4\pi j_\mu = J_{\mu\nu}a_\nu, \quad (31)$$

which is the source of the quanta which act on the second electron. The quantity

$$J_{\mu\nu}=-(e^2/\pi i)\int Sp[(p+q-m)^{-1} \times\gamma_\nu(p-m)^{-1}\gamma_\mu]d^4p, \quad (32)$$

is then characteristic for this problem of polarization of the vacuum.

One sees at once that $J_{\mu\nu}$ diverges badly. The modification of δ to f alters the amplitude with which the current j_μ will affect the scattered electron, but it can do nothing to prevent the divergence of the integral (32) and of its effects.

One way to avoid such difficulties is apparent. From one point of view we are considering all routes by which a given electron can get from one region of space-time to another, i.e., from the source of electrons to the apparatus which measures them. From this point of view the closed loop path leading to (32) is unnatural. It might be assumed that the only paths of meaning are those which start from the source and work their way in a continuous path (possibly containing many time reversals) to the detector. Closed loops would be excluded. We have already found that this may be done for electrons moving in a fixed potential.

Such a suggestion must meet several questions, however. The closed loops are a consequence of the usual hole theory in electrodynamics. Among other things, they are required to keep probability conserved. The probability that no pair is produced by a potential is

FIG. 6. Vacuum polarization effect on scattering, Eq. (30).

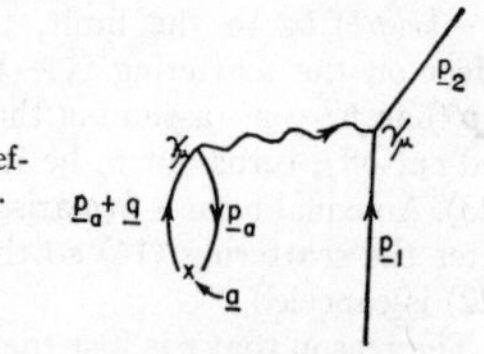

not unity and its deviation from unity arises from the imaginary part of $J_{\mu\nu}$. Again, with closed loops excluded, a pair of electrons once created cannot annihilate one another again, the scattering of light by light would be zero, etc. Although we are not experimentally sure of these phenomena, this does seem to indicate that the closed loops are necessary. To be sure, it is always possible that these matters of probability conservation, etc., will work themselves out as simply in the case of interacting particles as for those in a fixed potential. Lacking such a demonstration the presumption is that the difficulties of vacuum polarization are not so easily circumvented.[18]

An alternative procedure discussed in B is to assume that the function $K_+(2,1)$ used above is incorrect and is to be replaced by a modified function K_+' having no singularity on the light cone. The effect of this is to provide a convergence factor $C(p^2-m^2)$ for *every* integral over electron momenta.[19] This will multiply the integrand of (32) by $C(p^2-m^2)C((p+q)^2-m^2)$, since the integral was originally $\delta(p_a-p_b+q)d^4p_a d^4p_b$ and both p_a and p_b get convergence factors. The integral now converges but the result is unsatisfactory.[20]

One expects the current (31) to be conserved, that is $q_\mu j_\mu=0$ or $q_\mu J_{\mu\nu}=0$. Also one expects no current if a_ν is a gradient, or $a_\nu=q_\nu$ times a constant. This leads to the condition $J_{\mu\nu}q_\nu=0$ which is equivalent to $q_\mu J_{\mu\nu}=0$ since $J_{\mu\nu}$ is symmetrical. But when the expression (32) is integrated with such convergence factors it does not satisfy this condition. By altering the kernel from K to another, K', which does not satisfy the Dirac equation we have lost the gauge invariance, its consequent current conservation and the general consistency of the theory.

One can see this best by calculating $J_{\mu\nu}q_\nu$ directly from (32). The expression within the spur becomes $(p+q-m)^{-1}q(p-m)^{-1}\gamma_\mu$ which can be written as the difference of two terms: $(p-m)^{-1}\gamma_\mu-(p+q-m)^{-1}\gamma_\mu$. Each of these terms would give the same result if the integration d^4p were without a convergence factor, for

[18] It would be very interesting to calculate the Lamb shift accurately enough to be sure that the 20 megacycles expected from vacuum polarization are actually present.

[19] This technique also makes self-energy and radiationless scattering integrals finite even without the modification of δ_+ to f_+ for the radiation (and the consequent convergence factor $C(k^2)$ for the quanta). See B.

[20] Added to the terms given below (33) there is a term $\frac{1}{4}(\lambda^2-2\mu^2+\frac{1}{3}q^2)\delta_{\mu\nu}$ for $C(k^2)=-\lambda^2(k^2-\lambda^2)^{-1}$, which is not gauge invariant. (In addition the charge renormalization has $-7/6$ added to the logarithm.)

the first can be converted into the second by a shift of the origin of p, namely $p'=p+q$. This does not result in cancelation in (32) however, for the convergence factor is altered by the substitution.

A method of making (32) convergent without spoiling the gauge invariance has been found by Bethe and by Pauli. The convergence factor for light can be looked upon as the result of superposition of the effects of quanta of various masses (some contributing negatively). Likewise if we take the factor $C(p^2-m^2) = -\lambda^2(p^2-m^2-\lambda^2)^{-1}$ so that $(p^2-m^2)^{-1}C(p^2-m^2) = (p^2-m^2)^{-1}-(p^2-m^2-\lambda^2)^{-1}$ we are taking the difference of the result for electrons of mass m and mass $(\lambda^2+m^2)^{\frac{1}{2}}$. But we have taken this difference for *each* propagation between interactions with photons. They suggest instead that once created with a certain mass the electron should continue to propagate with this mass through all the potential interactions until it closes its loop. That is if the quantity (32), integrated over some finite range of p, is called $J_{\mu\nu}(m^2)$ and the corresponding quantity over the same range of p, but with m replaced by $(m^2+\lambda^2)^{\frac{1}{2}}$ is $J_{\mu\nu}(m^2+\lambda^2)$ we should calculate

$$J_{\mu\nu}{}^P=\int_0^\infty [J_{\mu\nu}(m^2)-J_{\mu\nu}(m^2+\lambda^2)]G(\lambda)d\lambda, \quad (32')$$

the function $G(\lambda)$ satisfying $\int_0^\infty G(\lambda)d\lambda=1$ and $\int_0^\infty G(\lambda)\lambda^2 d\lambda=0$. Then in the expression for $J_{\mu\nu}{}^P$ the range of p integration can be extended to infinity as the integral now converges. The result of the integration using this method is the integral on $d\lambda$ over $G(\lambda)$ of (see Appendix C)

$$J_{\mu\nu}{}^P=-\frac{e^2}{\pi}(q_\mu q_\nu-\delta_{\mu\nu}q^2)\left(-\frac{1}{3}\ln\frac{\lambda^2}{m^2} -\left[\frac{4m^2+2q^2}{3q^2}\left(1-\frac{\theta}{\tan\theta}\right)-\frac{1}{9}\right]\right), \quad (33)$$

with $q^2=4m^2\sin^2\theta$.

The gauge invariance is clear, since $q_\mu(q_\mu q_\nu-q^2\delta_{\mu\nu})=0$. Operating (as it always will) on a potential of zero divergence the $(q_\mu q_\nu-\delta_{\mu\nu}q^2)a_\nu$ is simply $-q^2a_\mu$, the D'Alembertian of the potential, that is, the current producing the potential. The term $-\frac{1}{3}(\ln(\lambda^2/m^2))(q_\mu q_\nu-q^2\delta_{\mu\nu})$ therefore gives a current proportional to the current producing the potential. This would have the same effect as a change in charge, so that we would have a difference $\Delta(e^2)$ between e^2 and the experimentally observed charge, $e^2+\Delta(e^2)$, analogous to the difference between m and the observed mass. This charge depends logarithmically on the cut-off, $\Delta(e^2)/e^2 = -(2e^2/3\pi)\ln(\lambda/m)$. After this renormalization of charge is made, no effects will be sensitive to the cut-off.

After this is done the final term remaining in (33), contains the usual effects[21] of polarization of the vacuum. It is zero for a free light quantum ($q^2=0$). For small q^2 it behaves as $(2/15)q^2$ (adding $-\frac{1}{5}$ to the logarithm in the Lamb effect). For $q^2>(2m)^2$ it is complex, the imaginary part representing the loss in amplitude required by the fact that the probability that no quanta are produced by a potential able to produce pairs ($(q^2)^{\frac{1}{2}}>2m$) decreases with time. (To make the necessary analytic continuation, imagine m to have a small negative imaginary part, so that $(1-q^2/4m^2)^{\frac{1}{2}}$ becomes $-i(q^2/4m^2-1)^{\frac{1}{2}}$ as q^2 goes from below to above $4m^2$. Then $\theta=\pi/2+iu$ where $\sinh u=+(q^2/4m^2-1)^{\frac{1}{2}}$, and $-1/\tan\theta=i\tanh u=+i(q^2-4m^2)^{\frac{1}{2}}(q^2)^{-\frac{1}{2}}$.)

Closed loops containing a number of quanta or potential interactions larger than two produce no trouble. Any loop with an odd number of interactions gives zero (I, reference 9). Four or more potential interactions give integrals which are convergent even without a convergence factor as is well known. The situation is analogous to that for self-energy. Once the simple problem of a single closed loop is solved there are no further divergence difficulties for more complex processes.[22]

8. LONGITUDINAL WAVES

In the usual form of quantum electrodynamics the longitudinal and transverse waves are given separate treatment. Alternately the condition $(\partial A_\mu/\partial x_\mu)\Psi=0$ is carried along as a supplementary condition. In the present form no such special considerations are necessary for we are dealing with the solutions of the equation $-\Box^2 A_\mu=4\pi j_\mu$ with a current j_μ which is conserved $\partial j_\mu/\partial x_\mu=0$. That means at least $\Box^2(\partial A_\mu/\partial x_\mu)=0$ and in fact our solution also satisfies $\partial A_\mu/\partial x_\mu=0$.

To show that this is the case we consider the amplitude for emission (real or virtual) of a photon and show that the divergence of this amplitude vanishes. The amplitude for emission for photons polarized in the μ direction involves matrix elements of γ_μ. Therefore what we have to show is that the corresponding matrix elements of $q_\mu\gamma_\mu=\mathbf{q}$ vanish. For example, for a first order effect we would require the matrix element of $\mathbf{q}$ between two states p_1 and $p_2=p_1+q$. But since $\mathbf{q}=\mathbf{p}_2-\mathbf{p}_1$ and $(\bar{u}_2\mathbf{p}_1u_1)=m(\bar{u}_2u_1)=(\bar{u}_2\mathbf{p}_2u_1)$ the matrix element vanishes, which proves the contention in this case. It also vanishes in more complex situations (essentially because of relation (34), below) (for example, try putting $\mathbf{e}_2=\mathbf{q}_2$ in the matrix (15) for the Compton Effect).

To prove this in general, suppose $\mathbf{a}_i$, $i=1$ to N are a set of plane wave disturbing potentials carrying momenta q_i (e.g., some may be emissions or absorptions of the same or different quanta) and consider a matrix for the transition from a state of momentum p_0 to p_N such

[21] E. A. Uehling, Phys. Rev. 48, 55 (1935), R. Serber, Phys. Rev. 48, 49 (1935).

[22] There are loops completely without external interactions. For example, a pair is created virtually along with a photon. Next they annihilate, absorbing this photon. Such loops are disregarded on the grounds that they do not interact with anything and are thereby completely unobservable. Any indirect effects they may have via the exclusion principle have already been included.

as $\boldsymbol{a}_N \prod_{i=1}^{N-1} (\boldsymbol{p}_i - m)^{-1}\boldsymbol{a}_i$ where $\boldsymbol{p}_i = \boldsymbol{p}_{i-1} + \boldsymbol{q}_i$ (and in the product, terms with larger i are written to the left). The most general matrix element is simply a linear combination of these. Next consider the matrix between states $\boldsymbol{p}_0$ and $\boldsymbol{p}_N + \boldsymbol{q}$ in a situation in which not only are the $\boldsymbol{a}_i$ acting but also another potential $\boldsymbol{a}\exp(-iq\cdot x)$ where $\boldsymbol{a} = \boldsymbol{q}$. This may act previous to all $\boldsymbol{a}_i$, in which case it gives $\boldsymbol{a}_N \prod (\boldsymbol{p}_i + \boldsymbol{q} - m)^{-1}\boldsymbol{a}_i(\boldsymbol{p}_0 + \boldsymbol{q} - m)^{-1}\boldsymbol{q}$ which is equivalent to $+\boldsymbol{a}_N \prod (\boldsymbol{p}_i + \boldsymbol{q} - m)^{-1}\boldsymbol{a}_i$ since $+(\boldsymbol{p}_0 + \boldsymbol{q} - m)^{-1}\boldsymbol{q}$ is equivalent to $(\boldsymbol{p}_0 + \boldsymbol{q} - m)^{-1} \times (\boldsymbol{p}_0 + \boldsymbol{q} - m)$ as $\boldsymbol{p}_0$ is equivalent to m acting on the initial state. Likewise if it acts after all the potentials it gives $\boldsymbol{q}(\boldsymbol{p}_N - m)^{-1}\boldsymbol{a}_N \prod (\boldsymbol{p}_i - m)^{-1}\boldsymbol{a}_i$ which is equivalent to $-\boldsymbol{a}_N \prod (\boldsymbol{p}_i - m)^{-1}\boldsymbol{a}_i$ since $\boldsymbol{p}_N + \boldsymbol{q} - m$ gives zero on the final state. Or again it may act between the potential $\boldsymbol{a}_k$ and $\boldsymbol{a}_{k+1}$ for each k. This gives

$$\sum_{k=1}^{N-1} \boldsymbol{a}_N \prod_{i=k+1}^{N-1} (\boldsymbol{p}_i + \boldsymbol{q} - m)^{-1}\boldsymbol{a}_i(\boldsymbol{p}_k + \boldsymbol{q} - m)^{-1} \times \boldsymbol{q}(\boldsymbol{p}_k - m)^{-1}\boldsymbol{a}_k \prod_{j=1}^{k-1} (\boldsymbol{p}_j - m)^{-1}\boldsymbol{a}_j.$$

However,

$$(\boldsymbol{p}_k + \boldsymbol{q} - m)^{-1}\boldsymbol{q}(\boldsymbol{p}_k - m)^{-1} = (\boldsymbol{p}_k - m)^{-1} - (\boldsymbol{p}_k + \boldsymbol{q} - m)^{-1}, \quad (34)$$

so that the sum breaks into the difference of two sums, the first of which may be converted to the other by the replacement of k by $k-1$. There remain only the terms from the ends of the range of summation,

$$+\boldsymbol{a}_N \prod_{i=1}^{N-1} (\boldsymbol{p}_i - m)^{-1}\boldsymbol{a}_i - \boldsymbol{a}_N \prod_{i=1}^{N-1} (\boldsymbol{p}_i + \boldsymbol{q} - m)^{-1}\boldsymbol{a}_i.$$

These cancel the two terms originally discussed so that the entire effect is zero. Hence any wave emitted will satisfy $\partial A_\mu/\partial x_\mu = 0$. Likewise longitudinal waves (that is, waves for which $A_\mu = \partial\phi/\partial x_\mu$ or $\boldsymbol{a} = \boldsymbol{q}$) cannot be absorbed and will have no effect, for the matrix elements for emission and absorption are similar. (We have said little more than that a potential $A_\mu = \partial\varphi/\partial x_\mu$ has no effect on a Dirac electron since a transformation $\psi' = \exp(-i\phi)\psi$ removes it. It is also easy to see in coordinate representation using integrations by parts.)

This has a useful practical consequence in that in computing probabilities for transition for unpolarized light one can sum the squared matrix over all four directions rather than just the two special polarization vectors. Thus suppose the matrix element for some process for light polarized in direction e_μ is $e_\mu M_\mu$. If the light has wave vector q_μ we know from the argument above that $q_\mu M_\mu = 0$. For unpolarized light progressing in the z direction we would ordinarily calculate $M_x^2 + M_y^2$. But we can as well sum $M_x^2 + M_y^2 + M_z^2 - M_t^2$ for $q_\mu M_\mu$ implies $M_t = M_z$ since $q_t = q_z$ for free quanta. This shows that unpolarized light is a relativistically invariant concept, and permits some simplification in computing cross sections for such light.

Incidentally, the virtual quanta interact through terms like $\gamma_\mu \cdots \gamma_\mu \boldsymbol{k}^{-2} d^4k$. Real processes correspond to poles in the formulae for virtual processes. The pole occurs when $\boldsymbol{k}^2 = 0$, but it looks at first as though in the sum on all four values of μ, of $\gamma_\mu \cdots \gamma_\mu$ we would have four kinds of polarization instead of two. Now it is clear that only two perpendicular to $\boldsymbol{k}$ are effective.

The usual elimination of longitudinal and scalar virtual photons (leading to an instantaneous Coulomb potential) can of course be performed here too (although it is not particularly useful). A typical term in a virtual transition is $\gamma_\mu \cdots \gamma_\mu \boldsymbol{k}^{-2} d^4k$ where the $\cdots$ represent some intervening matrices. Let us choose for the values of μ, the time t, the direction of vector part $\mathbf{K}$, of $\boldsymbol{k}$, and two perpendicular directions 1, 2. We shall not change the expression for these two 1, 2 for these are represented by transverse quanta. But we must find $(\gamma_t \cdots \gamma_t) - (\gamma_{\mathbf{K}} \cdots \gamma_{\mathbf{K}})$. Now $\boldsymbol{k} = k_4\gamma_t - K\gamma_{\mathbf{K}}$, where $K = (\mathbf{K}\cdot\mathbf{K})^{\frac{1}{2}}$, and we have shown above that $\boldsymbol{k}$ replacing the γ_μ gives zero.[23] Hence $K\gamma_{\mathbf{K}}$ is equivalent to $k_4\gamma_t$ and

$$(\gamma_t \cdots \gamma_t) - (\gamma_{\mathbf{K}} \cdots \gamma_{\mathbf{K}}) = ((K^2 - k_4^2)/K^2)(\gamma_t \cdots \gamma_t),$$

so that on multiplying by $\boldsymbol{k}^{-2}d^4k = d^4k(k_4^2 - K^2)^{-1}$ the net effect is $-(\gamma_t \cdots \gamma_t)d^4k/K^2$. The γ_t means just scalar waves, that is, potentials produced by charge density. The fact that $1/K^2$ does not contain k_4 means that k_4 can be integrated first, resulting in an instantaneous interaction, and the $d^3\mathbf{K}/K^2$ is just the momentum representation of the Coulomb potential, $1/r$.

9. KLEIN GORDON EQUATION

The methods may be readily extended to particles of spin zero satisfying the Klein Gordon equation,[24]

$$\Box^2\psi - m^2\psi = i\partial(A_\mu\psi)/\partial x_\mu + iA_\mu\partial\psi/\partial x_\mu - A_\mu A_\mu\psi. \quad (35)$$

[23] A little more care is required when both γ_μ's act on the same particle. Define $\boldsymbol{x} = k_4\gamma_t + K\gamma_{\mathbf{K}}$, and consider $(\boldsymbol{k}\cdots\boldsymbol{x}) + (\boldsymbol{x}\cdots\boldsymbol{k})$. Exactly this term would arise if a system, acted on by potential $\boldsymbol{x}$ carrying momentum $-\boldsymbol{k}$, is disturbed by an added potential $\boldsymbol{k}$ of momentum $+\boldsymbol{k}$ (the reversed sign of the momenta in the intermediate factors in the second term $\boldsymbol{x}\cdots\boldsymbol{k}$ has no effect since we will later integrate over all $\boldsymbol{k}$). Hence as shown above the result is zero, but since $(\boldsymbol{k}\cdots\boldsymbol{x}) + (\boldsymbol{x}\cdots\boldsymbol{k}) = k_4^2(\gamma_t\cdots\gamma_t) - K^2(\gamma_{\mathbf{K}}\cdots\gamma_{\mathbf{K}})$ we can still conclude $(\gamma_{\mathbf{K}}\cdots\gamma_{\mathbf{K}}) = k_4^2K^{-2}(\gamma_t\cdots\gamma_t)$.

[24] The equations discussed in this section were deduced from the formulation of the Klein Gordon equation given in reference 5, Section 14. The function ψ in this section has only one component and is not a spinor. An alternative formal method of making the equations valid for spin zero and also for spin 1 is (presumably) by use of the Kemmer-Duffin matrices β_μ, satisfying the commutation relation

$$\beta_\mu\beta_\nu\beta_\sigma + \beta_\sigma\beta_\nu\beta_\mu = \delta_{\mu\nu}\beta_\sigma + \delta_{\sigma\nu}\beta_\mu.$$

If we interpret $\boldsymbol{a}$ to mean $a_\mu\beta_\mu$, rather than $a_\mu\gamma_\mu$, for any a_μ, all of the equations in momentum space will remain formally identical to those for the spin 1/2; with the exception of those in which a denominator $(\boldsymbol{p}-m)^{-1}$ has been rationalized to $(\boldsymbol{p}+m)(\boldsymbol{p}^2-m^2)^{-1}$ since $\boldsymbol{p}^2$ is no longer equal to a number, $p\cdot p$. But $\boldsymbol{p}^3$ does equal $(p\cdot p)\boldsymbol{p}$ so that $(\boldsymbol{p}-m)^{-1}$ may now be interpreted as $(m\boldsymbol{p}+m^2+\boldsymbol{p}^2-p\cdot p)(p\cdot p-m^2)^{-1}m^{-1}$. This implies that equations in coordinate space will be valid of the function $K_+(2, 1)$ is given as $K_+(2,1) = [(i\nabla_2+m) - m^{-1}(\nabla_2^2 + \Box_2^2)]iI_+(2,1)$ with $\nabla_2 = \beta_\mu\partial/\partial x_{2\mu}$. This is all in virtue of the fact that the many component wave function ψ (5 components for spin 0, 10 for spin 1) satisfies $(i\nabla - m)\psi = A\psi$ which is formally identical to the Dirac Equation. See W. Pauli, Rev. Mod. Phys. 13, 203 (1940).

The important kernel is now $I_+(2,1)$ defined in (I, Eq. (32)). For a free particle, the wave function $\psi(2)$ satisfies $+\Box^2\psi - m^2\psi = 0$. At a point, 2, inside a space time region it is given by

$$\psi(2) = \int [\psi(1)\partial I_+(2,1)/\partial x_{1\mu} - (\partial\psi/\partial x_{1\mu})I_+(2,1)]N_\mu(1)d^3V_1,$$

(as is readily shown by the usual method of demonstrating Green's theorem) the integral being over an entire 3-surface boundary of the region (with normal vector N_μ). Only the positive frequency components of ψ contribute from the surface preceding the time corresponding to 2, and only negative frequencies from the surface future to 2. These can be interpreted as electrons and positrons in direct analogy to the Dirac case.

The right-hand side of (35) can be considered as a source of new waves and a series of terms written down to represent matrix elements for processes of increasing order. There is only one new point here, the term in $A_\mu A_\mu$ by which two quanta can act at the same time. As an example, suppose three quanta or potentials, $a_\mu \exp(-iq_a\cdot x)$, $b_\mu \exp(-iq_b\cdot x)$, and $c_\mu \exp(-iq_c\cdot x)$ are to act in that order on a particle of original momentum $p_{0\mu}$ so that $p_a = p_0 + q_a$ and $p_b = p_a + q_b$; the final momentum being $p_c = p_b + q_c$. The matrix element is the sum of three terms ($p^2 = p_\mu p_\mu$) (illustrated in Fig. 7)

$$\begin{aligned}&(p_c\cdot c + p_b\cdot c)(p_b^2 - m^2)^{-1}(p_b\cdot b + p_a\cdot b)\\ &\qquad\times(p_a^2 - m^2)^{-1}(p_a\cdot a + p_0\cdot a)\\ &-(p_c\cdot c + p_b\cdot c)(p_b^2 - m^2)^{-1}(b\cdot a)\\ &-(c\cdot b)(p_a^2 - m^2)^{-1}(p_a\cdot a + p_0\cdot a).\end{aligned} \tag{36}$$

The first comes when each potential acts through the perturbation $i\partial(A_\mu\psi)/\partial x_\mu + iA_\mu\partial\psi/\partial x_\mu$. These gradient operators in momentum space mean respectively the momentum after and before the potential A_μ operates. The second term comes from b_μ and a_μ acting at the same instant and arises from the $A_\mu A_\mu$ term in (a). Together b_μ and a_μ carry momentum $q_{b\mu} + q_{a\mu}$ so that after $b\cdot a$ operates the momentum is $p_0 + q_a + q_b$ or p_b. The final term comes from c_μ and b_μ operating together in a similar manner. The term $A_\mu A_\mu$ thus permits a new type of process in which two quanta can be emitted (or absorbed, or one absorbed, one emitted) at the same time. There is no $a\cdot c$ term for the order a, b, c we have assumed. In an actual problem there would be other terms like (36) but with alterations in the order in which the quanta a, b, c act. In these terms $a\cdot c$ would appear.

As a further example the self-energy of a particle of momentum p_μ is

$$(e^2/2\pi i m)\int [(2p-k)_\mu((p-k)^2 - m^2)^{-1} \times (2p-k)_\mu - \delta_{\mu\mu}]d^4k\,k^{-2}C(k^2),$$

where the $\delta_{\mu\mu} = 4$ comes from the $A_\mu A_\mu$ term and represents the possibility of the simultaneous emission and absorption of the same virtual quantum. This integral without the $C(k^2)$ diverges quadratically and would not converge if $C(k^2) = -\lambda^2/(k^2-\lambda^2)$. Since the interaction occurs through the gradients of the potential, we must use a stronger convergence factor, for example $C(k^2) = \lambda^4(k^2-\lambda^2)^{-2}$, or in general (17) with $\int_0^\infty \lambda^2 G(\lambda)d\lambda = 0$. In this case the self-energy converges but depends quadratically on the cut-off λ and is not necessarily small compared to m. The radiative corrections to scattering after mass renormalization are insensitive to the cut-off just as for the Dirac equation.

When there are several particles one can obtain Bose statistics by the rule that if two processes lead to the same state but with two electrons exchanged, their amplitudes are to be added (rather than subtracted as for Fermi statistics). In this case equivalence to the second quantization treatment of Pauli and Weisskopf should be demonstrable in a way very much like that given in I (appendix) for Dirac electrons. The Bose statistics mean that the sign of contribution of a closed loop to the vacuum polarization is the opposite of what it is for the Fermi case (see I). It is ($p_b = p_a + q$)

$$J_{\mu\nu} = \frac{e^2}{2\pi i m}\int [(p_{b\mu}+p_{a\mu})(p_{b\nu}+p_{a\nu})(p_a^2-m^2)^{-1} \times (p_b^2-m^2)^{-1} - \delta_{\mu\nu}(p_a^2-m^2)^{-1} - \delta_{\mu\nu}(p_b^2-m^2)^{-1}]d^4p_a$$

giving,

$$J_{\mu\nu}{}^P = \frac{e^2}{\pi}(q_\mu q_\nu - \delta_{\mu\nu}q^2)\left[\frac{1}{6}\ln\frac{\lambda^2}{m^2} + \frac{1}{9} - \frac{4m^2-q^2}{3q^2}\left(1 - \frac{\theta}{\tan\theta}\right)\right],$$

the notation as in (33). The imaginary part for $(q^2)^{\frac{1}{2}} > 2m$ is again positive representing the loss in the probability of finding the final state to be a vacuum, associated with the possibilities of pair production. Fermi statistics would give a gain in probability (and also a charge renormalization of opposite sign to that expected).

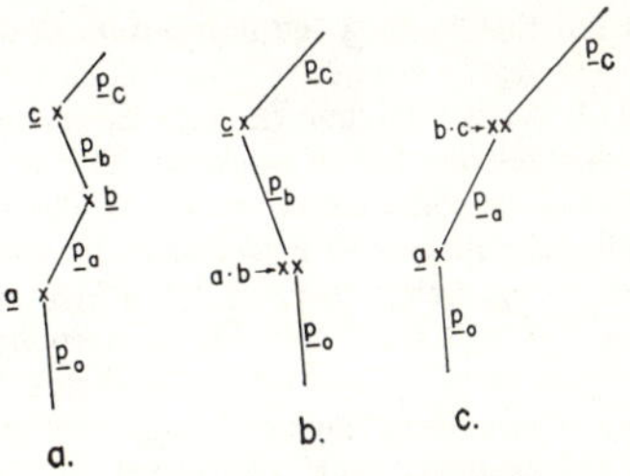

FIG. 7. Klein-Gordon particle in three potentials, Eq. (36). The coupling to the electromagnetic field is now, for example, $p_0\cdot a + p_a\cdot a$, and a new possibility arises, (b), of simultaneous interaction with two quanta $a\cdot b$. The propagation factor is now $(p\cdot p - m^2)^{-1}$ for a particle of momentum p_μ.

10. APPLICATION TO MESON THEORIES

The theories which have been developed to describe mesons and the interaction of nucleons can be easily expressed in the language used here. Calculations, to lowest order in the interactions can be made very easily for the various theories, but agreement with experimental results is not obtained. Most likely all of our present formulations are quantitatively unsatisfactory. We shall content ourselves therefore with a brief summary of the methods which can be used.

The nucleons are usually assumed to satisfy Dirac's equation so that the factor for propagation of a nucleon of momentum p is $(p-M)^{-1}$ where M is the mass of the nucleon (which implies that nucleons can be created in pairs). The nucleon is then assumed to interact with mesons, the various theories differing in the form assumed for this interaction.

First, we consider the case of neutral mesons. The theory closest to electrodynamics is the theory of vector mesons with vector coupling. Here the factor for emission or absorption of a meson is $g\gamma_\mu$ when this meson is "polarized" in the μ direction. The factor g, the "mesonic charge," replaces the electric charge e. The amplitude for propagation of a meson of momentum q in intermediate states is $(q^2-\mu^2)^{-1}$ (rather than q^{-2} as it is for light) where μ is the mass of the meson. The necessary integrals are made finite by convergence factors $C(q^2-\mu^2)$ as in electrodynamics. For scalar mesons with scalar coupling the only change is that one replaces the γ_μ by 1 in emission and absorption. There is no longer a direction of polarization, μ, to sum upon. For pseudoscalar mesons, pseudoscalar coupling replace γ_μ by $\gamma_5=i\gamma_x\gamma_y\gamma_z\gamma_t$. For example, the self-energy matrix of a nucleon of momentum p in this theory is

$$(g^2/\pi i)\int \gamma_5(p-k-M)^{-1}\gamma_5 d^4k(k^2-\mu^2)^{-1}C(k^2-\mu^2).$$

Other types of meson theory result from the replacement of γ_μ by other expressions (for example by $\frac{1}{2}(\gamma_\mu\gamma_\nu-\gamma_\nu\gamma_\mu)$ with a subsequent sum over all μ and ν for virtual mesons). Scalar mesons with vector coupling result from the replacement of γ_μ by $\mu^{-1}q$ where q is the final momentum of the nucleon minus its initial momentum, that is, it is the momentum of the meson if absorbed, or the negative of the momentum of a meson emitted. As is well known, this theory with neutral mesons gives zero for all processes, as is proved by our discussion on longitudinal waves in electrodynamics. Pseudoscalar mesons with pseudo-vector coupling corresponds to γ_μ being replaced by $\mu^{-1}\gamma_5 q$ while vector mesons with tensor coupling correspond to using $(2\mu)^{-1}(\gamma_\mu q-q\gamma_\mu)$. These extra gradients involve the danger of producing higher divergencies for real processes. For example, $\gamma_5 q$ gives a logarithmically divergent interaction of neutron and electron.[25] Although these divergencies can be held by strong enough convergence factors, the results then are sensitive to the method used for convergence and the size of the cut-off values of λ. For low order processes $\mu^{-1}\gamma_5 q$ is equivalent to the pseudoscalar interaction $2M\mu^{-1}\gamma_5$ because if taken between free particle wave functions of the nucleon of momenta p_1 and $p_2=p_1+q$, we have

$$(\bar{u}_2\gamma_5 q u_1)=(\bar{u}_2\gamma_5(p_2-p_1)u_1)=-(\bar{u}_2 p_2\gamma_5 u_1) \\ -(\bar{u}_2\gamma_5 p_1 u_1)=-2M(\bar{u}_2\gamma_5 u_1)$$

since γ_5 anticommutes with p_2 and p_2 operating on the state 2 equivalent to M as is p_1 on the state 1. This shows that the γ_5 interaction is unusually weak in the non-relativistic limit (for example the expected value of γ_5 for a free nucleon is zero), but since $\gamma_5^2=1$ is not small, pseudoscalar theory gives a more important interaction in second order than it does in first. Thus the pseudoscalar coupling constant should be chosen to fit nuclear forces including these important second order processes.[26] The equivalence of pseudoscalar and pseudovector coupling which holds for low order processes therefore does not hold when the pseudoscalar theory is giving its most important effects. These theories will therefore give quite different results in the majority of practical problems.

In calculating the corrections to scattering of a nucleon by a neutral vector meson field (γ_μ) due to the effects of virtual mesons, the situation is just as in electrodynamics, in that the result converges without need for a cut-off and depends only on gradients of the meson potential. With scalar (1) or pseudoscalar (γ_5) neutral mesons the result diverges logarithmically and so must be cut off. The part sensitive to the cut-off, however, is directly proportional to the meson potential. It may thereby be removed by a renormalization of mesonic charge g. After this renormalization the results depend only on gradients of the meson potential and are essentially independent of cut-off. This is in addition to the mesonic charge renormalization coming from the production of virtual nucleon pairs by a meson, analogous to the vacuum polarization in electrodynamics. But here there is a further difference from electrodynamics for scalar or pseudoscalar mesons in that the polarization also gives a term in the induced current proportional to the meson potential representing therefore an additional renormalization of the *mass of the meson* which usually depends quadratically on the cut-off.

Next consider charged mesons in the absence of an electromagnetic field. One can introduce isotopic spin operators in an obvious way. (Specifically replace the neutral γ_5, say, by $\tau_i\gamma_5$ and sum over $i=1$, 2 where $\tau_1=\tau_++\tau_-$, $\tau_2=i(\tau_+-\tau_-)$ and τ_+ changes neutron to proton (τ_+ on proton$=0$) and τ_- changes proton to neutron.) It is just as easy for practical problems simply to keep track of whether the particle is a proton or a neutron on a diagram drawn to help write down the

[25] M. Slotnick and W. Heitler, Phys. Rev. 75, 1645 (1949).

[26] H. A. Bethe, Bull. Am. Phys. Soc. 24, 3, Z3 (Washington, 1949).

matrix element. This excludes certain processes. For example in the scattering of a negative meson from $\mathbf{q}_1$ to $\mathbf{q}_2$ by a neutron, the meson $\mathbf{q}_2$ must be emitted first (in order of operators, not time) for the neutron cannot absorb the negative meson $\mathbf{q}_1$ until it becomes a proton. That is, in comparison to the Klein Nishina formula (15), only the analogue of second term (see Fig. 5(b)) would appear in the scattering of negative mesons by neutrons, and only the first term (Fig. 5(a)) in the neutron scattering of positive mesons.

The source of mesons of a given charge is not conserved, for a neutron capable of emitting negative mesons may (on emitting one, say) become a proton no longer able to do so. The proof that a perturbation $\mathbf{q}$ gives zero, discussed for longitudinal electromagnetic waves, fails. This has the consequence that vector mesons, if represented by the interaction γ_μ would not satisfy the condition that the divergence of the potential is zero. The interaction is to be taken[27] as $\gamma_\mu-\mu^{-2}q_\mu\mathbf{q}$ in emission and as γ_μ in absorption if the real emission of mesons with a non-zero divergence of potential is to be avoided. (The correction term $\mu^{-2}q_\mu\mathbf{q}$ gives zero in the neutral case.) The asymmetry in emission and absorption is only apparent, as this is clearly the same thing as subtracting from the original $\gamma_\mu\cdots\gamma_\mu$, a term $\mu^{-2}\mathbf{q}\cdots\mathbf{q}$. That is, if the term $-\mu^{-2}q_\mu\mathbf{q}$ is omitted the resulting theory describes a combination of mesons of spin one and spin zero. The spin zero mesons, coupled by vector coupling $\mathbf{q}$, are removed by subtracting the term $\mu^{-2}\mathbf{q}\cdots\mathbf{q}$.

The two extra gradients $\mathbf{q}\cdots\mathbf{q}$ make the problem of diverging integrals still more serious (for example the interaction between two protons corresponding to the exchange of two charged vector mesons depends quadratically on the cut-off if calculated in a straightforward way). One is tempted in this formulation to choose simply $\gamma_\mu\cdots\gamma_\mu$ and accept the admixture of spin zero mesons. But it appears that this leads in the conventional formalism to negative energies for the spin zero component. This shows one of the advantages of the method of second quantization of meson fields over the present formulation. There such errors of sign are obvious while here we seem to be able to write seemingly innocent expressions which can give absurd results. Pseudovector mesons with pseudovector coupling correspond to using $\gamma_5(\gamma_\mu-\mu^{-2}q_\mu\mathbf{q})$ for absorption and $\gamma_5\gamma_\mu$ for emission for both charged and neutral mesons.

In the presence of an electromagnetic field, whenever the nucleon is a proton it interacts with the field in the way described for electrons. The meson interacts in the scalar or pseudoscalar case as a particle obeying the Klein-Gordon equation. It is important here to use the method of calculation of Bethe and Pauli, that is, a virtual meson is assumed to have the same "mass" during all its interactions with the electromagnetic field. The result for mass μ and for $(\mu^2+\lambda^2)^{\frac{1}{2}}$ are subtracted and the difference integrated over the function $G(\lambda)d\lambda$. A separate convergence factor is not provided for each meson propagation between electromagnetic interactions, otherwise gauge invariance is not insured. When the coupling involves a gradient, such as $\gamma_5\mathbf{q}$ where $\mathbf{q}$ is the final minus the initial momentum of the nucleon, the vector potential $\mathbf{A}$ must be subtracted from the momentum of the proton. That is, there is an additional coupling $\pm\gamma_5\mathbf{A}$ (plus when going from proton to neutron, minus for the reverse) representing the new possibility of a simultaneous emission (or absorption) of meson and photon.

Emission of positive or absorption of negative virtual mesons are represented in the same term, the sign of the charge being determined by temporal relations as for electrons and positrons.

Calculations are very easily carried out in this way to lowest order in g^2 for the various theories for nucleon interaction, scattering of mesons by nucleons, meson production by nuclear collisions and by gamma-rays, nuclear magnetic moments, neutron electron scattering, etc., However, no good agreement with experiment results, when these are available, is obtained. Probably all of the formulations are incorrect. An uncertainty arises since the calculations are only to first order in g^2, and are not valid if $g^2/\hbar c$ is large.

The author is particularly indebted to Professor H. A. Bethe for his explanation of a method of obtaining finite and gauge invariant results for the problem of vacuum polarization. He is also grateful for Professor Bethe's criticisms of the manuscript, and for innumerable discussions during the development of this work. He wishes to thank Professor J. Ashkin for his careful reading of the manuscript.

APPENDIX

In this appendix a method will be illustrated by which the simpler integrals appearing in problems in electrodynamics can be directly evaluated. The integrals arising in more complex processes lead to rather complicated functions, but the study of the relations of one integral to another and their expression in terms of simpler integrals may be facilitated by the methods given here.

---

[27] The vector meson field potentials φ_μ satisfy

$$-\partial/\partial x_\nu(\partial\varphi_\mu/\partial x_\nu-\partial\varphi_\nu/\partial x_\mu)-\mu^2\varphi_\mu=-4\pi s_\mu,$$

where s_μ, the source for such mesons, is the matrix element of γ_μ between states of neutron and proton. By taking the divergence $\partial/\partial x_\mu$ of both sides, conclude that $\partial\varphi_\nu/\partial x_\nu=4\pi\mu^{-2}\partial s_\nu/\partial x_\nu$ so that the original equation can be rewritten as

$$\Box^2\varphi_\mu-\mu^2\varphi_\mu=-4\pi(s_\mu+\mu^{-2}\partial/\partial x_\mu(\partial s_\nu/\partial x_\nu)).$$

The right hand side gives in momentum representation $\gamma_\mu-\mu^{-2}q_\mu q_\nu\gamma_\nu$ the left yields the $(q^2-\mu^2)^{-1}$ and finally the interaction $s_\mu\varphi_\mu$ in the Lagrangian gives the γ_μ on absorption.

Proceeding in this way find generally that particles of spin one can be represented by a four-vector u_μ (which, for a free particle of momentum q satisfies $q\cdot u=0$). The propagation of virtual particles of momentum q from state ν to μ is represented by multiplication by the 4–4 matrix (or tensor) $P_{\mu\nu}=(\delta_{\mu\nu}-\mu^{-2}q_\mu q_\nu)\times(q^2-\mu^2)^{-1}$. The first-order interaction (from the Proca equation) with an electromagnetic potential $a\exp(-ik\cdot x)$ corresponds to multiplication by the matrix $E_{\mu\nu}=(q_2\cdot a+q_1\cdot a)\delta_{\mu\nu}-q_{2\nu}a_\mu-q_{1\mu}a_\nu$ where q_1 and $q_2=q_1+k$ are the momenta before and after the interaction. Finally, two potentials a, b may act simultaneously, with matrix $E'_{\mu\nu}=-(a\cdot b)\delta_{\mu\nu}+b_\mu a_\nu$.

As a typical problem consider the integral (12) appearing in the first order radiationless scattering problem:

$$\int \gamma_\mu(p_2-k-m)^{-1}a(p_1-k-m)^{-1}\gamma_\mu k^{-2}d^4kC(k^2), \tag{1a}$$

where we shall take $C(k^2)$ to be typically $-\lambda^2(k^2-\lambda^2)^{-1}$ and d^4k means $(2\pi)^{-2}dk_1dk_2dk_3dk_4$. We first rationalize the factors $(p-k-m)^{-1}=(p-k+m)((p-k)^2-m^2)^{-1}$ obtaining,

$$\int \gamma_\mu(p_2-k+m)a(p_1-k+m)\gamma_\mu k^{-2}d^4kC(k^2) \times((p_1-k)^2-m^2)^{-1}((p_2-k)^2-m^2)^{-1}. \tag{2a}$$

The matrix expression may be simplified. It appears to be best to do so *after* the integrations are performed. Since $AB=2A\cdot B-BA$ where $A\cdot B=A_\mu B_\mu$ is a number commuting with all matrices, find, if R is any expression, and A a vector, since $\gamma_\mu A=-A\gamma_\mu+2A_\mu$,

$$\gamma_\mu AR\gamma_\mu=-A\gamma_\mu R\gamma_\mu+2RA. \tag{3a}$$

Expressions between two γ_μ's can be thereby reduced by induction. Particularly useful are

$$\begin{aligned}\gamma_\mu\gamma_\mu&=4\\ \gamma_\mu A\gamma_\mu&=-2A\\ \gamma_\mu AB\gamma_\mu&=2(AB+BA)=4A\cdot B\\ \gamma_\mu ABC\gamma_\mu&=-2CBA\end{aligned} \tag{4a}$$

where A, B, C are any three vector-matrices (i.e., linear combinations of the four γ's).

In order to calculate the integral in (2a) the integral may be written as the sum of three terms (since $k=k_\sigma\gamma_\sigma$),

$$\gamma_\mu(p_2+m)a(p_1+m)\gamma_\mu J_1-[\gamma_\mu\gamma_\sigma a(p_1+m)\gamma_\mu +\gamma_\mu(p_2+m)a\gamma_\sigma\gamma_\mu]J_2+\gamma_\mu\gamma_\sigma a\gamma_\tau\gamma_\mu J_3, \tag{5a}$$

where

$$J_{(1;2;3)}=\int(1;k_\sigma;k_\sigma k_\tau)k^{-2}d^4kC(k^2) \times((p_2-k)^2-m^2)^{-1}((p_1-k)^2-m^2)^{-1}. \tag{6a}$$

That is for J_1 the $(1;k_\sigma;k_\sigma k_\tau)$ is replaced by 1, for J_2 by k_σ, and for J_3 by $k_\sigma k_\tau$.

More complex processes of the first order involve more factors like $((p_3-k)^2-m^2)^{-1}$ and a corresponding increase in the number of k's which may appear in the numerator, as $k_\sigma k_\tau k_\nu\cdots$. Higher order processes involving two or more virtual quanta involve similar integrals but with factors possibly involving $k+k'$ instead of just k, and the integral extending on $k^{-2}d^4kC(k^2)k'^{-2}d^4k'C(k'^2)$. They can be simplified by methods analogous to those used on the first order integrals.

The factors $(p-k)^2-m^2$ may be written

$$(p-k)^2-m^2=k^2-2p\cdot k-\Delta, \tag{7a}$$

where $\Delta=m^2-p^2$, $\Delta_1=m_1^2-p_1^2$, etc., and we can consider dealing with cases of greater generality in that the different denominators need not have the same value of the mass m. In our specific problem (6a), $p_1^2=m^2$ so that $\Delta_1=0$, but we desire to work with greater generality.

Now for the factor $C(k^2)/k^2$ we shall use $-\lambda^2(k^2-\lambda^2)^{-1}k^{-2}$. This can be written as

$$-\lambda^2/(k^2-\lambda^2)k^2=k^{-2}C(k^2)=-\int_0^{\lambda^2}dL(k^2-L)^{-2}. \tag{8a}$$

Thus we can replace $k^{-2}C(k^2)$ by $(k^2-L)^{-2}$ and at the end integrate the result with respect to L from zero to λ^2. We can for many practical purposes consider λ^2 very large relative to m^2 or p^2. When the original integral converges even without the convergence factor, it will be obvious since the L integration will then be convergent to infinity. If an infra-red catastrophe exists in the integral one can simply assume quanta have a small mass $\lambda_{\min}$ and extend the integral on L from $\lambda^2_{\min}$ to λ^2, rather than from zero to λ^2.

We then have to do integrals of the form

$$\int(1;k_\sigma;k_\sigma k_\tau)d^4k(k^2-L)^{-2}(k^2-2p_1\cdot k-\Delta_1)^{-1} \times(k^2-2p_2\cdot k-\Delta_2)^{-1}, \tag{9a}$$

where by $(1;k_\sigma;k_\sigma k_\tau)$ we mean that in the place of this symbol either 1, or k_σ, or $k_\sigma k_\tau$ may stand in different cases. In more complicated problems there may be more factors $(k^2-2p_i\cdot k-\Delta_i)^{-1}$ or other powers of these factors (the $(k^2-L)^{-2}$ may be considered as a special case of such a factor with $p_i=0$, $\Delta_i=L$) and further factors like $k_\sigma k_\tau k_\rho\cdots$ in the numerator. The poles in all the factors are made definite by the assumption that L, and the Δ's have infinitesimal negative imaginary parts.

We shall do the integrals of successive complexity by induction. We start with the simplest convergent one, and show

$$\int d^4k(k^2-L)^{-3}=(8iL)^{-1}. \tag{10a}$$

For this integral is $\int(2\pi)^{-2}dk_4d^3\mathbf{K}(k_4^2-\mathbf{K}\cdot\mathbf{K}-L)^{-3}$ where the vector $\mathbf{K}$, of magnitude $K=(\mathbf{K}\cdot\mathbf{K})^{\frac{1}{2}}$ is k_1, k_2, k_3. The integral on k_4 shows third order poles at $k_4=+(K^2+L)^{\frac{1}{2}}$ and $k_4=-(K^2+L)^{\frac{1}{2}}$. Imagining, in accordance with our definitions, that L has a small negative imaginary part only the first is below the real axis. The contour can be closed by an infinite semi-circle below this axis, without change of the value of the integral since the contribution from the semi-circle vanishes in the limit. Thus the contour can be shrunk about the pole $k_4=+(K^2+L)^{\frac{1}{2}}$ and the resulting k_4 integral is $-2\pi i$ times the residue at this pole. Writing $k_4=(K^2+L)^{\frac{1}{2}}+\epsilon$ and expanding $(k_4^2-K^2-L)^{-3}=\epsilon^{-3}(\epsilon+2(K^2+L)^{\frac{1}{2}})^{-3}$ in powers of ϵ, the residue, being the coefficient of the term ϵ^{-1}, is seen to be $6(2(K^2+L)^{\frac{1}{2}})^{-5}$ so our integral is

$$-(3i/32\pi)\int_0^\infty 4\pi K^2dK(K^2+L)^{-5/2}=(3/8i)(1/3L)$$

establishing (10a).

We also have $\int k_\sigma d^4k(k^2-L)^{-3}=0$ from the symmetry in the k space. We write these results as

$$(8i)\int(1;k_\sigma)d^4k(k^2-L)^{-3}=(1;0)L^{-1}, \tag{11a}$$

where in the brackets $(1;k_\sigma)$ and $(1;0)$ corresponding entries are to be used.

Substituting $k=k'-p$ in (11a), and calling $L-p^2=\Delta$ shows that

$$(8i)\int(1;k_\sigma)d^4k(k^2-2p\cdot k-\Delta)^{-3}=(1;p_\sigma)(p^2+\Delta)^{-1}. \tag{12a}$$

By differentiating both sides of (12a) with respect to Δ, or with respect to p_τ there follows directly

$$(24i)\int(1;k_\sigma;k_\sigma k_\tau)d^4k(k^2-2p\cdot k-\Delta)^{-4} =-(1;p_\sigma;p_\sigma p_\tau-\tfrac{1}{2}\delta_{\sigma\tau}(p^2+\Delta))(p^2+\Delta)^{-2}. \tag{13a}$$

Further differentiations give directly successive integrals including more k factors in the numerator and higher powers of $(k^2-2p\cdot k-\Delta)$ in the denominator.

The integrals so far only contain one factor in the denominator. To obtain results for two factors we make use of the identity

$$a^{-1}b^{-1}=\int_0^1 dx(ax+b(1-x))^{-2}, \tag{14a}$$

(suggested by some work of Schwinger's involving Gaussian integrals). This represents the product of two reciprocals as a parametric integral over one and will therefore permit integrals with two factors to be expressed in terms of one. For other powers of a, b, we make use of all of the identities, such as

$$a^{-2}b^{-1}=\int_0^1 2xdx(ax+b(1-x))^{-3}, \tag{15a}$$

deducible from (14a) by successive differentiations with respect to a or b.

To perform an integral, such as

$$(8i)\int(1;k_\sigma)d^4k(k^2-2p_1\cdot k-\Delta_1)^{-2}(k^2-2p_2\cdot k-\Delta_2)^{-1}, \tag{16a}$$

write, using (15a),

$$(\mathbf{k}^2-2p_1\cdot k-\Delta_1)^{-2}(\mathbf{k}^2-2p_2\cdot k-\Delta_2)^{-1}=\int_0^1 2xdx(\mathbf{k}^2-2p_x\cdot k-\Delta_x)^{-3},$$

where

$$\mathbf{p}_x=x\mathbf{p}_1+(1-x)\mathbf{p}_2 \quad\text{and}\quad \Delta_x=x\Delta_1+(1-x)\Delta_2, \tag{17a}$$

(note that Δ_x is *not* equal to $m^2-\mathbf{p}_x^2$) so that the expression (16a) is $(8i)\int_0^1 2xdx\int(1;k_\sigma)d^4k(\mathbf{k}^2-2p_x\cdot k-\Delta_x)^{-3}$ which may now be evaluated by (12a) and is

$$(16\text{a})=\int_0^1 (1;p_{x\sigma})2xdx(\mathbf{p}_x^2+\Delta_x)^{-1}, \tag{18a}$$

where $\mathbf{p}_x$, Δ_x are given in (17a). The integral in (18a) is elementary, being the integral of ratio of polynomials, the denominator of second degree in x. The general expression although readily obtained is a rather complicated combination of roots and logarithms.

Other integrals can be obtained again by parametric differentiation. For example differentiation of (16a), (18a) with respect to Δ_2 or $p_{2\tau}$ gives

$$(8i)\int(1;k_\sigma;k_\sigma k_\tau)d^4k(\mathbf{k}^2-2p_1\cdot k-\Delta_1)^{-2}(\mathbf{k}^2-2p_2\cdot k-\Delta_2)^{-2}$$
$$=-\int_0^1 (1;p_{x\sigma};p_{x\sigma}p_{x\tau}-\tfrac{1}{2}\delta_{\sigma\tau}(\mathbf{p}_x^2+\Delta_x))$$
$$\times 2x(1-x)dx(\mathbf{p}_x^2+\Delta_x)^{-2}, \tag{19a}$$

again leading to elementary integrals.

As an example, consider the case that the second factor is just $(\mathbf{k}^2-L)^{-2}$ and in the first put $\mathbf{p}_1=\mathbf{p}$, $\Delta_1=\Delta$. Then $\mathbf{p}_x=x\mathbf{p}$, $\Delta_x=x\Delta+(1-x)L$. There results

$$(8i)\int(1;k_\sigma;k_\sigma k_\tau)d^4k(\mathbf{k}^2-L)^{-2}(\mathbf{k}^2-2p\cdot k-\Delta)^{-2}$$
$$=-\int_0^1 (1;xp_\sigma;x^2p_\sigma p_\tau-\tfrac{1}{2}\delta_{\sigma\tau}(x^2\mathbf{p}^2+\Delta_x))$$
$$\times 2x(1-x)dx(x^2\mathbf{p}^2+\Delta_x)^{-2}. \tag{20a}$$

Integrals with three factors can be reduced to those involving two by using (14a) again. They, therefore, lead to integrals with two parameters (e.g., see application to radiative correction to scattering below).

The methods of calculation given in this paper are deceptively simple when applied to the lower order processes. For processes of increasingly higher orders the complexity and difficulty increases rapidly, and these methods soon become impractical in their present form.

A. Self-Energy

The self-energy integral (19) is

$$(e^2/\pi i)\int\gamma_\mu(\mathbf{p}-\mathbf{k}-m)^{-1}\gamma_\mu\mathbf{k}^{-2}d^4kC(\mathbf{k}^2), \tag{19}$$

so that it requires that we find (using the principle of (8a)) the integral on L from 0 to λ^2 of

$$\int\gamma_\mu(\mathbf{p}-\mathbf{k}+m)\gamma_\mu d^4k(\mathbf{k}^2-L)^{-2}(\mathbf{k}^2-2p\cdot k)^{-1},$$

since $(\mathbf{p}-\mathbf{k})^2-m^2=\mathbf{k}^2-2p\cdot k$, as $\mathbf{p}^2=m^2$. This is of the form (16a) with $\Delta_1=L$, $\mathbf{p}_1=0$, $\Delta_2=0$, $\mathbf{p}_2=\mathbf{p}$ so that (18a) gives, since $\mathbf{p}_x=(1-x)\mathbf{p}$, $\Delta_x=xL$,

$$(8i)\int(1;k_\sigma)d^4k(\mathbf{k}^2-L)^{-2}(\mathbf{k}^2-2p\cdot k)^{-1}$$
$$=\int_0^1 (1;(1-x)p_\sigma)2xdx((1-x)^2m^2+xL)^{-1},$$

or performing the integral on L, as in (8),

$$(8i)\int(1;k_\sigma)d^4k\mathbf{k}^{-2}C(\mathbf{k}^2)(\mathbf{k}^2-2p\cdot k)^{-1}$$
$$=\int_0^1 (1;(1-x)p_\sigma)2dx\ln\frac{x\lambda^2+(1-x)^2m^2.}{(1-x)^2m^2}$$

Assuming now that $\lambda^2\gg m^2$ we neglect $(1-x)^2m^2$ relative to $x\lambda^2$ in the argument of the logarithm, which then becomes $(\lambda^2/m^2)(x/(1-x)^2)$. Then since $\int_0^1dx\ln(x(1-x)^{-2})=1$ and $\int_0^1(1-x)dx\ln(x(1-x)^{-2})=-(1/4)$ find

$$(8i)\int(1;k_\sigma)\mathbf{k}^{-2}C(\mathbf{k}^2)d^4k(\mathbf{k}^2-2p\cdot k)^{-1}$$
$$=\left(2\ln\frac{\lambda^2}{m^2}+2;p_\sigma\left(\ln\frac{\lambda^2}{m^2}-\frac{1}{2}\right)\right),$$

so that substitution into (19) (after the $(\mathbf{p}-\mathbf{k}-\mathbf{m})^{-1}$ in (19) is replaced by $(\mathbf{p}-\mathbf{k}+m)(\mathbf{k}^2-2p\cdot k)^{-1}$) gives

$$\begin{aligned}(19)&=(e^2/8\pi)\gamma_\mu[(\mathbf{p}+m)(2\ln(\lambda^2/m^2)+2)\\&\qquad-\mathbf{p}(\ln(\lambda^2/m^2)-\tfrac{1}{2})]\gamma_\mu\\&=(e^2/8\pi)[8m(\ln(\lambda^2/m^2)+1)-\mathbf{p}(2\ln(\lambda^2/m^2)+5)],\end{aligned} \tag{20}$$

using (4a) to remove the γ_μ's. This agrees with Eq. (20) of the text, and gives the self-energy (21) when $\mathbf{p}$ is replaced by $\mathbf{m}$.

B. Corrections to Scattering

The term (12) in the radiationless scattering, after rationalizing the matrix denominators and using $\mathbf{p}_1^2=\mathbf{p}_2^2=m^2$ requires the integrals (9a), as we have discussed. This is an integral with three denominators which we do in two stages. First the factors $(\mathbf{k}^2-2p_1\cdot k)$ and $(\mathbf{k}^2-2p_2\cdot k)$ are combined by a parameter y;

$$(\mathbf{k}^2-2p_1\cdot k)^{-1}(\mathbf{k}^2-2p_2\cdot k)^{-1}=\int_0^1 dy(\mathbf{k}^2-2p_y\cdot k)^{-2},$$

from (14a) where

$$\mathbf{p}_y=y\mathbf{p}_1+(1-y)\mathbf{p}_2. \tag{21a}$$

We therefore need the integrals

$$(8i)\int(1;k_\sigma;k_\sigma k_\tau)d^4k(\mathbf{k}^2-L)^{-2}(\mathbf{k}^2-2p_y\cdot k)^{-2}, \tag{22a}$$

which we will then integrate with respect to y from 0 to 1. Next we do the integrals (22a) immediately from (20a) with $\mathbf{p}=\mathbf{p}_y$, $\Delta=0$:

$$(22\text{a})=-\int_0^1\int_0^1 (1;xp_{y\sigma};x^2p_{y\sigma}p_{y\tau}$$
$$-\tfrac{1}{2}\delta_{\sigma\tau}(x^2\mathbf{p}_y^2+(1-x)L))2x(1-x)dx(x^2\mathbf{p}_y^2+L(1-x))^{-2}dy.$$

We now turn to the integrals on L as required in (8a). The first term, (1), in $(1;k_\sigma;k_\sigma k_\tau)$ gives no trouble for large L, but if L is put equal to zero there results $x^{-2}\mathbf{p}_y^{-2}$ which leads to a diverging integral on x as $x\rightarrow 0$. This infra-red catastrophe is analyzed by using $\lambda_{\min}^2$ for the lower limit of the L integral. For the last term the upper limit of L must be kept as λ^2. Assuming $\lambda_{\min}^2\ll\mathbf{p}_y^2\ll\lambda^2$ the x integrals which remain are trivial, as in the self-energy case. One finds

$$-(8i)\int(\mathbf{k}^2-\lambda_{\min}^2)^{-1}d^4kC(\mathbf{k}^2-\lambda_{\min}^2)(\mathbf{k}^2-2p_1\cdot k)^{-1}(\mathbf{k}^2-2p_2\cdot k)^{-1}$$
$$=\int_0^1 \mathbf{p}_y^{-2}dy\ln(\mathbf{p}_y^2/\lambda_{\min}^2) \tag{23a}$$

$$-(8i)\int k_\sigma\mathbf{k}^{-2}d^4kC(\mathbf{k}^2)(\mathbf{k}^2-2p_1\cdot k)^{-1}(\mathbf{k}^2-2p_2\cdot k)^{-1}$$
$$=2\int_0^1 p_{y\sigma}\mathbf{p}_y^{-2}dy, \tag{24a}$$

$$-(8i)\int k_\sigma k_\tau\mathbf{k}^{-2}d^4kC(\mathbf{k}^2)(\mathbf{k}^2-2p_1\cdot k)^{-1}(\mathbf{k}^2-2p_2\cdot k)^{-1}$$
$$=\int_0^1 p_{y\sigma}p_{y\tau}\mathbf{p}_y^{-2}dy-\tfrac{1}{2}\delta_{\sigma\tau}\int_0^1 dy\ln(\lambda^2\mathbf{p}_y^{-2})+\tfrac{1}{4}\delta_{\sigma\tau}. \tag{25a}$$

The integrals on y give,

$$\int_0^1 \mathbf{p}_y^{-2}dy\ln(\mathbf{p}_y^2\lambda_{\min}^{-2})=4(m^2\sin 2\theta)^{-1}\left[\theta\ln(m\lambda_{\min}^{-1})-\int_0^\theta \alpha\tan\alpha d\alpha\right], \tag{26a}$$

$$\int_0^1 p_{y\sigma}\mathbf{p}_y^{-2}dy=\theta(m^2\sin 2\theta)^{-1}(p_{1\sigma}+p_{2\sigma}), \tag{27a}$$

$$\int_0^1 p_{y\sigma}p_{y\tau}\mathbf{p}_y^{-2}dy=\theta(2m^2\sin 2\theta)^{-1}(p_{1\sigma}+p_{1\tau})(p_{2\sigma}+p_{2\tau})$$
$$+\mathbf{q}^{-2}q_\sigma q_\tau(1-\theta\,\mathrm{ctn}\theta), \tag{28a}$$

$$\int_0^1 dy\ln(\lambda^2\mathbf{p}_y^{-2})=\ln(\lambda^2/m^2)+2(1-\theta\,\mathrm{ctn}\theta). \tag{29a}$$

These integrals on y were performed as follows. Since $p_2=p_1+q$ where q is the momentum carried by the potential, it follows from $p_2^2=p_1^2=m^2$ that $2p_1\cdot q=-q^2$ so that since $p_y=p_1+q(1-y)$, $p_y^2=m^2-q^2y(1-y)$. The substitution $2y-1=\tan\alpha/\tan\theta$ where θ is defined by $4m^2\sin^2\theta=q^2$ is useful for it means $p_y^2=m^2\sec^2\alpha/\sec^2\theta$ and $p_y^{-2}dy=(m^2\sin 2\theta)^{-1}d\alpha$ where α goes from $-\theta$ to $+\theta$.

These results are substituted into the original scattering formula (2a), giving (22). It has been simplified by frequent use of the fact that p_1 operating on the initial state is m, and likewise p_2 when it appears at the left is replacable by m. (Thus, to simplify:

$$\gamma_\mu p_2 a p_1 \gamma_\mu = -2p_1 a p_2 \text{ by (4a)},$$
$$= -2(p_2-q)a(p_1+q) = -2(m-q)a(m+q).$$

A term like $qaq=-q^2a+2(a\cdot q)q$ is equivalent to just $-q^2a$ since $q=p_2-p_1=m-m$ has zero matrix element.) The renormalization term requires the corresponding integrals for the special case $q=0$.

C. Vacuum Polarization

The expressions (32) and (32′) for $J_{\mu\nu}$ in the vacuum polarization problem require the calculation of the integral

$$J_{\mu\nu}(m^2) = -\frac{e^2}{\pi i}\int Sp[\gamma_\mu(p-\tfrac{1}{2}q+m)\gamma_\nu(p+\tfrac{1}{2}q+m)]d^4p$$
$$\times((p-\tfrac{1}{2}q)^2-m^2)^{-1}((p+\tfrac{1}{2}q)^2-m^2)^{-1}, \quad (32)$$

where we have replaced p by $p-\frac{1}{2}q$ to simplify the calculation somewhat. We shall indicate the method of calculation by studying the integral,

$$I(m^2)=\int p_\sigma p_\tau d^4p((p-\tfrac{1}{2}q)^2-m^2)^{-1}((p+\tfrac{1}{2}q)^2-m^2)^{-1}.$$

The factors in the denominator, $p^2-p\cdot q-m^2+\frac{1}{4}q^2$ and $p^2+p\cdot q-m^2+\frac{1}{4}q^2$ are combined as usual by (8a) but for symmetry we substitute $x=\frac{1}{2}(1+\eta)$, $(1-x)=\frac{1}{2}(1-\eta)$ and integrate η from -1 to $+1$:

$$I(m^2)=\int_{-1}^{+1} p_\sigma p_\tau d^4p(p^2-\eta p\cdot q-m^2+\tfrac{1}{4}q^2)^{-2}d\eta/2. \quad (30a)$$

But the integral on p will not be found in our list for it is badly divergent. However, as discussed in Section 7, Eq. (32′) we do not wish $I(m^2)$ but rather $\int_0^\infty[I(m^2)-I(m^2+\lambda^2)]G(\lambda)d\lambda$. We can calculate the difference $I(m^2)-I(m^2+\lambda^2)$ by first calculating the derivative $I'(m^2+L)$ of I with respect to m^2 at m^2+L and later integrating L from zero to λ^2. By differentiating (30a), with respect to m^2 find,

$$I'(m^2+L)=\int_{-1}^{+1} p_\sigma p_\tau d^4p(p^2-\eta p\cdot q-m^2-L+\tfrac{1}{4}q^2)^{-3}d\eta.$$

This still diverges, but we can differentiate again to get

$$I''(m^2+L)=3\int_{-1}^{+1} p_\sigma p_\tau d^4p(p^2-\eta p\cdot q-m^2-L+\tfrac{1}{4}q^2)^{-4}d\eta$$
$$=-(8i)^{-1}\int_{-1}^{+1}(\tfrac{1}{4}\eta^2 q_\sigma q_\tau D^{-2}-\tfrac{1}{2}\delta_{\sigma\tau}D^{-1})d\eta \quad (31a)$$

(where $D=\frac{1}{4}(\eta^2-1)q^2+m^2+L$), which now converges and has been evaluated by (13a) with $p=\frac{1}{2}\eta q$ and $\Delta=m^2+L-\frac{1}{4}q^2$. Now to get I' we may integrate I'' with respect to L as an indefinite integral and *we may choose any convenient arbitrary constant*. This is because a constant C in I' will mean a term $-C\lambda^2$ in $I(m^2)-I(m^2+\lambda^2)$ which vanishes since we will integrate the results times $G(\lambda)d\lambda$ and $\int_0^\infty \lambda^2 G(\lambda)d\lambda=0$. This means that the logarithm appearing on integrating L in (31a) presents no problem. We may take

$$I'(m^2+L)=(8i)^{-1}\int_{-1}^{+1}[\tfrac{1}{4}\eta^2 q_\sigma q_\tau D^{-1}+\tfrac{1}{2}\delta_{\sigma\tau}\ln D]d\eta+C\delta_{\sigma\tau},$$

a subsequent integral on L and finally on η presents no new problems. There results

$$-(8i)\int p_\sigma p_\tau d^4p((p-\tfrac{1}{2}q)^2-m^2)^{-1}((p+\tfrac{1}{2}q)^2-m^2)^{-1}$$
$$=(q_\sigma q_\tau-\delta_{\sigma\tau}q^2)\left[\frac{1}{9}-\frac{4m^2-q^2}{3q^2}\left(1-\frac{\theta}{\tan\theta}\right)+\frac{1}{6}\ln\frac{\lambda^2}{m^2}\right]$$
$$+\delta_{\sigma\tau}[(\lambda^2+m^2)\ln(\lambda^2m^{-2}+1)-C'\lambda^2], \quad (32a)$$

where we assume $\lambda^2 \gg m^2$ and have put some terms into the arbitrary constant C' which is independent of λ^2 (but in principle could depend on q^2) and which drops out in the integral on $G(\lambda)d\lambda$. We have set $q^2=4m^2\sin^2\theta$.

In a very similar way the integral with m^2 in the numerator can be worked out. It is, of course, necessary to differentiate this m^2 also when calculating I' and I''. There results

$$-(8i)\int m^2 d^4p((p-\tfrac{1}{2}q)^2-m^2)^{-1}((p+\tfrac{1}{2}q)^2-m^2)^{-1}$$
$$=4m^2(1-\theta\,\mathrm{ctn}\,\theta)-q^2/3+2(\lambda^2+m^2)\ln(\lambda^2m^{-2}+1)-C''\lambda^2), \quad (33a)$$

with another unimportant constant C''. The complete problem requires the further integral,

$$-(8i)\int(1;p_\sigma)d^4p((p-\tfrac{1}{2}q)^2-m^2)^{-1}((p+\tfrac{1}{2}q)^2-m^2)^{-1}$$
$$=(1,0)(4(1-\theta\,\mathrm{ctn}\,\theta)+2\ln(\lambda^2m^{-2})). \quad (34a)$$

The value of the integral (34a) times m^2 differs from (33a), of course, because the results on the right are not actually the integrals on the left, but rather equal their actual value minus their value for $m^2=m^2+\lambda^2$.

Combining these quantities, as required by (32), dropping the constants C', C'' and evaluating the spur gives (33). The spurs are evaluated in the usual way, noting that the spur of any odd number of γ matrices vanishes and $Sp(AB)=Sp(BA)$ for arbitrary A, B. The $Sp(1)=4$ and we also have

$$\tfrac{1}{4}Sp[(p_1+m_1)(p_2-m_2)]=p_1\cdot p_2-m_1m_2, \quad (35a)$$

$$\tfrac{1}{4}Sp[(p_1+m_1)(p_2-m_2)(p_3+m_3)(p_4-m_4)]$$
$$=(p_1\cdot p_2-m_1m_2)(p_3\cdot p_4-m_3m_4)$$
$$-(p_1\cdot p_3-m_1m_3)(p_2\cdot p_4-m_2m_4)$$
$$+(p_1\cdot p_4-m_1m_4)(p_2\cdot p_3-m_2m_3), \quad (36a)$$

where p_i, m_i are arbitrary four-vectors and constants.

It is interesting that the terms of order $\lambda^2\ln\lambda^2$ go out, so that the charge renormalization depends only logarithmically on λ^2. This is not true for some of the meson theories. Electrodynamics is suspiciously unique in the mildness of its divergence.

D. More Complex Problems

Matrix elements for complex problems can be set up in a manner analogous to that used for the simpler cases. We give three illustrations; higher order corrections to the Møller scatter-

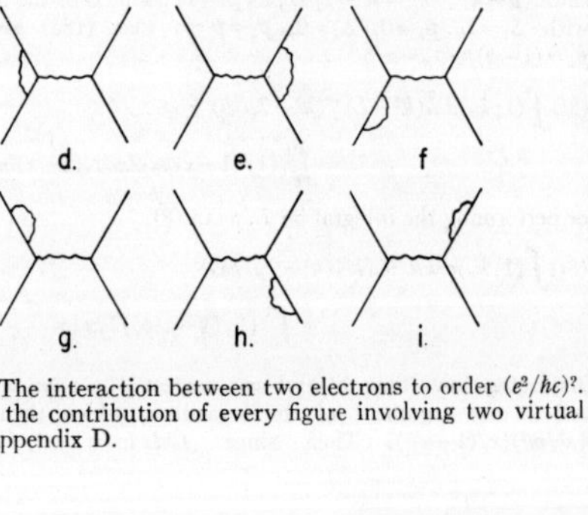

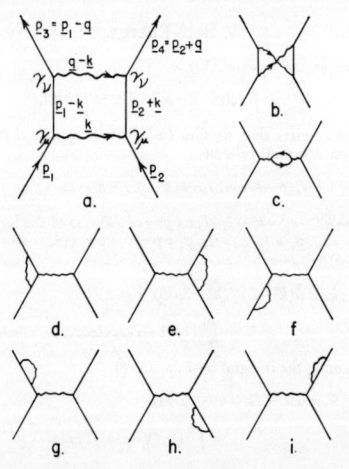

FIG. 8. The interaction between two electrons to order $(e^2/hc)^2$. One adds the contribution of every figure involving two virtual quanta, Appendix D.

ing, to the Compton scattering, and the interaction of a neutron with an electromagnetic field.

For the Møller scattering, consider two electrons, one in state u_1 of momentum p_1 and the other in state u_2 of momentum p_2. Later they are found in states u_3, p_3 and u_4, p_4. This may happen (first order in $e^2/\hbar c$) because they exchange a quantum of momentum $q=p_1-p_3=p_4-p_2$ in the manner of Eq. (4) and Fig. 1. The matrix element for this process is proportional to (translating (4) to momentum space)

$$(\bar{u}_4\gamma_\mu u_2)(\bar{u}_3\gamma_\mu u_1)q^{-2}. \tag{37a}$$

We shall discuss corrections to (37a) to the next order in $e^2/\hbar c$. (There is also the possibility that it is the electron at 2 which finally arrives at 3, the electron at 1 going to 4 through the exchange of quantum of momentum p_3-p_2. The amplitude for this process, $(\bar{u}_4\gamma_\mu u_1)(\bar{u}_3\gamma_\mu u_2)(p_3-p_2)^{-2}$, must be subtracted from (37a) in accordance with the exclusion principle. A similar situation exists to each order so that we need consider in detail only the corrections to (37a), reserving to the last the subtraction of the same terms with 3, 4 exchanged.)

One reason that (37a) is modified is that two quanta may be exchanged, in the manner of Fig. 8a. The total matrix element for all exchanges of this type is

$$(e^2/\pi i)\int(\bar{u}_3\gamma_\nu(p_1-k-m)^{-1}\gamma_\mu u_1)(\bar{u}_4\gamma_\nu(p_2+k-m)^{-1}\gamma_\mu u_2)\cdot k^{-2}(q-k)^{-2}d^4k, \tag{38a}$$

as is clear from the figure and the general rule that electrons of momentum p contribute in amplitude $(p-m)^{-1}$ between interactions γ_μ, and that quanta of momentum k contribute k^{-2}. In integrating on d^4k and summing over μ and ν, we add all alternatives of the type of Fig. 8a. If the time of absorption, γ_μ, of the quantum k by electron 2 is later than the absorption, γ_ν, of $q-k$, this corresponds to the virtual state p_2+k being a positron (so that (38a) contains over thirty terms of the conventional method of analysis).

In integrating over all these alternatives we have considered all possible distortions of Fig. 8a which preserve the order of events along the trajectories. We have not included the possibilities corresponding to Fig. 8b, however. Their contribution is

$$(e^2/\pi i)\int(\bar{u}_3\gamma_\nu(p_1-k-m)^{-1}\gamma_\mu u_1)\times(\bar{u}_4\gamma_\mu(p_2+q-k-m)^{-1}\gamma_\nu u_2)k^{-2}(q-k)^{-2}d^4k, \tag{39a}$$

as is readily verified by labeling the diagram. The contributions of all possible ways that an event can occur are to be added. This means that one adds with equal weight the integrals corresponding to each topologically distinct figure.

To this same order there are also the possibilities of Fig. 8d which give

$$(e^2/\pi i)\int(\bar{u}_3\gamma_\nu(p_3-k-m)^{-1}\gamma_\mu(p_1-k-m)^{-1}\gamma_\nu u_1)\times(\bar{u}_4\gamma_\mu u_2)k^{-2}q^{-2}d^4k.$$

This integral on k will be seen to be precisely the integral (12) for the radiative corrections to scattering, which we have worked out. The term may be combined with the renormalization terms resulting from the difference of the effects of mass change and the terms, Figs. 8f and 8g. Figures 8e, 8h, and 8i are similarly analyzed.

Finally the term Fig. 8c is clearly related to our vacuum polarization problem, and when integrated gives a term proportional to $(\bar{u}_4\gamma_\mu u_2)(\bar{u}_3\gamma_\nu u_1)J_{\mu\nu}q^{-4}$. If the charge is renormalized the term $\ln(\lambda/m)$ in $J_{\mu\nu}$ in (33) is omitted so there is no remaining dependence on the cut-off.

The only new integrals we require are the convergent integrals (38a) and (39a). They can be simplified by rationalizing the denominators and combining them by (14a). For example (38a) involves the factors $(k^2-2p_1\cdot k)^{-1}(k^2+2p_2\cdot k)^{-1}k^{-2}(q^2+k^2-2q\cdot k)^{-2}$. The first two may be combined by (14a) with a parameter x, and the second pair by an expression obtained by differentiation (15a) with respect to b and calling the parameter y. There results a factor $(k^2-2p_x\cdot k)^{-2}(k^2+yq^2-2yq\cdot k)^{-4}$ so that the integrals on d^4k now involve two factors and can be performed by the methods given earlier in the appendix. The subsequent integrals on the parameters x and y are complicated and have not been worked out in detail.

Working with charged mesons there is often a considerable reduction of the number of terms. For example, for the interaction between protons resulting from the exchange of two mesons only the term corresponding to Fig. 8b remains. Term 8a, for example, is impossible, for if the first proton emits a positive meson the second cannot absorb it directly for only neutrons can absorb positive mesons.

As a second example, consider the radiative correction to the Compton scattering. As seen from Eq. (15) and Fig. 5 this scattering is represented by two terms, so that we can consider the corrections to each one separately. Figure 9 shows the types of terms arising from corrections to the term of Fig. 5a. Calling k the momentum of the virtual quantum, Fig. 9a gives an integral

$$\int\gamma_\mu(p_2-k-m)^{-1}e_2(p_1+q_1-k-m)^{-1}e_1(p_1-k-m)^{-1}\gamma_\mu k^{-2}d^4k,$$

convergent without cut-off and reducible by the methods outlined in this appendix.

The other terms are relatively easy to evaluate. Terms b and c of Fig. 9 are closely related to radiative corrections (although somewhat more difficult to evaluate, for one of the states is not that of a free electron, $(p_1+q)^2\neq m^2$). Terms e, f are renormalization terms. From term d must be subtracted explicitly the effect of mass Δm, as analyzed in Eqs. (26) and (27) leading to (28) with $p'=p_1+q$, $a=e_2$, $b=e_1$. Terms g, h give zero since the vacuum polarization has zero effect on free light quanta, $q_1^2=0$, $q_2^2=0$. The total is insensitive to the cut-off λ.

The result shows an infra-red catastrophe, the largest part of the effect. When cut-off at $\lambda_{\min}$, the effect proportional to $\ln(m/\lambda_{\min})$ goes as

$$(e^2/\pi)\ln(m/\lambda_{\min})(1-2\theta\,\text{ctn}\,2\theta), \tag{40a}$$

times the uncorrected amplitude, where $(p_2-p_1)^2=4m^2\sin^2\theta$. This is the same as for the radiative correction to scattering for a deflection p_2-p_1. This is physically clear since the long wave quanta are not effected by short-lived intermediate states. The infra-red effects arise[28] from a final adjustment of the field from the asymptotic coulomb field characteristic of the electron of

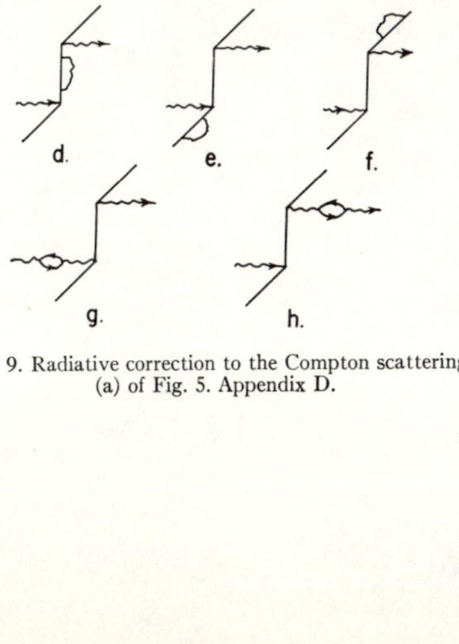

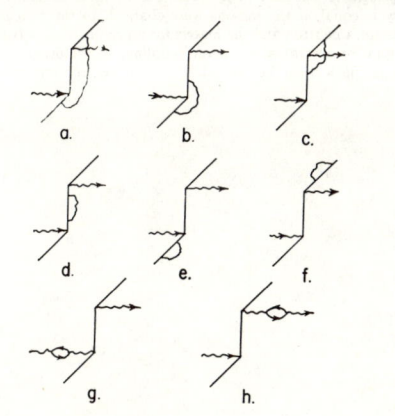

FIG. 9. Radiative correction to the Compton scattering term (a) of Fig. 5. Appendix D.

[28] F. Bloch and A. Nordsieck, Phys. Rev. 52, 54 (1937).

momentum $\boldsymbol{p}_1$ before the collision to that characteristic of an electron moving in a new direction $\boldsymbol{p}_2$ after the collision.

The complete expression for the correction is a very complicated expression involving transcendental integrals.

As a final example we consider the interaction of a neutron with an electromagnetic field in virtue of the fact that the neutron may emit a virtual negative meson. We choose the example of pseudoscalar mesons with pseudovector coupling. The change in amplitude due to an electromagnetic field $A=\boldsymbol{a}\exp(-iq\cdot x)$ determines the scattering of a neutron by such a field. In the limit of small $\boldsymbol{q}$ it will vary as $\boldsymbol{qa}-\boldsymbol{aq}$ which represents the interaction of a particle possessing a magnetic moment. The first-order interaction between an electron and a neutron is given by the same calculation by considering the exchange of a quantum between the electron and the nucleon. In this case a_μ is q^{-2} times the matrix element of γ_μ between the initial and final states of the electron, the states differing in momentum by $\boldsymbol{q}$.

The interaction may occur because the neutron of momentum $\boldsymbol{p}_1$ emits a negative meson becoming a proton which proton interacts with the field and then reabsorbs the meson (Fig. 10a). The matrix for this process is $(\boldsymbol{p}_2=\boldsymbol{p}_1+\boldsymbol{q})$,

$$\int(\gamma_5\boldsymbol{k})(\boldsymbol{p}_2-\boldsymbol{k}-M)^{-1}\boldsymbol{a}(\boldsymbol{p}_1-\boldsymbol{k}-M)^{-1}(\gamma_5\boldsymbol{k})(\boldsymbol{k}^2-\mu^2)^{-1}d^4k. \quad (41\text{a})$$

Alternatively it may be the meson which interacts with the field. We assume that it does this in the manner of a scalar potential satisfying the Klein Gordon Eq. (35), (Fig. 10b)

$$-\int(\gamma_5\boldsymbol{k}_2)(\boldsymbol{p}_1-\boldsymbol{k}_1-M)^{-1}(\gamma_5\boldsymbol{k}_1)(\boldsymbol{k}_2{}^2-\mu^2)^{-1}$$
$$\times(k_2\cdot a+k_1\cdot a)(\boldsymbol{k}_1{}^2-\mu^2)^{-1}d^4k_1, \quad (42\text{a})$$

where we have put $\boldsymbol{k}_2=\boldsymbol{k}_1+\boldsymbol{q}$. The change in sign arises because the virtual meson is negative. Finally there are two terms arising from the $\gamma_5\boldsymbol{a}$ part of the pseudovector coupling (Figs. 10c, 10d)

$$\int(\gamma_5\boldsymbol{k})(\boldsymbol{p}_2-\boldsymbol{k}-M)^{-1}(\gamma_5\boldsymbol{a})(\boldsymbol{k}^2-\mu^2)^{-1}d^4k, \quad (43\text{a})$$

and

$$\int(\gamma_5\boldsymbol{a})(\boldsymbol{p}_1-\boldsymbol{k}-M)^{-1}(\gamma_5\boldsymbol{k})(\boldsymbol{k}^2-\mu^2)^{-1}d^4k. \quad (44\text{a})$$

Using convergence factors in the manner discussed in the section on meson theories each integral can be evaluated and the results combined. Expanded in powers of q the first term gives the magnetic moment of the neutron and is insensitive to the cut-off, the next gives the scattering amplitude of slow electrons on neutrons, and depends logarithmically on the cut-off.

The expressions may be simplified and combined somewhat before integration. This makes the integrals a little easier and also shows the relation to the case of pseudoscalar coupling. For example in (41a) the final $\gamma_5\boldsymbol{k}$ can be written as $\gamma_5(\boldsymbol{k}-\boldsymbol{p}_1+M)$ since $\boldsymbol{p}_1=M$ when operating on the initial neutron state. This is $(\boldsymbol{p}_1-\boldsymbol{k}-M)\gamma_5+2M\gamma_5$ since γ_5 anticommutes with $\boldsymbol{p}_1$ and $\boldsymbol{k}$. The first term cancels the $(\boldsymbol{p}_1-\boldsymbol{k}-M)^{-1}$ and gives a term which just cancels (43a). In a like manner the leading factor $\gamma_5\boldsymbol{k}$ in (41a) is written as $-2M\gamma_5-\gamma_5(\boldsymbol{p}_2-\boldsymbol{k}-M)$, the second term leading to a simpler term containing no $(\boldsymbol{p}_2-\boldsymbol{k}-M)^{-1}$ factor and combining with a similar one from (44a). One simplifies the $\gamma_5\boldsymbol{k}_1$ and $\gamma_5\boldsymbol{k}_2$ in (42a) in an analogous way. There finally results terms like (41a), (42a) but with pseudoscalar coupling $2M\gamma_5$ instead of $\gamma_5\boldsymbol{k}$, no terms like (43a) or (44a) and a remainder, representing the difference in effects of pseudovector and pseudoscalar coupling. The pseudoscalar terms do not depend sensitively on the cut-off, but the difference term depends on it logarithmically. The difference term affects the electron-neutron interaction but not the magnetic moment of the neutron.

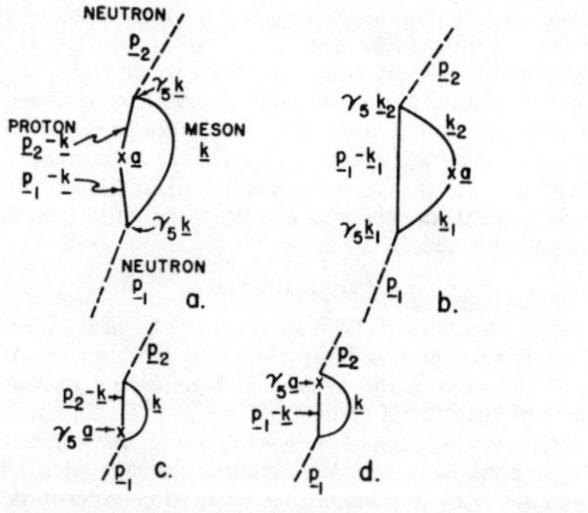

FIG. 10. According to the meson theory a neutron interacts with an electromagnetic potential $\boldsymbol{a}$ by first emitting a virtual charged meson. The figure illustrates the case for a pseudoscalar meson with pseudovector coupling. Appendix D.

Interaction of a proton with an electromagnetic potential can be similarly analyzed. There is an effect of virtual mesons on the electromagnetic properties of the proton even in the case that the mesons are neutral. It is analogous to the radiative corrections to the scattering of electrons due to virtual photons. The sum of the magnetic moments of neutron and proton for charged mesons is the same as the proton moment calculated for the corresponding neutral mesons. In fact it is readily seen by comparing diagrams, that for arbitrary $\boldsymbol{q}$, the scattering matrix to *first order in the electromagnetic potential* for a proton according to neutral meson theory is equal, if the mesons were charged, to the sum of the matrix for a neutron and the matrix for a proton. This is true, for any type or mixtures of meson coupling, to all orders in the coupling (neglecting the mass difference of neutron and proton).

Mathematical Formulation of the Quantum Theory of Electromagnetic Interaction

R. P. FEYNMAN*
Department of Physics, Cornell University, Ithaca, New York

(Received June 8, 1950)

The validity of the rules given in previous papers for the solution of problems in quantum electrodynamics is established. Starting with Fermi's formulation of the field as a set of harmonic oscillators, the effect of the oscillators is integrated out in the Lagrangian form of quantum mechanics. There results an expression for the effect of all virtual photons valid to all orders in $e^2/\hbar c$. It is shown that evaluation of this expression as a power series in $e^2/\hbar c$ gives just the terms expected by the aforementioned rules.

In addition, a relation is established between the amplitude for a given process in an arbitrary unquantized potential and in a quantum electrodynamical field. This relation permits a simple general statement of the laws of quantum electrodynamics.

A description, in Lagrangian quantum-mechanical form, of particles satisfying the Klein-Gordon equation is given in an Appendix. It involves the use of an extra parameter analogous to proper time to describe the trajectory of the particle in four dimensions.

A second Appendix discusses, in the special case of photons, the problem of finding what real processes are implied by the formula for virtual processes.

Problems of the divergences of electrodynamics are not discussed.

1. INTRODUCTION

IN two previous papers[1] rules were given for the calculation of the matrix element for any process in electrodynamics, to each order in $e^2/\hbar c$. No complete proof of the equivalence of these rules to the conventional electrodynamics was given in these papers. Secondly, no closed expression was given valid to all orders in $e^2/\hbar c$. In this paper these formal omissions will be remedied.[2]

In paper **II** it was pointed out that for many problems in electrodynamics the Hamiltonian method is not advantageous, and might be replaced by the over-all space-time point of view of a direct particle interaction. It was also mentioned that the Lagrangian form of quantum mechanics[3] was useful in this connection. The rules given in paper **II** were, in fact, first deduced in this form of quantum mechanics. We shall give this derivation here.

The advantage of a Lagrangian form of quantum mechanics is that in a system with interacting parts it permits a separation of the problem such that the motion of any part can be analyzed or solved first, and the results of this solution may then be used in the solution of the motion of the other parts. This separation is especially useful in quantum electrodynamics which represents the interaction of matter with the electromagnetic field. The electromagnetic field is an especially simple system and its behavior can be analyzed completely. What we shall show is that the net effect of the field is a delayed interaction of the particles. It is possible to do this easily only if it is not necessary at the same time to analyze completely the motion of the particles. The only advantage in our problems of the form of quantum mechanics in **C** is to permit one to separate these aspects of the problem. There are a number of disadvantages, however, such as a lack of familiarity, the apparent (but not real) necessity for dealing with matter in non-relativistic approximation, and at times a cumbersome mathematical notation and method, as well as the fact that a great deal of useful information that is known about operators cannot be directly applied.

It is also possible to separate the field and particle aspects of a problem in a manner which uses operators and Hamiltonians in a way that is much more familiar. One abandons the notation that the order of action of operators depends on their written position on the paper and substitutes some other convention (such that the order of operators is that of the time to which they refer). The increase in manipulative facility which accompanies this change in notation makes it easier to represent and to analyze the formal problems in electrodynamics. The method requires some discussion, however, and will be described in a succeeding paper. In this paper we shall give the derivations of the formulas of **II** by means of the form of quantum mechanics given in **C**.

The problem of interaction of matter and field will be analyzed by first solving for the behavior of the field in terms of the coordinates of the matter, and finally discussing the behavior of the matter (by matter is actually meant the electrons and positrons). That is to say, we shall first eliminate the field variables from the equations of motion of the electrons and then discuss the behavior of the electrons. In this way all of the rules given in the paper **II** will be derived.

Actually, the straightforward elimination of the field

*Now at the California Institute of Technology, Pasadena, California.

[1] R. P. Feynman, Phys. Rev. **76**, 749 (1949), hereafter called **I**, and Phys. Rev. **76**, 769 (1949), hereafter called **II**.

[2] See in this connection also the papers of S. Tomonaga, Phys. Rev. **74**, 224 (1948); S. Kanesawa and S. Tomonaga, Prog. Theoret. Phys. **3**, 101 (1948); J. Schwinger, Phys. Rev. **76**, 790 (1949); F. Dyson, Phys. Rev. **75**, 1736 (1949); W. Pauli and F. Villars, Rev. Mod. Phys. **21**, 434 (1949). The papers cited give references to previous work.

[3] R. P. Feynman, Rev. Mod. Phys. **20**, 367 (1948), hereafter called **C**.

variables will lead at first to an expression for the behavior of an arbitrary number of Dirac electrons. Since the number of electrons might be infinite, this can be used directly to find the behavior of the electrons according to hole theory by imagining that nearly all the negative energy states are occupied by electrons. But, at least in the case of motion in a fixed potential, it has been shown that this hole theory picture is equivalent to one in which a positron is represented as an electron whose space-time trajectory has had its time direction reversed. To show that this same picture may be used in quantum electrodynamics when the potentials are not fixed, a special argument is made based on a study of the relationship of quantum electrodynamics to motion in a fixed potential. Finally, it is pointed out that this relationship is quite general and might be used for a general statement of the laws of quantum electrodynamics.

Charges obeying the Klein-Gordon equation can be analyzed by a special formalism given in Appendix A. A fifth parameter is used to specify the four-dimensional trajectory so that the Lagrangian form of quantum mechanics can be used. Appendix B discusses in more detail the relation of real and virtual photon emission. An equation for the propagation of a self-interacting electron is given in Appendix C.

In the demonstration which follows we shall restrict ourselves temporarily to cases in which the particle's motion is non-relativistic, but the transition of the final formulas to the relativistic case is direct, and the proof could have been kept relativistic throughout.

The transverse part of the electromagnetic field will be represented as an assemblage of independent harmonic oscillators each interacting with the particles, as suggested by Fermi.[4] We use the notation of Heitler.[5]

2. QUANTUM ELECTRODYNAMICS IN LAGRANGIAN FORM

The Hamiltonian for a set of non-relativistic particles interacting with radiation is, classically, $H=H_p+H_I+H_c+H_{tr}$, where $H_p+H_I=\sum_n \frac{1}{2}m_n^{-1}(\mathbf{p}_n-e_n\mathbf{A}^{tr}(\mathbf{x}_n))^2$ is the Hamiltonian of the particles of mass m_n, charge e_n, coordinate $\mathbf{x}_n$ and momentum $\mathbf{p}_n$ and their interaction with the transverse part of the electromagnetic field. This field can be expanded into plane waves

$$\mathbf{A}^{tr}(\mathbf{x})=(8\pi)^{\frac{1}{2}}\sum_{\mathbf{K}}[\mathbf{e}_1(q_{\mathbf{K}}^{(1)}\cos(\mathbf{K}\cdot\mathbf{x})+q_{\mathbf{K}}^{(3)}\sin(\mathbf{K}\cdot\mathbf{x})) + \mathbf{e}_2(q_{\mathbf{K}}^{(2)}\cos(\mathbf{K}\cdot\mathbf{x})+q_{\mathbf{K}}^{(4)}\sin(\mathbf{K}\cdot\mathbf{x}))] \quad (1)$$

where $\mathbf{e}_1$ and $\mathbf{e}_2$ are two orthogonal polarization vectors at right angles to the propagation vector $\mathbf{K}$, magnitude k. The sum over $\mathbf{K}$ means, if normalized to unit volume, $\frac{1}{2}\int d^3\mathbf{K}/8\pi^3$, and each $q_{\mathbf{K}}^{(r)}$ can be considered as the coordinate of a harmonic oscillator. (The factor $\frac{1}{2}$ arises for the mode corresponding to $\mathbf{K}$ and to $-\mathbf{K}$ is the same.) The Hamiltonian of the transverse field represented as oscillators is

$$H_{tr}=\frac{1}{2}\sum_{\mathbf{K}}\sum_{r=1}^{4}((p_{\mathbf{K}}^{(r)})^2+k^2(q_{\mathbf{K}}^{(r)})^2)$$

where $p_{\mathbf{K}}^{(r)}$ is the momentum conjugate to $q_{\mathbf{K}}^{(r)}$. The longitudinal part of the field has been replaced by the Coulomb interaction,[6]

$$H_c=\tfrac{1}{2}\sum_n\sum_m e_ne_m/r_{nm}$$

where $r_{nm}^2=(\mathbf{x}_n-\mathbf{x}_m)^2$. As is well known,[4] when this Hamiltonian is quantized one arrives at the usual theory of quantum electrodynamics. To express these laws of quantum electrodynamics one can equally well use the Lagrangian form of quantum mechanics to describe this set of oscillators and particles. The classical Lagrangian equivalent to this Hamiltonian is $L=L_p+L_I+L_c+L_{tr}$ where

$$L_p=\tfrac{1}{2}\sum_n m_n\dot{\mathbf{x}}_n^2 \quad (2a)$$

$$L_I=\sum_n e_n\dot{\mathbf{x}}_n\cdot\mathbf{A}^{tr}(\mathbf{x}_n) \quad (2b)$$

$$L_{tr}=\tfrac{1}{2}\sum_{\mathbf{K}}\sum_r((\dot{q}_{\mathbf{K}}^{(r)})^2-k^2(q_{\mathbf{K}}^{(r)})^2) \quad (2c)$$

$$L_c=-\tfrac{1}{2}\sum_n\sum_m e_ne_m/r_{mn}. \quad (2d)$$

When this Lagrangian is used in the Lagrangian forms of quantum mechanics of **C**, what it leads to is, of course, mathematically equivalent to the result of using the Hamiltonian H in the ordinary way, and is therefore equivalent to the more usual forms of quantum electrodynamics (at least for non-relativistic particles). We may, therefore, proceed by using this Lagrangian form of quantum electrodynamics, with the assurance that the results obtained must agree with those obtained from the more usual Hamiltonian form.

The Lagrangian enters through the statement that the functional which carries the system from one state to another is $\exp(iS)$ where

$$S=\int Ldt=S_p+S_I+S_c+S_{tr}. \quad (3)$$

The time integrals must be written as Riemann sums with some care; for example,

$$S_I=\sum_n\int e_n\dot{\mathbf{x}}_n(t)\cdot\mathbf{A}^{tr}(\mathbf{x}_n(t))dt \quad (4)$$

becomes according to **C**, Eq. (19)

$$S_I=\sum_n\sum_i\tfrac{1}{2}e_n(\mathbf{x}_{n,i+1}-\mathbf{x}_{n,i})\cdot(\mathbf{A}^{tr}(\mathbf{x}_{n,i+1})+\mathbf{A}^{tr}(\mathbf{x}_{n,i})) \quad (5)$$

so that the velocity $\dot{\mathbf{x}}_{n,i}$ which multiplies $\mathbf{A}^{tr}(\mathbf{x}_{n,i})$ is

$$\dot{\mathbf{x}}_{n,i}=\tfrac{1}{2}\epsilon^{-1}(\mathbf{x}_{n,i+1}-\mathbf{x}_{n,i})+\tfrac{1}{2}\epsilon^{-1}(\mathbf{x}_{n,i}-\mathbf{x}_{n,i-1}). \quad (6)$$

[4] E. Fermi, Rev. Mod. Phys. 4, 87 (1932).

[5] W. Heitler, *The Quantum Theory of Radiation*, second edition (Oxford University Press, London, 1944).

[6] The term in the sum for $n=m$ is obviously infinite but must be included for relativistic invariance. Our problem here is to re-express the usual (and divergent) form of electrodynamics in the form given in **II**. Modifications for dealing with the divergences are discussed in **II** and we shall not discuss them further here.

In the Lagrangian form it is possible to eliminate the transverse oscillators as is discussed in **C**, Section 13. One must specify, however, the initial and final state of all oscillators. We shall first choose the special, simple case that all oscillators are in their ground states initially and finally, so that all photons are virtual. Later we do the more general case in which real quanta are present initially or finally. We ask, then, for the amplitude for finding no quanta present and the particles in state $\chi_{t''}$ at time t'', if at time t' the particles were in state $\psi_{t'}$ and no quanta were present.

The method of eliminating field oscillators is described in Section 13 of **C**. We shall simply carry out the elimination here using the notation and equations of **C**. To do this, for simplicity, we first consider in the next section the case of a particle or a system of particles interacting with a single oscillator, rather than the entire assemblage of the electromagnetic field.

3. FORCED HARMONIC OSCILLATOR

We consider a harmonic oscillator, coordinate q, Lagrangian $L=\frac{1}{2}(\dot{q}^2-\omega^2q^2)$ interacting with a particle or system of particles, action S_p, through a term in the Lagrangian $q(t)\gamma(t)$ where $\gamma(t)$ is a function of the coordinates (symbolized as x) of the particle. The precise form of $\gamma(t)$ for each oscillator of the electromagnetic field is given in the next section. We ask for the amplitude that at some time t'' the particles are in state $\chi_{t''}$ and the oscillator is in, say, an eigenstate m of energy $\omega(m+\frac{1}{2})$ (units are chosen such that $\hbar=c=1$) when it is given that at a previous time t' the particles were in state $\psi_{t'}$ and the oscillator in n. The amplitude for this is the transition amplitude [see **C**, Eq. (61)]

$$\langle\chi_{t''}\varphi_m|1|\psi_{t'}\varphi_n\rangle_{S_p+S_0+S_I}$$

$$=\int\int\chi_{t''}^*(x_{t''})\varphi_m^*(q_{t''})\exp i(S_p+S_0+S_I)$$

$$\cdot\varphi_n(q_{t'})\psi_{t'}(x_{t'})dx_{t''}dx_{t'}dq_{t''}dq_{t'}\mathfrak{D}x(t)\mathfrak{D}q(t)\quad(7)$$

where x represents the variables describing the particle, S_p is the action calculated classically for the particles for a given path going from coordinate $x_{t'}$ at t' to $x_{t''}$ at t'', S_0 is the action $\int\frac{1}{2}(\dot{q}^2-\omega^2q^2)dt$ for any path of the oscillator going from $q_{t'}$ at t' to $q_{t''}$ at t'', while

$$S_I=\int q(t)\gamma(t)dt,\quad(8)$$

the action of interaction, is a functional of both $q(t)$ and $x(t)$, the paths of oscillator and particles. The symbols $\mathfrak{D}x(t)$ and $\mathfrak{D}q(t)$ represent a summation over all possible paths of particles and oscillator which go between the given end points in the sense defined in **C**, Eq. (9). (That is, assuming time to proceed in infinitesimal steps, ϵ, an integral over all values of the coordinates x and q corresponding to each instant in time, suitably normalized.)

The problem may be broken in two. The result can be written as an integral over all paths of the particles only, of $(\exp iS_p)\cdot G_{mn}$:

$$\langle\chi_{t''}\varphi_m|1|\psi_{t'}\varphi_n\rangle_{S_p+S_0+S_I}=\langle\chi_{t''}|G_{mn}|\psi_{t'}\rangle_{S_p}\quad(9)$$

where G_{mn} is a functional of the path of the particles alone (since it depends on $\gamma(t)$) given by

$$G_{mn}=\left\langle\varphi_m\left|\exp i\int q(t)\gamma(t)dt\right|\varphi_n\right\rangle_{S_0}$$

$$=\int\varphi_m^*(q_{t''})\exp i(S_0+S_I)\varphi_n(q_{t'})dq_{t'}dq_{t''}\mathfrak{D}q(t)$$

$$=\int\varphi_m^*(q_j)\exp i\epsilon\sum_{i=0}^{j-1}[\tfrac{1}{2}\epsilon^{-2}(q_{i+1}-q_i)^2-\tfrac{1}{2}\omega^2q_i^2+q_i\gamma_i]$$

$$\cdot\varphi_n(q_0)dq_0a^{-1}dq_1a^{-1}dq_2\cdots a^{-1}dq_j\quad(10)$$

where we have written the $\mathfrak{D}q(t)$ out explicitly (and have set $a=(2\pi i\epsilon)^{\frac{1}{2}}$, $t''-t'=j\epsilon$, $q_{t'}=q_0$, $q_{t''}=q_j$). The last form can be written as

$$G_{mn}=\int\varphi_m^*(q_j)k(q_j,t'';q_0,t')\varphi_n(q_0)dq_0dq_j\quad(11)$$

where $k(q_j,t'';q_0,t')$ is the kernel [as in **I**, Eq. (2)] for a forced harmonic oscillator giving the amplitude for arrival at q_j at time t'' if at time t' it was known to be at q_0. According to **C** it is given by

$$k(q_j,t'';q_0,t')=(2\pi i\omega^{-1}\sin\omega(t''-t'))^{-\frac{1}{2}}$$
$$\times\exp iQ(q_j,t'';q_0,t')\quad(12)$$

where $Q(q_j,t'';q_0,t')$ is the action calculated along the classical path between the end points $q_j,t'';q_0,t'$, and is given explicitly in **C**.[7] It is

---

[7] That (12) is correct, at least insofar as it depends on q_0, can be seen directly as follows. Let $\bar{q}(t)$ be the classical path which satisfies the boundary condition $\bar{q}(t')=q_0$, $\bar{q}(t'')=q_j$. Then in the integral defining k replace each of the variables q_i by $q_i=\bar{q}_i+y_i$, $(\bar{q}_i=\bar{q}(t_i))$, that is, use the displacement y_i from the classical path $\bar{q}_i$ as the coordinate rather than the absolute position. With the substitution $q_i=\bar{q}_i+y_i$ in the action

$$S_0+S_I=\int(\tfrac{1}{2}\dot{q}^2-\tfrac{1}{2}\omega^2q^2+\gamma q)dt$$
$$=\int(\tfrac{1}{2}\dot{\bar{q}}^2-\tfrac{1}{2}\omega^2\bar{q}^2+\gamma\bar{q})dt+\int(\tfrac{1}{2}\dot{y}^2-\tfrac{1}{2}\omega^2y^2)dt$$

the terms linear in y drop out by integrations by parts using the equation of motion $\ddot{\bar{q}}=-\omega^2\bar{q}+\gamma(t)$ for the classical path, and the boundary conditions $y(t')=y(t'')=0$. That this should occur should occasion no surprise, for the action functional is an extremum at $q(t)=\bar{q}(t)$ so that it will only depend to second order in the displacements y from this extremal orbit $\bar{q}(t)$. Further, since the action functional is quadratic to begin with, it cannot depend on y more than quadratically. Hence

$$S_0+S_I=Q+\int(\tfrac{1}{2}\dot{y}^2-\tfrac{1}{2}\omega^2y^2)dt$$

so that since $dq_i=dy_i$,

$$k(q_j,t'';q_0,t')=\exp(iQ)\int\exp\left(i\int\frac{1}{2}(\dot{y}^2-\omega^2y^2)dt\right)\mathfrak{D}y(t).$$

The factor following the $\exp iQ$ is the amplitude for a free oscillator to proceed from $y=0$ at $t=t'$ to $y=0$ at $t=t''$ and does not there-

$$Q=\frac{\omega}{2\sin\omega(t''-t')}\Big[(q_j^2+q_0^2)\cos\omega(t''-t')-2q_jq_0$$
$$+\frac{2q_j}{\omega}\int_{t'}^{t''}\gamma(t)\sin\omega(t-t')dt$$
$$+\frac{2q_0}{\omega}\int_{t'}^{t''}\gamma(t)\sin\omega(t''-t)dt$$
$$-\frac{2}{\omega^2}\int_{t'}^{t''}\int_{t'}^{t}\gamma(t)\gamma(s)\sin\omega(t''-t)$$
$$\times\sin\omega(s-t')dsdt\Big]. \quad (13)$$

The solution of the motion of the oscillator can now be completed by substituting (12) and (13) into (11) and performing the integrals. The simplest case is for $m, n=0$ for which case[8]

$$\varphi_0(q_0)=(\omega/\pi)^{\frac{1}{4}}\exp(-\tfrac{1}{2}\omega q_0^2)\exp(-\tfrac{1}{2}i\omega t')$$

so that the integrals on q_0, q_j are just Gaussian integrals. There results

$$G_{00}=\exp\left(-\frac{1}{2}\omega^{-1}\int_{t'}^{t''}\int_{t'}^{t}\exp(-i\omega(t-s))\gamma(t)\gamma(s)dtds\right)$$

a result of fundamental importance in the succeeding developments. By replacing $t-s$ by its absolute value $|t-s|$ we may integrate both variables over the entire range and divide by 2. We will henceforth make the results more general by extending the limits on the integrals from $-\infty$ to $+\infty$. Thus if one wishes to study the effect on a particle of interaction with an oscillator for just the period t' to t'' one may use

$$G_{00}=\exp\left(-\frac{1}{4\omega}\int_{-\infty}^{\infty}\int_{-\infty}^{\infty}\right.$$
$$\left.\times\exp(-i\omega|t-s|)\gamma(t)\gamma(s)dtds\right) \quad (14)$$

imagining in this case that the interaction $\gamma(t)$ is zero outside these limits. We defer to a later section the discussion of other values of m, n.

Since G_{00} is simply an exponential, we can write it as $\exp(iI)$, consider that the complete "action" for the system of particles is $S=S_p+I$ and that one computes transition elements with this "action" instead of S_p (see **C**, Sec. 12). The functional I, which is given by

$$I=\tfrac{1}{4}i\omega^{-1}\int\int\exp(-i\omega|t-s|)\gamma(s)\gamma(t)dsdt \quad (15)$$

is complex, however; we shall speak of it as the complex action. It describes the fact that the system at one time can affect itself at a different time by means of a temporary storage of energy in the oscillator. When there are several independent oscillators with different interactions, the effect, if they are all in the lowest state at t' and t'', is the product of their separate G_{00} contributions. Thus the complex action is additive, being the sum of contributions like (15) for each of the several oscillators.

4. VIRTUAL TRANSITIONS IN THE ELECTROMAGNETIC FIELD

We can now apply these results to eliminate the transverse field oscillators of the Lagrangian (2). At first we can limit ourselves to the case of purely virtual transitions in the electromagnetic field, so that there is no photon in the field at t' and t''. That is, all of the field oscillators are making transitions from ground state to ground state.

The $\gamma_{\mathbf{K}}^{(r)}$ corresponding to each oscillator $q_{\mathbf{K}}^{(r)}$ is found from the interaction term L_I [Eq. (2b)], substituting the value of $\mathbf{A}^{tr}(\mathbf{x})$ given in (1). There results, for example,

$$\begin{aligned}\gamma_{\mathbf{K}}^{(1)}&=(8\pi)^{\frac{1}{2}}\textstyle\sum_n e_n(\mathbf{e}_1\cdot\dot{\mathbf{x}}_n)\cos(\mathbf{K}\cdot\mathbf{x}_n)\\ \gamma_{\mathbf{K}}^{(3)}&=(8\pi)^{\frac{1}{2}}\textstyle\sum_n e_n(\mathbf{e}_1\cdot\dot{\mathbf{x}}_n)\sin(\mathbf{K}\cdot\mathbf{x}_n)\end{aligned} \quad (16)$$

the corresponding results for $\gamma_{\mathbf{K}}^{(2)}$, $\gamma_{\mathbf{K}}^{(4)}$ replace $\mathbf{e}_1$ by $\mathbf{e}_2$.

The complex action resulting from oscillator of coordinate $q_{\mathbf{K}}^{(1)}$ is therefore

$$I_{\mathbf{K}}^{(1)}=\frac{8\pi i}{4k}\sum_n\sum_m\int\int e_ne_m\exp(-ik|t-s|)(\mathbf{e}_1\cdot\dot{\mathbf{x}}_n(t))$$
$$\times(\mathbf{e}_1\cdot\dot{\mathbf{x}}_m(s))\cdot\cos(\mathbf{K}\cdot\mathbf{x}_n(t))\cos(\mathbf{K}\cdot\mathbf{x}_m(s))dsdt.$$

The term $I_{\mathbf{K}}^{(3)}$ exchanges the cosines for sines, so in the sum $I_{\mathbf{K}}^{(1)}+I_{\mathbf{K}}^{(3)}$ the product of the two cosines, $\cos A\cdot\cos B$ is replaced by $(\cos A\cos B+\sin A\sin B)$ or $\cos(A-B)$. The terms $I_{\mathbf{K}}^{(2)}+I_{\mathbf{K}}^{(4)}$ give the same result with $\mathbf{e}_2$ replacing $\mathbf{e}_1$. The sum $(\mathbf{e}_1\cdot\mathbf{V})(\mathbf{e}_1\cdot\mathbf{V}')+(\mathbf{e}_2\cdot\mathbf{V})(\mathbf{e}_2\cdot\mathbf{V}')$ is $(\mathbf{V}\cdot\mathbf{V}')-k^{-2}(\mathbf{K}\cdot\mathbf{V})(\mathbf{K}\cdot\mathbf{V}')$ since it is the sum of the products of vector components in two orthogonal directions, so that if we add the product in the third direction (that of $\mathbf{K}$) we construct the complete scalar product. Summing over all $\mathbf{K}$ then, since $\sum_{\mathbf{K}}=\frac{1}{2}\int d^3\mathbf{K}/8\pi^3$ we find for the total complex action of all of the transverse oscillators,

$$I_{tr}=i\sum_n\sum_m\int_{t'}^{t''}dt\int_{t'}^{t''}ds\int e_ne_m\exp(-ik|t-s|)$$
$$\times[\dot{\mathbf{x}}_n(t)\cdot\dot{\mathbf{x}}_m(s)-k^{-2}(\mathbf{K}\cdot\dot{\mathbf{x}}_n(t))(\mathbf{K}\cdot\dot{\mathbf{x}}_m(s))]$$
$$\cdot\cos(\mathbf{K}\cdot(\mathbf{x}_n(t)-\mathbf{x}_m(s)))d^3\mathbf{K}/8\pi^2k. \quad (17)$$

fore depend on q_0, q_j, or $\gamma(t)$, being a function only of $t''-t'$. [That it is actually $(2\pi i\omega^{-1}\sin\omega(t''-t'))^{-\frac{1}{2}}$ can be demonstrated either by direct integration of the y variables or by using some normalizing property of the kernels k, for example that G_{00} for the case $\gamma=0$ must equal unity.] The expression for Q given in **C** on page 386 is in error, the quantities q_0 and q_j should be interchanged.

[8] It is most convenient to define the state φ_n with the phase factor $\exp[-i\omega(n+\frac{1}{2})t']$ and the final state with the factor $\exp[-i\omega(m+\frac{1}{2})t'']$ so that the results will not depend on the particular times t', t'' chosen.

This is to be added to S_p+S_c to obtain the complete action of the system with the oscillators removed.

The term in $(\mathbf{K}\cdot\dot{\mathbf{x}}_n(t))(\mathbf{K}\cdot\dot{\mathbf{x}}_m(s))$ can be simplified by integration by parts with respect to t and with respect to s [note that $\exp(-ik|t-s|)$ has a discontinuous slope at $t=s$, or break the integration up into two regions]. One finds

$$I_{tr}=R-I_c+I_{\text{transient}} \quad (18)$$

where

$$R=-i\sum_n\sum_m\int_{t'}^{t''}dt\int_{t'}^{t''}ds\int e_ne_m$$
$$\times\exp(-ik|t-s|)(1-\dot{\mathbf{x}}_n(t)\cdot\dot{\mathbf{x}}_m(s))$$
$$\cdot\cos\mathbf{K}\cdot(\mathbf{x}_n(t)-\mathbf{x}_m(s))d^3\mathbf{K}/8\pi^2k \quad (19)$$

and

$$I_c=-\sum_n\sum_m\int_{t'}^{t''}dt\int e_ne_m$$
$$\times\cos\mathbf{K}\cdot(\mathbf{x}_n(t)-\mathbf{x}_m(t))d^3\mathbf{K}/4\pi^2k^2 \quad (20)$$

comes from the discontinuity in slope of $\exp(-ik|t-s|)$ at $t=s$. Since

$$\int\cos(\mathbf{K}\cdot\mathbf{R})d^3\mathbf{K}/4\pi^2k^2=\int_0^\infty(kr)^{-1}\sin(kr)dk/\pi=(2r)^{-1}$$

this term I_c just cancels the Coulomb interaction term $S_c=\int L_cdt$. The term

$$I_{\text{transient}}=-\sum_n\sum_m e_ne_m\int\frac{d^3\mathbf{K}}{4\pi^2k^2}$$
$$\times\Big\{\int_{t'}^{t''}[\exp(-ik(t''-t))\cos\mathbf{K}\cdot(\mathbf{x}_n(t'')-\mathbf{x}_m(t))$$
$$+\exp(-ik(t-t'))\cos\mathbf{K}\cdot(\mathbf{x}_n(t)-\mathbf{x}_m(t'))]dt$$
$$+(2k)^{-1}i[\cos\mathbf{K}\cdot(\mathbf{x}_n(t'')-\mathbf{x}_m(t''))$$
$$+\cos\mathbf{K}\cdot(\mathbf{x}_n(t')-\mathbf{x}_m(t'))$$
$$-2\exp(-ik(t''-t'))\cos\mathbf{K}\cdot(\mathbf{x}_n(t')-\mathbf{x}_m(t''))]\Big\}. \quad (21)$$

is one which comes from the limits of integration at t' and t'', and involves the coordinates of the particle at either one of these times or the other. If t' and t'' are considered to be exceedingly far in the past and future, there is no correlation to be expected between these temporally distant coordinates and the present ones, so the effects of $I_{\text{transient}}$ will cancel out quantum mechanically by interference. This transient was produced by the sudden turning on of the interaction of field and particles at t' and its sudden removal at t''. Alternatively we can imagine the charges to be turned on after t' adiabatically and turned off slowly before t'' (in this case, in the term L_c, the charges should also be considered as varying with time). In this case, in the limit, $I_{\text{transient}}$ is zero.[9] Hereafter we shall drop the transient term and consider the range of integration of t to be from $-\infty$ to $+\infty$, imagining, if one needs a definition, that the charges vary with time and vanish in the direction of either limit.

To simplify R we need the integral

$$J=\int\exp(-ik|t|)\cos(\mathbf{K}\cdot\mathbf{R})d^3\mathbf{K}/8\pi^2k$$
$$=\int_0^\infty\exp(-ik|t|)\sin(kr)dk/2\pi r \quad (22)$$

where r is the length of the vector $\mathbf{R}$. Now

$$\int_0^\infty\exp(-ikx)dk=\lim_{\epsilon\to0}(-i(x-i\epsilon)^{-1})$$
$$=-ix^{-1}+\pi\delta(x)=\pi\delta_+(x)$$

where the equation serves to define $\delta_+(x)$ [as in **II**, Eq. (3)]. Hence, expanding $\sin(kr)$ in exponentials find

$$J=-(4\pi r)^{-1}((|t|-r)^{-1}-(|t|+r)^{-1})$$
$$+(4ir)^{-1}(\delta(|t|-r)-\delta(|t|+r))$$
$$=-(2\pi)^{-1}(t^2-r^2)^{-1}+(2i)^{-1}\delta(t^2-r^2)$$
$$=-\tfrac{1}{2}i\delta_+(t^2-r^2) \quad (23)$$

where we have used the fact that

$$\delta(t^2-r^2)=(2r)^{-1}(\delta(|t|-r)+\delta(|t|+r))$$

and that $\delta(|t|+r)=0$ since both $|t|$ and r are necessarily positive.

Substitution of these results into (19) gives finally,

$$R=-\frac{1}{2}\sum_n\sum_m\int_{-\infty}^{+\infty}\int_{-\infty}^{+\infty}e_ne_m(1-\dot{\mathbf{x}}_n(t)\cdot\dot{\mathbf{x}}_m(s))$$
$$\times\delta_+((t-s)^2-(\mathbf{x}_n(t)-\mathbf{x}_m(s))^2)dtds. \quad (24)$$

The total complex action of the system is then[10] S_p+R. Or, what amounts to the same thing; to obtain

[9] One can obtain the final result, that the total interaction is just R, in a formal manner starting from the Hamiltonian from which the longitudinal oscillators have not yet been eliminated. There are for each $\mathbf{K}$ and cos or sin, four oscillators $q_{\mu\mathbf{K}}$ corresponding to the three components of the vector potential ($\mu=1$, 2, 3) and the scalar potential ($\mu=4$). It must then be assumed that the wave functions of the initial and final state of the $\mathbf{K}$ oscillators is the function $(k/\pi)\exp[-\frac{1}{2}k(q_{1\mathbf{K}}^2+q_{2\mathbf{K}}^2+q_{3\mathbf{K}}^2-q_{4\mathbf{K}}^2)]$. The wave function suggested here has only formal significance, of course, because the dependence on $q_{4\mathbf{K}}$ is not square integrable, and cannot be normalized. If each oscillator were assumed actually in the ground state, the sign of the $q_{4\mathbf{K}}$ term would be changed to positive, and the sign of the frequency in the contribution of these oscillators would be reversed (they would have negative energy).

[10] The classical action for this problem is just $Sp+R'$ where R' is the real part of the expression (24). In view of the generalization of the Lagrangian formulation of quantum mechanics suggested in Section 12 of **C**, one might have anticipated that R would have been simply R'. This corresponds, however, to boundary conditions other than no quanta present in past and future. It is harder to interpret physically. For a system enclosed in a light tight box, however, it appears likely that both R and R' lead to the same results.

transition amplitudes including the effects of the field we must calculate the transition element of $\exp(iR)$:

$$\langle\chi_{t''}|\exp iR|\psi_{t'}\rangle_{S_p} \tag{25}$$

under the action S_p of the particles, excluding interaction. Expression (24) for R must be considered to be written in the usual manner as a Riemann sum and the expression (25) interpreted as defined in **C** [Eq. (39)]. Expression (6) must be used for $\dot{\mathbf{x}}_n$ at time t.

Expression (25), with (24), then contains all the effects of virtual quanta on a (at least non-relativistic) system according to quantum electrodynamics. It contains the effects to all orders in $e^2/\hbar c$ in a single expression. If expanded in a power series in $e^2/\hbar c$, the various terms give the expressions to the corresponding order obtained by the diagrams and methods of **II**. We illustrate this by an example in the next section.

5. EXAMPLE OF APPLICATION OF EXPRESSION (25)

We shall not be much concerned with the non-relativistic case here, as the relativistic case given below is as simple and more interesting. It is, however, very similar and at this stage it is worth giving an example to show how expressions resulting from (25) are to be interpreted according to the rules of **C**. For example, consider the case of a single electron, coordinate $\mathbf{x}$, either free or in an external given potential (contained for simplicity in S_p, not in[11] R). Its interaction with the field produces a reaction back on itself given by R as in (24) but in which we keep only a single term corresponding to $m=n$. Assume the effect of R to be small and expand $\exp(iR)$ as $1+iR$. Let us find the amplitude at time t'' of finding the electron in a state ψ with no quanta emitted, if at time t' it was in the same state. It is

$$\langle\psi_{t''}|1+iR|\psi_{t'}\rangle_{S_p}=\langle\psi_{t''}|1|\psi_{t'}\rangle_{S_p}+i\langle\psi_{t''}|R|\psi_{t'}\rangle_{S_p}$$

where $\langle\psi_{t''}|1|\psi_{t'}\rangle_{S_p}=\exp[-iE(t''-t')]$ if E is the energy of the state, and

$$\langle\psi_{t''}|R|\psi_{t'}\rangle_{S_p}=-\tfrac{1}{2}e^2\int_{t'}^{t''}dt\int_{t'}^{t''}ds\langle\psi_{t''}|(1-\dot{\mathbf{x}}_t\cdot\dot{\mathbf{x}}_s)$$
$$\times\delta_+((t-s)^2-(\mathbf{x}_t-\mathbf{x}_s)^2)|\psi_{t'}\rangle_{S_p}. \tag{26}$$

Here $\mathbf{x}_s=\mathbf{x}(s)$, etc. In (26) we shall limit the range of integrations by assuming $s<t$, and double the result.

The expression within the brackets $\langle\ \rangle_{S_p}$ on the right-hand side of (26) can be evaluated by the methods described in **C** [Eq. (29)]. An expression such as (26) can also be evaluated directly in terms of the propagation kernel $K(2, 1)$ [see **I**, Eq. (2)] for an electron moving in the given potential.

The term $\dot{\mathbf{x}}_s\cdot\dot{\mathbf{x}}_t$ in the non-relativistic case produces an interesting complication which does not have an analog for the relativistic case with the Dirac equation. We discuss it below, but for a moment consider in further detail expression (26) but with the factor $(1-\dot{\mathbf{x}}_s\cdot\dot{\mathbf{x}}_t)$ replaced simply by unity.

The kernel $K(2, 1)$ is defined and discussed in **I**. From its definition as the amplitude that the electron be found at $\mathbf{x}_2$ at time t_2, if at t_1 it was at $\mathbf{x}_1$, we have

$$K(\mathbf{x}_2, t_2; \mathbf{x}_1, t_1)=\langle\delta(\mathbf{x}-\mathbf{x}_2)_{t_2}|1|\delta(\mathbf{x}-\mathbf{x}_1)_{t_1}\rangle_{S_p} \tag{27}$$

that is, more simply $K(2, 1)$ is the sum of $\exp(iS_p)$ over all paths which go from space time point 1 to 2.

In the integrations over all paths implied by the symbol in (26) we can first integrate over all the $\mathbf{x}_i$ variables corresponding to times t_i from t' to s, not inclusive, the result being a factor $K(\mathbf{x}_s, s; \mathbf{x}_{t'}, t')$ according to (27). Next we integrate on the variables between s and t not inclusive, giving a factor $K(\mathbf{x}_t, t; \mathbf{x}_s, s)$ and finally on those between t and t'' giving $K(\mathbf{x}_{t''}, t''; \mathbf{x}_t, t)$. Hence the left-hand term in (26) excluding the $\dot{\mathbf{x}}_t\cdot\dot{\mathbf{x}}_s$ factor is

$$-e^2\int dt\int ds\int\psi^*(\mathbf{x}_{t''}, t'')K(\mathbf{x}_{t''}, t''; \mathbf{x}_t, t)\delta_+((t-s)^2$$
$$-(\mathbf{x}_t-\mathbf{x}_s)^2)\cdot K(\mathbf{x}_t, t; \mathbf{x}_s, s)K(\mathbf{x}_s, s; \mathbf{x}_{t'}, t')$$
$$\times\psi(\mathbf{x}_{t'}, t')d^3\mathbf{x}_{t''}d^3\mathbf{x}_td^3\mathbf{x}_sd^3\mathbf{x}_{t'} \tag{28}$$

which in improved notation and in the relativistic case is essentially the result given in **II**.

We have made use of a special case of a principle which may be stated more generally as

$$\langle\chi_{t''}|F(\mathbf{x}_1, t_1; \mathbf{x}_2, t_2; \cdots \mathbf{x}_k, t_k)|\psi_{t'}\rangle_{S_p}$$
$$=\int\chi^*(\mathbf{x}_{t''})K(\mathbf{x}_{t''}, t''; \mathbf{x}_1, t_1)\cdot K(\mathbf{x}_1, t_1; \mathbf{x}_2, t_2)\cdots$$
$$\times K(\mathbf{x}_{k-1}, t_{k-1}; \mathbf{x}_k, t_k)K(\mathbf{x}_k, t_k; \mathbf{x}_{t'}, t')$$
$$\cdot F(\mathbf{x}_1, t_1; \mathbf{x}_2, t_2; \cdots \mathbf{x}_k, t_k)\psi(\mathbf{x}_{t'})$$
$$\times d^3\mathbf{x}_{t''}d^3\mathbf{x}_1d^3\mathbf{x}_2\cdots d^3\mathbf{x}_kd^3\mathbf{x}_{t'} \tag{29}$$

where F is any function of the coordinate $\mathbf{x}_1$ at time t_1, $\mathbf{x}_2$ at t_2 up to $\mathbf{x}_k$, t_k, and, it is important to notice, we have assumed $t''>t_1>t_2>\cdots t_k>t'$.

Expressions of higher order arising for example from R^2 are more complicated as there are quantities referring to several different times mixed up, but they all can be interpreted readily. One simply breaks up the ranges of integrations of the time variables into parts such that in each the order of time of each variable is definite. One then interprets each part by formula (29).

[11] One can show from (25) how the correlated effect of many atoms at a distance produces on a given system the effects of an external potential. Formula (24) yields the result that this potential is that obtained from Liénard and Wiechert by retarded waves arising from the charges and currents resulting from the distant atoms making transitions. Assume the wave functions χ and ψ can be split into products of wave functions for system and distant atoms and expand $\exp(iR)$ assuming the effect of any individual distant atom is small. Coulomb potentials arise even from nearby particles if they are moving slowly.

As a simple example we may refer to the problem of the transition element

$$\left\langle \chi_{t''} \middle| \int U(\mathbf{x}(t), t)dt \int V(\mathbf{x}(s), s)ds \middle| \psi_{t'} \right\rangle$$

arising, say, in the cross term in U and V in an ordinary second order perturbation problem (disregarding radiation) with perturbation potential $U(\mathbf{x}, t)+V(\mathbf{x}, t)$. In the integration on s and t which should include the entire range of time for each, we can split the range of s into two parts, $s<t$ and $s>t$. In the first case, $s<t$, the potential V acts earlier than U, and in the other range, vice versa, so that

$$\left\langle \chi_{t''} \middle| \int U(\mathbf{x}_t, t)dt \int V(\mathbf{x}_s, s)ds \middle| \psi_{t'} \right\rangle$$

$$= \int_{t'}^{t''} dt \int_{t'}^{t} ds \int \chi^*(\mathbf{x}_{t''})K(\mathbf{x}_{t''}, t''; \mathbf{x}_t, t)$$

$$\times U(\mathbf{x}_t, t)K(\mathbf{x}_t, t; \mathbf{x}_s, s)V(\mathbf{x}_s, s)$$

$$\cdot K(\mathbf{x}_s, s; \mathbf{x}_{t'}, t')\psi(\mathbf{x}_{t'})d^3\mathbf{x}_{t''}d^3\mathbf{x}_t d^3\mathbf{x}_s d^3\mathbf{x}_{t'}$$

$$+\int_{t'}^{t''} dt \int_{t}^{t''} ds \int \chi^*(\mathbf{x}_{t''})K(\mathbf{x}_{t''}, t''; \mathbf{x}_s, s)$$

$$\times V(\mathbf{x}_s, s)K(\mathbf{x}_s, s; \mathbf{x}_t, t)U(\mathbf{x}_t, t)$$

$$\cdot K(\mathbf{x}_t, t; \mathbf{x}_{t'}, t')\psi(\mathbf{x}_{t'})d^3\mathbf{x}_{t''}d^3\mathbf{x}_s d^3\mathbf{x}_t d^3\mathbf{x}_{t'} \quad (30)$$

so that the single expression on the left is represented by two terms analogous to the two terms required in analyzing the Compton effect. It is in this way that the several terms and their corresponding diagrams corresponding to each process arise when an attempt is made to represent the transition elements of single expressions involving time integrals in terms of the propagation kernels K.

It remains to study in more detail the term in (26) arising from $\mathbf{x}^{\cdot}(t)\cdot\mathbf{x}^{\cdot}(s)$ in the interaction. The interpretation of such expressions is considered in detail in **C**, and we must refer to Eqs. (39) through (50) of that paper for a more thorough analysis. A similar type of term also arises in the Lagrangian formulation in simpler problems, for example the transition element

$$\left\langle \chi_{t''} \middle| \int \mathbf{x}^{\cdot}(t)\cdot\mathbf{A}(\mathbf{x}(t), t)dt \int \mathbf{x}^{\cdot}(s)\cdot\mathbf{B}(\mathbf{x}(s), s)ds \middle| \psi_{t'} \right\rangle$$

arising say, in the cross term in **A** and **B** in a second-order perturbation problem for a particle in a perturbing vector potential $\mathbf{A}(\mathbf{x}, t)+\mathbf{B}(\mathbf{x}, t)$. The time integrals must first be written as Riemannian sums, the velocity (see (6)) being replaced by $\mathbf{x}^{\cdot}=\frac{1}{2}\epsilon^{-1}(\mathbf{x}_{i+1}-\mathbf{x}_i)+\frac{1}{2}\epsilon^{-1}(\mathbf{x}_i-\mathbf{x}_{i-1})$ so that we ask for the transition element of

$$\sum_i\sum_j[\tfrac{1}{2}(\mathbf{x}_{i+1}-\mathbf{x}_i)+\tfrac{1}{2}(\mathbf{x}_i-\mathbf{x}_{i-1})]\cdot\mathbf{A}(\mathbf{x}_i, t_i)$$

$$\times[\tfrac{1}{2}(\mathbf{x}_{j+1}-\mathbf{x}_j)+\tfrac{1}{2}(\mathbf{x}_j-\mathbf{x}_{j-1})]\cdot\mathbf{B}(\mathbf{x}_j, t_j). \quad (31)$$

In **C** it is shown that when converted to operator notation the quantity $(x_{i-1}-x_i)/\epsilon$ is equivalent (nearly, see below) to an operator,

$$(x_{i+1}-x_i)/\epsilon \rightarrow i(Hx-xH) \quad (32)$$

operating in order indicated by the time index i (that is after x_l's for $l\leq i$ and before all x_l's for $l>i$). In nonrelativistic mechanics $i(Hx-xH)$ is the momentum operator p_x divided by the mass m. Thus in (31) the expression $[\frac{1}{2}(\mathbf{x}_{i+1}-\mathbf{x}_i)+\frac{1}{2}(\mathbf{x}_i-\mathbf{x}_{i-1})]\cdot\mathbf{A}(\mathbf{x}_i, t_i)$ becomes $\epsilon(\mathbf{p}\cdot\mathbf{A}+\mathbf{A}\cdot\mathbf{p})/2m$. Here again we must split the sum into two regions $j<i$ and $j>i$ so the quantities in the usual notation will operate in the right order such that eventually (31) becomes identical with the right-hand side of Eq. (30) but with $U(\mathbf{x}_t, t)$ replaced by the operator

$$\frac{1}{2m}\left(\frac{1}{i}\frac{\partial}{\partial\mathbf{x}_t}\cdot\mathbf{A}(\mathbf{x}_t, t)+\mathbf{A}(\mathbf{x}_t, t)\cdot\frac{1}{i}\frac{\partial}{\partial\mathbf{x}_t}\right)$$

standing in the same place, and with the operator

$$\frac{1}{2m}\left(\frac{1}{i}\frac{\partial}{\partial\mathbf{x}_s}\cdot\mathbf{B}(\mathbf{x}_s, s)+\frac{1}{i}\mathbf{B}(\mathbf{x}_s, s)\cdot\frac{\partial}{\partial\mathbf{x}_s}\right)$$

standing in the place of $V(\mathbf{x}_s, s)$. The sums and factors ϵ have now become $\int dt\int ds$.

This is nearly but not quite correct, however, as there is an additional term coming from the terms in the sum corresponding to the special values, $j=i$, $j=i+1$ and $j=i-1$. We have tacitly assumed from the appearance of the expression (31) that, for a given i, the contribution from just three such special terms is of order ϵ^2. But this is not true. Although the expected contribution of a term like $(x_{i+1}-x_i)(x_{j+1}-x_j)$ for $j\neq i$ is indeed of order ϵ^2, the expected contribution of $(x_{i+1}-x_i)^2$ is $+i\epsilon m^{-1}$ [**C**, Eq. (50)], that is, of order ϵ. In nonrelativistic mechanics the velocities are unlimited and in very short times ϵ the amplitude diffuses a distance proportional to the square root of the time. Making use of this equation then we see that the additional contribution from these terms is essentially

$$im^{-1}\epsilon\sum_i\mathbf{A}(\mathbf{x}_i, t_i)\cdot\mathbf{B}(\mathbf{x}_i, t_i)=im^{-1}\int\mathbf{A}(\mathbf{x}(t), t)\cdot\mathbf{B}(\mathbf{x}(t), t)dt$$

when summed on all i. This has the same effect as a first-order perturbation due to a potential $\mathbf{A}\cdot\mathbf{B}/m$. Added to the term involving the momentum operators we therefore have an additional term[12]

$$\frac{i}{m}\int_{t'}^{t''} dt \int \chi^*(\mathbf{x}_{t''})K(\mathbf{x}_{t''}, t''; \mathbf{x}_t, t)\mathbf{A}(\mathbf{x}_t, t)\cdot\mathbf{B}(\mathbf{x}_t, t) \cdot K(\mathbf{x}_t, t; \mathbf{x}_{t'}, t')\psi(\mathbf{x}_{t'})d^3\mathbf{x}_{t''}d^3\mathbf{x}_t d^3\mathbf{x}_{t'}. \quad (33)$$

In the usual Hamiltonian theory this term arises, of course, from the term $\mathbf{A}^2/2m$ in the expansion of the Hamiltonian

$$H=(2m)^{-1}(\mathbf{p}-\mathbf{A})^2=(2m)^{-1}(\mathbf{p}^2-\mathbf{p}\cdot\mathbf{A}-\mathbf{A}\cdot\mathbf{p}+\mathbf{A}^2)$$

while the other term arises from the second-order action of $\mathbf{p}\cdot\mathbf{A}+\mathbf{A}\cdot\mathbf{p}$. We shall not be interested in non-relativistic quantum electrodynamics in detail. The situation is simpler for Dirac electrons. For particles satisfying the Klein-Gordon equation (discussed in Appendix A) the situation is very similar to a four-dimensional analog of the non-relativistic case given here.

6. EXTENSION TO DIRAC PARTICLES

Expressions (24) and (25) and their proof can be readily generalized to the relativistic case according to the one electron theory of Dirac. We shall discuss the hole theory later. In the non-relativistic case we began with the proposition that the amplitude for a particle to proceed from one point to another is the sum over paths of $\exp(iS_p)$, that is, we have for example for a transition element

$$\langle\chi|1|\psi\rangle=\lim_{\epsilon\to 0}\int\cdots\int\chi^*(\mathbf{x}_N)\Phi_p(\mathbf{x}_N, \mathbf{x}_{N-1}, \cdots\mathbf{x}_0) \cdot\psi(\mathbf{x}_0)d^3\mathbf{x}_0 d^3\mathbf{x}_1\cdots d^3\mathbf{x}_N \quad (34)$$

where for $\exp(iS_p)$ we have written Φ_p, that is more precisely,

$$\Phi_p=\Pi_i A^{-1}\exp iS(\mathbf{x}_{i+1}, \mathbf{x}_i).$$

As discussed in C this form is related to the usual form of quantum mechanics through the observation that

$$(\mathbf{x}_{i+1}|\mathbf{x}_i)_\epsilon=A^{-1}\exp[iS(\mathbf{x}_{i+1}, \mathbf{x}_i)] \quad (35)$$

where $(\mathbf{x}_{i+1}|\mathbf{x}_i)_\epsilon$ is the transformation matrix from a representation in which $\mathbf{x}$ is diagonal at time t_i to one in which $\mathbf{x}$ is diagonal at time $t_{i+1}=t_i+\epsilon$ (so that it is identical to $K_0(\mathbf{x}_{i+1}, t_{i+1}; \mathbf{x}_i, t_i)$ for the small time interval ϵ). Hence the amplitude for a given path can also be written

$$\Phi_p=\Pi_i(\mathbf{x}_{i+1}|\mathbf{x}_i)_\epsilon \quad (36)$$

for which form, of course, (34) is exact irrespective of whether $(\mathbf{x}_{i+1}|\mathbf{x}_i)_\epsilon$ can be expressed in the simple form (35).

For a Dirac electron the $(\mathbf{x}_{i+1}|\mathbf{x}_i)_\epsilon$ is a 4×4 matrix (or $4^N\times 4^N$ if we deal with N electrons) but the expression (34) with (36) is still correct (as it is in fact for any quantum-mechanical system with a sufficiently general definition of the coordinate $\mathbf{x}_i$). The product (36) now involves operators, the order in which the factors are to be taken is the order in which the terms appear in time.

For a Dirac particle in a vector and scalar potential (times the electron charge e) $\mathbf{A}(\mathbf{x}, t)$, $A_4(\mathbf{x}, t)$, the quantity $(\mathbf{x}_{i+1}|\mathbf{x}_i)_\epsilon^{(A)}$ is related to that of a free particle to the first order in ϵ as

$$(\mathbf{x}_{i+1}|\mathbf{x}_i)_\epsilon^{(A)}=(\mathbf{x}_{i+1}|\mathbf{x}_i)_\epsilon^{(0)}\exp[-i(\epsilon A_4(\mathbf{x}_i, t_i) -(\mathbf{x}_{i+1}-\mathbf{x}_i)\cdot\mathbf{A}(\mathbf{x}_i, t_i))]. \quad (37)$$

This can be verified directly by substitution into the Dirac equation.[13] It neglects the variation of $\mathbf{A}$ and A_4 with time and space during the short interval ϵ. This produces errors only of order ϵ^2 in the Dirac case for the expected square velocity $(\mathbf{x}_{i+1}-\mathbf{x}_i)^2/\epsilon^2$ during the interval ϵ is finite (equaling the square of the velocity of light) rather than being of order $1/\epsilon$ as in the non-relativistic case. [This makes the relativistic case somewhat simpler in that it is not necessary to define the velocity as carefully as in (6); $(\mathbf{x}_{i+1}-\mathbf{x}_i)/\epsilon$ is sufficiently exact, and no term analogous to (33) arises.]

Thus $\Phi_p^{(A)}$ differs from that for a free particle, $\Phi_p^{(0)}$, by a factor $\Pi_i\exp-i(\epsilon A_4(\mathbf{x}_i, t_i)-(\mathbf{x}_{i+1}-\mathbf{x}_i)\cdot\mathbf{A}(\mathbf{x}_i, t_i))$ which in the limit can be written as

$$\exp\left\{-i\int[A_4(\mathbf{x}(t), t)-\dot{\mathbf{x}}(t)\cdot\mathbf{A}(\mathbf{x}(t), t)]dt\right\} \quad (38)$$

exactly as in the non-relativistic case.

The case of a Dirac particle interacting with the quantum-mechanical oscillators representing the field may now be studied. Since the dependence of $\Phi_p^{(A)}$ on $\mathbf{A}$, A_4 is through the same factor as in the non-relativistic case, when $\mathbf{A}$, A_4 are expressed in terms of the oscillator coordinates q, the dependence of Φ on the oscillator coordinates q is unchanged. Hence the entire analysis of the preceding sections which concern the results of the integration over oscillator coordinates can be carried through unchanged and the results will be expression (25) with formula (24) for R. Expression (25) is now interpreted as

$$\langle\chi_{t''}|\exp iR|\psi_{t'}\rangle=\lim_{\epsilon\to 0}\int\chi^*(\mathbf{x}_{t''}^{(1)}, \mathbf{x}_{t''}^{(2)}\cdots) \times\prod_n(\Phi_{p,n}^{(0)}d^3\mathbf{x}_{t''}^{(n)}d^3\mathbf{x}_{t''-\epsilon}^{(n)}\cdots d^3\mathbf{x}_{t'}^{(n)}) \cdot\exp(iR)\psi(\mathbf{x}_{t'}^{(1)}, \mathbf{x}_{t'}^{(2)}\cdots) \quad (39)$$

---

[12] The term corresponding to this for the self-energy expression (26) would give an integral over $\delta_+((t-t)^2-(\mathbf{x}_t-\mathbf{x}_t)^2)$ which is evidently infinite and leads to the quadratically divergent self-energy. There is no such term for the Dirac electron, but there is for Klein-Gordon particles. We shall not discuss the infinities in this paper as they have already been discussed in **II**.

[13] Alternatively, note that Eq. (37) is exact for arbitrarily large ϵ if the potential A_μ is constant. For if the potential in the Dirac equation is the gradient of a scalar function $A_\mu=\partial\chi/\partial x_\mu$ the potential may be removed by replacing the wave function by $\psi=\exp(-i\chi)\psi'$ (gauge transformation). This alters the kernel by a factor $\exp[-i(\chi(2)-\chi(1))]$ owing to the change in the initial and final wave functions. A constant potential A_μ is the gradient of $\chi=A_\mu x_\mu$ and can be completely removed by this gauge transformation, so that the kernel differs from that of a free particle by the factor $\exp[-i(A_\mu x_{\mu 2}-A_\mu x_{\mu 1})]$ as in (37).

where $\Phi_{p,n}^{(0)}$, the amplitude for a particular path for particle n is simply the expression (36) where $(\mathbf{x}_{i+1}|\mathbf{x}_i)_\epsilon$ is the kernel $K_{0,n}(\mathbf{x}_{i+1}^{(n)}, t_{i+1}; \mathbf{x}_i^{(n)}, t_i)$ for a free electron according to the one electron Dirac theory, with the matrices which appear operating on the spinor indices corresponding to particle (n) and the order of all operations being determined by the time indices.

For calculational purposes we can, as before, expand R as a power series and evaluate the various terms in the same manner as for the non-relativistic case. In such an expansion the quantity $\mathbf{x}^{\cdot}(t)$ is replaced, as we have seen in (32), by the operator $i(H\mathbf{x}-\mathbf{x}H)$, that is, in this case by $\boldsymbol{\alpha}$ operating at the corresponding time. There is no further complicated term analogous to (33) arising in this case, for the expected value of $(x_{i+1}-x_i)^2$ is now of order ϵ^2 rather than ϵ.

For example, for self-energy one sees that expression (28) will be (with other terms coming from those with $\mathbf{x}^{\cdot}(t)$ replaced by $\boldsymbol{\alpha}$ and with the usual β in back of each K_0 because of the definition of K_0 in relativity theory)

$$\langle\psi_{t''}|R|\psi_{t'}\rangle_{S_p}=-e^2\int\psi^*(\mathbf{x}_{t''})K_0(\mathbf{x}_{t''},t'';\mathbf{x}_t,t)\beta\alpha_\mu$$
$$\cdot\delta_+((t-s)^2-(\mathbf{x}_t-\mathbf{x}_s)^2)K_0(\mathbf{x}_t,t;\mathbf{x}_s,s)\beta\alpha_\mu$$
$$\cdot K_0(\mathbf{x}_s,s;\mathbf{x}_{t'},t')\beta\psi(\mathbf{x}_{t'})d^3\mathbf{x}_{t''}d^3\mathbf{x}_td^3\mathbf{x}_sd^3\mathbf{x}_{t'}dtds, \quad (40)$$

where $\alpha_4=1$, $\alpha_{1,2,3}=\alpha_{x,y,z}$ and a sum on the repeated index μ is implied in the usual way; $a_\mu b_\mu=a_4b_4-a_1b_1-a_2b_2-a_3b_3$. One can change $\beta\alpha_\mu$ to γ_μ and ψ^* to $\bar{\psi}\beta$. In this manner all of the rules referring to virtual photons discussed in **II** are deduced; but with the difference that K_0 is used instead of K_+ and we have the Dirac one electron theory with negative energy states (although we may have any number of such electrons).

7. EXTENSION TO POSITRON THEORY

Since in (39) we have an arbitrary number of electrons, we can deal with the hole theory in the usual manner by imagining that we have an infinite number of electrons in negative energy states.

On the other hand, in paper **I** on the theory of positrons, it was shown that the results of the hole theory in a system with a given external potential A_μ were equivalent to those of the Dirac one electron theory if one replaced the propagation kernel, K_0, by a different one, K_+, and multiplied the resultant amplitude by factor C_v involving A_μ. We must now see how this relation, derived in the case of external potentials, can also be carried over in electrodynamics to be useful in simplifying expressions involving the infinite sea of electrons.

To do this we study in greater detail the relation between a problem involving virtual photons and one involving purely external potentials. In using (25) we shall assume in accordance with the hole theory that the number of electrons is infinite, but that they all have the same charge, e. Let the states $\psi_{t'}$, $\chi_{t''}$, represent the vacuum plus perhaps a number of real electrons in positive energy states and perhaps also some empty negative energy states. Let us call the amplitude for the transition in an external potential B_μ, but *excluding virtual photons*, $T_0[B]$, a functional of $B_\mu(1)$. We have seen (38)

$$T_0[B]=\langle\chi_{t''}|\exp iP|\psi_{t'}\rangle \quad (41)$$

where

$$P=-\sum_n\int[B_4(\mathbf{x}^{(n)}(t),t)-\mathbf{x}^{\cdot(n)}(t)\cdot\mathbf{B}(\mathbf{x}^{(n)}(t),t)]dt$$

by (38). We can write this as

$$P=-\sum_n\int B_\mu(x_\nu^{(n)}(t))\dot{x}_\mu^{(n)}(t)dt$$

where $x_4(t)=t$ and $\dot{x}_4=1$, the other values of μ corresponding to space variables. The corresponding amplitude for the same process in the same potential, but *including* all the virtual photons we may call,

$$T_{e^2}[B]=\langle\chi_{t''}|\exp(iR)\exp(iP)|\psi_{t'}\rangle. \quad (42)$$

Now let us consider the effect on $T_{e^2}[B]$ of changing the coupling e^2 of the virtual photons. Differentiating (42) with respect to e^2 which appears only [14] in R we find

$$dT_{e^2}[B]/d(e^2)=\left\langle\chi_{t''}\left|-\frac{i}{2}\sum_n\sum_m\int\int dtds\dot{x}_\mu^{(n)}(t)\dot{x}_\mu^{(m)}(s)\right.\right.$$
$$\left.\left.\cdot\delta_+((x_\nu^{(n)}(t)-x_\nu^{(m)}(s))^2)\exp i(R+P)\right|\psi_{t'}\right\rangle. \quad (43)$$

We can also study the first-order effect of a change of B_μ:

$$\delta T_{e^2}[B]/\delta B_\mu(1)=-i\left\langle\chi_{t''}\left|\sum_n\int dt\dot{x}_\mu^{(n)}\delta^4(x_\alpha^{(n)}(t)-x_{\alpha,1})\right.\right.$$
$$\left.\left.\cdot\exp i(R+P)\right|\psi_{t'}\right\rangle \quad (44)$$

where $x_{\alpha,1}$ is the field point at which the derivative with respect to B_μ is taken[15] and the term (current density) $-\sum_n\int dt\dot{x}_\mu^{(n)}(t)\delta^4(x_\alpha^{(n)}(t)-x_{\alpha,1})$ is just $\delta P/\delta B_\mu(1)$. The function $\delta^4(x_\alpha^{(n)}-x_{\alpha,1})$ means $\delta(x_4^{(n)}-x_{4,1})$

[14] In changing the charge e^2 we mean to vary only the degree to which virtual photons are important. We do not contemplate changes in the influence of the external potentials. If one wishes, as e is raised the strength of the potential is decreased proportionally so that B_μ, the potential times the charge e, is held constant.

[15] The functional derivative is defined such that if $T[B]$ is a number depending on the functions $B_\mu(1)$, the first order variation in T produced by a change from B_μ to $B_\mu+\Delta B_\mu$ is given by

$$T[B+\Delta B]-T[B]=\int(\delta T[B]/\delta B_\mu(1))\Delta B_\mu(1)d\tau_1$$

the integral extending over all four-space $x_{\alpha,1}$.

$\times\delta(x_3^{(n)}-x_{3,1})\delta(x_2^{(n)}-x_{2,1})\delta(x_1^{(n)}-x_{1,1})$ that is, $\delta(2,1)$ with $x_{\alpha,2}=x_\alpha^{(n)}(t)$. A second variation of T gives, by differentiation of (44) with respect to $B_\nu(2)$,

$$\delta^2 T_{e^2}[B]/\delta B_\mu(1)\delta B_\nu(2)$$

$$=-\left\langle \chi_{t'}\middle|\sum_n\sum_m\int dt\int ds\dot{x}_\mu^{(n)}(t)\dot{x}_\nu^{(m)}(s)\cdot\delta^4(x_\alpha^{(n)}(t)-x_{\alpha,1})\delta^4(x_\beta^{(n)}(s)-x_{\beta,2})\times\exp i(R+P)\middle|\psi_{t'}\right\rangle.$$

Comparison of this with (43) shows that

$$dT_{e^2}[B]/d(e^2)=\tfrac{1}{2}i\int\int(\delta^2T_{e^2}[B]/\delta B_\mu(1)\delta B_\mu(2))\times\delta_+(s_{12}^2)d\tau_1 d\tau_2 \quad (45)$$

where $s_{12}^2=(x_{\mu,1}-x_{\mu,2})(x_{\mu,1}-x_{\mu,2})$.

We now proceed to use this equation to prove the validity of the rules given in **II** for electrodynamics. This we do by the following argument. The equation can be looked upon as a differential equation for $T_{e^2}[B]$. It determines $T_{e^2}[B]$ uniquely if $T_0[B]$ is known. We have shown it is valid for the hole theory of positrons. But in **I** we have given formulas for calculating $T_0[B]$ whose correctness relative to the hole theory we have there demonstrated. Hence we have shown that the $T_{e^2}[B]$ obtained by solving (45) with the initial condition $T_0[B]$ as given by the rules in **I** will be equal to that given for the same problem by the second quantization theory of the Dirac matter field coupled with the quantized electromagnetic field. But it is evident (the argument is given in the next paragraph) that the rules[16] given in **II** constitute a solution in power series in e^2 of the Eq. (45) [which for $e^2=0$ reduce to the $T_0[B]$ given in **I**]. Hence the rules in **II** must give, to each order in e^2, the matrix element for any process that would be calculated by the usual theory of second quantization of the matter and electromagnetic fields. This is what we aimed to prove.

That the rules of **II** represent, in a power series expansion, a solution of (45) is clear. For the rules there given may be stated as follows: Suppose that we have a process to order k in e^2 (i.e., having k virtual photons) and order n in the external potential B_μ. Then, *the matrix element for the process with one more virtual photon and two less potentials is that obtained from the previous matrix by choosing from the n potentials a pair, say $B_\mu(1)$ acting at 1 and $B_\nu(2)$ acting at 2, replacing them by $ie^2\delta_{\mu\nu}\delta_+(s_{12}^2)$, adding the results for each way of choosing the pair, and dividing by $k+1$, the present number of photons.* The matrix with no virtual photons ($k=0$) being given to any n by the rules of **I**, this permits terms to all orders in e^2 to be derived by recursion. It is evident that the rule in italics is that of **II**, and equally evident that it is a word expression of Eq. (45). [The factor $\frac{1}{2}$ in (45) arises since in integrating over all $d\tau_1$ and $d\tau_2$ we count each pair twice. The division by $k+1$ is required by the rules of **II** for, there, each diagram is to be taken only once, while in the rule given above we say what to do to add one extra virtual photon to k others. But which one of the $k+1$ is to be identified at the last photon added is irrelevant. It agrees with (45) of course for it is canceled on differentiating with respect to e^2 the factor $(e^2)^{k+1}$ for the $(k+1)$ photons.]

[16] That is, of course, those rules of **II** which apply to the unmodified electrodynamics of Dirac electrons. (The limitation excluding real photons in the initial and final states is removed in Sec. 8.) The same arguments clearly apply to nucleons interacting via neutral vector mesons, vector coupling. Other couplings require a minor extension of the argument. The modification to the $(\mathbf{x}_{i+1}|\mathbf{x}_i)_\epsilon$, as in (37), produced by some couplings cannot very easily be written without using operators in the exponents. These operators can be treated as numbers if their order of operation is maintained to be always their order in time. This idea will be discussed and applied more generally in a succeeding paper.

8. GENERALIZED FORMULATION OF QUANTUM ELECTRODYNAMICS

The relation implied by (45) between the formal solution for the amplitude for a process in an arbitrary unquantized external potential to that in a quantized field appears to be of much wider generality. We shall discuss the relation from a more general point of view here (still limiting ourselves to the case of no photons in initial or final state).

In earlier sections we pointed out that as a consequence of the Lagrangian form of quantum mechanics the aspects of the particles' motions and the behavior of the field could be analyzed separately. What we did was to integrate over the field oscillator coordinates first. We could, in principle, have integrated over the particle variables first. That is, we first solve the problem with the action of the particles and their interaction with the field and then multiply by the exponential of the action of the field and integrate over all the field oscillator coordinates. (For simplicity of discussion let us put aside from detailed special consideration the questions involving the separation of the longitudinal and transverse parts of the field.[9]) Now the integral over the particle coordinates for a given process is precisely the integral required for the analysis of the motion of the particles in an unquantized potential. With this observation we may suggest a generalization to all types of systems.

Let us suppose the formal solution for the amplitude for some given process with matter in an external potential $B_\mu(1)$ is some numerical quantity T_0. We mean matter in a more general sense now, for the motion of the matter may be described by the Dirac equation, or by the Klein-Gordon equation, or may involve charged or neutral particles other than electrons and positrons in any manner whatsoever. The quantity T_0 depends of course on the potential function $B_\mu(1)$; that is, it is a functional $T_0[B]$ of this potential. We assume we have some expression for it in terms of B_μ (exact, or to some desired degree of approximation in the strength of the potential).

Then the answer $T_{e^2}[B]$ to the corresponding problem in quantum electrodynamics is $T_0[A_\mu(1)+B_\mu(1)] \times\exp(iS_0)$ summed over all possible distributions of field $A_\mu(1)$, wherein S_0 is the action for the field $S_0=-(8\pi e^2)^{-1}\sum_\mu \int((\partial A_\mu/\partial t)^2-(\nabla A_\mu)^2)d^3\mathbf{x}dt$ the sum on μ carrying the usual minus sign for space components.

If $F[A]$ is any functional of $A_\mu(1)$ we shall represent by ${}_0|F[A]|_0$ this superposition of $F[A]\exp(iS_0)$ over distributions of A_μ for the case in which there are no photons in initial or final state. That is, we have

$$T_{e^2}[B]={}_0|T_0[A+B]|_0. \qquad (46)$$

The evaluation of ${}_0|F[A]|_0$ directly from the definition of the operation ${}_0|\quad|_0$ is not necessary. We can give the result in another way. We first note that the operation is linear,

$${}_0|F_1[A]+F_2[A]|_0={}_0|F_1[A]|_0+{}_0|F_2[A]|_0 \qquad (47)$$

so that if F is represented as a sum of terms each term can be analyzed separately. We have studied essentially the case in which $F[A]$ is an exponential function. In fact, what we have done in Section 4 may be repeated with slight modification to show that

$${}_0\left|\exp\left(-i\int j_\mu(1)A_\mu(1)d\tau_1\right)\right|_0$$

$$=\exp\left(-\tfrac{1}{2}ie^2\int\int j_\mu(1)j_\mu(2)\delta_+(s_{12}^2)d\tau_1d\tau_2\right) \qquad (48)$$

where $j_\mu(1)$ is an arbitrary function of position and time for each value of μ.

Although this gives the evaluation of ${}_0|\quad|_0$ for only a particular functional of A_μ the appearance of the arbitrary function $j_\mu(1)$ makes it sufficiently general to permit the evaluation for any other functional. For it is to be expected that any functional can be represented as a superposition of exponentials with different functions $j_\mu(1)$ (by analogy with the principle of Fourier integrals for ordinary functions). Then, by (47), the result of the operation is the corresponding superposition of expressions equal to the right-hand side of (48) with the various j's substituted for j_μ.

In many applications $F[A]$ can be given as a power series in A_μ:

$$F[A]=f_0+\int f_\mu(1)A_\mu(1)d\tau_1$$

$$+\int\int f_{\mu\nu}(1,2)A_\mu(1)A_\nu(2)d\tau_1d\tau_2+\cdots \qquad (49)$$

where $f_0, f_\mu(1), f_{\mu\nu}(1,2)\cdots$ are known numerical functions independent of A_μ. Then by (47)

$${}_0|F[A]|_0=f_0+\int f_\mu(1)\,{}_0|A_\mu(1)|_0d\tau_1$$

$$+\int\int f_{\mu\nu}(1,2)\,{}_0|A_\mu(1)A_\nu(2)|_0d\tau_1d\tau_2+\cdots \qquad (50)$$

where we set ${}_0|1|_0=1$ (from (48) with $j_\mu=0$). We can work out expressions for the successive powers of A_μ by differentiating both sides of (48) successively with respect to j_μ and setting $j_\mu=0$ in each derivative. For example, the first variation (derivative) of (48) with respect to $j_\mu(3)$ gives

$${}_0\left|-iA_\mu(3)\exp\left(-i\int j_\nu(1)A_\nu(1)d\tau_1\right)\right|_0$$

$$=-ie^2\int\delta_+(s_{34}^2)j_\mu(4)d\tau_4$$

$$\times\exp\left(-\tfrac{1}{2}ie^2\int\int j_\nu(1)j_\nu(2)\delta_+(s_{12}^2)d\tau_1d\tau_2\right). \qquad (51)$$

Setting $j_\mu=0$ gives

$${}_0|A_\mu(3)|_0=0.$$

Differentiating (51) again with respect to $j_\nu(4)$ and setting $j_\nu=0$ shows

$${}_0|A_\mu(3)A_\nu(4)|_0=ie^2\delta_{\mu\nu}\delta_+(s_{34}^2) \qquad (52)$$

and so on for higher powers. These results may be substituted into (50). Clearly therefore when $T_0[B+A]$ in (46) is expanded in a power series and the successive terms are computed in this way, we obtain the results given in II.

It is evident that (46), (47), (48) imply that $T_{e^2}[B]$ satisfies the differential equation (45) and conversely (45) with the definition (46) implies (47) and (48). For if $T_0[B]$ is an exponential

$$T_0[B]=\exp\left(-i\int j_\mu(1)B_\mu(1)d\tau_1\right) \qquad (53)$$

we have from (46), (48) that

$$T_{e^2}[B]=\exp\left[-\tfrac{1}{2}ie^2\int\int j_\mu(1)j_\mu(2)\delta_+(s_{12}^2)d\tau_1d\tau_2\right]$$

$$\cdot\exp\left[-i\int j_\nu(1)B_\nu(1)d\tau_1\right]. \qquad (54)$$

Direct substitution of this into Eq. (45) shows it to be a solution satisfying the boundary condition (53). Since the differential equation (45) is linear, if $T_0[B]$ is a superposition of exponentials, the corresponding superposition of solutions (54) is also a solution.

Many of the formal representations of the matter system (such as that of second quantization of Dirac electrons) represent the interaction with a fixed potential in a formal exponential form such as the left-hand side of (48), except that $j_\mu(1)$ is an operator instead of a numerical function. Equation (48) may still be used if care is exercised in defining the order of the operators on the right-hand side. The succeeding paper will discuss this in more detail.

Equation (45) or its solution (46), (47), (48) constitutes a very general and convenient formulation of the laws of quantum electrodynamics for virtual processes. Its relativistic invariance is evident if it is assumed that the unquantized theory giving $T_0[B]$ is invariant. It has been proved to be equivalent to the usual formulation for Dirac electrons and positrons (for Klein-Gordon particles see Appendix A). It is suggested that it is of wide generality. It is expressed in a form which has meaning even if it is impossible to express the matter system in Hamiltonian form; in fact, it only requires the existence of an amplitude for fixed potentials which obeys the principle of superposition of amplitudes. If $T_0[B]$ is known in power series in B, calculations of $T_{e^2}[B]$ in a power series of e^2 can be made directly using the italicized rule of Sec. 7. The limitation to virtual quanta is removed in the next section.

On the other hand, the formulation is unsatisfactory because for situations of importance it gives divergent results, even if $T_0[B]$ is finite. The modification proposed in **II** of replacing $\delta_+(s_{12}^2)$ in (45), (48) by $f_+(s_{12}^2)$ is not satisfactory owing to the loss of the theorems of conservation of energy or probability discussed in **II** at the end of Sec. 6. There is the additional difficulty in positron theory that even $T_0[B]$ is infinite to begin with (vacuum polarization). Computational ways of avoiding these troubles are given in **II** and in the references of footnote 2.

9. CASE OF REAL PHOTONS

The case in which there are real photons in the initial or the final state can be worked out from the beginning in the same manner.[17] We first consider the case of a system interacting with a single oscillator. From this result the generalization will be evident. This time we shall calculate the transition element between an initial state in which the particle is in state $\psi_{t'}$ and the oscillator is in its nth eigenstate (i.e., there are n photons in the field) to a final state with particle in $\chi_{t''}$, oscillator in mth level. As we have already discussed, when the coordinates of the oscillator are eliminated the result is the transition element $\langle\chi_{t''}|G_{mn}|\psi_{t'}\rangle$ where

$$G_{mn}=\int\varphi_m{}^*(q_j)k(q_j,t'';q_0,t')\varphi_n(q_0)dq_0dq_j \qquad (11)$$

where φ_m, φ_n are the wave functions[8] for the oscillator in state m, n and k is given in (12). The G_{mn} can be evaluated most easily by calculating the generating function

$$g(X,Y)=\textstyle\sum_m\sum_n G_{mn}X^mY^n(m!n!)^{-\frac{1}{2}} \qquad (55)$$

for arbitrary X, Y. If expression (11) is substituted in the left-hand side of (55), the expression can be simplified by use of the generating function relation for the eigenfunctions[8] of the harmonic oscillator

$$\sum_n\varphi_n(q_0)Y^n(n!)^{-\frac{1}{2}}=(\omega/\pi)^{\frac{1}{4}}\exp(-\tfrac{1}{2}i\omega t')$$
$$\times\exp\tfrac{1}{2}[\omega q_0{}^2-(Y\exp[-i\omega t']-(2\omega)^{\frac{1}{2}}q_0)^2]$$

Using a similar expansion for the $\varphi_m{}^*$ one is left with the exponential of a quadratic function of q_0 and q_j. The integration on q_0 and q_j is then easily performed to give

$$g(X,Y)=G_{00}\exp(XY+i\beta^*X+i\beta Y) \qquad (56)$$

from which expansion in powers of X and Y and comparison to (11) gives the final result

$$G_{mn}=G_{00}(m!n!)^{-\frac{1}{2}}\sum_r\frac{m!}{(m-r)!r!}\frac{n!}{(n-r)!r!}$$
$$\times r!(i\beta^*)^{m-r}(i\beta)^{n-r} \qquad (57)$$

where G_{00} is given in (14) and

$$\beta=(2\omega)^{-\frac{1}{2}}\int_{t'}^{t''}\gamma(t)\exp(-i\omega t)dt,$$
$$\beta^*=(2\omega)^{-\frac{1}{2}}\int_{t'}^{t''}\gamma(t)\exp(+i\omega t)dt, \qquad (58)$$

and the sum on r is to go from 0 to m or to n whichever is the smaller. (The sum can be expressed as a Laguerre polynomial but there is no advantage in this.)

Formula (57) is readily understandable. Consider first a simple case of absorption of one photon. Initially we have one photon and finally none. The amplitude for this is the transition element of $G_{01}=i\beta G_{00}$ or $\langle\chi_{t''}|i\beta G_{00}|\psi_{t'}\rangle$. This is the same as would result if we asked for the transition element for a problem in which all photons are virtual but there was present a perturbing potential $-(2\omega)^{-\frac{1}{2}}\gamma(t)\exp(-i\omega t)$ and we required the first-order effect of this potential. Hence photon absorption is like the first order action of a potential varying in time as $\gamma(t)\exp(-i\omega t)$ that is with a positive frequency (i.e., the sign of the coefficient of t in the exponential corresponds to positive energy). The amplitude for emission of one photon involves $G_{10}=i\beta^*G_{00}$, which is the same result except that the potential has negative frequency. Thus we begin by interpreting $i\beta^*$ as the amplitude for emission of one photon $i\beta$ as the amplitude for absorption of one.

Next for the general case of n photons initially and m finally we may understand (57) as follows. We first

[17] For an alternative method starting directly from the formula (24) for virtual photons, see Appendix B.

neglect Bose statistics and imagine the photons as individual distinct particles. If we start with n and end with m this process may occur in several different ways. The particle may absorb in total $n-r$ of the photons and the final m photons will represent r of the photons which were present originally plus $m-r$ new photons emitted by the particle. In this case the $n-r$ which are to be absorbed may be chosen from among the original n in $n!/(n-r)!r!$ different ways, and each contributes a factor $i\beta$, the amplitude for absorption of a photon. Which of the $m-r$ photons from among the m are emitted can be chosen in $m!/(m-r)!r!$ different ways and each photon contributes a factor $i\beta^*$ in amplitude. The initial r photons which do not interact with the particle can be re-arranged among the final r in $r!$ ways. We must sum over the alternatives corresponding to different values of r. Thus the form of G_{mn} can be understood. The remaining factor $(m!)^{-\frac{1}{2}}(n!)^{-\frac{1}{2}}$ may be interpreted as saying that in computing probabilities (which therefore involves the square of G_{mn}) the photons may be considered as independent but that if m are actually equal the statistical weight of each of the states which can be made by rearranging the m equal photons is only $1/m!$. This is the content of Bose statistics; that m equal particles in a given state represents just one state, i.e., has statistical weight unity, rather than the $m!$ statistical weight which would result if it is imagined that the particles and states can be identified and rearranged in $m!$ different ways. This holds for both the initial and final states of course. From this rule about the statistical weights of states the derivation of the blackbody distribution law follows.

The actual electromagnetic field is represented as a host of oscillators each of which behaves independently and produces its own factor such as G_{mn}. Initial or final states may also be linear combinations of states in which one or another oscillator is excited. The results for this case are of course the corresponding linear combination of transition elements.

For photons of a given direction of polarization and for sin or cos waves the explicit expression for β can be obtained directly from (58) by substituting the formulas (16) for the γ's for the corresponding oscillator. It is more convenient to use the linear combination corresponding to running waves. Thus we find the amplitude for absorption of a photon of momentum $\mathbf{K}$, frequency $k=(\mathbf{K}\cdot\mathbf{K})^{\frac{1}{2}}$ polarized in direction $\mathbf{e}$ is given by including a factor i times

$$\beta_{\mathbf{K},\mathbf{e}}=(4\pi)^{\frac{1}{2}}(2k)^{-\frac{1}{2}}\sum_n e_n\int_{t'}^{t''}\exp(-ikt)\times\exp(i\mathbf{K}\cdot\mathbf{x}_n(t))\mathbf{e}\cdot\mathbf{x}'_n(t)dt \quad (59)$$

in the transition element (25). The density of states in momentum space is now $(2\pi)^{-3}d^3\mathbf{K}$. The amplitude for emission is just i times the complex conjugate of this expression, or what amounts to the same thing, the same expression with the sign of the four vector k_μ reversed. Since the factor (59) is exactly the first-order effect of a vector potential

$$\mathbf{A}^{PH}=(2\pi/k)^{\frac{1}{2}}\mathbf{e}\exp(-i(kt-\mathbf{K}\cdot\mathbf{x}))$$

of the corresponding classical wave, we have derived the rules for handling real photons discussed in **II**.

We can express this directly in terms of the quantity $T_{e^2}[B]$, the amplitude for a given transition without emission of a photon. What we have said is that the amplitude for absorption of just one photon whose classical wave form is $A_\mu{}^{PH}(1)$ (time variation $\exp(-ikt_1)$ corresponding to positive energy k) is proportional to the first order (in ϵ) change produced in $T_{e^2}[B]$ on changing B to $B+\epsilon A^{PH}$. That is, more exactly,

$$\int(\delta T_{e^2}[B]/\delta B_\mu(1))A_\mu{}^{PH}(1)d\tau_1 \quad (60)$$

is the amplitude for absorption by the particle system of one photon, A^{PH}. (A superposition argument shows the expression to be valid not only for plane waves, but for spherical waves, etc., as given by the form of A^{PH}.) The amplitude for emission is the same expression but with the sign of the frequency reversed in A^{PH}. The amplitude that the system absorbs two photons with waves $A_\mu{}^{PH_1}$ and $A_\nu{}^{PH_2}$ is obtained from the next derivative,

$$\int\int(\delta^2T_{e^2}[B]/\delta B_\mu(1)\delta B_\nu(2))A_\mu{}^{PH_1}(1)A_\nu{}^{PH_2}(2)d\tau_1d\tau_2,$$

the same expression holding for the absorption of one and emission of the other, or emission of both depending on the sign of the time dependence of A^{PH_1} and A^{PH_2}. Larger photon numbers correspond to higher derivatives, absorption of l_1 emission of l_2 requiring the (l_1+l_2) derivaties. When two or more of the photons are exactly the same (e.g., $A^{PH_1}=A^{PH_2}$) the same expression holds for the amplitude that l_1 are absorbed by the system while l_2 are emitted. However, the statement that initially n of a kind are present and m of this kind are present finally, does not imply $l_1=n$ and $l_2=m$. It is possible that only $n-r=l_1$ were absorbed by the system and $m-r=l_2$ emitted, and that r remained from initial to final state without interaction. This term is weighed by the combinatorial coefficient $(m!n!)^{-\frac{1}{2}}\binom{m}{r}\binom{n}{r}r!$ and summed over the possibilities for r as explained in connection with (57). Thus once the amplitude for virtual processes is known, that for real photon processes can be obtained by differentiation.

It is possible, of course, to deal with situations in which the electromagnetic field is not in a definite state after the interaction. For example, we might ask for the total probability of a given process, such as a scattering, without regard for the number of photons emitted. This is done of course by squaring the amplitude for the emission of m photons of a given kind and summing on all m. Actually the sums and integrations over the oscillator momenta can usually easily be performed analytically. For example, the amplitude, starting from vacuum and ending with m photons of a given kind, is by (56) just

$$G_{m0}=(m!)^{-\frac{1}{2}}G_{00}(i\beta^*)^m. \tag{61}$$

The square of the amplitude summed on m requires the product of two such expressions (the $\gamma(t)$ in the β of one and in the other will have to be kept separately) summed on m:

$$\sum_m G_{m0}{}^*G_{m0}{}'=\sum_m G_{00}{}^*G_{00}{}'(m!)^{-1}\beta^m(\beta'^*)^m =G_{00}{}^*G_{00}{}'\exp(\beta\beta'^*).$$

In the resulting expression the sum over all oscillators is easily done. Such expressions can be of use in the analysis in a direct manner of problems of line width, of the Bloch-Nordsieck infra-red problem, and of statistical mechanical problems, but no such applications will be made here.

The author appreciates his opportunities to discuss these matters with Professor H. A. Bethe and Professer J. Ashkin, and the help of Mr. M. Baranger with the manuscript.

APPENDIX A. THE KLEIN-GORDON EQUATION

In this Appendix we describe a formulation of the equations for a particle of spin zero which was first used to obtain the rules given in **II** for such particles. The complete physical significance of the equations has not been analyzed thoroughly so that it may be preferable to derive the rules directly from the second quantization formulation of Pauli and Weisskopf. This can be done in a manner analogous to the derivation of the rules for the Dirac equation given in **I** or from the Schwinger-Tomonaga formulation[2] in a manner described, for example, by Rohrlich.[18] The formulation given here is therefore not necessary for a description of spin zero particles but is given only for its own interest as an alternative to the formulation of second quantization.

We start with the Klein-Gordon equation

$$(i\partial/\partial x_\mu-A_\mu)^2\psi=m^2\psi \tag{1A}$$

for the wave function ψ of a particle of mass m in a given external potential A_μ. We shall try to represent this in a manner analogous to the formulation of quantum mechanics in **C**. That is, we try to represent the amplitude for a particle to get from one point to another as a sum over all trajectories of an amplitude $\exp(iS)$ where S is the classical action for a given trajectory. To maintain the relativistic invariance in evidence the idea suggests itself of describing a trajectory in space-time by giving the four variables $x_\mu(u)$ as functions of some fifth parameter u (rather than expressing x_1, x_2, x_3 in terms of x_4). As we expect to represent paths which may reverse themselves in time (to represent pair production, etc., as in **I**) this is certainly a more convenient representation, for all four functions $x_\mu(u)$ may be considered as functions of a parameter u (somewhat analogous to proper time) which increase as we go along the trajectory, whether the trajectory is proceeding forward ($dx_4/du>0$) or backward ($dx_4/du<0$) in time.[19] We shall then have a new type of wave function $\varphi(x, u)$ a function of five variables, x standing for the four x_μ. It gives the amplitude for arrival at point x_μ with a certain value of the parameter u. We shall suppose that this wave function satisfies the equation

$$i\partial\varphi/\partial u=-\tfrac{1}{2}(i\partial/\partial x_\mu-A_\mu)^2\varphi \tag{2A}$$

which is seen to be analogous to the time-dependent Schrödinger equation, u replacing the time and the four coordinates of space-time x_μ replacing the usual three coordinates of space.

Since the potentials $A_\mu(x)$ are functions only of coordinates x_μ and are independent of u, the equation is separable in u and we can write a special solution in the form $\varphi=\exp(\frac{1}{2}im^2u)\psi(x)$ where $\psi(x)$, a function of the coordinates x_μ only, satisfies (1A) and the eigenvalue $\frac{1}{2}m^2$ conjugate to u is related to the mass m of the particle. Equation (2A) is therefore equivalent to the Klein-Gordon Eq. (1A) provided we ask in the end only for the solution of (1A) corresponding to the eigenvalue $\frac{1}{2}m^2$ for the quantity conjugate to u.

We may now proceed to represent Eq. (2A) in Lagrangian form in general and without regard to this eigenvalue condition. Only in the final solutions need we apply the eigenvalue condition. That is, if we have some special solution $\varphi(x, u)$ of (2A) we can select that part corresponding to the eigenvalue $\frac{1}{2}m^2$ by calculating

$$\psi(x)=\int_{-\infty}^{\infty}\exp(-\tfrac{1}{2}im^2u)\varphi(x, u)du$$

and thereby obtain a solution ψ of Eq. (1A).

Since (2A) is so closely analogous to the Schrödinger equation, it is easily written in the Lagrangian form described in **C**, simply by working by analogy. For example if $\varphi(x, u)$ is known at one value of u its value at a slightly larger value $u+\epsilon$ is given by

$$\varphi(x, u+\epsilon)=\int\exp i\epsilon\left[-\frac{(x_\mu-x_\mu')^2}{2\epsilon^2}-\frac{1}{2}\left(\frac{x_\mu-x_\mu'}{\epsilon}\right)(A_\mu(x)+A_\mu(x'))\right] \cdot\varphi(x', u)d^4\tau_{x'}(2\pi i\epsilon)^{-\frac{3}{2}}(-2\pi i\epsilon)^{-\frac{1}{2}} \tag{3A}$$

where $(x_\mu-x_\mu')^2$ means $(x_\mu-x_\mu')(x_\mu-x_\mu')$, $d^4\tau_{x'}=dx_1'dx_2'dx_3'dx_4'$ and the sign of the normalizing factor is changed for the x_4 component since the component has the reversed sign in its quadratic coefficient in the exponential, in accordance with our summation convention $a_\mu b_\mu=a_4b_4-a_1b_1-a_2b_2-a_3b_3$. Equation (3A), as can be verified readily as described in **C**, Sec. 6, is equivalent to first order in ϵ, to Eq. (2A). Hence, by repeated use of this equation the wave function at $u_0=n\epsilon$ can be represented in terms of that at $u=0$ by:

$$\varphi(x_{\nu,n}, u_0)=\int\exp-\frac{i\epsilon}{2}\sum_{i=1}^{n}\left[\left(\frac{x_{\mu,i}-x_{\mu,i-1}}{\epsilon}\right)^2 +\epsilon^{-1}(x_{\mu,i}-x_{\mu,i-1})(A_\mu(x_i)+A_\mu(x_{i-1}))\right] \cdot\varphi(x_{\nu,0}, 0)\prod_{i=0}^{n-1}(d^4\tau_i/4\pi^2\epsilon^2 i). \tag{4A}$$

That is, roughly, the amplitude for getting from one point to another with a given value of u_0 is the sum over all trajectories of $\exp(iS)$ where

$$S=-\int_0^{u_0}[\tfrac{1}{2}(dx_\mu/du)^2+(dx_\mu/du)A_\mu(x)]du, \tag{5A}$$

when sufficient care is taken to define the quantities, as in **C**. This completes the formulation for particles in a fixed potential but a few words of description may be in order.

In the first place in the special case of a free particle we can define a kernal $k^{(0)}(x, u_0; x', 0)$ for arrival from x_μ', 0 to x_μ at u_0 as the sum over all trajectories between these points of $\exp-i\int_0^{u_0}\frac{1}{2}(dx_\mu/du)^2du$. Then for this case we have

$$\varphi(x, u_0)=\int k^{(0)}(x, u_0; x', 0)\varphi(x', 0)d^4\tau_{x'} \tag{6A}$$

and it is easily verified that k_0 is given by

$$k^{(0)}(x, u_0; x', 0)=(4\pi^2u_0{}^2i)^{-1}\exp-i(x_\mu-x_\mu')^2/2u_0 \tag{7A}$$

for $u_0>0$ and by 0, by definition, for $u_0<0$. The corresponding

[18] F. Rohrlich (to be published).

[19] The physical ideas involved in such a description are discussed in detail by Y. Nambu, Prog. Theor. Phys. **5**, 82 (1950). An equation of type (2A) extended to the case of Dirac electrons has been studied by V. Fock, Physik Zeits. Sowjetunion **12**, 404 (1937).

kernel of importance when we select the eigenvalue $\frac{1}{2}m^2$ is[20]

$$2iI_+(x,x')=\int_{-\infty}^{\infty}k^{(0)}(x,u_0;x',0)\exp(-\tfrac{1}{2}im^2u_0)du_0$$
$$=\int_0^{\infty}du_0(4\pi^2u_0^2i)^{-1}\exp-\tfrac{1}{2}i(m^2u_0+u_0^{-1}(x_\mu-x_\mu')^2) \quad (8A)$$

(the last extends only from $u_0=0$ since k_0 is zero for negative u_0) which is identical to the I_+ defined in **II**.[21] This may be seen readily by studying the Fourier transform, for the transform of the integrand on the right-hand side is

$$\int(4\pi^2u_0^2i)^{-1}\exp(ip\cdot x)\exp-\tfrac{1}{2}i(m^2u_0+x_\mu^2/u_0)d^4\tau_x$$
$$=\exp-\tfrac{1}{2}iu_0(m^2-p_\mu^2)$$

so that the u_0 integration gives for the transform of I_+ just $1/(p_\mu^2-m^2)$ with the pole defined exactly as in **II**. Thus we are automatically representing the positrons as trajectories with the time sense reversed.

If $\Phi^{(0)}[x(u)]=\exp-i\int_0^{u_0}\frac{1}{2}(dx_\mu/du)^2du$ is the amplitude for a given trajectory $x_\nu(u)$ for a free particle, then the amplitude in a potential is

$$\Phi^{(A)}[x(u)]=\Phi^{(0)}[x(u)]\exp-i\int_0^{u_0}(dx_\mu/du)A_\mu(x)du. \quad (9A)$$

If desired this may be studied by perturbation methods by expanding the exponential in powers of A_μ.

For interpretation, the integral in (9A) must be written as a Riemann sum, and if a perturbation expansion is made, care must be taken with the terms quadratic in the velocity, for the effect of $(x_{\mu,i+1}-x_{\mu,i})(x_{\nu,i+1}-x_{\nu,i})$ is not of order ϵ^2 but is $-i\delta_{\mu\nu}\epsilon$. The "velocity" dx_μ/du becomes the momentum operator $p_\mu=+i\partial/\partial x_\mu$ operating half before and half after A_μ, just as in the non-relativistic Schrödinger equation discussed in Sec. 5. Furthermore, in exactly the same manner as in that case, but here in four dimensions, a term quadratic in A_μ arises in the second-order perturbation terms from the coincidence of two velocities for the same value of u.

As an example, the kernal $k^{(A)}(x,u_0;x',0)$ for proceeding from $x_\mu',0$ to x_μ,u_0 in a potential A_μ differs from $k^{(0)}$ to first order in A_μ by a term

$$-i\int_0^{u_0}duk^{(0)}(x,u_0;y,u)\tfrac{1}{2}(p_\mu A_\mu(y)+A_\mu(y)p_\mu)k^{(0)}(y,u;x',0)d\tau_y$$

the p_μ here meaning $+i\partial/\partial y_\mu$. The kernel of importance on selecting the eigenvalue $\frac{1}{2}m^2$ is obtained by multiplying this by $\exp(-\frac{1}{2}im^2u_0)$ and integrating u_0 from 0 to ∞. The kernel $k^{(0)}(x,u_0;y,u)$ depends only on $u'=u_0-u$ and in the integrals on u and u_0; $\int_0^\infty du_0\int_0^{u_0}du\exp(-\frac{1}{2}im^2u_0)\cdots$, can be written, on interchanging the order of integration and changing variables to u and u', $\int_0^\infty du\int_0^\infty du'\exp(-\frac{1}{2}im^2(u+u'))\cdots$. Now the integral on u' converts $k^{(0)}(x,u_0;y,u)$ to $2iI_+(x,y)$ by (8A), while that on u converts $k^{(0)}(y,u;x',0)$ to $2iI_+(y,x')$, so the result becomes

$$\int 2iI_+(x,y)(p_\mu A_\mu+A_\mu p_\mu)I_+(y,x')d^4\tau_y$$

as expected. The same principle works to any order so that the rules for a single Klein-Gordon particle in external potentials given in **II**, Section 9, are deduced.

The transition to quantum electrodynamics is simple for in (5A) we already have a transition amplitude represented as a sum (over trajectories, and eventually u_0) of terms, in each of which the potential appears in exponential form. We may make use of the general relation (54). Hence, for example, one finds for the case of no photons in the initial and final states, in the presence of an external potential B_μ, the amplitude that a particle proceeds from $(x_\mu',0)$ to (x_μ,u_0) is the sum over all trajectories of the quantity

$$\exp-i\left[\frac{1}{2}\int_0^{u_0}\left(\frac{dx_\mu}{du}\right)^2du+\int_0^{u_0}\frac{dx_\mu}{du}B_\mu(x(u))du\right.$$
$$\left.+\frac{e^2}{2}\int_0^{u_0}\int_0^{u_0}\frac{dx_\mu(u)}{du}\frac{dx_\nu(u')}{du'}\delta_+((x_\mu(u)-x_\mu(u'))^2)dudu'\right]. \quad (10A)$$

This result must be multiplied by $\exp(-\frac{1}{2}im^2u_0)$ and integrated on u_0 from zero to infinity to express the action of a Klein-Gordon particle acting on itself through virtual photons. The integrals are interpreted as Riemann sums, and if perturbation expansions are made, the necessary care is taken with the terms quadratic in velocity. When there are several particles (other than the virtual pairs already included) one use a separate u for each, and writes the amplitude for each set of trajectories as the exponental of $-i$ times

$$\frac{1}{2}\sum_n\int_0^{u_0^{(n)}}\left(\frac{dx_\mu^{(n)}}{du}\right)^2du+\sum_n\int_0^{u_0^{(n)}}\frac{dx_\mu^{(n)}}{du}B_\mu(x_\mu^{(n)}(u))du$$
$$+\frac{e^2}{2}\sum_n\sum_m\int_0^{u_0^{(n)}}\int_0^{u_0^{(m)}}\frac{dx_\nu^{(n)}(u)}{du}\frac{dx_\nu^{(m)}(u')}{du'}$$
$$\times\delta_+((x_\mu^{(n)}(u)-x_\mu^{(m)}(u'))^2)dudu', \quad (11A)$$

where $x_\mu^{(n)}(u)$ are the coordinates of the trajectory of the nth particle.[22] The solution should depend on the $u_0^{(n)}$ as $\exp(-\frac{1}{2}im^2\Sigma_n u_0^{(n)})$.

Actually, knowledge of the motion of a single charge implies a great deal about the behavior of several charges. For a pair which eventually may turn out to be a virtual pair may appear in the short run as two "other particles." As a virtual pair, that is, as the reverse section of a very long and complicated single track we know its behavior by (10A). We can assume that such a section can be looked at equally well, for a limited duration at least, as being due to other unconnected particles. This then implies a definite law of interaction of particles if the self-action (10A) of a single particle is known. (This is similar to the relation of real and virtual photon processes discussed in detail in Appendix B.) It is possible that a detailed analysis of this could show that (10A) implied that (11A) was correct for many particles. There is even reason to believe that the law of Bose-Einstein statistics and the expression for contributions from closed loops could be deduced by following this argument. This has not yet been analyzed completely, however, so we must leave this formulation in an incomplete form. The expression for closed loops should come out to be $C_v=\exp+L$ where L, the contribution from a

[20] The factor $2i$ in front of I_+ is simply to make the definition of I_+ here agree with that in **I** and **II**. In **II** it operates with $\mathbf{p}\cdot\mathbf{A}+\mathbf{A}\cdot\mathbf{p}$ as a perturbation. But the perturbation coming from (3A) in a natural way by expansion of the exponential is $-\frac{1}{2}i(\mathbf{p}\cdot\mathbf{A}+\mathbf{A}\cdot\mathbf{p})$.

[21] Expression (8A) is closely related to Schwinger's parametric integral representation of these functions. For example, (8A) becomes formula (45) of F. Dyson, Phys. Rev. **75**, 486 (1949) for $\Delta_F\equiv\Delta^{(1)}-2i\bar{\Delta}\equiv 2iI_+$ if $(2\alpha)^{-1}$ is substituted for u_0.

[22] The form (10A) suggests another interesting possibility for avoiding the divergences of quantum electrodynamics in this case. The divergences arise from the δ_+ function when $u=u'$. We might restrict the integration in the double integral such that $|u-u'|>\delta$ where δ is some finite quantity, very small compared with m^{-2}. More generally, we could keep the region $u=u'$ from contributing by including in the integrand a factor $F(u-u')$ where $F(x)\to 1$ for x large compared to some δ, and $F(0)=0$ (e.g., $F(x)$ acts qualitatively like $1-\exp(-x^2\delta^{-2})$. (Another way might be to replace u by a discontinuous variable, that is, we do not use the limit in (4A) as $\epsilon\to 0$ but set $\epsilon=\delta$.) The idea is that two interactions would contribute very little in amplitude if they followed one another too rapidly in u. It is easily verified that this makes the otherwise divergent integrals finite. But whether the resulting formulas make good physical sense is hard to see. The action of a potential would now depend on the value of u so that Eq. (2A), or its equivalent, would not be separable in u so that $\frac{1}{2}m^2$ would no longer be a strict eigenvalue for all disturbances. High energy potentials could excite states corresponding to other eigenvalues, possibly thereby corresponding to other masses. This note is meant only as a speculation, for not enough work has been done in this direction to make sure that a reasonable physical theory can be developed along these lines. (What little work has been done was not promising.) Analogous modifications can also be made for Dirac electrons.

single loop, is

$$L=2\int_0^\infty l(u_0)\exp(-\tfrac{1}{2}im^2u_0)du_0/u_0$$

where $l(u_0)$ is the sum over all trajectories which close on themselves ($x_\mu(u_0)=x_\mu(0)$) of $\exp(iS)$ with S given in (5A), and a final integration $d\tau_{x(0)}$ on $x_\mu(0)$ is made. This is equivalent to putting

$$l(u_0)=\int(k^{(A)}(x,u_0;x,0)-k^{(0)}(x,u_0;x,0))d\tau_x.$$

The term $k^{(0)}$ is subtracted only to simplify convergence problems (as adding a constant independent of A_μ to L has no effect).

APPENDIX B. THE RELATION OF REAL AND VIRTUAL PROCESSES

If one has a general formula for all virtual processes he should be able to find the formulas and states involved in real processes. That is to say, we should be able to deduce the formulas of Section 9 directly from the formulation (24), (25) (or its generalized equivalent such as (46), (48)) without having to go all the way back to the more usual formulation. We discuss this problem here.

That this possibility exists can be seen from the consideration that what looks like a real process from one point of view may appear as a virtual process occurring over a more extended time.

For example, if we wish to study a given real process, such as the scattering of light, we can, if we wish, include in principle the source, scatterer, and eventual absorber of the scattered light in our analysis. We may imagine that no photon is present initially, and that the source then emits light (the energy coming say from kinetic energy in the source). The light is then scattered and eventually absorbed (becoming kinetic energy in the absorber). From this point of view the process is virtual; that is, we start with no photons and end with none. Thus we can analyze the process by means of our formula for virtual processes, and obtain the formulas for real processes by attempting to break the analysis into parts corresponding to emission, scattering, and absorption.[23]

To put the problem in a more general way, consider the amplitude for some transition from a state empty of photons far in the past (time t') to a similar one far in the future ($t=t''$). Suppose the time interval to be split into three regions a, b, c in some convenient manner, so that region b is an interval $t_2>t>t_1$ around the present time that we wish to study. Region a, ($t_1>t>t'$), precedes b, and c, ($t''>t>t_2$), follows b. We want to see how it comes about that the phenomena during b can be analyzed by a study of transitions $g_{ji}(b)$ between some initial state i at time t_1 (which no longer need be photon-free), to some other final state j at time t_2. The states i and j are members of a large class which we will have to find out how to specify. (The single index i is used to represent a large number of quantum numbers, so that different values of i will correspond to having various numbers of various kinds of photons in the field, etc.) Our problem is to represent the over-all transition amplitude, $g(a,b,c)$, as a sum over various values of i, j of a product of three amplitudes,

$$g(a,b,c)=\sum_i\sum_j g_{0j}(c)g_{ji}(b)g_{i0}(a); \quad (1B)$$

first the amplitude that during the interval a the vacuum state makes transition to some state i, then the amplitude that during b the transition to j is made, and finally in c the amplitude that the transition from j to some photon-free state 0 is completed.

[23] The formulas for real processes deduced in this way are strictly limited to the case in which the light comes from sources which are originally dark, and that eventually all light emitted is absorbed again. We can only extend it to the case for which these restrictions do not hold by hypothesis, namely, that the details of the scattering process are independent of these characteristics of the light source and of the eventual disposition of the scattered light. The argument of the text gives a method for discovering formulas for real processes when no more than the formula for virtual processes is at hand. But with this method belief in the general validity of the resulting formulas must rest on the physical reasonableness of the above-mentioned hypothesis.

The mathematical problem of splitting $g(a,b,c)$ is made definite by the further condition that $g_{ji}(b)$ for given i, j must not involve the coordinates of the particles for times corresponding to regions a or c, $g_{i0}(a)$ must involve those only in region a, and $g_{0j}(c)$ only in c.

To become acquainted with what is involved, suppose first that we do not have a problem involving virtual photons, but just the transition of a one-dimensional Schrödinger particle going in a long time interval from, say, the origin o to the origin o, and ask what states i we shall need for intermediary time intervals. We must solve the problem (1B) where $g(a,b,c)$ is the sum over all trajectories going from o at t' to o at t'' of $\exp iS$ where $S=\int Ldt$. The integral may be split into three parts $S=S_a+S_b+S_c$ corresponding to the three ranges of time. Then $\exp(iS)=\exp(iS_c)\cdot\exp(iS_b)\cdot\exp(iS_a)$ and the separation (1B) is accomplished by taking for $g_{i0}(a)$ the sum over all trajectories lying in a from o to some end point x_{t_1} of $\exp(iS_a)$, for $g_{ji}(b)$ the sum over trajectories in b of $\exp(iS_b)$ between end points x_{t_1} and x_{t_2}, and for $g_{0j}(c)$ the sum of $\exp(iS_c)$ over the section of the trajectory lying in c and going from x_{t_2} to o. Then the sum on i and j can be taken to be the integrals on x_{t_1}, x_{t_2} respectively. Hence the various states i can be taken to correspond to particles being at various coordinates x. (Of course any other representation of the states in the sense of Dirac's transformation theory could be used equally well. Which, one, whether coordinate, momentum, or energy level representation, is of course just a matter of convenience and we cannot determine that simply from (1B).)

We can consider next the problem including virtual photons. That is, $g(a,b,c)$ now contains an additional factor $\exp(iR)$ where R involves a double integral $\int\int$ over all time. Those parts of the index i which correspond to the particle states can be taken in the same way as though R were absent. We study now the extra complexities in the states produced by splitting the R. Let us first (solely for simplicity of the argument) take the case that there are only two regions a, c separated by time t_0 and try to expand

$$g(a,c)=\sum_i g_{0i}(c)g_{i0}(a).$$

The factor $\exp(iR)$ involves R as a double integral which can be split into three parts $\int_a\int_a+\int_c\int_c+\int_a\int_c$ for the first of which both t, s are in a, for the second both are in c, for the third one is in a the other in c. Writing $\exp(iR)$ as $\exp(iR_{cc})\cdot\exp(iR_{aa})\cdot\exp(iR_{ac})$ shows that the factors R_{cc} and R_{aa} produce no new problems for they can be taken bodily into $g_{0i}(c)$ and $g_{i0}(a)$ respectively. However, we must disentangle the variables which are mixed up in $\exp(iR_{ac})$.

The expression for R_{ac} is just twice (24) but with the integral on s extending over the range a and that for t extending over c. Thus $\exp(iR_{ac})$ contains the variables for times in a and in c in a quite complicated mixture. Our problem is to write $\exp(iR_{ac})$ as a sum over possibly a vast class of states i of the product of two parts, like $h_i'(c)h_i(a)$, each of which involves the coordinates in one interval alone.

This separation may be made in many different ways, corresponding to various possible representations of the state of the electromagnetic field. We choose a particular one. First we can expand the exponential, $\exp(iR_{ac})$, in a power series, as $\sum_n i^n(n!)^{-1}(R_{ac})^n$. The states i can therefore be subdivided into subclasses corresponding to an integer n which we can interpret as the number of quanta in the field at time t_0. The amplitude for the case $n=0$ clearly just involves $\exp(iR_{aa})$ and $\exp(iR_{cc})$ in the way that it should if we interpret these as the amplitudes for regions a and c, respectively, of making a transition between a state of zero photons and another state of zero photons.

Next consider the case $n=1$. This implies an additional factor in the transitional element; the factor R_{ac}. The variables are still mixed up. But an easy way to perform the separation suggests itself. Namely, expand the $\delta_+((t-s)^2-(\mathbf{x}_n(t)-\mathbf{x}_m(s))^2)$ in R_{ac} as a Fourier integral as

$$i\int\exp(-ik|t-s|)\exp(-i\mathbf{K}\cdot(\mathbf{x}_n(t)-\mathbf{x}_m(s))d^3\mathbf{K}/4\pi^2k.$$

For the exponential can be written immediately as a product of $\exp+i(\mathbf{K}\cdot\mathbf{x}_m(s))$, a function only of coordinates for times s in a (suppose $s<t$), and $\exp-i\mathbf{K}\cdot\mathbf{x}_n(t)$ (a function only of coordinates during interval c). The integral on $d^3\mathbf{K}$ can be symbolized as a sum over states i characterized by the value of $\mathbf{K}$. Thus the state with $n=1$ must be further characterized by specifying a vector $\mathbf{K}$, interpreted as the momentum of the photon. Finally the factor $(1-\mathbf{x}'_n(t)\cdot\mathbf{x}'_m(s))$ in R_{ac} is simply the sum of four parts each of which is already split (namely 1, and each of the three components in the vector scalar product). Hence each photon of momentum $\mathbf{K}$ must still be characterized by specifying it as one of four varieties; that is, there are four polarizations.[24] Thus in trying to represent the effect of the past a on the future c we are lead to invent photons of four polarizations and characterized by a propagation vector $\mathbf{K}$.

The term for a given polarization and value of $\mathbf{K}$ (for $n=1$) is clearly just $-\beta_c\beta_a{}^*$ where the β_a is defined in (59) but with the time integral extending just over region a, while β_c is the same expression with the integration over region c. Hence the amplitude for transition during interval a from a state with no quanta to a state with one in a given state of polarization and momentum is calculated by inclusion of an extra factor $i\beta_a{}^*$ in the transition element. Absorption in region c corresponds to a factor $i\beta_c$.

We next turn to the case $n=2$. This requires analysis of $R_{ac}{}^2$. The δ_+ can be expanded again as a Fourier integral, but for each of the two δ_+ in $\frac{1}{2}R_{ac}{}^2$ we have a value of $\mathbf{K}$ which may be different. Thus we say, we have two photons, one of momentum $\mathbf{K}$ and one momentum $\mathbf{K}'$ and we sum over all values of $\mathbf{K}$ and $\mathbf{K}'$. (Similarly each photon is characterized by its own independent polarization index.) The factor $\frac{1}{2}$ can be taken into account neatly by asserting that we count each possible pair of photons as constituting just one state at time t_0. Then the $\frac{1}{2}$ arises for the sum over all $\mathbf{K}$, $\mathbf{K}'$ (and polarizations) counts each pair twice. On the other hand, for the terms representing two identical photons ($\mathbf{K}=\mathbf{K}'$) of like polarization, the $\frac{1}{2}$ cannot be so interpreted. Instead we invent the rule that a state of two like photons has statistical weight $\frac{1}{2}$ as great as that calculated as though the photons were different. This, generalized to n identical photons, is the rule of Bose statistics.

The higher values of n offer no problem. The $1/n!$ is interpreted combinatorially for different photons, and as a statistical factor when some are identical. For example, for all n identical one obtains a factor $(n!)^{-1}(-\beta_c\beta_a{}^*)^n$ so that $(n!)^{-\frac{1}{2}}(i\beta_a{}^*)^n$ can be interpreted as the amplitude for emission (from no initial photons) of n identical photons, in complete agreement with (61) for G_{m0}.

To obtain the amplitude for transitions in which neither the initial nor the final state is empty of photons we must consider the more general case of the division into three time regions (1B). This time we see that the factor which involves the coordinates in an entangled manner is $\exp i(R_{ab}+R_{bc}+R_{ac})$. It is to be expanded in the form $\sum_i\sum_j h_i''(c)h_{ij}'(b)h_j(a)$. Again the expansion in power series and development in Fourier series with a polarization sum will solve the problem. Thus the exponential is $\sum_r\sum_{l_1}\sum_{l_2}(iR_{ac})^r(iR_{ab})^{l_1}(iR_{bc})^{l_2}(l_1!)^{-1}(l_2!)^{-1}(r!)^{-1}$. Now the R are written as Fourier series, one of the terms containing l_1+l_2+r variables $\mathbf{K}$. Since l_1+r involve a, l_2+r involve c and l_1+l_2 involve b, this term will give the amplitude that l_1+r photons are emitted during the interval a, of those l_1 are absorbed during b but the remaining r, along with l_2 new ones emitted during b go on to be absorbed during the interval c. We have therefore $n=l_1+r$ photons in the state at time t_1 when b begins, and $m=l_2+r$ at t_2 when b is over. They each are characterized by momentum vectors and polarizations. When these are different the factors $(l_1!)^{-1}(l_2!)^{-1}(r!)^{-1}$ are absorbed combinatorially. When some are equal we must invoke the rule of the statistical weights. For example, suppose all l_1+l_2+r photons are identical. Then $R_{ab}=i\beta_b\beta_a{}^*$, $R_{bc}=i\beta_c\beta_b{}^*$, $R_{ac}=i\beta_c\beta_a{}^*$ so that our sum is

$$\sum_{l_1}\sum_{l_2}\sum_r(l_1!l_2!r!)^{-1}(i\beta_c)^{l_2+r}(i\beta_b)^{l_1}(i\beta_b{}^*)^{l_2}(i\beta_a{}^*)^{l_1+r}.$$

Putting $m=l_2+r$, $n=l_1+r$, this is the sum on n and m of

$$(i\beta_c)^m(m!)^{-\frac{1}{2}}[\textstyle\sum_r(m!n!)^{\frac{1}{2}}((m-r)!(n-r)!r!)^{-1}\times(i\beta_b{}^*)^{m-r}(i\beta_b)^{n-r}](n!)^{-\frac{1}{2}}(i\beta_a{}^*)^n.$$

The last factor we have seen is the amplitude for emission of n photons during interval a, while the first factor is the amplitude for absorption of m during c. The sum is therefore the factor for transition from n to m identical photons, in accordance with (57). We see the significance of the simple generating function (56).

We have therefore found rules for real photons in terms of those for virtual. The real photons are a way of representing and keeping track of those aspects of the past behavior which may influence the future.

If one starts from a theory involving an arbitrary modification of the direct interaction δ_+ (or in more general situations) it is possible in this way to discover what kinds of states and physical entities will be involved if one tries to represent in the present all the information needed to predict the future. With the Hamiltonian method, which begins by assuming such a representation, it is difficult to suggest modifications of a general kind, for one cannot formulate the problem without having a complete representation of the characteristics of the intermediate states, the particles involved in interaction, etc. It is quite possible (in the author's opinion, it is very likely) that we may discover that in nature the relation of past and future is so intimate for short durations that no simple representation of a present may exist. In such a case a theory could not find expression in Hamiltonian form.

An exactly similar analysis can be made just as easily starting with the general forms (46), (48). Also a coordinate representation of the photons could have been used instead of the familiar momentum one. One can deduce the rules (60), (61). Nothing essentially different is involved physically, however, so we shall not pursue the subject further here. Since they imply[23] all the rules for real photons, Eqs. (46), (47), (48) constitute a compact statement of all the laws of quantum electrodynamics. But they give divergent results. Can the result *after* charge and mass renormalization also be expressed to all orders in $e^2/\hbar c$ in a simple way?

APPENDIX C. DIFFERENTIAL EQUATION FOR ELECTRON PROPAGATION

An attempt has been made to find a differential wave equation for the propagation of an electron interacting with itself, analogous to the Dirac equation, but containing terms representing the self-action. Neglecting all effects of closed loops, one such equation has been found, but not much has been done with it. It is reported here for whatever value it may have.

An electron acting upon itself is, from one point of view, a complex system of a particle and a field of an indefinite number of photons. To find a differential law of propagation of such a system we must ask first what quantities known at one instant will permit the calculation of these same quantities an instant later. Clearly, a knowledge of the position of the particle is not enough. We should need to specify: (1) the amplitude that the electron is at x and there are no photons in the field, (2) the amplitude the electron is at x and there is one photon of such and such a kind in the field, (3) the amplitude there are two photons, etc. That is, a series of functions of ever increasing numbers of variables. Following this view, we shall be led to the wave equation of the theory of second quantization.

We may also take a different view. Suppose we know a quantity $\Phi_e2[B,x]$, a spinor function of x_μ, and functional of $B_\mu(1)$, defined as the amplitude that an electron arrives at x_μ with no photon in the field when it moves in an arbitrary external unquantized potential $B_\mu(1)$. We allow the electron also to interact with itself,

[24] Usually only two polarizations transverse to the propagation vector $\mathbf{K}$ are used. This can be accomplished by a further rearrangement of terms corresponding to the reverse of the steps leading from (17) to (19). We omit the details here as it is well-known that either formulation gives the same results. See **II**, Section 8.

but $\Phi_e 2$ is the amplitude at a given instant that there happens to be no photons present. As we have seen, a complete knowledge of this functional will also tell us the amplitude that the electron arrives at x and there is just one photon, of form $A_\mu^{PH}(1)$ present. It is, from (60), $\int(\delta\Phi_e 2[B,x]/\delta B_\mu(1))A_\mu^{PH}(1)d\tau_1$.

Higher numbers of photons correspond to higher functional derivatives of $\Phi_e 2$. Therefore, $\Phi_e 2[B,x]$ contains all the information requisite for describing the state of the electron-photon system, and we may expect to find a differential equation for it. Actually it satisfies ($\nabla=\gamma_\mu\partial/\partial x_\mu$, $B=\gamma_\mu B_\mu$),

$$(i\nabla-m)\Phi_e 2[B,x]=B(x)\Phi_e 2[B,x] + ie^2\gamma_\mu\int\delta_+(s_{x1}^2)(\delta\Phi_e 2[B,x]/\delta B_\mu(1))d\tau_1 \quad \text{(1C)}$$

as may be seen from a physical argument.[25] The operator $(i\nabla-m)$ operating on the x coordinate of $\Phi_e 2$ should equal, from Dirac's equation, the changes in $\Phi_e 2$ as we go from one position x to a neighboring position due to the action of vector potentials. The term $B(x)\Phi_e 2$ is the effect of the external potential. But $\Phi_e 2$ may also change for at the first position x we may have had a photon present (amplitude that it was emitted at another point 1 is $\delta\Phi_e 2/\delta B_\mu(1)$) which was absorbed at x (amplitude photon released at 1 gets to x is $\delta_+(s_{x1}^2)$ where s_{x1}^2 is the squared invariant distance from 1 to x) acting as a vector potential there (factor γ_μ). Effects of vacuum polarization are left out.

Expansion of the solution of (1C) in a power series in B and e^2 starting from a free particle solution for a single electron, produces a series of terms which agree with the rules of II for action of potentials and virtual photons to various orders. It is another matter to use such an equation for the practical solution of a problem to all orders in e^2. It might be possible to represent the self-energy problem as the variational problem for m, stemming from (1C). The δ_+ will first have to be modified to obtain a convergent result.

We are not in need of the general solution of (1C). (In fact, we have it in (46), (48) in terms of the solution $T_0[B]=\Phi_0[B,x]$ of the ordinary Dirac equation $(i\nabla-m)\Phi_0[B,x]=B\Phi_0[B,x]$. The general solution is too complicated, for complete knowledge of the motion of a self-acting electron in an arbitrary potential is essentially all of electrodynamics (because of the kind of relation of real and virtual processes discussed for photons in Appendix B, extended also to real and virtual pairs). Furthermore, it is easy to see that other quantities also satisfy (1C). Consider a system of many electrons, and single out some one for consideration, supposing all the others go from some definite initial state i to some definite final state f. Let $\Phi_e 2[B,x]$ be the amplitude that the special electron arrives at x, there are no photons present, and the other electrons go from i to f when there is an external potential B_μ present (which B_μ also acts on the other electrons). Then $\Phi_e 2$ also satisfies (1C). Likewise the amplitude with closed loops (all other electrons go vacuum to vacuum) also satisfies (1C) including all vacuum polarization effects. The various problems correspond to different assumptions as to the dependence of $\Phi_e 2[B,x]$ on B_μ in the limit of zero e^2. The Eq. (1C) without further boundary conditions is probably too general to be useful.

[25] Its general validity can also be demonstrated mathematically from (45). The amplitude for arriving at x with no photons in the field with virtual photon coupling e^2 is a transition amplitude. It must, therefore, satisfy (45) with $T_e 2[B]=\Phi_e 2[B,x]$ for any x. Hence show that the quantity

$$C_e 2[B,x]=(i\nabla-m-B(x))\Phi_e 2[B,x] - ie^2\gamma_\mu\int\delta_+(s_{x1}^2)(\delta\Phi_e 2[B,x]/\delta B_\mu(1))d\tau_1$$

also satisfies Eq. (45) by substituting $C_e 2[B,x]$ for $T_e 2[B]$ in (45) and using the fact that $\Phi_e 2[B,x]$ satisfies (45). Hence if $C_0[B,x]=0$ then $C_e 2[B,x]=0$ for all e^2. But $C_e 2[B,x]=0$ means that $\Phi_e 2[B,x]$ satisfies (1C). Therefore, that solution $\Phi_e^2[B,x]$ of (45) which also satisfies $(i\nabla-m-B(x))\Phi_0[B,x]=0$ (the propagation of a free electron without virtual photons) is a solution of (1C) as we wished to show. Equation (1C) may be more convenient than (45) for some purposes for it does not involve differentiation with respect to the coupling constant, and is more analogous to a wave equation.

The Radiation Theories of Tomonaga, Schwinger, and Feynman

F. J. Dyson
Institute for Advanced Study, Princeton, New Jersey

(Received October 6, 1948)

A unified development of the subject of quantum electrodynamics is outlined, embodying the main features both of the Tomonaga-Schwinger and of the Feynman radiation theory. The theory is carried to a point further than that reached by these authors, in the discussion of higher order radiative reactions and vacuum polarization phenomena. However, the theory of these higher order processes is a program rather than a definitive theory, since no general proof of the convergence of these effects is attempted.

The chief results obtained are (a) a demonstration of the equivalence of the Feynman and Schwinger theories, and (b) a considerable simplification of the procedure involved in applying the Schwinger theory to particular problems, the simplification being the greater the more complicated the problem.

I. INTRODUCTION

As a result of the recent and independent discoveries of Tomonaga,[1] Schwinger,[2] and Feynman,[3] the subject of quantum electrodynamics has made two very notable advances. On the one hand, both the foundations and the applications of the theory have been simplified by being presented in a completely relativistic way; on the other, the divergence difficulties have been at least partially overcome. In the reports so far published, emphasis has naturally been placed on the second of these advances; the magnitude of the first has been somewhat obscured by the fact that the new methods have been applied to problems which were beyond the range of the older theories, so that the simplicity of the methods was hidden by the complexity of the problems. Furthermore, the theory of Feynman differs so profoundly in its formulation from that of Tomonaga and Schwinger, and so little of it has been published, that its particular advantages have not hitherto been available to users of the other formulations. The advantages of the Feynman theory are simplicity and ease of application, while those of Tomonaga-Schwinger are generality and theoretical completeness.

The present paper aims to show how the Schwinger theory can be applied to specific problems in such a way as to incorporate the ideas of Feynman. To make the paper reasonably self-contained it is necessary to outline the foundations of the theory, following the method of Tomonaga; but this paper is not intended as a substitute for the complete account of the theory shortly to be published by Schwinger. Here the emphasis will be on the application of the theory, and the major theoretical problems of gauge-invariance and of the divergencies will not be considered in detail. The main results of the paper will be general formulas from which the radiative reactions on the motions of electrons can be calculated, treating the radiation interaction as a small perturbation, to any desired order of approximation. These formulas will be expressed in Schwinger's notation, but are in substance identical with results given previously by Feynman. The contribution of the present paper is thus intended to be twofold: first, to simplify the Schwinger theory for the benefit of those using it for calculations, and second, to demonstrate the equivalence of the various theories within their common domain of applicability.*

[1] Sin-itiro Tomonaga, Prog. Theoret. Phys. **1**, 27 (1946); Koba, Tati, and Tomonaga, Prog. Theoret. Phys. **2**, 101 198 (1947); S. Kanesawa and S. Tomonaga, Prog. Theoret. Phys. **3**, 1, 101 (1948); S. Tomonaga, Phys. Rev. **74**, 224 (1948).

[2] Julian Schwinger, Phys. Rev. **73**, 416 (1948); Phys. Rev. **74**, 1439 (1948). Several papers, giving a complete exposition of the theory, are in course of publication.

[3] R. P. Feynman, Rev. Mod. Phys. **20**, 367 (1948); Phys. Rev. **74**, 939, 1430 (1948); J. A. Wheeler and R. P. Feynman, Rev. Mod. Phys. **17**, 157 (1945). These articles describe early stages in the development of Feynman's theory, little of which is yet published.

* After this paper was written, the author was shown a letter, published in Progress of Theoretical Physics **3**, 205 (1948) by Z. Koba and G. Takeda. The letter is dated May 22, 1948, and briefly describes a method of treatment of radiative problems, similar to the method of this paper.

II. OUTLINE OF THEORETICAL FOUNDATIONS

Relativistic quantum mechanics is a special case of non-relativistic quantum mechanics, and it is convenient to use the usual non-relativistic terminology in order to make clear the relation between the mathematical theory and the results of physical measurements. In quantum electrodynamics the dynamical variables are the electromagnetic potentials $A_\mu(\mathbf{r})$ and the spinor electron-positron field $\psi_\alpha(\mathbf{r})$; each component of each field at each point $\mathbf{r}$ of space is a separate variable. Each dynamical variable is, in the Schrödinger representation of quantum mechanics, a time-independent operator operating on the state vector Φ of the system. The nature of Φ (wave function or abstract vector) need not be specified; its essential property is that, given the Φ of a system at a particular time, the results of all measurements made on the system at that time are statistically determined. The variation of Φ with time is given by the Schrödinger equation

$$i\hbar[\partial/\partial t]\Phi = \left\{\int H(\mathbf{r})d\tau\right\}\Phi, \quad (1)$$

where $H(\mathbf{r})$ is the operator representing the total energy-density of the system at the point $\mathbf{r}$. The general solution of (1) is

$$\Phi(t) = \exp\left\{[-it/\hbar]\int H(\mathbf{r})d\tau\right\}\Phi_0, \quad (2)$$

with Φ_0 any constant state vector.

Now in a relativistic system, the most general kind of measurement is not the simultaneous measurement of field quantities at different points of space. It is also possible to measure independently field quantities at different points of space at different times, provided that the points of space-time at which the measurements are made lie outside each other's light cones, so that the measurements do not interfere with each other. Thus the most comprehensive general type of measurement is a measurement of field quantities at each point $\mathbf{r}$ of space at a time $t(\mathbf{r})$, the locus of the points $(\mathbf{r}, t(\mathbf{r}))$ in space-time forming a 3-dimensional surface σ which is space-like (i.e., every pair of points on it is separated by a space-like interval). Such a measurement will be called "an observation of the system on σ." It is easy to see what the result of the measurement will be. At each point $\mathbf{r}'$ the field quantities will be measured for a state of the system with state vector $\Phi(t(\mathbf{r}'))$ given by (2). But all observable quantities at $\mathbf{r}'$ are operators which commute with the energy-density operator $H(\mathbf{r})$ at every point $\mathbf{r}$ different from $\mathbf{r}'$, and it is a general principle of quantum mechanics that if B is a unitary operator commuting with A, then for any state Φ the results of measurements of A are the same in the state Φ as in the state $B\Phi$. Therefore, the results of measurement of the field quantities at $\mathbf{r}'$ in the state $\Phi(t(\mathbf{r}'))$ are the same as if the state of the system were

$$\Phi(\sigma) = \exp\left\{-[i/\hbar]\int t(\mathbf{r})H(\mathbf{r})d\tau\right\}\Phi_0, \quad (3)$$

which differs from $\Phi(t(\mathbf{r}'))$ only by a unitary factor commuting with these field quantities. The important fact is that the state vector $\Phi(\sigma)$ depends only on σ and not on $\mathbf{r}'$. The conclusion reached is that observations of a system on σ give results which are completely determined by attributing to the system the state vector $\Phi(\sigma)$ given by (3).

The Tomonaga-Schwinger form of the Schrödinger equation is a differential form of (3). Suppose the surface σ to be deformed slightly near the point $\mathbf{r}$ into the surface σ', the volume of space-time separating the two surfaces being V. Then the quotient

$$[\Phi(\sigma') - \Phi(\sigma)]/V$$

tends to a limit as $V \to 0$, which we denote by $\partial\Phi/\partial\sigma(\mathbf{r})$ and call the functional derivative of Φ with respect to σ at the point $\mathbf{r}$. From (3) it follows that

$$i\hbar c[\partial\Phi/\partial\sigma(\mathbf{r})] = H(\mathbf{r})\Phi, \quad (4)$$

and (3) is, in fact, the general solution of (4).

The whole meaning of an equation such as (4) depends on the physical meaning which is attached to the statement "a system has a constant state vector Φ_0." In the present context, this statement means "results of measurements of

Results of the application of the method to a calculation of the second-order radiative correction to the Klein-Nishina formula are stated. All the papers of Professor Tomonaga and his associates which have yet been published were completed before the end of 1946. The isolation of these Japanese workers has undoubtedly constituted a serious loss to theoretical physics.

field quantities at any given point of space are independent of time." This statement is plainly non-relativistic, and so (4) is, in spite of appearances, a non-relativistic equation.

The simplest way to introduce a new state vector Ψ which shall be a relativistic invariant is to require that the statement "a system has a constant state vector Ψ" shall mean "a system consists of photons, electrons, and positrons, traveling freely through space without interaction or external disturbance." For this purpose, let

$$H(\mathbf{r}) = H_0(\mathbf{r}) + H_1(\mathbf{r}), \tag{5}$$

where H_0 is the energy-density of the free electromagnetic and electron fields, and H_1 is that of their interaction with each other and with any external disturbing forces that may be present. A system with constant Ψ is, then, one whose H_1 is identically zero; by (3) such a system corresponds to a Φ of the form

$$\Phi(\sigma) = T(\sigma)\Phi_0,$$

$$T(\sigma) = \exp\left\{-[i/\hbar]\int t(\mathbf{r})H_0(\mathbf{r})d\tau\right\}. \tag{6}$$

It is therefore consistent to write generally

$$\Phi(\sigma) = T(\sigma)\Psi(\sigma), \tag{7}$$

thus defining the new state vector Ψ of any system in terms of the old Φ. The differential equation satisfied by Ψ is obtained from (4), (5), (6), and (7) in the form

$$i\hbar c[\partial\Psi/\partial\sigma(\mathbf{r})] = (T(\sigma))^{-1}H_1(\mathbf{r})T(\sigma)\Psi. \tag{8}$$

Now if $q(\mathbf{r})$ is any time-independent field operator, the operator

$$q(x_0) = (T(\sigma))^{-1}q(\mathbf{r})T(\sigma)$$

is just the corresponding time-dependent operator as usually defined in quantum electrodynamics.[4] It is a function of the point x_0 of space-time whose coordinates are $(\mathbf{r}, ct(\mathbf{r}))$, but is the same for all surfaces σ passing through this point, by virtue of the commutation of $H_1(\mathbf{r})$ with $H_0(\mathbf{r}')$ for $\mathbf{r}' \neq \mathbf{r}$. Thus (8) may be written

$$i\hbar c[\partial\Psi/\partial\sigma(x_0)] = H_1(x_0)\Psi, \tag{9}$$

where $H_1(x_0)$ is the time-dependent form of the energy-density of interaction of the two fields with each other and with external forces. The left side of (9) represents the degree of departure of the system from a system of freely traveling particles and is a relativistic invariant; $H_1(x_0)$ is also an invariant, and thus is avoided one of the most unsatisfactory features of the old theories, in which the invariant H_1 was added to the non-invariant H_0. Equation (9) is the starting point of the Tomonaga-Schwinger theory.

III. INTRODUCTION OF PERTURBATION THEORY

Equation (9) can be solved explicitly. For this purpose it is convenient to introduce a one-parameter family of space-like surfaces filling the whole of space-time, so that one and only one member $\sigma(x)$ of the family passes through any given point x. Let $\sigma_0, \sigma_1, \sigma_2, \cdots$ be a sequence of surfaces of the family, starting with σ_0 and proceeding in small steps steadily into the past. By

$$\int_{\sigma_1}^{\sigma_0} H_1(x)dx$$

is denoted the integral of $H_1(x)$ over the 4-dimensional volume between the surfaces σ_1 and σ_0; similarly, by

$$\int_{-\infty}^{\sigma_0} H_1(x)dx, \quad \int_{\sigma_0}^{\infty} H_1(x)dx$$

are denoted integrals over the whole volume to the past of σ_0 and to the future of σ_0, respectively. Consider the operator

$$U = U(\sigma_0) = \left(1 - [i/\hbar c]\int_{\sigma_1}^{\sigma_0} H_1(x)dx\right) \times\left(1 - [i/\hbar c]\int_{\sigma_2}^{\sigma_1} H_1(x)dx\right)\cdots, \tag{10}$$

the product continuing to infinity and the surfaces $\sigma_0, \sigma_1, \cdots$ being taken in the limit infinitely close together. U satisfies the differential equation

$$i\hbar c[\partial U/\partial\sigma(x_0)] = H_1(x_0)U, \tag{11}$$

and the general solution of (9) is

$$\Psi(\sigma) = U(\sigma)\Psi_0, \tag{12}$$

with Ψ_0 any constant vector.

[4] See, for example, Gregor Wentzel, *Einführung in die Quantentheorie der Wellenfelder* (Franz Deuticke, Wien, 1943), pp. 18–26.

Expanding the product (10) in ascending powers of H_1 gives a series

$$U=1+(-i/\hbar c)\int_{-\infty}^{\sigma 0} H_1(x_1)dx_1+(-i/\hbar c)^2 \times\int_{-\infty}^{\sigma 0} dx_1\int_{-\infty}^{\sigma(x_1)} H_1(x_1)H_1(x_2)dx_2+\cdots. \quad (13)$$

Further, U is by (10) obviously unitary, and

$$U^{-1}=\bar{U}=1+(i/\hbar c)\int_{-\infty}^{\sigma 0} H_1(x_1)dx_1+(i/\hbar c)^2 \times\int_{-\infty}^{\sigma 0} dx_1\int_{-\infty}^{\sigma(x_1)} H_1(x_2)H_1(x_1)dx_2+\cdots. \quad (14)$$

It is not difficult to verify that U is a function of σ_0 alone and is independent of the family of surfaces of which σ_0 is one member. The use of a finite number of terms of the series (13) and (14), neglecting the higher terms, is the equivalent in the new theory of the use of perturbation theory in the older electrodynamics.

The operator $U(\infty)$, obtained from (10) by taking σ_0 in the infinite future, is a transformation operator transforming a state of the system in the infinite past (representing, say, converging streams of particles) into the same state in the infinite future (after the particles have interacted or been scattered into their final outgoing distribution). This operator has matrix elements corresponding only to real transitions of the system, i.e., transitions which conserve energy and momentum. It is identical with the Heisenberg S matrix.[5]

IV. ELIMINATION OF THE RADIATION INTERACTION

In most of the problem of electrodynamics, the energy-density $H_1(x_0)$ divides into two parts—

$$H_1(x_0)=H^i(x_0)+H^e(x_0), \quad (15)$$

$$H^i(x_0)=-[1/c]j_\mu(x_0)A_\mu(x_0), \quad (16)$$

the first part being the energy of interaction of the two fields with each other, and the second part the energy produced by external forces. It is usually not permissible to treat H^e as a small perturbation as was done in the last section. Instead, H^i alone is treated as a perturbation, the aim being to eliminate H^i but to leave H^e in its original place in the equation of motion of the system.

Operators $S(\sigma)$ and $S(\infty)$ are defined by replacing H_1 by H^i in the definitions of $U(\sigma)$ and $U(\infty)$. Thus $S(\sigma)$ satisfies the equation

$$i\hbar c[\partial S/\partial\sigma(x_0)]=H^i(x_0)S. \quad (17)$$

Suppose now a new type of state vector $\Omega(\sigma)$ to be introduced by the substitution

$$\Psi(\sigma)=S(\sigma)\Omega(\sigma). \quad (18)$$

By (9), (15), (17), and (18) the equation of motion for $\Omega(\sigma)$ is

$$i\hbar c[\partial\Omega/\partial\sigma(x_0)]=(S(\sigma))^{-1}H^e(x_0)S(\sigma)\Omega. \quad (19)$$

The elimination of the radiation interaction is hereby achieved; only the question, "How is the new state vector $\Omega(\sigma)$ to be interpreted?," remains.

It is clear from (19) that a system with a constant Ω is a system of electrons, positrons, and photons, moving under the influence of their mutual interactions, but in the absence of external fields. In a system where two or more particles are actually present, their interactions alone will, in general, cause real transitions and scattering processes to occur. For such a system, it is rather "unphysical" to represent a state of motion including the effects of the interactions by a constant state vector; hence, for such a system the new representation has no simple interpretation. However, the most important systems are those in which only one particle is actually present, and its interaction with the vacuum fields gives rise only to virtual processes. In this case the particle, including the effects of all its interactions with the vacuum, appears to move as a free particle in the absence of external fields, and it is eminently reasonable to represent such a state of motion by a constant state vector. Therefore, it may be said that the operator,

$$H_T(x_0)=(S(\sigma))^{-1}H^e(x_0)S(\sigma), \quad (20)$$

on the right of (19) represents the interaction of a physical particle with an external field, including radiative corrections. Equation (19) describes the extent to which the motion of a

[5] Werner Heisenberg, Zeits. f. Physik **120**, 513 (1943), **120**, 673 (1943), and Zeits. f. Naturforschung **1**, 608 (1946).

single physical particle deviates, in the external field, from the motion represented by a constant state-vector, i.e., from the motion of an observed "free" particle.

If the system whose state vector is constantly Ω undergoes no real transitions with the passage of time, then the state vector Ω is called "steady." More precisely, Ω is steady if, and only if, it satisfies the equation

$$S(\infty)\Omega = \Omega. \tag{21}$$

As a general rule, one-particle states are steady and many-particle states unsteady. There are, however, two important qualifications to this rule.

First, the interaction (20) itself will almost always cause transitions from steady to unsteady states. For example, if the initial state consists of one electron in the field of a proton, H_T will have matrix elements for transitions of the electron to a new state with emission of a photon, and such transitions are important in practice. Therefore, although the interpretation of the theory is simpler for steady states, it is not possible to exclude unsteady states from consideration.

Second, if a one-particle state as hitherto defined is to be steady, the definition of $S(\sigma)$ must be modified. This is because $S(\infty)$ includes the effects of the electromagnetic self-energy of the electron, and this self-energy gives an expectation value to $S(\infty)$ which is different from unity (and indeed infinite) in a one-electron state, so that Eq. (21) cannot be satisfied. The mistake that has been made occurred in trying to represent the observed electron with its electromagnetic self-energy by a wave field with the same characteristic rest-mass as that of the "bare" electron. To correct the mistake, let δm denote the electromagnetic mass of the electron, i.e., the difference in rest-mass between an observed and a "bare" electron. Instead of (5), the division of the energy-density $H(\mathbf{r})$ should have taken the form

$$H(\mathbf{r}) = (H_0(\mathbf{r}) + \delta mc^2\psi^*(\mathbf{r})\beta\psi(\mathbf{r})) + (H_1(\mathbf{r}) - \delta mc^2\psi^*(\mathbf{r})\beta\psi(\mathbf{r})).$$

The first bracket on the right here represents the energy-density of the free electromagnetic and electron fields with the observed electron rest-mass, and should have been used instead of $H_0(\mathbf{r})$ in the definition (6) of $T(\sigma)$. Consequently, the second bracket should have been used instead of $H_1(\mathbf{r})$ in Eq. (8).

The definition of $S(\sigma)$ has therefore to be altered by replacing $H^i(x_0)$ by[6]

$$H^I(x_0) = H^i(x_0) + H^S(x_0) = H^i(x_0) - \delta mc^2\bar{\psi}(x_0)\psi(x_0). \tag{22}$$

The value of δm can be adjusted so as to cancel out the self-energy effects in $S(\infty)$ (this is only a formal adjustment since the value is actually infinite), and then Eq. (21) will be valid for one-electron states. For the photon self-energy no such adjustment is needed since, as proved by Schwinger, the photon self-energy turns out to be identically zero.

The foregoing discussion of the self-energy problem is intentionally only a sketch, but it will be found to be sufficient for practical applications of the theory. A fuller discussion of the theoretical assumptions underlying this treatment of the problem will be given by Schwinger in his forthcoming papers. Moreover, it must be realized that the theory as a whole cannot be put into a finally satisfactory form so long as divergencies occur in it, however skilfully these divergencies are circumvented; therefore, the present treatment should be regarded as justified by its success in applications rather than by its theoretical derivation.

The important results of the present paper up to this point are Eq. (19) and the interpretation of the state vector Ω. The state vector Ψ of a system can be interpreted as a wave function giving the probability amplitude of finding any particular set of occupation numbers for the various possible states of free electrons, positrons, and photons. The state vector Ω of a system with a given Ψ on a given surface σ is, crudely speaking, the Ψ which the system would have had in the infinite past if it had arrived at the given Ψ on σ under the influence of the interaction $H^I(x_0)$ alone.

The definition of Ω being unsymmetrical between past and future, a new type of state vector Ω' can be defined by reversing the direction of time in the definition of Ω. Thus the Ω' of a system with a given Ψ on a given σ is the Ψ

[6] Here Schwinger's notation $\bar{\psi} = \psi^*\beta$ is used.

which the system would reach in the infinite future if it continued to move under the influence of $H^I(x_0)$ alone. More simply, Ω' can be defined by the equation

$$\Omega'(\sigma)=S(\infty)\Omega(\sigma). \tag{23}$$

Since $S(\infty)$ is a unitary operator independent of σ, the state vectors Ω and Ω' are really only the same vector in two different representations or coordinate systems. Moreover, for any steady state the two are identical by (21).

V. FUNDAMENTAL FORMULAS OF THE SCHWINGER AND FEYNMAN THEORIES

The Schwinger theory works directly from Eqs. (19) and (20), the aim being to calculate the matrix elements of the "effective external potential energy" H_T between states specified by their state vectors Ω. The states considered in practice always have Ω of some very simple kind, for example, Ω representing systems in which one or two free-particle states have occupation number one and the remaining free-particle states have occupation number zero. By analogy with (13), $S(\sigma_0)$ is given by

$$S(\sigma_0)=1+(-i/\hbar c)\int_{-\infty}^{\sigma_0} H^I(x_1)dx_1+(-i/\hbar c)^2 \times\int_{-\infty}^{\sigma_0} dx_1\int_{-\infty}^{\sigma(x_1)} H^I(x_1)H^I(x_2)dx_2+\cdots, \tag{24}$$

and $(S(\sigma_0))^{-1}$ by a corresponding expression analogous to (14). Substitution of these series into (20) gives at once

$$H_T(x_0)=\sum_{n=0}^{\infty}(i/\hbar c)^n\int_{-\infty}^{\sigma(x_0)} dx_1\int_{-\infty}^{\sigma(x_1)} dx_2\cdots \times\int_{-\infty}^{\sigma(x_{n-1})} dx_n\times[H^I(x_n),[\cdots,[H^I(x_2), [H^I(x_1), H^e(x_0)]]\cdots]]. \tag{25}$$

The repeated commutators in this formula are characteristic of the Schwinger theory, and their evaluation gives rise to long and rather difficult analysis. Using the first three terms of the series, Schwinger was able to calculate the second-order radiative corrections to the equations of motion of an electron in an external field, and obtained satisfactory agreement with experimental results. In this paper the development of the Schwinger theory will be carried no further; in principle the radiative corrections to the equations of motion of electrons could be calculated to any desired order of approximation from formula (25).

In the Feynman theory the basic principle is to preserve symmetry between past and future. Therefore, the matrix elements of the operator H_T are evaluated in a "mixed representation;" the matrix elements are calculated between an initial state specified by its state vector Ω_1 and a final state specified by its state vector Ω_2'. The matrix element of H_T between two such states in the Schwinger representation is

$$\Omega_2^*H_T\Omega_1=\Omega_2'^*S(\infty)H_T\Omega_1, \tag{26}$$

and therefore the operator which replaces H_T in the mixed representation is

$$H_F(x_0)=S(\infty)H_T(x_0) =S(\infty)(S(\sigma))^{-1}H^e(x_0)S(\sigma). \tag{27}$$

Going back to the original product definition of $S(\sigma)$ analogous to (10), it is clear that $S(\infty)\times(S(\sigma))^{-1}$ is simply the operator obtained from $S(\sigma)$ by interchanging past and future. Thus,

$$R(\sigma)=S(\infty)(S(\sigma))^{-1}=1+(-i/\hbar c)\times\int_{\sigma}^{\infty} H^I(x_1)dx_1+(-i/\hbar c)^2\int_{\sigma}^{\infty} dx_1 \times\int_{\sigma(x_1)}^{\infty} H^I(x_2)H^I(x_1)dx_2+\cdots. \tag{28}$$

The physical meaning of a mixed representation of this type is not at all recondite. In fact, a mixed representation is normally used to describe such a process as bremsstrahlung of an electron in the field of a nucleus when the Born approximation is not valid; the process of bremsstrahlung is a radiative transition of the electron from a state described by a Coulomb wave function, with a plane ingoing and a spherical outgoing wave, to a state described by a Coulomb wave function with a spherical ingoing and a plane outgoing wave. The initial and final states here belong to different orthogonal systems of wave functions, and so the transition matrix elements are calculated in a mixed representation. In the Feynman theory the situation is analogous; only the roles of the radiation interaction and the external (or Coulomb) field are interchanged; the radiation interaction is used instead of the Coulomb field to modify the state vectors (wave functions) of the initial and final states, and the external field instead of the radiation interaction causes transitions between these state vectors.

In the Feynman theory there is an additional simplification. For if matrix elements are being calculated between two states, either of which is steady (and this includes all cases so far considered), the mixed representation reduces to an ordinary representation. This occurs, for example, in treating a one-particle problem such as the radiative correction to the equations of motion of an electron in an external field; the operator $H_F(x_0)$, although in general it is not even Hermitian, can in this case be considered as an effective external potential energy acting on the particle, in the ordinary sense of the words.

This section will be concluded with the derivation of the fundamental formula (31) of the Feynman theory, which is the analog of formula (25) of the Schwinger theory. If

$$F_1(x_1), \quad \cdots, \quad F_n(x_n)$$

are any operators defined, respectively, at the points $x_1, \cdots, x_n$ of space-time, then

$$P(F_1(x_1), \quad \cdots, \quad F_n(x_n)) \tag{29}$$

will denote the product of these operators, taken in the order, reading from right to left, in which the surfaces $\sigma(x_1), \cdots, \sigma(x_n)$ occur in time. In most applications of this notation $F_i(x_i)$ will commute with $F_j(x_j)$ so long as x_i and x_j are outside each other's light cones; when this is the case, it is easy to see that (29) is a function of the points $x_1, \cdots, x_n$ only and is independent of the surfaces $\sigma(x_i)$. Consider now the integral

$$I_n=\int_{-\infty}^{\infty} dx_1\cdots\int_{-\infty}^{\infty} dx_n P(H^e(x_0), H^I(x_1), \cdots, H^I(x_n)).$$

Since the integrand is a symmetrical function of the points $x_1, \cdots, x_n$, the value of the integral is just $n!$ times the integral obtained by restricting the integration to sets of points $x_1, \cdots, x_n$ for which $\sigma(x_i)$ occurs after $\sigma(x_{i+1})$ for each i. The restricted integral can then be further divided into $(n+1)$ parts, the j'th part being the integral over those sets of points with the property that $\sigma(x_0)$ lies between $\sigma(x_{j-1})$ and $\sigma(x_j)$ (with obvious modifications for $j=1$ and $j=n+1$). Therefore,

$$I_n=n!\sum_{j=1}^{n+1}\int_{-\infty}^{\sigma(x_0)} dx_j\cdots\int_{-\infty}^{\sigma(x_{n-1})} dx_n \times\int_{\sigma(x_0)}^{\infty} dx_{j-1}\cdots\int_{\sigma(x_2)}^{\infty} dx_1 \times H^I(x_1)\cdots H^I(x_{j-1})H^e(x_0)H^I(x_j)\cdots H^I(x_n). \tag{30}$$

Now if the series (24) and (28) are substituted into (27), sums of integrals appear which are precisely of the form (30). Hence finally

$$H_F(x_0)=\sum_{n=0}^{\infty}(-i/\hbar c)^n[1/n!]I_n = \sum_{n=0}^{\infty}(-i/\hbar c)^n[1/n!]\int_{-\infty}^{\infty} dx_1\cdots\int_{-\infty}^{\infty} dx_n \times P(H^e(x_0), H^I(x_1), \cdots, H^I(x_n)). \tag{31}$$

By this formula the notation $H_F(x_0)$ is justified, for this operator now appears as a function of the point x_0 alone and not of the surface σ. The further development of the Feynman theory is mainly concerned with the calculation of matrix elements of (31) between various initial and final states.

As a special case of (31) obtained by replacing H^e by the unit matrix in (27),

$$S(\infty)=\sum_{n=0}^{\infty}(-i/\hbar c)^n[1/n!]\int_{-\infty}^{\infty} dx_1\cdots\int_{-\infty}^{\infty} dx_n \times P(H^I(x_1), \cdots, H^I(x_n)). \tag{32}$$

VI. CALCULATION OF MATRIX ELEMENTS

In this section the application of the foregoing theory to a general class of problems will be explained. The ultimate aim is to obtain a set of rules by which the matrix element of the operator (31) between two given states may be written down in a form suitable for numerical evaluation, immediately and automatically. The fact that such a set of rules exists is the basis of the Feynman radiation theory; the derivation in this section of the same rules from what is fundamentally the Tomonaga-Schwinger theory constitutes the proof of equivalence of the two theories.

To avoid excessive complication, the type of matrix element considered will be restricted in two ways. First, it will be assumed that the external potential energy is

$$H^e(x_0) = -[1/c] j_\mu(x_0) A_\mu^e(x_0), \quad (33)$$

that is to say, the interaction energy of the electron-positron field with electromagnetic potentials $A_\mu^e(x_0)$ which are given numerical functions of space and time. Second, matrix elements will be considered only for transitions from a state A, in which just one electron and no positron or photon is present, to another state B of the same character. These restrictions are not essential to the theory, and are introduced only for convenience, in order to illustrate clearly the principles involved.

The electron-positron field operator may be written

$$\psi_\alpha(x) = \sum_u \phi_{u\alpha}(x) a_u, \quad (34)$$

where the $\phi_{u\alpha}(x)$ are spinor wave functions of free electrons and positrons, and the a_u are annihilation operators of electrons and creation operators of positrons. Similarly, the adjoint operator

$$\bar{\psi}_\alpha(x) = \sum_u \bar{\phi}_{u\alpha}(x) \bar{a}_u, \quad (35)$$

where $\bar{a}_u$ are annihilation operators of positrons and creation operators of electrons. The electromagnetic field operator is

$$A_\mu(x) = \sum_v (A_{v\mu}(x) b_v + A_{v\mu}^*(x) \bar{b}_v), \quad (36)$$

where b_v and $\bar{b}_v$ are photon annihilation and creation operators, respectively. The charge-current 4-vector of the electron field is

$$j_\mu(x) = iec\bar{\psi}(x)\gamma_\mu\psi(x); \quad (37)$$

strictly speaking, this expression ought to be antisymmetrized to the form[7]

$$j_\mu(x) = \tfrac{1}{2} iec\{\bar{\psi}_\alpha(x)\psi_\beta(x) - \psi_\beta(x)\bar{\psi}_\alpha(x)\}(\gamma_\mu)_{\alpha\beta}, \quad (38)$$

but it will be seen later that this is not necessary in the present theory.

Consider the product P occurring in the n'th integral of (31); let it be denoted by P_n. From (16), (22), (33), and (37) it is seen that P_n is a sum of products of $(n+1)$ operators ψ_α, $(n+1)$ operators $\bar{\psi}_\alpha$, and not more than n operators A_μ, multiplied by various numerical factors. By Q_n may be denoted a typical product of factors ψ_α, $\bar{\psi}_\alpha$, and A_μ, not summed over the indices such as α and μ, so that P_n is a sum of terms such as Q_n. Then Q_n will be of the form (indices omitted)

$$Q_n = \bar{\psi}(x_{i_0})\psi(x_{i_0})\bar{\psi}(x_{i_1})\psi(x_{i_1})\cdots\bar{\psi}(x_{i_n})\psi(x_{i_n}) \times A(x_{j_1})\cdots A(x_{j_m}), \quad (39)$$

where $i_0, i_1, \cdots, i_n$ is some permutation of the integers $0, 1, \cdots, n$, and $j_1, \cdots, j_m$ are some, but not necessarily all, of the integers $1, \cdots, n$ in some order. Since none of the operators $\bar{\psi}$ and ψ commute with each other, it is especially important to preserve the order of these factors. Each factor of Q_n is a sum of creation and annihilation operators by virtue of (34), (35), and (36), and so Q_n itself is a sum of products of creation and annihilation operators.

Now consider under what conditions a product of creation and annihilation operators can give a non-zero matrix element for the transition $A \rightarrow B$. Clearly, one of the annihilation operators must annihilate the electron in state A, one of the creation operators must create the electron in state B, and the remaining operators must be divisible into pairs, the members of each pair respectively creating and annihilating the same particle. Creation and annihilation operators referring to different particles always commute or anticommute (the former if at least one is a photon operator, the latter if both are electron-positron operators). Therefore, if the two single operators and the various pairs of operators in the product all refer to different particles, the order of factors in the product can be altered so as to bring together the two single operators and the two members of each pair, without changing the value of the product except for a change of sign if the permutation made in the order of the electron and positron operators is odd. In the case when some of the single operators and pairs of operators refer to the same particle, it is not hard to verify that the same change in order of factors can be made, provided it is remembered that the division of the operators into pairs is no longer unique, and the change of order is to

[7] See Wolfgang Pauli, Rev. Mod. Phys. **13**, 203 (1941), Eq. (96), p. 224.

be made for each possible division into pairs and the results added together.

It follows from the above considerations that the matrix element of Q_n for the transition $A \rightarrow B$ is a sum of contributions, each contribution arising from a specific way of dividing the factors of Q_n into two single factors and pairs. A typical contribution of this kind will be denoted by M. The two factors of a pair must involve a creation and an annihilation operator for the same particle, and so must be either one $\bar{\psi}$ and one ψ or two A; the two single factors must be one $\bar{\psi}$ and one ψ. The term M is thus specified by fixing an integer k, and a permutation $r_0, r_1, \cdots, r_n$ of the integers $0, 1, \cdots, n$, and a division $(s_1,t_1), (s_2,t_2), \cdots, (s_h,t_h)$ of the integers $j_1, \cdots, j_m$ into pairs; clearly $m=2h$ has to be an even number; the term M is obtained by choosing for single factors $\bar{\psi}(x_k)$ and $\psi(x_{r_k})$, and for associated pairs of factors $(\bar{\psi}(x_i),\psi(x_{r_i}))$ for $i=0, 1, \cdots, k-1, k+1, \cdots, n$ and $(A(x_{s_i}),A(x_{t_i}))$ for $i=1, \cdots, h$. In evaluating the term M, the order of factors in Q_n is first to be permuted so as to bring together the two single factors and the two members of each pair, but without altering the order of factors within each pair; the result of this process is easily seen to be

$$Q_n' = \epsilon P(\bar{\psi}(x_0),\psi(x_{r_0})) \cdots P(\bar{\psi}(x_n),\psi(x_{r_n})) \times P(A(x_{s_1}),A(x_{t_1})) \cdots P(A(x_{s_h}),A(x_{t_h})), \quad (40)$$

a factor ϵ being inserted which takes the value ± 1 according to whether the permutation of $\bar{\psi}$ and ψ factors between (39) and (40) is even or odd. Then in (40) each product of two associated factors (but not the two single factors) is to be independently replaced by the sum of its matrix elements for processes involving the successive creation and annihilation of the same particle.

Given a bilinear operator such as $A_\mu(x)A_\nu(y)$, the sum of its matrix elements for processes involving the successive creation and annihilation of the same particle is just what is usually called the "vacuum expectation value" of the operator, and has been calculated by Schwinger. This quantity is, in fact (note that Heaviside units are being used)

$$\langle A_\mu(x)A_\nu(y)\rangle_0 = \tfrac{1}{2}\hbar c\delta_{\mu\nu}\{D^{(1)}+iD\}(x-y),$$

where $D^{(1)}$ and D are Schwinger's invariant D functions. The definitions of these functions will not be given here, because it turns out that the vacuum expectation value of $P(A_\mu(x),A_\nu(y))$ takes an even simpler form. Namely,

$$\langle P(A_\mu(x),A_\nu(y))\rangle_0 = \tfrac{1}{2}\hbar c\delta_{\mu\nu}D_F(x-y), \quad (41)$$

where D_F is the type of D function introduced by Feynman. $D_F(x)$ is an even function of x, with the integral expansion

$$D_F(x) = -[i/2\pi^2]\int_0^\infty \exp[i\alpha x^2]d\alpha, \quad (42)$$

where x^2 denotes the square of the invariant length of the 4-vector x. In a similar way it follows from Schwinger's results that

$$\langle P(\bar{\psi}_\alpha(x),\psi_\beta(y))\rangle_0 = \tfrac{1}{2}\eta(x,y)S_{F\beta\alpha}(x-y), \quad (43)$$

where

$$S_{F\beta\alpha}(x) = -(\gamma_\mu(\partial/\partial x_\mu)+\kappa_0)_{\beta\alpha}\Delta_F(x), \quad (44)$$

κ_0 is the reciprocal Compton wave-length of the electron, $\eta(x,y)$ is -1 or $+1$ according as $\sigma(x)$ is earlier or later than $\sigma(y)$ in time, and Δ_F is a function with the integral expansion

$$\Delta_F(x) = -[i/2\pi^2]\int_0^\infty \exp[i\alpha x^2 - i\kappa_0^2/4\alpha]d\alpha. \quad (45)$$

Substituting from (41) and (44) into (40), the matrix element M takes the form (still omitting the indices of the factors $\bar{\psi}$, ψ, and A of Q_n)

$$M = \epsilon \prod_{i\neq k} (\tfrac{1}{2}\eta(x_i,x_{r_i})S_F(x_i-x_{r_i})) \times \prod_j (\tfrac{1}{2}\hbar c D_F(x_{s_j}-x_{t_j}))P(\bar{\psi}(x_k),\psi(x_{r_k})). \quad (46)$$

The single factors $\bar{\psi}(x_k)$ and $\psi(x_{r_k})$ are conveniently left in the form of operators, since the matrix elements of these operators for effecting the transition $A\rightarrow B$ depend on the wave functions of the electron in the states A and B. Moreover, the order of the factors $\bar{\psi}(x_k)$ and $\psi(x_{r_k})$ is immaterial since they anticommute with each other; hence it is permissible to write

$$P(\bar{\psi}(x_k),\psi(x_{r_k})) = \eta(x_k,x_{r_k})\bar{\psi}(x_k)\psi(x_{r_k}).$$

Therefore (46) may be rewritten

$$M = \epsilon' \prod_{i\neq k} (\tfrac{1}{2}S_F(x_i-x_{r_i}))\prod_j (\tfrac{1}{2}\hbar c D_F(x_{s_j}-x_{t_j})) \times \bar{\psi}(x_k)\psi(x_{r_k}), \quad (47)$$

with

$$\epsilon' = \epsilon \prod_i \eta(x_i,x_{r_i}). \quad (48)$$

Now the product in (48) is $(-1)^p$, where p is the number of occasions in the expression (40) on which the ψ of a P bracket occurs to the left of the $\bar{\psi}$. Referring back to the definition of ϵ after Eq. (40), it follows that ϵ' takes the value $+1$ or -1 according to whether the permutation of $\bar{\psi}$ and ψ factors between (39) and the expression

$$\bar{\psi}(x_0)\psi(x_{r_0})\cdots\bar{\psi}(x_n)\psi(x_{r_n}) \qquad (49)$$

is even or odd. But (39) can be derived by an even permutation from the expression

$$\bar{\psi}(x_0)\psi(x_0)\cdots\bar{\psi}(x_n)\psi(x_n), \qquad (50)$$

and the permutation of factors between (49) and (50) is even or odd according to whether the permutation $r_0, \cdots, r_n$ of the integers $0, \cdots, n$ is even or odd. Hence, finally, ϵ' in (47) is $+1$ or -1 according to whether the permutation $r_0, \cdots, r_n$ is even or odd. It is important that ϵ' depends only on the type of matrix element M considered, and not on the points $x_0, \cdots, x_n$; therefore, it can be taken outside the integrals in (31).

One result of the foregoing analysis is to justify the use of (37), instead of the more correct (38), for the charge-current operator occurring in H^e and H^i. For it has been shown that in each matrix element such as M the factors $\bar{\psi}$ and ψ in (38) can be freely permuted, so that (38) can be replaced by (37), except in the case when the two factors form an associated pair. In the exceptional case, M contains as a factor the vacuum expectation value of the operator $j_\mu(x_i)$ at some point x_i; this expectation value is zero according to the correct formula (38), though it would be infinite according to (37); thus the matrix elements in the exceptional case are always zero. The conclusion is that only those matrix elements are to be calculated for which the integer r_i differs from i for every $i\neq k$, and in these elements the use of formula (37) is correct.

To write down the matrix elements of (31) for the transition $A\rightarrow B$, it is only necessary to take all the products Q_n, replace each by the sum of the corresponding matrix elements M given by (47), reassemble the terms into the form of the P_n from which they were derived, and finally substitute back into the series (31). The problem of calculating the matrix elements of (31) is thus in principle solved. However, in the following section it will be shown how this solution-in-principle can be reduced to a much simpler and more practical procedure.

VII. GRAPHICAL REPRESENTATION OF MATRIX ELEMENTS

Let an integer n and a product P_n occurring in (31) be temporarily fixed. The points $x_0, x_1, \cdots, x_n$ may be represented by $(n+1)$ points drawn on a piece of paper. A type of matrix element M as described in the last section will then be represented graphically as follows. For each associated pair of factors $(\bar{\psi}(x_i),\psi(x_{r_i}))$ with $i\neq k$, draw a line with a direction marked in it from the point x_i to the point x_{r_i}. For the single factors $\bar{\psi}(x_k)$, $\psi(x_{r_k})$, draw directed lines leading out from x_k to the edge of the diagram, and in from the edge of the diagram to x_{r_k}. For each pair of factors $(A(x_{s_i}),A(x_{t_i}))$, draw an undirected line joining the points x_{s_i} and x_{t_i}. The complete set of points and lines will be called the "graph" of M; clearly there is a one-to-one correspondence between types of matrix element and graphs, and the exclusion of matrix elements with $r_i=i$ for $i\neq k$ corresponds to the exclusion of graphs with lines joining a point to itself. The directed lines in a graph will be called "electron lines," the undirected lines "photon lines."

Through each point of a graph pass two electron lines, and therefore the electron lines together form one open polygon containing the vertices x_k and x_{r_k}, and possibly a number of closed polygons as well. The closed polygons will be called "closed loops," and their number denoted by l. Now the permutation $r_0, \cdots, r_n$ of the integers $0, \cdots, n$ is clearly composed of $(l+1)$ separate cyclic permutations. A cyclic permutation is even or odd according to whether the number of elements in it is odd or even. Hence the parity of the permutation $r_0, \cdots, r_n$ is the parity of the number of even-number cycles contained in it. But the parity of the number of odd-number cycles in it is obviously the same as the parity of the total number $(n+1)$ of elements. The total number of cycles being $(l+1)$, the parity of the number of even-number cycles is $(l-n)$. Since it was seen earlier that the ϵ' of Eq. (47) is determined just by the parity of the permutation $r_0, \cdots, r_n$, the above argument yields the simple formula

$$\epsilon' = (-1)^{l-n}. \tag{51}$$

This formula is one result of the present theory which can be much more easily obtained by intuitive considerations of the sort used by Feynman.

In Feynman's theory the graph corresponding to a particular matrix element is regarded, not merely as an aid to calculation, but as a picture of the physical process which gives rise to that matrix element. For example, an electron line joining x_1 to x_2 represents the possible creation of an electron at x_1 and its annihilation at x_2, together with the possible creation of a positron at x_2 and its annihilation at x_1. This interpretation of a graph is obviously consistent with the methods, and in Feynman's hands has been used as the basis for the derivation of most of the results, of the present paper. For reasons of space, these ideas of Feynman will not be discussed in further detail here.

To the product P_n correspond a finite number of graphs, one of which may be denoted by G; all possible G can be enumerated without difficulty for moderate values of n. To each G corresponds a contribution $C(G)$ to the matrix element of (31) which is being evaluated.

It may happen that the graph G is disconnected, so that it can be divided into subgraphs, each of which is connected, with no line joining a point of one subgraph to a point of another. In such a case it is clear from (47) that $C(G)$ is the product of factors derived from each subgraph separately. The subgraph G_1 containing the point x_0 is called the "essential part" of G, the remainder G_2 the "inessential part." There are now two cases to be considered, according to whether the points x_k and x_{r_k} lie in G_2 or in G_1 (they must clearly both lie in the same subgraph). In the first case, the factor $C(G_2)$ of $C(G)$ can be seen by a comparison of (31) and (32) to be a contribution to the matrix element of the operator $S(\infty)$ for the transition $A \rightarrow B$. Now letting G vary over all possible graphs with the same G_1 and different G_2, the sum of the contributions of all such G is a constant $C(G_1)$ multiplied by the total matrix element of $S(\infty)$ for the transition $A \rightarrow B$. But for one-particle states the operator $S(\infty)$ is by (21) equivalent to the identity operator and gives, accordingly, a zero matrix element for the transition $A \rightarrow B$. Consequently, the disconnected G for which x_k and x_{r_k} lie in G_2 give zero contribution to the matrix element of (31), and can be omitted from further consideration. When x_k and x_{r_k} lie in G_1, again the $C(G)$ may be summed over all G consisting of the given G_1 and all possible G_2; but this time the connected graph G_1 itself is to be included in the sum. The sum of all the $C(G)$ in this case turns out to be just $C(G_1)$ multiplied by the expectation value in the vacuum of the operator $S(\infty)$. But the vacuum state, being a steady state, satisfies (21), and so the expectation value in question is equal to unity. Therefore the sum of the $C(G)$ reduces to the single term $C(G_1)$, and again the disconnected graphs may be omitted from consideration.

The elimination of disconnected graphs is, from a physical point of view, somewhat trivial, since these graphs arise merely from the fact that meaningful physical processes proceed simultaneously with totally irrelevant fluctuations of fields in the vacuum. However, similar arguments will now be used to eliminate a much more important class of graphs, namely, those involving self-energy effects. A "self-energy part" of a graph G is defined as follows; it is a set of one or more vertices not including x_0, together with the lines joining them, which is connected with the remainder of G (or with the edge of the diagram) only by two electron lines or by one or two photon lines. For definiteness it may be supposed that G has a self-energy part F, which is connected with its surroundings only by one electron line entering F at x_1, and another leaving F at x_2; the case of photon lines can be treated in an entirely analogous way. The points x_1 and x_2 may or may not be identical. From G a "reduced graph" G_0 can be obtained by omitting F completely and joining the incoming line at x_1 with the outgoing line at x_2 to form a single electron line in G_0, the newly formed line being denoted by λ. Given G_0 and λ, there is conversely a well determined set Γ of graphs G which are associated with G_0 and λ in this way; G_0 itself is considered also to belong to Γ. It will now be shown that the sum $C(\Gamma)$ of the contributions $C(G)$ to the matrix element of (31) from all the graphs G of Γ reduces to a single term $C'(G_0)$.

Suppose, for example, that the line λ in G_0 leads from a point x_3 to the edge of the diagram. Then $C(G_0)$ is an integral containing in the integrand the matrix element of

$$\bar{\psi}_\alpha(x_3) \tag{52}$$

for creation of an electron into the state B. Let the momentum-energy 4-vector of the created electron be p; the matrix element of (52) is of the form

$$Y_\alpha(x_3)=a_\alpha \exp[-i(p\cdot x_3)/\hbar] \tag{53}$$

with a_α independent of x_3. Now consider the sum $C(\Gamma)$. It follows from an analysis of (31) that $C(\Gamma)$ is obtained from $C(G_0)$ by replacing the operator (52) by

$$\sum_{n=0}^{\infty}(-i/\hbar c)^n[1/n!]\int_{-\infty}^{\infty}dy_1\cdots\int_{-\infty}^{\infty}dy_n \times P(\bar{\psi}_\alpha(x_3), H^I(y_1), \cdots, H^I(y_n)). \tag{54}$$

(This is, of course, a consequence of the special character of the graphs of Γ.) It is required to calculate the matrix element of (54) for a transition from the vacuum state O to the state B, i.e., for the emission of an electron into state B. This matrix element will be denoted by Z_α; $C(\Gamma)$ involves Z_α in the same way that $C(G_0)$ involves (53). Now Z_α can be evaluated as a sum of terms of the same general character as (47); it will be of the form

$$Z_\alpha=\sum_i\int_{-\infty}^{\infty}K_i^{\alpha\beta}(y_i-x_3)Y_\beta(y_i)dy_i,$$

where the important fact is that K_i is a function only of the coordinate differences between y_i and x_3. By (53), this implies that

$$Z_\alpha=R_{\alpha\beta}(p)Y_\beta(x_3), \tag{55}$$

with R independent of x_3. From considerations of relativistic invariance, R must be of the form

$$\delta_{\beta\alpha}R_1(p^2)+(p_\mu\gamma_\mu)_{\beta\alpha}R_2(p^2),$$

where p^2 is the square of the invariant length of the 4-vector p. But since the matrix element (53) is a solution of the Dirac equation,

$$p^2=-\hbar^2\kappa_0^2, \quad (p_\mu\gamma_\mu)_{\beta\alpha}Y_\beta=i\hbar\kappa_0 Y_\alpha,$$

and so (55) reduces to

$$Z_\alpha=R_1Y_\alpha(x_3),$$

with R_1 an absolute constant. Therefore the sum $C(\Gamma)$ is in this case just $C'(G_0)$, where $C'(G_0)$ is obtained from $C(G_0)$by the replacement

$$\bar{\psi}(x_3)\rightarrow R_1\bar{\psi}(x_3). \tag{56}$$

In the case when the line λ leads into the graph G_0 from the edge of the diagram to the point x_3, it is clear that $C(\Gamma)$ will be similarly obtained from $C(G_0)$ by the replacement

$$\psi(x_3)\rightarrow R_1^*\psi(x_3). \tag{57}$$

There remains the case in which λ leads from one vertex x_3 to another x_4 of G_0. In this case $C(G_0)$ contains in its integrand the function

$$\tfrac{1}{2}\eta(x_3,x_4)S_{F\beta\alpha}(x_3-x_4), \tag{58}$$

which is the vacuum expectation value of the operator

$$P(\bar{\psi}_\alpha(x_3),\psi_\beta(x_4)) \tag{59}$$

according to (43). Now in analogy with (54), $C(\Gamma)$ is obtained from $C(G_0)$ by replacing (59) by

$$\sum_{n=0}^{\infty}(-i/\hbar c)^n[1/n!]\int_{-\infty}^{\infty}dy_1\cdots\int_{-\infty}^{\infty}dy_n \times P(\bar{\psi}_\alpha(x_3), \psi_\beta(x_4), H^I(y_1), \cdots, H^I(y_n)), \tag{60}$$

and the vacuum expectation value of this operator will be denoted by

$$\tfrac{1}{2}\eta(x_3,x_4)S'_{F\beta\alpha}(x_3-x_4). \tag{61}$$

By the methods of Section VI, (61) can be expanded as a series of terms of the same character as (47); this expansion will not be discussed in detail here, but it is easy to see that it leads to an expression of the form (61), with $S_F'(x)$ a certain universal function of the 4-vector x. It will not be possible to reduce (61) to a numerical multiple of (58), as Z_α was in the previous case reduced to a multiple of Y_α. Instead, there may be expected to be a series expansion of the form

$$S_{F\beta\alpha}(x)=(R_2+a_1(\Box^2-\kappa_0^2)+a_2(\Box^2-\kappa_0^2)^2+\cdots)S_{F\beta\alpha}(x)+(b_1+b_2(\Box^2-\kappa_0^2)+\cdots)\times(\gamma_\mu[\partial/\partial x_\mu]-\kappa_0)_{\beta\gamma}S_{F\gamma\alpha}(x), \tag{62}$$

where $\Box^2$ is the Dalembertian operator and the a, b are numerical coefficients. In this case $C(\Gamma)$ will be equal to the $C'(G_0)$ obtained from $C(G_0)$ by the replacement

$$S_F(x_3-x_4)\rightarrow S_F'(x_3-x_4). \tag{63}$$

Applying the same methods to a graph G with a self-energy part connected to its surroundings by two photon lines, the sum $C(\Gamma)$ will be obtained as a single contribution $C'(G_0)$ from the reduced graph G_0, $C'(G_0)$ being formed from $C(G_0)$ by the replacement

$$D_F(x_3-x_4)\rightarrow D_F'(x_3-x_4). \tag{64}$$

The function D_F' is defined by the condition that

$$\tfrac{1}{2}\hbar c\delta_{\mu\nu}D_F'(x_3-x_4) \tag{65}$$

is the vacuum expectation value of the operator

$$\sum_{n=0}^{\infty}(-i/\hbar c)^n[1/n!]\int_{-\infty}^{\infty}dy_1\cdots\int_{-\infty}^{\infty}dy_n$$
$$\times P(A_\mu(x_3), A_\nu(x_4), H^I(y_1), \cdots, H^I(y_n)), \tag{66}$$

and may be expanded in a series

$$D_F'(x)=(R_3+c_1\Box^2+c_2(\Box^2)^2+\cdots)D_F(x). \tag{67}$$

Finally, it is not difficult to see that for graphs G with self-energy parts connected to their surroundings by a single photon line, the sum $C(\Gamma)$ will be identically zero, and so such graphs may be omitted from consideration entirely.

As a result of the foregoing arguments, the contributions $C(G)$ of graphs with self-energy parts can always be replaced by modified contributions $C'(G_0)$ from a reduced graph G_0. A given G may be reducible in more than one way to give various G_0, but if the process of reduction is repeated a finite number of times a G_0 will be obtained which is "totally reduced," contains no self-energy part, and is uniquely determined by G. The contribution $C'(G_0)$ of a totally reduced graph to the matrix element of (31) is now to be calculated as a sum of integrals of expressions like (47), but with a replacement (56), (57), (63), or (64) made corresponding to every line in G_0. This having been done, the matrix element of (31) is correctly calculated by taking into consideration each totally reduced graph once and once only.

The elimination of graphs with self-energy parts is a most important simplification of the theory. For according to (22), H^I contains the subtracted part H^S, which will give rise to many additional terms in the expansion of (31). But if any such term is taken, say, containing the factor $H^S(x_i)$ in the integrand, every graph corresponding to that term will contain the point x_i joined to the rest of the graph only by two electron lines, and this point by itself constitutes a self-energy part of the graph. Therefore, all terms involving H^S are to be omitted from (31) in the calculation of matrix elements. The intuitive argument for omitting these terms is that they were only introduced in order to cancel out higher order self-energy terms arising from H^i, which are also to be omitted; the analysis of the foregoing paragraphs is a more precise form of this argument. In physical language, the argument can be stated still more simply; since δm is an unobservable quantity, it cannot appear in the final description of observable phenomena.

VIII. VACUUM POLARIZATION AND CHARGE RENORMALIZATION

The question now arises: What is the physical meaning of the new functions D_F' and S_F', and of the constant R_1? In general terms, the answer is clear. The physical processes represented by the self-energy parts of graphs have been pushed out of the calculations, but these processes do not consist entirely of unobservable interactions of single particles with their self-fields, and so cannot entirely be written off as "self-energy processes." In addition, these processes include the phenomenon of vacuum polarization, i.e., the modification of the field surrounding a charged particle by the charges which the particle induces in the vacuum. Therefore, the appearance of D_F', S_F', and R_1 in the calculations may be regarded as an explicit representation of the vacuum polarization phenomena which were implicitly contained in the processes now ignored.

In the present theory there are two kinds of vacuum polarization, one induced by the external field and the other by the quantized electron and photon fields themselves; these will be called "external" and "internal," respectively. It is only the internal polarization which is represented yet in explicit fashion by the substitutions (56), (57), (63), (64); the external will be included later.

To form a concrete picture of the function D_F', it may be observed that the function $D_F(y-z)$ represents in classical electrodynamics the retarded potential of a point charge at y acting upon a point charge at z, together with the retarded potential of the charge at z acting on the charge at y. Therefore, D_F may be spoken of loosely as "the electromagnetic interaction between two point charges." In this semiclassical picture, D_F' is then the electromagnetic interaction between two point charges, including the effects of the charge-distribution which each charge induces in the vacuum.

The complete phenomenon of vacuum polarization, as hitherto understood, is included in the above picture of the function D_F'. There is nothing left for S_F' to represent. Thus, one of the important conclusions of the present theory is that there is a second phenomenon occurring in nature, included in the term vacuum polarization as used in this paper, but additional to vacuum polarization in the usual sense of the word. The nature of the second phenomenon can best be explained by an example.

The scattering of one electron by another may be represented as caused by a potential energy (the Møller interaction) acting between them. If one electron is at y and the other at z, then, as explained above, the effect of vacuum polarization of the usual kind is to replace a factor D_F in this potential energy by D_F'. Now consider an analogous, but unorthodox, representation of the Compton effect, or the scattering of an electron by a photon. If the electron is at y and the photon at z, the scattering may be again represented by a potential energy, containing now the operator $S_F(y-z)$ as a factor; the potential is an exchange potential, because after the interaction the electron must be considered to be at z and the photon at y, but this does not detract from its usefulness. By analogy with the 4-vector charge-current density j_μ which interacts with the potential D_F, a spinor Compton-effect density u_α may be defined by the equation

$$u_\alpha(x)=A_\mu(x)(\gamma_\mu)_{\alpha\beta}\psi_\beta(x),$$

and an adjoint spinor by

$$\bar{u}_\alpha(x)=\bar{\psi}_\beta(x)(\gamma_\mu)_{\beta\alpha}A_\mu(x).$$

These spinors are not directly observable quantities, but the Compton effect can be adequately described as an exchange potential, of magnitude proportional to $S_F(y-z)$, acting between the Compton-effect density at any point y and the adjoint density at z. The second vacuum polarization phenomenon is described by a change in the form of this potential from S_F to S_F'. Therefore, the phenomenon may be pictured in physical terms as the inducing, by a given element of Compton-effect density at a given point, of additional Compton-effect density in the vacuum around it.

In both sorts of internal vacuum polarization, the functions D_F and S_F, in addition to being altered in shape, become multiplied by numerical (and actually divergent) factors R_3 and R_2; also the matrix elements of (31) become multiplied by numerical factors such as $R_1R_1^*$. However, it is believed (this has been verified only for second-order terms) that all n'th-order matrix elements of (31) will involve these factors only in the form of a multiplier

$$(eR_2R_3^{\frac{1}{2}})^n;$$

this statement includes the contributions from the higher terms of the series (62) and (67). Here e is defined as the constant occurring in the fundamental interaction (16) by virtue of (37). Now the only possible experimental determination of e is by means of measurements of the effects described by various matrix elements of (31), and so the directly measured quantity is not e but $eR_2R_3^{\frac{1}{2}}$. Therefore, in practice the letter e is used to denote this measured quantity, and the multipliers R no longer appear explicitly in the matrix elements of (31); the change in the meaning of the letter e is called "charge renormalization," and is essential if e is to be identified with the observed electronic charge. As a result of the renormalization, the divergent coefficients R_1, R_2, and R_3 in (56), (57), (62), and (67) are to be replaced by unity, and the higher coefficients a, b, and c by expressions involving only the renormalized charge e.

The external vacuum polarization induced by the potential A_μ^e is, physically speaking, only a special case of the first sort of internal polarization; it can be treated in a precisely similar manner. Graphs describing external polarization effects are those with an "external polarization part," namely, a part including the point x_0 and connected with the rest of the graph by only a single photon line. Such a graph is to be "reduced" by omitting the polarization part entirely and renaming with the label x_0 the point at the further end of the single photon line. A discussion similar to those of Section VII leads to the conclusion that only reduced graphs need be considered in the calculation of the matrix element of (31), and that the effect of external polarization is explicitly represented if in the contributions from these graphs a replacement

$$A_\mu^e(x) \to A_\mu^{e\prime}(x) \tag{68}$$

is made. After a renormalization of the unit of potential, similar to the renormalization of charge, the modified potential $A_\mu^{e\prime}$ takes the form

$$A_\mu^{e\prime}(x) = (1 + c_1\Box^2 + c_2(\Box^2)^2 + \cdots)A_\mu^e(x), \tag{69}$$

where the coefficients are the same as in (67).

It is necessary, in order to determine the functions D_F', S_F', and $A_\mu^{e\prime}$, to go back to formulas (60) and (66). The determination of the vacuum expectation values of the operators (60) and (66) is a problem of the same kind as the original problem of the calculation of matrix elements of (31), and the various terms in the operators (60) and (66) must again be split up, represented by graphs, and analyzed in detail. However, since D_F' and S_F' are universal functions, this further analysis has only to be carried out once to be applicable to all problems.

It is one of the major triumphs of the Schwinger theory that it enables an unambiguous interpretation to be given to the phenomenon of vacuum polarization (at least of the first kind), and to the vacuum expectation value of an operator such as (66). In making this interpretation, profound theoretical problems arise, particularly concerned with the gauge invariance of the theory, about which nothing will be said here. For Schwinger's solution of these problems, the reader must refer to his forthcoming papers. Schwinger's argument can be transferred without essential change into the framework of the present paper.

Having overcome the difficulties of principle, Schwinger proceeded to evaluate the function D_F' explicitly as far as terms of order $\alpha = (e^2/4\pi\hbar c)$ (heaviside units). In particular, he found for the coefficient c_1 in (67) and (69) the value $(-\alpha/15\pi\kappa_0^2)$ to this order.[8] It is hoped to publish in a sequel to the present paper a similar evaluation of the function S_F'; the analysis involved is too complicated to be summarized here.

IX. SUMMARY OF RESULTS

In this section the results of the preceding pages will be summarized, so far as they relate to the performance of practical calculations. In effect, this summary will consist of a set of rules for the application of the Feynman radiation theory to a certain class of problems.

Suppose an electron to be moving in an external field with interaction energy given by (33). Then the interaction energy to be used in calculating the motion of the electron, including radiative corrections of all orders, is

$$\begin{aligned} H_E(x_0) &= \sum_{n=0}^{\infty} (-i/\hbar c)^n [1/n!] J_n \\ &= \sum_{n=0}^{\infty} (-i/\hbar c)^n [1/n!] \int_{-\infty}^{\infty} dx_1 \cdots \int_{-\infty}^{\infty} dx_n \\ &\qquad \times P(H^e(x_0), H^i(x_1), \cdots, H^i(x_n)), \end{aligned} \tag{70}$$

with H^i given by (16), and the P notation as defined in (29).

To find the effective n'th-order radiative correction to the potential acting on the electron, it is necessary to calculate the matrix elements of J_n for transitions from one one-electron state to another. These matrix elements can be written down most conveniently in the form of an operator K_n bilinear in $\bar{\psi}$ and ψ, whose matrix elements for one-electron transitions are the same as those to be determined. In fact, the operator K_n itself is already the matrix element to be determined if the $\bar{\psi}$ and ψ contained in it are regarded as one-electron wave functions.

To write down K_n, the integrand P_n in J_n is first expressed in terms of its factors $\bar{\psi}$, ψ, and A, all suffixes being indicated explicitly, and the expression (37) used for j_μ. All possible graphs G with $(n+1)$ vertices are now drawn as described in Section VII, omitting disconnected graphs, graphs with self-energy parts, and graphs with external vacuum polarization parts as defined in Section VIII. It will be found that in each graph there are at each vertex two electron lines and one photon line, with the exception of x_0 at which there are two electron lines only; further,

[8] Schwinger's results agree with those of the earlier, theoretically unsatisfactory treatment of vacuum polarization. The best account of the earlier work is V. F. Weisskopf, Kgl. Danske Sels. Math.-Fys. Medd. 14, No. 6 (1936).

such graphs can exist only for even n. K_n is the sum of a contribution $K(G)$ from each G.

Given G, $K(G)$ is obtained from J_n by the following transformations. First, for each photon line joining x and y in G, replace two factors $A_\mu(x)A_\nu(y)$ in P_n (regardless of their positions) by

$$\tfrac{1}{2}\hbar c\delta_{\mu\nu}D_F'(x-y), \tag{71}$$

with D_F' given by (67) with $R_3=1$, the function D_F being defined by (42). Second, for each electron line joining x to y in G, replace two factors $\bar{\psi}_\alpha(x)\psi_\beta(y)$ in P_n (regardless of positions) by

$$\tfrac{1}{2}S'_{F\beta\alpha}(x-y) \tag{72}$$

with S_F' given by (62) with $R_2=1$, the function S_F being defined by (44) and (45). Third, replace the remaining two factors $P(\bar{\psi}_\gamma(z),\psi_\delta(w))$ in P_n by $\bar{\psi}_\gamma(z)\psi_\delta(w)$ in this order. Fourth, replace $A_\mu^e(x_0)$ by $A_\mu^{e\prime}(x_0)$ given by

$$A_\mu^{e\prime}(x)=A_\mu^e(x)-[\alpha/15\pi\kappa_0^2]\Box^2A_\mu^e(x) \tag{73}$$

or, more generally, by (69). Fifth, multiply the whole by $(-1)^l$, where l is the number of closed loops in G as defined in Section VII.

The above rules enable K_n to be written down very rapidly for small values of n. It should be observed that if K_n is being calculated, and if it is not desired to include effects of higher order than the n'th, then D_F', S_F', and $A_\mu^{e\prime}$ in (71), (72), and (73) reduce to the simple functions D_F, S_F, and A_μ^e. Also, the integrand in J_n is a symmetrical function of $x_1, \cdots, x_n$; therefore, graphs which differ only by a relabeling of the vertices $x_1, \cdots, x_n$ give identical contributions to K_n and need not be considered separately.

The extension of these rules to cover the calculation of matrix elements of (70) of a more general character than the one-electron transitions hitherto considered presents no essential difficulty. All that is necessary is to consider graphs with more than two "loose ends," representing processes in which more than one particle is involved. This extension is not treated in the present paper, chiefly because it would lead to unpleasantly cumbersome formulas.

X. EXAMPLE—SECOND-ORDER RADIATIVE CORRECTIONS

As an illustration of the rules of procedure of the previous section, these rules will be used for writing down the terms giving second-order

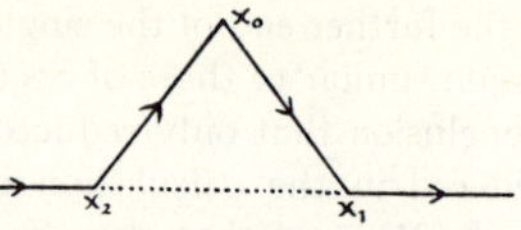

FIG. 1.

radiative corrections to the motion of an electron in an external field. Let the energy of the external field be

$$-[1/c]j_\mu(x_0)A_\mu^e(x_0). \tag{74}$$

Then there will be one second-order correction term

$$U=[\alpha/15\pi\kappa_0^2][1/c]j_\mu(x_0)\Box^2A_\mu^e(x_0)$$

arising from the substitution (73) in the zero-order term (74). This is the well-known vacuum polarization or Uehling term.[9]

The remaining second-order term arises from the second-order part J_2 of (70). Written in expanded form, J_2 is

$$J_2=ie^3\int_{-\infty}^{\infty}dx_1\int_{-\infty}^{\infty}dx_2P(\bar{\psi}_\alpha(x_0)(\gamma_\lambda)_{\alpha\beta}\psi_\beta(x_0)A_\lambda^e(x_0),$$
$$\bar{\psi}_\gamma(x_1)(\gamma_\mu)_{\gamma\delta}\psi_\delta(x_1)A_\mu(x_1),$$
$$\bar{\psi}_\epsilon(x_2)(\gamma_\nu)_{\epsilon\zeta}\psi_\zeta(x_2)A_\nu(x_2)).$$

Next, all admissable graphs with the three vertices x_0, x_1, x_2 are to be drawn. It is easy to see that there are only two such graphs, that G shown in Fig. 1, and the identical graph with x_1 and x_2 interchanged. The full lines are electron lines, the dotted line a photon line. The contribution $K(G)$ is obtained from J_2 by substituting according to the rules of Section IX; in this case $l=0$, and the primes can be omitted from (71), (72), (73) since only second-order terms are required. The integrand in $K(G)$ can be reassembled into the form of a matrix product, suppressing the suffixes $\alpha, \cdots, \zeta$. Then, multiplying by a factor 2 to allow for the second graph, the complete second-order correction to (74) arising from J_2 becomes

$$L=-i[e^3/8\hbar c]\int_{-\infty}^{\infty}dx_1\int_{-\infty}^{\infty}dx_2D_F(x_1-x_2)A_\mu^e(x_0)$$
$$\times\bar{\psi}(x_1)\gamma_\nu S_F(x_0-x_1)\gamma_\mu S_F(x_2-x_0)\gamma_\nu\psi(x_2).$$

[9] Robert Serber, Phys. Rev. **48**, 49 (1935); E. A. Uehling, Phys. Rev. **48**, 55 (1935).

This is the term which gives rise to the main part of the Lamb-Retherford line shift,[10] the anomalous magnetic moment of the electron,[11] and the anomalous hyperfine splitting of the ground state of hydrogen.[12]

The above expression L is formally simpler than the corresponding expression obtained by Schwinger, but the two are easily seen to be equivalent. In particular, the above expression does not lead to any great reduction in the labor involved in a numerical calculation of the Lamb shift. Its advantage lies rather in the ease with which it can be written down.

In conclusion, the author would like to express his thanks to the Commonwealth Fund of New York for financial support, and to Professors Schwinger and Feynman for the stimulating lectures in which they presented their respective theories.

Notes added in proof (To Section II). The argument of Section II is an over-simplification of the method of Tomonaga,[1] and is unsound. There is an error in the derivation of (3); derivatives occurring in $H(r)$ give rise to non-commutativity between $H(r)$ and field quantities at r' when r is a point on σ infinitesimally distant from r'. The argument should be amended as follows. Φ is defined only for flat surfaces $t(r)=t$, and for such surfaces (3) and (6) are correct. Ψ is defined for general surfaces by (12) and (10), and is verified to satisfy (9). For a flat surface, Φ and Ψ are then shown to be related by (7). Finally, since H_1 does not involve the derivatives in H, the argument leading to (3) can be correctly applied to prove that for general σ the state-vector $\Psi(\sigma)$ will completely describe results of observations of the system on σ.

(To Section III). A covariant perturbation theory similar to that of Section III has previously been developed by E. C G. Stueckelberg, Ann. d. Phys. **21**, 367 (1934); Nature, **153**, 143 (1944).

(To Section V). Schwinger's "effective potential" is not H_T given by (25), but is $H_T'=QH_TQ^{-1}$. Here Q is a "square-root" of $S(\infty)$ obtained by expanding $(S(\infty))^{\frac{1}{2}}$ by the binomial theorem. The physical meaning of this is that Schwinger specifies states neither by Ω nor by Ω', but by an intermediate state-vector $\Omega''=Q\Omega=Q^{-1}\Omega'$, whose definition is symmetrical between past and future. H_T' is also symmetrical between past and future. For one-particle states, H_T and H_T' are identical.

Equation (32) can most simply be obtained directly from the product expansion of $S(\infty)$.

(To Section VII). Equation (62) is incorrect. The function S_F' is well-behaved, but its fourier transform has a logarithmic dependence on frequency, which makes an expansion precisely of the form (62) impossible.

(To Section X). The term L still contains two divergent parts. One is an "infra-red catastrophe" removable by standard methods. The other is an "ultraviolet" divergence, and has to be interpreted as an additional charge-renormalization, or, better, cancelled by part of the charge-renormalization calculated in Section VIII.

[10] W. E. Lamb and R. C. Retherford, Phys. Rev. **72**, 241 (1947).
[11] P. Kusch and H. M. Foley, Phys. Rev. **74**, 250 (1948).
[12] J. E. Nafe and E. B. Nelson, Phys. Rev. **73**, 718 (1948); Aage Bohr, Phys. Rev. **73**, 1109 (1948).

The S Matrix in Quantum Electrodynamics

F. J. DYSON
Institute for Advanced Study, Princeton, New Jersey
(Received February 24, 1949)

The covariant quantum electrodynamics of Tomonaga, Schwinger, and Feynman is used as the basis for a general treatment of scattering problems involving electrons, positrons, and photons. Scattering processes, including the creation and annihilation of particles, are completely described by the S matrix of Heisenberg. It is shown that the elements of this matrix can be calculated, by a consistent use of perturbation theory, to any desired order in the fine-structure constant. Detailed rules are given for carrying out such calculations, and it is shown that divergences arising from higher order radiative corrections can be removed from the S matrix by a consistent use of the ideas of mass and charge renormalization.

Not considered in this paper are the problems of extending the treatment to include bound-state phenomena, and of proving the convergence of the theory as the order of perturbation itself tends to infinity.

I. INTRODUCTION

IN a previous paper[1] (to be referred to in what follows as I) the radiation theory of Tomonaga[2] and Schwinger[3] was applied in detail to the problem of the radiative corrections to the motion of a single electron in a given external field. It was shown that the rules of calculation for corrections of this kind were identical with those which had been derived by Feynman[4] from his own radiation theory. For the one-electron problem the radiative corrections were fully described by an operator H_T (Eq. (20) of I) which appeared as the "effective potential" acting upon the electron, after the interactions of the electron with its own self-field had been eliminated by a contact transformation. The difference between the Schwinger and Feynman theories lay only in the choice of a particular representation in which the matrix elements of H_T were calculated (Section V of I).

The present paper deals with the relation between the Schwinger and Feynman theories when the restriction to one-electron problems is removed. In these more general circumstances the two theories appear as complementary rather than identical. The Feynman method is essentially a set of rules for the calculation of the elements of the Heisenberg S matrix corresponding to any physical process, and can be applied with directness to all kinds of scattering problems.[5] The Schwinger method evaluates radiative corrections by exhibiting them as extra terms appearing in the Schrödinger equation of a system of particles and is suited especially to bound-state problems. In spite of the difference of principle, the two methods in practice involve the calculation of closely related expressions; moreover, the theory underlying them is in all cases the same. The systematic technique of Feynman, the exposition of which occupied the second half of I and occupies the major part of the present paper, is therefore now available for the evaluation not only of the S matrix but also of most of the operators occurring in the Schwinger theory.

The prominent part which the S matrix plays in this paper is due to its practical usefulness as the connecting link between the Feynman technique of calculation and the Hamiltonian formulation of quantum electrodynamics. This practical usefulness remains, whether or not one follows Heisenberg in believing that the S matrix may eventually replace the Hamiltonian altogether. It is still an unanswered question, whether the finiteness of the S matrix automatically implies the finiteness of all observable quantities, such as bound-state energy levels, optical transition probabilities, etc., occurring in electrodynamics. An affirmative answer to the question is in no way essential to the arguments of this paper. Even if a finite S matrix does not of itself imply finiteness of other observable quantities, it is probable that all such quantities will be finite; to verify this, it will be necessary to repeat the analysis of the present paper, keeping all the time closer to the original Schwinger theory than

[1] F. J. Dyson, Phys. Rev. **75**, 486 (1949).
[2] Sin-Itiro Tomonaga, Prog. Theor. Phys. **1**, 27 (1946); Koba, Tati, and Tomonaga, Prog. Theor. Phys. **2**, 101 (1947) and **2**, 198 (1947); S. Kanesawa and S. Tomonaga, Prog. Theor. Phys. **3**, 1 (1948) and **3**, 101 (1948); Sin-Itiro Tomonaga, Phys. Rev. **74**, 224 (1948); Ito, Koba, and Tomonaga, Prog. Theor. Phys. **3**, 276 (1948); Z. Koba and S. Tomonaga, Prog. Theor. Phys. **3**, 290 (1948).
[3] Julian Schwinger, Phys. Rev. **73**, 416 (1948); **74**, 1439 (1948); **75**, 651 (1949).
[4] Richard P. Feynman, Phys. Rev. **74**, 1430 (1948).
[5] The idea of using standard electrodynamics as a starting point for an explicit calculation of the S matrix has been previously developed by E. C. G. Stueckelberg, Helv. Phys. Acta, **14**, 51 (1941); **17**, 3 (1944); **18**, 195 (1945); **19**, 242 (1946); Nature, **153**, 143 (1944); Phys. Soc. Cambridge Conference Report, 199 (1947); E. C. G. Stueckelberg and D. Rivier, Phys. Rev. **74**, 218 (1948). Stueckelberg anticipated several features of the Feynman theory, in particular the use of the function D_F (in Stueckelberg's notation D^C) to represent retarded (i.e., causally transmitted) electromagnetic interactions. For a review of the earlier part of this work, see Gregor Wentzel, Rev. Mod. Phys. **19**, 1 (1947). The use of mass renormalization in scattering problems is due to H. W Lewis, Phys. Rev **73**, 173 (1948).

has here been possible. There is no reason for attributing a more fundamental significance to the S matrix than to other observable quantities, nor was it Heisenberg's intention to do so. In the last section of this paper, tentative suggestions are made for a synthesis of the Hamiltonian and Heisenberg philosophies.

II. THE FEYNMAN THEORY AS AN S MATRIX THEORY

The S matrix was originally defined by Heisenberg in terms of the stationary solutions of a scattering problem. A typical stationary solution is represented by a time-independent wave function Ψ' which has a part representing ingoing waves which are asymptotically of the form Ψ_1', and a part representing outgoing waves which are asymptotically of the form Ψ_2'. The S matrix is the transformation operator S with the property that

$$\Psi_2' = S\Psi_1' \quad (1)$$

for every stationary state Ψ'.

In Section III of I an operator $U(\infty)$ was defined and stated to be identical with the S matrix. Since $U(\infty)$ was defined in terms of time-dependent wave functions, a little care is needed in making the identification. In fact, the equation

$$\Psi_2 = U(\infty)\Psi_1 \quad (2)$$

held, where Ψ_1 and Ψ_2 were the asymptotic forms of the ingoing and outgoing parts of a wave function Ψ in the Ψ-representation of I (the "interaction representation" of Schwinger[3]). Now the time-independent wave function Ψ' corresponds to a time-dependent wave function

$$\exp[(-i/\hbar)Et]\Psi'$$

in the Schrödinger representation, where E is the total energy of the state; and this corresponds to a wave function in the interaction representation

$$\Psi = \exp[(+i/\hbar)t(H_0 - E)]\Psi', \quad (3)$$

where H_0 is the total free particle Hamiltonian. However, the asymptotic parts of the wave function Ψ', both ingoing and outgoing, represent freely traveling particles of total energy E, and are therefore eigenfunctions of H_0 with eigenvalue E. This implies, in virtue of (3), that the asymptotic parts Ψ_1 and Ψ_2 of Ψ are actually time-independent and equal, respectively, to Ψ_1' and Ψ_2'. Thus (1) and (2) are identical, and $U(\infty)$ is indeed the S matrix. Incidentally, $U(\infty)$ is also the "invariant collision operator" defined by Schwinger.[3]

There is a series expansion of $U(\infty)$ analogous to (32) of I, namely,

$$U(\infty) = \sum_{n=0}^{\infty} \left(\frac{-i}{\hbar c}\right)^n \frac{1}{n!} \int_{-\infty}^{\infty} dx_1 \cdots \int_{-\infty}^{\infty} dx_n \times P(H_1(x_1), \cdots, H_1(x_n)). \quad (4)$$

Here the P notation is as defined in Section V of I, and

$$H_1(x) = H^I(x) + H^e(x) \quad (5)$$

is the sum of the interaction energies of the electron field with the photon field and with the external potentials. The Feynman radiation theory provides a set of rules for the calculation of matrix elements of (4), between states composed of any number of ingoing and outgoing free particles. Also, quantities contained in (4) are the only ones with which the Feynman rules can deal directly. The Feynman theory is thus correctly characterized as an S matrix theory.

One particular way to analyze $U(\infty)$ is to use (5) to expand (4) in a series of terms of ascending order in H^e. Substitution from (5) into (4) gives

$$U(\infty) = \sum_{m=0}^{\infty} \sum_{n=0}^{\infty} \left(\frac{-i}{\hbar c}\right)^{m+n} \frac{1}{m!n!} \int_{-\infty}^{\infty} dx_1 \cdots \times \int_{-\infty}^{\infty} dx_{m+n} P(H^e(x_1), \cdots, H^e(x_m), \times H^I(x_{m+1}), \cdots, H^I(x_{m+n})). \quad (6)$$

In this double series, the term of zero order in H^e is $S(\infty)$, given by (32) of I. The term of first order is

$$U_1 = (-i/\hbar c) \int_{-\infty}^{\infty} H_F(x) dx, \quad (7)$$

where H_F is given by (31) of I. Clearly, $S(\infty)$ is the S matrix representing scattering of electrons and photons by each other in the absence of an external potential; U_1 is the S matrix representing the additional scattering produced by an external potential, when the external potential is treated in the first Born approximation; higher terms of the series (6) would correspond to treating the external potential in the second or higher Born approximation. The operator H_F played a prominent part in I, where it was in no way connected with a Born approximation; however, it was there introduced in a somewhat unnatural manner, and its physical meaning is made clearer by its appearance in (7). In fact, H_F may be defined by the statement that

$$(-i/\hbar)(\delta t)(\delta\omega)H_F(x)$$

is the contribution to the S matrix that would be produced by an external potential of strength H^e, acting for a small duration δt and over a small volume $\delta\omega$ in the neighborhood of the space-time point x.

The remainder of this section will be occupied with a statement of the Feynman rules for evaluating $U(\infty)$. Proofs will not be given, because the rules are only trivial generalizations of the rules which were given in I for the evaluation of matrix elements of H_F corresponding to one-electron transitions.

In evaluating $U(\infty)$ we shall not make any distinction between the external and radiative parts of the electromagnetic field; this is physically reasonable since it is to some extent a matter of convention how much of the field in a given situation is to be regarded as "external." The interaction energy occurring in (4) is then

$$H_1(x) = -ieA_\mu(x)\bar{\psi}(x)\gamma_\mu\psi(x) - \delta mc^2\bar{\psi}(x)\psi(x), \quad (8)$$

where A_μ is the total electromagnetic field, and the term in δm is included in order to allow for the fact that the interaction representation is defined in terms of the total mass of an electron including its "electromagnetic mass" δm (see Section IV of I). The first step in the evaluation of $U(\infty)$ is to substitute from (8) into (4), writing out in full the suffixes of the operators $\bar{\psi}_\alpha$, ψ_β which are concealed in the matrix product notation of (8). After such a substitution, (4) becomes

$$U(\infty) = \sum_{n=0}^{\infty} J_n, \quad (9)$$

where J_n is an n-fold integral with an integrand which is a polynomial in $\bar{\psi}_\alpha$, ψ_β and A_μ operators.

The most general matrix element of J_n is obtained by allowing some of the $\bar{\psi}_\alpha$, ψ_β and A_μ operators to annihilate particles in the initial state, some to create particles in the final state, while others are associated in pairs to perform a successive creation and annihilation of intermediate particles. The operators which are not associated in pairs, and which are available for the real creation and annihilation of particles, are called "free"; a particular type of matrix element of J_n is specified by enumerating which of the operators in the integrand are to be free and which are to be associated in pairs. As described more fully in Section VII of I, each type of matrix element of J_n is uniquely represented by a "graph" G consisting of n points (bearing the labels $x_1, \cdots, x_n$) and various lines terminating at these points.

The relation between a type of matrix element of J_n and its graph G is as follows. For every associated pair of operators $(\bar{\psi}(x), \psi(y))$, there is a directed line (electron line) joining x to y in G. For every associated pair of operators $(A(x), A(y))$, there is an undirected line (photon line) joining x and y in G. For every free operator $\bar{\psi}(x)$, there is a directed line in G leading from x to the edge of the diagram. For every free operator $\psi(x)$, there is a directed line in G leading to x from the edge of the diagram. For every free operator $A(x)$, there is an undirected line in G leading from x to the edge of the diagram. Finally, for a particular type of matrix element of J_n it is specified that at each point x_i either the part of $H_1(x_i)$ containing $A_\mu(x_i)$ or the part containing δm is operating; correspondingly, at each vertex x_i of G there are either two electron lines (one ingoing and one outgoing) and one photon line, or else two electron lines only. Lines joining one point to itself are always forbidden.

In every graph G, the electron lines form a finite number m of open polygonal arcs with ends at the edge of the diagram, and perhaps in addition a number l of closed polygonal loops. The corresponding type of matrix element of J_n has m free operators $\bar{\psi}$ and m free operators ψ; the two end segments of any one open arc correspond to two free operators, one $\bar{\psi}$ and one ψ, which will be called a "free pair." The matrix elements of J_n are now to be calculated by means of an operator $J(G)$, which is defined for each graph G of n vertices, and which is obtained from J_n by making the following five alterations.

First, at each point x_i, $H_1(x_i)$ is to be replaced by either the first or the second term on the right of (8), as indicated by the presence or absence of a photon line at the vertex x_i of G. Second, for every electron line joining a vertex x to a vertex y in G, two operators $\bar{\psi}_\alpha(x)$ and $\psi_\beta(y)$ in J_n, regardless of their positions, are to be replaced by the function

$$\tfrac{1}{2}S_{F\beta\alpha}(x-y), \quad (10)$$

as defined by (44) and (45) of I. Third, for every photon line joining two vertices x and y of G, two operators $A_\mu(x)$ and $A_\nu(y)$ in J_n, regardless of their positions, are to be replaced by the function

$$\tfrac{1}{2}\hbar c\delta_{\mu\nu}D_F(x-y), \quad (11)$$

defined by (42) of I. Fourth, all free operators in J_n are to be left unaltered, but the ordering by the P notation is to be dropped, and the order of the free $\bar{\psi}$ and ψ operators is to be arranged so that the two members of each free pair stand consecutively and in the order $\bar{\psi}\psi$; the order of the free pairs among themselves, and of all free A_μ operators, is left arbitrary. Fifth, the whole expression J_n is to be multiplied by

$$(-1)^{n-l-m}. \quad (12)$$

The Feynman rules for the evaluation of $U(\infty)$ are essentially contained in the above definition of the operators $J(G)$. To each value of n correspond only a finite number of graphs G, and all possible matrix elements of $U(\infty)$ are obtained by substituting into (9) for each J_n the sum of all the corresponding $J(G)$. It is necessary only to specify how the matrix element of a given $J(G)$ corresponding to a given scattering process may be written down.

The matrix element of $J(G)$ for a given process may be obtained, broadly speaking, by replacing each free operator in $J(G)$ by the wave function of the particle which it is supposed to create or annihilate. More specifically, each free $\bar{\psi}$ operator may either create an electron in the final state or annihilate a positron in the initial state, and the reverse processes are performed by a free ψ operator. Therefore, for a transition from a state involving A electrons and B positrons to a state involving C electrons and D positrons, only operators $J(G)$ containing $(A+D)=(B+C)$ free pairs contribute matrix elements. For each such $J(G)$, the $(A+D)$ free ψ operators are to be replaced in all possible combinations by the A initial electron wave functions and the D final positron wave functions, and the $(B+C)$ free $\bar{\psi}$ operators are to be similarly replaced by the initial positron and final electron wave functions, and the results of all such replacements added together, taking account of the antisymmetry of the total wave functions of the system in the individual particle wave functions. In the case of the free A_μ operators, the situation is rather different, since each such operator may either create a photon in the final state, or annihilate a photon in the initial state, or represent merely the external potential. Therefore, for a transition from a state with A photons to a state with B photons, any $J(G)$ with not less than $(A+B)$ free A_μ operators may give a matrix element. If the number of free A_μ operators in $J(G)$ is $(A+B+C)$, these operators are to be replaced in all possible combinations by the $(A+B)$ suitably normalized potentials corresponding to the initial and final photon states, and by the external potential taken C times, and the results of all such replacements added together, taking account now of the symmetry of the total wave functions in the individual photon states.

In practice cases are seldom likely to arise of scattering problems in which more than two similar particles are involved. The replacement of the free operators in $J(G)$ by wave functions can usually be carried out by inspection, and the enumeration of matrix-elements of $U(\infty)$ is practically complete as soon as the operators $J(G)$ have been written down.

The above rules for the calculation of $U(\infty)$ describe the state of affairs before any attempt has been made to identify and remove the various divergent parts of the expressions. In particular, contributions are included from all graphs G, even those which yield nothing but self-energy effects. For this reason, the rules here formulated are superficially different from those given for the one-electron problem in Section IX of I, which described the state of affairs after many divergencies had been removed. Needless to say, the rules are not complete until instructions have been supplied for the removal of all infinite quantities from the theory; in Sections V–VII of this paper it will be shown how the formal structure of the S matrix makes such a complete removal of infinities appear attainable.

Another essential limitation is introduced into the S matrix theory by the use of the expansion (4). All quantities discussed in this paper are expansions of this kind, in which it is assumed that not only the radiation interaction but also the external potential is small enough to be treated as a perturbation. It is well known that an expansion in powers of the external potential does not give a satisfactory approximation, either in problems involving bound states or in scattering problems at low energies. In particular, whenever a scattering problem allows the possibility of one of the incident particles being captured into a bound state, the capture process will not be represented in $U(\infty)$, since the initial and final states for processes described by $U(\infty)$ are always free-particle states. It is the expansion in powers of the external potential which breaks down when such a capture process is possible. Therefore it must be emphasized that the perturbation theory of this paper is applicable only to a restricted class of problems, and that in other situations the Schwinger theory will have to be used in its original form.

III. THE S MATRIX IN MOMENTUM SPACE

Both for practical applications to specific problems, and for general theoretical discussion, it is convenient to express the S matrix $U(\infty)$ in terms of momentum variables. For this purpose, it is enough to consider an expression which will be denoted by M, and which is a typical example of the units out of which all matrix elements of $U(\infty)$ are built up. A particular integer n and a particular graph G of n vertices being supposed fixed, the operator $J(G)$ is constructed as in the previous section, and M is defined as the number obtained by substituting for each of the free operators in $J(G)$ one particular free-particle wave function. More specifically, for each free operator $\psi(x)$ in $J(G)$ there is substituted

$$\psi(k)e^{ik_\mu x_\mu}, \tag{13}$$

where k_μ is some constant 4 vector representing the momentum and energy of an electron, or minus the momentum and energy of a positron, and where $\psi(k)$ is a constant spinor. For each free operator $\bar{\psi}(x)$ there is substituted

$$\bar{\psi}(k')e^{-ik_\mu' x_\mu}, \tag{14}$$

where $\bar{\psi}(k')$ is again a constant spinor. For each free operator $A_\mu(x)$ there is substituted

$$A_\mu(k'')e^{ik_\mu'' x_\mu}, \tag{15}$$

where $A_\mu(k'')$ is a constant 4 vector which may

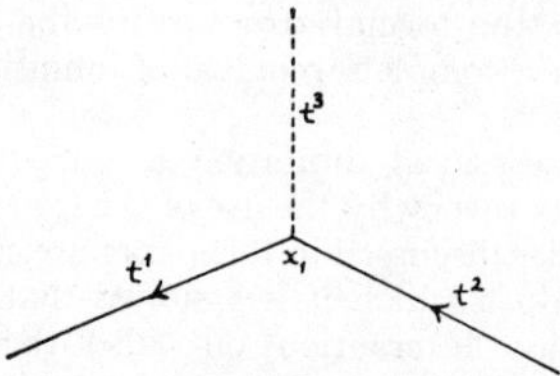

FIG. 1.

represent the polarization vector of a quantum whose momentum-energy 4-vector is either plus or minus k_μ''; alternatively, $A_\mu(k'')$ may represent the Fourier component of the external potential with a particular wave number and frequency specified by the 4 vector k''. There is no loss of generality in splitting up the external potential into Fourier components of the form (15). When the substitutions (13), (14), (15) are made in $J(G)$, the expression M which is obtained is still an n-fold integral over the whole of space-time, and in addition depends parametrically upon E constant 4 vectors in momentum-space, where E is the number of free operators in $J(G)$.

The graph G will contain E external lines, i.e., lines with one end at a vertex and the other end at the edge of the diagram. To each of these external lines corresponds one constant 4 vector, which may be denoted by k_μ^i, $i=1, \cdots, E$, and one constant spinor or polarization vector appearing in M, either $\psi(k^i)$ or $\bar{\psi}(k^i)$ or $A_\mu(k^i)$.

Suppose that G contains F internal lines, i.e., lines with both ends at vertices. To each of these lines corresponds a D_F or an S_F function in M, as specified by (11) or (10). These functions have been expressed by Feynman as 4-dimensional Fourier integrals of very simple form, namely

$$D_F(x)=\frac{1}{4\pi^3}\int e^{-ip_\mu x_\mu}\delta_+(p^2)dp, \qquad (16)$$

$$S_F(x)=\frac{1}{4\pi^3}\int e^{-ip_\mu x_\mu}[+ip_\mu\gamma_\mu-\kappa_0]\times\delta_+(p^2+\kappa_0^2)dp, \qquad (17)$$

where κ_0 is the electron reciprocal Compton wavelength,

$$p^2=p_\mu p_\mu=p_1^2+p_2^2+p_3^2-p_0^2, \qquad (18)$$

and the δ_+ function is defined by

$$\delta_+(a)=\tfrac{1}{2}\delta(a)+\frac{1}{2\pi ia}=\frac{1}{2\pi}\int_0^\infty e^{-iaz}dz. \qquad (19)$$

Substituting from (16) and (17) into M will introduce an F-fold integral over momentum space. Corresponding to each internal line of G, there will appear in M a 4 vector variable of integration, which may be denoted by p_μ^i, $i=1, \cdots, F$. However, after this substitution is made, the space-time variables $x_1, \cdots, x_n$ occur in M only in the exponential factors, and the integration over these variables can be performed. The result of the integration over x_j is to give

$$(2\pi)^4\delta(q_j), \qquad (20)$$

where the δ represents a simple 4-dimensional Dirac δ-function, and q_j is a 4 vector formed by taking an algebraic sum of the k^i and p^i 4 vectors corresponding to those lines of G which meet at x_j. The factor (20) in the integrand of M expresses the conservation of energy and momentum in the interaction occurring at the point x_j.

The transformation of M into terms of momentum variables is now complete. To summarize the results, M now appears as an F-fold integral over the variable 4 vectors p_μ^i in momentum space. In the integrand there appear, besides numerical factors;

(i) a constant spinor or polarization-vector, $\psi(k^i)$ or $\bar{\psi}(k^i)$ or $A_\mu(k^i)$, corresponding to each external line of G;

(ii) a factor

$$D_F(p^i)=\delta_+((p^i)^2) \qquad (21)$$

corresponding to each internal photon line of G;

(iii) a factor

$$S_F(p^i)=[+ip_\mu^i\gamma_\mu-\kappa_0]\delta_+((p^i)^2+\kappa_0^2) \qquad (22)$$

corresponding to each internal electron line of G;

(iv) a factor

$$\delta(q_j) \qquad (23)$$

corresponding to each vertex of G;

(v) a γ_μ operator, surviving from Eq. (8), corresponding to each vertex of G at which there is a photon line.

The important feature of the above analysis is that all the constituents of M are now localized and associated with individual lines and vertices in the graph G. It therefore becomes possible in an unambiguous manner to speak of "adding" or "subtracting" certain groups of factors in M, when G is modified by the addition or subtraction of certain lines and vertices. As an example of this method of analysis, we shall briefly discuss the treatment in the S matrix formalism of the "Lamb shift" and associated phenomena.

Suppose that a graph G, of any degree of complication, has a vertex x_1 at which two electron lines and a photon line meet. These three lines may be either internal or external, and the momentum 4 vectors associated with them in M may be either p^i or k^i; these 4 vectors are denoted by t^1, t^2, t^3 as indicated in Fig. 1. The factors in the integrand of

M arising from the vertex x_1 are

$$-ie\gamma_\mu(2\pi)^4\delta(t^1-t^2-t^3), \quad (24)$$

the two spinor indices of the γ_μ being available for matrix multiplication on both sides with the factors in M arising from the two electron lines at x_1.

Now suppose that G' is a graph identical with G, except that in the neighborhood of x_1 it is modified by the addition of two new vertices and three new lines, as indicated in Fig. 2. With the three new lines, which are all internal, are associated three 4 vector variables p^1, p^2, p^3, which occur as variables of integration in the expression M' formed from G' as M is from G. It can be proved, in view of Eqs. (21), (22), (23), that M' may be obtained from M simply by replacing the factor (24) in M by the expression

$$-\frac{ie^3}{\hbar c}(2\pi)^3\int\int\int dp^1dp^2dp^3$$
$$\delta(t^1-p^2+p^3)\delta(p^2-p^1-t^3)\delta(p^1-p^3-t^2)$$
$$\gamma_\lambda(+ip_\rho{}^2\gamma_\rho-\kappa_0)\gamma_\mu(+ip_\sigma{}^1\gamma_\sigma-\kappa_0)\gamma_\lambda$$
$$\delta_+((p^2)^2+\kappa_0{}^2)\delta_+((p^1)^2+\kappa_0{}^2)\delta_+((p^3)^2). \quad (25)$$

(The factorial coefficients appearing in (4) are just compensated by the fact that the two new vertices of G' may be labelled x_i, x_j in $(n+1)(n+2)$ ways, where n is the number of vertices in G.) In (25), two of the 4-dimensional δ-functions can be eliminated at once by integration over p^1 and p^2, and the third then reduces to the δ-function occurring in (24). Therefore M' can be obtained from M by replacing the operator γ_μ in (24) by an operator

$$L_\mu=L_\mu(t^1,t^2)$$
$$=2\alpha\int dp[\gamma_\lambda(+i(p_\rho+t_\rho{}^1)\gamma_\rho-\kappa_0)\gamma_\mu$$
$$\times(+i(p_\sigma+t_\sigma{}^2)\gamma_\sigma-\kappa_0)\gamma_\lambda]$$
$$\times\delta_+((p+t^1)^2+\kappa_0{}^2)$$
$$\times\delta_+((p+t^2)^2+\kappa_0{}^2)\delta_+(p^2). \quad (26)$$

Here α is the fine-structure constant, $(e^2/4\pi\hbar c)$ in Heaviside units. The operator L_μ can without great difficulty be calculated explicitly as a function of the 4 vectors t^1 and t^2, by methods developed by Feynman.

In the special case when Fig. 1 represents the graph G in its entirety, M is a matrix element for the scattering of a single electron by an external potential. Figure 2 then represents G' in its entirety, and M' is a second-order radiative correction to the scattering of the electron. In this case then the operator L_μ gives rise to what may be called "Lamb shift and associated phenomena." However, the above analysis applies equally to an expression M which may occur anywhere among the matrix elements of $U(\infty)$, and may represent any physical process whatever involving electrons, positrons and photons. There will always appear in $U(\infty)$, together with M, terms M' representing second-order radiative corrections to the same process; one term M' arises from each vertex of G at which a photon line ends; and M' is always to be obtained from M by substituting for an operator γ_μ the same operator L_μ. Furthermore, some higher radiative corrections to M will be obtained by substituting L_μ for γ_μ independently at two or more of the vertices of G.

By a "vertex part" of any graph will be meant a connected part of the graph, consisting of vertices and internal lines only, which touches precisely two electron lines and one photon line belonging to the remainder of the graph. The central triangle of Fig. 2 is an example of such a part. In other words, a vertex part of a graph is a part which can be substituted for the single vertex of Fig. 1 and give a physically meaningful result. Now the argument, by which the replacement of Fig. 1 by Fig. 2 was shown to be equivalent to the replacement of γ_μ by L_μ, can be used also when a more complicated vertex part is substituted for the vertex in Fig. 1. If G is any graph with a vertex x_1 as shown in Fig. 1, and G' is obtained from G by substituting for x_1 any vertex part V, and if M and M' are elements of $U(\infty)$ formed analogously from G and G', then M' can be obtained from M by replacing an operator γ_μ by an operator

$$\Lambda_\mu=\Lambda_\mu(V,t^1,t^2), \quad (27)$$

dependent only on V and the 4 vectors t^1, t^2 and independent of G.

To summarize the results of this section, it has been shown that the S matrix formalism allows a wide variety of higher order radiative processes to be calculated in the form of operators in momentum space. Such operators appear as radiative corrections to the fundamental interaction between the photon and electron-positron fields, and need only to be calculated once to be applicable to the various special problems of electrodynamics.

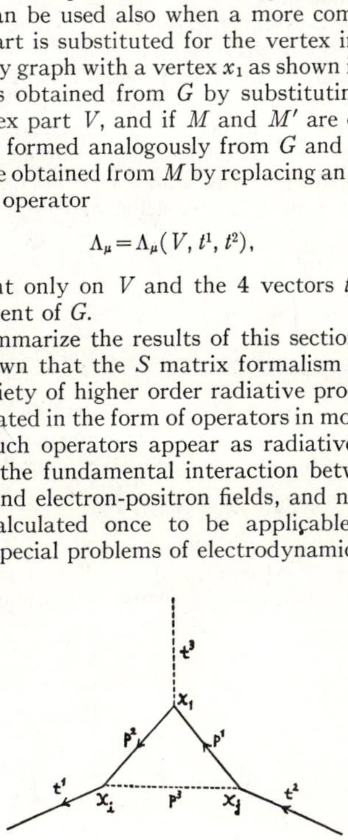

FIG. 2.

IV. FURTHER REDUCTION OF THE S MATRIX

It was shown in Section VII of I that, for the one-electron processes there considered, only connected graphs needed to be taken into account. In constructing the S matrix in general, this is no longer the case; disconnected graphs give matrix elements of $U(\infty)$ representing two or more collision processes occurring simultaneously among separate groups of particles, and such processes have physical reality. It is only permissable to omit a disconnected graph when one of its connected components is entirely lacking in external lines; such a component without external lines will give rise only to a constant multiplicative phase factor in every matrix element of $U(\infty)$ and is therefore devoid of physical significance.

On the other hand, the treatment in Section VII of I of graphs with "self-energy parts" applies almost without change to the general S matrix formalism. A "self-energy part" of a graph is a connected part, consisting of vertices and internal lines only, which can be inserted into the middle of a single line of a graph G so as to give a meaningful graph G'. In Fig. 3 is shown an example of such an insertion made in one of the lines of Fig. 1. Let M and M' be expressions derived in the manner of the previous section from the graphs G and G' of which parts are shown in Figs. 1 and 3. Suppose for definiteness that the line labelled t^1 is an internal line of G; then according to (22) it will contribute a factor $S_F(t^1)$ in M. By an argument similar to that leading to (26), it can be shown that M' may now be obtained from M by replacing $S_F(t^1)$ by

$$S_F(t^1)N(t^1)S_F(t^1)$$
$$=S_F(t^1)2\alpha\int dp[\gamma_\lambda(+i\gamma_\rho(p_\rho+t_\rho^1)-\kappa_0)\gamma_\lambda]$$
$$\times\delta_+((p+t^1)^2+\kappa_0^2)\delta_+(p^2)S_F(t^1). \quad (28)$$

In the same way, if G' were obtained from G by inserting in the t^1 line any self-energy part W, then M' would be obtained from M by replacing $S_F(t^1)$ by

$$S_F(t^1)\Sigma(W,t^1)S_F(t^1), \quad (29)$$

where Σ is an operator dependent only on W and t^1 and not on G. Moreover, if the t^1 line were an external line of G, then M' would be obtained from

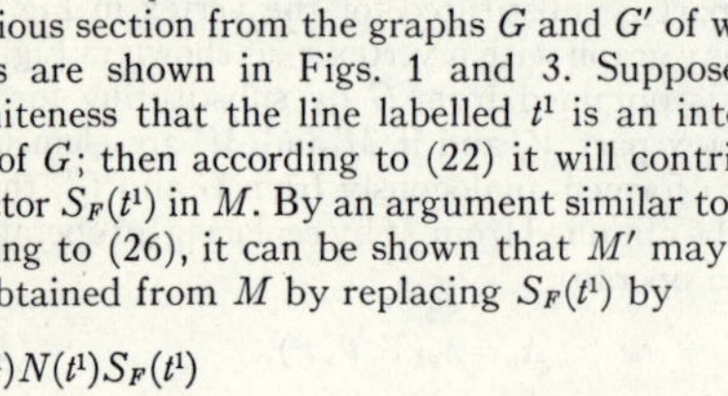

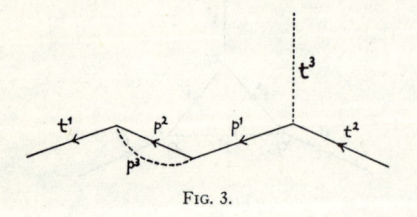

FIG. 3.

M by replacing a factor $\bar{\psi}(t^1)$by

$$\bar{\psi}(t^1)\Sigma(W,t^1)S_F(t^1). \quad (30)$$

As a special case, W may consist of a single point; then at this point it is the term in δm of the interaction (8) which is operating, and Σ reduces to a constant,

$$\Sigma(W,t^1)=-2\pi i(\delta mc/\hbar)=-2\pi i\delta\kappa_0. \quad (31)$$

The operator $N(t^1)$ in (28) describes in a general way the second-order contribution to the electron self-energy and to the phenomenon called "vacuum polarization of the second kind" in Section VIII of I. The self-energy contribution is supposed to be cancelled by (31); the constant $\delta\kappa_0$ being a power series in α, the linear term only is required to cancel the self-energy part of (28), and the higher terms are available for the cancellation of self-energy effects from operators $\Sigma(W,t^1)$ of higher order. The S matrix formalism makes clear the important fact that, since the operators $\Sigma(W,t^1)$ are universal operators independent of the graph G, the electron self-energy effects will be formally cancelled by a constant $\delta\kappa_0$ independent of the physical situation in which the effects occur.

When a self-energy part W' is inserted into a photon line of a graph G, for example the line labelled t^3 in Fig. 1, then the modification produced in M may be again described by a function $\Pi(W',t^3)$ independent of G. Specifically, if the t^3 line in G is internal, M' is obtained from M by replacing a factor $D_F(t^3)$ by

$$D_F(t^3)\Pi(W',t^3)D_F(t^3). \quad (32)$$

If the t^3 line is external, the replacement is of a factor $A_\mu(t^3)$ by

$$A_\mu(t^3)\Pi(W',t^3)D_F(t^3). \quad (33)$$

In addition to terms of the form (33), there will appear terms such as

$$A_\nu(t^3)t_\nu{}^3t_\mu{}^3\Pi'(W',t^3)D_F(t^3); \quad (34)$$

these however are zero in consequence of the gauge condition satisfied by A_ν. Similar terms in $t_\nu{}^3t_\mu{}^3$ will also appear with the expression (32); in this case the extra terms can be shown to vanish in consequence of the equation of conservation of charge satisfied by the electron-positron field. The functions $\Pi(W',t^3)$ describe the phenomenon of photon self-energy and the "vacuum polarization of the first kind" of Section VIII of I. Following Schwinger, one does not explicitly subtract away the divergent photon self-energy effects from the $\Pi(W',t^3)$, but one asserts that these effects are zero as a consequence of the gauge invariance of electrodynamics.

In Section VII of I, it was shown how self-energy parts could be systematically eliminated from all graphs, and their effects described by suitably modifying the functions D_F and S_F. The analysis was carried out in configuration space, and was confined to the one-electron problem. We are now in a position to extend this method to the whole S matrix formalism, working in momentum space, and furthermore to eliminate not only self-energy parts but also the "vertex parts" defined in the last section.

Every graph G has a uniquely defined "skeleton" G_0, which is the graph obtained by omitting all self-energy parts and vertex parts from G. A graph which is its own skeleton is called "irreducible;" all of its vertices will be of the kind depicted in Fig. 1. From every irreducible G_0, the G of which it is the skeleton can be built by inserting pieces in all possible ways at all vertices and lines of G_0; these G form a well-defined class Γ. The term "proper vertex part" must here be introduced, denoting a vertex part which is not divisible into two pieces joined only by a single line; thus a vertex part which is not proper is essentially redundant, being a proper vertex part plus one or more self-energy parts. The graphs of Γ are then accurately enumerated by inserting at some or all of the vertices of G_0 a proper vertex part, and in some or all of the lines a self-energy part, these insertions being made independently in all possible combinations.

Suppose that M is a constituent of a matrix element of $U(\infty)$, obtained from G_0 as described in Section III. Then every graph G in Γ will yield additional contributions to the same matrix element of $U(\infty)$; the sum of all such contributions, including M, is denoted by M_S. As a result of the analysis leading to (27), (29), and (32), and in view of the statistical independence of the insertions made at the different vertices and lines of G_0, the sum M_S will be obtained from M by the following substitutions. For every internal electron line of G_0, a factor $S_F(p^i)$ of M is replaced by

$$S_F'(p^i)=S_F(p^i)+S_F(p^i)\Sigma(p^i)S_F(p^i), \qquad (35)$$

where $\Sigma(p^i)$ is the sum of the $\Sigma(W, p^i)$ over all electron self-energy parts W. For every internal photon line, a factor $D_F(p^i)$ is replaced by

$$D_F'(p^i)=D_F(p^i)+D_F(p^i)\Pi(p^i)D_F(p^i), \qquad (36)$$

where $\Pi(p^i)$ is the sum of the $\Pi(W', p^i)$ over all photon self-energy parts W'. For every external line, a factor $\psi(k^i)$ or $\bar{\psi}(k^i)$ or $A_\mu(k^i)$ is replaced by

$$\begin{aligned}\psi'(k^i)&=S_F(k^i)\Sigma(k^i)\psi(k^i)+\psi(k^i),\\ \bar{\psi}'(k^i)&=\bar{\psi}(k^i)\Sigma(k^i)S_F(k^i)+\bar{\psi}(k^i),\\ A_\mu'(k^i)&=A_\mu(k^i)\Pi(k^i)D_F(k^i)+A_\mu(k^i),\end{aligned} \qquad (37)$$

respectively. For every vertex of G_0, where the incident lines carry momentum variables as shown in Fig. 1, an operator γ_μ is replaced by

$$\Gamma_\mu(t^1, t^2)=\gamma_\mu+\Lambda_\mu(t^1, t^2), \qquad (38)$$

where $\Lambda_\mu(t^1, t^2)$ is the sum of the $\Lambda_\mu(V, t^1, t^2)$ over all proper vertex parts V. The matrix elements of $U(\infty)$ will be correctly calculated, if one includes contributions only from irreducible graphs, after making in each contribution the replacements (35), (36), (37), (38).

To calculate the operators Λ_μ, Σ and Π, it is necessary to write down explicitly the integrals in momentum space, examples being (26) and (28), corresponding to every self-energy part W or proper vertex part V. When considering effects of higher order than the second, the parts W and V will themselves often be reducible, containing in their interior self-energy and vertex parts. It will again be convenient to omit such reducible V and W, and to include their effects by making the substitutions (35)–(38) in the integrals corresponding to irreducible V and W. In this way one obtains in general not explicit formulas, but integral equations, for Λ_μ, Σ and Π. For example,

$$\Lambda_\mu=\alpha I_\mu(\Lambda, \Sigma, \Pi) \qquad (39)$$

where I_μ is an integral in which Λ_μ, Σ and Π occur explicitly. Fortunately, the appearance of α on the right side of (39) makes it easy to solve such equations by a process of successive substitution, the forms of Λ_μ, Σ, and Π being obtained correct to order α^n when values correct to order α^{n-1} are substituted into the integrals.

The functions D_F' and S_F' of (35) and (36) are the Fourier transforms of the corresponding functions in I. The interpretation of these functions in Section VIII of I can be extended in an obvious way to include the operator Γ_μ. Since $\bar{\psi}\gamma_\mu\psi$ is the charge-current 4 vector of an electron without radiative corrections, $\bar{\psi}\Gamma_\mu\psi$ may be interpreted as an "effective current" carried by an electron, including the effects of exchange interactions between the electron and the electron-positron field around it.

An additional reduction in the number of graphs effectively contributing to $U(\infty)$ is obtained from a theorem of Furry.[6] The theorem was shown by Feynman to be an elegant consequence of his theory. In any graph G, a "closed loop" is a closed electron polygon, at the vertices of which a number p of photon lines originate; the loop is called odd or even according to the parity of p. If G contains a closed loop, then there will be another graph $\bar{G}$ also contributing to $U(\infty)$, obtained from G by reversing the sense of the electron lines in the loop. Now if M and $\bar{M}$ are contributions from G and $\bar{G}$, $\bar{M}$ is derived from M by interchanging the roles of electron and positron states in each of the interactions at the vertices of the loop; such an interchange is called "charge conjugation." It was shown by Schwinger that his theory is invariant under charge

[6] Wendell H. Furry, Phys. Rev. 51, 125 (1937).

conjugation, provided that the sign of e is at the same time reversed (this is the well-known charge symmetry of the Dirac hole theory). It is clear from (8) that the constant e appears once in M for each of the p loop vertices at which there is a photon line; at the remaining vertices only the constant δm is involved, and δm is an even function of e. Therefore the principle of charge-symmetry implies

$$\bar{M}=(-1)^p M. \qquad (40)$$

Taking p odd in (40) gives Furry's theorem; all contributions to $U(\infty)$ from graphs with one or more odd closed loops vanish identically.

By an "odd part" of a graph is meant any part, consisting only of vertices and internal lines, which touches no electron lines, and only an odd number of photon lines, belonging to the rest of the graph. The simplest type of odd part which can occur is a single odd closed loop. Conversely, it is easy to see that every odd part must include within itself at least one odd closed loop. Therefore, Furry's theorem allows all graphs with odd parts to be omitted from consideration in calculating $U(\infty)$.

V. INVESTIGATION OF DIVERGENCES IN THE S MATRIX

The δ_+ function defined by (19) has the property that, if b is real and $f(a)$ is any function analytic in the neighborhood of b, then

$$\int f(a)\delta_+(a-b)da=(1/2\pi i)\int f(a)(1/(a-b))da, \qquad (41)$$

where the first integral is along a stretch of the real axis including b, and the second integral is along the same stretch of the real axis but with a small detour into the complex plane passing underneath b. In the matrix elements of $U(\infty)$ there appear integrals of the form

$$\int dpF(p)\delta_+(p_1^2+p_2^2+p_3^2-p_0^2+c^2), \qquad (42)$$

integrated over all real values of p_1, p_2, p_3, p_0. By (41), one may write (42) in the form

$$\frac{1}{2\pi i}\int dp\frac{F(p)}{(p_1^2+p_2^2+p_3^2-p_0^2+c^2)}, \qquad (43)$$

in which it is understood that the integration is along the real axis for the variables p_1, p_2, p_3, and for p_0 is along the real axis with two small detours, one passing above the point $+(p_1^2+p_2^2+p_3^2+c^2)^{\frac{1}{2}}$, and one passing below the point $-(p_1^2+p_2^2+p_3^2+c^2)^{\frac{1}{2}}$. To equate (42) with (43) is certainly correct, when $F(p)$ is analytic at the critical values of p_0. In practice one has to deal with integrals (42) in which $F(p)$ itself involves δ_+ functions (see for example (26) and (28)); in these cases it is legitimate to replace each δ_+ function by a reciprocal, making a separate detour in the p_0 integration for each pole in the integrand, provided that no two poles coincide. Thus every constituent part M of $U(\infty)$ can be written as an integral of a rational algebraic function of momentum variables, by using instead of (21) and (22)

$$D_F(p^i)=\frac{1}{2\pi i(p^i)^2}, \qquad (44)$$

$$S_F(p^i)=\frac{(ip_\mu{}^i\gamma_\mu-\kappa_0)}{2\pi i((p^i)^2+\kappa_0^2)}. \qquad (45)$$

This representation of D_F and S_F as rational functions in momentum-space has been developed and extensively used by Feynman (unpublished).

There may appear in M infinities of three distinct kinds. These are (i) singularities caused by the coincidence of two or more poles of the integrand, (ii) divergences at small momenta caused by a factor (44) in the integrand, (iii) divergences at large momenta due to insufficiently rapid decrease of the whole integrand at infinity.

In this paper no attempt will be made to explore the singularities of type (i). Such singularities occur, for example, when a many-particle scattering process may for special values of the particle momenta be divided into independent processes involving separate groups of particles. It is probable that all singularities of type (i) have a similarly clear physical meaning; these singularities have long been known in the form of vanishing energy denominators in ordinary perturbation theory, and have never caused any serious trouble.

A divergence of type (ii) is the so-called "infrared catastrophe," and is well known to be caused by the failure of an expansion in powers of α to describe correctly the radiation of low momentum quanta. It would presumably be possible to eliminate this divergence from the theory by a suitable adaptation of the standard Bloch-Nordsieck[7] treatment; we shall not do this here. From a practical point of view, one may avoid the difficulty by arbitrarily writing instead of (44)

$$D_F(p^i)=\frac{1}{2\pi i((p^i)^2+\lambda^2)}, \qquad (46)$$

where λ is some non-zero momentum, smaller than any of the quantum momenta which are significant in the particular process under discussion.[8]

[7] F. Bloch and A. Nordsieck, Phys. Rev. **52**, 54 (1937).

[8] The device of introducing λ in order to avoid infra-red divergences must be used with circumspection. Schwinger (unpublished) has shown that a long standing discrepancy between two alternative calculations of the Lamb shift was due to careless use of λ in one of them.

It is the divergences of type (iii) which have always been the main obstacle to the construction of a consistent quantum electrodynamics, and which it is the purpose of the present theory to eliminate. In the following pages, attention will be confined to type (iii) divergences; when the word "convergent" is used, the proviso "except for possible singularities of types (i) and (ii)" should always be understood.

A divergent M is called "primitive" if, whenever one of the momentum 4 vectors in its integrand is held fixed, the integration over the remaining variables is convergent. Correspondingly, a primitive divergent graph is a connected graph G giving rise to divergent M, but such that, if any internal line is removed and replaced by two external lines, the modified G gives convergent M. To analyze the divergences of the theory, it is sufficient to enumerate the primitive divergent M and G and to examine their properties.

Let G be a primitive divergent graph, with n vertices, E external and F internal lines. A corresponding M will be an integral over F variable p^i of a product of F factors (44) and (45) and n factors (23). Since G is connected, the δ-functions (23) in the integrand enable $(n-1)$ of the variables p^i to be expressed in terms of the remaining $(F-n+1)$ p^i and the constants k^i, leaving one δ-function involving the k^i only and expressing conservation of momentum and energy for the whole system. An example of such integration over the δ-functions was the derivation of (26) from (25). After this, the integrations in M may be arranged as follows; the fourth components of the $(F-n+1)$ independent p^i are written

$$p_4{}^i = ip_0{}^i = i\alpha\pi_0{}^i, \qquad (47)$$

and the integration over α is performed first; subsequently, integration is carried out over the $3(F-n+1)$ independent $p_1{}^i$, $p_2{}^i$, $p_3{}^i$, and over the $(F-n)$ ratios of the $\pi_0{}^i$. M then has the form

$$M = \int dp_1{}^i dp_2{}^i dp_3{}^i d\pi_0{}^i \int_{-\infty}^{\infty} R\alpha^{F-n} d\alpha, \qquad (48)$$

where R is a rational function of α, the denominator of which is a product of F factors

$$(p_1{}^i)^2 + (p_2{}^i)^2 + (p_3{}^i)^2 + \mu^2 - (\alpha\pi_0{}^i + c^i)^2. \qquad (49)$$

Here the constants $\pi_0{}^i$, c^i are defined by the condition that

$$p_4{}^j = ip_0{}^j = i(\alpha\pi_0{}^j + c^j), \quad j = 1, 2, \cdots, F. \qquad (50)$$

Thus the c^i corresponding to the $(F-n+1)$ independent p^i are zero by (47), and the remainder are linear combinations of the k^i; also $(n-1)$ of the $\pi_0{}^i$ are linear combinations of the independent $\pi_0{}^i$ defined in (47). In view of (43), we take the integration variables in (48) to be real variables, with the exception of α which is to be integrated along a contour C deviating from the real axis at each of the $2F$ poles of R. As a general rule, C will detour above the real axis for $\alpha > 0$, and below it for $\alpha < 0$; the reverse will only occur at certain of the poles corresponding to denominators (49) for which

$$(p_1{}^i)^2 + (p_2{}^i)^2 + (p_3{}^i)^2 + \mu^2 < (c^i)^2. \qquad (51)$$

Such poles will be called "displaced." The integration over α alone will always be absolutely convergent. Therefore the contour C may be rotated in a counter-clockwise direction until it lies along the imaginary axis, and the value of M will be unchanged except for residues at the displaced poles.

Regarded as a function of the parameters k^i describing the incoming and outgoing particles, M will have a complicated behavior; the behavior will change abruptly whenever one of the c^i has a critical value for which (51) begins to be soluble for $p_1{}^i$, $p_2{}^i$, $p_3{}^i$, and a new displaced pole comes into existence. This behavior is explained by observing that displaced poles appear whenever there is sufficient energy available for one of the virtual particles involved in M to be actually emitted as a real particle. It is to be expected that the behavior of M should change when the process described by M begins to be in competition with other real processes. It is a feature of standard perturbation theory, that when a process A involves an intermediate state I which is variable over a continuous range, and in this range occurs a state II which is the final state of a competing process, then the matrix element for A involves an integral over I which has a singularity at the position II. In standard perturbation theory, this improper integral is always to be evaluated as a Cauchy principal value, and does not introduce any real divergence into the matrix element. In the theory of the present paper, the displaced poles give rise to similar improper integrals; these come under the heading of singularities of type (i) and will not be discussed further.

If $p_1{}^i$, $p_2{}^i$, $p_3{}^i$ satisfying (51) are held fixed, then the value of $p_4{}^i$ at the corresponding displaced pole is fixed by (50). The contribution to M from the displaced pole is just the expression obtained by holding the 4-vector p^i fixed in the original integral M, apart from bounded factors; since M is primitive divergent, this expression is convergent. The total contribution to M from the i'th displaced pole is the integral of this expression over the finite sphere (51) and is therefore finite. Strictly speaking, this argument requires not only the convergence of the expression, but uniform convergence in a finite region; however, it will be seen that the convergent integrals in this theory are convergent for large momenta by virtue of a sufficient preponderance of large denominators, and convergence produced in this way will always be uniform in a finite region.

M is thus, apart from finite parts, equal to the integral M' obtained by replacing α by $i\alpha$ in (48) and (49). Alternatively, M' is obtained from the original integral M by substituting for each $p_0{}^i$

$$ip_4{}^i+(1-i)c^i, \tag{52}$$

and then treating the $4(F-n+1)$ independent $p_\mu{}^i$, $\mu=1, 2, 3, 4$, as ordinary real variables. In M' the denominators of the integrand take the form

$$(p_1{}^i)^2+(p_2{}^i)^2+(p_3{}^i)^2+\mu^2+(p_4{}^i-(1+i)c^i)^2, \tag{53}$$

and are uniformly large for large values of $p_\mu{}^i$. The convergence of M' can now be estimated simply by counting powers of $p_\mu{}^i$ in numerator and denominator of the integrand. Since M' is known to converge whenever one of the p^i is held fixed and integration is carried out over the others, the convergence of the whole expression is assured provided that

$$K=2F-F_e-4[F-n+1]\geqslant 1. \tag{54}$$

Here $2F$ is the degree of the denominator, and F_e that of the numerator, which is by (44) and (45) equal to the number of internal electron lines in G. Let E_e and E_p be the numbers of external electron and photon lines in G, and let n_s be the number of vertices without photon lines incident. It follows from the structure of G that

$$2F=3n-n_s-E_e-E_p,$$
$$F_e=n-\tfrac{1}{2}E_e,$$

and so the convergence condition (52) is

$$K=\tfrac{3}{2}E_e+E_p+n_s-4\geqslant 1. \tag{55}$$

This gives the vital information that the only possible primitive divergent graphs are those with $E_e=2$, $E_p=0, 1$, and with $E_e=0$, $E_p=1, 2, 3, 4$. Further, the cases $E_e=0$, $E_p=1, 3$, do not arise, since these give graphs with odd parts which were shown to be harmless in Section IV. It should be observed that the course of the argument has been "if E_e and E_p do not have certain small values, then the integral M is convergent at infinity;" there is no objection to changing the order of integrations in M as was done in (48), since the argument requires that this be done only in cases when M is, in fact, absolutely convergent.

The possible primitive divergent graphs that have been found are all of a kind familiar to physicists. The case $E_e=2$, $E_p=0$ describes self-energy effects of a single electron; $E_e=0$, $E_p=2$ self-energy effects of a single photon; $E_e=2$, $E_p=1$ the scattering of a single electron in an electromagnetic field; and $E_e=0$, $E_p=4$ the "scattering of light by light" or the mutual scattering of two photons. Further, (55) shows that the divergence will never be more than logarithmic in the third and fourth cases, more than linear in the first, or more than quadratic in the second. Thus it appears that, however far quantum electrodynamics is developed in the discussion of many-particle interactions and higher order phenomena, no essentially new kinds of divergence will be encountered. This gives strong support to the view that "subtraction physics," of the kind used by Schwinger and Feynman, will be enough to make quantum electrodynamics into a consistent theory.

VI. SEPARATION OF DIVERGENCES IN THE S MATRIX

First it will be shown that the "scattering of light by light" does not in fact introduce any divergence into the theory. The possible primitive divergent M in the case $E_e=0$, $E_p=4$ will be of the form

$$\delta(k^1+k^2+k^3+k^4)A_\lambda(k^1)A_\mu(k^2)A_\nu(k^3)A_\rho(k^4)I_{\lambda\mu\nu\rho}, \tag{56}$$

where $I_{\lambda\mu\nu\rho}$ is an integral of the type

$$\int R_{\lambda\mu\nu\rho}(k^1, k^2, k^3, k^4, p^i)dp^i, \tag{57}$$

at most logarithmically divergent, and R is a certain rational function of the constant k^i and the variable p^i. In any physical situation where, for example, the $A(k)$ are the potentials corresponding to particular incident and outgoing photons, there will appear in $U(\infty)$ a matrix element which is the sum of (56) and the 23 similar expressions obtained by permuting the suffixes of $I_{\lambda\mu\nu\rho}$ in all possible ways. It may therefore be supposed that at the start $R_{\lambda\mu\nu\rho}$ has been symmetrized by summation over all permutations of suffixes; (56) is then a sum of contributions from 24 or fewer (according to the degree of symmetry existing) graphs G.

If, under the sign of integration in (57), the value $R(0)$ of R for $k^1=k^2=k^3=k^4=0$ is subtracted from R, the integrand acquires one extra power of $|p_\mu{}^i|^{-1}$ for large $|p_\mu{}^i|$, and the integral becomes absolutely convergent at infinity. Therefore

$$I_{\lambda\mu\nu\rho}=I_{\lambda\mu\nu\rho}(0)+J_{\lambda\mu\nu\rho}, \tag{58}$$

where $I(0)$ is a possibly divergent integral independent of the k^i, and J is a convergent integral vanishing when all k^i's are zero. To interpret this result physically, it is convenient to write (56) again in terms of space-time variables; this gives

$$M=\int I_{\lambda\mu\nu\rho}(0)A_\lambda(x)A_\mu(x)A_\nu(x)A_\rho(x)dx+N, \tag{59}$$

where N is a convergent expression involving derivatives of the $A(x)$ with respect to space and time. Now the first term in (59) is physically inadmissable; it is not gauge-invariant, and implies for example a scattering of light by an electric field depending on the absolute magnitude of the scalar potential, which has no physical meaning. Therefore $I(0)$ must vanish identically, and the whole expression (56) is convergent.

The fact that the scattering of light by light is finite in the lowest order in which it occurs has long been known.[9] It has also been verified by Feynman by direct calculation, using his own theory as described in this paper. The graphs which give rise to the lowest order scattering are shown in Fig. 4. It is found that the divergent parts of the corresponding M exactly cancel when the three contributions are added, or, what comes to the same thing, when the function $R_{\lambda\mu\nu\rho}$ is symmetrized. It is probable that the absence of divergence in the scattering of light by light is in all cases due to a similar cancellation, and it should not be difficult to prove this by calculation and thus avoid making an appeal to gauge-invariance.

The three remaining types of primitive divergent M are, in fact, divergent. However, these are just the expressions which have been studied in Sections III and IV and shown to be completely described by the operators Λ_μ, Σ, and Π. More specifically, when $E_e=2$, $E_p=0$, M will be of the form

$$\bar{\psi}(k^1)\Sigma(W, k^1)\psi(k^1), \tag{60}$$

where W is some electron self-energy part of a graph. When $E_e=0$, $E_p=2$, M will be

$$A_\mu(k^1)\Pi(W', k^1)A_\mu(k^1), \tag{61}$$

with W' some photon self-energy part. When $E_e=2$, $E_p=1$, M will be

$$\bar{\psi}(k^1)\Lambda_\mu(V, k^1, k^2)\psi(k^2)A_\mu(k^1-k^2), \tag{62}$$

with V some vertex part. Therefore, if some means can be found for isolating and removing the divergent parts from Λ_μ, Σ, and Π, the "irreducible" graphs defined in Section IV will not introduce any fresh divergences into the theory, and the rules of Section IV will lead to a divergence-free S matrix.

In considering Λ_μ, Σ, and Π in Section IV it was found convenient to divide vertex and self-energy parts themselves into the categories reducible and irreducible. An irreducible self-energy part W is required not only to have no vertex and self-energy parts inside itself; it is also required to be "proper," that is to say, it is not to be divisible into two pieces joined by a single line. In Section IV it was shown that to avoid redundancy the operator Λ_μ should be defined as a sum over proper vertex parts V only. By the same argument, in order to make (35), (36), (37) correct, it is essential to define Σ and Π as sums over both proper and improper self-energy parts. However, it is possible to define S_F' and D_F' in terms of proper self-energy parts only, at the cost of replacing the explicit definitions (35), (36) by implicit definitions. Let $\Sigma^*(p^i)$ be defined as the sum of the $\Sigma(W, p^i)$ over proper electron self-energy parts W, and let $\Pi^*(p^i)$ be defined similarly. Every W is either proper, or else it is a proper W joined by a single electron line to another self-energy part which may be proper or improper. Therefore, using (35), S_F' may be expressed in the two equivalent forms

$$\begin{aligned} S_F'(p^i) &= S_F(p^i)+S_F(p^i)\Sigma^*(p^i)S_F'(p^i) \\ &= S_F(p^i)+S_F'(p^i)\Sigma^*(p^i)S_F(p^i). \end{aligned} \tag{63}$$

Similarly,

$$\begin{aligned} D_F'(p^i) &= D_F(p^i)+D_F(p^i)\Pi^*(p^i)D_F'(p^i) \\ &= D_F(p^i)+D_F'(p^i)\Pi^*(p^i)D_F(p^i). \end{aligned} \tag{64}$$

It is sometimes convenient to work with the Σ and Π in the starred form, and sometimes in the unstarred form.

Consider the contribution $\Sigma(W, t^1)$ to the operator Σ^*, arising from an electron self-energy part W. It is supposed that W is irreducible, and the effects of possible insertions of self-energy and vertex parts inside W are for the time being neglected. Also it is supposed that W is not a single point, of which the contribution is given by (31). Then W has an even number $2l$ of vertices, at each of which a photon line is incident; and $\Sigma(W, t^1)$ will be of the form

$$e^{2l}\int R(t^1, p^i)dp^i, \tag{65}$$

where R is a certain rational function of the t^1 and p^i, and the integral is at most linearly divergent. The integrand in (65) is now written in the form

$$R(t^1, p^i) = R(0, p^i) + t_\mu^1\left(\frac{\partial R}{\partial t_\mu^1}(0, p^i)\right) + R_c(t^1, p^i), \tag{66}$$

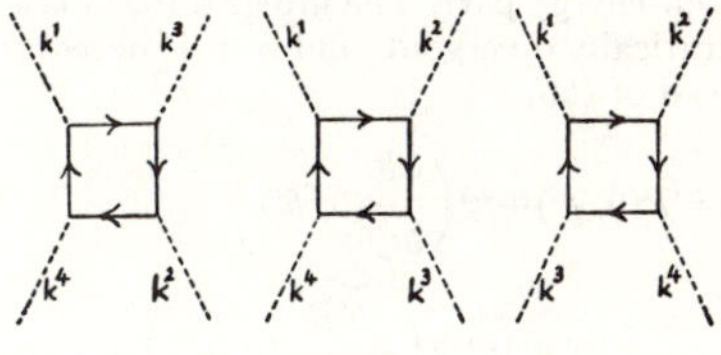

FIG. 4.

[9] H. Euler and B. Kockel, Naturwiss. **23**, 246 (1935); H. Euler, Ann. d. Phys. **26**, 398 (1936). In these early calculations of the scattering of light by light, the theory used is the Heisenberg electrodynamics, in which certain singularities are eliminated at the start by a procedure involving non-diagonal elements of the Dirac density matrix. In Feynman's calculation, on the other hand, a finite result is obtained without subtractions of any kind.

and for large values of the $|p_\mu{}^i|$ the remainder term R_c will tend to zero more rapidly by two powers of $|p_\mu{}^i|$ than R. Therefore, in complete analogy with (58),

$$\Sigma(W, t^1)=e^{2l}[A+B_\mu t_\mu{}^1+\Sigma_c(W, t^1)], \tag{67}$$

where A and B_μ are constant divergent operators, and $\Sigma_c(W, t^1)$ is defined by a covariant and absolutely convergent integral. $\Sigma_c(W, t^1)$ must, on grounds of covariance, be of the form

$$R_1((t^1)^2)+R_2((t^1)^2)t_\mu{}^1\gamma_\mu \tag{68}$$

with R_1 and R_2 particular functions of $(t^1)^2$; for the same reason, B_μ must be of the form $B\gamma_\mu$ with B a certain divergent integral. Now if t^1 happens to be the momentum-energy 4 vector of a free electron,

$$(t^1)^2=-\kappa_0^2, \quad t_\mu{}^1\gamma_\mu=i\kappa_0. \tag{69}$$

It is convenient to write

$$\Sigma_c(W, t^1)=A'+B'(t_\mu{}^1\gamma_\mu-i\kappa_0)+(t_\mu{}^1\gamma_\mu-i\kappa_0)S(W, t^1), \tag{70}$$

where $S(W, t^1)$ is zero for t^1 satisfying (69), and to include the first two terms in the constants A and B of (67); since all terms in (70) are finite, the separation of $S(W, t^1)$ is without ambiguity. Thus an equation of the form (67) is obtained, with

$$\Sigma_c(W, t^1)=(t_\mu{}^1\gamma_\mu-i\kappa_0)S(W, t^1). \tag{71}$$

Summing (67) over all irreducible W and including (31), gives for the operator Σ^*,

$$\Sigma^*(t^1)=A-2\pi i\delta\kappa_0+B(t_\mu{}^1\gamma_\mu-i\kappa_0)+(t_\mu{}^1\gamma_\mu-i\kappa_0)S_c(t^1). \tag{72}$$

Hence by (63) and (45)

$$S_F'(t^1)=(A-2\pi i\delta\kappa_0)S_F(t^1)S_F'(t^1)+\frac{1}{2\pi}BS_F'(t^1)+S_F(t^1)+\frac{1}{2\pi}S_c(t^1)S_F'(t^1). \tag{73}$$

In (72) and (73), A and B are infinite constants, and S_c a divergence-free operator which is zero when (69) holds; A, B, and S_c are power series in e starting with a term in e^2. In (72) and (73), however, effects of higher order corrections to the $\Sigma(W, t^1)$ themselves are not yet included.

A similar separation of divergent parts may be made for the $\Pi(W', t^1)$, when W' is an irreducible photon self-energy part. The integral (65) may now be quadratically divergent, and so it is necessary to use instead of (66)

$$R(t^1, p^i)=R(0, p^i)+t_\mu{}^1\left(\frac{\partial R}{\partial t_\mu{}^1}(0, p^i)\right)+\tfrac{1}{2}t_\mu{}^1t_\nu{}^1\left(\frac{\partial^2 R}{\partial t_\mu{}^1\partial t_\nu{}^1}(0, p^i)\right)+R_c(t^1, p^i),$$

and derive instead of (67)

$$\Pi(W', t^1)=e^{2l}[A+B_\mu t_\mu{}^1+C_{\mu\nu}t_\mu{}^1t_\nu{}^1+\Pi_c(W', t^1)]. \tag{74}$$

The A, B_μ, $C_{\mu\nu}$ are absolute constant numbers (not Dirac operators) and therefore covariance requires that $B_\mu=0$, $C_{\mu\nu}=C\delta_{\mu\nu}$. $\Pi_c(W', t^1)$ is defined by an absolutely convergent integral, and will be an invariant function of $(t^1)^2$ of a form

$$\Pi_c(W', t^1)=(t^1)^2D(W', t^1), \tag{75}$$

where $D(W', t^1)$ is zero for t^1 satisfying

$$(t^1)^2=0 \tag{76}$$

instead of (69). Summing (74) over all irreducible W''s will give

$$\Pi^*(t^1)=A'+C(t^1)^2+(t^1)^2D_c(t^1), \tag{77}$$

and hence by (64) and (44)

$$D_F'(t^1)=A'D_F(t^1)D_F'(t^1)+\frac{1}{2\pi i}CD_F'(t^1)+D_F(t^1)+\frac{1}{2\pi i}D_c(t^1)D_F'(t^1). \tag{78}$$

In (77) and (78), D_c is zero for t^1 satisfying (76), and is divergence free.

The constant A' in (77) is the quadratically divergent photon self-energy. It will give rise to matrix elements in $U(\infty)$ of the form

$$M=A'\int A_\mu(x)A_\mu(x)dx, \tag{79}$$

which are non-gauge invariant and inadmissable. Such matrix elements must be eliminated from the theory, as the first term of (59) was eliminated, by the statement that A' is zero. The verification of this statement, by direct calculation of the lowest order contribution to A', has been given by Schwinger.[3, 10]

The separation of the divergent part of Λ_μ again follows the lines laid down for Σ^*. Since the integral analogous to (65) is now only logarithmically divergent, no derivative term is required in (66), and the analog of (67) is

$$\Lambda_\mu(V, t^1, t^2)=e^{2l}[L_\mu+\Lambda_{\mu c}(V, t^1, t^2)], \tag{80}$$

where L_μ is a constant divergent operator, and $\Lambda_{\mu c}$ is convergent and zero for $t^1=t^2=0$. In (80), L_μ can only be of the form $L\gamma_\mu$. Also, if $t^1=t^2$ and t^1 satisfies (69), $\Lambda_{\mu c}$ will reduce to a finite multiple of γ_μ which can be included in the term $L\gamma_\mu$. Therefore it may be supposed that $\Lambda_{\mu c}$ in (80) is zero not for $t^1=t^2=0$ but for $t^1=t^2$ satisfying (69). The meaning of this

[10] Gregor Wentzel, Phys. Rev. **74**, 1070 (1948), presents the case against Schwinger's treatment of the photon self-energy.

physically is that $\Lambda_{\mu c}$ now gives zero contribution to the energy of a single electron in a constant electromagnetic potential, so that the whole measured static charge on an electron is included in the term $L\gamma_\mu$. Summing (80) over all irreducible vertex parts V, and using (38),

$$\Lambda_\mu(t^1, t^2) = L\gamma_\mu + \Lambda_{\mu c}(t^1, t^2), \tag{81}$$

$$\Gamma_\mu(t^1, t^2) = (1+L)\gamma_\mu + \Lambda_{\mu c}(t^1, t^2). \tag{82}$$

In (81) and (82), effects of higher order corrections to the $\Lambda_\mu(V, t^1, t^2)$ are again not yet included. Formally, (82) differs from (73) and (78) in not containing the unknown operator Γ_μ on both sides of the equation.

VII. REMOVAL OF DIVERGENCES FROM THE S MATRIX

The task remaining is to complete the formulas (73), (78), and (82), which show how the infinite parts can be separated from the operators Γ_μ, S_F', and D_F', and to include the corrections introduced into these operators by the radiative reactions which they themselves describe. In other words, we have to include radiative corrections to radiative corrections, and renormalizations of renormalizations, and so on *ad infinitum*. This task is not so formidable as it appears.

First, we observe that Λ_μ, Σ^*, and Π^* are defined by integral equations of the form (39), which will be referred to in the following pages as "the integral equations." More specifically, consider the contribution $\Lambda_\mu(V, t^1, t^2)$ to Λ_μ represented by (80), arising from a vertex part V with $(2l+1)$ vertices, l photon lines, and $2l$ electron lines. This contribution is defined by an integral analogous to (65), with an integrand which is a product of $(2l+1)$ operators γ_μ, l functions D_F, and $2l$ operators S_F. The exact $\Lambda_\mu(V, t^1, t^2)$ is to be obtained by replacing these factors, respectively, by Γ_μ, D_F', S_F', as described in Section IV. Now suppose that S_F' in the integrand is represented, to order e^{2n} say, by the sum of S_F and of a finite number of finite products of S_F with absolutely convergent operators $S(\bar{W}, t^1)$ such as appear in (71); similarly, let D_F' be represented by D_F plus a finite sum of finite products of D_F with functions $D(\bar{W}', t^1)$ appearing in (75); and let Γ_μ be represented by the sum of γ_μ and of a finite set of $\Lambda_{\mu c}(\bar{V}, t^1, t^2)$ from (80). Then the integral $\Lambda_\mu(V, t^1, t^2)$ will be determined to order e^{2n+2l}; and since the operators $S(\bar{W}, t^1)$, $D(\bar{W}', t^1)$, $\Lambda_{\mu c}(\bar{V}, t^1, t^2)$ always have a sufficiency of denominators for convergence, the theory of Section V can be applied to prove that this $\Lambda_\mu(V, t^1, t^2)$ will not be more than logarithmically divergent. Therefore the new $\Lambda_\mu(V, t^1, t^2)$ can be again separated into the form (80). The sum of these $\Lambda_\mu(V, t^1, t^2)$ will then be a $\Lambda_\mu(t^1, t^2)$ of the form (81), with constant L and convergent operator $\Lambda_{\mu c}$ determined to order e^{2n+2}. Thus (82) provides a new expression fo Γ_μ, determined to order e^{2n+2}.

FIG. 5.

The above procedure describes the general method for separating out the finite part from the contribution to Γ_μ arising from a reducible vertex part V_R. First, V_R is broken down into an irreducible vertex part V plus various inserted parts $\bar{W}$, $\bar{W}'$, $\bar{V}$; the contribution to Γ_μ from V_R is an integral $M(V_R)$ which is not only divergent as a whole, but also diverges when integrated over the variables belonging to one of the insertions $\bar{W}$, $\bar{W}'$, $\bar{V}$, the remaining variables being held fixed. The divergences are to be removed from $M(V_R)$ in succession, beginning with those arising from the inserted parts, and ending with those arising from V itself. This successive removal of divergences is a well-defined procedure, because any two of the insertions made in V are either completely non-overlapping or else arranged so that one is completely contained in the other.

In calculating the contribution to Σ^* or Π^* from reducible self-energy parts, additional complications arise. There is in fact only one irreducible photon self-energy part, the one denoted by W' in Fig. 5; and there is, besides the self-energy part consisting of a single point, just one irreducible electron self-energy part, denoted by W in Fig. 5. All other self-energy parts may be obtained by making various insertions in W or W'. However, reducible self-energy parts are to be enumerated by inserting vertex parts at only one, and not both, of the vertices of W or W'; otherwise the same self-energy part would appear more than once in the enumeration. And the contribution $M(W_R)$ to Σ^* arising from a reducible part W_R will be, in general, an integral which involves simultaneously divergences corresponding to each of the ways in which W_R might have been built up by insertions of vertex parts at either or both vertices of W. This complication arises because, in the special case when two vertex parts are both contained in a self-energy part and each contains one end-vertex of the self-energy part (and in no other case), it is possible for the two vertex parts to overlap without either being completely contained in the other.

The finite part of $M(W_R)$ is to be separated out as follows. In a unique way, W_R is obtained from W by inserting a vertex part $\bar{V}_a$ at a, and self-energy parts $\bar{W}_a$ and $\bar{W}_a'$ in the two lines of W. From $M(W_R)$ there are subtracted all divergences arising from $\bar{V}_a$, $\bar{W}_a$, $\bar{W}_a'$; let the remainder after this sub-

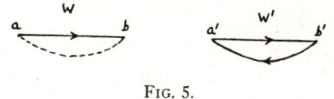

traction be $M'(W_R)$. Next, W_R is considered as built up from W by inserting some vertex part $\bar{V}_b$ at b, and self-energy parts $\bar{W}_b$ and $\bar{W}_b'$ in the two lines of W. The integral $M'(W_R)$ will still contain divergences arising from $\bar{V}_b$ (but none from $\bar{W}_b$ and $\bar{W}_b'$), and these divergences are to be subtracted, leaving a remainder $M''(W_R)$. The finite part of $M''(W_R)$ can finally be separated by applying to the whole integral the method of Section VI, which gives for $M''(W_R)$ an expression of the form (67), with Σ_c given by (71). Therefore the finite part of $M(W_R)$ is a well-determined quantity, and is an operator of the form (71).

The behavior of the higher order contributions to Σ^* and Π^* having now been qualitatively explained, we may describe the precise rules for the calculation of Σ^* and Π^* by the same kind of inductive scheme as was given for Λ_μ in the second paragraph of this Section. Apart from the constant term $(-2\pi i\delta\kappa_0)$, Σ^* is just the contribution $\Sigma(W, t^1)$ from the W of Fig. 5; and $\Sigma(W, t^1)$ is represented by an integral of the form (65) with $l=1$. The integrand in (65) was a product of two operators γ_μ, one operator D_F, and one operator S_F. The exact $\Sigma(W, t^1)$ is to be obtained by replacing D_F by D_F', S_F by S_F', and one only of the factors γ_μ by Γ_μ, say the γ_μ corresponding to the vertex a of W. Suppose that S_F' in the integrand is represented, to order e^{2n}, by the sum of S_F and of a finite number of finite products of S_F with operators $S(\bar{W}, t^1)$ such as appear in (71); and suppose that D_F' and Γ_μ are similarly represented. Then $\Sigma(W, t^1)$ will be determined to order e^{2n+2}. The new $\Sigma(W, t^1)$ will be a sum of integrals like the $M'(W_R)$ of the previous paragraph, still containing divergences arising from vertex parts at the vertex b of W, in addition to divergences arising from the graph W_R as a whole. When all these divergences are dropped, we have a $\Sigma_c(W, t^1)$ which is finite; substituting this $\Sigma_c(W, t^1)$ for Σ^* in (63) gives an S_F' which is also finite and determined to the order e^{2n+2}.

The above procedures start from given S_F', D_F' and Γ_μ represented to order e^{2n} by, respectively, S_F plus S_F multiplied by a finite sum of products of $S(\bar{W}, t^1)$, D_F plus D_F multiplied by a finite sum of products of $D(\bar{W}', t^1)$, and γ_μ plus a finite sum of $\Lambda_{\mu c}(\bar{V}, t^1, t^2)$. From these there are obtained new expressions for S_F', D_F', Γ_μ. In the new expressions there appear new convergent operators $S(W, t^1)$, $D(W', t^1)$, $\Lambda_{\mu c}(V, t^1, t^2)$, determined to order e^{2n+2}; in the divergent terms which are separated out and dropped from the new expressions, there appear divergent coefficients A, B, C, L, such as occur in (73), (78), (82), also now determined to order e^{2n+2}. After the dropping of the divergent terms, the new Γ_μ by (82) is a sum of γ_μ and a finite set of $\Lambda_{\mu c}(V, t^1, t^2)$; the new S_F' by (73) is S_F plus S_F multiplied by a finite sum of products of $S(W, t^1)$; and the new D_F' by (78) is D_F plus D_F multiplied by a finite sum of products of $D(W', t^1)$. That is to say, the new Γ_μ, S_F', D_F' can be substituted back into the integrals of the form (65), and so a third set of operators Γ_μ, S_F', D_F' is obtained, determined to order e^{2n+4}, and again with finite and divergent parts separated. In this way, always dropping the divergent terms before substituting back into the integral equations, the finite parts of Γ_μ, S_F', D_F', may be calculated by a process of successive approximation, starting with the zero-order values γ_μ, S_F, D_F. After n substitutions, the finite parts of Γ_μ, S_F', D_F' will be determined to order e^{2n}.

It is necessary finally to justify the dropping of the divergent terms. This will be done by showing that the "true" Γ_μ, S_F', D_F', which are obtained if the divergent terms are not dropped, are only numerical multiples of those obtained by dropping divergences, and that the numerical multiples can themselves be eliminated from the theory by a consistent use of the ideas of mass and charge renormalization. Let $\Gamma_{\mu 1}(e)$, $S_{F1}'(e)$, $D_{F1}'(e)$ be the operators obtained by the process of substitution dropping divergent terms; these operators are power series in e with finite operator coefficients (to avoid raising the question of the convergence of these power-series, all quantities are supposed defined only up to some finite order e^{2N}). Then we shall show that the true operators Γ_μ, S_F', D_F' are of the form

$$\Gamma_\mu = Z_1^{-1}\Gamma_{\mu 1}(e_1), \tag{83}$$

$$S_F' = Z_2 S_{F1}'(e_1), \tag{84}$$

$$D_F' = Z_3 D_{F1}'(e_1), \tag{85}$$

where Z_1, Z_2, Z_3 are constants to be determined, and e_1 is given by

$$e_1 = Z_1^{-1} Z_2 Z_3^{\frac{1}{2}} e. \tag{86}$$

This e_1 will turn out to be the "true" electronic charge. It has to be proved that the result of substituting (83), (84), (85) into the integral equations defining Γ_μ, S_F', D_F', is to reproduce these expressions exactly, when Z_1, Z_2, Z_3, and $\delta\kappa_0$ are suitably chosen.

Concerning the $\Gamma_{\mu 1}(e)$, $S_{F1}'(e)$, $D_{F1}'(e)$, it is known that, when these operators are substituted into the integral equations, they reproduce themselves with the addition of certain divergent terms. The additional divergent terms consist partly of the terms involving A, B, C, L, which are displayed in (73), (78), (82), and partly of terms arising (in the case of S_F' and D_F' only) from the peculiar behavior of the vertices b, b' in Fig. 5. The terms arising from b and b' have been discussed earlier; they may be called for brevity b-divergences. Originally, of course, there is no asymmetry between the divergences arising in Σ^* from vertex parts inserted at the two ends a and b of W; we have manufactured an asymmetry by including the divergences arising at a in the coefficient Z_1^{-1} of (83), while at b the operator γ_μ has not been replaced by Γ_μ and so the b divergences have not been so absorbed. It is thus to be expected that the effect of the b divergences, like that of the a divergences, will be merely to multiply all contributions to Σ^* by the constant Z_1^{-1}. Similarly, we expect that divergences at b' will multiply Π^* by the constant Z_1^{-1}. It can be shown, by a detailed argument too long to be given here, that these expectations are justified. (The interested reader is recommended to see for himself, by considering contributions to Σ^* arising from various self-energy parts, how it is that the finite terms of a given order are always reappearing in higher order multiplied by the same divergent coefficients.) Therefore, the complete expressions obtained by substituting $\Gamma_{\mu 1}(e)$, $S_{F1}'(e)$, $D_{F1}'(e)$, into the integral equations defining Λ_μ, Σ^*, Π^*, are

$$\Lambda_{\mu 1}(e)=\Lambda_{\mu c}(e)+L(e)\gamma_\mu, \tag{87}$$

$$S_F\Sigma_1^*(e)=-2\pi i\delta\kappa_0 S_F +Z_1^{-1}\left(A(e)S_F+\frac{1}{2\pi}B(e)+\frac{1}{2\pi}S_c(e)\right), \tag{88}$$

$$D_F\Pi_1^*(e)=Z_1^{-1}\left(\frac{1}{2\pi i}C(e)+\frac{1}{2\pi i}D_c(e)\right). \tag{89}$$

Here $A(e)$, $B(e)$, $C(e)$, $L(e)$ are well-defined power series in e, with coefficients which diverge never more strongly than as a power of a logarithm. The finite operators $\Lambda_{\mu c}(e)$, $S_c(e)$, $D_c(e)$, will, when all divergent terms are dropped, lead back to the $\Gamma_{\mu 1}(e)$, $S_{F1}'(e)$, $D_{F1}'(e)$, from which the substitution started; thus, according to (38), (63), (64),

$$\Gamma_{\mu 1}(e)=\gamma_\mu+\Lambda_{\mu c}(e), \tag{87'}$$

$$S_{F1}'(e)=S_F+\frac{1}{2\pi}S_c(e)S_{F1}'(e), \tag{88'}$$

$$D_{F1}'(e)=D_F+\frac{1}{2\pi i}D_c(e)D_{F1}'(e). \tag{89'}$$

Equations (87)–(89), (87')–(89'), describe precisely the way in which the $\Gamma_{\mu 1}(e)$, $S_{F1}'(e)$, $D_{F1}'(e)$, when substituted into the integral equations, reproduce themselves with the addition of divergent terms. And from these results it is easy to deduce the self-reproducing property of the operators (83)–(85), when substituted into the same equations.

Consider for example the effect of substituting from (83)–(85) into the term $\Sigma(W,t^1)$, given by (65) with $l=1$. The integrand of (65) is a product of one factor Γ_μ, one γ_μ, one S_F', and one D_F'. Therefore the substitution gives

$$Z_1^{-1}Z_2Z_3\Sigma_0(W), \tag{90}$$

where $\Sigma_0(W)$ is the expression (65) obtained by substituting $\Gamma_{\mu 1}(e_1)$, $S_{F1}'(e_1)$, $D_{F1}'(e_1)$, without the Z factors. Now the Z factors in (90) combine with the e^2 of (65) to give

$$Z_1Z_2^{-1}e_1^2,$$

and the remaining factor of $\Sigma_0(W)$ is explicitly a function of e_1 and not of e. Therefore (90) is

$$Z_1Z_2^{-1}\Sigma_1(W,e_1),$$

where $\Sigma_1(W,e)$ is the expression obtained by substituting the operators $\Gamma_{\mu 1}(e)$, $S_{F1}'(e)$, $D_{F1}'(e)$ into $\Sigma(W,t^1)$. Thus the $\Sigma^*(t^1)$, obtained by substituting from (83)–(85) into (65), is identical with the result of substituting the operators $\Gamma_{\mu 1}(e)$, $S_{F1}'(e)$, $D_{F1}'(e)$, and afterwards changing e to e_1 and multiplying the whole expression (except for the constant term in $\delta\kappa_0$) by $Z_1Z_2^{-1}$. More exactly, using (88), one can say that the Σ^* obtained by substituting from (83)–(85) is given by

$$S_F\Sigma^*=-2\pi i\delta\kappa_0 S_F +Z_2^{-1}\left(A(e_1)S_F+\frac{1}{2\pi}B(e_1)+\frac{1}{2\pi}S_c(e_1)\right). \tag{91}$$

Further, the S_F' obtained by substituting from (83)–(85) into the integral equations is given by (91) and

$$S_F'=S_F+S_F\Sigma^*S_F'. \tag{92}$$

It is now easy to verify, using (88'), that S_F' given by (91) and (92) will be identical with (84), provided that

$$Z_2=1+\frac{1}{2\pi}B(e_1), \tag{93}$$

$$\delta\kappa_0=\frac{1}{2\pi i}Z_2^{-1}A(e_1). \tag{94}$$

In a similar way, the D_F' obtained by substituting from (83)–(85) into the integral equations can be related with the $\Pi_1^*(e)$ of (89). This D_F' will be identical with (85) provided that

$$Z_3=1+\frac{1}{2\pi i}C(e_1). \tag{95}$$

Finally, the Γ_μ obtained by substituting from (83)–(85) can be shown to be

$$\Gamma_\mu=\gamma_\mu+Z_1^{-1}\Lambda_{\mu 1}(e_1),$$

with $\Lambda_{\mu 1}(e)$ given by (87). Using (87'), this Γ_μ will be identical with (83) provided that

$$Z_1=1-L(e_1). \tag{96}$$

Therefore, if Z_1, Z_2, Z_3, $\delta\kappa_0$ are defined by (96), (93), (95), (94), it is established that (83)–(85) give the correct forms of the operators Γ_μ, S_F', D_F', including all the effects of the radiative corrections which these operators introduce into themselves and into each other. The exact Eqs. (83)–(85) give a much simpler separation of the infinite from the finite parts of these operators than the approximate equations (73), (78), (82).

Consider now the result of using the exact operators (83)–(85) in calculating a constituent M of $U(\infty)$, where M is constructed from a certain irreducible graph G_0 according to the rules of Section IV. G_0 will have, say, F_e internal and E_e external electron lines, F_p internal and E_p external photon lines, and

$$n=F_e+\tfrac{1}{2}E_e=2F_p+E_p \tag{97}$$

vertices. In M there will be $\frac{1}{2}E_e$ factors $\psi'(k^i)$, $\frac{1}{2}E_e$ factors $\bar{\psi}'(k^i)$ and E_p factors $A_\mu'(k^i)$ given by (37). In $\psi'(k^i)$, k^i is the momentum-energy 4-vector of an electron, which satisfies (69), and the $S_c(k^i)$ in (73) are zero at every stage of the inductive definition of $S_{F1}'(e)$. Therefore (84), (35), (37) give in turn

$$\begin{aligned}
&S_F'(k^i)=Z_2S_F(k^i),\\
&\Sigma(k^i)=2\pi(Z_2-1)(k_\mu{}^i\gamma_\mu-i\kappa_0),\\
&\psi'(k^i)=\psi(k^i)+2\pi(Z_2-1)S_F(k^i)(k_\mu{}^i\gamma_\mu-i\kappa_0)\psi(k^i).
\end{aligned} \tag{98}$$

The expression (98) is indeterminate, since $(k_\mu{}^i\gamma_\mu-i\kappa_0)$ operating on $\psi(k^i)$ gives zero, while operating on $S_F(k^i)$ it gives the constant $(1/2\pi)$. Thus, according to the order in which the factors are evaluated, (98) will give for $\psi'(k^i)$ either the value $\psi(k^i)$ or the value $Z_2\psi(k^i)$. Similarly, $\bar{\psi}'(k^i)$ is indeterminate between $\bar{\psi}(k^i)$ and $Z_2\bar{\psi}(k^i)$, and, excluding for the moment $A_\mu(k^i)$ which are Fourier components of the external potential, $A_\mu'(k^i)$ is indeterminate between $A_\mu(k^i)$ and $Z_3A_\mu(k^i)$. In any case, considerations of covariance show that the $\psi'(k^i)$, $\bar{\psi}'(k^i)$, $A_\mu'(k^i)$ are numerical multiples of the $\psi(k^i)$, $\bar{\psi}(k^i)$, $A_\mu(k^i)$; thus the indeterminacy lies only in a constant factor multiplying the whole expression M.

There cannot be any indeterminacy in the magnitude of the matrix elements of $U(\infty)$, so long as this operator is restricted to be unitary. The indeterminacy in fact lies only in the normalization of the electron and photon wave functions $\psi(k^i)$, $\bar{\psi}(k^i)$, $A_\mu(k^i)$, which may or may not be regarded as altered by the continual interactions of these particles with the vacuum-fields around them. It can be shown that, if the wave functions are everywhere normalized in the usual way, the apparent indeterminacy is removed, and one must take

$$\begin{aligned}
\psi'(k^i)&=Z_2^{\frac{1}{2}}\psi(k^i),\\
\bar{\psi}'(k^i)&=Z_2^{\frac{1}{2}}\bar{\psi}(k^i),\\
A_\mu'(k^i)&=Z_3^{\frac{1}{2}}A_\mu(k^i).
\end{aligned} \tag{99}$$

It will be seen that (99) gives just the geometric mean of the two alternative values of $\psi'(k^i)$ obtained from (98).

When $A_\mu(k^i)$ is a Fourier component of the external potential, then in general $(k^i)^2\neq 0$, and $A_\mu'(k^i)$ is not indeterminate but is given by (37) and (85) in the form

$$A_\mu'(k^i)=2\pi iZ_3D_{F1}'(e_1)(k^i)^2A_\mu(k^i). \tag{100}$$

However, the unit in which external potentials are measured is defined by the dynamical effects which the potentials produce on known charges; and these dynamical effects are just the matrix elements of $U(\infty)$ in which (100) appears. Therefore the factor Z_3 in (100) has no physical significance, and will be changed when A_μ is measured in practical units. The correct constant which appears when practical units are used is $Z_3^{\frac{1}{2}}$; this is because the photon potentials A_μ in (99) were normalized in terms of practical units; and (100) should reduce to (99) when $(k^i)^2\to 0$, if the external A_μ and the photon A_μ are measured in the same units. Therefore the correct formula for A_μ', covering the cases both of photon and of external potentials, is

$$\left.\begin{aligned}
&A_\mu'(k^i)=2\pi iZ_3^{\frac{1}{2}}D_{F1}'(e_1)(k^i)^2A_\mu(k^i), && (k^i)^2\neq 0,\\
&A_\mu'(k^i)=Z_3^{\frac{1}{2}}A_\mu(k^i), && (k^i)^2=0.
\end{aligned}\right\} \tag{101}$$

In M there will appear F_e factors S_F', F_p factors D_F', and n factors Γ_μ, in addition to the factors of the type (99), (101). Hence by (97) the Z factors will occur in M only as the constant multiplier

$$Z_1^{-n}Z_2^{n}Z_3^{\frac{1}{2}n}.$$

By (86), this multiplier is exactly sufficient to convert the factor e^n, remaining in M from the original interaction (8), into a factor e_1^n. Thereby, both e and Z factors disappear from M, leaving only their combination e_1 in the operators $\Gamma_{\mu 1}(e_1)$, $S_{F1}'(e_1)$, $D_{F1}'(e_1)$, and in the factor e_1^n. If now e_1 is identified with the finite observed electronic charge, there no longer appear any divergent expressions in M. And since M is a completely general constituent of $U(\infty)$, the elimination of divergences from the S matrix is accomplished.

It hardly needs to be pointed out that the arguments of this section have involved extensive manipulations of infinite quantities. These manipulations have only a formal validity, and must be justified *a posteriori* by the fact that they ultimately lead to a clear separation of finite from infinite expressions. Such an *a posteriori* justification of dubious manipulations is an inevitable feature of any theory which aims to extract meaningful results from not completely consistent premises.

We conclude with two disconnected remarks. First, it is probable that $Z_1 = Z_2$ identically, though this has been proved so far only up to the order e^2. If this conjecture is correct, then all charge-renormalization effects arise according to (86) from the coefficient Z_3 alone, and the arguments of this paper can be somewhat simplified. Second, Eqs. (88′), (89′), which define the fundamental operators S_{F1}', D_{F1}', may be solved for these operators. Thus

$$S_{F1}'(e) = \left[1 - \frac{1}{2\pi} S_c(e)\right]^{-1} S_F, \tag{88''}$$

$$D_{F1}'(e) = \left[1 - \frac{1}{2\pi i} D_c(e)\right]^{-1} D_F. \tag{89''}$$

In electrodynamics, the S_c and D_c are small radiative corrections, and it will always be legitimate and convenient to expand (88″) and (89″) by the binomial theorem. If, however, the methods of the present paper are to be applied to meson fields, with coupling constants which are not small, then it will be desirable not to expand these expressions; in this way one may hope to escape partially from the limitations which the use of weak-coupling approximations imposes on the theory.

VIII. SUMMARY OF RESULTS

The results of the preceding sections divide themselves into two groups. On the one hand, there is a set of rules by which the element of the S matrix corresponding to any given scattering process may be calculated, without mentioning the divergent expressions occurring in the theory. On the other hand, there is the specification of the divergent expressions, and the interpretation of these expressions as mass and charge renormalization factors.

The first group of results may be summarized as follows. Given a particular scattering problem, with specified initial and final states, the corresponding matrix element of $U(\infty)$ is a sum of contributions from various graphs G as described in Section II. A particular contribution M from a particular G is to be written down as an integral over momentum variables according to the rules of Section III; the integrand is a product of factors $\psi(k^i)$, $\bar{\psi}(k^i)$, $A_\mu(k^i)$, $S_F(p^i)$, $D_F(p^i)$, $\delta(q_j)$, γ_μ, the factors corresponding in a prescribed way to the lines and vertices of G. According to Section IV, contributions M are only to be admitted from irreducible G; the effects of reducible graphs are included by replacing in M the factors ψ, $\bar{\psi}$, A_μ, S_F, D_F, γ_μ, by the corresponding expressions (37), (35), (36), (38). These replacements are then shown in Section VII to be equivalent to the following: each factor S_F in M is replaced by $S_{F1}'(e)$, each factor D_F by $D_{F1}'(e)$, each factor γ_μ by $\Gamma_{\mu 1}(e)$, each factor A_μ when it represents an external potential is replaced by

$$A_{\mu 1}(k^i) = 2\pi i D_{F1}'(e)(k^i)^2 A_\mu(k^i), \tag{102}$$

factors ψ, $\bar{\psi}$, A_μ representing particle wave-functions are left unchanged, and finally e wherever it occurs in M is replaced by e_1. The definition of M is completed by the specification of $S_{F1}'(e)$, $D_{F1}'(e)$, $\Gamma_{\mu 1}(e)$; it is in the calculation of these operators that the main difficulty of the theory lies. The method of obtaining these operators is the process of successive substitution and integration explained in the first part of Section VII; the operators so calculated are divergence-free, the divergent parts at every stage of the calculation being explicitly dropped after being separated from the finite parts by the method of Section VI.

The above rules determine each contribution M to $U(\infty)$ as a divergence-free expression, which is a function of the observed mass m and the observed charge e_1 of the electron, both of which quantities are taken to have their empirical values. The divergent parts of the theory are irrelevant to the calculation of $U(\infty)$, being absorbed into the unobservable constants δm and e occurring in (8). A place where some ambiguity might appear in M is in the calculation of the operators $S_{F1}'(e)$, $D_{F1}'(e)$, $\Gamma_{\mu 1}(e)$, when the method of Section VI is used to separate out the finite parts $S(W, t^1)$, $D(W', t^1)$, $\Lambda_{\mu c}(V, t^1, t^2)$, from the expressions (67), (74), (80). Even in this place the rules of Section VI give unambiguous directions for making the separation; only there is a question whether some alternative directions might be equally reasonable. For example, it is possible to separate out a finite part from $\Sigma(W, t^1)$ according to (67), and not to make the further step of using (70) to separate out a finite part $S(W, t^1)$ which vanishes when (69) holds. Actually it is easy to verify that such an alternative procedure will not change the value of M, but will only make its evaluation more complicated; it will lead to an expression for M in which one (infinite) part of the mass and charge renormalizations is absorbed into the constants δm and e, while other finite mass and charge renormalizations are left explicitly in the formulas. It is just these finite renormalization effects which the second step in the separation of $S(W, t^1)$ and $\Lambda_{\mu c}(V, t^1, t^2)$ is designed to avoid. Therefore it may be concluded that the rules of calculation of $U(\infty)$ are not only divergence-free but unambiguous.

As anyone acquainted with the history of the Lamb shift[11] knows, the utmost care is required

[11] H. A. Bethe, *Electromagnetic Shift of Energy Levels*, Report to Solvay Conference, Brussels (1948).

before it can be said that any particular rule of calculation is unambiguous. The rules given in this paper are unambiguous, in the sense that each quantity to be calculated is an integral in momentum-space which is absolutely convergent at infinity; such an integral has always a well-defined value. However, the rules would not be unambiguous if it were allowed to split the integrand into several parts and to evaluate the integral by integrating the parts separately and then adding the results; ambiguities would arise if ever the partial integrals were not absolutely convergent. A splitting of the integrals into conditionally convergent parts may seem unnatural in the context of the present paper, but occurs in a natural way when calculations are based upon a perturbation theory in which electron and positron states are considered separately from each other. The absolute convergence of the integrals in the present theory is essentially connected with the fact that the electron and positron parts of the electron-positron field are never separated; this finds its algebraic expression in the statement that the quadratic denominator in (45) is never to be separated into partial fractions. Therefore the absence of ambiguity in the rules of calculation of $U(\infty)$ is achieved by introducing into the theory what is really a new physical hypothesis, namely that the electron-positron field always acts as a unit and not as a combination of two separate fields. A similar hypothesis is made for the electromagnetic field, namely that this field also acts as a unit and not as a sum of one part representing photon emission and another part representing photon absorption.

Finally, it must be said that the proof of the finiteness and unambiguity of $U(\infty)$ given in this paper makes no pretence of being complete and rigorous. It is most desirable that these general arguments should as soon as possible be supplemented by an explicit calculation of at least one fourth-order radiative effect, to make sure that no unforeseen difficulties arise in that order.

The second group of results of the theory is the identification of δm and e by (94) and (86). Although these two equations are strictly meaningless, both sides being infinite, yet it is a satisfactory feature of the theory that it determines the unobservable constants δm and e formally as power series in the observable e_1, and not vice versa. There is thus no objection in principle to identifying e_1 with the observed electronic charge and writing

$$(e_1^2/4\pi\hbar c)=\alpha=1/137. \qquad (103)$$

The constants appearing in (8) are then, by (94) and (86),

$$\delta m=m(A_1\alpha+A_2\alpha^2+\cdots), \qquad (104)$$

$$e=e_1(1+B_1\alpha+B_2\alpha^2+\cdots), \qquad (105)$$

where the A_i and B_i are logarithmically divergent numerical coefficients, independent of m and e_1.

IX. DISCUSSION OF FURTHER OUTLOOK

The surprising feature of the S matrix theory, as outlined in this paper, is its success in avoiding difficulties. Starting from the methods of Tomonaga, Schwinger and Feynman, and using no new ideas or techniques, one arrives at an S matrix from which the well-known divergences seem to have conspired to eliminate themselves. This automatic disappearance of divergences is an empirical fact, which must be given due weight in considering the future prospects of electrodynamics. Paradoxically opposed to the finiteness of the S matrix is the second fact, that the whole theory is built upon a Hamiltonian formalism with an interaction-function (8) which is infinite and therefore physically meaningless.

The arguments of this paper have been essentially mathematical in character, being concerned with the consequences of a particular mathematical formalism. In attempting to assess their significance for the future, one must pass from the language of mathematics to the language of physics. One must assume provisionally that the mathematical formalism corresponds to something existing in nature, and then enquire to what extent the paradoxical results of the formalism can be reconciled with such an assumption. In accordance with this program, we interpret the contrast between the divergent Hamiltonian formalism and the finite S matrix as a contrast between two pictures of the world, seen by two observers having a different choice of measuring equipment at their disposal. The first picture is of a collection of quantized fields with localizable interactions, and is seen by a fictitious observer whose apparatus has no atomic structure and whose measurements are limited in accuracy only by the existence of the fundamental constants c and h. This observer is able to make with complete freedom on a sub-microscopic scale the kind of observations which Bohr and Rosenfeld[12] employ in a more restricted domain in their classic discussion of the measurability of field-quantities; and he will be referred to in what follows as the "ideal" observer. The second picture is of a collection of observable quantities (in the terminology of Heisenberg), and is the picture seen by a real observer, whose apparatus consists of atoms and elementary particles and whose measurements are limited in accuracy not only by c and h but also by other constants such as α and m. The real observer

[12] N. Bohr and L. Rosenfeld, Kgl. Dansk. Vid. Sels. Math.-Phys. Medd. **12**, No. 8 (1933). A second paper by Bohr and Rosenfeld is to be published later, and is abstracted in a booklet by A. Pais, *Developments in the Theory of the Electron* (Princeton University Press, Princeton, 1948).

makes spectroscopic observations, and performs experiments involving bombardments of atomic systems with various types of mutually interacting subatomic projectiles, but to the best of our knowledge he cannot measure the strength of a single field undisturbed by the interaction of that field with others. The ideal observer, utilizing his apparatus in the manner described in the analysis of the Hamiltonian formalism by Bohr and Rosenfeld,[12] makes measurements of precisely this last kind, and it is in terms of such measurements that the commutation-relations of the fields are interpreted. The interaction-function (8) will presumably always remain unobservable to the real observer, who is able to determine positions of particles only with limited accuracy, and who must always obtain finite results from his measurements. The ideal observer, however, using non-atomic apparatus whose location in space and time is known with infinite precision, is imagined to be able to disentangle a single field from its interactions with others, and to measure the interaction (8). In conformity with the Heisenberg uncertainty principle, it can perhaps be considered a physical consequence of the infinitely precise knowledge of location allowed to the ideal observer, that the value obtained by him when he measures (8) is infinite.

If the above analysis is correct, the divergences of electrodynamics are directly attributable to the fact that the Hamiltonian formalism is based upon an idealized conception of measurability. The paradoxical feature of the present situation does not then lie in the mere coexistence of a finite S matrix with an infinite interaction-function. The empirically found correlation, between expressions which are unobservable to a real observer and expressions which are infinite, is a physically intelligible and acceptable feature of the theory. The paradox is the fact that it is necessary in the present paper to start from the infinite expressions in order to deduce the finite ones. Accordingly, what is to be looked for in a future theory is not so much a modification of the present theory which will make all infinite quantities finite, but rather a turning-round of the theory so that the finite quantities shall become primary and the infinite quantities secondary.

One may expect that in the future a consistent formulation of electrodynamics will be possible, itself free from infinities and involving only the physical constants m and e_1, and such that a Hamiltonian formalism with interaction (8), with divergent coefficients δm and e, may in suitably idealized circumstances be deduced from it. The Hamiltonian formalism should appear as a limiting form of a description of the world as seen by a certain type of observer, the limit being approached more and more closely as the precision of measurement allowed to the observer tends to infinity.

The nature of a future theory is not a profitable subject for theoretical speculation. The future theory will be built, first of all upon the results of future experiments, and secondly upon an understanding of the interrelations between electrodynamics and mesonic and nucleonic phenomena. The purpose of the foregoing remarks is merely to point out that there is now no longer, as there has seemed to be in the past, a compelling necessity for a future theory to abandon some essential features of the present electrodynamics. The present electrodynamics is certainly incomplete, but is no longer certainly incorrect.

In conclusion, the author would like to express his profound indebtedness to Professor Feynman for many of the ideas upon which this paper is built, to Professor Oppenheimer for valuable discussions, and to the Commonwealth Fund of New York for financial support.

THE LAGRANGIAN IN QUANTUM MECHANICS.

By P. A. M. Dirac.

(Received November 19, 1932).

Quantum mechanics was built up on a foundation of analogy with the Hamiltonian theory of classical mechanics. This is because the classical notion of canonical coordinates and momenta was found to be one with a very simple quantum analogue, as a result of which the whole of the classical Hamiltonian theory, which is just a structure built up on this notion, could be taken over in all its details into quantum mechanics.

Now there is an alternative formulation for classical dynamics, provided by the Lagrangian. This requires one to work in terms of coordinates and velocities instead of coordinates and momenta. The two formulations are, of course, closely related, but there are reasons for believing that the Lagrangian one is the more fundamental.

In the first place the Lagrangian method allows one to collect together all the equations of motion and express them as the stationary property of a certain action function. (This action function is just the time-integral of the Lagrangian). There is no corresponding action principle in terms of the coordinates and momenta of the Hamiltonian theory. Secondly the Lagrangian method can easily be expressed relativistically, on account of the action function being a relativistic invariant; while the Hamiltonian method is essentially non-relativistic in form, since it marks out a particular time variable as the canonical conjugate of the Hamiltonian function.

For these reasons it would seem desirable to take up the question of what corresponds in the quantum theory to the Lagrangian method of the classical theory. A little consideration shows, however, that one cannot expect to be able to take over the classical Lagrangian equations in any very direct way. These equations involve partial derivatives of the Lagrangian with respect to the coordinates and velocities and no meaning can be given to such derivatives in quantum mechanics. The only differentiation process that can be carried out with respect to the dynamical variables of quantum mechanics is that of forming Poisson brackets and this process leads to the Hamiltonian theory.[1]

We must therefore seek our quantum Lagrangian theory in an indirect way. We must try to take over the ideas of the classical Lagrangian theory, not the equations of the classical Lagrangian theory.

Contact Transformations.

Lagrangian theory is closely connected with the theory of contact transformations. We shall therefore begin with a discussion of the analogy between classical and quantum contact transformations. Let the two sets of variables be p_r, q_r and P_r, Q_r, $(r = 1, 2 \ldots n)$ and suppose the q's and Q's to be all independent, so that any function of the dynamical variables can be expressed in terms of them. It is well known that in the classical theory the transformation equations for this case can be put in the form

$$p_r = \frac{\partial S}{\partial q_r}, \quad P_r = -\frac{\partial S}{\partial Q_r}, \tag{1}$$

where S is some function of the q's and Q's.

[1] Processes for partial differentiation with respect to matrices have been given by Born, Heisenberg and Jordan (ZS. f. Physik **35**, 561, 1926) but these processes do not give us means of differentiation with respect to dynamical variables, since they are not independent of the representation chosen. As an example of the difficulties involved in differentiation with respect to quantum dynamical variables, consider the three components of an angular momentum, satisfying

$$m_x m_y - m_y m_x = ihm_z.$$

We have here m_z expressed explicitly as a function of m_x and m_y, but we can give no meaning to its partial derivative with respect to m_x or m_y.

In the quantum theory we may take a representation in which the q's are diagonal, and a second representation in which the Q's are diagonal. There will be a transformation function $(q'|Q')$ connecting the two representations. We shall now show that this transformation function is the quantum analogue of $e^{iS/h}$.

If α is any function of the dynamical variables in the quantum theory, it will have a „mixed" representative $(q'|\alpha|Q')$, which may be defined in terms of either of the usual representatives $(q'|\alpha|q'')$, $(Q'|\alpha|Q'')$ by

$$(q'|\alpha|Q') = \int (q'|\alpha|q'')\,dq''\,(q''|Q') = \int (q'|Q'')\,dQ''\,(Q''|\alpha|Q').$$

From the first of these definitions we obtain

$$(q'|q_r|Q') = q'_r\,(q'|Q') \tag{2}$$

$$(q'|p_r|Q') = -ih\frac{\partial}{\partial q'_r}(q'|Q') \tag{3}$$

and from the second

$$(q'|Q_r|Q') = Q'_r\,(q'|Q') \tag{4}$$

$$(q'|P_r|Q') = ih\frac{\partial}{\partial Q'_r}(q'|Q'). \tag{5}$$

Note the difference in sign in (3) and (5).

Equations (2) and (4) may be generalised as follows. Let $f(q)$ be any function of the q's and $g(Q)$ any function of the Q's. Then

$$(q'|f(q)\,g(Q)|Q') = \iint (q'|f(q)|q'')\,dq''\,(q''|Q'')\,dQ''\,(Q''|g(Q)|Q')$$
$$= f(q')\,g(Q')\,(q'|Q').$$

Further, if $f_k(q)$ and $g_k(Q)$, $(k = 1, 2 \ldots, m)$ denote two sets of functions of the q's and Q's respectively,

$$(q'|\Sigma_k f_k(q) g_k(Q)|Q') = \Sigma_k f_k(q')\,g_k(Q')\cdot(q'|Q').$$

Thus if α is any function of the dynamical variables and we suppose it to be expressed as a function $\alpha(qQ)$ of the q's and Q's in a „well-ordered" way, that is, so that it consists of a sum of terms of the form $f(q)\,g(Q)$, we shall have

$$(q'|\alpha(qQ)|Q') = \alpha(q'Q')\,(q'|Q'). \tag{6}$$

This is a rather remarkable equation, giving us a connection between $\alpha(qQ)$, which is a function of operators, and $\alpha(q'Q')$, which is a function of numerical variables.

Let us apply this result for $\alpha = p_r$. Putting

$$(q' \mid Q') = e^{iU/h}, \tag{7}$$

where U is a new function of the q''s and Q''s we get from (3)

$$(q' \mid p_r \mid Q') = \frac{\partial U(q'Q')}{\partial q'_r} (q' \mid Q').$$

By comparing this with (6) we obtain

$$p_r = \frac{\partial U(qQ)}{\partial q_r}$$

as an equation between operators or dynamical variables, which holds provided $\partial U / \partial q_r$ is well-ordered. Similarly, by applying the result (6) for $\alpha = P_r$ and using (5), we get

$$P_r = -\frac{\partial U(qQ)}{\partial Q_r},$$

provided $\partial U / \partial Q_r$ is well-ordered. These equations are of the same form as (1) and show that the U defined by (7) is the analogue of the classical function S, which is what we had to prove.

Incidentally, we have obtained another theorem at the same time, namely that equations (1) hold also in the quantum theory provided the right-hand sides are suitably interpreted, the variables being treated classically for the purpose of the differentiations and the derivatives being then well-ordered. This theorem has been previously proved by Jordan by a different method.[1]

The Lagrangian and the Action Principle.

The equations of motion of the classical theory cause the dynamical variables to vary in such a way that their values q_t, p_t at any time t are connected with their values q_T, p_T at any other time T by a contact transformation, which may be put into the form (1) with $q, p = q_t, p_t$; $Q, P = q_T, p_T$ and S equal to the time integral of the Lagrangian over the range

[1] Jordan, ZS. f. Phys. **38**, 513, 1926.

T to t. In the quantum theory the q_t, p_t will still be connected with the q_T, p_T by a contact transformation and there will be a transformation function $(q_t | q_T)$ connecting the two representations in which the q_t and the q_T are diagonal respectively. The work of the preceding section now shows that

$$(q_t | q_T) \text{ corresponds to } \exp\left[i\int_T^t Ldt/h\right], \tag{8}$$

where L is the Lagrangian. If we take T to differ only infinitely little from t, we get the result

$$(q_{t+dt} | q_t) \text{ corresponds to } \exp\,[iL\,dt/h]. \tag{9}$$

The transformation functions in (8) and (9) are very fundamental things in the quantum theory and it is satisfactory to find that they have their classical analogues, expressible simply in terms of the Lagrangian. We have here the natural extension of the well-known result that the phase of the wave function corresponds to Hamilton's principle function in classical theory. The analogy (9) suggests that we ought to consider the classical Lagrangian, not as a function of the coordinates and velocities, but rather as a function of the coordinates at time t and the coordinates at time $t+dt$.

For simplicity in the further discussion in this section we shall take the case of a single degree of freedom, although the argument applies also to the general case. We shall use the notation

$$\exp\left[i\int_T^t L\,dt/h\right] = A\,(tT),$$

so that $A(tT)$ is the classical analogue of $(q_t | q_T)$.

Suppose we divide up the time interval $T \to t$ into a large number of small sections $T \to t_1$, $t_1 \to t_2$, ..., $t_{m-1} \to t_m$, $t_m \to t$ by the introduction of a sequence of intermediate times t_1, t_2, ... t_m. Then

$$A\,(tT) = A\,(tt_m)\,A\,(t_m t_{m-1}) \ldots A\,(t_2 t_1)\,A\,(t_1 T). \tag{10}$$

Now in the quantum theory we have

$$(q_t | q_T) = \int (q_t | q_m)\,dq_m (q_m | q_{m-1})\,dq_{m-1} \ldots (q_2 | q_1) dq_1 (q_1 | q_T), \tag{11}$$

where q_k denotes q at the intermediate time t_k, $(k=1, 2 \ldots m)$. Equation (11) at first sight does not seem to correspond properly to equation (10), since on the right-hand side of (11) we must integrate after doing the multiplication while on the right-hand side of (10) there is no integration.

Let us examine this discrepancy by seeing what becomes of (11) when we regard t as extremely small. From the results (8) and (9) we see that the integrand in (11) must be of the form $e^{iF/h}$ where F is a function of $q_T, q_1, q_2 \ldots q_m, q_t$ which remains finite as h tends to zero. Let us now picture one of the intermediate q's, say q_k, as varying continuously while the others are fixed. Owing to the smallness of h, we shall then in general have F/h varying extremely rapidly. This means that $e^{iF/h}$ will vary periodically with a very high frequency about the value zero, as a result of which its integral will be practically zero. The only important part in the domain of integration of q_k is thus that for which a comparatively large variation in q_k produces only a very small variation in F. This part is the neighbourhood of a point for which F is stationary with respect to small variations in q_k.

We can apply this argument to each of the variables of integration in the right-hand side of (11) and obtain the result that the only important part in the domain of integration is that for which F is stationary for small variations in all the intermediate q's. But, by applying (8) to each of the small time sections, we see that F has for its classical analogue

$$\int_{t_m}^{t} L\,dt + \int_{t_{m-1}}^{t_m} L\,dt + \ldots + \int_{t_1}^{t_2} L\,dt + \int_{T}^{t_1} L\,dt = \int_{T}^{t} L\,dt,$$

which is just the action function which classical mechanics requires to be stationary for small variations in all the intermediate q's. This shows the way in which equation (11) goes over into classical results when h becomes extremely small.

We now return to the general case when h is not small. We see that, for comparison with the quantum theory, equation (10) must be interpreted in the following way. Each of the quantities A must be considered as a function of the q's at the two times to which it refers. The right-hand side is then a function, not only of q_T and q_t, but also of q_1, q_2, ... q_m, and in order to get from it a function of q_T and q_t only, which we can equate to the left-hand side, we must substitute for $q_1, q_2 \ldots q_m$ their values given by the action principle. This process of substitution for the intermediate q's then corresponds to the process of integration over all values of these q's in (11).

Equation (11) contains the quantum analogue of the action principle, as may be seen more explicitly from the following argument. From equation (11) we can extract the statement (a rather trivial one) that, if we take specified values for q_T and q_t, then the importance of our considering any set of values for the intermediate q's is determined by the importance of this set of values in the integration on the right-hand side of (11). If we now make h tend to zero, this statement goes over into the classical statement that, if we take specified values for q_T and q_t, then the importance of our considering any set of values for the intermediate q's is zero unless these values make the action function stationary. This statement is one way of formulating the classical action principle.

Application to Field Dynamics.

We may treat the problem of a vibrating medium in the classical theory by Lagrangian methods which form a natural generalisation of those for particles. We choose as our coordinates suitable field quantities or potentials. Each coordinate is then a function of the four space-time variables x, y, z, t, corresponding to the fact that in particle theory it is a function of just the one variable t. Thus the one independent variable t of particle theory is to be generalised to four independent variables x, y, z, t.[1]

[1] It is customary in field dynamics to regard the values of a field quantity for two different values of (x, y, z) but the same value of t as two different coordinates, instead of as two values of the same coordi-

We introduce at each point of space-time a Lagrangian density, which must be a function of the coordinates and their first derivatives with respect to x, y, z and t, corresponding to the Lagrangian in particle theory being a function of coordinates and velocities. The integral of the Lagrangian density over any (four-dimensional) region of space-time must then be stationary for all small variations of the coordinates inside the region, provided the coordinates on the boundary remain invariant.

It is now easy to see what the quantum analogue of all this must be. If S denotes the integral of the classical Lagrangian density over a particular region of space-time, we should expect there to be a quantum analogue of $e^{iS/h}$ corresponding to the $(q_t | q_T)$ of particle theory. This $(q_t | q_T)$ is a function of the values of the coordinates at the ends of the time interval to which it refers and so we should expect the quantum analogue of $e^{iS/h}$ to be a function (really a functional) of the values of the coordinates on the boundary of the space-time region. This quantum analogue will be a sort of „generalized transformation function". It cannot in general be interpreted, like $(q_t | q_T)$, as giving a transformation between one set of dynamical variables and another, but it is a four-dimensional generalization of $(q_t | q_T)$ in the following sense.

Corresponding to the composition law for $(q_t | q_T)$

$$(q_t | q_T) = \int (q_t | q_1) \, dq_1 \, (q_1 | q_T), \tag{12}$$

the generalized transformation function (g.t.f.) will have the following composition law. Take a given region of space-time and divide it up into two parts. Then the g.t.f. for the whole region will equal the product of the g.t.f.'s for the two parts, integrated over all values for the coordinates on the common boundary of the two parts.

Repeated application of (12) gives us (11) and repeated application of the corresponding law for g.t.f.'s will enable

nate for two different points in the domain of independent variables, and in this way to keep to the idea of a single independent variable t. This point of view is necessary for the Hamiltonian treatment, but for the Lagrangian treatment the point of view adopted in the text seems preferable on account of its greater space-time symmetry.

us in a similar way to connect the g.t.f. for any region with the g.t.f.'s for the very small sub-regions into which that region may be divided. This connection will contain the quantum analogue of the action principle applied to fields.

The square of the modulus of the transformation function $(q_t | q_T)$ can be interpreted as the probability of an observation of the coordinates at the later time t giving the result q_t for a state for which an observation of the coordinates at the earlier time T is certain to give the result q_T. A corresponding meaning for the square of the modulus of the g.t.f. will exist only when the g.t.f. refers to a region of space-time bounded by two separate (three-dimensional) surfaces, each extending to infinity in the space directions and lying entirely outside any light-cone having its vertex on the surface. The square of the modulus of the g. t. f. then gives the probability of the coordinates having specified values at all points on the later surface for a state for which they are given to have definite values at all points on the earlier surface. The g.t.f. may in this case be considered as a transformation function connecting the values of the coordinates and momenta on one of the surfaces with their values on the other.

We can alternatively consider $|(q_t | q_T)|^2$ as giving the relative a priori probability of any state yielding the results q_T and q_t when observations of the q's are made at time T and at time t (account being taken of the fact that the earlier observation will alter the state and affect the later observation). Correspondingly we can consider the square of the modulus of the g.t.f. for any space-time region as giving the relative a priori probability of specified results being obtained when observations are made of the coordinates at all points on the boundary. This interpretation is more general than the preceding one, since it does not require a restriction on the shape of the space-time region.

St John's College, Cambridge.

Space-Time Approach to Non-Relativistic Quantum Mechanics

R. P. FEYNMAN

Cornell University, Ithaca, New York

Non-relativistic quantum mechanics is formulated here in a different way. It is, however, mathematically equivalent to the familiar formulation. In quantum mechanics the probability of an event which can happen in several different ways is the absolute square of a sum of complex contributions, one from each alternative way. The probability that a particle will be found to have a path $x(t)$ lying somewhere within a region of space time is the square of a sum of contributions, one from each path in the region. The contribution from a single path is postulated to be an exponential whose (imaginary) phase is the classical action (in units of $\hbar$) for the path in question. The total contribution from all paths reaching x, t from the past is the wave function $\psi(x, t)$. This is shown to satisfy Schroedinger's equation. The relation to matrix and operator algebra is discussed. Applications are indicated, in particular to eliminate the coordinates of the field oscillators from the equations of quantum electrodynamics.

1. INTRODUCTION

IT is a curious historical fact that modern quantum mechanics began with two quite different mathematical formulations: the differential equation of Schroedinger, and the matrix algebra of Heisenberg. The two, apparently dissimilar approaches, were proved to be mathematically equivalent. These two points of view were destined to complement one another and to be ultimately synthesized in Dirac's transformation theory.

This paper will describe what is essentially a third formulation of non-relativistic quantum theory. This formulation was suggested by some of Dirac's[1,2] remarks concerning the relation of classical action[3] to quantum mechanics. A probability amplitude is associated with an entire motion of a particle as a function of time, rather than simply with a position of the particle at a particular time.

The formulation is mathematically equivalent to the more usual formulations. There are, therefore, no fundamentally new results. However, there is a pleasure in recognizing old things from a new point of view. Also, there are problems for which the new point of view offers a distinct advantage. For example, if two systems A and B interact, the coordinates of one of the systems, say B, may be eliminated from the equations describing the motion of A. The inter-

[1] P. A. M. Dirac, *The Principles of Quantum Mechanics* (The Clarendon Press, Oxford, 1935), second edition, Section 33; also, Physik. Zeits. Sowjetunion **3**, 64 (1933).

[2] P. A. M. Dirac, Rev. Mod. Phys. **17**, 195 (1945).

[3] Throughout this paper the term "action" will be used for the time integral of the Lagrangian along a path. When this path is the one actually taken by a particle, moving classically, the integral should more properly be called Hamilton's first principle function.

action with B is represented by a change in the formula for the probability amplitude associated with a motion of A. It is analogous to the classical situation in which the effect of B can be represented by a change in the equations of motion of A (by the introduction of terms representing forces acting on A). In this way the coordinates of the transverse, as well as of the longitudinal field oscillators, may be eliminated from the equations of quantum electrodynamics.

In addition, there is always the hope that the new point of view will inspire an idea for the modification of present theories, a modification necessary to encompass present experiments.

We first discuss the general concept of the superposition of probability amplitudes in quantum mechanics. We then show how this concept can be directly extended to define a probability amplitude for any motion or path (position *vs.* time) in space-time. The ordinary quantum mechanics is shown to result from the postulate that this probability amplitude has a phase proportional to the action, computed classically, for this path. This is true when the action is the time integral of a quadratic function of velocity. The relation to matrix and operator algebra is discussed in a way that stays as close to the language of the new formulation as possible. There is no practical advantage to this, but the formulae are very suggestive if a generalization to a wider class of action functionals is contemplated. Finally, we discuss applications of the formulation. As a particular illustration, we show how the coordinates of a harmonic oscillator may be eliminated from the equations of motion of a system with which it interacts. This can be extended directly for application to quantum electrodynamics. A formal extension which includes the effects of spin and relativity is described.

2. THE SUPERPOSITION OF PROBABILITY AMPLITUDES

The formulation to be presented contains as its essential idea the concept of a probability amplitude associated with a completely specified motion as a function of time. It is, therefore, worthwhile to review in detail the quantum-mechanical concept of the superposition of probability amplitudes. We shall examine the essential changes in physical outlook required by the transition from classical to quantum physics.

For this purpose, consider an imaginary experiment in which we can make three measurements successive in time: first of a quantity A, then of B, and then of C. There is really no need for these to be of different quantities, and it will do just as well if the example of three successive position measurements is kept in mind. Suppose that a is one of a number of possible results which could come from measurement A, b is a result that could arise from B, and c is a result possible from the third measurement C.[4] We shall assume that the measurements A, B, and C are the type of measurements that completely specify a state in the quantum-mechanical case. That is, for example, the state for which B has the value b is not degenerate.

It is well known that quantum mechanics deals with probabilities, but naturally this is not the whole picture. In order to exhibit, even more clearly, the relationship between classical and quantum theory, we could suppose that classically we are also dealing with probabilities but that all probabilities either are zero or one. A better alternative is to imagine in the classical case that the probabilities are in the sense of classical statistical mechanics (where, possibly, internal coordinates are not completely specified).

We define P_{ab} as the probability that if measurement A gave the result a, then measurement B will give the result b. Similarly, P_{bc} is the probability that if measurement B gives the result b, then measurement C gives c. Further, let P_{ac} be the chance that if A gives a, then C gives c. Finally, denote by P_{abc} the probability of all three, i.e., if A gives a, then B gives b, and C gives c. If the events between a and b are independent of those between b and c, then

$$P_{abc} = P_{ab}P_{bc}. \tag{1}$$

This is true according to quantum mechanics when the statement that B is b is a complete specification of the state.

[4] For our discussion it is not important that certain values of a, b, or c might be excluded by quantum mechanics but not by classical mechanics. For simplicity, assume the values are the same for both but that the probability of certain values may be zero.

In any event, we expect the relation

$$P_{ac}=\sum_b P_{abc}. \quad (2)$$

This is because, if initially measurement A gives a and the system is later found to give the result c to measurement C, the quantity B must have had some value at the time intermediate to A and C. The probability that it was b is P_{abc}. We sum, or integrate, over all the mutually exclusive alternatives for b (symbolized by $\sum_b$).

Now, the essential difference between classical and quantum physics lies in Eq. (2). In classical mechanics it is always true. In quantum mechanics it is often false. We shall denote the quantum-mechanical probability that a measurement of C results in c when it follows a measurement of A giving a by $P_{ac}{}^q$. Equation (2) is replaced in quantum mechanics by this remarkable law:[5] There exist complex numbers φ_{ab}, φ_{bc}, φ_{ac} such that

$$P_{ab}=|\varphi_{ab}|^2,\quad P_{bc}=|\varphi_{bc}|^2,\quad \text{and}\quad P_{ac}{}^q=|\varphi_{ac}|^2. \quad (3)$$

The classical law, obtained by combining (1) and (2),

$$P_{ac}=\sum_b P_{ab}P_{bc} \quad (4)$$

is replaced by

$$\varphi_{ac}=\sum_b \varphi_{ab}\varphi_{bc}. \quad (5)$$

If (5) is correct, ordinarily (4) is incorrect. The logical error made in deducing (4) consisted, of course, in assuming that to get from a to c the system had to go through a condition such that B had to have some definite value, b.

If an attempt is made to verify this, i.e., if B is measured between the experiments A and C, then formula (4) is, in fact, correct. More precisely, if the apparatus to measure B is set up and used, but no attempt is made to utilize the results of the B measurement in the sense that only the A to C correlation is recorded and studied, then (4) is correct. This is because the B measuring machine has done its job; if we wish, we could read the meters at any time without disturbing the situation any further. The experiments which gave a and c can, therefore, be separated into groups depending on the value of b.

Looking at probability from a frequency point of view (4) simply results from the statement that in each experiment giving a and c, B had some value. The only way (4) could be wrong is the statement, "B had some value," must sometimes be meaningless. Noting that (5) replaces (4) only under the circumstance that we make no attempt to measure B, we are led to say that the statement, "B had some value," may be meaningless whenever we make no attempt to measure B.[6]

Hence, we have different results for the correlation of a and c, namely, Eq. (4) or Eq. (5), depending upon whether we do or do not attempt to measure B. No matter how subtly one tries, the attempt to measure B must disturb the system, at least enough to change the results from those given by (5) to those of (4).[7] That measurements do, in fact, cause the necessary disturbances, and that, essentially, (4) could be false was first clearly enunciated by Heisenberg in his uncertainty principle. The law (5) is a result of the work of Schroedinger, the statistical interpretation of Born and Jordan, and the transformation theory of Dirac.[8]

Equation (5) is a typical representation of the wave nature of matter. Here, the chance of finding a particle going from a to c through several different routes (values of b) may, if no attempt is made to determine the route, be represented as the square of a sum of several complex quantities—one for each available route.

[5] We have assumed b is a non-degenerate state, and that therefore (1) is true. Presumably, if in some generalization of quantum mechanics (1) were not true, even for pure states b, (2) could be expected to be replaced by: There are complex numbers φ_{abc} such that $P_{abc}=|\varphi_{abc}|^2$. The analog of (5) is then $\varphi_{ac}=\sum_b \varphi_{abc}$.

[6] It does not help to point out that we *could* have measured B had we wished. The fact is that we did not.

[7] How (4) actually results from (5) when measurements disturb the system has been studied particularly by J. von Neumann (*Mathematische Grundlagen der Quantenmechanik* (Dover Publications, New York, 1943)). The effect of perturbation of the measuring equipment is effectively to change the phase of the interfering components, by θ_b, say, so that (5) becomes $\varphi_{ac}=\sum_b e^{i\theta_b}\varphi_{ab}\varphi_{bc}$. However, as von Neumann shows, the phase shifts must remain unknown if B is measured so that the resulting probability P_{ac} is the square of φ_{ac} averaged over all phases, θ_b. This results in (4).

[8] If $\mathbf{A}$ and $\mathbf{B}$ are the operators corresponding to measurements A and B, and if ψ_a and ψ_b are solutions of $\mathbf{A}\psi_a=a\psi_a$ and $\mathbf{B}\chi_b=b\chi_b$, then $\varphi_{ab}=\int\chi_b{}^*\psi_a dx=(\chi_b{}^*,\psi_a)$. Thus, φ_{ab} is an element $(a|b)$ of the transformation matrix for the transformation from a representation in which $\mathbf{A}$ is diagonal to one in which $\mathbf{B}$ is diagonal.

Probability can show the typical phenomena of interference, usually associated with waves, whose intensity is given by the square of the sum of contributions from different sources. The electron acts as a wave, (5), so to speak, as long as no attempt is made to verify that it is a particle; yet one can determine, if one wishes, by what route it travels just as though it were a particle; but when one does that, (4) applies and it does act like a particle.

These things are, of course, well known. They have already been explained many times.[9] However, it seems worth while to emphasize the fact that they are all simply direct consequences of Eq. (5), for it is essentially Eq. (5) that is fundamental in my formulation of quantum mechanics.

The generalization of Eqs. (4) and (5) to a large number of measurements, say $A, B, C, D, \cdots, K$, is, of course, that the probability of the sequence $a, b, c, d, \cdots, k$ is

$$P_{abcd\cdots k} = |\varphi_{abcd\cdots k}|^2.$$

The probability of the result a, c, k, for example, if $b, d, \cdots$ are measured, is the classical formula:

$$P_{ack} = \sum_b \sum_d \cdots P_{abcd\cdots k}, \qquad (6)$$

while the probability of the same sequence a, c, k if no measurements are made between A and C and between C and K is

$$P_{ack}{}^q = \Big|\sum_b \sum_d \cdots \varphi_{abcd\cdots k}\Big|^2. \qquad (7)$$

The quantity $\varphi_{abcd\cdots k}$ we can call the probability amplitude for the condition $A=a$, $B=b$, $C=c$, $D=d, \cdots, K=k$. (It is, of course, expressible as a product $\varphi_{ab}\varphi_{bc}\varphi_{cd}\cdots\varphi_{jk}$.)

3. THE PROBABILITY AMPLITUDE FOR A SPACE-TIME PATH

The physical ideas of the last section may be readily extended to define a probability amplitude for a particular completely specified space-time path. To explain how this may be done, we shall limit ourselves to a one-dimensional problem, as the generalization to several dimensions is obvious.

Assume that we have a particle which can take up various values of a coordinate x. Imagine that we make an enormous number of successive position measurements, let us say separated by a small time interval ϵ. Then a succession of measurements such as $A, B, C, \cdots$ might be the succession of measurements of the coordinate x at successive times $t_1, t_2, t_3, \cdots$, where $t_{i+1} = t_i + \epsilon$. Let the value, which might result from measurement of the coordinate at time t_i, be x_i. Thus, if A is a measurement of x at t_1 then x_1 is what we previously denoted by a. From a classical point of view, the successive values, $x_1, x_2, x_3, \cdots$ of the coordinate practically define a path $x(t)$. Eventually, we expect to go the limit $\epsilon \rightarrow 0$.

The probability of such a path is a function of $x_1, x_2, \cdots, x_i, \cdots$, say $P(\cdots x_i, x_{i+1}, \cdots)$. The probability that the path lies in a particular region R of space-time is obtained classically by integrating P over that region. Thus, the probability that x_i lies between a_i and b_i, and x_{i+1} lies between a_{i+1} and b_{i+1}, etc., is

$$\cdots \int_{a_i}^{b_i} \int_{a_{i+1}}^{b_{i+1}} \cdots P(\cdots x_i, x_{i+1}, \cdots) \cdots dx_i dx_{i+1} \cdots$$
$$= \int_R P(\cdots x_i, x_{i+1}, \cdots) \cdots dx_i dx_{i+1} \cdots, \qquad (8)$$

the symbol $\int_R$ meaning that the integration is to be taken over those ranges of the variables which lie within the region R. This is simply Eq. (6) with $a, b, \cdots$ replaced by $x_1, x_2, \cdots$ and integration replacing summation.

In quantum mechanics this is the correct formula for the case that $x_1, x_2, \cdots, x_i, \cdots$ were actually all measured, and then only those paths lying within R were taken. We would expect the result to be different if no such detailed measurements had been performed. Suppose a measurement is made which is capable only of determining that the path lies somewhere within R.

The measurement is to be what we might call an "ideal measurement." We suppose that no further details could be obtained from the same measurement without further disturbance to the system. I have not been able to find a precise definition. We are trying to avoid the extra uncertainties that must be averaged over if, for example, more information were measured but

[9] See, for example, W. Heisenberg, *The Physical Principles of the Quantum Theory* (University of Chicago Press, Chicago, 1930), particularly Chapter IV.

not utilized. We wish to use Eq. (5) or (7) for all x_i and have no residual part to sum over in the manner of Eq. (4).

We expect that the probability that the particle is found by our "ideal measurement" to be, indeed, in the region R is the square of a complex number $|\varphi(R)|^2$. The number $\varphi(R)$, which we may call the probability amplitude for region R is given by Eq. (7) with $a, b, \cdots$ replaced by $x_i, x_{i+1}, \cdots$ and summation replaced by integration:

$$\varphi(R) = \lim_{\epsilon\to 0} \int_R \times \Phi(\cdots x_i, x_{i+1}\cdots)\cdots dx_i dx_{i+1}\cdots. \quad (9)$$

The complex number $\Phi(\cdots x_i, x_{i+1}\cdots)$ is a function of the variables x_i defining the path. Actually, we imagine that the time spacing ϵ approaches zero so that Φ essentially depends on the entire path $x(t)$ rather than only on just the values of x_i at the particular times t_i, $x_i = x(t_i)$. We might call Φ the probability amplitude functional of paths $x(t)$.

We may summarize these ideas in our first postulate:

I. If an ideal measurement is performed to determine whether a particle has a path lying in a region of space-time, then the probability that the result will be affirmative is the absolute square of a sum of complex contributions, one from each path in the region.

The statement of the postulate is incomplete. The meaning of a sum of terms one for "each" path is ambiguous. The precise meaning given in Eq. (9) is this: A path is first defined only by the positions x_i through which it goes at a sequence of equally spaced times,[10] $t_i = t_{i-1} + \epsilon$. Then all values of the coordinates within R have an equal weight. The actual magnitude of the weight depends upon ϵ and can be so chosen that the probability of an event which is certain shall be normalized to unity. It may not be best to do so, but we have left this weight factor in a proportionality constant in the second postulate. The limit $\epsilon \to 0$ must be taken at the end of a calculation.

When the system has several degrees of freedom the coordinate space x has several dimensions so that the symbol x will represent a set of coordinates $(x^{(1)}, x^{(2)}, \cdots, x^{(k)})$ for a system with k degrees of freedom. A path is a sequence of configurations for successive times and is described by giving the configuration x_i or $(x_i^{(1)}, x_i^{(2)}, \cdots, x_i^{(k)})$, i.e., the value of each of the k coordinates for each time t_i. The symbol dx_i will be understood to mean the volume element in k dimensional configuration space (at time t_i). The statement of the postulates is independent of the coordinate system which is used.

The postulate is limited to defining the results of position measurements. It does not say what must be done to define the result of a momentum measurement, for example. This is not a real limitation, however, because in principle the measurement of momentum of one particle can be performed in terms of position measurements of other particles, e.g., meter indicators. Thus, an analysis of such an experiment will determine what it is about the first particle which determines its momentum.

4. THE CALCULATION OF THE PROBABILITY AMPLITUDE FOR A PATH

The first postulate prescribes the type of mathematical framework required by quantum mechanics for the calculation of probabilities. The second postulate gives a particular content to this framework by prescribing how to compute the important quantity Φ for each path:

II. The paths contribute equally in magnitude, but the phase of their contribution is the classical action (in units of $\hbar$); i.e., the time integral of the Lagrangian taken along the path.

That is to say, the contribution $\Phi[x(t)]$ from a given path $x(t)$ is proportional to $\exp(i/\hbar)S[x(t)]$, where the action $S[x(t)] = \int L(\dot{x}(t), x(t))dt$ is the time integral of the classical Lagrangian $L(\dot{x}, x)$ taken along the path in question. The Lagrangian, which may be an explicit function of the time, is a function of position and velocity. If we suppose it to be a quadratic function of the

[10] There are very interesting mathematical problems involved in the attempt to avoid the subdivision and limiting processes. Some sort of complex measure is being associated with the space of functions $x(t)$. Finite results can be obtained under unexpected circumstances because the measure is not positive everywhere, but the contributions from most of the paths largely cancel out. These curious mathematical problems are sidestepped by the subdivision process. However, one feels as Cavalieri must have felt calculating the volume of a pyramid before the invention of calculus.

velocities, we can show the mathematical equivalence of the postulates here and the more usual formulation of quantum mechanics.

To interpret the first postulate it was necessary to define a path by giving only the succession of points x_i through which the path passes at successive times t_i. To compute $S=\int L(\dot{x}, x)dt$ we need to know the path at all points, not just at x_i. We shall assume that the function $x(t)$ in the interval between t_i and t_{i+1} is the path followed by a classical particle, with the Lagrangian L, which starting from x_i at t_i reaches x_{i+1} at t_{i+1}. This assumption is required to interpret the second postulate for discontinuous paths. The quantity $\Phi(\cdots x_i, x_{i+1}, \cdots)$ can be normalized (for various ϵ) if desired, so that the probability of an event which is certain is normalized to unity as $\epsilon\rightarrow 0$.

There is no difficulty in carrying out the action integral because of the sudden changes of velocity encountered at the times t_i as long as L does not depend upon any higher time derivatives of the position than the first. Furthermore, unless L is restricted in this way the end points are not sufficient to define the classical path. Since the classical path is the one which makes the action a minimum, we can write

$$S=\sum_i S(x_{i+1}, x_i), \qquad (10)$$

where

$$S(x_{i+1}, x_i)=\text{Min.}\int_{t_i}^{t_{i+1}} L(\dot{x}(t), x(t))dt. \qquad (11)$$

Written in this way, the only appeal to classical mechanics is to supply us with a Lagrangian function. Indeed, one could consider postulate two as simply saying, "Φ is the exponential of i times the integral of a real function of $x(t)$ and its first time derivative." Then the classical equations of motion might be derived later as the limit for large dimensions. The function of x and $\dot{x}$ then could be shown to be the classical Lagrangian within a constant factor.

Actually, the sum in (10), even for finite ϵ, is infinite and hence meaningless (because of the infinite extent of time). This reflects a further incompleteness of the postulates. We shall have to restrict ourselves to a finite, but arbitrarily long, time interval.

Combining the two postulates and using Eq. (10), we find

$$\varphi(R)=\lim_{\epsilon\rightarrow 0}\int_R \times\exp\left[\frac{i}{\hbar}\sum_i S(x_{i+1}, x_i)\right]\cdots\frac{dx_{i+1}}{A}\frac{dx_i}{A}\cdots, \qquad (12)$$

where we have let the normalization factor be split into a factor $1/A$ (whose exact value we shall presently determine) for each instant of time. The integration is just over those values $x_i, x_{i+1}, \cdots$ which lie in the region R. This equation, the definition (11) of $S(x_{i+1}, x_i)$, and the physical interpretation of $|\varphi(R)|^2$ as the probability that the particle will be found in R, complete our formulation of quantum mechanics.

5. DEFINITION OF THE WAVE FUNCTION

We now proceed to show the equivalence of these postulates to the ordinary formulation of quantum mechanics. This we do in two steps. We show in this section how the wave function may be defined from the new point of view. In the next section we shall show that this function satisfies Schroedinger's differential wave equation.

We shall see that it is the possibility, (10), of expressing S as a sum, and hence Φ as a product, of contributions from successive sections of the path, which leads to the possibility of defining a quantity having the properties of a wave function.

To make this clear, let us imagine that we choose a particular time t and divide the region R in Eq. (12) into pieces, future and past relative to t. We imagine that R can be split into: (a) a region R', restricted in any way in space, but lying entirely earlier in time than some t', such that $t'<t$; (b) a region R'' arbitrarily restricted in space but lying entirely later in time than t'', such that $t''>t$; (c) the region between t' and t'' in which all the values of x coordinates are unrestricted, i.e., all of space-time between t' and t''. The region (c) is not absolutely necessary. It can be taken as narrow in time as desired. However, it is convenient in letting us consider varying t a little without having to redefine R' and R''. Then $|\varphi(R', R'')|^2$ is the probability that the path occupies R' and R''. Because R' is entirely previous to R'', considering the time t as the present, we can express this as the probability that the path had been in region R' and will be in region R''. If we divide by a factor, the probability that the path is in R', to renormalize the probability we find: $|\varphi(R', R'')|^2$ is the (relative) probability that if the system were in region R' it will be found later in R''.

This is, of course, the important quantity in predicting the results of many experiments. We prepare the system in a certain way (e.g., it was in region R') and then measure some other property (e.g., will it be found in region R''?). What does (12) say about computing this quantity, or rather the quantity $\varphi(R', R'')$ of which it is the square?

Let us suppose in Eq. (12) that the time t corresponds to one particular point k of the subdivision of time into steps ϵ, i.e., assume $t=t_k$, the index k, of course, depending upon the subdivision ϵ. Then, the exponential being the exponential of a sum may be split into a product of two factors

$$\exp\left[\frac{i}{\hbar}\sum_{i=k}^{\infty} S(x_{i+1}, x_i)\right] \cdot \exp\left[\frac{i}{\hbar}\sum_{i=-\infty}^{k-1} S(x_{i+1}, x_i)\right]. \qquad (13)$$

The first factor contains only coordinates with index k or higher, while the second contains only coordinates with index k or lower. This split is possible because of Eq. (10), which results essentially from the fact that the Lagrangian is a function only of positions and velocities. First, the integration on all variables x_i for $i>k$ can be performed on the first factor resulting in a function of x_k (times the second factor). Next, the integration on all variables x_i for $i<k$ can be performed on the second factor also, giving a function of x_k. Finally, the integration on x_k can be performed. That is, $\varphi(R', R'')$ can be written as the integral over x_k of the product of two factors. We will call these $\chi^*(x_k, t)$ and $\psi(x_k, t)$:

$$\varphi(R', R'') = \int \chi^*(x, t)\psi(x, t)dx, \qquad (14)$$

where

$$\psi(x_k, t) = \lim_{\epsilon\to 0}\int_{R'} \times\exp\left[\frac{i}{\hbar}\sum_{i=-\infty}^{k-1} S(x_{i+1}, x_i)\right]\frac{dx_{k-1}}{A}\frac{dx_{k-2}}{A}\cdots, \qquad (15)$$

and

$$\chi^*(x_k, t) = \lim_{\epsilon\to 0}\int_{R''} \exp\left[\frac{i}{\hbar}\sum_{i=k}^{\infty} S(x_{i+1}, x_i)\right] \cdot\frac{1}{A}\frac{dx_{k+1}}{A}\frac{dx_{k+2}}{A}\cdots. \qquad (16)$$

The symbol R' is placed on the integral for ψ to indicate that the coordinates are integrated over the region R', and, for t_i between t' and t, over all space. In like manner, the integral for χ^* is over R'' and over all space for those coordinates corresponding to times between t and t''. The asterisk on χ^* denotes complex conjugate, as it will be found more convenient to define (16) as the complex conjugate of some quantity, χ.

The quantity ψ depends only upon the region R' previous to t, and is completely defined if that region is known. It does not depend, in any way, upon what will be done to the system after time t. This latter information is contained in χ. Thus, with ψ and χ we have separated the past history from the future experiences of the system. This permits us to speak of the relation of past and future in the conventional manner. Thus, if a particle has been in a region of space-time R' it may at time t be said to be in a certain condition, or state, determined only by its past and described by the so-called wave function $\psi(x, t)$. This function contains all that is needed to predict future probabilities. For, suppose, in another situation, the region R' were different, say r', and possibly the Lagrangian for times before t were also altered. But, nevertheless, suppose the quantity from Eq. (15) turned out to be the same. Then, according to (14) the probability of ending in any region R'' is the same for R' as for r'. Therefore, future measurements will not distinguish whether the system had occupied R' or r'. Thus, the wave function $\psi(x, t)$ is sufficient to define those attributes which are left from past history which determine future behavior.

Likewise, the function $\chi^*(x, t)$ characterizes the experience, or, let us say, experiment to which the system is to be subjected. If a different region, r'' and different Lagrangian after t, were to give the same $\chi^*(x, t)$ *via* Eq. (16), as does region R'', then no matter what the preparation, ψ, Eq. (14) says that the chance of finding the system in R'' is always the same as finding it in r''. The two "experiments" R'' and r'' are equivalent, as they yield the same results. We shall say loosely that these experiments are to determine with what probability the system is in state χ. Actually, this terminology is poor. The system is really in state ψ. The reason we can associate a state with an experiment is, of course, that for an ideal experiment there turns out to be a unique state (whose wave function is $\chi(x, t)$) for which the experiment succeeds with certainty.

Thus, we can say: the probability that a system in state ψ will be found by an experiment whose characteristic state is χ (or, more loosely, the chance that a system in state ψ will appear to be in χ) is

$$\left|\int \chi^*(x, t)\psi(x, t)dx\right|^2. \quad (17)$$

These results agree, of course, with the principles of ordinary quantum mechanics. They are a consequence of the fact that the Lagrangian is a function of position, velocity, and time only.

6. THE WAVE EQUATION

To complete the proof of the equivalence with the ordinary formulation we shall have to show that the wave function defined in the previous section by Eq. (15) actually satisfies the Schroedinger wave equation. Actually, we shall only succeed in doing this when the Lagrangian L in (11) is a quadratic, but perhaps inhomogeneous, form in the velocities $\dot{x}(t)$. This is not a limitation, however, as it includes all the cases for which the Schroedinger equation has been verified by experiment.

The wave equation describes the development of the wave function with time. We may expect to approach it by noting that, for finite ϵ, Eq. (15) permits a simple recursive relation to be developed. Consider the appearance of Eq. (15) if we were to compute ψ at the next instant of time:

$$\psi(x_{k+1}, t+\epsilon) = \int_{R'} \exp\left[\frac{i}{\hbar}\sum_{i=-\infty}^{k} S(x_{i+1}, x_i)\right] \times\frac{dx_k}{A}\frac{dx_{k-1}}{A}\cdots. \quad (15')$$

This is similar to (15) except for the integration over the additional variable x_k and the extra term in the sum in the exponent. This term means that the integral of (15′) is the same as the integral of (15) except for the factor $(1/A)\exp(i/\hbar)S(x_{k+1}, x_k)$. Since this does not contain any of the variables x_i for i less than k, all of the integrations on dx_i up to dx_{k-1} can be performed with this factor left out. However, the result of these integrations is by (15) simply $\psi(x_k, t)$. Hence, we find from (15′) the relation

$$\psi(x_{k+1}, t+\epsilon) = \int \exp\left[\frac{i}{\hbar}S(x_{k+1}, x_k)\right]\psi(x_k, t)dx_k/A. \quad (18)$$

This relation giving the development of ψ with time will be shown, for simple examples, with suitable choice of A, to be equivalent to Schroedinger's equation. Actually, Eq. (18) is not exact, but is only true in the limit $\epsilon\rightarrow 0$ and we shall derive the Schroedinger equation by assuming (18) is valid to first order in ϵ. The Eq. (18) *need* only be true for small ϵ to the first order in ϵ. For if we consider the factors in (15) which carry us over a finite interval of time, T, the number of factors is T/ϵ. If an error of order ϵ^2 is made in each, the resulting error will not accumulate beyond the order $\epsilon^2(T/\epsilon)$ or $T\epsilon$, which vanishes in the limit.

We shall illustrate the relation of (18) to Schroedinger's equation by applying it to the simple case of a particle moving in one dimension in a potential $V(x)$. Before we do this, however, we would like to discuss some approximations to the value $S(x_{i+1}, x_i)$ given in (11) which will be sufficient for expression (18).

The expression defined in (11) for $S(x_{i+1}, x_i)$ is difficult to calculate exactly for arbitrary ϵ from classical mechanics. Actually, it is only necessary that an approximate expression for $S(x_{i+1}, x_i)$ be used in (18), provided the error of the approximation be of an order smaller than the first in ϵ. We limit ourselves to the case that the Lagrangian is a quadratic, but perhaps inhomogeneous, form in the velocities $\dot{x}(t)$. As we shall see later, the paths which are important are those for which $x_{i+1}-x_i$ is of order $\epsilon^{\frac{1}{2}}$. Under these circumstances, it is sufficient to calculate the integral in (11) over the classical path taken by a *free* particle.[11]

In *Cartesian coordinates*[12] the path of a free particle is a straight line so the integral of (11) can be taken along a straight line. Under these circumstances it is sufficiently accurate to replace the integral by the trapezoidal rule

$$S(x_{i+1}, x_i) = \frac{\epsilon}{2} L\left(\frac{x_{i+1}-x_i}{\epsilon}, x_{i+1}\right) + \frac{\epsilon}{2} L\left(\frac{x_{i+1}-x_i}{\epsilon}, x_i\right) \quad (19)$$

or, if it proves more convenient,

$$S(x_{i+1}, x_i) = \epsilon L\left(\frac{x_{i+1}-x_i}{\epsilon}, \frac{x_{i+1}+x_i}{2}\right). \quad (20)$$

These are not valid in a general coordinate system, e.g., spherical. An even simpler approximation may be used if, in addition, there is no vector potential or other terms linear in the velocity (see page 376):

$$S(x_{i+1}, x_i) = \epsilon L\left(\frac{x_{i+1}-x_i}{\epsilon}, x_{i+1}\right). \quad (21)$$

Thus, for the simple example of a particle of mass m moving in one dimension under a potential $V(x)$, we can set

$$S(x_{i+1}, x_i) = \frac{m\epsilon}{2}\left(\frac{x_{i+1}-x_i}{\epsilon}\right)^2 - \epsilon V(x_{i+1}). \quad (22)$$

[11] It is assumed that the "forces" enter through a scalar and vector potential and not in terms involving the square of the velocity. More generally, what is meant by a free particle is one for which the Lagrangian is altered by omission of the terms linear in, and those independent of, the velocities.

[12] More generally, coordinates for which the terms quadratic in the velocity in $L(\dot{x}, x)$ appear with constant coefficients.

For this example, then, Eq. (18) becomes

$$\psi(x_{k+1}, t+\epsilon) = \int \exp\left[\frac{i\epsilon}{\hbar}\left\{\frac{m}{2}\left(\frac{x_{k+1}-x_k}{\epsilon}\right)^2 - V(x_{k+1})\right\}\right]\psi(x_k, t)dx_k/A. \quad (23)$$

Let us call $x_{k+1}=x$ and $x_{k+1}-x_k=\xi$ so that $x_k=x-\xi$. Then (23) becomes

$$\psi(x, t+\epsilon) = \int \exp\frac{im\xi^2}{\epsilon\cdot 2\hbar} \cdot \exp\frac{-i\epsilon V(x)}{\hbar} \cdot \psi(x-\xi, t)\frac{d\xi}{A}. \quad (24)$$

The integral on ξ will converge if $\psi(x, t)$ falls off sufficiently for large x (certainly if $\int\psi^*(x)\psi(x)dx=1$). In the integration on ξ, since ϵ is very small, the exponential of $im\xi^2/2\hbar\epsilon$ oscillates extremely rapidly except in the region about $\xi=0$ (ξ of order $(\hbar\epsilon/m)^{\frac{1}{2}}$). Since the function $\psi(x-\xi, t)$ is a relatively smooth function of ξ (since ϵ may be taken as small as desired), the region where the exponential oscillates rapidly will contribute very little because of the almost complete cancelation of positive and negative contributions. Since only small ξ are effective, $\psi(x-\xi, t)$ may be expanded as a Taylor series. Hence,

$$\psi(x, t+\epsilon) = \exp\left(\frac{-i\epsilon V(x)}{\hbar}\right) \cdot \times \int \exp\left(\frac{im\xi^2}{2\hbar\epsilon}\right)\left[\psi(x, t) - \xi\frac{\partial\psi(x, t)}{\partial x} + \frac{\xi^2}{2}\frac{\partial^2\psi(x, t)}{\partial x^2} - \cdots\right]d\xi/A. \quad (25)$$

Now

$$\int_{-\infty}^{\infty} \exp(im\xi^2/2\hbar\epsilon)d\xi = (2\pi\hbar\epsilon i/m)^{\frac{1}{2}},$$

$$\int_{-\infty}^{\infty} \exp(im\xi^2/2\hbar\epsilon)\xi d\xi = 0, \quad (26)$$

$$\int_{-\infty}^{\infty} \exp(im\xi^2/2\hbar\epsilon)\xi^2 d\xi = (\hbar\epsilon i/m)(2\pi\hbar\epsilon i/m)^{\frac{1}{2}},$$

while the integral containing ξ^3 is zero, for like

the one with ξ it possesses an odd integrand, and the ones with ξ^4 are of at least the order ϵ smaller than the ones kept here.[13] If we expand the left-hand side to first order in ϵ, (25) becomes

$$\psi(x,t)+\epsilon\frac{\partial\psi(x,t)}{\partial t}$$

$$=\exp\left(\frac{-i\epsilon V(x)}{\hbar}\right)\frac{(2\pi\hbar\epsilon i/m)^{\frac{1}{2}}}{A}\times\left[\psi(x,t)+\frac{\hbar\epsilon i}{m}\frac{\partial^2\psi(x,t)}{\partial x^2}+\cdots\right]. \quad (27)$$

In order that both sides may agree to *zero* order in ϵ, we must set

$$A=(2\pi\hbar\epsilon i/m)^{\frac{1}{2}}. \quad (28)$$

Then expanding the exponential containing $V(x)$, we get

$$\psi(x,t)+\epsilon\frac{\partial\psi}{\partial t}=\left(1-\frac{i\epsilon}{\hbar}V(x)\right)\times\left(\psi(x,t)+\frac{\hbar\epsilon i}{2m}\frac{\partial^2\psi}{\partial x^2}\right). \quad (29)$$

Canceling $\psi(x,t)$ from both sides, and comparing terms to first order in ϵ and multiplying by $-\hbar/i$ one obtains

$$-\frac{\hbar}{i}\frac{\partial\psi}{\partial t}=\frac{1}{2m}\left(\frac{\hbar}{i}\frac{\partial}{\partial x}\right)^2\psi+V(x)\psi, \quad (30)$$

which is Schroedinger's equation for the problem in question.

The equation for χ^* can be developed in the same way, but adding a factor *decreases* the time by one step, i.e., χ^* satisfies an equation like (30) but with the sign of the time reversed. By taking complex conjugates we can conclude that χ satisfies the same equation as ψ, i.e., an experiment can be defined by the particular state χ to which it corresponds.[14]

This example shows that most of the contribution to $\psi(x_{k+1},t+\epsilon)$ comes from values of x_k in $\psi(x_k,t)$ which are quite close to x_{k+1} (distant of order $\epsilon^{\frac{1}{2}}$) so that the integral equation (23) can, in the limit, be replaced by a differential equation. The "velocities," $(x_{k+1}-x_k)/\epsilon$ which are important are very high, being of order $(\hbar/m\epsilon)^{\frac{1}{2}}$ which diverges as $\epsilon\to 0$. The paths involved are, therefore, continuous but possess no derivative. They are of a type familiar from study of Brownian motion.

It is these large velocities which make it so necessary to be careful in approximating $S(x_{k+1},x_k)$ from Eq. (11).[15] To replace $V(x_{k+1})$ by $V(x_k)$ would, of course, change the exponent in (18) by $i\epsilon[V(x_k)-V(x_{k+1})]/\hbar$ which is of order $\epsilon(x_{k+1}-x_k)$, and thus lead to unimportant terms of higher order than ϵ on the right-hand side of (29). It is for this reason that (20) and (21) are equally satisfactory approximations to $S(x_{i+1},x_i)$ when there is no vector potential. A term, linear in velocity, however, arising from a vector potential, as $A\dot{x}dt$ must be handled more carefully. Here a term in $S(x_{k+1},x_k)$ such as $A(x_{k+1})\times(x_{k+1}-x_k)$ differs from $A(x_k)(x_{k+1}-x_k)$ by a term of order $(x_{k+1}-x_k)^2$, and, therefore, of order ϵ. Such a term would lead to a change in the resulting wave equation. For this reason the approximation (21) is not a sufficiently accurate approximation to (11) and one like (20), (or (19) from which (20) differs by terms of order higher than ϵ) must be used. If $\mathbf{A}$ represents the vector potential and $\mathbf{p}=(\hbar/i)\nabla$, the momentum operator, then (20) gives, in the Hamiltonian operator, a term $(1/2m)(\mathbf{p}-(e/c)\mathbf{A})\cdot(\mathbf{p}-(e/c)\mathbf{A})$, while (21) gives $(1/2m)(\mathbf{p}\cdot\mathbf{p}-(2e/c)\mathbf{A}\cdot\mathbf{p}+(e^2/c^2)\mathbf{A}\cdot\mathbf{A})$. These two expressions differ by $(\hbar e/2imc)\nabla\cdot\mathbf{A}$

[13] Really, these integrals are oscillatory and not defined, but they may be defined by using a convergence factor. Such a factor is automatically provided by $\psi(x-\xi,t)$ in (24). If a more formal procedure is desired replace $\hbar$ by $\hbar(1-i\delta)$, for example, where δ is a small positive number, and then let $\delta\to 0$.

[14] Dr. Hartland Snyder has pointed out to me, in private conversation, the very interesting possibility that there may be a generalization of quantum mechanics in which the states measured by experiment cannot be prepared; that is, there would be no state into which a system may be put for which a particular experiment gives certainty for a result. The class of functions χ is not identical to the class of available states ψ. This would result if, for example, χ satisfied a different equation than ψ.

[15] Equation (18) is actually exact when (11) is used for $S(x_{i+1},x_i)$ for arbitrary ϵ for cases in which the potential does not involve x to higher powers than the second (e.g., free particle, harmonic oscillator). It is necessary, however, to use a more accurate value of A. One can define A in this way. Assume classical particles with k degrees of freedom start from the point x_i, t_i with uniform density in momentum space. Write the number of particles having a given component of momentum in range dp as dp/p_0 with p_0 constant. Then $A=(2\pi\hbar i/p_0)^{k/2}\rho^{-\frac{1}{2}}$, where ρ is the density in k dimensional coordinate space x_{i+1} of these particles at time t_{i+1}.

which may not be zero. The question is still more important in the coefficient of terms which are quadratic in the velocities. In these terms (19) and (20) are not sufficiently accurate representations of (11) in general. It is when the coefficients are constant that (19) or (20) can be substituted for (11). If an expression such as (19) is used, say for spherical coordinates, when it is not a valid approximation to (11), one obtains a Schroedinger equation in which the Hamiltonian operator has some of the momentum operators and coordinates in the wrong order. Equation (11) then resolves the ambiguity in the usual rule to replace p and q by the non-commuting quantities $(\hbar/i)(\partial/\partial q)$ and q in the classical Hamiltonian $H(p, q)$.

It is clear that the statement (11) is independent of the coordinate system. Therefore, to find the differential wave equation it gives in any coordinate system, the easiest procedure is first to find the equations in Cartesian coordinates and then to transform the coordinate system to the one desired. It suffices, therefore, to show the relation of the postulates and Schroedinger's equation in rectangular coordinates.

The derivation given here for one dimension can be extended directly to the case of three-dimensional Cartesian coordinates for any number, K, of particles interacting through potentials with one another, and in a magnetic field, described by a vector potential. The terms in the vector potential require completing the square in the exponent in the usual way for Gaussian integrals. The variable x must be replaced by the set $x^{(1)}$ to $x^{(3K)}$ where $x^{(1)}$, $x^{(2)}$, $x^{(3)}$ are the coordinates of the first particle of mass m_1, $x^{(4)}$, $x^{(5)}$, $x^{(6)}$ of the second of mass m_2, etc. The symbol dx is replaced by $dx^{(1)}dx^{(2)}\cdots dx^{(3K)}$, and the integration over dx is replaced by a $3K$-fold integral. The constant A has, in this case, the value $A=(2\pi\hbar\epsilon i/m_1)^{\frac{3}{2}}(2\pi\hbar\epsilon i/m_2)^{\frac{3}{2}}\cdots(2\pi\hbar\epsilon i/m_K)^{\frac{3}{2}}$. The Lagrangian is the classical Lagrangian for the same problem, and the Schroedinger equation resulting will be that which corresponds to the classical Hamiltonian, derived from this Lagrangian. The equations in any other coordinate system may be obtained by transformation. Since this includes all cases for which Schroedinger's equation has been checked with experiment, we may say our postulates are able to describe what can be described by non-relativistic quantum mechanics, neglecting spin.

7. DISCUSSION OF THE WAVE EQUATION

The Classical Limit

This completes the demonstration of the equivalence of the new and old formulations. We should like to include in this section a few remarks about the important equation (18).

This equation gives the development of the wave function during a small time interval. It is easily interpreted physically as the expression of Huygens' principle for matter waves.• In geometrical optics the rays in an inhomogeneous medium satisfy Fermat's principle of least *time*. We may state Huygens' principle in wave optics in this way: If the amplitude of the wave is known on a given surface, the amplitude at a near by point can be considered as a sum of contributions from all points of the surface. Each contribution is delayed in phase by an amount proportional to the *time* it would take the light to get from the surface to the point along the ray of least *time* of geometrical optics. We can consider (22) in an analogous manner starting with Hamilton's first principle of least *action* for classical or "geometrical" mechanics. If the amplitude of the wave ψ is known on a given "surface," in particular the "surface" consisting of all x at time t, its value at a particular nearby point at time $t+\epsilon$, is a sum of contributions from all points of the surface at t. Each contribution is delayed in phase by an amount proportional to the *action* it would require to get from the surface to the point along the path of least *action* of classical mechanics.[16]

Actually Huygens' principle is not correct in optics. It is replaced by Kirchoff's modification which requires that both the amplitude and its derivative must be known on the adjacent surface. This is a consequence of the fact that the wave equation in optics is second order in the time. The wave equation of quantum mechanics is first order in the time; therefore, Huygens' principle *is* correct for matter waves, action replacing time.

[16] See in this connection the very interesting remarks of Schroedinger, Ann. d. Physik **79**, 489 (1926).

The equation can also be compared mathematically to quantities appearing in the usual formulations. In Schroedinger's method the development of the wave function with time is given by

$$-\frac{\hbar}{i}\frac{\partial\psi}{\partial t}=\mathbf{H}\psi, \tag{31}$$

which has the solution (for any ϵ if $\mathbf{H}$ is time independent)

$$\psi(x, t+\epsilon)=\exp(-i\epsilon\mathbf{H}/\hbar)\psi(x, t). \tag{32}$$

Therefore, Eq. (18) expresses the operator $\exp(-i\epsilon\mathbf{H}/\hbar)$ by an approximate integral operator for small ϵ.

From the point of view of Heisenberg one considers the position at time t, for example, as an operator $\mathbf{x}$. The position $\mathbf{x}'$ at a later time $t+\epsilon$ can be expressed in terms of that at time t by the operator equation

$$\mathbf{x}'=\exp(i\epsilon\mathbf{H}/\hbar)\mathbf{x}\exp-(i\epsilon\mathbf{H}/\hbar). \tag{33}$$

The transformation theory of Dirac allows us to consider the wave function at time $t+\epsilon$, $\psi(x', t+\epsilon)$, as representing a state in a representation in which $\mathbf{x}'$ is diagonal, while $\psi(x, t)$ represents the same state in a representation in which $\mathbf{x}$ is diagonal. They are, therefore, related through the transformation function $(x'|x)_\epsilon$ which relates these representations:

$$\psi(x', t+\epsilon)=\int (x'|x)_\epsilon\psi(x,t)\,dx.$$

Therefore, the content of Eq. (18) is to show that for small ϵ we can set

$$(x'|x)_\epsilon=(1/A)\exp(iS(x', x)/\hbar) \tag{34}$$

with $S(x', x)$ defined as in (11).

The close analogy between $(x'|x)_\epsilon$ and the quantity $\exp(iS(x', x)/\hbar)$ has been pointed out on several occasions by Dirac.[1] In fact, we now see that to sufficient approximations the two quantities may be taken to be proportional to each other. Dirac's remarks were the starting point of the present development. The points he makes concerning the passage to the classical limit $\hbar\rightarrow 0$ are very beautiful, and I may perhaps be excused for briefly reviewing them here.

First we note that the wave function at x'' at time t'' can be obtained from that at x' at time t' by

$$\psi(x'', t'')=\lim_{\epsilon\rightarrow 0}\int\cdots\int \times\exp\left[\frac{i}{\hbar}\sum_{i=0}^{j-1}S(x_{i+1}, x_i)\right] \times\psi(x', t')\frac{dx_0}{A}\frac{dx_1}{A}\cdots\frac{dx_{j-1}}{A}, \tag{35}$$

where we put $x_0\equiv x'$ and $x_j\equiv x''$ where $j\epsilon=t''-t'$ (between the times t' and t'' we assume no restriction is being put on the region of integration). This can be seen either by repeated applications of (18) or directly from Eq. (15). Now we ask, as $\hbar\rightarrow 0$ what values of the intermediate coordinates x_i contribute most strongly to the integral? These will be the values most likely to be found by experiment and therefore will determine, in the limit, the classical path. If $\hbar$ is very small, the exponent will be a very rapidly varying function of any of its variables x_i. As x_i varies, the positive and negative contributions of the exponent nearly cancel. The region at which x_i contributes most strongly is that at which the phase of the exponent varies least rapidly with x_i (method of stationary phase). Call the sum in the exponent S;

$$S=\sum_{i=0}^{j-1}S(x_{i+1}, x_i). \tag{36}$$

Then the classical orbit passes, approximately, through those points x_i at which the rate of change of S with x_i is small, or in the limit of small $\hbar$, zero, i.e., the classical orbit passes through the points at which $\partial S/\partial x_i=0$ for all x_i. Taking the limit $\epsilon\rightarrow 0$, (36) becomes in view of (11)

$$S=\int_{t'}^{t''}L(\dot{x}(t), x(t))dt. \tag{37}$$

We see then that the classical path is that for which the integral (37) suffers no first-order change on varying the path. This is Hamilton's principle and leads directly to the Lagrangian equations of motion.

8. OPERATOR ALGEBRA

Matrix Elements

Given the wave function and Schroedinger's equation, of course all of the machinery of operator or matrix algebra can be developed. It is, however, rather interesting to express these concepts in a somewhat different language more closely related to that used in stating the postulates. Little will be gained by this in elucidating operator algebra. In fact, the results are simply a translation of simple operator equations into a somewhat more cumbersome notation. On the other hand, the new notation and point of view are very useful in certain applications described in the introduction. Furthermore, the form of the equations permits natural extension to a wider class of operators than is usually considered (e.g., ones involving quantities referring to two or more different times). If any generalization to a wider class of action functionals is possible, the formulae to be developed will play an important role.

We discuss these points in the next three sections. This section is concerned mainly with definitions. We shall define a quantity which we call a transition element between two states. It is essentially a matrix element. But instead of being the matrix element between a state ψ and another χ corresponding to the *same* time, these two states will refer to different times. In the following section a fundamental relation between transition elements will be developed from which the usual commutation rules between coordinate and momentum may be deduced. The same relation also yields Newton's equation of motion in matrix form. Finally, in Section 10 we discuss the relation of the Hamiltonian to the operation of displacement in time.

We begin by defining a transition element in terms of the probability of transition from one state to another. More precisely, suppose we have a situation similar to that described in deriving (17). The region R consists of a region R' previous to t', all space between t' and t'' and the region R'' after t''. We shall study the probability that a system in region R' is later found in region R''. This is given by (17). We shall discuss in this section how it changes with changes in the form of the Lagrangian between t' and t''. In Section 10 we discuss how it changes with changes in the preparation R' or the experiment R''.

The state at time t' is defined completely by the preparation R'. It can be specified by a wave function $\psi(x', t')$ obtained as in (15), but containing only integrals up to the time t'. Likewise, the state characteristic of the experiment (region R'') can be defined by a function $\chi(x'', t'')$ obtained from (16) with integrals only beyond t''. The wave function $\psi(x'', t'')$ at time t'' can, of course, also be gotten by appropriate use of (15). It can also be gotten from $\psi(x', t')$ by (35). According to (17) with t'' used instead of t, the probability of being found in χ if prepared in ψ is the square of what we shall call the transition amplitude $\int \chi^*(x'', t'')\psi(x'', t'')dx''$. We wish to express this in terms of χ at t'' and ψ at t'. This we can do with the aid of (35). Thus, the chance that a system prepared in state $\psi_{t'}$ at time t' will be found after t'' to be in a state $\chi_{t''}$ is the square of the transition amplitude

$$\langle\chi_{t''}|1|\psi_{t'}\rangle_S = \lim_{\epsilon\to 0}\int\cdots\int \chi^*(x'', t'') \times\exp(iS/\hbar)\psi(x', t')\frac{dx_0}{A}\cdots\frac{dx_{j-1}}{A}dx_j, \quad (38)$$

where we have used the abbreviation (36).

In the language of ordinary quantum mechanics if the Hamiltonian, $\mathbf{H}$, is constant, $\psi(x, t'') = \exp[-i(t''-t')\mathbf{H}/\hbar]\psi(x, t')$ so that (38) is the matrix element of $\exp[-i(t''-t')\mathbf{H}/\hbar]$ between states $\chi_{t''}$ and $\psi_{t'}$.

If F is any function of the coordinates x_i for $t' < t_i < t''$, we shall define the transition element of F between the states ψ at t' and χ at t'' for the action S as $(x'' \equiv x_j, x' \equiv x_0)$:

$$\langle\chi_{t''}|F|\psi_{t'}\rangle_S = \lim_{\epsilon\to 0}\int\cdots\int \times\chi^*(x'', t'')F(x_0, x_1, \cdots x_j) \cdot\exp\left[\frac{i}{\hbar}\sum_{i=0}^{j-1}S(x_{i+1}, x_i)\right]\psi(x', t')\frac{dx_0}{A}\cdots\frac{dx_{j-1}}{A}dx_j. \quad (39)$$

In the limit $\epsilon\to 0$, F is a functional of the path $x(t)$.

We shall see presently why such quantities are important. It will be easier to understand if we stop for a moment to find out what the quantities correspond to in conventional notation. Suppose F is simply x_k where k corresponds to some time $t=t_k$. Then on the right-hand side of (39) the integrals from x_0 to x_{k-1} may be performed to produce $\psi(x_k, t)$ or $\exp[-i(t-t')\mathbf{H}/\hbar]\psi_{t'}$. In like manner the integrals on x_i for $j \geqslant i > k$ give $\chi^*(x_k, t)$ or $\{\exp[-i(t''-t)\mathbf{H}/\hbar]\chi_{t''}\}^*$. Thus, the transition element of x_k,

$$\langle \chi_{t''} | F | \psi_{t'} \rangle_S$$

$$= \int \chi_{t''}{}^* e^{-(i/\hbar)\mathbf{H}(t''-t)} x e^{-(i/\hbar)\mathbf{H}(t-t')} \psi_{t'} dx$$

$$= \int \chi^*(x, t) x \psi(x, t) dx \quad (40)$$

is the matrix element of $\mathbf{x}$ at time $t=t_k$ between the state which would develop at time t from $\psi_{t'}$ at t' and the state which will develop from time t to $\chi_{t''}$ at t''. It is, therefore, the matrix element of $\mathbf{x}(t)$ between these states.

Likewise, according to (39) with $F=x_{k+1}$, the transition element of x_{k+1} is the matrix element of $\mathbf{x}(t+\epsilon)$. The transition element of $F=(x_{k+1}-x_k)/\epsilon$ is the matrix element of $(\mathbf{x}(t+\epsilon)-\mathbf{x}(t))/\epsilon$ or of $i(\mathbf{Hx}-\mathbf{xH})/\hbar$, as is easily shown from (40). We can call this the matrix element of velocity $\dot{x}(t)$.

Suppose we consider a second problem which differs from the first because, for example, the potential is augmented by a small amount $U(, \mathbf{x}t)$. Then in the new problem the quantity replacing S is $S'=S+\sum_i \epsilon U(x_i, t_i)$. Substitution into (38) leads directly to

$$\langle \chi_{t''} | 1 | \psi_{t'} \rangle_{S'}$$

$$= \left\langle \chi_{t''} \left| \exp \frac{i\epsilon}{\hbar} \sum_{i=1}^{j} U(x_i, t_i) \right| \psi_{t'} \right\rangle_S. \quad (41)$$

Thus, transition elements such as (39) are important insofar as F may arise in some way from a change δS in an action expression. We denote, by observable functionals, those functionals F which can be defined, (possibly indirectly) in terms of the changes which are produced by possible changes in the action S. The condition that a functional be observable is somewhat similar to the condition that an operator be Hermitian. The observable functionals are a restricted class because the action must remain a quadratic function of velocities. From one observable functional others may be derived, for example, by

$$\langle \chi_{t''} | F | \psi_{t'} \rangle_{S'}$$

$$= \left\langle \chi_{t''} \left| F \exp \frac{i\epsilon}{\hbar} \sum_{i=1}^{j} U(x_i, t_i) \right| \psi_{t'} \right\rangle_S \quad (42)$$

which is obtained from (39).

Incidentally, (41) leads directly to an important perturbation formula. If the effect of U is small the exponential can be expanded to first order in U and we find

$$\langle \chi_{t''} | 1 | \psi_{t'} \rangle_{S'} = \langle \chi_{t''} | 1 | \psi_{t'} \rangle_S + \frac{i}{\hbar} \langle \chi_{t''} | \sum_i \epsilon U(x_i, t_i) | \psi_{t'} \rangle. \quad (43)$$

Of particular importance is the case that $\chi_{t''}$ is a state in which $\psi_{t'}$ would not be found at all were it not for the disturbance, U (i.e., $\langle \chi_{t''} | 1 | \psi_{t'} \rangle_S = 0$). Then

$$\frac{1}{\hbar^2} |\langle \chi_{t''} | \sum_i \epsilon U(x_i, t_i) | \psi_{t'} \rangle_S|^2 \quad (44)$$

is the probability of transition as induced to first order by the perturbation. In ordinary notation,

$$\langle \chi_{t''} | \sum_i \epsilon U(x_i, t_i) | \psi_{t'} \rangle_S$$

$$= \int \left\{ \int \chi_{t''}{}^* e^{-(i/\hbar)\mathbf{H}(t''-t)} \mathbf{U} e^{-(i/\hbar)\mathbf{H}(t-t')} \psi_{t'} dx \right\} dt$$

so that (44) reduces to the usual expression[17] for time dependent perturbations.

9. NEWTON'S EQUATIONS

The Commutation Relation

In this section we find that different functionals may give identical results when taken between any two states. This equivalence between functionals is the statement of operator equations in the new language.

If F depends on the various coordinates, we can, of course, define a new functional $\partial F/\partial x_k$

[17] P. A. M. Dirac, *The Principles of Quantum Mechanics* (The Clarendon Press, Oxford, 1935), second edition, Section 47, Eq. (20).

by differentiating it with respect to one of its variables, say $x_k (0<k<j)$. If we calculate $\langle\chi_{t''}|\partial F/\partial x_k|\psi_{t'}\rangle_S$ by (39) the integral on the right-hand side will contain $\partial F/\partial x_k$. The only other place that the variable x_k appears is in S. Thus, the integration on x_k can be performed by parts. The integrated part vanishes (assuming wave functions vanish at infinity) and we are left with the quantity $-F(\partial/\partial x_k)\exp(iS/\hbar)$ in the integral. However, $(\partial/\partial x_k)\exp(iS/\hbar) = (i/\hbar)(\partial S/\partial x_k)\exp(iS/\hbar)$, so the right side represents the transition element of $-(i/\hbar)F(\partial S/\partial x_k)$, i.e.,

$$\left\langle\chi_{t''}\left|\frac{\partial F}{\partial x_k}\right|\psi_{t'}\right\rangle_S = -\frac{i}{\hbar}\left\langle\chi_{t''}\left|F\frac{\partial S}{\partial x_k}\right|\psi_{t'}\right\rangle_S. \quad (45)$$

This very important relation shows that two different functionals may give the same result for the transition element between any two states. We say they are equivalent and symbolize the relation by

$$\frac{\hbar}{i}\frac{\partial F}{\partial x_k} \underset{S}{\leftrightarrow} F\frac{\partial S}{\partial x_k}, \quad (46)$$

the symbol $\underset{S}{\leftrightarrow}$ emphasizing the fact that functionals equivalent under one action may not be equivalent under another. The quantities in (46) need not be observable. The equivalence is, nevertheless, true. Making use of (36) one can write

$$\frac{\hbar}{i}\frac{\partial F}{\partial x_k} \underset{S}{\leftrightarrow} F\left[\frac{\partial S(x_{k+1}, x_k)}{\partial x_k} + \frac{\partial S(x_k, x_{k-1})}{\partial x_k}\right]. \quad (47)$$

This equation is true to zero and first order in ϵ and has as consequences the commutation relations of momentum and coordinate, as well as the Newtonian equations of motion in matrix form.

In the case of our simple one-dimensional problem, $S(x_{i+1}, x_i)$ is given by the expression (15), so that

$$\partial S(x_{k+1}, x_k)/\partial x_k = -m(x_{k+1} - x_k)/\epsilon,$$

and

$$\partial S(x_k, x_{k-1})/\partial x_k = +m(x_k - x_{k-1})/\epsilon - \epsilon V'(x_k);$$

where we write $V'(x)$ for the derivative of the potential, or force. Then (47) becomes

$$\frac{\hbar}{i}\frac{\partial F}{\partial x_k} \underset{S}{\leftrightarrow} F\left[-m\left(\frac{x_{k+1}-x_k}{\epsilon} - \frac{x_k - x_{k-1}}{\epsilon}\right) - \epsilon V'(x_k)\right]. \quad (48)$$

If F does not depend on the variable x_k, this gives Newton's equations of motion. For example, if F is constant, say unity, (48) just gives (dividing by ϵ)

$$0 \underset{S}{\leftrightarrow} -\frac{m}{\epsilon}\left(\frac{x_{k+1}-x_k}{\epsilon} - \frac{x_k - x_{k-1}}{\epsilon}\right) - V'(x_k).$$

Thus, the transition element of mass times acceleration $[(x_{k+1}-x_k)/\epsilon - (x_k - x_{k-1})/\epsilon]/\epsilon$ between any two states is equal to the transition element of force $-V'(x_k)$ between the same states. This is the matrix expression of Newton's law which holds in quantum mechanics.

What happens if F does depend upon x_k? For example, let $F = x_k$. Then (48) gives, since $\partial F/\partial x_k = 1$,

$$\frac{\hbar}{i} \underset{S}{\leftrightarrow} x_k\left[-m\left(\frac{x_{k+1}-x_k}{\epsilon} - \frac{x_k - x_{k-1}}{\epsilon}\right) - \epsilon V'(x_k)\right]$$

or, neglecting terms of order ϵ,

$$m\left(\frac{x_{k+1}-x_k}{\epsilon}\right)x_k - m\left(\frac{x_k - x_{k-1}}{\epsilon}\right)x_k \underset{S}{\leftrightarrow} \frac{\hbar}{i}. \quad (49)$$

In order to transfer an equation such as (49) into conventional notation, we shall have to discover what matrix corresponds to a quantity such as $x_k x_{k+1}$. It is clear from a study of (39) that if F is set equal to, say, $f(x_k)g(x_{k+1})$, the corresponding operator in (40) is

$$e^{-(i/\hbar)(t''-t-\epsilon)\mathbf{H}}g(\mathbf{x})e^{-(i/\hbar)\epsilon\mathbf{H}}f(\mathbf{x})e^{-(i/\hbar)(t-t')\mathbf{H}},$$

the matrix element being taken between the states $\chi_{t''}$ and $\psi_{t'}$. The operators corresponding to functions of x_{k+1} will appear to the left of the operators corresponding to functions of x_k, i.e., *the order of terms in a matrix operator product corresponds to an order in time of the corresponding factors in a functional.* Thus, if the functional can and is written in such a way that in each term factors corresponding to later times appear to the left of factors corresponding to earlier terms, the corresponding operator can immediately be written down if the order of the operators is kept the same as in the functional.[18] Obviously, the order of factors in a functional is of no consequence. The ordering just facilitates translation into conventional operator notation. To write Eq. (49) in the way desired for easy translation would require the factors in the second term on the left to be reversed in order. We see, therefore, that it corresponds to

$$\mathbf{px}-\mathbf{xp}=\hbar/i$$

where we have written $\mathbf{p}$ for the operator $m\dot{\mathbf{x}}$.

The relation between functionals and the corresponding operators is defined above in terms of the order of the factors in time. It should be remarked that this rule must be especially carefully adhered to when quantities involving velocities or higher derivatives are involved. The correct functional to represent the operator $(\dot{x})^2$ is actually $(x_{k+1}-x_k)/\epsilon\cdot(x_k-x_{k-1})/\epsilon$ rather than $[(x_{k+1}-x_k)/\epsilon]^2$. The latter quantity diverges as $1/\epsilon$ as $\epsilon\rightarrow 0$. This may be seen by replacing the second term in (49) by its value $x_{k+1}\cdot m(x_{k+1}-x_k)/\epsilon$ calculated an instant ϵ later in time. This does not change the equation to zero order in ϵ. We then obtain (dividing by ϵ)

$$\left(\frac{x_{k+1}-x_k}{\epsilon}\right)^2 \underset{S}{\leftrightarrow} -\frac{\hbar}{im\epsilon}. \qquad (50)$$

This gives the result expressed earlier that the root mean square of the "velocity" $(x_{k+1}-x_k)/\epsilon$ between two successive positions of the path is of order $\epsilon^{-\frac{1}{2}}$.

It will not do then to write the functional for kinetic energy, say, simply as

$$\tfrac{1}{2}m[(x_{k+1}-x_k)/\epsilon]^2 \qquad (51)$$

for this quantity is infinite as $\epsilon\rightarrow 0$. In fact, it is not an observable functional.

One can obtain the kinetic energy as an observable functional by considering the first-order change in transition amplitude occasioned by a change in the mass of the particle. Let m be changed to $m(1+\delta)$ for a short time, say ϵ, around t_k. The change in the action is $\frac{1}{2}\delta\epsilon m[(x_{k+1}-x_k)/\epsilon]^2$ the derivative of which gives an expression like (51). But the change in m changes the normalization constant $1/A$ corresponding to dx_k as well as the action. The constant is changed from $(2\pi\hbar\epsilon i/m)^{-\frac{1}{2}}$ to $(2\pi\hbar\epsilon i/m(1+\delta))^{-\frac{1}{2}}$ or by $\frac{1}{2}\delta(2\pi\hbar\epsilon i/m)^{-\frac{1}{2}}$ to first order in δ. The total effect of the change in mass in Eq. (38) to the first order in δ is

$$\langle\chi_{t''}|\tfrac{1}{2}\delta\epsilon im[(x_{k+1}-x_k)/\epsilon]^2/\hbar+\tfrac{1}{2}\delta|\psi_{t'}\rangle.$$

We expect the change of order δ lasting for a time ϵ to be of order $\delta\epsilon$. Hence, dividing by $\delta\epsilon i/\hbar$, we can define the kinetic energy functional as

$$\text{K.E.}=\tfrac{1}{2}m[(x_{k+1}-x_k)/\epsilon]^2+\hbar/2\epsilon i. \qquad (52)$$

This is finite as $\epsilon\rightarrow 0$ in view of (50). By making use of an equation which results from substituting $m(x_{k+1}-x_k)/\epsilon$ for F in (48) we can also show that the expression (52) is equal (to order ϵ) to

$$\text{K.E.}=\tfrac{1}{2}m\left(\frac{x_{k+1}-x_k}{\epsilon}\right)\left(\frac{x_k-x_{k-1}}{\epsilon}\right). \qquad (53)$$

That is, the easiest way to produce observable functionals involving powers of the velocities is to replace these powers by a product of velocities, each factor of which is taken at a slightly different time.

10. THE HAMILTONIAN

Momentum

The Hamiltonian operator is of central importance in the usual formulation of quantum mechanics. We shall study in this section the functional corresponding to this operator. We could immediately define the Hamiltonian functional by adding the kinetic energy functional (52) or (53) to the potential energy. This method is artificial and does not exhibit the important relationship of the Hamiltonian to time. We shall define the Hamiltonian functional by the changes made in a state when it is displaced in time.

To do this we shall have to digress a moment to point out that the subdivision of time into *equal* intervals is not necessary. Clearly, any subdivision into instants t_i will be satisfactory; the limits are to be taken as the largest spacing, $t_{i+1}-t_i$, approaches zero. The total action S must

[18] Dirac has also studied operators containing quantities referring to different times. See reference 2.

now be represented as a sum

$$S=\sum_i S(x_{i+1},\, t_{i+1};\, x_i,\, t_i), \tag{54}$$

where

$$S(x_{i+1},\, t_{i+1};\, x_i,\, t_i)=\int_{t_i}^{t_{i+1}} L(\dot{x}(t),\, x(t))dt, \tag{55}$$

the integral being taken along the classical path between x_i at t_i and x_{i+1} at t_{i+1}. For the simple one-dimensional example this becomes, with sufficient accuracy,

$$S(x_{i+1},\, t_{i+1};\, x_i,\, t_i) = \left\{\frac{m}{2}\left(\frac{x_{i+1}-x_i}{t_{i+1}-t_i}\right)^2 - V(x_{i+1})\right\}(t_{i+1}-t_i); \tag{56}$$

the corresponding normalization constant for integration on dx_i is $A=(2\pi\hbar i(t_{i+1}-t_i)/m)^{-\frac{1}{2}}$.

The relation of H to the change in a state with displacement in time can now be studied. Consider a state $\psi(t)$ defined by a space-time region R'. Now imagine that we consider another state at time t, $\psi_\delta(t)$, defined by another region R_δ'. Suppose the region R_δ' is exactly the same as R' except that it is earlier by a time δ, i.e., displaced bodily toward the past by a time δ. All the apparatus to prepare the system for R_δ' is identical to that for R' but is operated a time δ sooner. If L depends explicitly on time, it, too, is to be displaced, i.e., the state ψ_δ is obtained from the L used for state ψ except that the time t in L_δ is replaced by $t+\delta$. We ask how does the state ψ_δ differ from ψ? In any measurement the chance of finding the system in a fixed region R'' is different for R' and R_δ'. Consider the change in the transition element $\langle\chi|1|\psi_\delta\rangle_{S_\delta}$ produced by the shift δ. We can consider this shift as effected by decreasing all values of t_i by δ for $i\leqslant k$ and leaving all t_i fixed for $i>k$, where the time t lies in the interval between t_{k+1} and t_k.[19] This change will have no effect on $S(x_{i+1},\, t_{i+1};\, x_i,\, t_i)$ as defined by (55) as long as both t_{i+1} and t_i are changed by the same amount. On the other hand, $S(x_{k+1},\, t_{k+1};\, x_k,\, t_k)$ is changed to $S(x_{k+1},\, t_{k+1};\, x_k,\, t_k-\delta)$. The constant $1/A$ for the integration on dx_k is also altered to $(2\pi\hbar i(t_{k+1}-t_k+\delta)/m)^{-\frac{1}{2}}$. The effect of these changes on the transition element is given to the first order in δ by

$$\langle\chi|1|\psi\rangle_S-\langle\chi|1|\psi_\delta\rangle_{S_\delta}=\frac{i\delta}{\hbar}\langle\chi|H_k|\psi\rangle_S, \tag{57}$$

here the Hamiltonian functional H_k is defined by

$$H_k=\frac{\partial S(x_{k+1},\, t_{k+1};\, x_k,\, t_k)}{\partial t_k}+\frac{\hbar}{2i(t_{k+1}-t_k)}. \tag{58}$$

The last term is due to the change in $1/A$ and serves to keep H_k finite as $\epsilon\rightarrow 0$. For example, for the expression (56) this becomes

$$H_k=\frac{m}{2}\left(\frac{x_{k+1}-x_k}{t_{k+1}-t_k}\right)^2+\frac{\hbar}{2i(t_{k+1}-t_k)}+V(x_{k+1}),$$

which is just the sum of the kinetic energy functional (52) and that of the potential energy $V(x_{k+1})$.

The wave function $\psi_\delta(x,t)$ represents, of course, the same state as $\psi(x,t)$ will be after time δ, i.e., $\psi(x,t+\delta)$. Hence, (57) is intimately related to the operator equation (31).

One could also consider changes occasioned by a time shift in the final state χ. Of course, nothing new results in this way for it is only the relative shift of χ and ψ which counts. One obtains an alternative expression

$$H_k=-\frac{\partial S(x_{k+1},\, t_{k+1};\, x_k,\, t_k)}{\partial t_{k+1}}+\frac{\hbar}{2i(t_{k+1}-t_k)}. \tag{59}$$

This differs from (58) only by terms of order ϵ.

The time rate of change of a functional can be computed by considering the effect of shifting both initial and final state together. This has the same effect as calculating the transition element of the functional referring to a later time. What results is the analog of the operator equation

$$\frac{\hbar}{i}\dot{\mathbf{f}}=\mathbf{Hf}-\mathbf{fH}.$$

The momentum functional p_k can be defined in an analagous way by considering the changes

[19] From the point of view of mathematical rigor, if δ is finite, as $\epsilon\rightarrow 0$ one gets into difficulty in that, for example, the interval $t_{k+1}-t_k$ is kept finite. This can be straightened out by assuming δ to vary with time and to be turned on smoothly before $t=t_k$ and turned off smoothly after $t=t_k$. Then keeping the time variation of δ fixed, let $\epsilon\rightarrow 0$. Then seek the first-order change as $\delta\rightarrow 0$. The result is essentially the same as that of the crude procedure used above.

made by displacements of position:

$$\langle\chi|1|\psi\rangle_S-\langle\chi|1|\psi_\Delta\rangle_{S_\Delta}=\frac{i\Delta}{\hbar}\langle\chi|p_k|\psi\rangle_S.$$

The state ψ_Δ is prepared from a region R_Δ' which is identical to region R' except that it is moved a distance Δ in space. (The Lagrangian, if it depends explicitly on x, must be altered to $L_\Delta=L(\dot{x}, x-\Delta)$ for times previous to t.) One finds[20]

$$p_k=\frac{\partial S(x_{k+1}, x_k)}{\partial x_{k+1}}=-\frac{\partial S(x_{k+1}, x_k)}{\partial x_k}. \qquad (60)$$

Since $\psi_\Delta(x, t)$ is equal to $\psi(x-\Delta, t)$, the close connection between p_k and the x-derivative of the wave function is established.

Angular momentum operators are related in an analogous way to rotations.

The derivative with respect to t_{i+1} of $S(x_{i+1}, t_{i+1}; x_i, t_i)$ appears in the definition of H_i. The derivative with respect to x_{i+1} defines p_i. But the derivative with respect to t_{i+1} of $S(x_{i+1}, t_{i+1}; x_i, t_i)$ is related to the derivative with respect to x_{i+1}, for the function $S(x_{i+1}, t_{i+1}; x_i, t_i)$ defined by (55) satisfies the Hamilton-Jacobi equation. Thus, the Hamilton-Jacobi equation is an equation expressing H_i in terms of the p_i. In other words, it expresses the fact that time displacements of states are related to space displacements of the same states. This idea leads directly to a derivation of the Schroedinger equation which is far more elegant than the one exhibited in deriving Eq. (30).

11. INADEQUACIES OF THE FORMULATION

The formulation given here suffers from a serious drawback. The mathematical concepts needed are new. At present, it requires an unnatural and cumbersome subdivision of the time interval to make the meaning of the equations clear. Considerable improvement can be made through the use of the notation and concepts of the mathematics of functionals. However, it was thought best to avoid this in a first presentation. One needs, in addition, an appropriate measure for the space of the argument functions $x(t)$ of the functionals.[10]

It is also incomplete from the physical standpoint. One of the most important characteristics of quantum mechanics is its invariance under unitary transformations. These correspond to the canonical transformations of classical mechanics. Of course, the present formulation, being equivalent to ordinary formulations, can be mathematically demonstrated to be invariant under these transformations. However, it has not been formulated in such a way that it is *physically* obvious that it is invariant. This incompleteness shows itself in a definite way. No direct procedure has been outlined to describe measurements of quantities other than position. Measurements of momentum, for example, of one particle, can be defined in terms of measurements of positions of other particles. The result of the analysis of such a situation does show the connection of momentum measurements to the Fourier transform of the wave function. But this is a rather roundabout method to obtain such an important physical result. It is to be expected that the postulates can be generalized by the replacement of the idea of "paths in a region of space-time R" to "paths of class R," or "paths having property R." But which properties correspond to which physical measurements has not been formulated in a general way.

12. A POSSIBLE GENERALIZATION

The formulation suggests an obvious generalization. There are interesting classical problems which satisfy a principle of least action but for which the action cannot be written as an integral of a function of positions and velocities. The action may involve accelerations, for example. Or, again, if interactions are not instantaneous, it may involve the product of coordinates at two different times, such as $\int x(t)x(t+T)dt$. The action, then, cannot be broken up into a sum of small contributions as in (10). As a consequence, no wave function is available to describe a state. Nevertheless, a transition probability can be defined for getting from a region R' into another R''. Most of the theory of the transition elements $\langle\chi_{t''}|F|\psi_{t'}\rangle_S$ can be carried over. One simply invents a symbol, such as $\langle R''|F|R'\rangle_S$ by an

[20] We did not immediately substitute p_i from (60) into (47) because (47) would then no longer have been valid to both zero order and the first order in ϵ. We could derive the commutation relations, but not the equations of motion. The two expressions in (60) represent the momenta at each end of the interval t_i to t_{i+1}. They differ by $\epsilon V'(x_{k+1})$ because of the force acting during the time ϵ.

equation such as (39) but with the expressions (19) and (20) for ψ and χ substituted, and the more general action substituted for S. Hamiltonian and momentum functionals can be defined as in section (10). Further details may be found in a thesis by the author.[21]

13. APPLICATION TO ELIMINATE FIELD OSCILLATORS

One characteristic of the present formulation is that it can give one a sort of bird's-eye view of the space-time relationships in a given situation. Before the integrations on the x_i are performed in an expression such as (39) one has a sort of format into which various F functionals may be inserted. One can study how what goes on in the quantum-mechanical system at different times is interrelated. To make these vague remarks somewhat more definite, we discuss an example.

In classical electrodynamics the fields describing, for instance, the interaction of two particles can be represented as a set of oscillators. The equations of motion of these oscillators may be solved and the oscillators essentially eliminated (Lienard and Wiechert potentials). The interactions which result involve relationships of the motion of one particle at one time, and of the other particle at another time. In quantum electrodynamics the field is again represented as a set of oscillators. But the motion of the oscillators cannot be worked out and the oscillators eliminated. It is true that the oscillators representing longitudinal waves may be eliminated. The result is instantaneous electrostatic interaction. The electrostatic elimination is very instructive as it shows up the difficulty of self-interaction very distinctly. In fact, it shows it up so clearly that there is no ambiguity in deciding what term is incorrect and should be omitted. This entire process is not relativistically invariant, nor is the omitted term. It would seem to be very desirable if the oscillators, representing transverse waves, could also be eliminated. This presents an almost insurmountable problem in the conventional quantum mechanics. We expect that the motion of a particle a at one time depends upon the motion of b at a previous time, and *vice versa*. A wave function $\psi(x_a, x_b; t)$, however, can only describe the behavior of both particles at one time. There is no way to keep track of what b did in the past in order to determine the behavior of a. The only way is to specify the state of the set of oscillators at t, which serve to "remember" what b (and a) had been doing.

The present formulation permits the solution of the motion of all the oscillators and their complete elimination from the equations describing the particles. This is easily done. One must simply solve for the motion of the oscillators before one integrates over the various variables x_i for the particles. It is the integration over x_i which tries to condense the past history into a single state function. This we wish to avoid. Of course, the result depends upon the initial and final states of the oscillator. If they are specified, the result is an equation for $\langle\chi_{t''}|1|\psi_{t'}\rangle$ like (38), but containing as a factor, besides $\exp(iS/\hbar)$ another functional G depending only on the coordinates describing the paths of the particles.

We illustrate briefly how this is done in a very simple case. Suppose a particle, coordinate $x(t)$, Lagrangian $L(\dot{x}, x)$ interacts with an oscillator, coordinate $q(t)$, Lagrangian $\frac{1}{2}(\dot{q}^2-\omega^2q^2)$, through a term $\gamma(x, t)q(t)$ in the Lagrangian for the system. Here $\gamma(x, t)$ is any function of the coordinate $x(t)$ of the particle and the time.[22] Suppose we desire the probability of a transition from a state at time t', in which the particle's wave function is $\psi_{t'}$ and the oscillator is in energy level n, to a state at t'' with the particle in $\chi_{t''}$ and oscillator in level m. This is the square of

$$\langle\chi_{t''}\varphi_m|1|\psi_{t'}\varphi_n\rangle_{S_p+S_0+S_I}$$

$$=\int\cdots\int\varphi_m^*(q_j)\chi_{t''}^*(x_j)$$

$$\times\exp\frac{i}{\hbar}(S_p+S_0+S_I)\psi_{t'}(x_0)\varphi_n(q_0)$$

$$\cdot\frac{dx_0}{A}\frac{dq_0}{a}\cdots\frac{dx_{j-1}}{A}\frac{dq_{j-1}}{a}dx_jdq_j. \quad (61)$$

[21] The theory of electromagnetism described by J. A. Wheeler and R. P. Feynman, Rev. Mod. Phys. 17, 157 (1945) can be expressed in a principle of least action involving the coordinates of particles alone. It was an attempt to quantize this theory, without reference to the fields, which led the author to study the formulation of quantum mechanics given here. The extension of the ideas to cover the case of more general action functions was developed in his Ph.D. thesis, "The principle of least action in quantum mechanics" submitted to Princeton University, 1942.

[22] The generalization to the case that γ depends on the velocity, $\dot{x}$, of the particle presents no problem.

Here $\varphi_n(q)$ is the wave function for the oscillator in state n, S_p is the action

$$\sum_{i=0}^{j-1} S_p(x_{i+1}, x_i)$$

calculated for the particle as though the oscillator were absent,

$$S_0=\sum_{i=0}^{j-1}\left[\frac{\epsilon}{2}\left(\frac{q_{i+1}-q_i}{\epsilon}\right)^2-\frac{\epsilon\omega^2}{2}q_{i+1}{}^2\right]$$

that of the oscillator alone, and

$$S_I=\sum_{i=0}^{j-1}\gamma_i q_i$$

(where $\gamma_i=\gamma(x_i,t_i)$) is the action of interaction between the particle and the oscillator. The normalizing constant, a, for the oscillator is $(2\pi\epsilon i/\hbar)^{-\frac{1}{2}}$. Now the exponential depends quadratically upon all the q_i. Hence, the integrations over all the variables q_i for $0<i<j$ can easily be performed. One is integrating a sequence of Gaussian integrals.

The result of these integrations is, writing $T=t''-t'$, $(2\pi i\hbar\sin\omega T/\omega)^{-\frac{1}{2}}\exp i(S_p+Q(q_j,q_0))/\hbar$, where $Q(q_j,q_0)$ turns out to be just the classical action for the forced harmonic oscillator (see reference 15). Explicitly it is

$$\begin{aligned}Q(q_j,q_0)=\frac{\omega}{2\sin\omega T}\bigg[&(\cos\omega T)(q_j^2+q_0^2)-2q_jq_0\\&+\frac{2q_0}{\omega}\int_{t'}^{t''}\gamma(t)\sin\omega(t-t')dt\\&+\frac{2q_j}{\omega}\int_{t'}^{t''}\gamma(t)\sin\omega(t''-t)dt\\&-\frac{2}{\omega^2}\int_{t'}^{t''}\int_{t'}^{t}\gamma(t)\gamma(s)\sin\omega(t''-t)\\&\times\sin\omega(s-t')dsdt\bigg].\end{aligned}$$

It has been written as though $\gamma(t)$ were a continuous function of time. The integrals really should be split into Riemann sums and the quantity $\gamma(x_i,t_i)$ substituted for $\gamma(t_i)$. Thus, Q depends on the coordinates of the particle at all times through the $\gamma(x_i,t_i)$ and on that of the oscillator at times t' and t'' only. Thus, the quantity (61) becomes

$$\begin{aligned}\langle\chi_{t''}\varphi_m|1|\psi_{t'}\varphi_n\rangle_{S_p+S_0+S_I}&=\int\cdots\int\chi_{t''}{}^*(x_j)G_{mn}\\&\cdot\exp\left(\frac{iS_p}{\hbar}\right)\psi_{t'}(x_0)\frac{dx_0}{A}\cdots\frac{dx_{j-1}}{A}dx_j\\&=\langle\chi_{t''}|G_{mn}|\psi_{t'}\rangle_{S_p}\end{aligned}$$

which now contains the coordinates of the particle only, the quantity G_{mn} being given by

$$G_{mn}=(2\pi i\hbar\sin\omega T/\omega)^{-\frac{1}{2}}\int\int\varphi_m{}^*(q_j)\times\exp(iQ(q_j,q_0)/\hbar)\varphi_n(q_0)dq_jdq_0.$$

Proceeding in an analogous manner one finds that all of the oscillators of the electromagnetic field can be eliminated from a description of the motion of the charges.

14. STATISTICAL MECHANICS

Spin and Relativity

Problems in the theory of measurement and statistical quantum mechanics are often simplified when set up from the point of view described here. For example, the influence of a perturbing measuring instrument can be integrated out in principle as we did in detail for the oscillator. The statistical density matrix has a fairly obvious and useful generalization. It results from considering the square of (38). It is an expression similar to (38) but containing integrations over two sets of variables dx_i and dx_i'. The exponential is replaced by $\exp i(S-S')/\hbar$, where S' is the same function of the x_i' as S is of x_i. It is required, for example, to describe the result of the elimination of the field oscillators where, say, the final state of the oscillators is unspecified and one desires only the sum over all final states m.

Spin may be included in a formal way. The Pauli spin equation can be obtained in this way:

One replaces the vector potential interaction term in $S(x_{i+1}, x_i)$,

$$\frac{e}{2c}(\mathbf{x}_{i+1}-\mathbf{x}_i)\cdot\mathbf{A}(\mathbf{x}_i)+\frac{e}{2c}(\mathbf{x}_{i+1}-\mathbf{x}_i)\cdot\mathbf{A}(\mathbf{x}_{i+1})$$

arising from expression (13) by the expression

$$\frac{e}{2c}(\boldsymbol{\sigma}\cdot(\mathbf{x}_{i+1}-\mathbf{x}_i))(\boldsymbol{\sigma}\cdot\mathbf{A}(\mathbf{x}_i)) + \frac{e}{2c}(\boldsymbol{\sigma}\cdot\mathbf{A}(\mathbf{x}_{i+1}))(\boldsymbol{\sigma}\cdot(\mathbf{x}_{i+1}-\mathbf{x}_i)).$$

Here $\mathbf{A}$ is the vector potential, $\mathbf{x}_{i+1}$ and $\mathbf{x}_i$ the vector positions of a particle at times t_{i+1} and t_i and $\boldsymbol{\sigma}$ is Pauli's spin vector matrix. The quantity Φ must now be expressed as $\prod_i \exp iS(x_{i+1}, x_i)/\hbar$ for this differs from the exponential of the sum of $S(x_{i+1}, x_i)$. Thus, Φ is now a spin matrix.

The Klein Gordon relativistic equation can also be obtained formally by adding a fourth coordinate to specify a path. One considers a "path" as being specified by four functions $x^{(\mu)}(\tau)$ of a parameter τ. The parameter τ now goes in steps ϵ as the variable t went previously. The quantities $x^{(1)}(t)$, $x^{(2)}(t)$, $x^{(3)}(t)$ are the space coordinates of a particle and $x^{(4)}(t)$ is a corresponding time. The Lagrangian used is

$$\sum_{\mu=1}^{4}{}' \left[(dx^\mu/d\tau)^2+(e/c)(dx^\mu/d\tau)A_\mu\right],$$

where A_μ is the 4-vector potential and the terms in the sum for $\mu=1, 2, 3$ are taken with reversed sign. If one seeks a wave function which depends upon τ periodically, one can show this must satisfy the Klein Gordon equation. The Dirac equation results from a modification of the Lagrangian used for the Klein Gordon equation, which is analagous to the modification of the non-relativistic Lagrangian required for the Pauli equation. What results directly is the square of the usual Dirac operator.

These results for spin and relativity are purely formal and add nothing to the understanding of these equations. There are other ways of obtaining the Dirac equation which offer some promise of giving a clearer physical interpretation to that important and beautiful equation.

The author sincerely appreciates the helpful advice of Professor and Mrs. H. C. Corben and of Professor H. A. Bethe. He wishes to thank Professor J. A. Wheeler for very many discussions during the early stages of the work.

The Theory of Quantized Fields. I*

JULIAN SCHWINGER
Harvard University, Cambridge, Massachusetts
(Received March 2, 1951)

The conventional correspondence basis for quantum dynamics is here replaced by a self-contained quantum dynamical principle from which the equations of motion and the commutation relations can be deduced. The theory is developed in terms of the model supplied by localizable fields. A short review is first presented of the general quantum-mechanical scheme of operators and eigenvectors, in which emphasis is placed on the differential characterization of representatives and transformation functions by means of infinitesimal unitary transformations. The fundamental dynamical principle is stated as a variational equation for the transformation function connecting eigenvectors associated with different spacelike surfaces, which describes the temporal development of the system. The generator of the infinitesimal transformation is the variation of the action integral operator, the space-time volume integral of the invariant lagrange function operator. The invariance of the lagrange function preserves the form of the dynamical principle under coordinate transformations, with the exception of those transformations which include a reversal in the positive sense of time, where a separate discussion is necessary. It will be shown in Sec. III that the requirement of invariance under time reflection imposes a restriction upon the operator properties of fields, which is simply the connection between the spin and statistics of particles. For a given dynamical system, changes in the transformation function arise only from alterations of the eigenvectors associated with the two surfaces, as generated by operators constructed from field variables attached to those surfaces. This yields the operator principle of stationary action, from which the equations of motion are obtained. Commutation relations are derived from the generating operator associated with a given surface. In particular, canonical commutation relations are obtained for those field components that are not restricted by equations of constraint. The surface generating operator also leads to generalized Schrödinger equations for the representative of an arbitrary state. Action integral variations which correspond to changing the dynamical system are discussed briefly. A method for constructing the transformation function is described, in a form appropriate to an integral spin field, which involves solving Hamilton-Jacobi equations for ordered operators. In Sec. III, the exceptional nature of time reflection is indicated by the remark that the charge and the energy-momentum vector behave as a pseudoscalar and pseudovector, respectively, for time reflection transformations. This shows, incidentally, that positive and negative charge must occur symmetrically in a completely covariant theory. The contrast between the pseudo energy-momentum vector and the proper displacement vector then indicates that time reflection cannot be described within the unitary transformation framework. This appears most fundamentally in the basic dynamical principle. It is important to recognize here that the contributions to the lagrange function of half-integral spin fields behave like pseudoscalars with respect to time reflection. The non-unitary transformation required to represent time reflection is found to be the replacement of a state vector by its dual, or complex conjugate vector, together with the transposition of all operators. The fundamental dynamical principle is then invariant under time reflection if inverting the order of all operators in the lagrange function leaves an integral spin contribution unaltered, and reverses the sign of a half-integral spin contribution. This implies the essential commutativity, or anti-commutativity, of integral and half-integral field components, respectively, which is the connection between spin and statistics.

I. INTRODUCTION

DESPITE extensive developments in the concepts and techniques of the theory of quantized fields, quantitative success has been achieved thus far only in the restricted domain of quantum electrodynamics. Furthermore, the existence of divergences, whether concealed or explicit, serves to emphasize that the present quantum theory of fields must, in some respect, be incomplete. It is not our purpose to propose a solution of this basic problem, but rather to present a general theory of quantum field dynamics which unifies several independently developed procedures and which may provide a framework capable of admitting fundamentally new physical ideas.

Quantum mechanics involves two distinct sets of hypotheses—the general mathematical scheme of linear operators and state vectors with its associated probability interpretation and the commutation relations and equations of motion for specific dynamical systems. It is the latter aspect that we wish to develop, by substituting a single quantum dynamical principle for the conventional array of assumptions based on classical hamiltonian dynamics and the correspondence principle.[1] We shall find it useful, however, first to review briefly some aspects of the mathematical formalism that find repeated application in the construction of our theory.

The simultaneous eigenvectors of some complete set of commuting hermitian operators, $\Psi(\alpha')$, provide a description of the arbitrary state Ψ by means of the representative

$$(\alpha'|)=(\Psi(\alpha'),\Psi), \tag{1.1}$$

which has the interpretation of a probability amplitude. Two such representations, associated with different complete sets of commuting operators, are related by

$$(\alpha'|)=\int(\alpha'|\beta')d\beta'(\beta'|), \tag{1.2}$$

where $\int d\beta'$ indicates integration and summation over

* The author wishes to acknowledge the hospitality of the Brookhaven National Laboratory, which is under the auspices of the AEC. The general program of this series was initiated there during the early summer of 1949, and the present paper was largely written at this Laboratory during the summer of 1950.

[1] Although our attention will be focused on field dynamics, the analogous development of particle quantum dynamics should be evident.

the totality of eigenvalues β', and

$$(\alpha'|\beta')=(\Psi(\alpha'),\Psi(\beta')) \tag{1.3}$$

is the transformation function. As a special example of Eq. (1.2), we have

$$(\alpha'|\gamma')=\int(\alpha'|\beta')d\beta'(\beta'|\gamma'), \tag{1.4}$$

the multiplicative composition law of transformation functions.

The set of commuting hermitian operators

$$\bar{\alpha}=U\alpha U^{-1}, \tag{1.5}$$

which is obtained from α with the aid of the arbitrary unitary operator U, has the property that its eigenvalues are identical with those of α, and that its eigenvectors are given by

$$\Psi(\bar{\alpha}')=U\Psi(\alpha'), \tag{1.6}$$

where $\bar{\alpha}'$ and α' are the same set of eigenvalues. Conversely, two sets of operators that possess the same eigenvalue spectrum are related by a unitary transformation. Note that the transformation function $(\bar{\alpha}'|\alpha'')$ may also be viewed as the matrix of U^{-1} in the original eigenvector system,

$$\begin{aligned}(\bar{\alpha}'|\alpha'')&=(U\Psi(\alpha'),\Psi(\alpha''))=(\Psi(\alpha'),U^{\dagger}\Psi(\alpha''))\\&=(\alpha'|U^{-1}|\alpha'').\end{aligned} \tag{1.7}$$

The unitary operator

$$U=1-(i/\hbar)F,\quad U^{-1}=1+(i/\hbar)F, \tag{1.8}$$

in which F is an infinitesimal hermitian operator, induces an infinitesimal transformation in the commuting set of operators,

$$\bar{\alpha}=U\alpha U^{-1}=\alpha-\delta\alpha, \tag{1.9}$$

where

$$i\hbar\delta\alpha=\alpha F-F\alpha=[\alpha,F]. \tag{1.10}$$

If the system is such that it is possible to obtain operators $\delta\alpha$ that commute with the complete set α, one can treat the $\delta\alpha$ as arbitrary, infinitesimal numbers, and $\Psi(\bar{\alpha}')$ provides an eigenvector of α with the eigenvalue set $\alpha'+\delta\alpha$. This evidently corresponds to the special circumstance of α having a continuous eigenvalue spectrum.

The concept of infinitesimal unitary transformation can be used to provide a differential characterization for the representative of a state, or for a transformation function. The change in the representative $(\alpha'|)$ when the commuting set of operators is altered by the unitary transformation generated by the infinitesimal hermitian operator F, is given by

$$\delta(\alpha'|)=((\alpha-\delta\alpha)'|)-(\alpha'|)=(\delta\Psi(\alpha'),\Psi), \tag{1.11}$$

where

$$\delta\Psi(\alpha')=U\Psi(\alpha')-\Psi(\alpha')=-(i/\hbar)F\Psi(\alpha'). \tag{1.12}$$

Therefore,

$$\delta(\alpha'|)=(i/\hbar)(\Psi(\alpha'),F\Psi)=(i/\hbar)(\alpha'|F|), \tag{1.13}$$

or

$$(\hbar/i)\delta(\alpha'|)=\int(\alpha'|F|\alpha'')d\alpha''(\alpha''|), \tag{1.14}$$

which is a differential equation for the representative $(\alpha'|)$. In a similar manner, we can characterize the transformation function $(\alpha'|\beta')$ by the effect of altering the two commuting sets α and β into $\alpha-\delta\alpha$ and $\beta-\delta\beta$, as induced by the two infinitesimal generating operators F_α, and F_β. Thus,

$$\begin{aligned}\delta(\alpha'|\beta')&=(\delta\Psi(\alpha'),\Psi(\beta'))+(\Psi(\alpha'),\delta\Psi(\beta'))\\&=(i/\hbar)(\alpha'|(F_\alpha-F_\beta)|\beta'),\end{aligned} \tag{1.15}$$

or

$$(\hbar/i)\delta(\alpha'|\beta')=\int(\alpha'|F_\alpha|\alpha'')d\alpha''(\alpha''|\beta')-\int(\alpha'|\beta'')d\beta''(\beta''|F_\beta|\beta'). \tag{1.16}$$

II. QUANTUM DYNAMICS OF LOCALIZABLE FIELDS

A localizable field is a dynamical system characterized by one or more operator functions of the space-time coordinates, $\phi^\alpha(x)$. Contained in this statement are the assumptions that the operators x_μ, representing position measurements, are commutative,

$$[x_\mu,x_\nu]=0, \tag{2.1}$$

and furthermore, that they commute with the field operators,

$$[x_\mu,\phi^\alpha]=0, \tag{2.2}$$

so that

$$(x|\phi^\alpha|x')=\delta(x-x')\phi^\alpha(x). \tag{2.3}$$

The difficulties associated with current field theories may be attributable to the implicit hypothesis of localizability. However, our development of quantum field dynamics will be confined to such fields. It remains to be seen whether other systems can be included within its scope.

The problem of constructing a complete set of commuting operators, that is, of simultaneously measurable physical quantities, necessarily involves specific properties of the fields. Nevertheless, as a general principle associated with relativistic requirements, we must expect such mutually commuting operators to be formed from field quantities at physically independent space-time points, that is, points which cannot be connected even by light signals. A continuous set of such points form a spacelike surface, which is a geometrical concept independent of the coordinate system. Therefore, a base vector system, $\Psi(\zeta',\sigma)$, will be specified by a spacelike surface σ and by the eigenvalues ζ' of a complete set of commuting operators constructed from field quantities attached to that surface. A change of representation will correspond, in general, to the introduction of another set of commuting operators on a different spacelike surface. Of particular importance is the transformation $\zeta_2, \sigma_2 \to \zeta_1, \sigma_1$, in which ζ_1 and ζ_2 are similarly constructed operator sets which possess the same eigenvalue spectrum and are therefore related by a unitary transformation [Eqs. (1.5) and (1.6)],

$$\zeta_1 = U_{12}\zeta_2 U_{12}^{-1},$$
$$\Psi(\zeta_1', \sigma_1) = U_{12}\Psi(\zeta_2', \sigma_2), \quad \zeta_1' = \zeta_2', \tag{2.4}$$

so that [Eq. (1.7)]

$$(\zeta_1', \sigma_1 | \zeta_2'', \sigma_2) = (\zeta_2', \sigma_2 | U_{12}^{-1} | \zeta_2'', \sigma_2). \tag{2.5}$$

A description of the temporal development of a system is evidently accomplished by stating the relationship between eigenvectors associated with different spacelike surfaces, or, in other words, by exhibiting the transformation function (2.5). Accordingly, we may expect that the quantum dynamical laws will find their proper expression in terms of the transformation function. A differential formulation of this type will now be constructed.

The operator U_{12}^{-1} describes the development of the system from σ_2 to σ_1 and involves, not only the detailed dynamical characteristics of the system in this space-time region, but also the choice of commuting operators, ζ_1 and ζ_2, on the surfaces σ_1 and σ_2. Any infinitesimal change in the quantities on which the transformation function depends induces a corresponding alteration in U_{12}^{-1},

$$\delta(\zeta_1', \sigma_1 | \zeta_2'', \sigma_2) = (\zeta_2', \sigma_2 | \delta U_{12}^{-1} | \zeta_2'', \sigma_2). \tag{2.6}$$

Now it is a consequence of the unitary property that $iU_{12}\delta U_{12}^{-1}$ must be hermitian. Accordingly, we write

$$\delta U_{12}^{-1} = (i/\hbar) U_{12}^{-1} \delta W_{12}, \tag{2.7}$$

where δW_{12} is an infinitesimal hermitian operator, and obtain

$$\delta(\zeta_1', \sigma_1 | \zeta_2'', \sigma_2) = (i/\hbar)(\zeta_1', \sigma_1 | \delta W_{12} | \zeta_2'', \sigma_2). \tag{2.8}$$

The composition law of transformation functions [Eq. (1.4)],

$$(\zeta_1', \sigma_1 | \zeta_3''', \sigma_3)$$
$$= \int (\zeta_1', \sigma_1 | \zeta_2'', \sigma_2) d\zeta''(\zeta_2'', \sigma_2 | \zeta_3''', \sigma_3), \tag{2.9}$$

imposes a restriction on δW, the generating operator of infinitesimal transformations. Thus,

$$(\zeta_1', \sigma_1 | \delta W_{13} | \zeta_3''', \sigma_3)$$
$$= \int (\zeta_1', \sigma_1 | \delta W_{12} | \zeta_2'', \sigma_2) d\zeta''(\zeta_2'', \sigma_2 | \zeta_3''', \sigma_3)$$
$$+ \int (\zeta_1', \sigma_1 | \zeta_2'', \sigma_2) d\zeta''(\zeta_2'', \sigma_2 | \delta W_{23} | \zeta_3''', \sigma_3), \tag{2.10}$$

or

$$\delta W_{13} = \delta W_{12} + \delta W_{23}; \tag{2.11}$$

the infinitesimal generating operators satisfy an additive law of composition.

Our basic assumption is that δW_{12} is obtained by variation of the quantities contained in a hermitian operator W_{12}, which must have the general form

$$W_{12} = (1/c) \int_{\sigma_2}^{\sigma_1} (dx) \mathfrak{L}[x], \tag{2.12}$$

according to the additive requirement (2.11). Individual systems are described by stating $\mathfrak{L}$ as an invariant hermitian function of the fields and their coordinate derivatives,

$$\mathfrak{L}[x] = \mathfrak{L}(\phi^\alpha(x), \phi_\mu{}^\alpha(x)), \quad \phi_\mu{}^\alpha(x) = \partial_\mu \phi^\alpha(x). \tag{2.13}$$

In conformity with their classical analogs, we shall call W and $\mathfrak{L}$ the action integral and lagrange function operators, respectively. The invariance of the lagrange function, and therefore of the action integral, guarantees that our fundamental dynamical principle,

$$\delta(\zeta_1', \sigma_1 | \zeta_2'', \sigma_2)$$
$$= (i/\hbar)(\zeta_1', \sigma_1 | \delta W_{12} | \zeta_2'', \sigma_2)$$
$$= (i/\hbar c)(\zeta_1', \sigma_1 | \delta \int_{\sigma_2}^{\sigma_1} (dx) \mathfrak{L} | \zeta_2'', \sigma_2), \tag{2.14}$$

is unaltered in form by a change in the coordinate system. An exception must be made, however, for those coordinate transformations that include a reversal in the positive sense of time, which require a separate discussion. We shall see that the requirement of invariance under time reflection imposes a general restriction upon the commutation properties of fields, which is simply the connection between the spin and statistics of elementary particles.

If the parameters of the system are not altered, the variation of the transformation function in Eq. (2.14) arises only from infinitesimal changes of ζ_1, σ_1 and ζ_2, σ_2. Such transformations may be characterized by infinitesimal generating operators, $F(\sigma_1)$ and $F(\sigma_2)$, which act on the eigenvectors $\Psi(\zeta_1', \sigma_1)$ and $\Psi(\zeta_2'', \sigma_2)$, and are therefore expressed in terms of operators associated with the surfaces σ_1 and σ_2, respectively. On referring to Eq. (1.15), we obtain for such variations,

$$\delta W_{12} = F(\sigma_1) - F(\sigma_2). \tag{2.15}$$

This is the operator principle of stationary action, for it states that the action integral operator is unaltered by infinitesimal variations of the field quantities in the interior of the region bounded by σ_1 and σ_2, being dependent only on operators attached to the boundary surfaces. The equations of motion for the field are contained in this principle.[2]

[2] In the following discussions, one should keep in mind that the lagrange functions of the simple systems usually considered are no more than quadratic in the components of individual fields.

The evaluation of δW_{12} involves adding the independent effects of changing the field components at each point by $\delta_0\phi^\alpha(x)$, and of altering the region of integration by a displacement δx_μ of the points on the boundary surfaces. Thus,

$$\delta W_{12}=(1/c)\int_{\sigma_2}^{\sigma_1}(dx)\delta_0\mathcal{L}$$
$$+(1/c)\left(\int_{\sigma_1}-\int_{\sigma_2}\right)d\sigma_\mu\delta x_\mu\mathcal{L}, \quad (2.16)$$

where

$$\delta_0\mathcal{L}=(\partial\mathcal{L}/\partial\phi^\alpha)\delta_0\phi^\alpha+(\partial\mathcal{L}/\partial\phi_\mu{}^\alpha)\partial_\mu\delta_0\phi^\alpha$$
$$=[(\partial\mathcal{L}/\partial\phi^\alpha)-\partial_\mu(\partial\mathcal{L}/\partial\phi_\mu{}^\alpha)]\delta_0\phi^\alpha$$
$$+\partial_\mu[(\partial\mathcal{L}/\partial\phi_\mu{}^\alpha)\delta_0\phi^\alpha]. \quad (2.17)$$

This expression for $\delta_0\mathcal{L}$ is to be understood symbolically, since the order of the operators in $\mathcal{L}$ must not be altered in the course of effecting the variation. Accordingly, the commutation properties of $\delta_0\phi^\alpha$ are involved in obtaining the consequences of the stationary requirement on the action integral. For simplicity, we shall introduce here the explicit assumption that the commutation properties of $\delta_0\phi^\alpha$ and the structure of the lagrange function must be so related that identical contributions are produced by terms that differ fundamentally only in the position of $\delta_0\phi^\alpha$. We may now infer the equations of motion

$$\partial_\mu(\partial\mathcal{L}/\partial\phi_\mu{}^\alpha)=\partial\mathcal{L}/\partial\phi^\alpha. \quad (2.18)$$

From the resulting form of δW_{12} we obtain the infinitesimal generating operator $F(\sigma)$, which acts on eigenvectors associated with the surface σ,

$$F(\sigma)=(1/c)\int_\sigma d\sigma_\mu[(\partial\mathcal{L}/\partial\phi_\mu{}^\alpha)\delta_0\phi^\alpha+\mathcal{L}\delta x_\mu]. \quad (2.19)$$

The total variation, $\delta\phi^\alpha(x)$, is composed additively of the variation $\delta_0\phi^\alpha(x)$ at the point x, and of the change in $\phi^\alpha(x)$ produced by moving from the point x on σ to $x+\delta x$ on $\sigma+\delta\sigma$. In evaluating the latter, we shall take into account that the field components $\phi^\alpha(x)$, although stated in terms of some fixed coordinate system, are most advantageously considered in relation to the local coordinate system provided by σ at the point x. Only such motions are contemplated that correspond to a local rigid displacement of the surface σ. This restriction is expressed by

$$\partial_\mu\delta x_\nu=-\partial_\nu\delta x_\mu, \quad (2.20)$$

being the condition that an infinitesimal space vector on σ be mapped into one of equal length on $\sigma+\delta\sigma$. The displacement induced change in $\phi^\alpha(x)$ may be obtained by an alteration in the coordinate system that reduces, in the neighborhood of x, to the equivalent local coordinate transformation. Thus, under the infinitesimal coordinate transformation

$$x_\mu'-x_\mu=-\delta x_\mu, \quad (2.21)$$

where

$$\delta x_\mu=\epsilon_\mu-\epsilon_{\mu\nu}x_\nu, \quad \epsilon_{\mu\nu}=-\epsilon_{\nu\mu}=\partial_\mu\delta x_\nu, \quad (2.22)$$

the field components suffer a linear transformation, as expressed by

$$\phi^{\alpha\prime}(x')-\phi^\alpha(x)=(i/\hbar)\tfrac{1}{2}\epsilon_{\mu\nu}S_{\mu\nu}{}^{\alpha\beta}\phi^\beta(x). \quad (2.23)$$

Therefore,

$$\phi^{\alpha\prime}(x)-\phi^\alpha(x)=\partial_\mu\phi^\alpha(x)\delta x_\mu+(i/\hbar)\tfrac{1}{2}\epsilon_{\mu\nu}S_{\mu\nu}{}^{\alpha\beta}\phi^\beta(x), \quad (2.24)$$

and

$$\delta\phi^\alpha(x)=\delta_0\phi^\alpha(x)+\phi_\mu{}^\alpha(x)\delta x_\mu$$
$$+(i/\hbar)\tfrac{1}{2}\partial_\mu\delta x_\nu S_{\mu\nu}{}^{\alpha\beta}\phi^\beta(x). \quad (2.25)$$

With the introduction of the total variation, the infinitesimal generating operator $F(\sigma)$ assumes the form

$$F(\sigma)=\int_\sigma d\sigma_\mu[\Pi_\mu{}^\alpha\delta\phi^\alpha+(1/c)\mathcal{L}\delta x_\mu-\Pi_\mu{}^\alpha\phi_\nu{}^\alpha\delta x_\nu$$
$$-(i/2\hbar)\Pi_\mu{}^\alpha S_{\lambda\nu}{}^{\alpha\beta}\phi^\beta\partial_\lambda\delta x_\nu], \quad (2.26)$$

where

$$c\Pi_\mu{}^\alpha=\partial\mathcal{L}/\partial\phi_\mu{}^\alpha. \quad (2.27)$$

To simplify the last term of Eq. (2.26), we define

$$f_{\mu\lambda\nu}=-f_{\lambda\mu\nu}=(i/2\hbar)[\Pi_\mu{}^\alpha S_{\lambda\nu}{}^{\alpha\beta}\phi^\beta+\Pi_\nu{}^\alpha S_{\lambda\mu}{}^{\alpha\beta}\phi^\beta$$
$$+\Pi_\lambda{}^\alpha S_{\nu\mu}{}^{\alpha\beta}\phi^\beta], \quad (2.28)$$

and obtain

$$(i/2\hbar)\Pi_\mu{}^\alpha S_{\lambda\nu}{}^{\alpha\beta}\phi^\beta\partial_\lambda\delta x_\nu=f_{\mu\lambda\nu}\partial_\lambda\delta x_\nu$$
$$=\partial_\lambda(f_{\mu\lambda\nu}\delta x_\nu)+\partial_\lambda f_{\lambda\mu\nu}\delta x_\nu, \quad (2.29)$$

since the last two terms of $f_{\mu\lambda\nu}$ are symmetrical in λ and ν, and therefore do not contribute to Eq. (2.29), in view of Eq. (2.20). We now remark that, in virtue of $f_{\mu\lambda\nu}=-f_{\lambda\mu\nu}$,

$$\int_\sigma d\sigma_\mu\partial_\lambda(f_{\mu\lambda\nu}\delta x_\nu)=0, \quad (2.30)$$

provided $f_{\mu\lambda\nu}\delta x_\nu$ effectively approaches zero, with sufficient rapidity, at infinitely remote points[3] on σ. Finally then,

$$F(\sigma)=\int_\sigma d\sigma_\mu[\Pi_\mu{}^\alpha\delta\phi^\alpha+(1/c)T_{\mu\nu}\delta x_\nu], \quad (2.31)$$

where

$$(1/c)T_{\mu\nu}=(1/c)\mathcal{L}\delta_{\mu\nu}-\Pi_\mu{}^\alpha\phi_\nu{}^\alpha-\partial_\lambda f_{\lambda\mu\nu} \quad (2.32)$$

is the stress tensor operator. As we shall demonstrate, this tensor has the property of being symmetrical.

$$T_{\mu\nu}=T_{\nu\mu}, \quad (2.33)$$

[3] All such characterizations of a spatially closed system, in terms of an operator approaching zero at infinity, are to be understood as a restriction to states for which the matrix elements of the operator have this property.

as an expression of the conservation of angular momentum.

Conservation laws are associated with variations that leave the action integral unchanged, since

$$\delta W_{12}=F(\sigma_1)-F(\sigma_2)=0 \qquad (2.34)$$

implies the constancy of the corresponding generating operator. The mechanical conservation laws for an isolated system are derived by considering a rigid displacement of the entire field, or equivalently, of the coordinate system, which is described by a common infinitesimal translation and rotation of the surfaces σ_1 and σ_2,

$$\delta x_\mu=\epsilon_\mu-\epsilon_{\mu\nu}x_\nu,\quad \epsilon_{\mu\nu}=-\epsilon_{\nu\mu}, \qquad (2.35)$$

combined with the field variation $\delta\phi^\alpha=0$. The displacement generating operator is then given by

$$F_{\delta x}(\sigma)=\epsilon_\mu P_\mu(\sigma)+\tfrac{1}{2}\epsilon_{\mu\nu}J_{\mu\nu}(\sigma), \qquad (2.36)$$

where

$$P_\nu(\sigma)=(1/c)\int_\sigma d\sigma_\mu T_{\mu\nu}, \qquad (2.37)$$

and

$$J_{\mu\nu}(\sigma)=(1/c)\int d\sigma_\lambda M_{\lambda\mu\nu},$$
$$M_{\lambda\mu\nu}=x_\mu T_{\lambda\nu}-x_\nu T_{\lambda\mu}. \qquad (2.38)$$

Accordingly,

$$P_\nu(\sigma_1)-P_\nu(\sigma_2)=0, \qquad (2.39)$$

and

$$J_{\mu\nu}(\sigma_1)-J_{\mu\nu}(\sigma_2)=0, \qquad (2.40)$$

which are the conservation laws for the energy-momentum vector, and the angular momentum tensor, respectively. Since the surfaces σ_1 and σ_2 are arbitrary, we infer the corresponding differential conservation laws,

$$\partial_\mu T_{\mu\nu}=0, \qquad (2.41)$$

and

$$\partial_\lambda M_{\lambda\mu\nu}=0, \qquad (2.42)$$

which, in conjunction, imply the symmetry of the stress tensor:

$$\partial_\lambda M_{\lambda\mu\nu}=T_{\mu\nu}-T_{\nu\mu}=0. \qquad (2.43)$$

The conservation law of charge can be obtained from the required invariance of the hermitian lagrange function under constant phase transformations—the multiplication of mutually hermitian conjugate pairs of field components by $\exp(\pm i\gamma)$. We consider infinitesimal phase transformations and, for convenience, write

$$\gamma=(e/\hbar c)\delta\lambda. \qquad (2.44)$$

Thus, we postulate the invariance of $\mathfrak{L}$ under the infinitesimal transformation

$$\delta\phi^\alpha=-(ie/\hbar c)\epsilon^\alpha\delta\lambda\phi^\alpha, \qquad (2.45)$$

where ϵ^α is characteristic of the field component ϕ^α, and may assume the values 0, or ±1. The associated generating operator is

$$F_{\delta\lambda}(\sigma)=-(ie/\hbar c)\int_\sigma d\sigma_\mu\Pi_\mu{}^\alpha\epsilon^\alpha\phi^\alpha\delta\lambda$$
$$=(1/c)Q(\sigma)\delta\lambda, \qquad (2.46)$$

where

$$Q(\sigma)=(1/c)\int_\sigma d\sigma_\mu j_\mu \qquad (2.47)$$

and

$$j_\mu=-(iec/\hbar)\Pi_\mu{}^\alpha\epsilon^\alpha\phi^\alpha. \qquad (2.48)$$

The implied conservation law,

$$Q(\sigma_1)-Q(\sigma_2)=0, \qquad (2.49)$$

is that of the total charge in the system.

It is important to notice the ambiguity in the lagrange function that is associated with given equations of motion. Thus, two lagrange functions that are related by

$$\bar{\mathfrak{L}}(\phi^\alpha,\phi_\mu{}^\alpha)=\mathfrak{L}(\phi^\alpha,\phi_\mu{}^\alpha)+c\partial_\nu f_\nu(\phi^\alpha,\phi_\mu{}^\alpha) \qquad (2.50)$$

provide action integral operators that differ by surface integrals:

$$\bar{W}_{12}=W_{12}+\left(\int_{\sigma_1}-\int_{\sigma_2}\right)d\sigma_\nu f_\nu. \qquad (2.51)$$

Therefore, the principle of stationary action for $\bar{W}_{12}$ is automatically satisfied by the equations of motion deduced from W_{12}, and

$$\delta\bar{W}_{12}=\bar{F}(\sigma_1)-\bar{F}(\sigma_2), \qquad (2.52)$$

where

$$\bar{F}=F+\delta w,\quad w=\int_\sigma d\sigma_\nu f_\nu. \qquad (2.53)$$

Hence, augmenting a lagrange function by the divergence of an arbitrary vector does not affect the equations of motion, but modifies the infinitesimal generating operator associated with a given surface σ. However, this ambiguity of the lagrange function corresponds precisely to the possibility of subjecting the commuting set of operators on σ to an arbitrary unitary transformation.

We verify this statement by specializing the general transformation theory to unitary transformations on a given surface. Let us introduce $\bar{\zeta}$, a new set of commuting operators on σ, which are obtained from ζ by a unitary transformation,

$$\Psi(\bar{\zeta}',\sigma)=\mathfrak{U}\Psi(\zeta',\sigma), \qquad (2.54)$$

where $\mathfrak{U}$ is characterized by an infinitesimal hermitian generating operator δw, according to

$$\delta\mathfrak{U}^{-1}=(i/\hbar)\mathfrak{U}^{-1}\delta w. \qquad (2.55)$$

As the analog of Eq. (2.8) we have, therefore,

$$\delta(\bar{\zeta}',\sigma|\zeta'',\sigma)=(i/\hbar)(\bar{\zeta}',\sigma|\delta w|\zeta'',\sigma); \qquad (2.56)$$

but

$$\delta w = \bar{F} - F, \tag{2.57}$$

where $\bar{F}$ and F are, respectively, the operators generating infinitesimal transformations of $\bar{\zeta}$ and ζ. This is just of the form (2.53); and conversely, by employing a particular w we obtain from Eq. (2.56) a differential equation to determine the transformation function that defines the new representation.

The commutation relations of our theory are implicit in the significance of F as an infinitesimal generating operator. We shall consider first those transformations that do not alter the surface σ, so that $\delta x_\nu = 0$. It is convenient to write

$$d\sigma_\mu = n_\mu d\sigma, \tag{2.58}$$

where n_μ is a unit timelike vector and $d\sigma$ is the numerical measure of the surface element. To avoid irrelevant geometrical complications in the following discussion, we shall henceforth restrict σ to be a plane surface, so that n_μ is constant on σ. Note, incidentally, that coordinate derivatives can be decomposed into components normal and tangential to σ,

$$\begin{gathered}\partial_\mu = n_\mu \partial_n + \partial_{t\mu}, \\ \partial_n = -n_\mu \partial_\mu, \quad \partial_{t\mu} = (\delta_{\mu\nu} + n_\mu n_\nu)\partial_\nu,\end{gathered} \tag{2.59}$$

and that the equations of motion read

$$\partial_n \Pi^\alpha = (1/c)(\partial \mathfrak{L}/\partial \phi^\alpha) - \partial_{t\mu} \Pi_\mu{}^\alpha. \tag{2.60}$$

We have here introduced the notation

$$\Pi^\alpha = n_\mu \Pi_\mu{}^\alpha \tag{2.61}$$

for a quantity which, more precisely, should be written $\Pi^\alpha(x, \sigma)$.

The generating operator F now becomes

$$F_{\delta\phi} = \int d\sigma \Pi^\alpha \delta\phi^\alpha. \tag{2.62}$$

Another significant form, associated with a different base vector system, is obtained from Eq. (2.53) with

$$f_\nu = -\Pi_\nu{}^\alpha \phi^\alpha. \tag{2.63}$$

Indeed, we have

$$\begin{aligned}\delta \int d\sigma_\nu f_\nu &= -\delta \int d\sigma \Pi^\alpha \phi^\alpha \\ &= -\int d\sigma (\Pi^\alpha \delta\phi^\alpha + \delta\Pi^\alpha \phi^\alpha),\end{aligned} \tag{2.64}$$

so that

$$\bar{F} = F_{\delta\Pi} = -\int d\sigma \delta\Pi^\alpha \phi^\alpha. \tag{2.65}$$

It should be emphasized again that these operator expressions are symbolic in the sense that the actual positions in which $\delta\phi^\alpha$ and $\delta\Pi^\alpha$ appear depend upon the structure of the lagrange function.

To obtain the proper interpretation of $F_{\delta\phi}$ or $F_{\delta\Pi}$, it is necessary to recognize that some of the Π^α can be identically equal to zero. This expresses the possibility that derivatives in timelike directions of some of the ϕ^α may not occur in the lagrange function. Accordingly, we shall divide the quantities ϕ^α and Π^α into two sets: ϕ^a and Π^a, called the canonical variables, and ϕ^A, Π^A, termed the constraint variables, in which the second set is characterized by

$$\Pi^A \equiv 0. \tag{2.66}$$

The name ascribed to the ϕ^A refers to the fact that, for these quantities, the equations of motion (2.60) degenerate into equations of constraint,

$$(1/c)(\partial \mathfrak{L}/\partial \phi^A) = \partial_{t\mu} \Pi_\mu{}^A, \tag{2.67}$$

that is, relations among the variables on σ. The nature of these relations can be made more apparent by exploiting the requirement that Eq. (2 66) be independent of the coordinate system. We shall later show that the implied restriction on the structure of $\mathfrak{L}$ is expressed by

$$n_\mu \Pi_\nu{}^A - n_\nu \Pi_\mu{}^A = (i/h)\Pi^a S_{\mu\nu}{}^{aA}. \tag{2.68}$$

On multiplication with n_ν, we obtain from this equation that

$$\Pi_\mu{}^A = (i/h)\Pi^a S_{\mu\nu}{}^{aA} n_\nu, \tag{2.69}$$

which enables the constraint equations to be written

$$(1/c)(\partial \mathfrak{L}/\partial \phi^A) = (i/h)\partial_\mu \Pi^a S_{\mu\nu}{}^{aA} n_\nu. \tag{2.70}$$

We shall now assume that it is possible to solve the left side of Eq. (2.70) for ϕ^A, thus exhibiting explicitly the constraint variables as functions of the canonical variables. This excludes systems for which the ϕ^A are fundamentally ambiguous in consequence of the existence of gauge transformations. The latter situation will be discussed subsequently in terms of the familiar example provided by the electromagnetic field.

It is evident from these considerations, and from the structure of the generating operators,

$$\begin{aligned}F_{\delta\phi} &= \int d\sigma \Pi^a \delta\phi^a, \\ F_{\delta\Pi} &= -\int d\sigma \delta\Pi^a \phi^a,\end{aligned} \tag{2.71}$$

that only the canonical variables are dynamically independent on σ. Accordingly, $F_{\delta\phi}$ is to be interpreted as the generator of that infinitesimal transformation of the commuting operator set ζ on σ which is produced by changing ϕ^a into $\phi^a - \delta\phi^a$. Similarly, $F_{\delta\Pi}$ is regarded as generating the infinitesimal transformation of ζ in which Π^a is replaced by $\Pi^a - \delta\Pi^a$. Thus, ϕ^a and Π^a are special examples of a set of independent field coordinates, and the most general possibility is implicit in the transformation (2.53). Associated with any such set of operators is the conjugate set appearing in F, as Π^a is conjugate to ϕ^a, and $-\phi^a$ to Π^a.

We shall now examine the change in the matrix of G, an arbitrary function of field variables on σ, which is produced by the infinitesimal transformation generated by $F_{\delta\phi}$, say. Thus, we have

$$\begin{aligned}\delta(\zeta',\sigma|G|\zeta'',\sigma)&=(\Psi(\zeta',\sigma),G\delta\Psi(\zeta'',\sigma))\\&\quad+(\delta\Psi(\zeta',\sigma),G\Psi(\zeta'',\sigma))\\&=-(i/\hbar)(\zeta',\sigma|[G,F_{\delta\phi}]|\zeta'',\sigma).\end{aligned}\quad(2.72)$$

On the other hand, we have

$$\delta(\zeta',\sigma|G|\zeta'',\sigma)=(\zeta',\sigma|\delta_\phi G|\zeta'',\sigma),\quad(2.73)$$

where $\delta_\phi G$ means the change in G produced on increasing ϕ^a by $\delta\phi^a$. This simply expresses the fact that replacing ϕ^a by $\phi^a-\delta\phi^a$ in both G and ζ leaves the relation between them, and therefore the matrix, unaltered. We thereby infer the commutation relation,

$$[G,F_{\delta\phi}]=i\hbar\delta_\phi G,\quad(2.74)$$

with its evident generalization in regard to the field coordinates, including, in particular,

$$[G,F_{\delta\Pi}]=i\hbar\delta_\Pi G.\quad(2.75)$$

Of special importance are the results obtained from Eqs. (2.74) and (2.75) with $G=\phi^a$ and Π^a:

$$\begin{aligned}&\left[\phi^a(x),\int_\sigma d\sigma'\Pi^b(x')\delta\phi^b(x')\right]=i\hbar\delta\phi^a(x),\\&\left[\Pi^a(x),\int_\sigma d\sigma'\Pi^b(x')\delta\phi^b(x')\right]=0,\end{aligned}\quad(2.76)$$

and

$$\begin{aligned}&\left[\int_\sigma d\sigma'\delta\Pi^b(x')\phi^b(x'),\Pi^a(x)\right]=i\hbar\delta\Pi^a(x),\\&\left[\int_\sigma d\sigma'\delta\Pi^b(x')\phi^b(x'),\phi^a(x)\right]=0,\end{aligned}\quad(2.77)$$

in which we have invoked the dynamical independence of the ϕ^a and the Π^a.

To extract explicit commutation relations among the ϕ^a and Π^a we must know the operator properties of $\delta\phi^a$ and $\delta\Pi^a$. The requirement that the formalism be invariant with respect to time reflection supplies the desired information. It will be shown in Sec. III that $\delta\phi^b(x')$ and $\delta\Pi^b(x')$ commute with all field quantities $\phi^a(x)$ and $\Pi^a(x)$, on σ, except when both a and b designate components of fields that possess half-integral spin, in which event they anti-commute. Accordingly, the commutation relations of Eqs. (2.76) and (2.77) become

$$\begin{aligned}&\int_\sigma d\sigma'[\phi^a(x),\Pi^b(x')]_\pm\delta\phi^b(x')=i\hbar\delta\phi^a(x),\\&\int_\sigma d\sigma'[\Pi^a(x),\Pi^b(x')]_\pm\delta\phi^b(x')=0,\\&\int_\sigma d\sigma'\delta\Pi^b(x')[\phi^b(x'),\Pi^a(x)]_\pm=i\hbar\delta\Pi^a(x),\\&\int_\sigma d\sigma'\delta\Pi^b(x')[\phi^b(x'),\phi^a(x)]_\pm=0,\end{aligned}\quad(2.78)$$

where

$$[A,B]_-=AB-BA\quad(2.79)$$

and

$$[A,B]_+=AB+BA.\quad(2.80)$$

Since the $\delta\phi^a$ and $\delta\Pi^a$ are quite arbitrary, we have derived the fundamental commutation relations,

$$\begin{aligned}&[\phi^a(x),\Pi^b(x')]_\pm=i\hbar\delta_{ab}\delta_\sigma(x-x'),\\&[\phi^a(x),\phi^b(x')]_\pm=[\Pi^a(x),\Pi^b(x')]_\pm=0.\end{aligned}\quad(2.81)$$

Here $\delta_\sigma(x-x')$ denotes the three-dimensional delta-function, which is defined by

$$\int_\sigma d\sigma'\delta_\sigma(x-x')f(x')=f(x),\quad(2.82)$$

where $f(x)$ is an arbitrary function. The commutation properties of the ϕ^A can then be obtained from their explicit expression in terms of the canonical variables. Thus, according to Eq. (2.70),

$$(1/c)[\partial\mathfrak{L}/\partial\phi^A(x),\phi^a(x')]_\pm=\partial_\mu\delta_\sigma(x-x')S_{\mu\nu}{}^{aA}n_\nu,\quad(2.83)$$

and

$$[\partial\mathfrak{L}/\partial\phi^A(x),\Pi^a(x')]_\pm=0.\quad(2.84)$$

In the requirement that commutators be employed, for components of an integral spin field, and anti-commutators for components of a half-integral spin field, we have the connection between the spin and statistics of particles. We shall note here that the commutation properties of a Bose-Einstein system, that is, an integral spin field, can be represented by means of differential operators. According to Eq. (1.13), a suitable representative of an arbitrary state obeys

$$\begin{aligned}\delta(\zeta',\sigma|)&=(i/\hbar)(\zeta',\sigma|F_{\delta\phi}|)\\&=(i/\hbar)\left(\zeta',\sigma\left|\int_\sigma d\sigma\Pi^a\delta\phi^a\right|\right),\end{aligned}\quad(2.85)$$

in which σ is not altered. Now the characteristic property of an integral spin field is that $\delta\phi^a$ commutes with all dynamical variables and can therefore be treated as a number. The representation involved in

Eq. (2.85) is evidently that which is labeled by the continuous eigenvalues of the $\phi^a(x)$ at all points of σ. In terms of the notation

$$\delta(\phi', \sigma|) = \int_\sigma d\sigma \delta\phi^{a'}(x)(\delta/\delta\phi^{a'}(x))(\phi', \sigma|), \quad (2.86)$$

we obtain

$$(\hbar/i)(\delta/\delta\phi^{a'}(x))(\phi', \sigma|) = (\phi', \sigma|\Pi^a(x)|), \quad (2.87)$$

and similarly,

$$i\hbar(\delta/\delta\Pi^{a'}(x))(\Pi', \sigma|) = (\Pi', \sigma|\phi^a(x)|). \quad (2.88)$$

We shall make further application of the general commutation relations (2.74) and (2.75) by successively placing $G = P_\nu$, $J_{\mu\nu}$, and Q. According to Eq. (2.30), the last term of Eq. (2.32) makes no contribution to P_ν, so that

$$P_\nu = -\int d\sigma \Pi^a \partial_\nu \phi^a + (1/c)\int d\sigma_\nu \mathfrak{L}. \quad (2.89)$$

In the evaluation of δP_ν we encounter

$$(1/c)\int d\sigma_\nu \delta\mathfrak{L} = \int d\sigma_\nu \partial_\mu(\Pi_\mu{}^a \delta\phi^a) = \int d\sigma_\mu \partial_\nu(\Pi_\mu{}^a\delta\phi^a)$$

$$= \int d\sigma \partial_\nu(\Pi^a\delta\phi^a), \quad (2.90)$$

whence

$$\delta P_\nu = \int d\sigma(\partial_\nu \Pi^a \delta\phi^a - \delta\Pi^a \partial_\nu\phi^a). \quad (2.91)$$

The rearrangements of Eq. (2.90) have involved Eq. (2.17), as simplified by the equations of motion, and the assumption that the system is spatially closed. We thereby obtain

$$[\phi^a(x), P_\nu] = (\hbar/i)\partial_\nu\phi^a(x),$$
$$[\Pi^a(x), P_\nu] = (\hbar/i)\partial_\nu\Pi^a(x), \quad (2.92)$$

in virtue of the commutativity of P_ν with $\delta\phi^a$ and $\delta\Pi^a$, which is a consequence of the fact that half-integral spin field components must appear paired in the vector P_ν. Incidentally, a commutator $[F^{(1)}, F^{(2)}]$, which has been evaluated by considering the effect on $F^{(2)}$ of the transformation generated by $F^{(1)}$, can equally well be viewed from the reverse standpoint. Thus, the relations (2.92) also exhibit P_ν in the role of the translation generator.

The angular momentum tensor $J_{\mu\nu}$ is easily brought into a form analogous to (2.89),

$$J_{\mu\nu} = -\int d\sigma \Pi^a[(x_\mu\partial_\nu - x_\nu\partial_\mu)\phi^a + (i/\hbar)S_{\mu\nu}{}^{a\beta}\phi^\beta]$$

$$+ (1/c)\int (d\sigma_\nu x_\mu \mathfrak{L} - d\sigma_\mu x_\nu \mathfrak{L}). \quad (2.93)$$

The contribution of the second term to $\delta J_{\mu\nu}$ is evaluated as follows,

$$\delta(1/c)\int (d\sigma_\nu x_\mu \mathfrak{L} - d\sigma_\mu x_\nu \mathfrak{L})$$

$$= \int [d\sigma_\nu x_\mu \partial_\lambda(\Pi_\lambda{}^a\delta\phi^a) - d\sigma_\mu x_\nu\partial_\lambda(\Pi_\lambda{}^a\delta\phi^a)]$$

$$= \int d\sigma_\lambda (x_\mu\partial_\nu - x_\nu\partial_\mu)(\Pi_\lambda{}^a\delta\phi^a)$$

$$+ \int (d\sigma_\mu \Pi_\nu{}^a - d\sigma_\nu \Pi_\mu{}^a)\delta\phi^a$$

$$= \int d\sigma[(x_\mu\partial_\nu - x_\nu\partial_\mu)(\Pi^a\delta\phi^a)$$

$$+ (n_\mu\Pi_\nu{}^a - n_\nu\Pi_\mu{}^a)\delta\phi^a]. \quad (2.94)$$

Therefore,

$$\delta J_{\mu\nu} = -\int d\sigma\delta\Pi^a[(x_\mu\partial_\nu - x_\nu\partial_\mu)\phi^a + (i/\hbar)S_{\mu\nu}{}^{a\beta}\phi^\beta]$$

$$+ \int d\sigma[(x_\mu\partial_\nu - x_\nu\partial_\mu)\Pi^a - (i/\hbar)\Pi^b S_{\mu\nu}{}^{ba}$$

$$+ n_\mu\Pi_\nu{}^a - n_\nu\Pi_\mu{}^a]\delta\phi^a. \quad (2.95)$$

We have thus derived the commutation relations

$$[\phi^a, J_{\mu\nu}] = (x_\mu(\hbar/i)\partial_\nu - x_\nu(\hbar/i)\partial_\mu)\phi^a + S_{\mu\nu}{}^{a\beta}\phi^\beta,$$
$$[\Pi^a, J_{\mu\nu}] = (x_\mu(\hbar/i)\partial_\nu - x_\nu(\hbar/i)\partial_\mu)\Pi^a - \Pi^b S_{\mu\nu}{}^{ba}$$
$$+ (\hbar/i)(n_\mu\Pi_\nu{}^a - n_\nu\Pi_\mu{}^a), \quad (2.96)$$
$$0 = -\Pi^a S_{\mu\nu}{}^{aA} + (\hbar/i)(n_\mu\Pi_\nu{}^A - n_\nu\Pi_\mu{}^A),$$

which exhibit $J_{\mu\nu}$ in the role of the rotation generator and illustrate the formation of $J_{\mu\nu}$ as the superposition of orbital and spin angular momenta. The third equation of Eq. (2.96), the statement that $\Pi^A \equiv 0$ is a property independent of the coordinate system, has already been employed in Eq. (2.68).

According to Eqs. (2.47) and (2.48), the charge operator is given by

$$Q = -(ie/\hbar)\int d\sigma \Pi^a \epsilon^a \phi^a. \quad (2.97)$$

Therefore, we have

$$\delta Q = -(ie/\hbar)\int d\sigma(\delta\Pi^a\epsilon^a\phi^a + \Pi^a\epsilon^a\delta\phi^a), \quad (2.98)$$

from which we obtain the commutation relations

$$[\phi^a, Q] = e\epsilon^a\phi^a, \quad [\Pi^a, Q] = -e\epsilon^a\Pi^a. \quad (2.99)$$

These indicate the significance of e as the elementary charge, and exhibit Q in the role of the phase transformation generator. Note, however, that the derivation of Eq. (2.99) from the latter viewpoint is not restricted to the canonical variables, although nothing new is obtained thereby.

The general infinitesimal operator (2.31) describes the transformation from the commuting operator set ζ, σ to $\zeta-\delta\zeta, \sigma+\delta\sigma$, as indicated by

$$\Psi((\zeta-\delta\zeta)', \sigma+\delta\sigma)=[1-(i/\hbar)F]\Psi(\zeta', \sigma). \quad (2.100)$$

The operator F is additively composed of two parts,

$$F=F_{\delta\phi}+F_{\delta x}, \quad (2.101)$$

where $F_{\delta\phi}$ induces, via $\delta\phi^a$, a change in the commuting operator set defined in relation to the local coordinate system provided by a fixed σ,

$$\Psi((\zeta-\delta\zeta)', \sigma)=[1-(i/\hbar)F_{\delta\phi}]\Psi(\zeta', \sigma), \quad (2.102)$$

while

$$F_{\delta x}=(1/c)\int d\sigma_\mu T_{\mu\nu}\delta x_\nu \quad (2.103)$$

generates the change in σ described by δx_ν, for a fixed set of commuting operators defined relative to σ,

$$\Psi(\zeta', \sigma+\delta\sigma)=[1-(i/\hbar)F_{\delta x}]\Psi(\zeta', \sigma). \quad (2.104)$$

Consistent with our restriction to plane surfaces, we consider only rigid displacements of σ, for which the generating operator has already been given in Eq. (2.36).

Differential equations that describe the change in the representative of an arbitrary state, as produced by rigid displacements, are inferred from

$$\delta_x(\zeta', \sigma|)=(\delta_x\Psi(\zeta', \sigma), \Psi)=(i/\hbar)(\zeta', \sigma|F_{\delta x}|). \quad (2.105)$$

In terms of the notation

$$\delta_x(\zeta', \sigma|)=\epsilon_\mu\delta_\mu(\zeta', \sigma|)+\tfrac{1}{2}\epsilon_{\mu\nu}\delta_{\mu\nu}(\zeta', \sigma|), \quad (2.106)$$

we obtain generalized Schrödinger equations[4] for translations,

$$(\hbar/i)\delta_\mu(\zeta', \sigma|)=(\zeta', \sigma|P_\mu(\sigma)|)$$
$$=\int(\zeta', \sigma|P_\mu(\sigma)|\zeta'', \sigma)d\zeta''(\zeta'', \sigma|), \quad (2.107)$$

and rotations,

$$(\hbar/i)\delta_{\mu\nu}(\zeta', \sigma|)=(\zeta', \sigma|J_{\mu\nu}(\sigma)|)$$
$$=\int(\zeta', \sigma|J_{\mu\nu}(\sigma)|\zeta'', \sigma)d\zeta''(\zeta'', \sigma|). \quad (2.108)$$

An operator $G(\sigma)$, which is constructed from field quantities on σ, has a matrix $(\zeta', \sigma|G(\sigma)|\zeta'', \sigma)$ that is independent of σ, since the relation between $G(\sigma)$ and the ζ on σ is unchanged by an alteration of the surface. The components of $P_\mu(\sigma)$ referred to axes based on σ are of this nature; and, consequently, the matrix of $P_\mu(\sigma)$ in Eq. (2.107) involves the orientation of σ relative to the coordinate system, but is otherwise independent of σ. Commutation relations between P_μ, $J_{\mu\nu}$ and $G(\sigma)$ follow from this property of the $G(\sigma)$ matrix. Thus, we have

$$0=\delta_x G(\sigma)-(i/\hbar)[G(\sigma), F_{\delta x}], \quad (2.109)$$

whence

$$[G(\sigma), P_\mu]=(\hbar/i)\delta_\mu G(\sigma), \quad (2.110)$$

and

$$[G(\sigma), J_{\mu\nu}]=(\hbar/i)\delta_{\mu\nu}G(\sigma). \quad (2.111)$$

As the first of several illustrations of these commutation relations, we choose $G(\sigma)=\phi^\alpha(x)$. According to Eq. (2.25), we have

$$[\phi^\alpha(x), P_\mu]=(\hbar/i)\partial_\mu\phi^\alpha(x), \quad (2.112)$$

and

$$[\phi^\alpha(x), J_{\mu\nu}]=(x_\mu(\hbar/i)\partial_\nu-x_\nu(\hbar/i)\partial_\mu)\phi^\alpha(x)$$
$$+S_{\mu\nu}{}^{\alpha\beta}\phi^\beta(x), \quad (2.113)$$

which is in agreement with Eqs. (2.92) and (2.96), but without the restriction of the latter to the components ϕ^a. A particularly simple example is provided by $G(\sigma)=Q$, the total charge. Since this operator is independent of σ, we have

$$[Q, P_\mu]=[Q, J_{\mu\nu}]=0, \quad (2.114)$$

which state, inversely, that P_μ and $J_{\mu\nu}$ are unaffected by phase transformations. The effect of a displacement of σ on the quantity $G(\sigma)=P_\lambda e_\lambda(\sigma)$, where $e_\lambda(\sigma)$ is an arbitrary vector that is rigidly attached to σ, comes entirely from the rotation of the vector $e_\lambda(\sigma)$,

$$\delta_x(P_\lambda e_\lambda(\sigma))=-\epsilon_{\mu\nu}P_\mu e_\nu(\sigma). \quad (2.115)$$

Therefore, we have

$$[P_\mu, P_\nu]=0, \quad (2.116)$$

and

$$[P_\lambda, J_{\mu\nu}]=i\hbar(\delta_{\nu\lambda}P_\mu-\delta_{\mu\lambda}P_\nu). \quad (2.117)$$

Our last example, $G(\sigma)=J_{\lambda\kappa}e_\lambda^{(1)}(\sigma)e_\kappa^{(2)}(\sigma)$, where both $e_\lambda^{(1)}(\sigma)$ and $e_\kappa^{(2)}(\sigma)$ are arbitrary vectors rigidly attached to σ, is actually an extension of the type of operator under consideration, since

$$J_{\lambda\kappa}e_\lambda^{(1)}e_\kappa^{(2)}=(1/c)\int d\sigma_\mu[x_\lambda e_\lambda^{(1)}T_{\mu\kappa}e_\kappa^{(2)}$$
$$-x_\kappa e_\kappa^{(2)}T_{\mu\lambda}e_\lambda^{(1)}] \quad (2.118)$$

involves space-time coordinates, in addition to field variables. The necessary revision of Eq. (2.109) is

$$\delta_x G(\sigma)=(i/\hbar)[G(\sigma), F_{\delta x}]+\partial_x G(\sigma), \quad (2.119)$$

where $\partial_x G(\sigma)$ denotes the displacement induced change in $G(\sigma)$, associated with the explicit appearance of space-time coordinates. In the example provided by

[4] Note that these Schrödinger equations have been obtained from the Heisenberg picture, in which the arbitrary state vector is fixed. P. A. M. Dirac, *The Principles of Quantum Mechanics* (The Clarendon Press, Oxford, 1947), third edition, Sec. 32.

Eq. (2.118), $\partial_x G(\sigma)$ arises from the translation (but not rotation) of σ,

$$\partial_x(J_{\lambda\kappa}e_\lambda^{(1)}e_\kappa^{(2)})=(\epsilon_\lambda P_\kappa-\epsilon_\kappa P_\lambda)e_\lambda^{(1)}e_\kappa^{(2)}. \quad (2.120)$$

On combining this with

$$\delta_x(J_{\lambda\kappa}e_\lambda^{(1)}e_\kappa^{(2)})=-\epsilon_{\mu\nu}J_{\mu\kappa}e_\nu^{(1)}e_\kappa^{(2)}-\epsilon_{\mu\nu}J_{\lambda\mu}e_\lambda^{(1)}e_\nu^{(2)}, \quad (2.121)$$

we again obtain Eq. (2.117), and

$$[J_{\lambda\kappa}, J_{\mu\nu}]=i\hbar(\delta_{\lambda\nu}J_{\mu\kappa}+\delta_{\lambda\mu}J_{\kappa\nu}+\delta_{\kappa\nu}J_{\lambda\mu}+\delta_{\kappa\mu}J_{\nu\lambda}). \quad (2.122)$$

We may remark, as an example of a general procedure for constructing representations of operator commutation properties, that the identity

$$[[\phi^a, J_{\lambda\kappa}], J_{\mu\nu}]-[[\phi^a, J_{\mu\nu}], J_{\lambda\kappa}] =[\phi^a, [J_{\lambda\kappa}, J_{\mu\nu}]] \quad (2.123)$$

leads to analogous commutation relations for the representatives of orbital and spin angular momentum in Eq. (2.113).

In a final comment concerning commutation relations, we observe that the commutators of generating operators are of significance in connection with integrability conditions for the infinitesimal transformations generated by these operators.[5] If $F^{(1)}$ and $F^{(2)}$ are two such generators of infinitesimal transformations, we have

$$\begin{aligned}\delta^{(1)}\Psi(\zeta', \sigma)&=\Psi((\zeta-\delta^{(1)}\zeta)', \sigma+\delta^{(1)}\sigma)-\Psi(\zeta', \sigma)\\&=-(i/\hbar)F^{(1)}\Psi(\zeta', \sigma),\\ \delta^{(2)}\Psi(\zeta', \sigma)&=\Psi((\zeta-\delta^{(2)}\zeta)', \sigma+\delta^{(2)}\sigma)-\Psi(\zeta', \sigma)\\&=-(i/\hbar)F^{(2)}\Psi(\zeta', \sigma).\end{aligned} \quad (2.124)$$

Now, the difference between the results of the two ways in which these transformations can be successively applied may be regarded as the effect of a third, related transformation,

$$\begin{aligned}(\delta^{(1)}\delta^{(2)}-\delta^{(2)}\delta^{(1)})\Psi(\zeta', \sigma)&=\Psi((\zeta+(\delta^{(1)}\delta^{(2)}-\delta^{(2)}\delta^{(1)})\zeta)', \sigma+(\delta^{(1)}\delta^{(2)}-\delta^{(2)}\delta^{(1)})\sigma)\\&\quad-\Psi(\zeta', \sigma)\\&=-(i/\hbar)F^{[12]}\Psi(\zeta', \sigma).\end{aligned} \quad (2.125)$$

Therefore,

$$[F^{(1)}, F^{(2)}]=i\hbar F^{[12]} \quad (2.126)$$

is a condition necessary to the integrability of Eq. (2.124). A simple illustration of this viewpoint is provided by rigid displacements:

$$\begin{aligned}\delta^{(1,2)}x_\mu&=\epsilon_\mu^{(1,2)}-\epsilon_{\mu\nu}^{(1,2)}x_\nu,\\ F_{\delta x}^{(1,2)}&=\epsilon_\mu^{(1,2)}P_\mu+\tfrac{1}{2}\epsilon_{\mu\nu}^{(1,2)}J_{\mu\nu},\end{aligned} \quad (2.127)$$

since

$$\begin{aligned}(\delta^{(1)}\delta^{(2)}-\delta^{(2)}\delta^{(1)})x_\mu&=-\epsilon_{\mu\nu}^{(1)}\epsilon_\nu^{(2)}+\epsilon_{\mu\nu}^{(2)}\epsilon_\nu^{(1)}\\&\quad-(-\epsilon_{\mu\lambda}^{(1)}\epsilon_{\lambda\nu}^{(2)}+\epsilon_{\mu\lambda}^{(2)}\epsilon_{\lambda\nu}^{(1)})x_\nu\\&=\epsilon_\mu^{[12]}-\epsilon_{\mu\nu}^{[12]}x_\nu\end{aligned} \quad (2.128)$$

is another rigid displacement. The ensuing commutation relations are just Eqs. (2.116), (2.117), and (2.122).

In our discussions of the variational principle (2.14), we have dealt with the properties of a given dynamical system. The principle is also applicable, however, to variations in which the system is altered, as characterized by a change in the structure of the lagrange function. For a variation of this type, we have

$$\delta(\zeta_1', \sigma_1|\zeta_2'', \sigma_2)=(i/\hbar c)\int_{\sigma_2}^{\sigma_1}(dx)(\zeta_1', \sigma_1|\delta\mathfrak{L}[x]|\zeta_2'', \sigma_2), \quad (2.129)$$

or

$$\delta(\zeta_1', \sigma_1|\zeta_2'', \sigma_2)=(i/\hbar c)\int_{\sigma_2}^{\sigma_1}(dx)\int(\zeta_1', \sigma_1|\zeta^a, \sigma)d\zeta^a(\zeta^a, \sigma|\delta\mathfrak{L}[x]|\zeta^b, \sigma)d\zeta^b(\zeta^b, \sigma|\zeta_2'', \sigma_2), \quad (2.130)$$

where the surface σ contains the point x. If two independent variations of this nature are applied successively we obtain

$$\begin{aligned}\delta^{(2)}\delta^{(1)}(\zeta_1', \sigma_1|\zeta_2'', \sigma_2)&=(i/\hbar c)\int_{\sigma_2}^{\sigma_1}(dx)\bigg[\int\delta^{(2)}(\zeta_1', \sigma_1|\zeta^a, \sigma)d\zeta^a(\zeta^a, \sigma|\delta^{(1)}\mathfrak{L}[x]|\zeta_2'', \sigma_2)\\&\qquad+\int(\zeta_1', \sigma_1|\delta^{(1)}\mathfrak{L}[x]|\zeta^b, \sigma)d\zeta^b\delta^{(2)}(\zeta^b, \sigma|\delta_2'', \sigma_2)\bigg]\\&=(i/\hbar c)^2\int_{\sigma_2}^{\sigma_1}(dx)\bigg[\int_\sigma^{\sigma_1}(dx')(\zeta_1', \sigma_1|\delta^{(2)}\mathfrak{L}[x']\delta^{(1)}\mathfrak{L}[x]|\zeta_2'', \sigma_2)\\&\qquad+\int_{\sigma_2}^{\sigma}(dx')(\zeta_1', \sigma_1|\delta^{(1)}\mathfrak{L}[x]\delta^{(2)}\mathfrak{L}[x']|\zeta_2'', \sigma_2)\bigg].\end{aligned} \quad (2.131)$$

We shall introduce here a notation for chronologically ordered operators,

$$(A(x)B(x'))_+=\begin{cases}A(x)B(x'), & x_0>x_0'\\ B(x')A(x), & x_0'>x_0,\end{cases} \quad (2.132)$$

[5] See, for example, H. Weyl, *The Theory of Group and Quantum Mechanics* (E. P. Dutton and Company, Inc., New York, 1931), p. 177.

which is an invariant concept provided that the operators involved commute when $x-x'$ is a spacelike interval and that the positive sense of time is preserved. Thus we may write Eq. (2.131) more compactly as

$$\delta^{(1)}\delta^{(2)}(\zeta_1',\sigma_1|\zeta_2'',\sigma_2)=\delta^{(2)}\delta^{(1)}(\zeta_1',\sigma_1|\zeta_2'',\sigma_2)$$
$$=(i/\hbar c)^2\int_{\sigma_2}^{\sigma_1}(dx)\int_{\sigma_2}^{\sigma_1}(dx')(\zeta_1',\sigma_1|(\delta^{(1)}\mathcal{L}[x]\delta^{(2)}\mathcal{L}[x'])_+|\zeta_2'',\sigma_2). \quad (2.133)$$

These results will find frequent application in later work.

We shall conclude this section by indicating, in connection with a Bose-Einstein field, a method for constructing the transformation function $(\zeta_1',\sigma_1|\zeta_2'',\sigma_2)$, which has as its classical analog the Hamilton-Jacobi theory of field mechanics. The actual motion of the system is implicit in the form assumed by the variation of the action integral,

$$\delta W_{12}=\left[\int_\sigma d\sigma\Pi^a\delta\phi^a+\epsilon_\mu P_\mu(\sigma)+\tfrac{1}{2}\epsilon_{\mu\nu}J_{\mu\nu}(\sigma)\right]_{\sigma_2}^{\sigma_1}, \quad (2.134)$$

in which we continue the restriction to plane spacelike surfaces. It is implied by Eq. (2.134) that W_{12} can be exhibited as a function of σ_1, σ_2, and of the ϕ^a on these surfaces, and therefore that the Π^a, P_μ and $J_{\mu\nu}$ associated with each surface can also be so exhibited. With the aid of commutation relations between the ϕ^a on σ_1 and on σ_2, it will be possible to order the operators in Eq. (2.134) so that the ϕ^a on σ_1 everywhere stand to the left of the ϕ^a on σ_2. The differential expression, thus ordered, shall be denoted by $\delta\mathcal{W}(\phi_1,\sigma_1;\phi_2,\sigma_2)$, from which we obtain differential equations connecting the various ordered operators,

$$(\delta/\delta\phi^a(x_1))\mathcal{W}=\Pi^a(x_1),\quad (\delta/\delta\phi^a(x_2))\mathcal{W}=-\Pi^a(x_2),$$
$$\delta_\mu^{(1)}\mathcal{W}=P_\mu(\sigma_1),\quad \delta_\mu^{(2)}\mathcal{W}=-P_\mu(\sigma_2),\quad (2.135)$$
$$\delta_{\mu\nu}^{(1)}\mathcal{W}=J_{\mu\nu}(\sigma_1),\quad \delta_{\mu\nu}^{(2)}\mathcal{W}=-J_{\mu\nu}(\sigma_2),$$

where x_1 and x_2 are arbitrary points on σ_1 and σ_2, respectively. In conjunction with the commutation relations (2.81), these Hamilton-Jacobi operator equations serve to determine the ordered operator $\mathcal{W}(\phi_1,\sigma_1;\phi_2,\sigma_2)$, to within an additive constant.

It is important to recognize that $\mathcal{W}\neq W_{12}$, and indeed, that $\mathcal{W}$ is a non-hermitian operator.[6] This is a consequence of the noncommutativity of the ϕ^a on σ_1 and on σ_2 in a manner which depends upon the location of these surfaces. Thus, if the operator W_{12} is first ordered and then varied, the result will differ from what is obtained by ordering δW_{12}. We now turn to the differential characterization of the transformation function labelled by eigenvalues of ϕ^a on σ_1 and σ_2,

$$\delta(\phi',\sigma_1|\phi'',\sigma_2)$$
$$=(i/\hbar)(\phi',\sigma_1|\delta\mathcal{W}(\phi_1,\sigma_1;\phi_2,\sigma_2)|\phi'',\sigma_2), \quad (2.136)$$

and observe that, in virtue of the ordering in $\delta\mathcal{W}$, the operators ϕ^a on σ_1 and on σ_2 act directly on their respective eigenvectors and can be replaced by the associated eigenvalues:

$$\delta(\phi',\sigma_1|\phi'',\sigma_2)$$
$$=(i/\hbar)\delta\mathcal{W}(\phi',\sigma_1;\phi'',\sigma_2)(\phi',\sigma_1|\phi'',\sigma_2). \quad (2.137)$$

The transformation function is thereby obtained as[7]

$$(\phi',\sigma_1|\phi'',\sigma_2)=\exp[(i/\hbar)\mathcal{W}(\phi',\sigma_1;\phi'',\sigma_2)], \quad (2.138)$$

where the constant of integration, which is additively contained in $\mathcal{W}$, can be determined from the condition

$$\lim_{\sigma_1\to\sigma_2}(\phi',\sigma_1|\phi'',\sigma_2)=\delta(\phi'-\phi''). \quad (2.139)$$

III. TIME REFLECTION

The general physical requirement of invariance with respect to coordinate transformations applies not only to translations and rotations of the coordinate system, but also to reflections of the coordinate axes. Among

[6] The elementary example of a one-dimensional free particle will suffice to illustrate this. The Hamilton-Jacobi equations for the construction of $\mathcal{W}(x(t_1),x(t_2),t)$, $t=t_1-t_2$, are

$$(\partial/\partial x(t_1))\mathcal{W}=-(\partial/\partial x(t_2))\mathcal{W}=p,\quad -(\partial/\partial t)\mathcal{W}=p^2/2m.$$

According to the solution of the equations of motion,

$$x(t_1)-x(t_2)=(t/m)p,$$

we have

$$[x(t_1),x(t_2)]=-i\hbar t/m,$$

whence

$$-(\partial/\partial t)\mathcal{W}=(m/2t^2)(x(t_1)-x(t_2))^2$$
$$=(m/2t^2)[x^2(t_1)-2x(t_1)x(t_2)+x^2(t_2)]-i\hbar/2t.$$

The solution of the Hamilton-Jacobi operator equations is

$$\mathcal{W}=(m/2t)[x^2(t_1)-2x(t_1)x(t_2)+x^2(t_2)]+\tfrac{1}{2}i\hbar\log(At),$$

which should be compared with the hermitian action integral

$$W_{12}=\tfrac{1}{2}mv^2t=(m/2t)(x(t_1)-x(t_2))^2$$
$$=(m/2t)[x^2(t_1)-2x(t_1)x(t_2)+x^2(t_2)]-\tfrac{1}{2}i\hbar.$$

Incidentally, the analog of Eq. (2.138) is

$$(x',t_1|x'',t_2)=\exp[(i/\hbar)\mathcal{W}(x',x'',t)]$$
$$=(At)^{-\frac{1}{2}}\exp[(im/2\hbar t)(x'-x'')^2],$$

where the constant A is determined to be

$$A=2\pi i\hbar/m$$

from the analog of Eq. (2.139),

$$\lim_{t\to 0}(x',t_1|x'',t_2)=\delta(x'-x'').$$

[7] The exponential form of Eq. (2.138) is familiar as a basis for establishing a correspondence connection with classical Hamilton-Jacobi particle mechanics. Dirac employed this form in a discussion of unitary transformations and recognized, in part, that the Hamilton-Jacobi equations are rigorous as relations among ordered operators (see the end of the section quoted in reference 4). In Feynman's version of quantum mechanics [R. P. Feynman, Revs. Modern Phys. **20**, 367 (1948)], the exponential form is employed for infinitesimal time intervals, with the real part of $\mathcal{W}$ defined as the classical action integral.

the latter transformations, time reflection has a singular position. Its special nature can be indicated by the transformation properties of some integrated physical quantities. Thus, the expectation value of the energy-momentum vector,

$$\langle P_\nu \rangle = (1/c) \int_\sigma d\sigma_\mu \langle T_{\mu\nu} \rangle, \tag{3.1}$$

is actually a pseudovector with respect to time reflection. With the plane surface σ chosen perpendicular to the time axis, the components of $\langle P_\nu \rangle$ are obtained as three-dimensional volume integrals,

$$\begin{aligned} \langle P_0 \rangle &= (1/c) \int d\sigma \langle T_{00} \rangle, \\ \langle P_k \rangle &= (1/c) \int d\sigma \langle T_{0k} \rangle, \quad k = 1, 2, 3, \end{aligned} \tag{3.2}$$

and the time reflection $x_0 \to -x_0$, $x_k \to x_k$ induces $\langle P_0 \rangle \to \langle P_0 \rangle$, $\langle P_k \rangle \to -\langle P_k \rangle$, according to the transformation properties of tensors. This differs in sign from a proper vector transformation. In particular, the energy does not reverse sign under time reflection. More generally, this property of $\langle P_\nu \rangle$ is obtained from the pseudovector character of $d\sigma_\mu$, which expresses the pseudoscalar nature of a four-dimensional volume element with respect to time reflection. Similarly, the expectation value of the charge

$$\langle Q \rangle = (1/c) \int d\sigma_\mu \langle j_\mu \rangle = (1/c) \int d\sigma \langle j_0 \rangle \tag{3.3}$$

behaves as a pseudoscalar under time reflection. Hence, this transformation interchanges positive and negative charge, and both signs must occur symmetrically in a covariant theory. Indeed, for some purposes the requirement of charge symmetry can be substituted for the more incisive demand of invariance under time reflection.

The significant implication of these properties is that time reflection cannot be included within the general framework of unitary transformations. Thus, on referring to the Schrödinger equation for translations (2.107), or the analogous operator equation (2.110), we encounter a contradiction between the transformation properties of the proper vector translation operator δ_μ and of the pseudovector P_μ. This difficulty appears most fundamentally in our basic variational principle (2.14). With $\mathfrak{L}$ behaving as a scalar and (dx) as a pseudoscalar, reflection of the time axis introduces a minus sign on the right side of this equation. However, it is important to notice that the scalar nature of $\mathfrak{L}$ cannot be maintained for that part of the lagrange function which describes half-integral spin fields. Indeed, such contributions to $\mathfrak{L}$ behave like pseudoscalars with respect to time reflection.[8] If we were to consider only such a half-integral spin field, the basic dynamical equation would preserve its structure under time reversal, but at the expense of violating the general transformation properties of all physical quantities; charge would remain unaltered, and energy would reverse sign under time reflection. The latter difficulty simply indicates that, on inclusion of the contributions of integral spin fields, the various parts of $\mathfrak{L}$ would transform differently, thus emphasizing again the general failure of Eq. (2.14) to admit time reflection as a unitary transformation.

To aid in investigating the extended class of transformations that is required to include time reflection we shall introduce some notational developments. The scalar product of two vectors, Ψ_a and Ψ_b, can be written

$$(a|b) = \Psi_a^* \Psi_b = \Psi_b \Psi_a^*, \tag{3.4}$$

thereby being regarded as the invariant combination of a vector Ψ_b with the dual, complex conjugate vector Ψ_a^* We allow operators to act both on the left and on the right of vectors, Ψ and Ψ^*. Thus, an operator associated with A, the transposed operator A^T, is defined by[9]

$$A\Psi = \Psi A^T, \quad \Psi^* A = A^T \Psi^*, \tag{3.5}$$

or by

$$(a|A|b) = \Psi_a^* A \Psi_b = \Psi_b A^T \Psi_a^*. \tag{3.6}$$

We also define the associated complex conjugate operator A^*,

$$(A\Psi)^* = A^* \Psi^*. \tag{3.7}$$

The connection with the hermitian conjugate operator $A^\dagger$ is obtained from the definition of the latter,

$$(A\Psi)^* = \Psi^* A^\dagger, \tag{3.8}$$

---

[8] The fundamental invariant of a spin $\frac{1}{2}$ field is $\bar{\psi}\psi = \psi^\dagger \gamma_0 \psi$. The transformation that represents time reflection, $\psi' = R\psi$, can be obtained from its equivalence with a rotation through the angle π in the (45) plane; $R = \exp[i\pi\frac{1}{2}\sigma_{45}] = i\sigma_{45}$. Accordingly,

$$\bar{\psi}'\psi' = \psi^\dagger R^{-1} \gamma_0 R \psi = -\bar{\psi}\psi,$$

which indicates the pseudoscalar character of the spin $\frac{1}{2}$ field lagrange function, with respect to time reflection. The corresponding behavior of fields with other spin values can be obtained from the observation that a spinor of rank n contains fields of spin $\frac{1}{2}n$, $\frac{1}{2}n - 1$, $\cdots$. The basic invariant and time reflection operator for a spinor of rank n are

$$\bar{\psi}\psi = \psi^\dagger \prod_{k=1}^{n} \gamma_0^{(k)} \psi,$$

and

$$R = \exp\left[i\pi\tfrac{1}{2} \sum_{k=1}^{n} \sigma_{45}^{(k)} \right] = \prod_{k=1}^{n} i\sigma_{45}^{(k)}$$

Therefore,

$$\bar{\psi}'\psi' = \psi^\dagger R^{-1} \prod_{k=1}^{n} \gamma_0^{(k)} R \psi = (-1)^n \bar{\psi}\psi,$$

which shows the pseudoscalar nature of the lagrange function for all half-integral spin fields.

[9] Note how the familiar property of transposition, $(AB)^T = B^T A^T$, follows from this definition: $AB\Psi = A(\Psi B^T) = \Psi B^T A^T$.

namely,

$$A^{\dagger}=A^{*T}. \tag{3.9}$$

Conventional quantum mechanics contemplates transformations only within the Ψ vector space, and contragradient transformations within the dual Ψ^* space. We shall now consider transformations that interchange the two spaces, as in

$$\Psi_a \rightarrow \Psi_{\bar{a}} = \Psi_a^*. \tag{3.10}$$

The effect of Eq. (3.10) is indicated by

$$(a|b)=\Psi_a^*\Psi_b=\Psi_{\bar{a}}\Psi_{\bar{b}}^*=(\bar{b}|\bar{a}), \tag{3.11}$$

and

$$(a|A|b)=\Psi_a^*A\Psi_b=\Psi_{\bar{a}}A\Psi_{b}^*=(\bar{b}|A^T|\bar{a}). \tag{3.12}$$

More generally, if

$$\Psi_{\bar{a}}^*=R\Psi_a, \tag{3.13}$$

where R is a unitary operator, we have

$$(a|b)=(\bar{b}|\bar{a}), \quad (a|A|b)=(\bar{b}|\bar{A}|\bar{a}), \tag{3.14}$$

in which

$$\bar{A}=(RAR^{-1})^T. \tag{3.15}$$

Now, we have

$$\overline{AB}=(RABR^{-1})^T=(RBR^{-1})^T(RAR^{-1})^T=\bar{B}\bar{A}, \tag{3.16}$$

and therefore

$$(a|[A,B]|b)=-(\bar{b}|[\bar{A},\bar{B}]|\bar{a}). \tag{3.17}$$

We have here precisely the sign change that is required to preserve the structure of equations like Eq. (2.110) under time reflection.

We now examine whether it is possible to satisfy the requirement of invariance under time reflection by means of transformations of the type (3.13). When we introduce the coordinate transformation

$$\bar{x}_0=-x_0, \quad \bar{x}_k=x_k, \quad k=1,2,3, \tag{3.18}$$

in conjunction with the eigenvector transformation

$$\Psi^*(\bar{\zeta}', \sigma)=R\Psi(\zeta', \sigma), \tag{3.19}$$

the fundamental dynamical equation (2.14) becomes

$$\delta(\bar{\zeta}_2'', \sigma_2|\bar{\zeta}_1', \sigma_1)$$

$$=(i/\hbar c)(\bar{\zeta}_2'', \sigma_2|\delta\int_{\sigma_1}^{\sigma_2}(d\bar{x})\bar{\mathfrak{L}}|\bar{\zeta}_1', \sigma_1), \tag{3.20}$$

where

$$\bar{\mathfrak{L}}=(R\mathfrak{L}R^{-1})^T=\mathfrak{L}^T((R\phi^\alpha R^{-1})^T, \pm\bar{\partial}_\mu(R\phi^\alpha R^{-1})^T). \tag{3.21}$$

In the last statement, the $\pm$ sign indicates the effect of the coordinate transformation (3.18) on the components of the gradient vector, while the notation $\mathfrak{L}^T(\ )$ symbolizes the reversal in the order of all factors induced by the operation of transposition. The operator R will now be chosen to produce that linear transformation of the ϕ^α,

$$R\phi^\alpha R^{-1}=R^{\alpha\beta}\phi^\beta, \tag{3.22}$$

which compensates the effect of the gradient vector transformation. Thus. we have

$$\bar{\mathfrak{L}}=(\pm)\mathfrak{L}^T(\phi^{\alpha T}, \bar{\partial}_\mu\phi^{\alpha T}), \tag{3.23}$$

where the $(\pm)$ sign here refers to the fact that the structure of the lagrange function, for half-integral spin fields, can be maintained only at the expense of a change in sign. We now see that if

$$\bar{\mathfrak{L}}=\mathfrak{L}(\phi^{\alpha T}, \bar{\partial}_\mu\phi^{\alpha T}), \tag{3.24}$$

the form of our fundamental dynamical equation will have been preserved under time reflection, since Eq. (3.20) will then differ from Eq. (2.14) only in the substitution of $\phi^{\alpha T}$ for ϕ^α as the appropriate field variable, and in the interchange of σ_1 and σ_2, which simply reflects the reversed temporal sense in which the dynamical development of the system is to be traced.

Invariance under time reflection thus requires that inverting the order of all factors in the lagrange function leave a scalar term unchanged, and reverse the sign of a pseudoscalar term. This can be satisfied, of course, by an explicit symmetrization or antisymmetrization of the various terms in $\mathfrak{L}$. When the lagrange function, thus arranged, is employed in the principle of stationary action, the variations $\delta_0\phi^\alpha$ will likewise be disposed in a symmetrical or antisymmetrical manner. We must now recall that the equations of motion (2.18), which do not depend explicitly on the nature of the field commutation properties, have been obtained by postulating the equality of terms in $\delta_0\mathfrak{L}$ that differ basically only in the location of $\delta_0\phi^\alpha$. Since such terms appear with the same sign in scalar components of $\mathfrak{L}$, and with opposite signs in pseudoscalar components, we deduce a corresponding commutativity, or anticommutativity, between $\delta_0\phi^\alpha$ and the other operators in the individual terms of $\delta_0\mathfrak{L}$.

The information concerning commutation properties that has thus been obtained is restricted to operators at common space-time points, since this is the nature of the terms in $\mathfrak{L}$. Commutation relations between field quantities located at distinct points of a space-like surface are implied by the general compatibility requirement for physical quantities attached to points with a spacelike interval. Components of integral spin fields, and bilinear combinations of the components of half-integral spin fields, are the basic physicial quantities to which this compatibility condition applies. By considering the general possibilities of coupling between the various fields, we may draw from these two expressions of relativistic invariance the consequence that the variations $\delta\phi^b(x')$, and therefore the conjugate variations $\delta\Pi^b(x')$, commute or anticommute with $\phi^a(x)$, $\Pi^a(x)$ for all x and x' on a given σ, where the relation of anticommutativity holds when both a and b refer to components of half-integral spin fields. The consistency of this statement with the general commutation relations that have already been deduced from it is easily verified. By subjecting the canonical variables in Eq. (2.81) to independent variations, we obtain

$$[\phi^a(x), \delta\phi^b(x')]_\pm = [\Pi^a(x), \delta\phi^b(x')]_\pm = 0,$$
$$[\phi^a(x), \delta\Pi^b(x')]_\pm = [\Pi^a(x), \delta\Pi^b(x')]_\pm = 0, \quad (3.25)$$

which is valid for all x, x' on σ. In addition, Eq. (2.81) properly states that all physical quantities commute at distinct points of σ.

We conclude that the connection between the spin and statistics of particles is implicit in the requirement of invariance under coordinate transformations.[10]

[10] The discussion of the spin and statistics connection by W. Pauli [Phys. Rev. 58, 716 (1940)] is somewhat more negative in character, although based on closely related physical requirements. Thus, Pauli remarks that Bose-Einstein quantization of a half-integral spin field implies an energy that possesses no lower bound, and that Fermi-Dirac quantization of an integral spin field leads to an algebraic contradiction with the commutativity of physical quantities located at points with a spacelike interval. Another postulate which has been employed, that of charge symmetry [W. Pauli and F. J. Belinfante, Physica 7, 177 (1940)], suffices to determine the nature of the commutation relations for sufficiently simple systems. As we have noticed, it is a consequence of time reflection invariance. The comments of Feynman on vacuum polarization and statistics [Phys. Rev. 76, 749 (1949)] appear to be an illustration of the charge symmetry requirement, since a contradiction is established when the charge symmetrical concept of the vacuum is applied to a Bose-Einstein spin ½ field, or to a Fermi-Dirac spin 0 field.

The Theory of Quantized Fields. II

JULIAN SCHWINGER
Harvard University, Cambridge, Massachusetts
(Received February 19, 1953)

The arguments leading to the formulation of the action principle for a general field are presented. In association with the complete reduction of all numerical matrices into symmetrical and antisymmetrical parts, the general field is decomposed into two sets, which are identified with Bose-Einstein and Fermi-Dirac fields. The spin restriction on the two kinds of fields is inferred from the time reflection invariance requirement. The consistency of the theory is verified in terms of a criterion involving the various generators of infinitesimal transformations. Following a discussion of charged fields, the electromagnetic field is introduced to satisfy the postulate of general gauge invariance. As an aspect of the latter, it is recognized that the electromagnetic field and charged fields are not kinematically independent. After a discussion of the field strength commutation relations, the independent dynamical variables of the electromagnetic field are exhibited in terms of a special gauge.

THE general program of this series[1] is the construction of a theory of quantized fields in terms of a single fundamental dynamical principle. We shall first present a revised account of the developments contained in the initial paper.

THE DYNAMICAL PRINCIPLE

The transformation functions connecting various representations have the two fundamental properties

$$(\alpha'|\gamma')=\int(\alpha'|\beta')d\beta'(\beta'|\gamma'),$$

$$(\alpha'|\beta')^*=(\beta'|\alpha'),$$

where $\int d\beta'$ symbolizes both integration and summation over the eigenvalue spectrum. If $\delta(\alpha'|\beta')$ is any infinitesimal alteration of the transformation function, we may write

$$\delta(\alpha'|\beta')=i(\alpha'|\delta W_{\alpha\beta}|\beta'), \tag{1}$$

which serves as the definition of the infinitesimal operator $\delta W_{\alpha\beta}$. The requirement that any infinitesimal alteration maintain the multiplicative composition law of transformation functions implies an additive composition law for the infinitesimal operators,

$$\delta W_{\alpha\gamma}=\delta W_{\alpha\beta}+\delta W_{\beta\gamma}. \tag{2}$$

If the α and β representations are identical, we infer that

$$\delta W_{\alpha\alpha}=0,$$

which expresses the fixed orthonormality requirements on the eigenvectors of a given representation. On identifying the α and γ representations, we learn that

$$\delta W_{\beta\alpha}=-\delta W_{\alpha\beta}.$$

The second property of transformation functions implies that

$$-i(\alpha'|\delta W_{\alpha\beta}|\beta')^*=-i(\beta'|\delta W_{\alpha\beta}{}^\dagger|\alpha')$$
$$=i(\beta'|\delta W_{\beta\alpha}|\alpha'),$$

or

$$\delta W_{\alpha\beta}{}^\dagger=\delta W_{\alpha\beta};$$

the infinitesimal operators $\delta W_{\alpha\beta}$ are Hermitian.

The $\delta W_{\alpha\beta}$ possess another additivity property referring to the composition of two dynamically independent systems. Thus, if I and II designate such systems,

$$(\alpha_{\mathrm{I}}'\alpha_{\mathrm{II}}'|\beta_{\mathrm{I}}'\beta_{\mathrm{II}}')=(\alpha_{\mathrm{I}}'|\beta_{\mathrm{I}}')(\alpha_{\mathrm{II}}'|\beta_{\mathrm{II}}'),$$

and if $\delta W_{\alpha\beta}{}^{\mathrm{I}}$ and $\delta W_{\alpha\beta}{}^{\mathrm{II}}$ are the operators characterizing infinitesimal changes of the separate transformation functions, that of the composite system is

$$\delta W_{\alpha\beta}=\delta W_{\alpha\beta}{}^{\mathrm{I}}+\delta W_{\alpha\beta}{}^{\mathrm{II}}.$$

Infinitesimal alterations of eigenvectors that preserve the orthonormality properties have the form

$$\delta\Psi(\alpha')=-iG_\alpha\Psi(\alpha'),$$
$$\delta\Psi(\alpha')^\dagger=i\Psi(\alpha')^\dagger G_\alpha,$$

where the generator G_α is an infinitesimal Hermitian operator which possesses an additivity property for the composition of dynamically independent systems. If the two eigenvectors of a transformation function are varied independently, the resulting change of the transformation function has the general structure (1), with

$$\delta W_{\alpha\beta}=G_\alpha-G_\beta.$$

The vector

$$\Psi(\alpha')+\delta\Psi(\alpha')=(1-iG_\alpha)\Psi(\alpha'),$$

can be characterized as an eigenvector of the operator set

$$\bar{\alpha}=(1-iG_\alpha)\alpha(1+iG_\alpha)=\alpha-\delta\alpha,$$

with the eigenvalues α'. Here

$$\delta\alpha=-i[\alpha,G_\alpha].$$

This infinitesimal unitary transformation of the eigenvector $\Psi(\alpha')$ induces a transformation of any operator F such that

$$(\alpha'|F|\alpha'')=(\bar{\alpha}'|\bar{F}|\bar{\alpha}'').$$

[1] J. Schwinger, Phys. Rev. 82, 914 (1951), Part I.

We write this in the form

$$(\bar{\alpha}'|F|\bar{\alpha}'')-(\alpha'|F|\alpha'')=(\bar{\alpha}'|(F-\bar{F})|\bar{\alpha}''),$$

or, in virtue of the infinitesimal nature of the transformation,

$$\delta(\alpha'|F|\alpha'')=(\alpha'|\delta F|\alpha''),$$

where the left side refers to the change in the eigenvectors for a fixed F, while the right side provides an equivalent variation of the operator F, given by

$$\delta F=F-\bar{F}=-i[F,G_\alpha].$$

If the change consists in the alteration of some parameter τ, upon which the dynamical variables depend, and which may occur explicitly in F, we have

$$\begin{aligned}\bar{F}&=F-(\delta F)_\tau\\&=F+\delta_\tau F-\partial_\tau F,\end{aligned}$$

where $\delta_\tau F$ is the total alteration in F, from which is subtracted $\partial_\tau F$, the change in F associated with the explicit appearance of τ, since the latter cannot be produced by an operator transformation. We thereby obtain the "equation of motion" with respect to the parameter τ,

$$\delta_\tau F=\partial_\tau F+i[F,G_\tau]. \tag{3}$$

For dynamical systems obeying the postulate of local action, complete descriptions are provided by sets of physical quantities, ζ, associated with space-like surfaces, σ. An infinitesimal alteration of the general transformation function $(\zeta_1'\sigma_1|\zeta_2''\sigma_2)$ is characterized by

$$\delta(\zeta_1'\sigma_1|\zeta_2''\sigma_2)=i(\zeta_1'\sigma_1|\delta W_{12}|\zeta_2''\sigma_2). \tag{4}$$

Here the indices 1 and 2 refer both to the choice of complete set of commuting operators ζ, and to the space-like surface σ. We can, in particular, consider transformations between the same set of operators on different surfaces, or between different sets of commuting operators on the same surface, as in

$$\delta(\zeta'\sigma|\bar{\zeta}'\sigma)=i(\zeta'\sigma|\delta W|\bar{\zeta}'\sigma). \tag{5}$$

One type of change of the general transformation function consists in the introduction, independently on σ_1 and on σ_2, of infinitesimal unitary transformations of the operators, including displacements of these surfaces. The transformations will be generated by operators G_1 and G_2, constructed from dynamical variables on σ_1 and σ_2, respectively, and

$$\delta W_{12}=G_1-G_2. \tag{6}$$

When the transformation function connects two different sets of operators on the same surface, which are subjected to infinitesimal transformations generated by G and $\bar{G}$, respectively, we have, referring to (5),

$$\delta W=G-\bar{G}. \tag{7}$$

Since physical phenomena at distinct points on a space-like surface are dynamically independent, a generator G must have the additive form

$$G=\int_\sigma d\sigma G_{(0)}(x)=\int_\sigma d\sigma_\mu G_\mu(x),$$

where $d\sigma$ is the numerical measure of an element of space-like area and $G_{(0)}(x)$ is to be regarded as the time-like component of a vector in a local coordinate system based on σ in order to give the surface integral an invariant form. If one can interpret $G_\mu(x)$ on σ_1, and on σ_2, as the values of a vector defined at all points, the difference of surface integrals in (6) can be transformed into the volume integral

$$\delta W_{12}=\int_{\sigma_2}^{\sigma_1}(dx)\partial_\mu G_\mu(x),$$

$$(\partial_\mu=\partial/\partial x_\mu).$$

A second type of transformation function alteration is obtained on considering that the transformation connecting ζ_1, σ_1, and ζ_2, σ_2 can be constructed through the intermediary of an infinite succession of transformations relating operators on infinitesimally neighboring surfaces. According to the general additivity property (2),

$$\delta W_{12}=\sum_{\sigma_2}^{\sigma_1}\delta W_{\sigma+d\sigma,\,\sigma},$$

where $\delta W_{\sigma+d\sigma,\,\sigma}$ characterizes a modification of the transformation function connecting infinitesimally differing complete sets of operators on the infinitesimally separated surfaces σ and $\sigma+d\sigma$. If the choice of intermediate operators depends continuously upon the surface, we shall have

$$\delta W_{\sigma,\,\sigma}=0,$$

and, referring again to the dynamical independence of phenomena at points separated by a space-like interval, with the consequent additivity property, we see that $\delta W_{\sigma+d\sigma,\,\sigma}$ will have the general form

$$\delta W_{\sigma+d\sigma,\,\sigma}=\int_\sigma^{\sigma+d\sigma}(dx)\delta\mathfrak{L}(x).$$

Therefore

$$\delta W_{12}=\int_{\sigma_2}^{\sigma_1}(dx)\delta\mathfrak{L}(x). \tag{8}$$

The combination of these two types of modifications is described by

$$\delta W_{12}=G_1-G_2+\int_{\sigma_2}^{\sigma_1}(dx)\delta\mathfrak{L}(x),$$

which involves dynamical variables on the surfaces σ_1, σ_2, and in the interior of the volume bounded by these surfaces. On the other hand, we can write this as the volume integral

$$\delta W_{12}=\int_{\sigma_2}^{\sigma_1}(dx)[\partial_\mu G_\mu(x)+\delta\mathfrak{L}(x)],$$

which indicates, conversely, that any part of $\delta\mathfrak{L}(x)$, possessing the form of a divergence, contributes only to the generation of unitary transformations on σ_1 and σ_2.

The fundamental dynamical principle is contained in the postulate that there exists a class of transformation function alterations for which the characterizing operators δW_{12} are obtained by appropriate variation of a single operator W_{12},

$$\delta W_{12}=\delta(W_{12}).$$

Of course, this principle must be implemented by the explicit specification of that class.

The operator W_{12}, the action integral operator, evidently possess the form

$$W_{12}=\int_{\sigma_2}^{\sigma_1}(dx)\mathfrak{L}(x).$$

The Hermitian requirement on δW_{12} is satisfied if W_{12} is Hermitian, which implies the same property for $\mathfrak{L}(x)$, the Lagrange function operator. In order that relations between states on σ_1 and σ_2 be invariantly characterized, the Lagrange function must be a scalar with respect to the transformations of the orthochronous[2] Lorentz group, which preserve the temporal order of σ_1 and σ_2. A dynamical system is specified by exhibiting the Lagrange function in terms of a set of fundamental dynamical variables in the infinitesimal neighborhood of the point x. Contained in this Lagrange function will be certain numerical parameters, which may be functions of x. Any change of these parameters modifies the structure of the Lagrange function and is thus an alteration of the dynamical system. Accordingly, infinitesimal changes of the dynamical system are described by

$$\delta W_{12}=\int_{\sigma_2}^{\sigma_1}(dx)\delta\mathfrak{L}(x),$$

where $\delta\mathfrak{L}=\delta(\mathfrak{L})$, and the numerical parameters are the object of variation. This form is in agreement with (8). For a fixed dynamical system, W_{12} can be altered by displacing the surfaces σ_1, σ_2 and by varying the dynamical variables contained in the Lagrange function. The transformation function $(\zeta_1'\sigma_1|\zeta_2''\sigma_2)$ describes the relation between two states of the given system so that a change in the transformation function can only arise from alterations of the states on σ_1 and σ_2. Hence, for a fixed dynamical system we must have

$$\delta W_{12}=G_1-G_2,$$

[2] This name was suggested by H. J. Bhabha, Revs. Modern Phys. **21**, 451 (1949).

where $\delta W_{12}=\delta(W_{12})$ and the objects of variation here are σ_1, σ_2, and the dynamical variables of which $\mathfrak{L}$ is a function.

The latter statement is the operator principle of stationary action. It asserts that W_{12} must be stationary with respect to variations of the dynamical variables in the interior of the region defined by σ_1 and σ_2, since G_1 and G_2 only contain dynamical variables associated with the boundaries of the region. This principle implies equations of motion for the dynamical variables, that is to say, field equations, and provides expressions for the generators G_1 and G_2. The class of variations to which our postulate refers can now be defined through the requirement that this information concerning field equations and infinitesimal unitary transformations be self-consistent.

There exists much freedom within this class, as may be inferred from the remark that two Lagrange functions, differing by the divergence of a vector, describe the same dynamical system. Thus

$$\bar{\mathfrak{L}}(x)=\mathfrak{L}(x)-\partial_\mu f_\mu(x),$$

yields

$$\bar{W}_{12}=W_{12}-(W_1-W_2), \tag{9}$$

where, on each surface,

$$W=\int_\sigma d\sigma_\mu f_\mu=\int_\sigma d\sigma f_{(0)}.$$

Accordingly, the stationary action principle for $\bar{W}_{12}$ is satisfied if it is obeyed by W_{12}, since

$$\delta\bar{W}_{12}=\bar{G}_1-\bar{G}_2.$$

Here

$$\delta W_1=G_1-\bar{G}_1,\quad \delta W_2=G_2-\bar{G}_2,$$

define $\bar{G}_1$ and $\bar{G}_2$, which are new generators of infinitesimal unitary transformations on σ_1 and σ_2, respectively. The latter equations possess the form (7), and thus characterize transformation functions connecting two different representations on a common surface. Indeed, with a suitably elaborate notation, we recognize in (9) the additivity property of action operators,

$$W(\bar{\zeta}_1\sigma_1,\bar{\zeta}_2\sigma_2)=W(\bar{\zeta}_1\sigma_1,\zeta_1\sigma_1)+W(\zeta_1\sigma_1,\zeta_2\sigma_2)+W(\zeta_2\sigma_2,\bar{\zeta}_2\sigma_2),$$

where, for example,

$$W_1=-W(\bar{\zeta}_1\sigma_1,\zeta_1\sigma_1)=W(\zeta_1\sigma_1,\bar{\zeta}_1\sigma_1),$$

and

$$W_2=W(\zeta_2\sigma_2,\bar{\zeta}_2\sigma_2).$$

To be consistent with the postulate of local action, the field equations must be differential equations of finite order. One can always convert such equations into systems of first order equations by suitable adjunction of variables. We shall designate the fundamental dynamical variables that obey first-order field equations by $\chi_r(x)$, which form the components of the general field operator $\chi(x)$. With no loss in generality, we take

$\chi(x)$ to be a Hermitian operator,

$$\chi_r(x)^\dagger=\chi_r(x).$$

If the Lagrange function is to yield field equations of the desired structure, it must be linear in the first derivatives of the field operators with respect to the space-time coordinates. Furthermore, if these field equations are to emerge as explicit equations of motion for field components, that part of the Lagrange function containing first coordinate derivatives must be bilinear in the field components. With these preliminary remarks, we write the following general expression for the Lagrange function,

$$\mathfrak{L}=\tfrac{1}{2}(\chi\mathfrak{A}_\mu\partial_\mu\chi-\partial_\mu\chi\mathfrak{A}_\mu\chi)-\mathfrak{H}(\chi), \tag{10}$$

in which a matrix notation is employed,

$$\chi\mathfrak{A}_\mu\partial_\mu\chi=\chi_r(\mathfrak{A}_\mu)_{rs}\partial_\mu\chi_s.$$

The derivative terms have been symmetrized with respect to the operation of integration by parts, a process which adds a divergence to the Lagrange function, and is thus without effect on the structure of the dynamical system. In order that $\mathfrak{L}$ be a Hermitian operator, the general function $\mathfrak{H}$ must possess this character,

$$\mathfrak{H}(\chi)^\dagger=\mathfrak{H}(\chi),$$

and the numerical matrices $\mathfrak{A}_\mu$; $\mu=0, 1, 2, 3$ $(x_4=ix_0, \mathfrak{A}_4=i\mathfrak{A}_0)$ must be skew-Hermitian,

$$\mathfrak{A}_\mu{}^\dagger=\mathfrak{A}_\mu{}^{\mathrm{tr}*}=-\mathfrak{A}_\mu;\quad \mu=0, 1, 2, 3.$$

Although we are interested in complete dynamical systems, it is advantageous mathematically to employ devices based upon the properties of external sources. Accordingly, we add to (10) a term designed to describe the generation of the field $\chi(x)$ by an external source $\xi(x)$, which is to be regarded as a field quantity of the same general nature as $\chi(x)$,

$$\mathfrak{L}_{\mathrm{source}}=\tfrac{1}{2}(\xi\mathfrak{B}\chi+\chi\mathfrak{B}\xi). \tag{11}$$

This is a Hermitian operator if $\mathfrak{B}$ is a Hermitian matrix,

$$\mathfrak{B}^\dagger=\mathfrak{B}.$$

For the source concept to be meaningful, all components of χ must occur coupled with the source components in (11), which requires that $\mathfrak{B}$ be a nonsingular numerical matrix.

An orthochronous Lorentz transformation

$$'x_\mu=r_{\mu\nu}x_\nu+t_\mu,$$

$$r^{\mathrm{tr}}r=1,\quad r_{44}>0,$$

induces a linear transformation on the field components,

$$'\chi=L\chi=\chi L^{\mathrm{tr}},$$

where L must be a real matrix,

$$L^*=L,$$

to maintain the Hermiticity of $'\chi$. The scalar requirement on $\mathfrak{L}$ is satisfied if $\mathfrak{H}$ is a scalar,

$$\mathfrak{H}(L\chi)=\mathfrak{H}(\chi),$$

and if

$$L^{\mathrm{tr}}\mathfrak{A}_\mu L=r_{\mu\nu}\mathfrak{A}_\nu. \tag{12}$$

We shall suppose that the source possesses the same transformation properties as the field. The condition for the source term of the Lagrange function to be a scalar is then given by

$$L^{\mathrm{tr}}\mathfrak{B}L=\mathfrak{B}. \tag{13}$$

Note that $\mathfrak{A}_\mu{}^{\mathrm{tr}}$ and $\mathfrak{B}^{\mathrm{tr}}$ also obey Eqs. (12) and (13), respectively, and that these equations can be combined into

$$L^{-1}(\mathfrak{B}^{-1}\mathfrak{A}_\mu)L=r_{\mu\nu}(\mathfrak{B}^{-1}\mathfrak{A}_\nu),$$

in view of the nonsingular character of $\mathfrak{B}$.

For an infinitesimal Lorentz transformation,

$$'x_\mu=x_\mu-\epsilon_{\mu\nu}x_\nu+\epsilon_\mu,\quad \epsilon_{\mu\nu}=-\epsilon_{\nu\mu},$$

the matrix L can be written

$$L=1-i\tfrac{1}{2}\epsilon_{\mu\nu}S_{\mu\nu}, \tag{14}$$

where

$$S_{\mu\nu}{}^*=-S_{\mu\nu};\quad \mu, \nu=0, \cdots 3. \tag{15}$$

The infinitesimal version of (13) is

$$-S_{\mu\nu}{}^{\mathrm{tr}}=\mathfrak{B}S_{\mu\nu}\mathfrak{B}^{-1}=S_{\mu\nu}{}^\dagger$$

or

$$(\mathfrak{B}S_{\mu\nu})^\dagger=(\mathfrak{B}S_{\mu\nu}),$$

in which the complex conjugate statements refer to the components indicated in (15). Similarly,

$$\mathfrak{A}_\mu S_{\nu\lambda}-S_{\nu\lambda}{}^\dagger\mathfrak{A}_\mu=i(\delta_{\mu\lambda}\mathfrak{A}_\nu-\delta_{\mu\nu}\mathfrak{A}_\lambda) \tag{16}$$

and

$$[\mathfrak{B}^{-1}\mathfrak{A}_\mu, S_{\nu\lambda}]=i(\delta_{\mu\lambda}\mathfrak{B}^{-1}\mathfrak{A}_\nu-\delta_{\mu\nu}\mathfrak{B}^{-1}\mathfrak{A}_\lambda).$$

If one views $'\chi=(1-i\frac{1}{2}\epsilon_{\mu\nu}S_{\mu\nu})\chi$ as a field in the original coordinate system and thus subject to the same dependence upon that coordinate system as χ, it is inferred that

$$L^{-1}S_{\mu\nu}L=r_{\mu\lambda}r_{\nu\kappa}S_{\lambda\kappa}.$$

For infinitesimal transformations, this reads

$$i[S_{\mu\nu}, S_{\lambda\kappa}]=\delta_{\mu\kappa}S_{\nu\lambda}-\delta_{\nu\kappa}S_{\mu\lambda}+\delta_{\nu\lambda}S_{\mu\kappa}-\delta_{\mu\lambda}S_{\nu\kappa}.$$

In performing the variation of the action integral, we shall treat the two types of quantities, coordinates and field variables, on somewhat the same footing, although the former are numbers and the latter operators. We introduce an arbitrary variation of the coordinates, δx_μ, throughout the interior of the region, but subject to the condition that the boundaries remain plane surfaces,

$$\partial_\mu\delta x_\nu+\partial_\nu\delta x_\mu=0, \tag{17}$$

on σ_1 and σ_2. The field components $\chi_r(x)$ are dependent both upon the coordinate system and the "intrinsic field." Under a rotation of the coordinate system, the field components are altered in the manner described by (14). Accordingly, we write the general variation of the field as the sum of an intrinsic field variation, and of the variation induced by the local rotation of the coordinate system,

$$\delta(\chi)=\delta\chi-i\tfrac{1}{2}(\partial_\mu\delta x_\nu)S_{\mu\nu}\chi,$$

where the antisymmetry of $S_{\mu\nu}$ ensures that only the rotation part of the coordinate displacement is effective. For the source field, a prescribed function of the coordinates, we have

$$\delta(\xi)=\delta x_\mu\partial_\mu\xi. \tag{18}$$

We also remark that

$$\delta(dx)=(dx)\partial_\mu\delta x_\mu,$$

and

$$\delta(\partial_\mu)=-(\partial_\mu\delta x_\nu)\partial_\nu,$$

whence

$$\delta(\partial_\mu\chi)=\partial_\mu\delta(\chi)-(\partial_\mu\delta x_\nu)\partial_\nu\chi. \tag{19}$$

The Lorentz invariance of $\mathfrak{L}$ produces a significant simplification, in computing the contribution to $\delta(\mathfrak{L})$ from the coordinate induced variation of χ. Thus, if $\partial_\mu\delta x_\nu$ were antisymmetrical and constant, its coefficient in the variation of the Lagrange function would vanish identically, save for the source term since the rotation induced change of ξ is not present in (18). Accordingly, for the general coordinate variation of (10), there remains only those terms in which $\partial_\mu\delta x_\nu$ is differentiated, or occurs in the dilation combination, $\partial_\mu\delta x_\nu+\partial_\nu\delta x_\mu$. Both types are contained entirely in (19), which leads to

$$\begin{aligned}\delta(\mathfrak{L})=\delta\mathfrak{L}-\tfrac{1}{2}(\partial_\mu\delta x_\nu+\partial_\nu\delta x_\mu)\tfrac{1}{2}(\chi\mathfrak{A}_\mu\partial_\nu\chi-\partial_\nu\chi\mathfrak{A}_\mu\chi)\\ -i\tfrac{1}{2}(\partial_\mu\partial_\nu\delta x_\lambda)\tfrac{1}{2}\chi(\mathfrak{A}_\mu S_{\nu\lambda}+S_{\nu\lambda}{}^\dagger\mathfrak{A}_\mu)\chi\\ -i\tfrac{1}{4}(\partial_\mu\delta x_\nu)(\xi\mathfrak{B}S_{\mu\nu}\chi-\chi S_{\mu\nu}{}^\dagger\mathfrak{B}\xi).\end{aligned}$$

In virtue of the symmetry of the second derivative,

$$\begin{aligned}(\partial_\mu\partial_\nu\delta x_\lambda)\chi(\mathfrak{A}_\mu S_{\nu\lambda}+S_{\nu\lambda}{}^\dagger\mathfrak{A}_\mu)\chi\\ =(\partial_\mu(\partial_\nu\delta x_\lambda+\partial_\lambda\delta x_\nu))\chi(\mathfrak{A}_\nu S_{\mu\lambda}+S_{\mu\lambda}{}^\dagger\mathfrak{A}_\nu)\chi\\ \to-(\partial_\nu\delta x_\lambda+\partial_\lambda\delta x_\nu)\partial_\mu[\chi(\mathfrak{A}_\nu S_{\mu\lambda}+S_{\mu\lambda}{}^\dagger\mathfrak{A}_\nu)\chi],\end{aligned}$$

where the last step expresses the result of an integration by parts, for which the integrated term vanishes, since the dilation tensor is zero on the boundaries (Eq. (17)). Collecting the coefficients of $\partial_\mu\delta x_\nu$ into the tensor $T_{\mu\nu}$, we have

$$\begin{aligned}\delta(W_{12})&=\int_{\sigma_2}^{\sigma_1}(dx)[\delta\mathfrak{L}+(\partial_\mu\delta x_\nu)T_{\mu\nu}]\\ &=\int_{\sigma_2}^{\sigma_1}(dx)[\delta\mathfrak{L}-\delta x_\nu\partial_\mu T_{\mu\nu}+\partial_\mu(T_{\mu\nu}\delta x_\nu)],\end{aligned}$$

where

$$\begin{aligned}T_{\mu\nu}=\mathfrak{L}\delta_{\mu\nu}-\tfrac{1}{2}(\chi\mathfrak{A}_{(\mu}\partial_{\nu)}\chi-\partial_{(\nu}\chi\mathfrak{A}_{\mu)}\chi)\\ -i\tfrac{1}{4}(\xi\mathfrak{B}S_{\mu\nu}\chi-\chi S_{\mu\nu}{}^\dagger\mathfrak{B}\xi)\\ +i\tfrac{1}{2}\partial_\lambda[\chi(\mathfrak{A}_{(\mu}S_{\lambda\nu)}+S_{\lambda(\nu}{}^\dagger\mathfrak{A}_{\mu)})\chi],\end{aligned} \tag{20}$$

and we have employed a notation for the symmetrical part of a tensor,

$$\mathfrak{A}_{(\mu}\partial_{\nu)}=\tfrac{1}{2}(\mathfrak{A}_\mu\partial_\nu+\mathfrak{A}_\nu\partial_\mu).$$

The expression for $\delta\mathfrak{L}$ is

$$\begin{aligned}\delta\mathfrak{L}=\delta\chi\mathfrak{A}_\mu\partial_\mu\chi-\partial_\mu\chi\mathfrak{A}_\mu\delta\chi-\delta\mathfrak{H}+\tfrac{1}{2}(\delta\chi\mathfrak{B}\xi+\xi\mathfrak{B}\delta\chi)\\ +\delta x_\mu\tfrac{1}{2}(\chi\mathfrak{B}\partial_\mu\xi+\partial_\mu\xi\mathfrak{B}\chi)+\partial_\mu[\tfrac{1}{2}(\chi\mathfrak{A}_\mu\delta\chi-\delta\chi\mathfrak{A}_\mu\chi)].\end{aligned}$$

Hence, on applying the principle of stationary action to coordinate and field variations, separately, we obtain

$$\partial_\mu T_{\mu\nu}=\tfrac{1}{2}(\chi\mathfrak{B}\partial_\mu\xi+\partial_\mu\xi\mathfrak{B}\chi),$$

and

$$\delta\mathfrak{H}=\delta\chi\mathfrak{A}_\mu\partial_\mu\chi-\partial_\mu\chi\mathfrak{A}_\mu\delta\chi+\tfrac{1}{2}(\delta\chi\mathfrak{B}\xi+\xi\mathfrak{B}\delta\chi), \tag{21}$$

while the surface terms yield, on σ_1 and σ_2, the infinitesimal generator

$$G=\int_\sigma d\sigma_\mu[\tfrac{1}{2}(\chi\mathfrak{A}_\mu\delta\chi-\delta\chi\mathfrak{A}_\mu\chi)+T_{\mu\nu}\delta x_\nu].$$

The operator $\mathfrak{H}$ is an arbitrary, invariant function of the field χ. If its variation is to possess the form (21), with $\delta\chi$ appearing on the left and on the right, the latter must possess elementary operator properties, characterizing the class of variations to which the action principle refers. Thus, we should be able to displace $\delta\chi$ entirely to the left, or to the right, in the structure of $\delta\mathfrak{H}$,

$$\delta\mathfrak{H}=\delta\chi(\partial_l\mathfrak{H}/\partial\chi)=(\partial_r\mathfrak{H}/\partial\chi)\delta\chi,$$

which defines the left and right derivatives of $\mathfrak{H}$ with respect to χ. In view of the complete symmetry between left and right in the process of multiplication, we infer that the expressions with $\delta\chi$ on the left and on the right are, in fact, identical. The field equations, therefore, possess the two equivalent forms

$$2\mathfrak{A}_\mu\partial_\mu\chi=(\partial_l\mathfrak{H}/\partial\chi)-\mathfrak{B}\xi,$$
$$-\partial_\mu\chi 2\mathfrak{A}_\mu=(\partial_r\mathfrak{H}/\partial\chi)-\xi\mathfrak{B},$$

and G can be equivalently written

$$\begin{aligned}G&=\int_\sigma d\sigma_\mu[\chi\mathfrak{A}_\mu\delta\chi+T_{\mu\nu}\delta x_\nu]\\ &=\int_\sigma d\sigma_\mu[-\delta\chi\mathfrak{A}_\mu\chi+T_{\mu\nu}\delta x_\nu].\end{aligned} \tag{22}$$

In keeping with the restriction of the stationary action principle to fixed dynamical systems, the external source has not been altered. If we now introduce an infinitesimal variation of ξ, and extend the argument of the previous paragraph to $\delta\xi$, we obtain the two equivalent expressions for the change induced in W_{12},

$$\delta_\xi W_{12}=\int_{\sigma_2}^{\sigma_1}(dx)\delta\xi\mathfrak{B}\chi=\int_{\sigma_2}^{\sigma_1}(dx)\chi\mathfrak{B}\delta\xi.$$

The corresponding modification in the relation between

states on σ_1 and on σ_2 can be ascribed to the individual states only if one introduces a convention, of the nature of a boundary condition. Thus, we may suppose that the state on σ_2 is unaffected by varying the external source in the region between σ_1 and σ_2. In this "retarded" description, $\delta_\xi W_{12}$ generates the infinitesimal transformation of the state on σ_1. An alternative, "advanced" description corresponds to $-\delta_\xi W_{12}$ generating the change in the state on σ_2, with a fixed state on σ_1. These are just the simplest of possible boundary conditions.

The suitability of the designations, retarded and advanced, can be seen by considering the matrix of an operator constructed from dynamical variables on some surface σ, intermediate between σ_1 and σ_2,

$$(\zeta_1'\sigma_1|F(\sigma)|\zeta_2''\sigma_2) = \int (\zeta_1'\sigma_1|\zeta'\sigma)d\zeta'(\zeta'\sigma|F(\sigma)|\zeta''\sigma)d\zeta''(\zeta''\sigma|\zeta_2''\sigma_2).$$

An infinitesimal change of the source ξ produces the following change in the matrix element,

$$\begin{aligned}\delta_\xi(\zeta_1'\sigma_1|F(\sigma)|\zeta_2''\sigma_2) &= (\zeta_1'\sigma_1|(\partial_\xi F(\sigma)+i\delta_\xi W_{1\sigma}F(\sigma)+iF(\sigma)\delta_\xi W_{\sigma 2})|\zeta_2''\sigma_2)\\ &= (\zeta_1'\sigma_1|(\partial_\xi F(\sigma)+i(F(\sigma)\delta_\xi W_{12})_+)|\zeta_2''\sigma_2),\end{aligned}$$

in which we have allowed for the possibility that $F(\sigma)$ may be explicitly dependent upon the source, and introduced a notation for temporally ordered products. The matrix element depends upon the external source through the operator $F(\sigma)$, and the eigenvectors on σ_1 and σ_2. One thereby gets various expressions for $\delta_\xi F(\sigma)$, depending upon the boundary conditions that are adopted. Thus, if the state on σ_2 is prescribed, we find

$$\begin{aligned}\delta_\xi F(\sigma)]_{\text{ret}} &= \partial_\xi F(\sigma)+i(F(\sigma)\delta_\xi W_{12})_+ - i\delta W_{12}F(\sigma)\\ &= \partial_\xi F(\sigma)+i[F(\sigma), \delta_\xi W_{\sigma 2}],\end{aligned} \tag{23}$$

which only involves changes in the source prior to, or on σ. The opposite convention yields the analogous result

$$\begin{aligned}\delta_\xi F(\sigma)]_{\text{adv}} &= \partial_\xi F(\sigma)+i(F(\sigma)\delta_\xi W_{12})_+ - iF(\sigma)\delta W_{12}\\ &= \partial_\xi F(\sigma)-i[F(\sigma), \delta_\xi W_{1\sigma}].\end{aligned}$$

Note that

$$\delta_\xi F(\sigma)]_{\text{ret}} - \delta_\xi F(\sigma)]_{\text{adv}} = i[F(\sigma), \delta_\xi W_{12}].$$

The operator G of Eq. (22) consists of two parts,

$$G = G_\chi + G_x,$$

where

$$G_\chi = \int_\sigma d\sigma_\mu \chi \mathfrak{A}_\mu \delta\chi = -\int_\sigma d\sigma_\mu \delta\chi \mathfrak{A}_\mu \chi,$$

and

$$G_x = \int d\sigma_\mu T_{\mu\nu}\delta x_\nu = \epsilon_\nu P_\nu + \tfrac{1}{2}\epsilon_{\mu\nu}J_{\mu\nu}.$$

The latter form of G_x is a consequence of the restriction to plane space-like surfaces, limiting displacements to infinitesimal translations and rotations,

$$\delta x_\nu = \epsilon_\nu + \epsilon_{\mu\nu}x_\mu,$$

with the associated operators, the energy-momentum vector

$$P_\nu = \int d\sigma_\mu T_{\mu\nu},$$

and angular momentum tensor

$$J_{\mu\nu} = \int d\sigma_\lambda M_{\lambda\mu\nu},$$

$$M_{\lambda\mu\nu} = x_\mu T_{\lambda\nu} - x_\nu T_{\lambda\mu}.$$

The operator G_x evidently generates the infinitesimal transformation of an eigenvector, produced by the displacement of the surface to which it refers. With the notation

$$\delta_x\Psi(\zeta'\sigma) = (\epsilon_\nu\delta_\nu + \tfrac{1}{2}\epsilon_{\mu\nu}\delta_{\mu\nu})\Psi(\zeta'\sigma),$$

we have

$$i\delta_\nu\Psi(\zeta'\sigma) = P_\nu\Psi(\zeta'\sigma), \quad -i\delta_\nu\Psi(\zeta'\sigma)^\dagger = \Psi(\zeta'\sigma)^\dagger P_\nu,$$

and

$$i\delta_{\mu\nu}\Psi(\zeta'\sigma) = J_{\mu\nu}\Psi(\zeta'\sigma), \quad -i\delta_{\mu\nu}\Psi(\zeta'\sigma)^\dagger = \Psi(\zeta'\sigma)^\dagger J_{\mu\nu}.$$

If $F(\sigma)$ is an arbitrary function of dynamical variables on σ, and possibly of nondynamical parameters dependent on σ, we use the notation

$$\begin{aligned}\delta_x F(\sigma) &= (\epsilon_\nu\delta_\nu + \tfrac{1}{2}\epsilon_{\mu\nu}\delta_{\mu\nu})F(\sigma),\\ \partial_x F(\sigma) &= (\epsilon_\nu\partial_\nu + \tfrac{1}{2}\epsilon_{\mu\nu}\partial_{\mu\nu})F(\sigma),\end{aligned}$$

to distinguish between the total change on displacement, and that occasioned by the explicit appearance of nondynamical parameters. On referring to Eq. (3), we see that

$$\begin{aligned}\delta_\nu F(\sigma) &= \partial_\nu F(\sigma) + i[F(\sigma), P_\nu],\\ \delta_{\mu\nu}F(\sigma) &= \partial_{\mu\nu}F(\sigma) + i[F(\sigma), J_{\mu\nu}].\end{aligned}$$

The proper interpretation of the generating operator G_χ can be obtained by noting its equivalence with an appropriately chosen infinitesimal variation of the external source. Consider the following infinitesimal surface distribution on the negative side of σ,

$$\mathfrak{B}\delta\xi = \mathfrak{A}_{(0)}\delta\chi\delta(x_{(0)}), \tag{24}$$

which is not incompatible with the operator properties of these variations. We have assumed, for simplicity, that the equation of the surface σ is $x_{(0)}=0$. With this choice,

$$\delta_\xi W_{12} = \int_\sigma d\sigma\chi\mathfrak{A}_{(0)}\delta\chi = G_\chi.$$

The change that is produced in χ can be deduced from

the variation of the field equatons,

$$2\mathfrak{A}_\mu\partial_\mu\delta_\xi\chi-\delta_\xi(\partial_l\mathcal{H}/\partial\chi)=-\mathfrak{B}\delta\xi$$
$$=-\mathfrak{A}_{(0)}\delta\chi\delta(x_{(0)}).$$

Evidently there is a discontinuity in $\delta_\xi\chi$, on crossing the surface distribution $\delta\xi$, which is given by

$$2\mathfrak{A}_{(0)}\delta_\xi\chi]=-\mathfrak{A}_{(0)}\delta\chi.$$

In the retarded description, say, $\delta_\xi\chi$ is zero prior to the source bearing surface, so that the discontinuity in $\delta_\xi\chi$ is the change induced in χ on (the positive side of) σ. Thus, the surface variation of the external source simulates the transformation generated by G_χ, in which $\mathfrak{A}_{(0)}\chi$ on σ is replaced by

$$\mathfrak{A}_{(0)}\bar{\chi}=\mathfrak{A}_{(0)}\chi+\mathfrak{A}_{(0)}\delta_\xi\chi$$
$$=\mathfrak{A}_{(0)}\chi-\tfrac{1}{2}\mathfrak{A}_{(0)}\delta\chi. \qquad (25)$$

The matrix $\mathfrak{A}_{(0)}$ has been retained in this statement since it is a singular matrix, in general. The number of components of χ that appear independently in (25) equals the rank of the matrix $\mathfrak{A}_{(0)}$, and this is the number of independent component field equations that are equations of motion, in that they contain time-like derivatives. The expression of (25) in terms of the generator G_χ is

$$[\mathfrak{A}_{(0)}\chi, G_\chi]=i\tfrac{1}{2}\mathfrak{A}_{(0)}\delta\chi. \qquad (26)$$

The factor of $\frac{1}{2}$ that appears in this result stems from the treatment of all components of $\mathfrak{A}_{(0)}\chi$ on the same footing; we have not divided them into two sets of which one is fixed and the other varied.* If F is an arbitrary function of $\mathfrak{A}_{(0)}\chi$ on σ, we write

$$[F, G_\chi]=i(\delta F)_\chi=i\tfrac{1}{2}\delta F,$$

in which the components of $\mathfrak{A}_{(0)}\chi$ are the objects of variation. When the field equations that are equations of constraint prove sufficient to express all components of χ in terms of $\mathfrak{A}_{(0)}\chi$, we can extend (26) into

$$[\chi, G_\chi]=i\tfrac{1}{2}\delta\chi.$$

Of course, one must distinguish between these variations, in which only the $\mathfrak{A}_{(0)}\chi$ are independent, and the independent variations of all components of χ which produce the equations of constraint from the action principle.

In order to facilitate the explicit construction of the field commutation relations, we shall introduce a reducibility hypothesis, which is associated with the Lorentz invariant process of separating the matrices $\mathfrak{A}_\mu$, $\mathfrak{B}$ into symmetrical and antisymmetrical parts. We require that the field and the source decompose into two sets, of the first kind $\chi^{(1)}=\phi$, $\xi^{(1)}=\zeta$, and of the second kind, $\chi^{(2)}=\psi$, $\xi^{(2)}=\eta$, as a concomitant of the decomposition

$$\mathfrak{A}_\mu=\mathfrak{A}_\mu^{(1)}+\mathfrak{A}_\mu^{(2)}, \qquad \mathfrak{B}=\mathfrak{B}^{(1)}+\mathfrak{B}^{(2)},$$
$$\mathfrak{A}_\mu^{(1)\,\mathrm{tr}}=-\mathfrak{A}_\mu^{(1)}, \qquad \mathfrak{B}^{(1)\,\mathrm{tr}}=\mathfrak{B}^{(1)},$$
$$\mathfrak{A}_\mu^{(2)\,\mathrm{tr}}=\mathfrak{A}_\mu^{(2)}, \qquad \mathfrak{B}^{(2)\,\mathrm{tr}}=-\mathfrak{B}^{(2)}.$$

The matrices of the first kind are real ($\mu=0,\cdots 3$), and those of the second kind are imaginary. We shall not write the distinguishing index when no confusion is possible.

According to this reducibility hypothesis, the field equations in the two equivalent forms

$$2\mathfrak{A}_\mu\partial_\mu\chi=(\partial_l\mathcal{H}/\partial\chi)-\mathfrak{B}\xi,$$
$$-2\mathfrak{A}_\mu^{\mathrm{tr}}\partial_\mu\chi=(\partial_r\mathcal{H}/\partial\chi)-\mathfrak{B}^{\mathrm{tr}}\xi,$$

separate into the two sets

$$2\mathfrak{A}_\mu\partial_\mu\phi=(\partial\mathcal{H}/\partial\phi)-\mathfrak{B}\zeta, \quad (\partial_l\mathcal{H}/\partial\phi)=(\partial_r\mathcal{H}/\partial\phi),$$

and

$$2\mathfrak{A}_\mu\partial_\mu\psi=(\partial_l\mathcal{H}/\partial\psi)-\mathfrak{B}\eta, \quad (\partial_l\mathcal{H}/\partial\psi)=-(\partial_r\mathcal{H}/\partial\psi).$$

Furthermore, the generator

$$G_\chi=\int d\sigma\chi\mathfrak{A}_{(0)}\delta\chi=\int d\sigma(-\mathfrak{A}_{(0)}{}^{\mathrm{tr}}\delta\chi)\chi,$$

decomposes into $G_\phi+G_\psi$, where

$$G_\phi=\int d\sigma\phi\mathfrak{A}_{(0)}\delta\phi=\int d\sigma(\mathfrak{A}_{(0)}\delta\phi)\phi,$$

and (27)

$$G_\psi=\int d\sigma\psi\mathfrak{A}_{(0)}\delta\psi=\int d\sigma(-\mathfrak{A}_{(0)}\delta\psi)\psi.$$

These results reflect the form assumed by the Lagrange function,

$$\mathcal{L}=\tfrac{1}{2}\{\phi\mathfrak{A}_\mu, \partial_\mu\phi\}+\tfrac{1}{2}[\psi\mathfrak{A}_\mu, \partial_\mu\psi]-\mathcal{H}(\phi,\psi)$$
$$+\tfrac{1}{2}\{\zeta\mathfrak{B}, \phi\}+\tfrac{1}{2}[\eta\mathfrak{B}, \psi].$$

The equivalence between left and right derivatives of the arbitrary function $\mathcal{H}$, with respect to field components of the first kind, and of the two expressions for G_ϕ, shows that $\delta\phi$ commutes with all fields at the same point. It is compatible with the field equations to extend this statement to fields at arbitrary points,

$$[\phi(x), \delta\phi(x')]=[\psi(x), \delta\phi(x')]=0,$$

provided the source components are included,

$$[\zeta(x), \delta\phi(x')]=[\eta(x), \delta\phi(x')]=0.$$

It follows from (27) that the relation between ψ and $\delta\psi$ is one of anticommutivity. The opposite signs of the left and right derivatives of $\mathcal{H}$ with respect to ψ is then accounted for by

$$[\phi(x), \delta\psi(x')]=\{\psi(x), \delta\psi(x')\}=0,$$

provided only that $\mathcal{H}$ is an even function of the vari-

* *Note added in proof:*—Further discussion of this point will be found in a paper submitted to the *Philosophical Magazine.*

ables of the second kind. The inclusion of the source components

$$[\zeta(x), \delta\psi(x')] = \{\eta(x), \delta\psi(x')\} = 0,$$

insures compatibility with the field equations. We have now obtained the explicit characterization of the class of variations to which our fundamental postulate refers.

Let us also notice that

$$\delta_\xi W_{12} = \int_{\sigma_2}^{\sigma_1} (dx)\chi\mathfrak{B}\delta\xi = \int_{\sigma_2}^{\sigma_1} (dx)(\mathfrak{B}^{\mathrm{tr}}\delta\xi)\chi,$$

decomposes into $\delta_\zeta W_{12} + \delta_\eta W_{12}$, where

$$\delta_\zeta W_{12} = \int_{\sigma_2}^{\sigma_1} (dx)\phi\mathfrak{B}\delta\zeta = \int_{\sigma_2}^{\sigma_1} (dx)(\mathfrak{B}\delta\zeta)\phi,$$

and

$$\delta_\eta W_{12} = \int_{\sigma_2}^{\sigma_1} (dx)\psi\mathfrak{B}\delta\eta = \int_{\sigma_2}^{\sigma_1} (dx)(-\mathfrak{B}\delta\eta)\psi.$$

We can conclude that source variations have the same operator properties as field variations, as already exploited in Eq. (24).

The operator properties of $\mathfrak{A}_{(0)}\chi$ on a given σ can now be deduced from (26), with the results

$$\begin{aligned}
&[\mathfrak{A}_{(0)}\phi(x), \phi(x')\mathfrak{A}_{(0)}] = i\tfrac{1}{2}\mathfrak{A}_{(0)}\delta_\sigma(x-x'),\\
&[\mathfrak{A}_{(0)}\phi(x), \psi(x')\mathfrak{A}_{(0)}] = 0, \qquad (28)\\
&\{\mathfrak{A}_{(0)}\psi(x), \psi(x')\mathfrak{A}_{(0)}\} = i\tfrac{1}{2}\mathfrak{A}_{(0)}\delta_\sigma(x-x'),
\end{aligned}$$

in which $\delta_\sigma(x-x')$ is the three-dimensional delta function appropriate to the surface σ. The numerical forms of these commutators and anticommutators insures their consistency with the operator properties of $\delta\mathfrak{A}_{(0)}\phi$ and $\delta\mathfrak{A}_{(0)}\psi$. The dynamical variables of the first and second kind thus describe Bose-Einstein and Fermi-Dirac fields, respectively, which are unified in the general field χ.

Since the rank of the antisymmetrical matrix $\mathfrak{A}_{(0)}^{(1)}$ is necessarily even, there are an even number of independent field components of the first kind, say $2n^{(1)}$. One can always arrange the matrix $\mathfrak{A}_{(0)}^{(1)}$ so that all elements are zero beyond the first $2n^{(1)}$ rows and columns. We shall denote this nonsingular submatrix of dimensionality $2n^{(1)}$ by $\mathfrak{A}_{(0)}^{(1)}$, and the associated independent components of ϕ by $\boldsymbol{\phi}$. The first commutation relation of (28) can then be written

$$[\boldsymbol{\phi}(x), \boldsymbol{\phi}(x')] = i\tfrac{1}{2}\mathfrak{A}_{(0)}^{-1}\delta_\sigma(x-x').$$

The matrix $\mathfrak{B}^{(2)}$, associated with Fermi-Dirac fields, is antisymmetrical and nonsingular. Hence the total number of field components of the second kind is even. If we allow for the possibility that $\mathfrak{A}_{(0)}^{(2)}$ may be singular, and arrange the rows and columns so that the nonsingular submatrix $\mathfrak{A}_{(0)}^{(2)}$ is associated with the independent components $\boldsymbol{\psi}$, we obtain

$$\{\boldsymbol{\psi}(x), \boldsymbol{\psi}(x')\} = i\tfrac{1}{2}\mathfrak{A}_{(0)}^{-1}\delta_\sigma(x-x'),$$

which requires that the real, symmetrical matrix $i\mathfrak{A}_{(0)}^{(2)-1}$ be positive definite.

We shall argue that the number of independent field components of the second kind, the dimensionality of $\mathfrak{A}_{(0)}^{(2)}$, must be even, $2n^{(2)}$. Let us imagine that, by a suitable real transformation, $\mathfrak{A}_{(0)}^{(2)}$ is brought into diagonal form. If the number of components in $\boldsymbol{\psi}$ is odd, the product of all these components at a given point commutes with $\boldsymbol{\psi}$ at that point. Thus, as far as the algebra of operators at a given point is concerned, this product is a multiple of the unit operator (the necessary commutivity with $\boldsymbol{\psi}$ at other points on σ can always be achieved), which contradicts the assumption that all components of $\boldsymbol{\psi}$ are independent.

The relation between invariance under time reflection, and the connection between spin and statistics, may be noted here. The time reflection transformation

$$'x_4 = -x_4, \quad 'x_k = x_k,$$

induces a transformation of the field

$$'\chi = L_4\chi,$$

such that

$$L_4^{\mathrm{tr}}\mathfrak{A}_4L_4 = -\mathfrak{A}_4, \quad L_4^{\mathrm{tr}}\mathfrak{A}_kL_4 = \mathfrak{A}_k, \qquad (29)$$

and

$$L_4^{\mathrm{tr}}\mathfrak{B}L_4 = \mathfrak{B}, \quad \mathfrak{H}(L_4\chi) = \mathfrak{H}(\chi).$$

However, this preservation of the form of the Lagrange function is only apparent, for fields of the second kind. Since $-i\mathfrak{A}_{(0)}^{(2)}$ is a non-negative matrix, one can only satisfy the first equation of (29) with an imaginary $L_4^{(2)}$ which produces skew-Hermitian field components $'\chi^{(2)}$. But the invariance of the Lagrange function is not the correct criterion for invariance under time reflection. The reversal of the time sense inverts the order of σ_1 and σ_2, and thus introduces a minus sign in the action integral, which can only be compensated by changing the sign of i in (4). We shall describe this as a transformation from the algebra of the operators χ to the complex conjugate algebra of operators χ^*. Since the linear transformation designed to maintain the form of $\mathfrak{L}(\phi, \partial_\mu\phi; \psi, \partial_\mu\psi)$ has effectively replaced $\mathfrak{L}$ with $\mathfrak{L}(\phi, \partial_\mu\phi; i\psi, i\partial_\mu\psi)$, the criterion for invariance reads

$$\mathfrak{L}(\phi, \partial_\mu\phi; i\psi, i\partial_\mu\psi)^* = \mathfrak{L}(\phi^*, \partial_\mu\phi^*; \psi^*, \partial_\mu\psi^*).$$

The derivative term in $\mathfrak{L}$ is indeed invariant since the matrices $\mathfrak{A}_\mu^{(1)}$ and $\mathfrak{A}_\mu^{(2)}$ are real and imaginary, respectively. We describe this by saying that the theory is kinematically invariant under time reflection. In order that it be dynamically invariant, $\mathfrak{H}$ must be such that

$$\mathfrak{H}(\phi, i\psi)^* = \mathfrak{H}(\phi^*, \psi^*).$$

Since $\mathfrak{H}$ is an even function of the components of ψ, the latter are to be paired with the aid of imaginary matrices, characteristic of the variables of the second kind. The source term is invariant if source and field transform in the same way.

The correlation between spin and statistics enters on

observing that an imaginary L_4 is characteristic of half-integral spin fields. We can prove this by remarking that all the transformation properties of L_4 are satisfied by

$$L_4=\exp(-\tfrac{1}{2}\pi i S_{14})L_1\exp(\tfrac{1}{2}\pi i S_{14})=\exp(-\pi i S_{14})L_1,$$

where L_1 is the matrix describing the reflection of the first space axis. The latter form is a consequence of

$$L_1^{-1}S_{14}L_1=-S_{14}.$$

The essential point with regard to the reality of L_4 is that $S_{14}=iS_{10}$ is a real matrix, whence

$$L_4^*=\exp(\pi i S_{14})L_1=\exp(2\pi i S_{14})L_4.$$

Now S_{14} must possess the same eigenvalues as S_{12}, say, which implies that L_4 is real for an integral spin field, and imaginary for a half-integral spin field. The requirement of time reflection invariance thus restricts fields of the first (B.E.) and second (F.D.) kind to integral and half-integral spins, respectively. This correlation is also satisfactory in that it identifies the double-valued, half-integral spin fields with fields of the second kind, of which $\mathfrak{L}$ is an even function.

We have introduced several kinds of generators of infinitesimal transformations. A criterion for consistency is obtained from the alternative evaluations of the commutator of two such generators,

$$[G_a,G_b]=i(\delta G_a)_b=-i(\delta G_b)_a,$$

namely

$$(\delta G_a)_b+(\delta G_b)_a=0.$$

As a first example, we consider the two generators

$$G_x=\epsilon_\nu P_\nu(\sigma_1)+\tfrac{1}{2}\epsilon_{\mu\nu}J_{\mu\nu}(\sigma_1),$$

and

$$G_\xi=\int_{\sigma_2}^{\sigma_1}(dx)\chi\mathfrak{B}\delta\xi,$$

in the retarded description. In preparation for the test, we remark that

$$P_\nu(\sigma_1)-P_\nu(\sigma_2)=\int_{\sigma_2}^{\sigma_1}(dx)\partial_\mu T_{\mu\nu}$$
$$=\int_{\sigma_2}^{\sigma_1}(dx)\tfrac{1}{2}(\chi\mathfrak{B}\partial_\nu\xi+\partial_\nu\xi\mathfrak{B}\chi),$$

and that

$$J_{\mu\nu}(\sigma_1)-J_{\mu\nu}(\sigma_2)=\int_{\sigma_2}^{\sigma_1}(dx)\partial_\lambda M_{\lambda\mu\nu}$$
$$=\int_{\sigma_2}^{\sigma_1}(dx)[x_\mu\partial_\lambda T_{\lambda\nu}-x_\nu\partial_\lambda T_{\lambda\mu}+T_{\mu\nu}-T_{\nu\mu}].$$

Since

$$T_{\mu\nu}-T_{\nu\mu}=-i\tfrac{1}{2}(\xi\mathfrak{B}S_{\mu\nu}\chi-\chi S_{\mu\nu}{}^\dagger\mathfrak{B}\xi),$$

we have

$$J_{\mu\nu}(\sigma_1)-J_{\mu\nu}(\sigma_2)=\int_{\sigma_2}^{\sigma_1}(dx)\tfrac{1}{2}[\chi\mathfrak{B}(x_\mu\partial_\nu-x_\nu\partial_\mu+iS_{\mu\nu})\xi+(x_\mu\partial_\nu-x_\nu\partial_\mu+iS_{\mu\nu})\xi\mathfrak{B}\chi].$$

In the absence of an external source, $T_{\mu\nu}$ is symmetrical and divergenceless, and P_ν, $J_{\mu\nu}$ are conserved. For simplicity, we shall confine our verification to the situation of no source, in which the infinitesimal $\delta\xi$ is distributed in the region between σ_1 and σ_2. Hence

$$\delta_\xi P_\nu(\sigma_1)=-\int_{\sigma_2}^{\sigma_1}(dx)\partial_\nu\chi\mathfrak{B}\delta\xi,$$

and

$$\delta_\xi J_{\mu\nu}(\sigma_1)=-\int_{\sigma_2}^{\sigma_1}(dx)(x_\mu\partial_\nu-x_\nu\partial_\mu+iS_{\mu\nu})\chi\mathfrak{B}\delta\xi.$$

The consistency requirement

$$(\delta G_\xi)_x=\int_{\sigma_2}^{\sigma_1}(dx)(\delta\chi)_x\mathfrak{B}\delta\xi=\delta_\xi G_x,$$

then demands that

$$-(\delta\chi)_x=\epsilon_\nu\partial_\nu\chi+\tfrac{1}{2}\epsilon_{\mu\nu}(x_\mu\partial_\nu-x_\nu\partial_\mu+iS_{\mu\nu})\chi, \tag{30}$$

which is indeed true in virtue of the equivalence between $(\delta\chi(x))_x$, induced by the displacement δx_μ, and $'\chi(x)-\chi(x)$, induced by the coordinate transformation $'x_\mu=x_\mu+\delta x_\mu$.

Alternative forms of P_ν and $J_{\mu\nu}$ are convenient for testing the consistency of G_x and G_χ. The following relations derived from (16),

$$\chi\mathfrak{A}_\nu\partial_\mu\chi-\chi\mathfrak{A}_\mu\partial_\nu\chi=i\chi(\mathfrak{A}_\lambda S_{\mu\nu}-S_{\mu\nu}{}^\dagger\mathfrak{A}_\lambda)\partial_\lambda\chi,$$
$$\partial_\mu\chi\mathfrak{A}_\nu\chi-\partial_\nu\chi\mathfrak{A}_\mu\chi=i\partial_\lambda\chi(\mathfrak{A}_\lambda S_{\mu\nu}-S_{\mu\nu}{}^\dagger\mathfrak{A}_\lambda)\chi,$$

enable us to write $T_{\mu\nu}$ as

$$T_{\mu\nu}=\mathfrak{L}\delta_{\mu\nu}-\tfrac{1}{2}(\chi\mathfrak{A}_\mu\partial_\nu\chi-\partial_\nu\chi\mathfrak{A}_\mu\chi)+\partial_\lambda s_{\lambda\mu\nu}+\rho_{\mu\nu},$$

where

$$s_{\lambda\mu\nu}=-s_{\mu\lambda\nu}=i\tfrac{1}{4}\chi(2\mathfrak{A}_{(\mu}S_{\lambda\nu)}+2S_{\lambda(\nu}{}^\dagger\mathfrak{A}_{\mu)}-\mathfrak{A}_\lambda S_{\mu\nu}-S_{\mu\nu}{}^\dagger\mathfrak{A}_\lambda)\chi,$$

and

$$\rho_{\mu\nu}=-i\tfrac{1}{4}[S_{\mu\nu}\chi(\partial_l\mathfrak{H}/\partial\chi)+(\partial_r\mathfrak{H}/\partial\chi)S_{\mu\nu}\chi].$$

In virtue of the antisymmetry of $s_{\lambda\mu\nu}$ in the first two indices, $\partial_\lambda s_{\lambda\mu\nu}$ is automatically divergenceless and does not contribute to the energy-momentum vector P_ν,

$$P_\nu=\int d\sigma_\mu[\mathfrak{L}\delta_{\mu\nu}-\tfrac{1}{2}(\chi\mathfrak{A}_\mu\partial_\nu\chi-\partial_\nu\chi\mathfrak{A}_\mu\chi)+\rho_{\mu\nu}],$$

but does enter in

$$J_{\mu\nu}=\int d\sigma_\lambda[-\tfrac{1}{2}\chi\mathfrak{A}_\lambda(x_\mu\partial_\nu-x_\nu\partial_\mu+iS_{\mu\nu})\chi+\tfrac{1}{2}(x_\mu\partial_\nu-x_\nu\partial_\mu+iS_{\mu\nu})\chi\mathfrak{A}_\lambda\chi+x_\mu\rho_{\lambda\nu}-x_\nu\rho_{\lambda\mu}]+\int(d\sigma_\nu x_\mu-d\sigma_\mu x_\nu)\mathfrak{L}.$$

The components of P_ν in a local coordinate system are

$$P_{(0)}=\int d\sigma[\mathfrak{H}-\chi\mathfrak{A}_{(k)}\partial_{(k)}\chi-\tfrac{1}{2}(\xi\mathfrak{B}\chi+\chi\mathfrak{B}\xi)],$$
$$P_{(k)}=\int d\sigma[-\chi\mathfrak{A}_{(0)}\partial_{(k)}\chi+\rho_{(0)(k)}], \tag{31}$$

while those of $J_{\mu\nu}$ are

$$J_{(0)(k)}=x_{(0)}P_{(k)}-\int d\sigma x_{(k)}[\mathfrak{H}-\tfrac{1}{2}(\chi\mathfrak{A}_{(l)}\partial_{(l)}\chi-\partial_{(l)}\chi\mathfrak{A}_{(l)}\chi)-\tfrac{1}{2}(\xi\mathfrak{B}\chi+\chi\mathfrak{B}\xi)]-\tfrac{1}{2}i\int d\sigma\chi(\mathfrak{A}_{(0)}S_{(0)(k)}+S_{(0)(k)}{}^{\dagger}\mathfrak{A}_{(0)})\chi, \tag{32}$$

$$J_{(k)(l)}=\int d\sigma[-\chi\mathfrak{A}_{(0)}(x_{(k)}\partial_{(l)}-x_{(l)}\partial_{(k)}+iS_{(k)(l)})\chi+x_{(k)}\rho_{(0)(l)}-x_{(l)}\rho_{(0)(k)}].$$

The quantity $\rho_{\mu\nu}$ is closely related to the infinitesimal expression of the scalar character of $\mathfrak{H}$,

$$\mathfrak{H}(\chi-i\tfrac{1}{2}\epsilon_{\mu\nu}S_{\mu\nu}\chi)-\mathfrak{H}(\chi)=0.$$

We can, indeed, conclude that

$$\rho_{\mu\nu}=0,$$

if $\mathfrak{H}$ is no more than quadratic in the components of various independent fields. We shall also prove this without the latter restriction, but, for simplicity, with the limitation that there are no equations of constraint. The commutation relations equivalent to (30),

$$[\chi,P_\nu]=-i\partial_\nu\chi,$$
$$[\chi,J_{\mu\nu}]=-i(x_\mu\partial_\nu-x_\nu\partial_\mu+iS_{\mu\nu})\chi,$$

imply that

$$[\chi,N_{\mu\nu}]=S_{\mu\nu}\chi,$$

where

$$N_{\mu\nu}=J_{\mu\nu}-x_\mu P_\nu+x_\nu P_\mu.$$

This enables one to express the scalar requirement on $\mathfrak{H}$ in the form

$$[\mathfrak{H},N_{\mu\nu}]=0.$$

The components

$$N_{(0)(k)}=\int d\sigma'(x_{(k)}-x_{(k)}')[\mathfrak{H}(x')-\tfrac{1}{2}(\chi\mathfrak{A}_{(l)}\partial_{(l)}'\chi-\partial_{(l)}'\chi\mathfrak{A}_{(l)}\chi)-\tfrac{1}{2}(\xi\mathfrak{B}\chi+\chi\mathfrak{B}\xi)-\tfrac{1}{2}i\int d\sigma\chi(\mathfrak{A}_{(0)}S_{(0)(k)}+S_{(0)(k)}{}^{\dagger}\mathfrak{A}_{(0)})\chi,$$

do not involve the unknown $\rho_{(0)(k)}$. According to our simplifying assumption of no constraint equations, the commutators (anticommutators) of all field components at x and x' contain the three-dimensional delta function $\delta_\sigma(x-x')$ and therefore vanish when multiplied by $x_{(k)}-x_{(k)}'$. Furthermore,

$$[\mathfrak{H}(x),\chi(x')]\mathfrak{A}_{(0)}=\tfrac{1}{2}i(\partial_r\mathfrak{H}/\partial\chi)\delta_\sigma(x-x'),$$

and

$$\mathfrak{A}_{(0)}[\chi(x'),\mathfrak{H}(x)]=\tfrac{1}{2}i(\partial_l\mathfrak{H}/\partial\chi)\delta_\sigma(x-x'),$$

from which we obtain

$$[\mathfrak{H},N_{(0)(k)}]=2i\rho_{(0)(k)}=0.$$

With this information, the proof is easily extended to all components of $\rho_{\mu\nu}$.

The consistency of the generators G_x and G_χ requires that

$$\tfrac{1}{2}\delta_\chi(\epsilon_\nu P_\nu+\tfrac{1}{2}\epsilon_{\mu\nu}J_{\mu\nu})=-\int d\sigma(\delta\chi)_x\mathfrak{A}_{(0)}\delta\chi,$$

or

$$\delta_\chi P_\nu=\int d\sigma\partial_\nu\chi 2\mathfrak{A}_{(0)}\delta\chi,$$

$$\delta_\chi J_{\mu\nu}=\int d\sigma(x_\mu\partial_\nu-x_\nu\partial_\mu+iS_{\mu\nu})\chi 2\mathfrak{A}_{(0)}\delta\chi,$$

which can now be verified from the expressions (31) and (32), with $\rho_{(0)(k)}=0$.

CHARGED FIELDS

Our considerations thus far specifically exclude the electromagnetic field (and the gravitational field). We introduce the concept of charge by requiring that the Lagrange function be invariant under constant phase (special gauge) transformations, the infinitesimal version of which is

$$'\chi=(1-i\delta\lambda\mathcal{E})\chi.$$

Here $\delta\lambda$ is a constant, and $\mathcal{E}$ is an imaginary matrix which can be viewed as a rotation matrix referring to a space other than the four-dimensional world. The invariance requirement implies that

$$\mathcal{E}^\dagger=\mathfrak{B}\mathcal{E}\mathfrak{B}^{-1},$$

or

$$(\mathfrak{B}\mathcal{E})^\dagger=\mathfrak{B}\mathcal{E},$$

and

$$[\mathcal{E},\mathfrak{B}^{-1}\mathfrak{A}_\mu]=[\mathcal{E},S_{\mu\nu}]=0,$$

and that

$$\mathfrak{H}(\chi-i\delta\lambda\mathcal{E}\chi)-\mathfrak{H}(\chi)=0.$$

We now write the general variation as

$$\delta(\chi)=\delta\chi-i\tfrac{1}{2}(\partial_\mu\delta x_\nu)S_{\mu\nu}\chi-i\delta\lambda\mathcal{E}\chi,$$

where $\delta\lambda$, characterizing a local phase transformation, is an arbitrary function of x, consistent with constant values on σ_1 and on σ_2. The additional contribution to $\delta(\mathfrak{L})$ thereby produced is

$$j_\mu\partial_\mu\delta\lambda-i\tfrac{1}{2}(\xi\mathfrak{B}\mathcal{E}\chi-\chi\mathfrak{B}\mathcal{E}\xi)\delta\lambda,$$

where

$$j_\mu=-i\chi\mathfrak{A}_\mu\mathcal{E}\chi$$

is the charge-current vector. The stationary action principle requires that

$$\partial_\mu j_\mu = -i\tfrac{1}{2}(\xi\mathfrak{B}\mathcal{E}\chi - \chi\mathfrak{B}\mathcal{E}\xi), \tag{33}$$

and yields as the phase transformation generator

$$G_\lambda = \int_\sigma d\sigma_\mu j_\mu \delta\lambda = Q\delta\lambda,$$

where Q is the charge operator.

The integral statement derived from (33),

$$Q(\sigma_1) - Q(\sigma_2) = \int_{\sigma_2}^{\sigma_1} (dx) i\tfrac{1}{2}(\chi\mathfrak{B}\mathcal{E}\xi - \xi\mathfrak{B}\mathcal{E}\chi),$$

becomes the conservation of charge in the absence of an external source. If an infinitesimal source is introduced in the region bounded by σ_1 and σ_2, we then have, in the retarded description,

$$\delta_\xi Q(\sigma_1) = -i\int_{\sigma_2}^{\sigma_1} (dx)\delta\xi\mathfrak{B}\mathcal{E}\chi$$

$$= i[Q(\sigma_1), G_\xi],$$

whence

$$[\chi, Q] = \mathcal{E}\chi.$$

This commutation relation also follows directly from the significance of G_λ, indicating the consistency of the latter with G_ξ.

We shall suppose that the matrix $\mathfrak{B}$ is an element of the algebra generated by $\mathfrak{B}^{-1}\mathfrak{A}_\mu$ and $S_{\mu\nu}$. It follows that $\mathfrak{B}$ commutes with $\mathcal{E}$, and therefore that the latter is explicitly Hermitian,

$$\mathcal{E}^\dagger = \mathcal{E}.$$

Such an antisymmetrical, imaginary matrix possesses real eigenvalues which are symmetrically distributed about zero; nonvanishing eigenvalues occur in oppositely signed pairs. Since $\mathcal{E}$ commutes with all members of the above-mentioned algebra, the charge-bearing character of a given field depends upon the reducibility of this algebra. Thus, if the algebra for a certain kind of field is irreducible, the only matrix commuting with all members of the algebra is the symmetrical unit matrix. Hence $\mathcal{E}=0$, and the field is electrically neutral. If, however, the matrix algebra is reducible to two similar algebras, as in

$$\mathfrak{A}_\mu = \begin{pmatrix} \mathfrak{A}_\mu & 0 \\ 0 & \mathfrak{A}_\mu \end{pmatrix}, \tag{34}$$

the matrix $\mathcal{E}$ exists and has the form (with the same partitioning)

$$\mathcal{E} = e\begin{pmatrix} 0 & -i \\ i & 0 \end{pmatrix}. \tag{35}$$

This describes a charged field, composed of particles with charges $\pm e$, the eigenvalues of $\mathcal{E}$. If three similar algebras are involved, the field contains particles with charges 0, $\pm e$.

To present $\mathcal{E}$ as a diagonal matrix, we must forego the choice of Hermitian field components. Thus, for the example of a charged F.D. field, where the field components decompose into $\psi_{(1)}$, $\psi_{(2)}$, corresponding to the structures (34) and (35), the mutually Hermitian conjugate operators

$$\psi_{(+)} = \psi_{(1)} - i\psi_{(2)}, \quad \psi_{(-)} = \psi_{(1)} + i\psi_{(2)},$$

are associated with eigenvalues $+e$ and $-e$, respectively. On introducing these field components, the derivative term in the Lagrange function, the electric current vector, and the commutation relations, respectively, read

$$\tfrac{1}{4}[\psi_{(-)}\mathfrak{A}_\mu, \partial_\mu\psi_{(+)}] + \tfrac{1}{4}[\psi_{(+)}\mathfrak{A}_\mu, \partial_\mu\psi_{(-)}], \tag{36}$$

$$-ie\tfrac{1}{2}(\psi_{(-)}\mathfrak{A}_\mu\psi_{(+)} - \psi_{(+)}\mathfrak{A}_\mu\psi_{(-)}), \tag{37}$$

and

$$\begin{aligned} &\{\mathfrak{A}_{(0)}\psi_{(+)}(x), \psi_{(+)}(x')\mathfrak{A}_{(0)}\} \\ &\qquad = \{\mathfrak{A}_{(0)}\psi_{(-)}(x), \psi_{(-)}(x')\mathfrak{A}_{(0)}\} = 0, \\ &\{\mathfrak{A}_{(0)}\psi_{(+)}(x), \psi_{(-)}(x')\mathfrak{A}_{(0)}\} \\ &\qquad = \{\mathfrak{A}_{(0)}\psi_{(-)}(x), \psi_{(+)}(x')\mathfrak{A}_{(0)}\} = i\mathfrak{A}_{(0)}\delta_\sigma(x-x'). \end{aligned} \tag{38}$$

There is evident symmetry with respect to the substitution $\psi_{(+)} \leftrightarrow \psi_{(-)}$, $e \leftrightarrow -e$.

Since $\psi_{(+)}$ and $\psi_{(-)}$ are Hermitian conjugate operators, we can arbitrarily select one as the primary non-Hermitian field. We shall write

$$\mathfrak{B}^{-1}\mathfrak{A}_\mu = i\gamma_\mu,$$

and

$$\psi_{(+)} = \psi, \quad \psi_{(-)}\mathfrak{B} = \psi^\dagger\mathfrak{B} = \bar{\psi}.$$

This yields the following forms for (36), (37), and (38):

$$\begin{aligned} &\tfrac{1}{4}[\bar{\psi}\gamma_\mu, i\partial_\mu\psi] - \tfrac{1}{4}[i\partial_\mu\bar{\psi}\gamma_\mu, \psi], \\ &e\tfrac{1}{2}[\bar{\psi}\gamma_\mu, \psi], \end{aligned} \tag{39}$$

and

$$\begin{aligned} &\{\gamma_{(0)}\psi(x), \gamma_{(0)}\psi(x')\} = \{\bar{\psi}(x)\gamma_{(0)}, \bar{\psi}(x')\gamma_{(0)}\} = 0, \\ &\{\gamma_{(0)}\psi(x), \bar{\psi}(x')\gamma_{(0)}\} = \gamma_{(0)}\delta_\sigma(x-x'). \end{aligned} \tag{40}$$

To express the now slightly obscured symmetry between positive and negative charge, we call $\psi_{(-)}$ the charge conjugate field

$$\psi^c = (-\mathfrak{B}^{-1})\bar{\psi}, \tag{41}$$

and state this symmetry as invariance under the substitution $\psi \leftrightarrow \psi^c$, $e \leftrightarrow -e$.

The matrices γ_μ; $\mu = 0, \cdots 3$, obey

$$\gamma_\mu^\dagger = \mathfrak{B}\gamma_\mu\mathfrak{B}^{-1},$$

and

$$\gamma_\mu^{\text{tr}} = -\mathfrak{B}\gamma_\mu\mathfrak{B}^{-1}, \tag{42}$$

since they are purely imaginary matrices. One should also recall that $\mathfrak{B}$ is an antisymmetrical, imaginary matrix. If we were to depart from these special structures by subjecting all matrices to an arbitrary unitary transformation, we should find that the only formal changes occur in (41) and (42), where the matrix $\mathfrak{B}$ appears modified by an orthogonal, rather than a unitary transformation. Hence, in a general representation these equations read

$$\psi^c = C\bar{\psi},$$

$$\gamma_\mu{}^{tr} = -C^{-1}\gamma_\mu C,$$

where C still exhibits the symmetry of $\mathfrak{B}$, appropriate to the example of a half-integral spin field,

$$C^{tr} = -C.$$

The commutation relations (40) are in the canonical form which corresponds to the division of the independent field components into two sets, such that one has vanishing anticommutators (commutators, for an integral spin field) among members of the same set. The generator of changes in ψ and $\bar{\psi}$, Eq. (27) in the notation of the charged half-integral spin field example, is

$$G(\psi, \bar{\psi}) = \tfrac{1}{2} i \int d\sigma (\bar{\psi}\gamma_{(0)}\delta\psi - \delta\bar{\psi}\gamma_{(0)}\psi),$$

which can be deduced directly from the Lagrange function derivative term (39). Associated with the freedom of altering the Lagrange function by the addition of a divergence, are various expressions for generating operators of changes in the field components. Thus, we have the following two simple possibilities for the derivative term and the associated generating operator,

$$\tfrac{1}{2}[\bar{\psi}\gamma_\mu, i\partial_\mu\psi],$$

$$G(\psi) = i \int d\sigma \bar{\psi}\gamma_{(0)}\delta\psi,$$

and

$$-\tfrac{1}{2}[i\partial_\mu\bar{\psi}\gamma_\mu, \psi],$$

$$G(\bar{\psi}) = -i \int d\sigma \delta\bar{\psi}\gamma_{(0)}\psi.$$

Evidently $G(\psi)$, for example, in the generator of alterations in the components $\gamma_{(0)}\psi$, with no change in $\bar{\psi}\gamma_{(0)}$. The associated commutation relations,

$$[\gamma_{(0)}\psi, G(\psi)] = i\gamma_{(0)}\delta\psi,$$

$$[\bar{\psi}\gamma_{(0)}, G(\psi)] = 0,$$

are satisfied in virtue of (40), and, conversely, in conjunction with the analogous statements for $G(\bar{\psi})$, imply these operator properties of the field components. The connection with the generator in the symmetrical treatment of all field components is given by

$$G(\psi, \bar{\psi}) = \tfrac{1}{2}G(\psi) + \tfrac{1}{2}G(\bar{\psi}),$$

which indicates the origin of the factor (1/2) in the general Eq. (26).

THE ELECTROMAGNETIC FIELD

The postulate of general gauge invariance motivates the introduction of the electromagnetic field. If all fields and sources are subjected to the general gauge transformation,

$$'\chi = \exp(-i\lambda(x)\mathcal{E})\chi = \chi \exp(i\lambda(x)\mathcal{E}),$$

the Lagrange function we have been considering alters in the following manner,

$$'\mathfrak{L} = \mathfrak{L} + j_\mu\partial_\mu\lambda.$$

The addition of the electromagnetic field Lagrange function,

$$\mathfrak{L}_{\text{emf}} = \tfrac{1}{2}\{j_\mu, A_\mu\} - \tfrac{1}{4}\{F_{\mu\nu}, \partial_\mu A_\nu - \partial_\nu A_\mu\} + \tfrac{1}{4}F_{\mu\nu}{}^2 + J_\mu A_\mu, \quad (43)$$

provides a compensating quantity through the associated gauge transformation

$$'A_\mu = A_\mu - \partial_\mu\lambda.$$

The term involving the external current J_μ is effectively gauge invariant if

$$\partial_\mu J_\mu = 0,$$

since the modification is in the form of a divergence. In the same sense, there is no objection to employing a form of the Lagrange function in which the second term of (43) is replaced by

$$\tfrac{1}{2}\{\partial_\mu F_{\mu\nu}, A_\nu\}. \quad (44)$$

We write the general variation of A_μ in the form

$$\delta(A_\mu) = \delta A_\mu - (\partial_\mu\delta x_\nu)A_\nu$$
$$= \delta A_\mu - \tfrac{1}{2}(\partial_\mu\delta x_\nu - \partial_\nu\delta x_\mu)A_\nu - \tfrac{1}{2}(\partial_\mu\delta x_\nu + \partial_\nu\delta x_\mu)A_\nu,$$

which ascribes to A_μ the same transformation properties as the gradient of a scalar, thus preserving the possibility of gauge transformations under arbitrary coordinate deformations. In a similar way,

$$\delta(F_{\mu\nu}) = \delta F_{\mu\nu} - (\partial_\mu\delta x_\lambda)F_{\lambda\nu} - (\partial_\nu\delta x_\lambda)F_{\mu\lambda}.$$

With regard to the derivation of the electromagnetic field equations from the action principle, it should be noted that general gauge invariance requires that the sources of charged fields depend implicitly upon the vector potential A_μ. We express this dependence by

$$\delta_A\xi(x') = \int (dx)(\delta\xi(x')/\delta A_\mu(x))\delta A_\mu(x).$$

Since the infinitesimal gauge transformation, $\delta A_\mu = -\partial_\mu\delta\lambda$, must induce the change $\delta\xi = -i\delta\lambda\mathcal{E}\xi$, we learn that

$$\partial_\mu(\delta\xi(x')/\delta A_\mu(x)) = -i\mathcal{E}\xi(x)\delta(x-x'). \quad (45)$$

One obtains the following field equations on varying

$F_{\mu\nu}$ and A_ν in the complete Lagrange function,

$$F_{\mu\nu}=\partial_\mu A_\nu-\partial_\nu A_\mu, \quad (46)$$

$$\partial_\nu F_{\mu\nu}=j_\mu+k_\mu+J_\mu, \quad (47)$$

where

$$k_\mu(x)=\tfrac{1}{2}\int(dx')[(\delta\xi(x')/\delta A_\mu(x))\mathfrak{B}\chi(x') +\chi(x')\mathfrak{B}(\delta\xi(x')/\delta A_\mu(x))],$$

is the contribution to the total current vector associated with charged field sources. We derive from (45) that

$$\partial_\mu k_\mu=i\tfrac{1}{2}(\xi\mathfrak{B}\mathcal{E}\chi-\chi\mathfrak{B}\mathcal{E}\xi).$$

But the total current vector is divergenceless in consequence of the electromagnetic field equations. Therefore

$$\partial_\mu j_\mu=-i\tfrac{1}{2}(\xi\mathfrak{B}\mathcal{E}\chi-\chi\mathfrak{B}\mathcal{E}\xi),$$

which is in agreement with (33).

After removing the terms in $\delta(\mathfrak{L}_{\text{emf}})$ that contribute to the field equations, we are left with

$$\delta(\mathfrak{L}_{\text{emf}})=\tfrac{1}{2}\{\delta j_\mu, A_\mu\}-\partial_\mu(F_{\mu\nu}\delta A_\nu)+A_\mu\partial_\nu J_\mu\delta x_\nu -\tfrac{1}{2}(\partial_\mu\delta x_\nu+\partial_\nu\delta x_\mu)(\tfrac{1}{2}\{j_\mu, A_\nu\}-\tfrac{1}{2}\{F_{\mu\lambda}, F_{\nu\lambda}\}) -(\partial_\mu\delta x_\nu)J_\mu A_\nu, \quad (48)$$

in which

$$\tfrac{1}{2}\{\delta j_\mu, A_\mu\}=-i\delta\chi\mathfrak{A}_\mu\mathcal{E}\{\chi, A_\mu\}=-i\{A_\mu,\chi\}\mathfrak{A}_\mu\mathcal{E}\delta\chi.$$

This term alters the field equations of charged fields,

$$2\mathfrak{A}_\mu(\partial_\mu\chi-i\mathcal{E}\tfrac{1}{2}\{A_\mu,\chi\})=(\partial_l\mathcal{H}/\partial\chi)-\mathfrak{B}\xi,$$

$$-(\partial_\mu\chi+i\tfrac{1}{2}\{\chi, A_\mu\}\mathcal{E})2\mathfrak{A}_\mu=(\partial_r\mathcal{H}/\partial\chi)-\xi\mathfrak{B}.$$

We have anticipated that not all components of A_μ commute with χ. The tensor $T_{\mu\nu}$ is now obtained as

$$T_{\mu\nu}=\cdots+\tfrac{1}{2}\{F_{\mu\lambda}, F_{\nu\lambda}\}-\tfrac{1}{2}\{j_{(\mu}, A_{\nu)}\}-J_\mu A_\nu, \quad (49)$$

where $\cdots$ stands for (20), but with $\mathfrak{L}$ the complete Lagrange function. The action principle supplies the differential equation

$$\partial_\mu T_{\mu\nu}=\tfrac{1}{2}(\chi\mathfrak{B}\partial_\nu\xi+\partial_\nu\xi\mathfrak{B}\chi)+A_\mu\partial_\nu J_\mu. \quad (50)$$

The divergence term in (48) yields the infinitesimal generator

$$G_A=-\int d\sigma_\mu F_{\mu\nu}\delta A_\nu=-\int d\sigma F_{(0)(k)}\delta A_{(k)}, \quad (51)$$

while the Lagrange function with the derivative term (44) would give

$$G_F=\int d\sigma_\mu\delta F_{\mu\nu}A_\nu=\int d\sigma\delta F_{(0)(k)}A_{(k)}. \quad (52)$$

The change in the action integral produced by a variation of the external current J_μ is given by

$$\delta_J W_{12}=\int_{\sigma_2}^{\sigma_1}(dx)\delta J_\mu A_\mu.$$

If δJ_μ has the explicitly divergenceless form

$$\delta J_\mu=\partial_\nu\delta M_{\mu\nu}, \quad M_{\mu\nu}=-M_{\nu\mu}, \quad (53)$$

where $\delta M_{\mu\nu}$ vanishes on σ_1 and σ_2, we find that

$$\delta_J W_{12}=\int_{\sigma_2}^{\sigma_1}(dx)\tfrac{1}{2}\delta M_{\mu\nu}F_{\mu\nu},$$

which makes it unnecessary to introduce an external source that is directly coupled to the field strength tensor $F_{\mu\nu}$.

The special nature of the electromagnetic field[3] is apparent in the form of the operator (52) generating changes in the local electric field components. Since one of the field equations is the equation of constraint

$$\partial_{(k)}F_{(0)(k)}=j_{(0)}+k_{(0)}+J_{(0)}, \quad (54)$$

the three variations $\delta F_{(0)(k)}$ cannot be arbitrarily assigned; the electromagnetic field and charged fields are not kinematically independent. This is evidently an aspect of the gauge invariance that links the two types of fields. Alternatively, we see from (51) that $A_{(0)}$ is not a dynamical variable subject to independent variations. But there is no field equation that expresses $A_{(0)}$ in terms of independent dynamical variables, in virtue of the arbitrariness associated with the existence of gauge transformations. Furthermore, a variation of $A_{(k)}$ in the form of a gradient, that is, a gauge transformation, yields a generating operator which, in consequence of (54), no longer contains electromagnetic field dynamical variables. Thus, in either form, (51) or (52), there are only two kinematically independent variations of the electromagnetic field quantities.

We now apply these generators to deduce commutation properties for the gauge invariant field strength components. According to the effect of a variation $\delta A_{(k)}$, upon the local components of $F_{\mu\nu}$ we have

$$[F_{(0)(k)}, G_A]=0,$$

$$[F_{(k)(l)}, G_A]=i(\partial_{(k)}\delta A_{(l)}-\partial_{(l)}\delta A_{(k)}),$$

whence

$$[F_{(0)(k)}(x), F_{(0)(l)}(x')]=0, \quad (55)$$

and

$$[F_{(k)(l)}(x), F_{(0)(m)}(x')] =i(\delta_{(k)(m)}\partial_{(l)}-\delta_{(l)(m)}\partial_{(k)})\delta_\sigma(x-x'). \quad (56)$$

In using G_F, we must restrict the electric field variation according to

$$\partial_{(k)}\delta F_{(0)(k)}=0,$$

which is identically satisfied on writing

$$\delta F_{(0)(k)}=\partial_{(l)}\delta Z_{(k)(l)}, \quad Z_{(k)(l)}=-Z_{(l)(k)}.$$

[3] Papers dealing with the situation peculiar to the electromagnetic field are legion. Of the older literature, the closest in spirit to our procedure is that of W. Pauli, *Handbuch der Physik* (Edwards Brothers, Ann Arbor, 1943), Vol. 24.

This yields the form

$$G_F=\int d\sigma\tfrac{1}{2}F_{(k)(l)}\delta Z_{(k)(l)}.$$

The expression of changes induced by $\delta F_{(0)(k)}$,

$$[F_{(k)(l)}, G_F]=0,$$
$$[F_{(0)(k)}, G_F]=i\delta F_{(0)(k)},$$

then provides the commutation properties

$$[F_{(k)(l)}(x), F_{(m)(n)}(x')]=0,$$
$$[F_{(0)(k)}(x), F_{(l)(m)}(x')] = i(\delta_{(k)(l)}\partial_{(m)}-\delta_{(k)(m)}\partial_{(l)})\delta_\sigma(x-x'). \tag{57}$$

where the latter is equivalent to (56).

An alternative derivation employs an infinitesimal change in the external source, distributed on (the negative side of) σ, $x_{(0)}=0$,

$$\delta M_{\mu\nu}=\delta m_{\mu\nu}\delta(x_{(0)}),$$

for which the associated generator is

$$G_m=\int d\sigma[\tfrac{1}{2}\delta m_{(k)(l)}F_{(k)(l)}-\delta m_{(0)(k)}F_{(0)(k)}].$$

The alteration produced in the field components follows from the field Eq. (47), and the form of (46) given by

$$\partial_\mu F_{\nu\lambda}+\partial_\nu F_{\lambda\mu}+\partial_\lambda F_{\mu\nu}=0. \tag{58}$$

Thus,

$$\partial_{(0)}(\delta F_{(0)(l)}-\delta M_{(0)(l)})=-\partial_{(k)}\delta F_{(k)(l)}+\partial_{(k)}\delta M_{(k)(l)},$$
$$\partial_{(0)}\delta F_{(k)(l)}=\partial_{(l)}\delta F_{(0)(k)}-\partial_{(k)}\delta F_{(0)(l)},$$

which yields the following discontinuities in $\delta F_{\mu\nu}$ on crossing the surface,

$$\delta F_{(0)(l)}]=\partial_{(k)}\delta m_{(k)(l)},$$
$$\delta F_{(k)(l)}]=\partial_{(l)}\delta m_{(0)(k)}-\partial_{(k)}\delta m_{(0)(l)}.$$

In the retarded description, these discontinuities are the actual changes in the field components on σ. On referring to the general formula (23), we obtain

$$\partial_{(k)}\delta m_{(k)(l)}=i[F_{(0)(l)}, G_m],$$
$$\partial_{(l)}\delta m_{(0)(k)}-\partial_{(k)}\delta m_{(0)(l)}=i[F_{(k)(l)}, G_m].$$

In view of the arbitrary values of $\delta m_{\mu\nu}$ on σ, these equations imply field strength commutation relations, which are identical with (55) and (57).

We give a related procedure which also illustrates the possibility of evaluating commutators of field quantities at points in time-like relation. The two field Eqs. (47) and (58) can be combined into (we incorporate k_ν with j_ν)

$$-\partial_\lambda^2 F_{\mu\nu}=\partial_\mu(j_\nu+J_\nu)-\partial_\nu(j_\mu+J_\mu).$$

A change in the external current, of the form (53), yields

$$-\partial_\lambda^2\delta_M F_{\mu\nu}-\partial_\mu\delta_M j_\nu+\partial_\nu\delta_M j_\mu =\partial_\mu\partial_\lambda\delta M_{\nu\lambda}-\partial_\nu\partial_\lambda\delta M_{\mu\lambda}, \tag{59}$$

where, in the retarded description

$$\delta_M F_{\mu\nu}(x)=i\left[F_{\mu\nu}(x), \int_{\sigma_2}^{\sigma}(dx')\tfrac{1}{2}\delta M_{\lambda\kappa}(x')F_{\lambda\kappa}(x')\right]$$
$$=\int_{\sigma_2}^{\sigma_1}(dx')\tfrac{1}{2}\delta M_{\lambda\kappa}(x')\eta_+(x-x')i \times[F_{\mu\nu}(x), F_{\lambda\kappa}(x')],$$

and η_+ is the discontinuous function

$$\eta_+(x-x')=1,\quad x_0>x_0'$$
$$=0,\quad x_0<x_0'.$$

We have a similar expression for $\delta_M j_\nu(x)$. On comparing the coefficients of $\delta M_{\lambda\kappa}(x')$ in (59) (our two treatments employing external sources are thus distinguished by surface and volume distributions of $\delta M_{\mu\nu}$, respectively), we find

$$-\partial_\lambda^2\eta_+(x-x')i[F_{\mu\nu}(x), F_{\lambda\kappa}(x')]$$
$$-\partial_\mu\eta_+(x-x')i[j_\nu(x), F_{\lambda\kappa}(x')]$$
$$+\partial_\nu\eta_+(x-x')i[j_\mu(x), F_{\lambda\kappa}(x')]$$
$$=(\delta_{\nu\lambda}\partial_\mu\partial_\kappa-\delta_{\nu\kappa}\partial_\mu\partial_\lambda-\delta_{\mu\lambda}\partial_\nu\partial_\kappa +\delta_{\mu\kappa}\partial_\nu\partial_\lambda)\delta(x-x'). \tag{60}$$

The value of $i[F_{\mu\nu}(x), F_{\lambda\kappa}(x')]$, for equal times, is then obtained from the coefficient of the differentiated delta function of the time coordinate, with the anticipated result.

In the approximation that neglects the dynamical relation between currents and fields at points in time-like relation, the differential Eq. (60) has the solution

$$\eta_+(x-x')i[F_{\mu\nu}(x), F_{\lambda\kappa}(x')] =(\delta_{\nu\lambda}\partial_\mu\partial_\kappa-\delta_{\nu\kappa}\partial_\mu\partial_\lambda-\delta_{\mu\lambda}\partial_\nu\partial_\kappa+\delta_{\mu\kappa}\partial_\nu\partial_\lambda)D_{\text{ret}}(x-x'),$$

where $D_{\text{ret}}(x-x')$ is the familiar retarded solution of

$$-\partial_\lambda^2 D_{\text{ret}}=\delta(x-x'). \tag{61}$$

Had we employed the advanced description, η_+ would be replaced by $-\eta_-$, where

$$\eta_-(x-x')=0,\quad x_0>x_0'$$
$$=1,\quad x_0<x_0',$$

and the advanced solution of (61) would appear. Subtracting these two results, we find

$$i[F_{\mu\nu}(x), F_{\lambda\kappa}(x')] =(\delta_{\nu\lambda}\partial_\mu\partial_\kappa-\delta_{\nu\kappa}\partial_\mu\partial_\lambda-\delta_{\mu\lambda}\partial_\nu\partial_\kappa+\delta_{\mu\kappa}\partial_\nu\partial_\lambda)D(x-x'),$$

in which $D(x-x')$ is the homogeneous solution of (61) provided by

$$D=D_{\text{ret}}-D_{\text{adv}}.$$

The kinematical relation between the electromagnetic field and charged fields, on a given σ, is most clearly indicated in a special choice of gauge, the so-called radiation gauge,

$$\partial_{(k)}A_{(k)}=0. \tag{62}$$

With this choice, the constraint equation for the electric field reads

$$\partial_{(k)}F_{(0)(k)}=-\partial_{(k)}{}^2A_{(0)}=j_{(0)}+J_{(0)},$$

so that the scalar potential is completely determined by the charge density,

$$A_{(0)}(x)=\int_\sigma d\sigma'\mathfrak{D}_\sigma(x-x')(j_{(0)}(x')+J_{(0)}(x')),$$

where

$$\mathfrak{D}_\sigma(x-x')=(1/4\pi)[(x_{(k)}-x_{(k)}')^2]^{-\frac{1}{2}}.$$

Evidently, $A_{(0)}$ does not commute with the components of charged fields. In this gauge, then, the dependence of the electric field upon the charged fields is made explicit through the decomposition of the electric field into transverse and longitudinal parts,

$$\begin{aligned}F_{(0)(k)}&=-\partial_{(0)}A_{(k)}-\partial_{(k)}A_{(0)}\\&=F_{(0)(k)}{}^{(T)}+F_{(0)(k)}{}^{(L)}.\end{aligned}$$

The inference that the transverse fields are the independent dynamical variables of the electromagnetic field in this gauge is confirmed on examining the generators G_A and G_F. Indeed,

$$G_A=-\int d\sigma F_{(0)(k)}\delta A_{(k)}=-\int d\sigma F_{(0)(k)}{}^{(T)}\delta A_{(k)},$$

and

$$G_F=\int d\sigma\delta F_{(0)(k)}A_{(k)}=\int d\sigma\delta F_{(0)(k)}{}^{(T)}A_{(k)},$$

in view of the transverse nature of $A_{(k)}$, Eq. (62). We can now derive the commutation properties of these dynamical variables from

$$[A_{(k)},G_A]=i\delta A_{(k)},\qquad [F_{(0)(k)}{}^{(T)},G_A]=0,$$
$$[F_{(0)(k)}{}^{(T)},G_F]=i\delta F_{(0)(k)}{}^{(T)},\qquad [A_{(k)},G_F]=0,$$

on taking into account the restrictions

$$\partial_{(k)}\delta A_{(k)}=\partial_{(k)}\delta F_{(0)(k)}{}^{(T)}=0,$$

produced by the transverse nature of these quantities. The Lagrange multiplier device permits us to deduce that

$$i[A_{(k)}(x),F_{(0)(l)}{}^{(T)}(x')]=\delta_{(k)(l)}\delta_\sigma(x-x')+\partial_{(l)}'\lambda_{(k)}.$$

The divergenceless character of the transverse electric field supplied the information

$$\partial_{(l)}'^2\lambda_{(k)}=\partial_{(k)}\delta_\sigma(x-x'),$$

whence

$$\lambda_{(k)}=-\partial_{(k)}\mathfrak{D}_\sigma(x-x').$$

The resulting commutator

$$\begin{aligned}i[A_{(k)}(x),F_{(0)(l)}{}^{(T)}(x')]&=\delta_{(k)(l)}\delta_\sigma(x-x')-\partial_{(k)}\partial_{(l)}'\mathfrak{D}_\sigma(x-x')\\&=(\delta_{(k)(l)}\delta_\sigma(x-x'))^{(T)},\end{aligned}$$

is also consistent with the transverse nature of $A_{(k)}$. The remaining commutation relations are

$$[A_{(k)}(x),A_{(l)}(x')]=[F_{(0)(k)}{}^{(T)}(x),F_{(0)(l)}{}^{(T)}(x')]=0.$$

We shall use the device of the external current to derive the commutation relations between the electromagnetic field tensor and the displacement generators P_ν, $J_{\mu\nu}$. According to (49) and (50),

$$P_\nu(\sigma_1)-P_\nu(\sigma_2)=\int_{\sigma_2}^{\sigma_1}(dx)[\cdots+A_\lambda\partial_\nu J_\lambda],$$

$$J_{\mu\nu}(\sigma_1)-J_{\mu\nu}(\sigma_2)=\int_{\sigma_2}^{\sigma_1}(dx)[\cdots+A_\lambda(x_\mu\partial_\nu-x_\nu\partial_\mu)J_\lambda+A_\mu J_\nu-A_\nu J_\mu],$$

in which we have indicated only the terms containing the external current. We consider an infinitesimal change in the latter possessing the form (53). In the retarded description, the resulting changes of P_ν and $J_{\mu\nu}$ on σ_1 are

$$\delta_M P_\nu(\sigma_1)=-\int_{\sigma_2}^{\sigma_1}(dx)\tfrac{1}{2}\delta M_{\lambda\kappa}\partial_\nu F_{\lambda\kappa},$$

$$\delta_M J_{\mu\nu}(\sigma_1)=-\int_{\sigma_2}^{\sigma_1}(dx)[\tfrac{1}{2}\delta M_{\lambda\kappa}(x_\mu\partial_\nu-x_\nu\partial_\mu)F_{\lambda\kappa}+\delta M_{\lambda\nu}F_{\mu\nu}-\delta M_{\lambda\mu}F_{\nu\lambda}].$$

When expressed in terms of the generator

$$G_M=\int_{\sigma_2}^{\sigma_1}(dx)\tfrac{1}{2}\delta M_{\lambda\kappa}F_{\lambda\kappa},$$

the following commutators are encountered,

$$\begin{aligned}&i[F_{\lambda\kappa},P_\nu]=\partial_\nu F_{\lambda\kappa},\\&i[F_{\lambda\kappa},J_{\mu\nu}]=(x_\mu\partial_\nu-x_\nu\partial_\mu)F_{\lambda\kappa}+\delta_{\nu\kappa}F_{\mu\lambda}-\delta_{\mu\kappa}F_{\nu\lambda}+\delta_{\mu\lambda}F_{\nu\kappa}-\delta_{\nu\lambda}F_{\mu\kappa}.\end{aligned}$$

Finally, we remark that the extension of (31) to include the electromagnetic field, in the radiation gauge, is

$$\begin{aligned}P_{(0)}=\int d\sigma[&\tfrac{1}{2}(F_{(0)(k)}{}^{(T)})^2+\tfrac{1}{4}(F_{(k)(l)})^2\\&+\mathcal{H}-\chi\mathfrak{A}_{(k)}(\partial_{(k)}-i\mathcal{E}A_{(k)})\chi\\&-J_{(k)}A_{(k)}+\tfrac{1}{2}(j_{(0)}+J_{(0)})A_{(0)}-\tfrac{1}{2}(\xi\mathfrak{B}\chi+\chi\mathfrak{B}\xi)],\end{aligned}$$

and

$$P_{(k)}=\int d\sigma\left[\tfrac{1}{2}\{F_{(0)(l)}{}^{(T)}, F_{(k)(l)}\}-\chi\mathfrak{A}_{(0)}\partial_{(k)}\chi\right].$$

In arriving at the expression for $P_{(0)}$, the noncommutivity of $A_{(0)}$ with χ must be taken into consideration, but produces no actual contribution. A variation of each of the independent fields yields

$$\delta P_\mu=\int d\sigma\left[\delta F_{(0)(k)}{}^{(T)}\partial_\mu A_{(k)}-\delta A_{(k)}\partial_\mu F_{(0)(k)}{}^{(T)}-\delta\chi 2\mathfrak{A}_{(0)}\partial_\mu\chi\right],$$

which confirms the consistency of the translation generator with the various field variation generators.

The Connection Between Spin and Statistics[1]

W. PAULI
Physikalisches Institut, Eidg. Technischen Hochschule, Zürich, Switzerland and Institute for Advanced Study, Princeton, New Jersey

(Received August 19, 1940)

In the following paper we conclude for the relativistically invariant wave equation for free particles: From postulate (I), according to which the energy must be positive, the necessity of *Fermi-Dirac* statistics for particles with arbitrary half-integral spin; from postulate (II), according to which observables on different space-time points with a space-like distance are commutable, the necessity of *Einstein-Bose* statistics for particles with arbitrary integral spin. It has been found useful to divide the quantities which are irreducible against Lorentz transformations into four symmetry classes which have a commutable multiplication like $+1$, -1, $+\epsilon$, $-\epsilon$ with $\epsilon^2=1$.

§1. UNITS AND NOTATIONS

SINCE the requirements of the relativity theory and the quantum theory are fundamental for every theory, it is natural to use as units the vacuum velocity of light c, and Planck's constant divided by 2π which we shall simply denote by $\hbar$. This convention means that all quantities are brought to the dimension of the power of a length by multiplication with powers of $\hbar$ and c. The reciprocal length corresponding to the rest mass m is denoted by $\kappa=mc/\hbar$.

As time coordinate we use accordingly the length of the light path. In specific cases, however, we do not wish to give up the use of the imaginary time coordinate. Accordingly, a tensor index denoted by small Latin letters i, refers to the imaginary time coordinate and runs from 1 to 4. A special convention for denoting the complex conjugate seems desirable. Whereas for quantities with the index 0 an asterisk signifies the complex-conjugate in the ordinary sense (e.g., for the current vector S_i the quantity S_0^* is the complex conjugate of the charge density S_0), in general $U^*_{ik\cdots}$ signifies: the complex-conjugate of $U_{ik\cdots}$ multiplied with $(-1)^n$, where n is the number of occurrences of the digit 4 among the i, k, $\cdots$ (e.g. $S_4=iS_0$, $S_4^*=iS_0^*$).

Dirac's spinors u_ρ with $\rho=1,\cdots,4$ have always a Greek index running from 1 to 4, and u_ρ^* means the complex-conjugate of u_ρ in the ordinary sense.

Wave functions, insofar as they are ordinary vectors or tensors, are denoted in general with capital letters, U_i, $U_{ik}\ldots$. The symmetry character of these tensors must in general be added explicitly. As classical fields the electromagnetic and the gravitational fields, as well as fields with rest mass zero, take a special place, and are therefore denoted with the usual letters φ_i, $f_{ik}=-f_{ki}$, and $g_{ik}=g_{ki}$, respectively.

The energy-momentum tensor T_{ik} is so defined, that the energy-density W and the momentum density G_k are given in natural units by $W=-T_{44}$ and $G_k=-iT_{k4}$ with $k=1, 2, 3$.

§2. IRREDUCIBLE TENSORS. DEFINITION OF SPINS

We shall use only a few general properties of those quantities which transform according to irreducible representations of the Lorentz group.[2] The proper Lorentz group is that continuous linear group the transformations of which leave the form

$$\sum_{k=1}^{4} x_k^2=\mathbf{x}^2-x_0^2$$

invariant and in addition to that satisfy the condition that they have the determinant $+1$

[1] This paper is part of a report which was prepared by the author for the Solvay Congress 1939 and in which slight improvements have since been made. In view of the unfavorable times, the Congress did not take place, and the publication of the reports has been postponed for an indefinite length of time. The relation between the present discussion of the connection between spin and statistics, and the somewhat less general one of Belinfante, based on the concept of charge invariance, has been cleared up by W. Pauli and F. J. Belinfante, Physica 7, 177 (1940).

[2] See B. L. v. d. Waerden, *Die gruppentheoretische Methode in der Quantentheorie* (Berlin, 1932).

and do not reverse the time. A tensor or spinor which transforms irreducibly under this group can be characterized by two integral positive numbers (p, q). (The corresponding "angular momentum quantum numbers" (j, k) are then given by $p=2j+1$, $q=2k+1$, with integral or half-integral j and k.)* The quantity $U(j, k)$ characterized by (j, k) has $p\cdot q=(2j+1)(2k+1)$ independent components. Hence to $(0, 0)$ corresponds the scalar, to $(\frac{1}{2}, \frac{1}{2})$ the vector, to $(1, 0)$ the self-dual skew-symmetrical tensor, to $(1, 1)$ the symmetrical tensor with vanishing spur, etc. Dirac's spinor u_ρ reduces to two irreducible quantities $(\frac{1}{2}, 0)$ and $(0, \frac{1}{2})$ each of which consists of two components. If $U(j, k)$ transforms according to the representation

$$U_r' = \sum_{s=1}^{(2j+1)(2k+1)} \Lambda_{rs} U_s,$$

then $U^*(k, j)$ transforms according to the complex-conjugate representation Λ^*. Thus for $k=j$, $\Lambda^*=\Lambda$. This is true only if the components of $U(j, k)$ and $U(k, j)$ are suitably ordered. For an arbitrary choice of the components, a similarity transformation of Λ and Λ^* would have to be added. In view of §1 we represent generally with U^* the quantity the transformation of which is equivalent to Λ^* if the transformation of U is equivalent to Λ.

The most important operation is the reduction of the product of two quantities

$$U_1(j_1, k_1)\cdot U_2(j_2, k_2)$$

which, according to the well-known rule of the composition of angular momenta, decompose into several $U(j, k)$ where, independently of each other j, k run through the values

$$j=j_1+j_2,\ j_1+j_2-1,\ \cdots,\ |j_1-j_2|$$
$$k=k_1+k_2,\ k_1+k_2-1,\ \cdots,\ |k_1-k_2|.$$

By limiting the transformations to the subgroup of space rotations alone, the distinction between the two numbers j and k disappears and $U(j, k)$ behaves under this group just like the product of two irreducible quantities $U(j)\,U(k)$ which in turn reduces into several irreducible $U(l)$ each having $2l+1$ components, with

$$l=j+k,\ j+k-1,\ \cdots,\ |j-k|.$$

Under the space rotations the $U(l)$ with integral l transform according to single-valued representation, whereas those with half-integral l transform according to double-valued representations. Thus the unreduced quantities $T(j, k)$ with integral (half-integral) $j+k$ are single-valued (double-valued).

If we now want to determine the spin value of the particles which belong to a given field it seems at first that these are given by $l=j+k$. Such a definition would, however, not correspond to the physical facts, for there then exists no relation of the spin value with the number of independent plane waves, which are possible in the absence of interaction) for given values of the components k_i in the phase factor $\exp i(\mathbf{kx})$. In order to define the spin in an appropriate fashion,[3] we want to consider first the case in which the rest mass m of all the particles is different from zero. In this case we make a transformation to the rest system of the particle, where all the space components of k_i are zero, and the wave function depends only on the time. In this system we reduce the field components, which according to the field equations do not necessarily vanish, into parts irreducible against space rotations. To each such part, with $r=2s+1$ components, belong r different eigenfunctions which under space rotations transform among themselves and which belong to a particle with spin s. If the field equations describe particles with only one spin value there then exists in the rest system only one such irreducible group of components. From the Lorentz invariance, it follows, for an arbitrary system of reference, that r or $\sum r$ eigenfunctions always belong to a given arbitrary k_i. The number of quantities $U(j, k)$ which enter the theory is, however, in a general coordinate system more complicated, since these quantities together with the vector k_i have to satisfy several conditions.

In the case of zero rest mass there is a special degeneracy because, as has been shown by Fierz, this case permits a gauge transformation of the

* In the spinor calculus this is a spinor with $2j$ undotted and $2k$ dotted indices.

[3] See M. Fierz, Helv. Phys. Acta **12**, 3 (1939); also L. de Broglie, Comptes rendus **208**, 1697 (1939); **209**, 265 (1939).

second kind.* If the field now describes only one kind of particle with the rest mass zero and a certain spin value, then there are for a given value of k_i only two states, which cannot be transformed into each other by a gauge transformation. The definition of spin may, in this case, not be determined so far as the physical point of view is concerned because the total angular momentum of the field cannot be divided up into orbital and spin angular momentum by measurements. But it is possible to use the following property for a definition of the spin. If we consider, in the q number theory, states where only one particle is present, then not all the eigenvalues $j(j+1)$ of the square of the angular momentum are possible. But j begins with a certain minimum value s and takes then the values $s, s+1, \cdots$.[4] This is only the case for $m=0$. For photons, $s=1$; $j=0$ is not possible for one single photon.[5] For gravitational quanta $s=2$ and the values $j=0$ and $j=1$ do not occur.

In an arbitrary system of reference and for arbitrary rest masses, the quantities U all of which transform according to double-valued (single-valued) representations with half-integral (integral) $j+k$ describe only particles with half-integral (integral) spin. A special investigation is required only when it is necessary to decide whether the theory describes particles with one single spin value or with several spin values.

§3. Proof of the Indefinite Character of the Charge in Case of Integral and of the Energy in Case of Half-Integral Spin

We consider first a theory which contains only U with integral $j+k$, i.e., which describes particles with integral spins only. It is not assumed that only particles with one single spin value will be described, but all particles shall have integral spin.

We divide the quantities U into two classes: (1) the "$+1$ class" with j integral, k integral; (2) the "-1 class" with j half-integral, k half-integral.

The notation is justified because, according to the indicated rules about the reduction of a product into the irreducible constituents under the Lorentz group, the product of two quantities of the $+1$ class or two quantities of the -1 class contains only quantities of the $+1$ class, whereas the product of a quantity of the $+1$ class with a quantity of the -1 class contains only quantities of the -1 class. It is important that the complex conjugate U^* for which j and k are interchanged belong to the same class as U. As can be seen easily from the multiplication rule, tensors with even (odd) number of indices reduce only to quantities of the $+1$ class (-1 class). The propagation vector k_i we consider as belonging to the -1 class, since it behaves after multiplication with other quantities like a quantity of the -1 class.

We consider now a homogeneous and linear equation in the quantities U which, however, does not necessarily have to be of the first order. Assuming a plane wave, we may put k_l for $-i\partial/\partial x_l$. Solely on account of the invariance against the *proper* Lorentz group it must be of the typical form

$$\sum kU^+=\sum U^-, \quad \sum kU^-=\sum U^+. \qquad (1)$$

This typical form shall mean that there may be as many different terms of the same type present, as there are quantities U^+ and U^-. Furthermore, among the U^+ may occur the U^+ as well as the $(U^+)^*$, whereas other U may satisfy reality conditions $U=U^*$. Finally we have omitted an *even* number of k factors. These may be present in arbitrary number in the term of the sum on the left- or right-hand side of these equations. It is now evident that these equations remain invariant under the substitution

$$k_i\to -k_i;\quad U^+\to U^+,\quad [(U^+)^*\to (U^+)^*];$$
$$U^-\to -U^-,\quad [(U^-)^*\to -(U^-)^*]. \qquad (2)$$

Let us consider now tensors T of even rank (scalars, skew-symmetrical or symmetrical tensors of the 2nd rank, etc.), which are composed quadratically or bilinearly of the U's. They are then composed solely of quantities with even j

---

* By "gauge-transformation of the first kind" we understand a transformation $U\to Ue^{i\alpha}$ $U^*\to U^*e^{-i\alpha}$ with an arbitrary space and time function α. By "gauge-transformation of the second kind" we understand a transformation of the type

$$\varphi_k\to\varphi_k - \frac{1}{\epsilon} i \frac{\partial\alpha}{\partial x_k}$$

as for those of the electromagnetic potentials.

[4] The general proof for this has been given by M. Fierz, Helv. Phys. Acta **13**, 45 (1940).

[5] See for instance W. Pauli in the article "Wellenmechanik" in the *Handbuch der Physik*, Vol. **24**/2, p. 260.

and even k and thus are of the typical form

$$T\sim\sum U^+U^+ + \sum U^-U^- + \sum U^+kU^-, \quad (3)$$

where again a possible even number of k factors is omitted and no distinction between U and U^* is made. Under the substitution (2) they remain unchanged, $T\rightarrow T$.

The situation is different for tensors of odd rank S (vectors, etc.) which consist of quantities with half-integral j and half-integral k. These are of the typical form

$$S\sim\sum U^+kU^+ + \sum U^-kU^- + \sum U^- \quad (4)$$

and hence change the sign under the substitution (2), $S\rightarrow -S$. Particularly is this the case for the current vector s_i. To the transformation $k_i\rightarrow -k_i$ belongs for arbitrary wave packets the transformation $x_i\rightarrow -x_i$ and it is remarkable that from the invariance of Eq. (1) against the proper Lorentz group alone there follows an invariance property for the change of sign of all the coordinates. In particular, the indefinite character of the current density and the total charge for even spin follows, since to every solution of the field equations belongs another solution for which the components of s_k change their sign. The definition of a definite particle density for even spin which transforms like the 4-component of a vector is therefore impossible.

We now proceed to a discussion of the somewhat less simple case of half-integral spins. Here we divide the quantities U, which have half-integral $j+k$, in the following fashion: (3) the "$+\epsilon$ class" with j integral k half-integral, (4) the "$-\epsilon$ class" with j half-integral k integral.

The multiplication of the classes (1), $\cdots$, (4), follows from the rule $\epsilon^2=1$ and the commutability of the multiplication. This law remains unchanged if ϵ is replaced by $-\epsilon$.

We can summarize the multiplication law between the different classes in the following multiplication table:

	1	-1	ϵ	$-\epsilon$
1	1	-1	ϵ	$-\epsilon$
-1	-1	$+1$	$-\epsilon$	$+\epsilon$
ϵ	ϵ	$-\epsilon$	$+1$	-1
$-\epsilon$	$-\epsilon$	ϵ	-1	$+1$

We notice that these classes have the multiplication law of Klein's "four-group."

It is important that here the complex-conjugate quantities for which j and k are interchanged do not belong to the same class, so that

$$\begin{aligned} &U^{+\epsilon},\ (U^{-\epsilon})^* \text{ belong to the } +\epsilon \text{ class}\\ &U^{-\epsilon},\ (U^{+\epsilon})^* \qquad\qquad\quad -\epsilon \text{ class.}\end{aligned}$$

We shall therefore cite the complex-conjugate quantities explicitly. (One could even choose the $U^{+\epsilon}$ suitably so that *all* quantities of the $-\epsilon$ class are of the form $(U^{+\epsilon})^*$.)

Instead of (1) we obtain now as typical form

$$\begin{aligned}\sum kU^{+\epsilon}+\sum k(U^{-\epsilon})^*&=\sum U^{-\epsilon}+\sum (U^{+\epsilon})^*\\ \sum kU^{-\epsilon}+\sum k(U^{+\epsilon})^*&=\sum U^{+\epsilon}+\sum (U^{-\epsilon})^*,\end{aligned} \quad (5)$$

since a factor k or $-i\partial/\partial x$ always changes the expression from one of the classes $+\epsilon$ or $-\epsilon$ into the other. As above, an even number of k factors have been omitted.

Now we consider instead of (2) the substitution

$$\begin{aligned}&k_i\rightarrow -k_i;\quad U^{+\epsilon}\rightarrow iU^{+\epsilon};\quad (U^{-\epsilon})^*\rightarrow i(U^{-\epsilon})^*;\\ &(U^{+\epsilon})^*\rightarrow -i(U^{+\epsilon})^*;\quad U^{-\epsilon}\rightarrow -iU^{-\epsilon}.\end{aligned} \quad (6)$$

This is in accord with the algebraic requirement of the passing over to the complex conjugate, as well as with the requirement that quantities of the same class as $U^{+\epsilon}$, $(U^{-\epsilon})^*$ transform in the same way. Furthermore, it does not interfere with possible reality conditions of the type $U^{+\epsilon}=(U^{-\epsilon})^*$ or $U^{-\epsilon}=(U^{+\epsilon})^*$. Equations (5) remain unchanged under the substitution (6).

We consider again tensors of even rank (scalars, tensors of 2nd rank, etc.), which are composed bilinearly or quadratically of the U and their complex-conjugate. For reasons similar to the above they must be of the form

$$\begin{aligned}T\sim&\sum U^{+\epsilon}U^{+\epsilon}+\sum U^{-\epsilon}U^{-\epsilon}+\sum U^{+\epsilon}kU^{-\epsilon}+\sum U^{+\epsilon}(U^{-\epsilon})^*+\sum U^{-\epsilon}(U^{+\epsilon})^*+\sum (U^{-\epsilon})^*kU^{-\epsilon}\\ &+\sum (U^{+\epsilon})^*kU^{+\epsilon}+\sum (U^{-\epsilon})^*k(U^{+\epsilon})^*+\sum (U^{-\epsilon})^*(U^{-\epsilon})^*+\sum (U^{+\epsilon})^*(U^{+\epsilon})^*.\end{aligned} \quad (7)$$

Furthermore, the tensors of odd rank (vectors, etc.) must be of the form

$$\begin{aligned}S\sim&\sum U^{+\epsilon}kU^{+\epsilon}+\sum U^{-\epsilon}kU^{-\epsilon}+\sum U^{+\epsilon}U^{-\epsilon}+\sum U^{+\epsilon}k(U^{-\epsilon})^*+\sum U^{-\epsilon}k(U^{+\epsilon})^*+\sum U^{-\epsilon}(U^{-\epsilon})^*\\ &+\sum U^{+\epsilon}(U^{+\epsilon})^*+\sum (U^{-\epsilon})^*k(U^{-\epsilon})^*+\sum (U^{+\epsilon})^*k(U^{+\epsilon})^*+\sum (U^{-\epsilon})^*(U^{+\epsilon})^*.\end{aligned} \quad (8)$$

The result of the substitution (6) is now the opposite of the result of the substitution (2): the tensors of even rank change their sign, the tensors of odd rank remain unchanged:

$$T \to -T; \quad S \to +S. \tag{9}$$

In case of half-integral spin, therefore, a positive definite energy density, as well as a positive definite total energy, is impossible. The latter follows from the fact, that, under the above substitution, the energy density in every space-time point changes its sign as a result of which the total energy changes also its sign.

It may be emphasized that it was not only unnecessary to assume that the wave equation is of the first order,* but also that the question is left open whether the theory is also invariant with respect to space reflections ($\mathbf{x}' = -\mathbf{x}, x_0' = x_0$). This scheme covers therefore also Dirac's two component wave equations (with rest mass zero).

These considerations do not prove that for integral spins there always exists a definite energy density and for half-integral spins a definite charge density. In fact, it has been shown by Fierz[6] that this is not the case for spin >1 for the densities. There exists, however (in the c number theory), a definite total charge for half-integral spins and a definite total energy for the integral spins. The spin value $\frac{1}{2}$ is discriminated through the possibility of a definite charge density, and the spin values 0 and 1 are discriminated through the possibility of defining a definite energy density. Nevertheless, the present theory permits arbitrary values of the spin quantum numbers of elementary particles as well as arbitrary values of the rest mass, the electric charge, and the magnetic moments of the particles.

§4. Quantization of the Fields in the Absence of Interactions. Connection Between Spin and Statistics

The impossibility of defining in a physically satisfactory way the particle density in the case of integral spin and the energy density in the case of half-integral spins in the c-number theory is an indication that a satisfactory interpretation of the theory within the limits of the one-body problem is not possible.* In fact, all relativistically invariant theories lead to particles, which in external fields can be emitted and absorbed in pairs of opposite charge for electrical particles and singly for neutral particles. The fields must, therefore, undergo a second quantization. For this we do not wish to apply here the canonical formalism, in which time is unnecessarily sharply distinguished from space, and which is only suitable if there are no supplementary conditions between the canonical variables.[7] Instead, we shall apply here a generalization of this method which was applied for the first time by Jordan and Pauli to the electromagnetic field.[8] This method is especially convenient in the absence of interaction, where all fields $U^{(r)}$ satisfy the wave equation of the second order

$$\Box U^{(r)} - \kappa^2 U^{(r)} = 0,$$

where

$$\Box \equiv \sum_{k=1}^{4} \frac{\partial^2}{\partial x_k^2} = \Delta - \frac{\partial^2}{\partial x_0^2}$$

and κ is the rest mass of the particles in units $\hbar/c$.

An important tool for the second quantization is the invariant D function, which satisfies the wave equation (9) and is given in a periodicity volume V of the eigenfunctions by

$$D(\mathbf{x}, x_0) = \frac{1}{V} \sum \exp\left[i(\mathbf{kx})\right] \frac{\sin k_0 x_0}{k_0} \tag{10}$$

or in the limit $V \to \infty$

$$D(\mathbf{x}, x_0) = \frac{1}{(2\pi)^3} \int d^3k \exp\left[i(\mathbf{kx})\right] \frac{\sin k_0 x_0}{k_0}. \tag{11}$$

* But we exclude operations like $(k^2+\kappa^2)^{\frac{1}{2}}$, which operate at finite distances in the coordinate space.

[6] M. Fierz, Helv. Phys. Acta **12**, 3 (1939).

* The author therefore considers as not conclusive the original argument of Dirac, according to which the field equation must be of the first order.

[7] On account of the existence of such conditions the canonical formalism is not applicable for spin >1 and therefore the discussion about the connection between spin and statistics by J. S. de Wet, Phys. Rev. **57**, 646 (1940), which is based on that formalism is not general enough.

[8] The consistent development of this method leads to the "many-time formalism" of Dirac, which has been given by P. A. M. Dirac, *Quantum Mechanics* (Oxford, second edition, 1935).

By k_0 we understand the positive root

$$k_0=+(k^2+\kappa^2)^{\frac{1}{2}}. \quad (12)$$

The D function is uniquely determined by the conditions:

$$\Box D-\kappa^2 D=0;\quad D(\mathbf{x},0)=0;\quad \left(\frac{\partial D}{\partial x_0}\right)_{x_0=0}=\delta(\mathbf{x}). \quad (13)$$

For $\kappa=0$ we have simply

$$D(\mathbf{x},x_0)=\{\delta(r-x_0)-\delta(r+x_0)\}/4\pi r. \quad (14)$$

This expression also determines the singularity of $D(\mathbf{x}, x_0)$ on the light cone for $\kappa\neq 0$. But in the latter case D is no longer different from zero in the inner part of the cone. One finds for this region[9]

$$D(\mathbf{x},x_0)=-\frac{1}{4\pi r}\frac{\partial}{\partial r}F(r,x_0)$$

with

$$F(r,x_0)=\begin{cases} J_0[\kappa(x_0^2-r^2)^{\frac{1}{2}}] & \text{for } x_0>r\\ 0 & \text{for } r>x_0>-r\\ -J_0[\kappa(x_0^2-r^2)^{\frac{1}{2}}] & \text{for } -r>x_0.\end{cases} \quad (15)$$

The jump from $+$ to $-$ of the function F on the light cone corresponds to the δ singularity of D on this cone. For the following it will be of decisive importance that D vanish in the exterior of the cone (i.e., for $r>x_0>-r$).

The form of the factor d^3k/k_0 is determined by the fact that d^3k/k_0 is invariant on the hyperboloid (k) of the four-dimensional momentum space $(\mathbf{k}, k_0)$. It is for this reason that, apart from D, there exists just one more function which is invariant and which satisfies the wave equation (9), namely,

$$D_1(\mathbf{x},x_0)=\frac{1}{(2\pi)^3}\int d^3k\exp[i(\mathbf{kx})]\frac{\cos k_0x_0}{k_0}. \quad (16)$$

For $\kappa=0$ one finds

$$D_1(\mathbf{x},x_0)=\frac{1}{2\pi^2}\frac{1}{r^2-x_0^2}. \quad (17)$$

In general it follows

$$D_1(\mathbf{x},x_0)=\frac{1}{4\pi}\frac{1}{r}\frac{\partial}{\partial r}F_1(r,x_0)$$

$$F_1(r,x_0)=\begin{cases} N_0[\kappa(x_0^2-r^2)^{\frac{1}{2}}] & \text{for } x_0>r\\ -iH_0^{(1)}[i\kappa(r^2-x_0^2)^{\frac{1}{2}}] & \text{for } r>x_0>-r\\ N_0[\kappa(x_0^2-r^2)^{\frac{1}{2}}] & \text{for } -r>x_0.\end{cases} \quad (18)$$

Here N_0 stands for Neumann's function and $H_0^{(1)}$ for the first Hankel cylinder function. The strongest singularity of D, on the surface of the light cone is in general determined by (17).

We shall, however, expressively postulate in the following *that all physical quantities at finite distances exterior to the light cone* (*for* $|x_0'-x_0''|<|\mathbf{x}'-\mathbf{x}''|$) *are commutable.** It follows from this that the bracket expressions of all quantities which satisfy the force-free wave equation (9) can be expressed by the function D and (a finite number) of derivatives of it without using the function D_1. This is also true for brackets with the $+$ sign, since otherwise it would follow that gauge invariant quantities, which are constructed bilinearly from the $U^{(r)}$, as for example the charge density, are noncommutable in two points with a space-like distance.[10]

The justification for our postulate lies in the fact that measurements at two space points with a space-like distance can never disturb each other, since no signals can be transmitted with velocities greater than that of light. Theories which would make use of the D_1 function in their quantization would be very much different from the known theories in their consequences.

At once we are able to draw further conclusions about the number of derivatives of D function which can occur in the bracket expressions, if we take into account the invariance of the theories under the transformations of the restricted Lorentz group and if we use the results of the preceding section on the class division of the tensors. We assume the quantities $U^{(r)}$ to be ordered in such a way that each field component is composed only of quantities of the same class.

[9] See P. A. M. Dirac, Proc. Camb. Phil. Soc. **30**, 150 (1934).

* For the canonical quantization formalism this postulate is satisfied implicitly. But this postulate is much more general than the canonical formalism.

[10] See W. Pauli, Ann. de l'Inst. H. Poincaré **6**, 137 (1936), esp. §3.

We consider especially the bracket expression of a field component $U^{(r)}$ with its own complex conjugate

$$[U^{(r)}(\mathbf{x}', x_0'),\ U^{*(r)}(\mathbf{x}'', x_0'')].$$

We distinguish now the two cases of half-integral and integral spin. In the former case this expression transforms according to (8) under Lorentz transformations as a tensor of odd rank. In the second case, however, it transforms as a tensor of even rank. Hence we have for half-integral spin

$$[U^{(r)}(\mathbf{x}', x_0'),\ U^{*(r)}(\mathbf{x}'', x_0'')] = \text{odd number of derivatives of the function } D(\mathbf{x}'-\mathbf{x}'', x_0'-x_0'') \tag{19a}$$

and similarly for integral spin

$$[U^{(r)}(\mathbf{x}', x_0'),\ U^{*(r)}(\mathbf{x}'', x_0'')] = \text{even number of derivatives of the function } D(\mathbf{x}'-\mathbf{x}'', x_0'-x_0''). \tag{19b}$$

This must be understood in such a way that on the right-hand side there may occur a complicated sum of expressions of the type indicated. We consider now the following expression, which is symmetrical in the two points

$$X \equiv [U^{(r)}(\mathbf{x}', x_0'),\ U^{*(r)}(\mathbf{x}'', x_0'')] + [U^{(r)}(\mathbf{x}'', x_0''),\ U^{*(r)}(\mathbf{x}', x_0')]. \tag{20}$$

Since the D function is even in the space coordinates odd in the time coordinate, which can be seen at once from Eqs. (11) or (15), it follows from the symmetry of X that X = even number of space-like times odd numbers of time-like derivatives of $D(\mathbf{x}'-\mathbf{x}'', x_0'-x_0'')$. This is fully consistent with the postulate (19a) for half-integral spin, but in contradiction with (19b) for integral spin unless X vanishes. We have therefore the result for integral spin

$$[U^{(r)}(\mathbf{x}', x_0'),\ U^{*(r)}(\mathbf{x}'', x_0'')] + [U^{(r)}(\mathbf{x}'', x_0''),\ U^{*(r)}(\mathbf{x}', x_0')] = 0. \tag{21}$$

So far we have not distinguished between the two cases of Bose statistics and the exclusion principle. In the former case, one has the ordinary bracket with the $-$ sign, in the latter case, according to Jordan and Wigner, the bracket

$$[A, B]_+ = AB + BA$$

with the $+$ sign. *By inserting the brackets with the $+$ sign into (20) we have an algebraic contradiction,* since the left-hand side is essentially positive for $x' = x''$ and cannot vanish unless both $U^{(r)}$ and $U^{*(r)}$ vanish.*

Hence we come to the result: *For integral spin the quantization according to the exclusion principle is not possible. For this result it is essential, that the use of the D_1 function in place of the D function be, for general reasons, discarded.*

On the other hand, it is formally possible to quantize the theory for half-integral spins according to Einstein-Bose-statistics, *but according to the general result of the preceding section the energy of the system would not be positive.* Since for physical reasons it is necessary to postulate this, we must apply the exclusion principle in connection with Dirac's hole theory.

For the positive proof that a theory with a positive total energy is possible by quantization according to Bose-statistics (exclusion principle) for integral (half-integral) spins, we must refer to the already mentioned paper by Fierz. In another paper by Fierz and Pauli[11] the case of an external electromagnetic field and also the connection between the special case of spin 2 and the gravitational theory of Einstein has been discussed.

In conclusion we wish to state, that according to our opinion the connection between spin and statistics is one of the most important applications of the special relativity theory.

---

* This contradiction may be seen also by resolving $U^{(r)}$ into eigenvibrations according to

$$U^{*(r)}(\mathbf{x}, x_0) = V^{-\frac{1}{2}} \Sigma_k \{ U_+^*(k) \exp [i\{-(\mathbf{kx}) + k_0 x_0\}] + U_-(k) \exp [i\{(\mathbf{kx}) - k_0 x_0\}]\}$$
$$U^{(r)}(\mathbf{x}, x_0) = V^{-\frac{1}{2}} \Sigma_k \{ U_+(k) \exp [i\{(\mathbf{kx}) - k_0 x_0\}] + U_-^*(k) \exp [i\{-(\mathbf{kx}) + k_0 x_0\}]\}.$$

The equation (21) leads then, among others, to the relation

$$[U_+^*(k), U_+(k)] + [U_-(k), U_-^*(k)] = 0,$$

a relation, which is not possible for brackets with the $+$ sign unless $U_\pm(k)$ and $U_\pm^*(k)$ vanish.

[11] M. Fierz and W. Pauli, Proc. Roy. Soc. **A173**, 211 (1939).

ON THE GREEN'S FUNCTIONS OF QUANTIZED FIELDS. I

BY JULIAN SCHWINGER

HARVARD UNIVERSITY

Communicated May 22, 1951

The temporal development of quantized fields, in its particle aspect, is described by propagation functions, or Green's functions. The construction of these functions for coupled fields is usually considered from the viewpoint of perturbation theory. Although the latter may be resorted to for detailed calculations, it is desirable to avoid founding the formal theory of the Green's functions on the restricted basis provided by the assumption of expandability in powers of coupling constants. These notes are a preliminary account of a general theory of Green's functions, in which the defining property is taken to be the representation of the fields of prescribed sources.

We employ a quantum dynamical principle for fields which has been described elsewhere.[1] This principle is a differential characterization of the function that produces a transformation from eigenvalues of a complete set of commuting operators on one space-like surface to eigenvalues of another set on a different surface,[2]

$$\delta(\zeta_1', \sigma_1|\zeta_2'', \sigma_2) = i(\zeta_1', \sigma_1|\delta\int_{\sigma_2}^{\sigma_1}(dx)\mathfrak{L}|\zeta_2'', \sigma_2). \qquad (1)$$

Here $\mathfrak{L}$ is the Lagrange function operator of the system. For the example of coupled Dirac and Maxwell fields, with external sources for each field, the Lagrange function may be taken as

$$\mathfrak{L} = -{}^1/_4[\bar{\psi}, \gamma_\mu(-i\partial_\mu - eA_\mu)\psi + m\psi] + {}^1/_2[\bar{\psi}, \eta] + \text{Herm. conj.} + {}^1/_4F_{\mu\nu}{}^2 - {}^1/_4\{F_{\mu\nu}, \partial_\mu A_\nu - \partial_\nu A_\mu\} + J_\mu A_\mu, \qquad (2)$$

which implies the equations of motion

$$\gamma_\mu(-i\partial_\mu - eA_\mu)\psi + m\psi = \eta,$$
$$F_{\mu\nu} = \partial_\mu A_\nu - \partial_\nu A_\mu, \quad \partial_\nu F_{\mu\nu} = J_\mu + j_\mu, \qquad (3)$$

where

$$j_\mu = e{}^1/_2[\bar{\psi}, \gamma_\mu\psi]. \qquad (4)$$

With regard to commutation relations, we need only note the anticommutativity of the source spinors with the Dirac field components.

We shall restrict our attention to changes in the transformation function that arise from variations of the external sources. In terms of the notation

$$(\zeta_1', \sigma_1|\zeta_2'', \sigma_2) = \exp i\mathfrak{W},$$
$$(\zeta_1', \sigma_1|F(x)|\zeta_2'', \sigma_2)/(\zeta_1', \sigma_1|\zeta_2'', \sigma_2) = \langle F(x)\rangle, \qquad (5)$$

the dynamical principle can then be written

$$\delta \mathcal{W} = \int_{\sigma_2}^{\sigma_1} (dx)\langle \delta \mathcal{L}(x)\rangle, \tag{6}$$

where

$$\langle \delta \mathcal{L}(x)\rangle = \langle \bar{\psi}(x)\rangle \delta\eta(x) + \delta\bar{\eta}(x)\langle \psi(x)\rangle + \langle A_\mu(x)\rangle \delta J_\mu(x). \tag{7}$$

The effect of a second, independent variation is described by

$$\delta'\langle \delta \mathcal{L}(x)\rangle = i \int_{\sigma_2}^{\sigma_1} (dx')[\langle(\delta \mathcal{L}(x)\delta'\mathcal{L}(x'))_+\rangle - \langle \delta\mathcal{L}(x)\rangle\langle \delta'\mathcal{L}(x')\rangle], \tag{8}$$

in which the notation $(\ )_+$ indicates temporal ordering of the operators. As examples we have

$$\delta_\eta\langle \psi(x)\rangle = i \int_{\sigma_{2'}}^{\sigma_1} (dx')[\langle(\psi(x)\bar{\psi}(x')\delta\eta(x'))_+\rangle - \langle \psi(x)\rangle\langle \bar{\psi}(x')\delta\eta(x')\rangle], \tag{9}$$

and

$$\delta_J\langle \psi(x)\rangle = i \int_{\sigma_2}^{\sigma_1} (dx')[\langle \psi(x)A_\mu(x'))_+\rangle - \langle \psi(x)\rangle\langle A_\mu(x')\rangle]\delta J_\mu(x'). \tag{10}$$

The latter result can be expressed in the notation

$$-i(\delta/\delta J_\mu(x'))\langle \psi(x)\rangle = \langle(\psi(x)A_\mu(x'))_+\rangle - \langle \psi(x)\rangle\langle A_\mu(x')\rangle, \tag{11}$$

although one may supplement the right side with an arbitrary gradient. This consequence of the charge conservation condition, $\partial_\mu J_\mu = 0$, corresponds to the gauge invariance of the theory.

A Green's function for the Dirac field, in the absence of an actual spinor source, is defined by

$$\delta_\eta\langle \psi(x)\rangle]_{\eta=0} = \int_{\sigma_2}^{\sigma_1} (dx')G(x, x')\delta\eta(x'). \tag{12}$$

According to (9), and the anticommutativity of $\delta\eta(x')$ with $\psi(x)$, we have

$$G(x, x') = i\langle(\psi(x)\bar{\psi}(x'))_+\rangle\epsilon(x, x'), \tag{13}$$

where $\epsilon(x, x') = (x_0 - x_0')/|x_0 - x_0'|$. On combining the differential equation for $\langle \psi(x)\rangle$ with (11), we obtain the functional differential equation

$$[\gamma_\mu(-i\partial_\mu - e\langle A_\mu(x)\rangle + ie\delta/\delta J_\mu(x)) + m]G(x, x') = \delta(x - x'). \tag{14}$$

An accompanying equation for $\langle A_\mu(x)\rangle$ is obtained by noting that

$$\langle j_\mu(x)\rangle = ie \text{ tr } \gamma_\mu G(x, x')_{x' \to x}, \tag{15}$$

in which the trace refers to the spinor indices, and an average is to be taken of the forms obtained with $x_0' \to x_0 \pm 0$. Thus, with the special choice of gauge, $\partial_\nu\langle A_\nu(x)\rangle = 0$, we have

$$-\partial_\nu^2\langle A_\mu(x)\rangle = J_\mu(x) + ie \text{ tr } \gamma_\mu G(x, x). \tag{16}$$

The simultaneous equations (14) and (16) provide a rigorous description of $G(x, x')$ and $\langle A_\mu(x)\rangle$.

A Maxwell field Green's function is defined by

$$\mathcal{G}_{\mu\nu}(x, x') = (\delta/\delta J_\nu(x'))\langle A_\mu(x)\rangle = (\delta/\delta J_\mu(x))\langle A_\nu(x')\rangle = i[\langle (A_\mu(x)A_\nu(x'))_+\rangle - \langle A_\mu(x)\rangle\langle A_\nu(x')\rangle]. \quad (17)$$

The differential equations obtained from (16) and the gauge condition are

$$-\partial_\lambda^2\mathcal{G}_{\mu\nu}(x, x') = \delta_{\mu\nu}\delta(x - x') + ie \operatorname{tr} \gamma_\mu(\delta/\delta J_\nu(x'))G(x, x),$$
$$\partial_\mu\mathcal{G}_{\mu\nu}(x, x') = 0 \ (= \partial_\nu\chi). \quad (18)$$

More complicated Green's functions can be discussed in an analogous manner. The Dirac field Green's function defined by

$$\delta_\eta^2\langle(\psi(x_1)\psi(x_2))_+\rangle\epsilon(x_1, x_2)_{\eta = 0} = \int_{\sigma_2}^{\sigma_1} (dx_1') \int_{\sigma_2}^{\sigma_1} (dx_2')G(x_1, x_2; x_1', x_2')\delta\bar{\eta}(x_1')\delta\bar{\eta}(x_2'), \quad (19)$$

may be called a "two-particle" Green's function, as distinguished from the "one-particle" $G(x, x')$. It is given explicitly by

$$G(x_1, x_2; x_1', x_2') = \langle(\psi(x_1)\psi(x_2)\bar{\psi}(x_1')\bar{\psi}(x_2'))_+\rangle\ \epsilon,$$
$$\epsilon = \epsilon(x_1, x_2)\epsilon(x_1', x_2')\epsilon(x_1, x_1')\epsilon(x_1, x_2')\epsilon(x_2, x_1')\epsilon(x_2, x_2'). \quad (20)$$

This function is antisymmetrical with respect to the interchange of x_1 and x_2, and of x_1' and x_2' (including the suppressed spinor indices). It obeys the differential equation

$$\mathfrak{F}_1G(x_1, x_2; x_1', x_2') = \delta(x_1 - x_1')G(x_2, x_2') - \delta(x_1 - x_2')G(x_2, x_1'), \quad (21)$$

where $\mathfrak{F}$ is the functional differential operator of (14). More symmetrically written, this equation reads

$$\mathfrak{F}_1\mathfrak{F}_2G(x_1, x_2; x_1', x_2') = \delta(x_1 - x_1')\delta(x_2 - x_2') - \delta(x_1 - x_2')\delta(x_2 - x_1'), \quad (22)$$

in which the two differential operators are commutative.

The replacement of the Dirac field by a Kemmer field involves alterations beyond those implied by the change in statistics. Not all components of the Kemmer field are dynamically independent. Thus, if 0 refers to some arbitrary time-like direction, we have

$$m(1 - \beta_0^2)\psi = (1 - \beta_0^2)\eta - \beta_k(-i\partial_k - eA_k)\beta_0^2\psi, \quad k = 1, 2, 3, \quad (23)$$

which is an equation of constraint expressing $(1 - \beta_0^2)\psi$ in terms of the independent field components $\beta_0^2\psi$, and of the external source. Accordingly, in computing $\delta_\eta\langle\psi(x)\rangle$ we must take into account the change induced in $(1 - \beta_0^2)\psi(x)$, whence

$$G(x, x') = i\langle(\psi(x)\bar{\psi}(x'))_+\rangle + (1/m)(1 - \beta_0^2)\delta(x - x'). \quad (24)$$

The temporal ordering is with respect to the arbitrary time-like direction

The Green's function is independent of this direction, however, and satisfies equations which are of the same form as (14) and (16), save for a sign change in the last term of the latter equation which arises from the different statistics associated with the integral spin field.

[1] Schwinger, J., *Phys. Rev.*, June 15, 1951 issue.
[2] We employ units in which $\hbar = c = 1$.

ON THE GREEN'S FUNCTIONS OF QUANTIZED FIELDS. II

By Julian Schwinger

Harvard University

Communicated May 22, 1951

In all of the work of the preceding note there has been no explicit reference to the particular states on σ_1 and σ_2 that enter in the definitions of the Green's functions. This information must be contained in boundary conditions that supplement the differential equations. We shall determine these boundary conditions for the Green's functions associated with vacuum states on both σ_1 and σ_2. The vacuum, as the lowest energy state of the system, can be defined only if, in the neighborhood of σ_1 and σ_2, the actual external electromagnetic field is constant in some time-like direction (which need not be the same for σ_1 and σ_2). In the Dirac one-particle Green's function, for example,

$$\begin{aligned} G(x, x') &= i\langle\psi(x)\bar{\psi}(x')\rangle,\ x_0 > x_0', \\ &= -i\langle\bar{\psi}(x')\,\psi(x)\rangle,\ x_0 < x_0', \end{aligned} \tag{25}$$

the temporal variation of $\psi(x)$ in the vicinity of σ_1 can then be represented by

$$\psi(x) = \exp\,[iP_0(x_0 - X_0)]\psi(X)\exp\,[-iP_0(x_0 - X_0)], \tag{26}$$

where P_0 is the energy operator and X is some fixed point. Therefore,

$$x \sim \sigma_1\colon G(x, x') = i\langle\psi(X)\exp\,[-i(P_0 - P_0^{\text{vac}})(x_0 - X_0)]\bar{\psi}(x')\rangle, \tag{27}$$

in which P_0^{vac} is the vacuum energy eigenvalue. Now $P_0 - P_0^{\text{vac}}$ has no negative eigenvalues, and accordingly $G(x, x')$, as a function of x_0 in the vicinity of σ_1, contains only positive frequencies, which are energy values for states of unit positive charge. The statement is true of every time-like direction, if the external field vanishes in this neighborhood.

A representation similar to (26) for the vicinity of σ_2 yields

$$x \sim \sigma_2\colon G(x, x') = -i\langle\bar{\psi}(x')\exp\,[i(P_0 - P_0^{\text{vac}})(x_0 - X_0)]\psi(X)\rangle, \tag{28}$$

which contains only negative frequencies. In absolute value, these are the energies of unit negative charge states. We thus encounter Green's functions that obey the temporal analog of the boundary condition characteristic of a source radiating into space.[1] In keeping with this analogy, such Green's functions can be derived from a retarded proper time Green's function by a Fourier decomposition with respect to the mass.

The boundary condition that characterizes the Green's functions associated with vacuum states on σ_1 and σ_2 involves these surfaces only to the extent that they must be in the region of outgoing waves. Accordingly, the domain of these functions may conveniently be taken as the entire four-dimensional space. Thus, if the Green's function $G_+(x, x')$, defined by (14), (16), and the outgoing wave boundary condition, is represented by the integro-differential equation,

$$\gamma_\mu(-i\partial_\mu - eA_{+\mu}(x))G_+(x, x') + \int (dx'')M(x, x'')G_+(x'', x') = \delta(x - x'), \tag{29}$$

the integration is to be extended over all space-time. This equation can be more compactly written as

$$[\gamma(p - eA_+) + M]G_+ = 1, \tag{30}$$

by regarding the space-time coordinates as matrix indices. The mass operator M is then symbolically defined by

$$MG_+ = mG_+ + ie\gamma(\delta/\delta J)G_+. \tag{31}$$

In these formulae, A_+ and $\delta/\delta J$ are considered to be diagonal matrices,

$$(x|A_{+\mu}|x') = \delta(x - x')A_{+\mu}(x). \tag{32}$$

There is some advantage, however, in introducing "photon coordinates" explicitly (while continuing to employ matrix notation for the "particle coordinates"). Thus

$$\gamma A_+ \rightarrow \int (d\xi)\gamma(\xi)A_+(\xi), \tag{33}$$

where $\gamma(\xi)$ is defined by

$$(x|\,\gamma_\mu(\xi)\,|x') = \gamma_\mu\delta(x - \xi)\delta(x - x'). \tag{34}$$

The differential equation for $A_+(\xi)$ can then be written

$$-\partial_\xi^2 A_+(\xi) = J(\xi) + ie \operatorname{Tr}[\gamma(\xi)G_+], \tag{35}$$

where Tr denotes diagonal summation with respect to spinor indices and particle coordinates. The associated photon Green's function differential equation is

$$-\partial_\xi^2 \mathcal{G}_+(\xi, \xi') = \delta(\xi - \xi') + ie \operatorname{Tr}[\gamma(\xi)(\delta/\delta J(\xi'))G_+]. \tag{36}$$

To express the variational derivatives that occur in (31) and (36) we introduce an auxiliary quantity defined by

$$\Gamma(\xi) = -(\delta/\delta eA_+(\xi))G_+^{-1} = \gamma(\xi) - (\delta/\delta eA_+(\xi))M. \tag{37}$$

Thus

$$(\delta/\delta J(\xi))G_+ = e\int(d\xi')G_+\Gamma(\xi')G_+\mathcal{G}_+(\xi', \xi), \tag{38}$$

from which we obtain

$$M = m + ie^2\int(d\xi)(d\xi')\gamma(\xi)G_+\Gamma(\xi')\mathcal{G}_+(\xi', \xi), \tag{39}$$

and

$$-\partial_\xi^2\mathcal{G}_+(\xi, \xi') + \int(d\xi'')P(\xi, \xi'')\mathcal{G}_+(\xi'', \xi') = \delta(\xi - \xi'),$$
$$P(\xi, \xi') = -ie^2 \operatorname{Tr}[\gamma(\xi)G_+\Gamma(\xi')G_+] \tag{40}$$

With the introduction of matrix notation for the photon coordinates, this Green's function equation becomes

$$(k^2 + P)\mathcal{G}_+ = 1, \; [\xi_\mu, k_\nu] = i\delta_{\mu\nu}, \tag{41}$$

and the polarization operator P is given by

$$P = -ie^2 \operatorname{Tr}[\gamma G_+\Gamma G_+]. \tag{42}$$

In this notation, the mass operator expression reads

$$M = m + ie^2 \operatorname{T}\rho\,[\gamma G_+\Gamma\mathcal{G}_+], \tag{43}$$

where Tρ denotes diagonal summation with respect to the photon coordinates, including the vector indices.

The two-particle Green's function

$$G_+(x_1, x_2; x_1', x_2') = (x_1, x_2|G_{12}|x_1', x_2'), \tag{44}$$

can be represented by the integro-differential equation

$$[(\gamma\pi + M)_1(\gamma\pi + M)_2 - I_{12}]G_{12} = 1_{12},$$
$$\pi = p - eA_+, \tag{45}$$

thereby introducing the interaction operator I_{12}. The unit operator 1_{12} is defined by the matrix representation

$$(x_1, x_2|1_{12}|x_1', x_2') = \delta(x_1 - x_1')\delta(x_2 - x_2') - \delta(x_1 - x_2')\delta(x_2 - x_1'). \tag{46}$$

On comparison with (21) we find that the interaction operator can be characterized symbolically by

$$I_{12}G_{12} = -ie^2 \operatorname{T}\rho[\gamma_1\Gamma_2\mathcal{G}_+]G_{12} - ie^2 \operatorname{T}\rho[\gamma_1 G_1\delta/\delta J](I_{12}G_{12})$$
$$= -ie^2 \operatorname{T}\rho[\gamma_2\Gamma_1\mathcal{G}_+]G_{12} - ie^2 \operatorname{T}\rho[\gamma_2 G_2\delta/\delta J](I_{12}G_{12}), \quad (47)$$

where G_1 and G_2 are the one-particle Green's functions of the indicated particle coordinates.

The various operators that enter in the Green's function equations, the mass operator M, the polarization operator P, the interaction operator I_{12}, can be constructed by successive approximation. Thus, in the first approximation,

$$M(x, x') = m\delta(x - x') + ie^2\gamma_\mu G_+(x, x')\gamma_\mu D_+(x, x'),$$
$$P_{\mu\nu}(\xi, \xi') = -ie^2 \operatorname{tr}[\gamma_\mu G_+(\xi, \xi')\gamma_\nu G_+(\xi', \xi)],$$
$$I(x_1, x_2; x_1', x_2') = -ie^2\gamma_{1\mu}\gamma_{2\mu}D_+(x_1, x_2)(x_1, x_2| 1_{12}| x_1', x_2'), \quad (48)$$

where

$$\mathcal{G}_{\mu\nu}(\xi, \xi') = \delta_{\mu\nu}D_+(\xi, \xi'), \quad (49)$$

and the Green's functions that appear in these formulae refer to the 0th approximation ($M = m$, $P = 0$). We also have, in the first approximation,

$$\Gamma_\mu(\xi; x, x') = \gamma_\mu\delta(\xi - x)\delta(x - x')$$
$$-ie^2\gamma_\nu G_+(x, \xi)\gamma_\mu G_+(\xi, x')\gamma_\nu D_+(x, x') \quad (50)$$

Perturbation theory, as applied in this manner, must not be confused with the expansion of the Green's functions in powers of the charge. The latter procedure is restricted to the treatment of scattering problems.

The solutions of the homogeneous Green's function equations constitute the wave functions that describe the various states of the system. Thus, we have the one-particle wave equation

$$(\gamma\pi + M)\psi = 0, \quad (51)$$

and the two particle wave equation

$$[(\gamma\pi + M)_1(\gamma\pi + M)_2 - I_{12}]\psi_{12} = 0, \quad (52)$$

which are applicable equally to the discussion of scattering and to the properties of bound states. In particular, the total energy and momentum eigenfunctions of two particles in isolated interaction are obtained as the solutions of (52) which are eigenfunctions for a common displacement of the two space-time coordinates. It is necessary to recognize, however, that the mass operator, for example, can be largely represented in its effect by an alteration in the mass constant and by a scale change of the Green's function. Similarly, the major effect of the polarization operator is to multiply the photon Green's function by a factor, which everywhere appears associated with the charge. It is only after these renormalizations have been performed that we deal with wave equations that involve the empirical mass and charge, and are thus of immediate physical applicability.

The details of this theory will be published elsewhere, in a series of articles entitled "The Theory of Quantized Fields."

[1] Green's functions of this variety have been discussed by Stueckelberg, E. C. G., *Helv. Phys. Acta*, **19**, 242 (1946), and by Feynman, R. P., *Phys. Rev.*, **76**, 749 (1949).

Electrodynamic Displacement of Atomic Energy Levels. III. The Hyperfine Structure of Positronium

ROBERT KARPLUS AND ABRAHAM KLEIN
Harvard University, Cambridge, Massachusetts
(Received May 13, 1952)

A functional integro-differential equation for the electron-positron Green's function is derived from a consideration of the effect of sources of the Dirac field. This equation contains an electron-positron interaction operator from which functional derivatives may be eliminated by an iteration procedure. The operator is evaluated so as to include the effects of one and two virtual quanta. It contains an interaction resulting from quantum exchange as well as one resulting from virtual annihilation of the pair. The wave functions of the electron-positron system are the solutions of the homogeneous equation related to the Green's function equation. The eigenvalues of the total energy of the system may be found by a four-dimensional perturbation technique. The system bound by the Coulomb interaction is here treated as the unperturbed situation. Numerical values for the spin-dependent change of the energy from the Coulomb value in the ground state are finally obtained accurate to order α relative to the hyperfine structure α^2 Ry. The result for the singlet-triplet energy difference is

$$\Delta W_{ts}=\tfrac{1}{2}\alpha^2\,\mathrm{Ry}_\infty[7/3-(32/9+2\ln 2)\alpha/\pi]=2.0337\times 10^5\ \mathrm{Mc/sec}.$$

Theory and experiment are in agreement.

I. INTRODUCTION

THE investigation to be described in this paper was suggested by the current theoretical interest in the quantum-mechanical two-body problem[1–3] and the recent accurate measurement of the ground state hyperfine structure of positronium.[4,5] The system composed of one electron and one positron in interaction is the simplest accessible to calculation because it is purely electrodynamic in nature. Moreover, the success of quantum electrodynamics in predicting with great accuracy the properties of a single particle in an external field indicates the absence of fundamental difficulties from the theory in the range of energies that are significant in positronium.

The discussion of the bound states of the electron-positron system is based upon a rigorous functional differential equation for the Green's function of that system, derived in Sec. II by the method described by Schwinger.[1] In order to obtain a useful approximate form of this equation (and of the associated homogeneous equation) we have iterated the implicitly defined interaction operator, in this way automatically generating to any required order the interaction kernel obtained from scattering considerations by Bethe and Salpeter.[3] In the present case we have included all interaction terms involving the emission and absorption of one or two quanta. The latter include self-energy and vacuum polarization corrections to one-photon exchange processes as well as two-photon exchange terms. The particle-antiparticle relationship of electron and positron is represented by terms describing one- and two-photon virtual annihilation of the pair.[6–8] In contrast to the case of scattering, only the irreducible[3] interactions appear explicitly.

Our subsequent concern is with the solution of the associated homogeneous equation. It should be emphasized at the outset that we shall be silent (out of ignorance) on the question of the fundamental interpretation of a wave function which refers to individual times for each of the particles. The possibility, nevertheless, of obtaining a solution to our problem entirely within the framework of the present formalism depends on two conditions. The first of these is that most of the binding is accounted for by the instantaneous Coulomb interaction. Salpeter[9] has shown that when the interaction is instantaneous, the wave equation can be rigorously reduced to one involving only equal times for the two particles. Moreover, the wave function for arbitrary individual time coordinates can be expressed in terms of that for equal times. This last circumstance can also be exploited in the development of a perturbation theory which yields the contribution to the energy levels of a small non-instantaneous interaction.[9] The relevant results of this treatment are given in Sec. III.

The second condition is that the free particle approximation for all intermediate states shall be an adequate one. The essential point here is that whether one derives an explicit interaction operator by the iteration procedure adopted in the present paper (tantamount to an expansion of the intrinsic nonlinearity in terms of free particle properties) or by a partial summation of a scattering kernel, the propagation which naturally enters in intermediate states is that of free particles. In the treatment of fine-structure effects, the contribu-

[1] J. Schwinger, Proc. Nat. Acad. Sci. US **37**, 452, 455 (1951).
[2] M. Gell-Mann and F. Low, Phys. Rev. **84**, 350 (1951).
[3] H. A. Bethe and E. E. Salpeter, Phys. Rev. **84**, 1232 (1951).
[4] M. Deutsch and S. C. Brown, Phys. Rev. **85**, 1047 (1952).
[5] M. Deutsch, latest result reported at the Washington Meeting of the American Physical Society, May, 1952. Phys. Rev. **87**, 212(T) (1952).
[6] J. Pirenne, Arch. sci. phys. et nat. **28**, 233 (1946); **29**, 121, 207, and 265 (1947).
[7] V. B. Berestetski and L. D. Landau, J. Exptl. Theoret. Phys. (U.S.S.R.) **19**, 673 (1949). See also V. B. Berestetski, J. Exptl. Theoret. Phys. (U.S.S.R.) **19**, 1130 (1949).
[8] R. A. Ferrell, Phys. Rev. **84**, 858 (1951) and Ph.D. thesis (Princeton, 1951). Dr. Ferrell kindly sent us a copy of his thesis.
[9] E. E. Salpeter, Phys. Rev. **87**, 328 (1952). We are indebted to Dr. Salpeter for making available to us a copy of his paper prior to publication. We have found his ideas very helpful in our work.

tion of nonrelativistic intermediate states, where the Coulomb binding cannot be ignored, must then be obtained in a manner reminiscent of the first treatments of the Lamb shift.[9] This will not be necessary in the present paper since we shall be concerned with the hyperfine (spin-spin) type of interaction to which only relativistic intermediate states contribute to the required precision.[10]

The practical goal of this work is to obtain the splitting of the singlet-triplet ground-state doublet of positronium correct to order α^3 Ry. Previous calculations,[6-8] accurate to order α^2 Ry, have included the lowest order contributions of the ordinary spin-spin coupling arising from the Breit[11] interaction (the analog of which in hydrogen is responsible for its hyperfine structure) and of the one-photon virtual annihilation force, characteristic of the system of particle-antiparticle. The expression for the energy shift given in Sec. III, Eq. (3.6) yields these again in lowest approximation and contains as well the matrix elements of all interactions which can contribute to the required accuracy.

Section IV is devoted to the detailed evaluation of all the matrix elements that may be looked upon as generalized Breit interactions because they depend purely on the exchange of photons between the two particles. In Sec. V we consider the annihilation interaction peculiar to the electron-positron system. Finally, the comparison with experiment is given in Sec. VI.

II. THE WAVE EQUATION

A discussion of the one-particle electron and positron Green's function associated with the vacuum state will serve as an introduction to this section. If the notation of reference 1 is extended to include the positron field variables $\psi'(x)$, $\bar{\psi}'(x)$, and their sources that are related to the electron variables $\psi(x)$, $\bar{\psi}(x)$, and their sources by the usual charge conjugating matrix C,

$$C^\dagger C=1, \quad C=-\tilde{C}, \quad C\tilde{\gamma}C^{-1}=-\gamma, \qquad \psi'=C\bar{\psi}, \quad \bar{\psi}'=C^{-1}\psi, \quad \eta'=C\bar{\eta}, \quad \bar{\eta}'=C^{-1}\eta, \tag{2.1}$$

the Green's functions are defined by the vacuum expectation values

$$\delta_\eta\langle\psi(x)\rangle_0|_{\eta=0}=\int_{\sigma_2}^{\sigma_1} d^4x' G^-(x,x')\delta\eta(x') \tag{2.2a}$$

and

$$\delta_{\eta'}\langle\psi'(x)\rangle_0|_{\eta'=0}=\int_{\sigma_2}^{\sigma_1} d^4x' G^+(x,x')\delta\eta(x'), \tag{2.2b}$$

where $\delta\eta$ and $\delta\eta'$ are arbitrary variations of the electron and positron sources, respectively. The Green's functions can be expressed in terms of expectation values by

$$G^-(x,x')=i\langle(\psi(x)\bar{\psi}(x'))_+\rangle_0\epsilon(x,x') \tag{2.3a}$$

and

$$G^+(x,x')=i\langle(\psi'(x)\bar{\psi}'(x'))_+\rangle_0\epsilon(x,x') \tag{2.3b}$$

and satisfy the differential equations

$$[\gamma_\mu(-i\partial_\mu-eA_{+\mu}(x)+ie\delta/\delta J_\mu(x))+m] \times G^-(x,x')=\delta(x-x') \tag{2.4a}$$

and

$$[\gamma_\mu(-i\partial_\mu+eA_{+\mu}(x)-ie\delta/\delta J_\mu(x))+m] \times G^+(x,x')=\delta(x-x'), \tag{2.4b}$$

with the outgoing wave boundary condition. They are, of course, related by the matrix C:

$$G_{\alpha\beta}{}^+(x,x')=-C_{\alpha\alpha'}C^{-1}{}_{\beta\beta'}G_{\beta'\alpha'}{}^-(x',x). \tag{2.5}$$

We shall now introduce matrix notation for the combined particle coordinates and spinor indices, and the combined photon coordinates and vector indices. Because the formulas will get quite involved, the matrix indices will be expressed as arguments, by numbers for the particles and by ξ, ξ', $\cdots$ for the photons, and the summation convention will be understood. Functions of one coordinate are to be diagonal matrices; quantities affixed with only one matrix index are to be vectors with respect to that index. The arguments of the Dirac matrices will refer only to the vector and spinor indices of these quantities; they will be unit matrices in the coordinates. Similarly, functions of the coordinates alone must be understood as multiples of the Dirac unit matrix.

As an example, Eqs. (2.4) and (2.5) will be transcribed with the symbols $\mathfrak{F}^-$ and $\mathfrak{F}^+$ standing for the functional differential operators in Eq. (2.4):

$$\mathfrak{F}^-(12)G^-(23)=\delta(13); \tag{2.4'a}$$

$$\mathfrak{F}^+(12)G^+(23)=\delta(13); \tag{2.4'b}$$

$$G^+(12)=-C(11')C^{-1}(22')G^-(2'1'). \tag{2.5'}$$

If the mass operator $M(12)$ is defined in the usual way,

$$M^\pm(12)G^\pm(23)=\mathfrak{M}^\pm(12)G^\pm(23), \tag{2.6}$$

where $\mathfrak{M}$ is the functional differential operator

$$\mathfrak{M}^\pm(12)=m\delta(12)\mp ie\gamma(\xi,12)\delta/\delta J(\xi), \tag{2.7}$$

then the Green's function equations (2.4) can be written in terms of integro-differential operators F that are obtained from the $\mathfrak{F}$ by the replacement of $\mathfrak{M}$ by M.

A vertex operator $\Gamma(\xi,12)$ must now be defined for each Green's function,

$$\Gamma^+(\xi,12)=(\delta/\delta eA_+(\xi))(G^+(12))^{-1} =(\delta/\delta eA_+(\xi))F^+(12) \tag{2.8a}$$

and

$$\Gamma^-(\xi,12)=-(\delta/\delta eA_+(\xi))(G^-(12))^{-1} =-(\delta/\delta eA_+(\xi))F^-(12). \tag{2.8b}$$

In the absence of an external field these two quantities become equal because then the charge occurs always

[10] R. Karplus and A. Klein, Phys. Rev. 85, 972 (1952).
[11] G. Breit, Phys. Rev. 34, 553 (1929); 36, 383 (1930); 39, 616 (1932).

to an even power only, and the two differ just in the sign of the charge.

We now proceed to the two-particle system. The electron-positron Green's function for the vacuum state is defined by the relation

$$\delta_\eta \delta_{\eta'} \langle (\psi(x_1)\psi'(x_2))_+\rangle_0 |_{\eta=\eta'=0} \epsilon(x_1, x_2)$$

$$= \int_{\sigma_2}^{\sigma_1} d^4x_1' \int_{\sigma_2}^{\sigma_1} d^4x_2' G^{-+}(x_1x_2, x_1'x_2') \times \delta\eta(x_1')\delta\eta'(x_2'). \quad (2.9)$$

Evaluation of the variations with the help of Eq. (9), reference 1, leads to the explicit expression

$$G^{-+}(x_1x_2, x_1'x_2')$$
$$= \langle (\psi(x_1)\psi'(x_2)\bar{\psi}(x_1')\bar{\psi}'(x_2'))_+\rangle_0 \epsilon$$
$$- \langle (\psi(x_1)\psi'(x_2))_+\rangle_0 \epsilon(x_1, x_2)$$
$$\times \langle (\bar{\psi}(x_1')\bar{\psi}'(x_2'))_+\rangle_0 \epsilon(x_1', x_2'). \quad (2.10)$$

As might be expected, this Green's function is related to a charge conjugate of the two-electron Green's function with arguments interchanged properly, by Eqs. (13, 20), reference 1:

$$G_{\alpha\beta\gamma\delta}{}^{-+}(x_1x_2, x_1'x_2')$$
$$= -C_{\beta\beta'}C^{-1}{}_{\delta\delta'}G_{\alpha\delta'\gamma\beta'}{}^{--}(x_1x_2', x_1'x_2)$$
$$- C_{\beta\beta'}C^{-1}{}_{\delta\delta'}G_{\alpha\beta'}{}^{-}(x_1x_2)G_{\delta'\gamma}{}^{-}(x_2'x_1'). \quad (2.11)$$

The antisymmetry of the two-electron Green's function assures that both direct and exchange processes are contained in the electron-positron Green's function; the second term merely corrects for the fact that the uncoupled electron-positron system cannot undergo an exchange process. In this case,

$$G_{\alpha\delta'\gamma\beta'}{}^{--}(x_1x_2', x_1'x_2) \to G_{\alpha\gamma}{}^{-}(x_1x_1')G_{\delta'\beta'}{}^{-}(x_2'x_2)$$
$$- G_{\alpha\beta'}{}^{-}(x_1x_2)G_{\delta'\gamma}{}^{-}(x_2'x_1'), \quad (2.12)$$

whence

$$G_{\alpha\beta\gamma\delta}{}^{-+}(x_1x_2, x_1'x_2')$$
$$\to -C_{\beta\beta'}C^{-1}{}_{\delta\delta'}G_{\alpha\gamma}{}^{-}(x_1x_1')G_{\delta'\beta'}{}^{-}(x_2'x_2)$$
$$= G_{\alpha\gamma}{}^{-}(x_1x_1')G_{\beta\delta}{}^{+}(x_2x_2'), \quad (2.13)$$

the proper description for noninteracting particles.

The differential equation for G^{-+} may be obtained with the help of that for G^{--}, Eq. (21), reference 1, and of Eq. (2.4'). They yield

$$\mathfrak{F}^-(11')G^{-+}(1'2, 34) = \delta(13)G^+(24)$$
$$+ ie\gamma(\xi, 11')C(1'2')C^{-1}(44')$$
$$\times G^+(22')(\delta/\delta J(\xi))G^-(4'3) \quad (2.14)$$

and

$$F^+(22')\mathfrak{F}^-(11')G^{-+}(1'2', 34) = \delta(13)\delta(24)$$
$$+ ie\gamma(\xi, 11')C(1'2)C^{-1}(44')(\delta/\delta J(\xi))G^-(4'3). \quad (2.15)$$

Finally, the equation may be written in the form

$$[F^-(11')F^+(22') - I(12, 1'2')]G^{-+}(1'2', 34)$$
$$= \delta(13)\delta(24), \quad (2.16)$$

where the interaction operator $I(1234)$ is defined by

$$I(12, 1'2')G^{-+}(1'2', 34)$$
$$= -F^+(22')[\mathfrak{M}^-(11') - M^-(11')]G^{-+}(1'2', 34)$$
$$+ ie\gamma(\xi, 13')C(3'2)C^{-1}(44')(\delta/\delta J(\xi))G^-(4'3),$$
$$= -F^-(11')[\mathfrak{M}^+(22') - M^+(22')]G^{-+}(1'2', 34)$$
$$- ie\gamma(\xi, 22')C(2'1)C^{-1}(33')(\delta/\delta J(\xi))G^+(3'4). \quad (2.17)$$

The second expression arises when $\mathfrak{F}^+$ and then F^- are applied to the Green's function. These expressions must now be rearranged so as to yield the interaction operator explicitly as an integral operator up to the desired order of accuracy. In other words, the functional derivatives may occur only in terms that contribute negligibly to the effect that is being investigated. The subsequent operations will be directed at finding an expression that is suitable for the purposes of this paper. (For other effects, such as the Lamb shift in positronium, a different form of the interaction operator is necessary.)

With the help of the definition of the vertex operator, Eq. (2.8), the lowest order interaction may be separated as follows:

$$I(12, 1'2')G^{-+}(1'2', 34)$$
$$= ie^2\gamma(\xi, 11')\mathcal{G}_+(\xi, \xi')\Gamma^+(\xi', 22')G^{-+}(1'2', 34)$$
$$- [\mathfrak{M}^-(11') - M^-(11')]F^+(22')G^{-+}(1'2', 34)$$
$$+ ie^2\gamma(\xi, 13')C(3'2)\mathcal{G}_+(\xi, \xi')C^{-1}(2'4')$$
$$\times \Gamma^-(\xi', 4'1')G^-(1'3)G^+(2'4). \quad (2.18)$$

The second term in Eq. (2.18) can be simplified by the use of Eqs. (2.16) and (2.6), whence it becomes

$$- ie^2\gamma(\xi, 11')G^-(1'1'')[\delta/\delta eJ(\xi)]$$
$$\times I(1''2, 3'4')G^{-+}(3'4', 34). \quad (2.19)$$

The last term, finally, is brought into more useful form with the help of the identity

$$\mathcal{G}_+(\xi, \xi')C^{-1}(2'4')\Gamma^-(\xi', 4'1')G^-(1'3)G^+(2'4)$$
$$= D_+(\xi, \xi')C^{-1}(2'4')\gamma(\xi', 4'1')G^{-+}(1'2', 34), \quad (2.20)$$

which may be verified by iteration of both sides. The interaction operator therefore is given by

$$I(12, 34) = ie^2\gamma(\xi, 13)\mathcal{G}_+(\xi, \xi')\Gamma^+(\xi', 24)$$
$$+ ie^2\gamma(\xi, 11')C(1'2)D_+(\xi, \xi')C^{-1}(44')\gamma(\xi', 4'3)$$
$$- ie^2\gamma(\xi, 11')G(1'1'')[(\delta/\delta eJ(\xi))I(1''2, 3'4')$$
$$\times G^{-+}(3'4', 3''4'')][G^{-+}(3''4'', 34)]^{-1}. \quad (2.21)$$

This, and a corresponding expression obtained from the alternative form of Eq. (2.17) correspond to Eq. (47), reference 1; the only difference lies in the second term above, which represents the interaction due to the virtual annihilation of the electron-positron pair. The last term contains the effects of higher order electrodynamic processes involving more than one virtual photon, such as multiple photon exchanges and the corrections that symmetrize the first term in the interaction so that it depends on the vertex operator of both the electron and the positron.

We are interested in the effects of one and two virtual quanta, terms of order e^4 in the interaction. For this reason, the functional derivative in Eq. (2.21) needs be evaluated only to the lowest order,

$$\begin{aligned}
&[(\delta/\delta eJ(\xi))I(1''2,3'4')G^{-+}(3'4',3''4'')] \\
&\quad\times[G^{-+}(3''4'',34)]^{-1}\cong -I(1''2,3'4') \\
&\quad\times G^{-+}(3'4',3''4'')[\delta/\delta eJ(\xi)] \\
&\quad\times[F^{-}(3''3)F^{+}(4''4)]\cong -ie^2[\gamma(\bar{\xi},1''3') \\
&\quad\times\gamma(\bar{\xi}',24')+\gamma(\bar{\xi},1''2')C(2'2)C^{-1}(4''2'') \\
&\quad\times\gamma(\bar{\xi}',2''3')]D_+(\bar{\xi},\bar{\xi}')G^{-}(3'3'')G^{+}(4'4'') \\
&\quad\times D_+(\xi,\xi')[-F^{+}(4''4)\gamma(\xi',3''3) \\
&\qquad\qquad +F^{-}(3''3)\gamma(\xi',4''4)]. \qquad (2.22)
\end{aligned}$$

When this expression is multiplied out, the first of the four terms is conveniently included in a symmetrical lowest order interaction, and the $(\pm)$ superscripts can be dropped in the limit of vanishing external field. This form of the approximate interaction operator,

$$\begin{aligned}
&I(12,34)\cong ie^2\Gamma(\xi,13)\mathcal{G}_+(\xi,\xi')\Gamma(\xi',24) \\
&\quad+ie^2\gamma(\xi,11')C(1'2)D_+(\xi,\xi')C^{-1}(44')\gamma(\xi',4'3) \\
&\quad+(ie^2)^2\gamma(\xi,11')G(1'1'')\gamma(\bar{\xi},1''3)\gamma(\bar{\xi}',24') \\
&\quad\times G(4'4'')\gamma(\xi',4''4)D_+(\xi,\xi')D_+(\bar{\xi},\bar{\xi}') \\
&\quad+(ie^2)^2\gamma(\xi,11')G(1'1'')\gamma(\bar{\xi},1''2')C(2'2)D_+(\xi\xi') \\
&\quad\times D_+(\bar{\xi},\bar{\xi}')[C^{-1}(33')\gamma(\bar{\xi}',3'4')G(4'4'')\gamma(\xi',4''4) \\
&\qquad\qquad +C^{-1}(44')\gamma(\bar{\xi}',4'3')G(3'3'')\gamma(\bar{\xi}',3''3)], \qquad (2.23)
\end{aligned}$$

can be easily understood in terms of the equivalent Feynman diagram.

The wave functions $\psi(12)$ of the electron-positron system are solutions of the homogeneous equation,

$$[F^{-}(11')F^{+}(22')-I(12,1'2')]\psi(1'2')=0, \qquad (2.24)$$

related to Eq. (2.16). It is important to realize that the operators $F(12)$ also contain electrodynamic corrections. These may be obtained from the corrections to the one-particle Green's function $G(12)$, of which $F(12)$ is the inverse.[12] For the nonrelativistic states in which we are interested, the operator $F(12)$ is a multiple of the Dirac operator $\bar{F}(12)$ that depends on the experimental mass m of the electron,

$$F^{\pm}(12)=(1-\alpha B/2\pi)^{-1}\bar{F}^{\pm}(12), \qquad (2.25)$$

with

$$\bar{F}^{\pm}(x,x')=\delta(x-x')[\gamma_\mu(-i\partial_\mu'\pm eA_{+\mu}(x'))+m]. \qquad (2.26)$$

We may now introduce the interaction

$$\bar{I}(12,34)=(1-\alpha B/\pi)I(12,34), \qquad (2.27)$$

which enters the equation of the usual form for the wave function,

$$[\bar{F}^{-}(11')\bar{F}^{+}(22')-\bar{I}(12,1'2')]\psi(1'2')=0. \qquad (2.28)$$

To find the energy levels of the system, we seek solutions of the form

$$\psi(x_1x_2)=e^{iKX}\varphi_K(x);\quad X=\tfrac{1}{2}(x_1+x_2),\quad x=x_1-x_2, \qquad (2.29)$$

that are eigenfunctions of the total momentum operator with eigenvalue K. This eigenvalue is the goal of the calculation. In the absence of an external field, the interaction operator conserves the total momentum, so that it is possible to write an equation for the function $\varphi_K(x)$ of the relative coordinate x,

$$[F_K(xx')-I_K(x,x')]\varphi_K(x')=0, \qquad (2.30)$$

where

$$\begin{aligned}
&e^{iKX}[F_K(xx')]_{\alpha\beta\gamma\delta} \\
&\quad=\int\bar{F}_{\alpha\gamma}(X+\tfrac{1}{2}x,X'+\tfrac{1}{2}x')\bar{F}_{\beta\delta}(X-\tfrac{1}{2}x,X'-\tfrac{1}{2}x') \\
&\qquad\qquad\times e^{iK'X'}d^4X', \qquad (2.31)
\end{aligned}$$

and $I_K(x,x')$ is similarly related to $\bar{I}(1234)$. The Dirac indices in Eq. (2.30) are summed in the same way as those in Eq. (2.24); φ_K still has two sets of Dirac indices even through it has but one four-vector argument. To avoid complications in the notation, this matrix notation will be continued; where necessary, superscripts 1 and 2 will distinguish Dirac matrices that operate, respectively, on the first and second particle index of the wave function $\varphi_K(x)$.

Before we proceed to solve Eq. (2.30), we shall decompose the first two contributions to $\bar{I}(1234)$, Eqs. (2.23) and (2.27). With the help of the expressions[12]

$$\begin{aligned}
&\Gamma_\mu(\xi,13)=\gamma_\mu(\xi,13)(1+\alpha B/2\pi) \\
&\qquad\qquad +\bar{\Lambda}_\mu^{(2)}(1-\xi,\xi-3) \qquad (2.32)
\end{aligned}$$

and[12,13]

$$\begin{aligned}
&\mathcal{G}_{+\mu\nu}(\xi,\xi')=(1+\alpha A/2\pi)D_+(\xi,\xi')\delta_{\mu\nu} \\
&\qquad\qquad +\bar{D}_+^{(2)}(\xi,\xi')\delta_{\mu\nu}, \qquad (2.33)
\end{aligned}$$

[12] R. Karplus and N. M. Kroll, Phys. Rev. **77**, 536 (1950).

[13] Note that

$$g_+=\tfrac{1}{2}iD_F',\quad D_+=\tfrac{1}{2}iD_F,\quad \bar{D}_+^{(2)}=\tfrac{1}{2}i\bar{D}_F^{(2)}.$$

they become

$$4\pi i\alpha\gamma_\mu(\xi,13)D_+(\xi,\xi')\gamma_\mu(\xi',24)$$
$$+ie^2\gamma_\mu(\xi,11')C(1'2)D_+(\xi\xi')C^{-1}(43')\gamma_\mu(\xi',3'3)$$
$$\times(1-\alpha B/\pi)+4\pi i\alpha\gamma_\mu(\xi,13)D_+(\xi,\xi')$$
$$\times\bar{\Lambda}_\mu^{(2)}(2-\xi',\xi'-4)+4\pi i\alpha\bar{\Lambda}_\mu^{(2)}(1-\xi,\xi-3)$$
$$\times D_+(\xi,\xi')\gamma_\mu(\xi',24)+4\pi i\alpha\gamma_\mu(\xi,13)$$
$$\bar{D}_+^{(2)}(\xi\xi')\gamma_\mu(\xi',24), \quad (2.34)$$

up to terms involving two virtual photons. The experimental value of the fine structure constant α has been written to absorb the charge renormalization factor in Eq. (2.33),[12]

$$4\pi\alpha=e^2(1+\alpha A/2\pi)=4\pi/137.03\cdots. \quad (2.35)$$

III. PERTURBATION THEORY

Salpeter[9] has discussed a method for finding the eigenvalues of the total energy of a two-particle system described by an equation like Eq. (2.30) if the interaction function does not differ greatly from a local instantaneous interaction of the form

$$\delta(x-x')\delta(t)f(\mathbf{r}) \quad (x_\mu=\mathbf{r},t;\quad i=1,2,3). \quad (3.1)$$

Such a term can indeed be separated from the center-of-mass transform of the first two contributions of Eq. (2.34), which may be written

$$I^C(x,x')+I_{K1}(x,x')=I^C(x,x')$$
$$+I_{K1B}(x,x')+I_{K1A}(x,x'), \quad (3.2)$$

where

$$I^C(x,x')=-i\alpha\delta(x-x')\gamma_0{}^1\gamma_0{}^2\delta(t)/r, \quad (3.3)$$

the Coulomb interaction, and

$$I_{K1B}=2i\alpha(2\pi)^{-3}\delta(x-x')$$
$$\times\int d^4ke^{ikx}\left[\frac{\gamma^1\cdot\gamma^2}{k_\mu{}^2}-\frac{\gamma_0{}^1\gamma_0{}^2k_0{}^2}{k_i{}^2k_\mu{}^2}\right], \quad (3.4)$$

$$I_{K1A}=ie^2(\gamma_\mu C)\delta(x)\delta(x')(C^{-1}\gamma_\mu)(1-\alpha B/\pi)/K_\nu{}^2. \quad (3.5)$$

These include the Breit[11] interaction, retardation effects, and the virtual annihilation exchange interaction. All the contributions derivable from Eqs. (2.23) and (2.34) that are not included in Eqs. (3.2–5) depend on the appearance of two virtual quanta. The two-quantum terms that are included in Eq. (2.34) will be denoted by $I_{K2B}{}^{(1)}$, while those that are explicit in Eq. (2.23) will be denoted by $I_{K2A}(x,x')$ or $I_{K2B}{}^{(2)}(x,x')$ depending on whether they are exchange or direct interactions.

The change in energy levels produced by the perturbations I_{K1} and I_{K2} acting on the electron-positron system bound by the Coulomb interaction Eq. (3.3) is then given to a sufficient approximation by[9]

$$\Delta E=-i\int d^4xd^4x'\,\bar{\varphi}_C(x)$$
$$\times\Big\{I_{K1}(x,x')+I_{K2A}(x,x')+I_{K2B}{}^{(1)}(x,x')$$
$$+I_{K2B}{}^{(2)}(x,x')+\int d^4x''d^4x'''I_{K1}(x,x'')$$
$$\times[F_{KC}(x'',x''')]^{-1}I_{K1}(x''',x')\Big\}\varphi_C(x'), \quad (3.6)$$

measured in the reference frame in which the total spatial momentum vanishes,

$$K_\mu=(\mathbf{0},K_0). \quad (3.7)$$

The function $\varphi_C(x)$ is the relativistic Coulomb wave function that is a good approximation to the actual wave function of the state whose energy level is sought. It is a solution of

$$[F_{KC}(x,x')-I^C(x,x')]\varphi_C(x')=0, \quad (3.8)$$

whence

$$\Delta E=K_0-K_0{}^C. \quad (3.9)$$

The expression Eq. (3.6) is accurate to order α relative to the fine structure contribution I_{K1} and further presupposes that the intermediate states in the second-order perturbation term, the last in Eq. (3.6), can be replaced by free particle states. This is the case for the spin-spin interaction under investigation.

Before closing this section, we must briefly discuss the wave function $\varphi_C(x)$ that enters into Eq. (3.6). As is the case with the electrodynamic corrections to the magnetic interactions in hydrogen, the contributions to ΔE come mostly from the vicinity of the relative coordinate origin. The two-photon contributions, therefore, will be at most of the order $\alpha^2|\varphi_0(0)|^2$, where $\varphi_0(r)$ is the Pauli wave function for the ground state of positronium. Since this is the smallest magnitude that is being considered, contributions to these terms that are proportional to the relative momentum can be neglected. It therefore suffices to approximate $\bar{\varphi}_C(x)[\;]\varphi_C(x')$ by the product of $|\varphi_0(0)|^2=(\frac{1}{2}\alpha m)^3/\pi$ and the appropriate spin matrix element, which will be denoted by $\langle\;\rangle$. In calculating the effect of I_{K1}, which contains contributions of order $\alpha|\varphi_0(0)|^2$ due to the exchange of one virtual photon, the relative momentum can no longer be neglected. Indeed, corrections of relative order α that arise from the large momentum components of the wave function must not be omitted. As Salpeter[9] has pointed out, an improvement over the Pauli wave functions is obtained when the integral equation,

$$\varphi_C(x)=-i\alpha\int[F_{KC}(x,x')]^{-1}\varphi_C(\mathbf{r}',0)d\mathbf{r}'/r', \quad (3.10)$$

is used for an iteration procedure based on the Pauli wave function,

$$\varphi_C(x)\cong -i\alpha\int [F_{KC}(x, x')]^{-1}\varphi_0(\mathbf{r}')d\mathbf{r}'/r'. \quad (3.11)$$

IV. THE DIRECT INTERACTION

We turn now to the evaluation of the matrix elements for the energy shift that was obtained in the previous section. We shall consider first the contributions ΔE_B of those terms which arise from direct interaction, namely, those in which an electron-positron pair is present in each intermediate state. According to Eq. (3.6) and the definition preceding this equation,

$$\begin{aligned}\Delta E_B = &-i\int d^4x d^4x'\,\bar{\varphi}_C(x) I_{K1B}(x, x')\varphi_C(x')\\ &-i|\varphi_0(0)|^2\int d^4x d^4x'\Big\langle I_{K2B}^{(2)}(x, x')\\ &+\int d^4x'' d^4x''' I_{K1B}(x, x'')\\ &\times[F_{KC}(x'', x''')]^{-1} I_{K1B}(x''', x')\Big\rangle\\ &-i|\varphi_0(0)|^2\int d^4x d^4x'\langle I_{K2B}^{(1)}(x, x')\rangle. \end{aligned} \quad (4.1)$$

The one-photon part of the interaction,

$$\begin{aligned}\Delta E_{B1} &= -i\int d^4x d^4x'\,\bar{\varphi}_C(x) I_{K1B}(x, x')\varphi_C(x')\\ &= \frac{2\alpha}{(2\pi)^3}\int d^4x d^4k\,\bar{\varphi}_C(x)e^{ikx}\\ &\qquad\times\left[\frac{\gamma^1\cdot\gamma^2}{k_\mu^2}-\frac{\gamma_0^1\gamma_0^2k_0^2}{k_i^2k_\mu^2}\right]\varphi_C(x), \end{aligned} \quad (4.2)$$

presents the greatest complication because it contains the lowest order hyperfine structure as leading term. When the approximate Coulomb wave function evaluated in the appendix is inserted here, one obtains a spin matrix element and multiple momentum integral which is multiplied by the explicit factor $\alpha^3|\varphi_0(0)|^2$:

$$\begin{aligned}\Delta E_{B1} = &\frac{8\alpha^3|\varphi_0(0)|^2}{(2\pi)^4m^2}\int d^4k d\mathbf{k}' d\mathbf{k}'' dt e^{-ik_0t}\\ &\times\frac{m^2}{(k'^2+\frac{1}{4}\alpha^2m^2)^2}\frac{m^2}{(k''^2+\frac{1}{4}\alpha^2m^2)^2}\delta(\mathbf{k}-\mathbf{k}'+\mathbf{k}'')\\ &\times\left\langle F_{k'}(-t)\left[\frac{\gamma^1\cdot\gamma^2}{k_\mu^2}-\frac{\gamma_0^1\gamma_0^2k_0^2}{k_i^2k_\mu^2}\right]F_{k''}(t)\right\rangle. \end{aligned} \quad (4.3)$$

The following observations can now be made about that part of the energy change which depends on the spin of both particles. Only large contributions of magnitude α^{-2} and α^{-1} will be important in the integral. It can be seen that only small values of the momenta k', $k''\lesssim\alpha m$ make such large contributions. The important region of integration, therefore, extends over small values of either or of both these momenta. When both momenta are large, k' and $k''\gtrsim m$, the integral becomes negligible for the purposes of the present calculation. A term proportional to k'^2 and k''^2 in the spin matrix element, for instance, is negligible because in its evaluation one may neglect $(\alpha m)^2$ compared to k'^2 and k''^2, so that the integral in Eq. (4.3) becomes effectively independent of α.[14] One may now see that the spin-dependent contribution of the retarded Coulomb interaction involves one of the $\boldsymbol{\alpha}^1\cdot\mathbf{k}'\boldsymbol{\alpha}^2\cdot\mathbf{k}'$ terms of both $F(t)$ operators and is therefore a negligible large momentum effect. The Breit interaction, of course, is important and contributes in conjunction with only one factor $\boldsymbol{\alpha}^1\cdot\mathbf{k}'\boldsymbol{\alpha}^2\cdot\mathbf{k}'$. Since corrections that involve an additional factor k''^2 are too small, one may use an approximate expression

$$\begin{aligned}F_k(t)\cong&\tfrac{1}{2}(1+\boldsymbol{\alpha}^1\cdot\mathbf{k}/2m)(1-\boldsymbol{\alpha}^2\cdot\mathbf{k}/2m)\\ &\times[(m/E)(e^{-i(E-m)|t|}+e^{-i(E+m)|t|})\\ &+(e^{-i(E-m)|t|}-e^{-i(E+m)|t|})] \end{aligned} \quad (4.4)$$

to evaluate ΔE_{B1}.

The spin matrix element has now become quite simple,

$$\begin{aligned}&\langle(1+\boldsymbol{\alpha}^1\cdot\mathbf{k}'/2m)(1-\boldsymbol{\alpha}^2\cdot\mathbf{k}'/2m)\\ &\quad\times\boldsymbol{\alpha}^1\cdot\boldsymbol{\alpha}^2(1+\boldsymbol{\alpha}^1\cdot\mathbf{k}''/2m)(1-\boldsymbol{\alpha}^2\cdot\mathbf{k}''/2\mathrm{m})\rangle\\ &\quad\rightarrow\langle\boldsymbol{\sigma}^1\cdot\boldsymbol{\sigma}^2k^2-\boldsymbol{\sigma}^1\cdot\mathbf{k}\boldsymbol{\sigma}^2\cdot\mathbf{k}\rangle\rightarrow\tfrac{2}{3}\langle\boldsymbol{\sigma}^1\cdot\boldsymbol{\sigma}^2\rangle k^2, \end{aligned} \quad (4.5)$$

since the δ-function implies that $\mathbf{k}'-\mathbf{k}''=\mathbf{k}$, and the integrand has the necessary spherical symmetry. The k_0 integration with the usual treatment of the poles yields

$$\int_{-\infty}^{\infty} e^{-ik_0t}dk_0(k^2-k_0^2-i\epsilon)^{-1}=\pi ik^{-1}e^{-ik|t|}(\epsilon>0). \quad (4.6)$$

The function of time in Eq. (4.3) is therefore even, so that the time integration may be carried out only over positive values if a factor of two is supplied. The integrals encountered are of the form

$$\int_0^{\infty} dte^{-i(E'\pm m)t}e^{-ikt}e^{-i(E''\pm m)t} = -i(k+E'+E''\pm m\pm m)^{-1}, \quad (4.7)$$

since the denominator never vanishes. The energy

[14] Detailed examination shows that the integral actually is proportional to $\log\alpha$ in this case. This dependence, however, is still negligible for our purposes.

change has now been reduced to

$$\Delta E_{B1}=\frac{4}{3}\langle\sigma^1\cdot\sigma^2\rangle\alpha^3(2\pi)^{-3}|\varphi_0(0)|^2$$

$$\times\int d\mathbf{k}d\mathbf{k}'d\mathbf{k}''\delta(\mathbf{k}-\mathbf{k}'+\mathbf{k}'')(k'^2+\tfrac{1}{4}\alpha^2m^2)^{-2}$$

$$\times(k''^2+\tfrac{1}{4}\alpha^2m^2)^{-2}\left\{\frac{(E'+m)(E''+m)}{4E'E''}\right.$$

$$\times\frac{k}{k+E'+E''-2m}+\frac{m^2-E'E''}{2E'E''}\frac{k}{k+E'+E''}$$

$$\left.+\frac{(E'-m)(E''-m)}{4E'E''}\frac{k}{k+E'+E''+2m}\right\}.\quad(4.8)$$

As it stands, the integral in Eq. (4.8) is quite difficult to carry out. We must remember, however, that at least one of the two variables k', k'' must be small compared to m, a fact which permits replacement of the corresponding kinetic energy by the rest energy. Furthermore, the occurrence of a factor $(E'-m)$ implies that the particular term contributes only for large $k'\sim m$, whence k'' must be small, and *vice versa*. In such a case, the small momentum may also be neglected in the argument of the δ-function. The remaining integration can then be carried out:

$$\Delta E_{B1}=\frac{4}{3}\langle\sigma^1\cdot\sigma^2\rangle\alpha^3(2\pi)^{-3}|\varphi_0(0)|^2$$

$$\times\int d\mathbf{k}d\mathbf{k}'d\mathbf{k}''\{(m^2/E'E'')\delta(\mathbf{k}+\mathbf{k}''-\mathbf{k}')$$

$$\times(k'^2+\tfrac{1}{4}\alpha^2m^2)^{-2}(k''^2+\tfrac{1}{4}\alpha^2m^2)^{-2}$$

$$+((k-E')/2mE'kk'^2)\delta(\mathbf{k}-\mathbf{k}')(k''^2+\tfrac{1}{4}\alpha^2m^2)^{-2}$$

$$+((k-E'')/2mE''kk''^2)\delta(\mathbf{k}+\mathbf{k}'')(k'^2+\tfrac{1}{4}\alpha^2m^2)^{-2}\}$$

$$=\frac{2\pi}{3}\langle\sigma^1\cdot\sigma^2\rangle\frac{\alpha}{m^2}|\varphi_0(0)|^2\left\{1-\frac{4\alpha}{\pi}-\frac{2\alpha}{\pi}\ln\frac{m}{2k_m}\right\}.\quad(4.9)$$

In the first term both k' and k'' are of the order αm, in the second $k'\sim m$, and in the third $k''\sim m$. A cutoff k_m has here been introduced as a lower limit on the final momentum integration. Its presence shows that some contributions of order $\alpha^2|\varphi_0(0)|^2$ to ΔE_{B1} do arise from small values of momentum, contrary to expectation. It will be seen, however, that the direct interaction $I_{K2B}^{(2)}$ and the second-order effect of I_{K1B} also contain contributions from small values of the momentum as represented by the appearance of $\ln(m/2k_m)$. Just as here, these are being treated incorrectly because of the assumption of free intermediate states that is implicit in the derivation of the interaction operator. The justification of this treatment lies in the fact that the sum of the direct interactions is independent of the cutoff; that a cutoff need not have been introduced at all if the terms had been grouped properly according to the photon momentum that makes the contribution rather than according to the physical process that is represented.

The evaluation of the remainder of Eq. (4.1) is relatively simple. The second line contributes

$$\Delta E_{B2}^{(2)}=-i|\varphi_0(0)|^2(4\pi i\alpha)^2$$

$$\times\int d^4xd^4x'e^{-iK^C(X-X')}d^4X'$$

$$\times\{\langle(\gamma_\mu{}^1G^1(X+\tfrac{1}{2}x,X'+\tfrac{1}{2}x')\gamma_\nu{}^1)$$

$$\times(\gamma_\nu{}^2G^2(X-\tfrac{1}{2}x,X'-\tfrac{1}{2}x')\gamma_\mu{}^2)\rangle$$

$$\times D_+(X-X'+\tfrac{1}{2}(x-x'))D_+(X'-X+\tfrac{1}{2}(x+x'))$$

$$+(2\pi)^{-8}\int d^4kd^4k'e^{ikx}e^{ik'x'}/k_\mu{}^2k_\mu'^2$$

$$\times\langle(\gamma^1\cdot\gamma^2-\gamma_0{}^1\gamma_0{}^2(k_0{}^2/k_i{}^2))G^1(X+\tfrac{1}{2}x,X'+\tfrac{1}{2}x')$$

$$\times G^2(X-\tfrac{1}{2}x,X'-\tfrac{1}{2}x')(\gamma^1\cdot\gamma^2-\gamma_0{}^1\gamma_0{}^2(k_0{}^2/k_i{}^2))\rangle\},\quad(4.10)$$

an expression derived from Eqs. (2.23), (2.31), and (3.4). When Fourier transforms are introduced for the Green's functions, the energy may be written

$$\Delta E_{B2}^{(2)}=\frac{4\alpha^2}{(2\pi)^2}|\varphi_0(0)|^2\int d^4k(k_\mu{}^2)^{-2}$$

$$\times\left\langle\left(\gamma_\mu{}^1\frac{\gamma^1(\tfrac{1}{2}K^C-k)-m}{E^2-(m-k_0)^2}\gamma_\nu{}^1\right)\right.$$

$$\times\left(\gamma_\nu{}^2\frac{\gamma^2(\tfrac{1}{2}K^C-k)-m}{E^2-(m-k_0)^2}\gamma_\mu{}^2\right)$$

$$+(\gamma^1\cdot\gamma^2-\gamma_0{}^1\gamma_0{}^2k_0{}^2/k_i{}^2)\frac{\gamma^1(\tfrac{1}{2}K^C-k)-m}{E^2-(m-k_0)^2}$$

$$\left.\cdot\frac{\gamma^2(\tfrac{1}{2}K^C+k)-m}{E^2-(m+k_0)^2}(\gamma^1\cdot\gamma^2-\gamma_0{}^1\gamma_0{}^2k_0{}^2/k_i{}^2)\right\rangle,\quad(4.11)$$

where, as before,

$$E^2=k^2+m^2-i\epsilon\quad(\epsilon>0),\quad\tfrac{1}{2}K_0{}^C\simeq m\quad(4.12a)$$

and

$$k_\mu{}^2=k^2-k_0{}^2-i\epsilon\quad(4.12b)$$

define the treatment of the poles.

Explicit display of the spin matrix element and spherical averaging wisely precede the momentum integration,

$$\Delta E_{B2}^{(2)}=(8\alpha^2/\pi)|\varphi_0(0)|^2\langle\boldsymbol{\sigma}^1\cdot\boldsymbol{\sigma}^2\rangle$$

$$\times\int_0^\infty k^2dk\int_{-\infty}^{\infty}dk_0(k_\mu^2)^{-2}$$

$$\times\{(k_0^2-\tfrac{2}{3}k^2)[E^2-(m-k_0)^2]^{-2}$$

$$+\tfrac{1}{3}k_0^2[E^2-(m-k_0)^2]^{-1}[E^2-(m+k_0)^2]^{-1}\};\quad(4.13)$$

it depends on the identities

$$\gamma_i\gamma_j=-\delta_{ij}-i\sigma_{ij};\quad\langle\sigma_{ij}^1\sigma_{ij}^2\rangle=2\langle\boldsymbol{\sigma}^1\cdot\boldsymbol{\sigma}^2\rangle.\quad(4.14)$$

The evaluation of the integrals is straightforward, except that the same cutoff k_m for small momentum values must be introduced. The result is

$$\Delta E_{B2}^{(2)}=\frac{2\pi}{3}\frac{\alpha}{m^2}\langle\boldsymbol{\sigma}^1\cdot\boldsymbol{\sigma}^2\rangle|\varphi_0(0)|^2\left\{\frac{5}{2}\frac{\alpha}{\pi}+\frac{2\alpha}{\pi}\ln\frac{m}{2k_m}\right\},\quad(4.15)$$

and gives the total effect independent of k_m of processes where all quanta are exchanged between the two particles,

$$\Delta E_{B1}+\Delta E_{B2}^{(2)}=\frac{2\pi}{3}\frac{\alpha}{m^2}\langle\boldsymbol{\sigma}^1\cdot\boldsymbol{\sigma}^2\rangle|\varphi_0(0)|^2\left\{1-\frac{3}{2}\frac{\alpha}{\pi}\right\}.\quad(4.16)$$

The perturbation $\Delta E_{B2}^{(1)}$ includes effects of vacuum fluctuations on the exchange of a single quantum. The spin-dependent corrections to the vertex operator are contained in the anomalous-magnetic moment $(\alpha/2\pi)(e/2m)$ of each particle while the vacuum polarization has no effect on the singlet-triplet separation. The added contribution is therefore

$$\Delta E_{B2}^{(1)}=\frac{2\pi}{3}\frac{\alpha}{m^2}\langle\boldsymbol{\sigma}^1\cdot\boldsymbol{\sigma}^2\rangle|\varphi_0(0)|^2\{\alpha/\pi\}.\quad(4.17)$$

V. EXCHANGE INTERACTION

In this section we shall evaluate the matrix elements of the exchange energy, embracing all processes in which there is an intermediate state with no pairs present. The energy change, according to Eq. (3.6), is

$$\Delta E_A=-i\int d^4xd^4x'\,\bar{\varphi}_C(x)I_{K1A}(x,x')\varphi_C(x')(1-\alpha B/\pi)$$

$$-i|\varphi_0(0)|^2\int d^4xd^4x'\langle I_{K2A}(x,x')\rangle$$

$$-i|\varphi_0(0)|^2\int d^4xd^4x'd^4x''d^4x'''$$

$$\times\langle I_{K1A}(x,x'')[F_{KC}(x'',x''')]^{-1}I_{K1A}(x''',x')$$

$$+I_{K1B}(x,x'')[F_{KC}(x'',x''')]^{-1}I_{K1A}(x''',x')$$

$$+I_{K1A}(x,x'')[F_{KC}(x'',x''')]^{-1}I_{K1B}(x''',x')\rangle.\quad(5.1)$$

Consideration of the virtual two-quantum annihilation I_{K2A} and of the second-order single-quantum annihilation will be postponed to the end of this section. We only anticipate the result [see Appendix, Eq. (A.3)] that the latter will contribute a term that renormalizes the charge occurring in the first-order virtual annihilation from its uncorrected value e^2 to the measured value $4\pi\alpha$ [see Eq. (2.35)]; to the order considered in this paper, therefore, all quantities depend on α from here on.

The first one and last two terms in Eq. (5.1) present some complications since the quantity B is actually a divergent integral.[12] We expect that other divergent integrals will make the complete result finite, but we must exercise great care to obtain the correct finite result. For purposes of orientation it is instructive to consider briefly the matrix element in Eq. (5.1) for noninteracting, nonrelativistic initial and final states, because the high energy divergences may be expected to be the same in this simpler case as in the positronium atom. The wave function $\varphi_C(x)$ then represents the initial state plus a correction due to one Coulomb scattering, while $\int[F(x,x')]^{-1}I_{K1B}(x',x'')\varphi_0(0)d^4x'd^4x''$ represents the correction to the initial state due to the Breit interaction and retardation effects. The three terms we are now considering, therefore, comprise the matrix element of the virtual annihilation in the initial

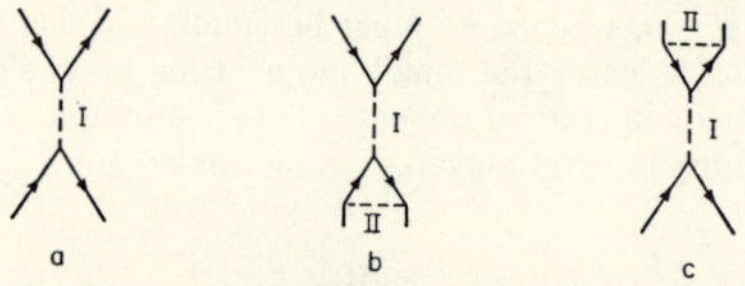

FIG. 1. Feynman diagrams for virtual annihilation electron-positron scattering.

state plus a correction due to the four-dimensional interaction represented by one quantum exchange. The Feynman diagrams for these processes, Fig. 1, show that the electrodynamic corrections, Fig. 1b, 1c, are just the correction to the vertex operator, and therefore contain each a contribution $(\alpha/2\pi)B$ multiplying the basic interaction Fig. 1a.[12,15] To our order of accuracy, the divergent integrals disappear.

With this understanding we can attempt to evaluate the actual matrix element in Eq. (5.1). In order to keep track of the infinite quantities, it is very convenient to regulate the interaction brought about by photon II in Fig. 1 with a heavy photon of mass Λ.[16] The integral B can be evaluated to B_Λ,[12]

$$B_\Lambda=(i\pi^2)^{-1}\int_0^1 udu\int d^4k$$

$$\times\{(k^2+m^2u^2)^{-2}-(k^2+m^2u^2+\Lambda^2(1-u))^{-2}$$

$$-4m^2(1-u-\tfrac{1}{4}u^2)(k^2+m^2u^2)^{-3}\}$$

$$=\ln(\Lambda/m)+\tfrac{1}{4}-\ln(m/2k_m),\quad(5.2)$$

[15] J. C. Ward, Phys. Rev. 78, 182 (1950).

[16] R. P. Feynman, Phys. Rev. 74, 1430 (1948); W. Pauli and F. Villars, Revs. Modern Phys. 21, 434 (1949).

where quantities depending inversely on Λ have been omitted and the low energy cutoff k_m has been introduced [see text following Eq. (4.9)].

The structure of the exchange interaction I_{K1A} implies that the energy change corresponding to Fig. 1 can be written

$$\Delta E_{A1} = -\pi\alpha m^{-2}\bar{\varphi}_\Lambda(0)_{\alpha\beta}(\gamma_j C)_{\alpha\beta} \times (C^{-1}\gamma_j)_{\beta'\alpha'}\varphi_\Lambda(0)_{\alpha'\beta'}, \quad (5.3)$$

where

$$(C^{-1}\gamma_j)_{\beta'\alpha'}\varphi_\Lambda(0)_{\alpha'\beta'} = \mathrm{Tr}[C^{-1}\gamma_j\varphi_\Lambda(0)]$$

$$= [1-(\alpha/2\pi)B_\Lambda]2\alpha(2\pi)^{-2}\,\mathrm{Tr}\bigg[\int d\mathbf{k}(m^2/E)$$

$$\times(k^2+\tfrac{1}{4}\alpha^2m^2)^{-2}\bigg\{(1-\tilde{\gamma}\cdot\mathbf{k}/2m)$$

$$\times C^{-1}\gamma_j(1+\gamma\cdot\mathbf{k}/2m)+C^{-1}\gamma_j k^2/4m^2\bigg\}\varphi_0(0)$$

$$+(i/2\pi)\int d^4k\{\tilde{\gamma}_i(m-\tilde{\gamma}(\tfrac{1}{2}K^C+k))$$

$$\times C^{-1}\gamma_j(m-\gamma(\tfrac{1}{2}K^C-k))\gamma_i-(k_0{}^2/k_i{}^2)$$

$$\times\tilde{\gamma}_0(m-\tilde{\gamma}(\tfrac{1}{2}K^C+k))C^{-1}\gamma_j(m-\gamma(\tfrac{1}{2}K^C-k))\gamma_0\}$$

$$\times[k_\mu{}^2]^{-1}[E^2-(m+k_0)^2]^{-1}[E^2-(m-k_0)^2]^{-1}\varphi_0(0)$$

$$-(i/2\pi)\int d^4k\{\tilde{\gamma}_\mu(m-\tilde{\gamma}(\tfrac{1}{2}K^C+k))$$

$$\times C^{-1}\gamma_j(m-\gamma(\tfrac{1}{2}K^C-k))\gamma_\mu\}[k_\mu{}^2+\Lambda^2]^{-1}$$

$$\times[E^2-(m+k_0)^2]^{-1}[E^2-(m-k_0)^2]^{-1}\varphi_0(0)\bigg]. \quad (5.4)$$

In writing the contribution of the regulating term, the last in Eq. (5.4), we have taken advantage of the fact that a very short range potential has no bound state so that the scattering picture described by Fig. 1 is applicable. The total energy has been approximated by $2m$ everywhere except in the correction to the Coulomb wave function, which comes from Eq. (A.9) evaluated at the origin. Only a space-like pair-producing Dirac matrix need be taken in Eq. (5.3). The trace is evaluated with the help of the facts that the Pauli wave function has only large components and that the charge-conjugating matrix C is an odd Dirac matrix. After integrating over k_0 with the usual treatment of the poles and after spherical averaging of the momentum integral, Eq. (5.4) becomes

$$\mathrm{Tr}[C^{-1}\gamma_j\varphi_\Lambda(0)]$$

$$=[1-(\alpha/2\pi)B_\Lambda]\,\mathrm{Tr}[C^{-1}\gamma_j\varphi_0(0)](-\alpha/\pi m^2)$$

$$\times\int_0^\infty k^2dk\{-m^2E^{-1}(4k^2/3+2m^2)$$

$$\times(k^2+\tfrac{1}{4}\alpha^2m^2)^{-2}+\tfrac{1}{3}(E^{-1}-k^{-1})+m^2k^{-3}$$

$$-k^2(k^2+\Lambda^2-\Lambda^4/4m^2)^{-1}[-(m^2-\tfrac{1}{4}\Lambda^2)$$

$$\times E^{-1}(4k^2/3+2m^2)k^{-4}+\tfrac{1}{3}(E^{-1}-E'^{-1})$$

$$+(m^2-\tfrac{1}{2}\Lambda^2)k^{-2}E'^{-1}]\}. \quad (5.5)$$

Here

$$E'=(k^2+\Lambda^2)^{\frac{1}{2}}, \quad (5.6)$$

in the second set of terms, which came from the regulating expression. One can observe that these reduce to the first set when $\Lambda=0$ if αm there is neglected with respect to k. The integrations are similar to the ones encountered in connection with Eq. (4.9) but made more complicated by the regulator. If one expands the result in powers of $(m/\Lambda)^2$ and keeps only the leading term, one obtains

$$\mathrm{Tr}[C^{-1}\gamma_j\varphi_\Lambda(0)]$$

$$=[1-(\alpha/2\pi)B_\Lambda]\,\mathrm{Tr}[C^{-1}\gamma_j\varphi_0(0)]$$

$$\times\{1+(\alpha/2\pi)[\ln(\Lambda/m)]-4+\tfrac{1}{4}-\ln(m/2k_m)\}$$

$$=(1-2\alpha/\pi)\,\mathrm{Tr}[C^{-1}\gamma_j\varphi_0(0)], \quad (5.7)$$

with B_Λ given by Eq. (5.2), whence

$$\Delta E_{A1}=-(\pi\alpha/m^2)(1-4\alpha/\pi)\times\mathrm{Tr}[\bar{\varphi}_0(0)\gamma_jC]\,\mathrm{Tr}[C^{-1}\gamma_j\varphi_0(0)], \quad (5.8)$$

because γ_jC is a symmetrical matrix. When the usual representation of the charge conjugating matrix

$$C=C^{-1}=\gamma_0\gamma_2 \quad (5.9)$$

is inserted, the direct product of the Dirac matrices in Eq. (5.8) can be relabeled so that the operators refer to the spins of the individual particles:[6–8]

$$\bar{\varphi}_0(0)_{\alpha\beta}(\gamma_jC)_{\beta\alpha}(C^{-1}\gamma_j)_{\beta'\alpha'}\varphi_0(0)_{\alpha'\beta'}\rightarrow-\varphi_0{}^*(0)_{\alpha\beta}$$

$$\times[\tfrac{3}{2}\delta_{\alpha\alpha'}\delta_{\beta\beta'}+\tfrac{1}{2}\boldsymbol{\sigma}_{\alpha\alpha'}\cdot\boldsymbol{\sigma}_{\beta\beta'}]\varphi_0(0)_{\alpha'\beta'}$$

$$=-|\varphi_0(0)|^2\langle S^2\rangle, \quad (5.10)$$

where S is the total spin of the system,

$$\mathbf{S}=\tfrac{1}{2}(\boldsymbol{\sigma}^1+\boldsymbol{\sigma}^2). \quad (5.11)$$

We then obtain the known effect of the virtual annihilation[6–8] plus a large correction of relative order α,

$$\Delta E_{A1}=(\pi\alpha/m^2)\langle S^2\rangle|\varphi_0(0)|^2\{1-4\alpha/\pi\}. \quad (5.12)$$

We now turn to the contribution of the second-order single quantum annihilation,

$$\Delta E_{A2}^{(1)} = i\pi^2\alpha^2 m^{-4}\,\mathrm{Tr}[\varphi_0(0)\gamma_i C](C^{-1}\gamma_i)$$
$$\times[F_{KC}(0,0)]^{-1}(\gamma_j C)\,\mathrm{Tr}[C^{-1}\gamma_j\varphi_0(0)], \quad (5.13)$$

with the spin sums as inferred from the derivation of this expression. In the appendix this effect is interpreted in terms of the polarization of the vacuum by the photon produced in the virtual annihilation. The evaluation given there together with Eq. (5.10) shows that the effect on the singlet-triplet splitting is[8]

$$\Delta E_{A2}^{(1)} = (\pi\alpha/m^2)\langle S^2\rangle|\varphi_0(0)|^2\{-8\alpha/9\pi\}, \quad (5.14)$$

since the renormalization constant A, Eq. (A.3) has been incorporated already.

The final item to be discussed in this section is the energy shift associated with two-quantum virtual annihilation given by

$$\Delta E_{A2}^{(2)} = \frac{i\alpha^2}{\pi^2}|\varphi_0(0)|^2$$
$$\times[k_\rho^2(K^C-k)_\sigma^2[(\tfrac{1}{2}K^C-k)_\lambda^2+m^2]^2]^{-1}$$
$$\times\langle(\gamma_\mu[\gamma(\tfrac{1}{2}K^C-k)-m]\gamma_\nu C)$$
$$\times(-C^{-1}\gamma_\mu[\gamma(\tfrac{1}{2}K^C-k)m]\gamma_\nu$$
$$+C^{-1}\gamma_\nu(\gamma(\tfrac{1}{2}K^C-k)-m]\gamma_\mu)\rangle, \quad (5.15)$$

when Fourier representations are introduced for the Green's functions. The spin matrix elements of the two parentheses is to be taken as trace with the final and initial state wave functions, as in Eqs. (5.3) and (5.4). The momentum integration is simplified by the usual procedure of combining the three distinct denominators according to the formula

$$\int d^4k[k_\mu^2(K^C-k)_\nu^2[(\tfrac{1}{2}K^C-k)_\lambda^2+m^2]^2]^{-1}$$
$$=6\int_0^1 x^2dx\int_0^1 ydy\int d^4k$$
$$\times[[k-\tfrac{1}{2}K^C(xy+2(1-x))]^2$$
$$+xm^2(xy^2-4(1-x)(1-y))-i\epsilon]^{-4}. \quad (5.16)$$

The displacement $k_0\rightarrow k_0+m(y+2(1-x))$ brings the denominator into the form

$$k^2+xm^2[x(2-y)^2-4(1-y)] \quad (5.17)$$

and leaves the numerator proportional to

$$\tfrac{1}{4}k_\lambda^2\langle(\gamma_\mu\gamma_\rho\gamma_\nu C)(-C^{-1}\gamma_\mu\gamma_\rho\gamma_\nu+C^{-1}\gamma_\nu\gamma_\rho\gamma_\mu)\rangle \quad (5.18)$$

after hyperspherical averaging and the discarding of some terms whose dependence on the Dirac matrices prevents them from contributing. Since the wave functions in which the spin matrix elements are evaluated have only large components, Eq. (5.18) can be simplified to

$$-3k_\mu^2\langle(\gamma_0\gamma_5 C)(C^{-1}\gamma_0\gamma_5)\rangle = 3k_\mu^2\langle(\gamma_2\gamma_5)(\gamma_2\gamma_5)\rangle, \quad (5.19)$$

where

$$\gamma_5=\gamma_1\gamma_2\gamma_3\gamma_0,\quad \gamma_5^2=-1. \quad (5.20)$$

Rearrangement of indices according to Eq. (5.10) finally produces the ordinary spin matrix element of a function of the total spin, Eq. (5.11),

$$3k_\mu^2\langle 2-S^2\rangle. \quad (5.21)$$

The momentum integration,

$$\int d^4k k_\mu^2[k_\nu^2+\Delta^2]^{-4} = i\pi^2/3\Delta^2, \quad (5.22)$$

brings the energy perturbation into the form

$$\Delta E_{A2}^{(2)} = -6\alpha^2 m^{-2}\langle 2-S^2\rangle|\varphi_0(0)|^2$$
$$\times\int_0^1 ydy\int_0^1 xdx[x(2-y)^2-4(1-y)-i\epsilon]^{-1}$$
$$=-(\alpha^2/m^2)\langle 2-S^2\rangle|\varphi_0(0)|^2(2-2\ln 2+\pi i). \quad (5.23)$$

The real part of this expression corresponds to the energy change of the level while the imaginary part corresponds to the well-known[17] decay rate of the singlet state by two-photon annihilation,

$$\tau^{-1}=\alpha^3\,\mathrm{Ry}_\infty = 0.804\times 10^{10}\ \mathrm{sec}^{-1}. \quad (5.24)$$

The total contribution of the virtual annihilation interaction may be collected from Eqs. (5.12), (5.13), and (5.23),

$$\Delta E_A = (\pi\alpha/m^2)|\varphi_0(0)|^2\{\langle S^2\rangle(1-4\alpha/\pi-8\alpha/9\pi)$$
$$+2\langle S^2-2\rangle(1-\ln 2)\}. \quad (5.25)$$

VI. SUMMARY

The dependence of the 1^1S and 1^3S states in positronium on the spin of the system is obtained by the addition of Eqs. (4.16), (4.17), and (5.25):

$$\Delta E = (2\pi\alpha/m^2)|\varphi_0(0)|^2\{\tfrac{1}{3}\langle\sigma^1\cdot\sigma^2\rangle[1-\tfrac{1}{2}\alpha/\pi]$$
$$+\tfrac{1}{2}\langle S^2\rangle[1-(26/9+2\ln 2)\alpha/\pi]\}.$$

By taking the difference of the value of these operators in the singlet and triplet states, one arrives at the hyperfine splitting

$$\Delta W_{ts} = (2\pi\alpha/m^2)|\varphi_0(0)|^2\{7/3-(32/9+2\ln 2)\alpha/\pi\}$$
$$=\tfrac{1}{2}\alpha^2\,\mathrm{Ry}_\infty\{7/3-(32/9+2\ln 2)\alpha/\pi\}$$
$$=2.0337\times 10^5\ \mathrm{Mc/sec}.$$

[17] J. A. Wheeler, Ann. N. Y. Acad. Sci. 48, 219 (1946).

The singlet state is the lower one. It can be seen that most of the rather large negative electrodynamic correction comes from the virtual annihilation interaction.

When the experiment of Deutsch and Brown[4] is interpreted on the basis of a Zeeman effect that depends on the total magnetic moment $(e\hbar/2mc)(1+\alpha/2\pi)$ of each particle, the value of the separation obtained by them is[5]

$$\Delta W_{ts}=(2.035\pm0.003)10^5 \text{ Mc/sec.}$$

Theory and experiment are thus in satisfactory agreement.

We are grateful to V. F. Weisskopf for calling this problem to our attention. The authors are also indebted to the members of the Institute for Advanced Study, Princeton, for an informative discussion.

APPENDIX

The operator $[F^-(13)F^+(24)]^{-1}=G^-(13)G^+(24)$, the noninteracting two-particle Green's function, and its Fourier transforms appear so frequently that this appendix will be devoted to a discussion of some of its properties.

In connection with the second-order effect of the virtual annihilation, there appears the tensor

$$\begin{aligned}(C^{-1}\gamma_i)&[F_{KC}(0,0)]^{-1}(\gamma_j C)\\ &=\int d^4X' e^{-iK(X-X')}(C^{-1}\gamma_i)_{\alpha\alpha'}G_{\alpha'\beta'}^-(X,X')\\ &\times G_{\alpha\beta}^+(X,X')(\gamma_j C)_{\beta'\beta}=\int d^4X' e^{-iK(X-X')}\\ &\times \mathrm{Tr}[\gamma_i G^-(X,X')\gamma_j G^-(X',X)],\end{aligned} \tag{A.1}$$

by Eq. (2.5). This is, however, precisely the quantity that appears in the vacuum polarization tensor.[12] It is equal to

$$-\frac{i}{(4\pi)^2}(\delta_{ij}(K^C)^2-K_i^C K_j^C)\times\left[2A+\int_0^1 dV\frac{V^2(1-\frac{1}{3}V^2)(K^C)^2}{m^2+\frac{1}{4}(K^C)^2(1-V^2)}\right]. \tag{A.2}$$

In the frame where $K_i^C=0$, $(K^C)^2\cong-4m^2$, the tensor becomes

$$\frac{im^2}{\pi\alpha}\delta_{ij}\left(\frac{\alpha}{2\pi}A-\frac{8}{9}\frac{\alpha}{\pi}\right). \tag{A.3}$$

The center-of-mass transform of the noninteracting Green's function appears in the integral equation (3.10). Its Fourier representation is

$$[F_{KC}(x,x')]^{-1}=\frac{1}{(2\pi)^4}\int d^4k e^{ik(x-x')}\times\frac{[m-\gamma^1(\frac{1}{2}K^C+k)][m-\gamma^2(\frac{1}{2}K^C-k)]}{[E^2-(\frac{1}{2}K_0{}^C+k_0)^2][E^2-(\frac{1}{2}K_0{}^C-k_0)^2]}, \tag{A.4}$$

when the center of mass is at rest. The quantity E^2 is defined by

$$E^2=k^2+m^2-i\epsilon,\quad \epsilon>0, \tag{A.5}$$

since Eq. (A.4) represents the noninteracting two-particle Green's function for outgoing waves. The integration over the fourth component of the momentum can be carried out with the help of the prescription Eq. (A.5). It gives the explicit function of the relative time coordinates,

$$[F_{KC}(x,x')]^{-1}=i(2\pi)^{-3}\int d\mathbf{k}e^{i\mathbf{k}\cdot(\mathbf{r}-\mathbf{r}')}\times(k^2+\tfrac{1}{4}\alpha^2m^2)F_k(t-t'), \tag{A.6}$$

where

$$\begin{aligned}F_k(t)=&\frac{m}{2E}[e^{-i(E-m)|t|}+e^{-i(E+m)|t|}]\\ &\times\left[\left(1+\frac{\boldsymbol{\alpha}^1\cdot\mathbf{k}}{2m}\right)\left(1-\frac{\boldsymbol{\alpha}^2\cdot\mathbf{k}}{2m}\right)+\frac{k^2}{4m^2}\right]\\ &+\tfrac{1}{2}[e^{-i(E-m)|t|}-e^{-i(E+m)|t|}]\\ &\times\left[\left(1+\frac{\boldsymbol{\alpha}^1\cdot\mathbf{k}}{2m}\right)\left(1-\frac{\boldsymbol{\alpha}^2\cdot\mathbf{k}}{2m}\right)\right.\\ &\left.-\frac{k^2}{4m^2}\left(1+\frac{t}{|t|}\frac{\boldsymbol{\alpha}^1\cdot\mathbf{k}-\boldsymbol{\alpha}^2\cdot\mathbf{k}}{E}\right)\right],\end{aligned} \tag{A.7}$$

and the total energy of the $1S$ state has been inserted,

$$K_0{}^C=2m-\tfrac{1}{2}\alpha^2m. \tag{A.8}$$

The wave function derived from Eq. (3.11) with the help of the operator just obtained is

$$\varphi_C(x)=(2\alpha/(2\pi)^2)\int d\mathbf{k}e^{i\mathbf{k}\cdot\mathbf{r}}m\times(k^2+\tfrac{1}{4}\alpha^2m^2)^{-2}F_k(t)\varphi_0(0). \tag{A.9}$$

ON THE MAGNITUDE OF THE RENORMALIZATION CONSTANTS IN QUANTUM ELECTRODYNAMICS

BY

GUNNAR KÄLLEN

With the aid of an exact formulation of the renormalization method in quantum electrodynamics which has been developed earlier, it is shown that not all of the renormalization constants can be finite quantities. It must be stressed that this statement is here made without any reference to perturbation theory.

Introduction.

In a previous paper[1], the author has given a formulation of quantum electrodynamics in terms of the renormalized Heisenberg operators and the experimental mass and charge of the electron. The consistency of the renormalization method was there shown to depend upon the behaviour of certain functions ($\Pi(p^2)$, $\Sigma_1(p^2)$ and $\Sigma_2(p^2)$) for large, negative values of the argument p^2. If the integrals

$$\int^{\infty} \frac{\Pi(-a)}{a}\, da, \quad \int^{\infty} \frac{\Sigma_i(-a)}{a}\, da \quad (i = 1, 2) \tag{1}$$

converge, quantum electrodynamics is a completely consistent theory, and the renormalization constants themselves are finite quantities. This would seem to contradict what has appeared to be a well-established fact for more than twenty years, but it must be remembered that all calculations of self-energies etc. have been made with the aid of expansions in the coupling constant e. Thus what we know is really only that, for example, the self-energy of the electron, considered as a function of e, is not analytic at the origin. It has even been suggested[2] that a different scheme of approximation may drastically alter the results obtained with the aid of a straightforward application of perturbation theory. It is the aim of the present paper to show—without any attempt at extreme mathematical rigour—that this is actually not the case in present quantum electrodynamics. The best we can

[1] G. KÄLLÉN, Helv. Phys. Acta **25**, 417 (1952), here quoted as I.

[2] *Cf., e. g.*, W. THIRRING, Z. f. Naturf. **6a** 462 (1951). N. HU, Phys. Rev. **80**, 1109 (1950).

hope for is that the renormalized theory is finite or, in other words, that the integrals

$$\int^{\infty} \frac{\Pi(-a)}{a^2}\, da, \quad \int^{\infty} \frac{\Sigma_i(-a)}{a^2}\, da, \tag{2}$$

appearing in the renormalized operators, do converge. No discussion of this point, however, will be given here.

General Outline of the Method.

We start our investigation with the assumption that all the quantities K, $(1-L)^{-1}$ and $\frac{1}{N}$ (for notations, cf. I) are finite or that the integrals (1) converge. This will be shown to lead to a lower bound for $\Pi(p^2)$ which has a finite limit for $-p^2 \to \infty$, thus contradicting our assumption. In this way it is proved that not all of the three quantities above can be finite. Our lower bound for $\Pi(p^2)$ is obtained from the formula (cf. I, Eqs. (32) and (32 a))

$$\Pi(p^2) = \frac{V}{-3p^2} \sum_{p^{(z)}=p}{}' |\langle 0 | j_\nu | z\rangle|^2 (-1)^{N_4^{(z)}}.^{1)} \tag{3}$$

It was shown in I that, in spite of the signs appearing in (3), the sum for $\Pi(p^2)$ could be written as a sum over only positive terms. Thus we get a lower bound for $\Pi(p^2)$, if we consider the following expression

$$\Pi(p^2) > \frac{V}{-3p^2} \sum_{q+q'=p}{}' |\langle 0 | j_\nu | q, q'\rangle|^2. \tag{4}$$

In Eq. (4), $\langle 0 | j_\nu | q, q'\rangle$ denotes a matrix element of the current (defined in I, Eq. (3)) between the vacuum and a state with one electron-positron pair (for $x_0 \to -\infty$). The energy-momentum vector of the electron is equal to q and of the positron is equal to q'. The sum is to be extended over all states for which $q + q' = p$. We can note here that, if we develop the function $\Pi(p^2)$ in powers of e^2 and consider just the first term in this expansion, only the states included in (4) will give a contribution. For this case, the sum is easily computed, *e. g.* in the following way:

---

[1]) $\Sigma' |\langle 0 | j_\nu | z\rangle|^2 = \sum \left(\sum_{k=1}^{3} |\langle 0 | j_k | z\rangle|^2 - |\langle 0 | j_4 | z\rangle|^2 \right)$

$$\left.\begin{aligned}\Pi^{(0)}(p^2) &= \frac{V}{-3p^2}\sum_{q+q'=p}{}' |\langle 0|j_\nu^{(0)}|q,q'\rangle|^2 \\ &= \frac{Ve^2}{-3p^2}\sum_{q+q'=p}\langle 0|\overline{\psi}^{(0)}|q'\rangle\gamma_\nu\langle 0|\psi^{(0)}|q\rangle\langle q|\overline{\psi}^{(0)}|0\rangle\gamma_\nu\langle q'|\psi^{(0)}|0\rangle \\ &= \frac{e^2}{12\pi^2}\left(1-\frac{2m^2}{p^2}\right)\sqrt{1+\frac{4m^2}{p^2}}\,\frac{1}{2}\left[1-\frac{p^2+4m^2}{|p^2+4m^2|}\right].\end{aligned}\right\} \quad (5)$$

The function $\Pi^{(0)}(p^2)$ has the constant limit $\frac{e^2}{12\pi^2}$ for large values of $-p^2$. This corresponds, of course, to the well-known divergence for the first-order charge-renormalization. We shall see, however, that with the assumptions we have made here the lower bound for the complete $\Pi(p^2)$, obtained from (4), is rather similar to $\Pi^{(0)}(p^2)$.

An Exact Expression for the Matrix Element of the Current.

Our next problem is to obtain a formula for $\langle 0|j_\nu|q,q'\rangle$ with which we can estimate the matrix element for large values of $-(q+q')^2$. For this purpose we first compute

$$\left.\begin{aligned}[j_\mu(x), \psi^{(0)}(x')] &= -N\int_{-\infty}^{x} S(13)\,[j_\mu(x), f(3)]\,dx''' \\ &\quad - iN\int_{x_0'''=x_0} S(13)\,\gamma_4\,[j_\mu(x), \psi(3)]\,d^3x'''.\end{aligned}\right\} \quad (6)$$

(Cf. I, Eq. (54).) The last commutator can be computed without difficulty if we introduce the following formula for $j_\mu(x)$

$$j_\mu(x) = \frac{ieN^2}{1-L}\xi_{\mu\lambda}s_\lambda(x) + \frac{L}{1-L}\xi_{\mu\lambda}\frac{\partial^2 A_\nu(x)}{\partial x_\lambda \partial x_\nu} - L\delta_{\mu 4}\Box A_4(x) \quad (7)$$

with

$$\xi_{\mu\lambda} = \delta_{\mu\lambda} - L\delta_{\mu 4}\delta_{\lambda 4} \quad (7\text{a})$$

and

$$s_\lambda(x) = \frac{1}{2}[\overline{\psi}(x), \gamma_\lambda\psi(x)]. \quad (7\text{b})$$

The expression (7) is written in such a way that the second time-derivatives of all the A_μ's drop out. With the aid of I, Eqs. (4)—(7) we now get

$$\left.\begin{aligned}[j_\mu(x), \psi(3)]_{x_0'''=x_0} &= \frac{ieN^2}{1-L}\xi_{\mu\lambda}[s_\lambda(x), \psi(3)] \\ &= -\frac{ie}{1-L}\xi_{\mu\lambda}\gamma_4\gamma_\lambda\psi(x)\delta(\bar{x}-\bar{x}''').\end{aligned}\right\} \quad (8)$$

It thus follows that

$$\left.\begin{aligned}[j_\mu(x), \psi^{(0)}(x')] = &-N\int_{-\infty}^{x} S(13)[j_\mu(x), f(3)]\,dx''' \\ &-\frac{eN}{1-L}\xi_{\mu\lambda}S(1x)\gamma_\lambda\psi(x).\end{aligned}\right\} \quad (9)$$

We then proceed by computing

$$\left.\begin{aligned}&\langle 0|\{[j_\mu(x), \psi^{(0)}(x')], \bar{\psi}^{(0)}(x'')\}|0\rangle \\ &= \frac{ieN}{1-L}\xi_{\mu\lambda}S(1x)\gamma_\lambda S(x2) - N\int_{-\infty}^{x} S(13)\,dx''' \\ &\times[\langle 0|[j_\mu(x), \{\bar{\psi}^{(0)}(2), f(3)\}]|0\rangle - \langle 0|\{[j_\mu(x), \bar{\psi}^{(0)}(2)], f(3)\}|0\rangle].\end{aligned}\right\} \quad (10)$$

If this expression is considered as an identity in x' and x'' it will obviously give us a formula for $\langle 0|j_\mu|q, q'\rangle$ and for $\langle q|j_\mu|q'\rangle$. (Cf. I, Eqs. (68) and (77).) We transform the right-hand side of (10) in the following way:

$$\{\bar{\psi}^{(0)}(2), f(3)\} = N\int_{-\infty}^{x'''}\{f(3), \bar{f}(4)\}S(42)\,dx^{\mathrm{IV}} - \frac{i}{N}[ie\gamma A(3) + K]S(32) \quad (11)$$

and, hence,

$$\left.\begin{aligned}\langle 0|[j_\mu(x), \{\bar{\psi}^{(0)}(2), f(3)\}]|0\rangle &= \frac{e}{N}\gamma_\lambda S(32)\langle 0|[j_\mu(x), A_\lambda(3)]|0\rangle \\ &+ N\int_{-\infty}^{x'''} dx^{\mathrm{IV}}\langle 0|[j_\mu(x), \{f(3), \bar{f}(4)\}]|0\rangle S(42).\end{aligned}\right\} \quad (12)$$

The last term in (10) can be treated in a similar way:

$$[j_\mu(x), \bar{\psi}^{(0)}(2)] = N\int_{-\infty}^{x}[j_\mu(x), \bar{f}(4)]S(42)\,dx^{\mathrm{IV}} + \frac{eN}{1-L}\bar{\psi}(x)\gamma_\lambda S(x2)\xi_{\lambda\mu} \quad (13)$$

and

$$\left.\begin{aligned}N\int_{-\infty}^{x} S(13)\,dx'''\langle 0|\{\bar{\psi}(x), f(3)\}|0\rangle &= -\langle 0|\{\bar{\psi}(x), \psi^{(0)}(x') \\ + iN\int_{x_0'''=x_0} S(13)\gamma_4\psi(3)\,d^3x'''\}|0\rangle &= iS(1x)\left[1-\frac{1}{N}\right].\end{aligned}\right\} \quad (14)$$

Collecting (12), (13) and (14) we get

$$\left.\begin{aligned} &\langle 0 | \{[j_\mu(x), \psi^{(0)}(x')], \overline{\psi}^{(0)}(x'')\} | 0 \rangle \\ &= \frac{ie}{1-L}[1+2(N-1)]\,\xi_{\mu\lambda} S(1x)\gamma_\lambda S(x2) \\ &- e\int_{-\infty}^{x} S(13)\gamma_\lambda S(32)\, dx''' \langle 0 | [j_\mu(x), A_\lambda(3)] | 0 \rangle \\ &- N^2 \int_{-\infty}^{x} dx''' \int_{-\infty}^{x'''} dx^{\mathrm{IV}} S(13) \langle 0 | [j_\mu(x), \{f(3), \bar{f}(4)\}] | 0 \rangle S(42) \\ &+ N^2 \int_{-\infty}^{x} dx''' \int_{-\infty}^{x} dx^{\mathrm{IV}} S(13) \langle 0 | \{f(3), [j_\mu(x), \bar{f}(4)]\} | 0 \rangle S(42). \end{aligned}\right\} \quad (15)$$

The second term in (15) can be rewritten with the aid of the functions $\Pi(p^2)$ and $\overline{\Pi}(p^2)$.

$$\left.\begin{aligned} \langle 0 | [j_\mu(x), A_\lambda(3)] | 0 \rangle &= \int D_R(34) \langle 0 | [j_\mu(x), j_\lambda(4)] | 0 \rangle\, dx^{\mathrm{IV}} \\ &= \frac{-1}{(2\pi)^3}\int dp e^{ip(3x)}\, \varepsilon(p)\, [p_\mu p_\lambda - p^2 \delta_{\mu\lambda}] \frac{\Pi(p^2)}{p^2}. \end{aligned}\right\} \quad (16)$$

We are, however, more interested in the expression

$$\left.\begin{aligned} \frac{1}{2}[1+\varepsilon(x3)] \langle 0 | [j_\mu(x), A_\lambda(3)] | 0 \rangle &= \frac{i\delta_{\mu\lambda}}{(2\pi)^4}\int dp e^{ip(x3)} [\overline{\Pi}(p^2) \\ &+ i\pi\varepsilon(p)\Pi(p^2)] + \frac{1}{2}[1+\varepsilon(x3)] \frac{\partial^2 \Phi(3x)}{\partial x_\mu \partial x_\lambda}, \end{aligned}\right\} \quad (17)$$

where

$$\Phi(x) = \frac{1}{(2\pi)^3}\int dp e^{ipx}\, \varepsilon(p) \frac{\Pi(p^2)}{p^2}. \quad (17\,\mathrm{a})$$

Obviously, we have

$$\Phi(3x) = 0 \quad (18\,\mathrm{a})$$

$$\frac{\partial \Phi(3x)}{\partial x_0'''} = -i\overline{\Pi}(0)\,\delta(\overline{x} - \overline{x}''') \quad (18\,\mathrm{b})$$

for $x_0''' = x_0$. It thus follows

$$\varepsilon(x3) \frac{\partial^2 \Phi(3x)}{\partial x_\mu \partial x_\lambda} = \frac{\partial^2}{\partial x_\mu \partial x_\lambda}[\varepsilon(x3)\Phi(3x)] + 2i\overline{\Pi}(0)\,\delta_{\mu 4}\delta_{\lambda 4}\delta(x3). \quad (19)$$

Using the equation

$$\frac{\partial}{\partial x_\lambda'''} S(13)\gamma_\lambda S(32) = 0, \tag{20}$$

we get

$$\left.\begin{aligned} &-e\int_{-\infty}^{+\infty}\frac{1}{2}[1+\varepsilon(x3)]\,S(13)\gamma_\lambda S(32)\langle 0|[j_\mu(x), A_\lambda(3)]|0\rangle\,dx''' \\ &= -\frac{ie}{(2\pi)^4}\int dx'''\int dp e^{ip(x3)} S(13)\gamma_\mu S(32)\,[\bar{\Pi}(p^2)+i\pi\varepsilon(p)\,\Pi(p^2)] \\ &\qquad + i\delta_{\mu 4}\frac{L}{1-L}S(1x)\gamma_4 S(x2). \end{aligned}\right\} \tag{21}$$

Introducing (21) into (15) we obtain

$$\left.\begin{aligned} &\langle 0|\{[j_\mu(x), \psi^{(0)}(x')], \bar{\psi}^{(0)}(x'')\}|0\rangle \\ &= ie\int dx'''\int\frac{dp}{(2\pi)^4}e^{ip(x3)}S(13)\gamma_\mu S(32)\,[1-\bar{\Pi}(p^2) \\ &\qquad +\bar{\Pi}(0)-i\pi\varepsilon(p)\,\Pi(p^2)] \\ &-N^2\int_{-\infty}^{x}dx'''\int_{-\infty}^{x'''}dx^{\mathrm{IV}}\,S(13)\langle 0|[j_\mu(x), \{f(3), \bar{f}(4)\}]|0\rangle S(42) \\ &+N^2\int_{-\infty}^{x}dx'''\int_{-\infty}^{x}dx^{\mathrm{IV}}\,S(13)\langle 0|\{f(3), [j_\mu(x), \bar{f}(4)]\}|0\rangle S(42) \\ &\qquad +\frac{2\,ie\,(N-1)}{1-L}\xi_{\mu\lambda}S(1x)\gamma_\lambda S(x2). \end{aligned}\right\} \tag{22}$$

The first term in (22) describes the vacuum polarization and is quite similar to the corresponding expression for a weak external field (cf. I, Appendix). The remaining terms contain the anomalous magnetic moment, the main contribution to the Lamb shift etc. Introducing the notation

$$\left.\begin{aligned} &-N^2\theta(x3)\,\theta(34)\langle 0|[j_\mu(x), \{f(3), \bar{f}(4)\}]|0\rangle \\ &+N^2\theta(x3)\,\theta(x4)\langle 0|\{f(3), [j_\mu(x), \bar{f}(4)]\}|0\rangle \\ &\qquad -\frac{2\,ie\,(N-1)}{1-L}L\delta_{\mu 4}\gamma_4\,\delta(x3)\,\delta(34) \\ &= \frac{ie}{(2\pi)^8}\iint dp\,dp'\,e^{ip'(3x)+ip(x4)}\Lambda_\mu(p',p) \end{aligned}\right\} \tag{23}$$

$$\theta(x) = \frac{1}{2}[1 + \varepsilon(x)], \tag{23 a}$$

we obtain from (22)

$$\left.\begin{aligned} &\langle 0 | j_\mu | q, q' \rangle \\ &= \langle 0 | j_\mu^{(0)} | q, q' \rangle \Big[1 - \bar{\Pi}((q+q')^2) + \bar{\Pi}(0) - i\pi\Pi((q+q')^2) \\ &+ 2\frac{N'-1}{1-L} \Big] + ie \langle 0 | \bar{\psi}^{(0)} | q' \rangle \Lambda_\mu(-q', q) \langle 0 | \psi^{(0)} | q \rangle . \end{aligned}\right\} \tag{24}$$

This is the desired formula for the matrix element of the current.

Analysis of the Function $\Lambda\mu(p', p)$.

We now want to investigate the function $\Lambda\mu(p', p)$ in some detail, especially studying its behaviour for large values of $-(q+q')^2$ in (24). For simplicity, we put $\mu = k+4$ and study

$$\left.\begin{aligned} ie\Lambda_k(p', p) = &\iint dx''' dx^{\mathrm{IV}} e^{-ip'(3x) - ip(x4)} N^2 \{ \theta(x3)\theta(x4) \langle 0 | \{ f(3), \\ &[j_k(x), \bar{f}(4)] \} | 0 \rangle - \theta(x3)\theta(34) \langle 0 | [j_k(x), \{ f(3), \bar{f}(4) \}] | 0 \rangle \} . \end{aligned}\right\} \tag{25}$$

We treat the two terms in (25) separately. The first vacuum expectation value can be transformed to momentum space with the aid of the functions

$$A_k^{(+)}(p', p) = V^2 \sum_{\substack{p^{(z)} = p \\ p^{(z')} = p'}} \langle 0 | f | z' \rangle \langle z' | j_k | z \rangle \langle z | \bar{f} | 0 \rangle \tag{26}$$

$$A_k^{(-)}(p', p) = V^2 \sum_{\substack{\cdots\cdots \\ \cdots\cdots}} \langle 0 | \bar{f} | z' \rangle \langle z' | j_k | z \rangle \langle z | f | 0 \rangle \tag{27}$$

$$B_k^{(+)}(p', p) = V^2 \sum_{\substack{\cdots\cdots \\ \cdots\cdots}} \langle 0 | f | z' \rangle \langle z' | \bar{f} | z \rangle \langle z | j_k | 0 \rangle \tag{28}$$

$$B_k^{(-)}(p', p) = V^2 \sum_{\substack{\cdots\cdots \\ \cdots\cdots}} \langle 0 | j_k | z' \rangle \langle z' | \bar{f} | z \rangle \langle z | f | 0 \rangle . \tag{29}$$

It then follows that

$$\left.\begin{aligned} \langle 0 | \{ f(3), [j_k(x), \bar{f}(4)] \} | 0 \rangle = \frac{1}{V^2} \sum_{p, p'} \{ & e^{ip'(3x) + ip(x4)} A_k^{(+)}(p', p) \\ - e^{ip'(34) + ip(4x)} B_k^{(+)}(p', p) + & e^{ip'(x4) + ip(43)} B_k^{(-)}(p', p) \\ - & e^{ip'(4x) + ip(x3)} A_k^{(-)}(p', p) \} . \end{aligned}\right\} \tag{30}$$

Our discussion started with the assumption that all the renormalization constants and, of course, all the matrix elements of the operators $j_\mu(x)$ and $f(x)$ are finite. As this is a condition on the behaviour of, for example, the function $\Pi(p^2)$ for large values of $-p^2$, and as this function is defined as a sum of matrix elements, it is clear that we also have a condition on the matrix elements themselves, *i. e.* on the functions A and B defined in (26)—(29) for large values of $-p^2, -p'^2$ and $-(p-p')^2$. To get more detailed information on this point we consider the expression

$$\left.\begin{aligned} &\langle z|[j_\mu(x), A_\nu^{(0)}(x')]|z\rangle \\ &= -i\frac{L}{1-L}\frac{\partial^2 D(x'-x)}{\partial x_\mu \partial x_\nu} + \int dx'' F_{\mu\nu}(x-x'')D(x'-x'') \end{aligned}\right\} \quad (31)$$

with

$$F_{\mu\nu}(x-x'') = \theta(x-x'')\langle z|[j_\mu(x), j_\nu(x'')]|z\rangle \quad (32)$$

(cf. I, Eq. (A. 8) and the equation of motion for $A_\mu(x)$). Supposing, for simplicity, that $|z\rangle$ does not contain a photon with energy-momentum vector k, we have

$$\left.\begin{aligned} &\langle z|j_\mu(x)|z, k\rangle \\ &= -\frac{L}{1-L}k_\mu k_\nu\langle 0|A_\nu^{(0)}(x)|k\rangle + i\int dx'' F_{\mu\nu}(x-x'')\langle 0|A_\nu^{(0)}(x'')|k\rangle. \end{aligned}\right\} \quad (33)$$

Writing

$$F_{\mu\lambda}(x-x'') = \theta(x-x'')\frac{-1}{(2\pi)^3}\int dp e^{ip(x-x'')} F_{\mu\lambda}(p) \quad (34)$$

and using the formula

$$\varepsilon(x-x'') = \frac{1}{i\pi}P\int\frac{d\tau}{\tau}e^{i\tau(x_0-x_0'')} \quad (35)$$

we get

$$iF_{\mu\lambda}(x-x'') = \frac{-1}{(2\pi)^4}\int dp e^{ip(x-x'')}\{\overline{F}_{\mu\lambda}(p) + i\pi F_{\mu\lambda}(p)\} \quad (36)$$

with

$$\overline{F}_{\mu\lambda}(p) = P\int\frac{d\tau}{\tau}F_{\mu\lambda}(\overline{p}, p_0+\tau). \quad (37)$$

We further note that from (34) it follows that

$$F_{\mu\lambda}(p) = V \sum_{p^{(z')} = p^{(z)} + p} \langle z | j_\lambda | z' \rangle \langle z' | j_\mu | z \rangle - V \sum_{p^{(z')} = p^{(z)} - p} \langle z | j_\mu | z' \rangle \langle z' | j_\lambda | z \rangle. \quad (38)$$

If every expression appearing in our formalism is finite, the integral in (37) must converge. This means that[1])

$$\lim_{p_0 \to \pm\infty} F_{\mu\lambda}(\bar{p}, p_0) = 0. \quad (39)$$

Putting $\mu = \lambda = k$ we then get from (38) and (39)

$$\lim_{p_0 \to \infty} \sum_{p^{(z')} = p^{(z)} + p} |\langle z | j_k | z' \rangle|^2 (-1)^{N_4^{(z)} + N_4^{(z')}} = 0 \quad (40\,a)$$

and

$$\lim_{p_0 \to -\infty} \sum_{p^{(z')} = p^{(z)} - p} |\langle z | j_k | z' \rangle|^2 (-1)^{N_4^{(z)} + N_4^{(z')}} = 0. \quad (40\,b)$$

If we first consider a state $|z\rangle$ with no scalar or longitudinal photons, it can be shown with the aid of the gauge-invariance of the current operator (cf. I, p. 426. Eq. (47) there can be verified explicitly with the aid of (32) and (33) above) that only states $|z'\rangle$ with transversal photons will give a non-vanishing contribution to (40 a) and (40 b), and these contributions are all positive. We thus obtain the result

$$\lim_{|p_0^{(z)} - p_0^{(z')}| \to \infty} |\langle z | j_k | z' \rangle|^2 = 0 \quad (41)$$

if none of the states $|z\rangle$ and $|z'\rangle$ contains a scalar or a longitudinal photon. Because of Lorentz invariance which requires that Eq. (41) is valid in every coordinate system, it follows, however, that (41) must be valid for all kinds of states. If we make a Lorentz transformation, the "transversal" states in the new coordinate system will in general be a mixture of all kinds of states in the old system. If (41) were not valid also for the scalar and longitudinal states in the old system, it could not hold for the transversal states in the new system.

---

[1]) The case in which the integrals converge without the functions vanishing will be discussed in the Appendix.

From equation (41) we conclude that

$$\lim_{-(p-p')^2\to\infty} A_k^{(\pm)}(p',p) = 0 \qquad (42\,a)$$

$$\lim_{-p^2\to\infty} B_k^{(+)}(p',p) = 0 \qquad (42\,b)$$

$$\lim_{-p'^2\to\infty} B_k^{(-)}(p',p) = 0. \qquad (42\,c)$$

It is, of course, not immediately clear that the sum over all the terms in (26)—(29) must vanish because every term vanishes. What really follows from (40) is, however, that the sum of all the absolute values of $\langle z | j_\mu | z' \rangle$ must vanish. If the limits in A and B are then performed in such a way that p^2 and p'^2 are kept fixed for A and $(p-p')^2$ and one of the p^2's are kept fixed for the B's, equations (42) will follow.

To summarize the argument so far, we have shown that if we write

$$\langle 0 | \{f(3), [j_k(x), \bar{f}(4)]\} | 0 \rangle = \frac{1}{(2\pi)^6} \iint dp dp' e^{ip'(3x)+ip(x4)} F_k(p',p) \qquad (43)$$

we have

$$\lim_{-(p-p')^2\to\infty} F_k(p',p) = 0. \qquad (44)$$

Introducing the notations

$$\bar{F}_k(p',p) = \int \frac{d\tau}{\tau} F_k(p'-\varepsilon\tau, p) \qquad (45\,a)$$

and

$$\tilde{F}_k(p',p) = \int \frac{d\tau}{\tau} F_k(p', p+\varepsilon\tau) \qquad (45\,b)$$

(ε is a "vector" with the components $\varepsilon_k = 0$ for $k \neq 4$ and $\varepsilon_0 = 1$) we find from (44) and the assumption that the integrals in (45) converge that

$$\lim_{-(p-p')^2\to\infty} \bar{F}_k(p',p) = \lim_{-(p-p')^2\to\infty} \tilde{F}_k(p',p) = 0 \qquad (46)$$

(cf. the Appendix). With the aid of the notations (45) we can now write

$$
\left.\begin{aligned}
&\theta(x3)\,\theta(x4)\langle 0|\{f(3),[j_k(x),\bar{f}(4)]\}|0\rangle\\
&=\frac{-1}{(2\pi)^8}\iint dpdp'\,e^{ip'(3x)+ip(x4)}\,[\tilde{\bar{F}}_k(p',p)\\
&-\pi^2 F_k(p',p)+i\pi(\bar{F}_k(p',p)+\tilde{F}_k(p',p))].
\end{aligned}\right\}\quad(47)
$$

In quite a similar way it can be shown that the second term in (25) can be written in a form analogous to (47) with the aid of a function $G_k(p',p)$ which also has the properties (44) and (46). It thus follows

$$
\lim_{-(p-p')^2\to\infty} \Lambda_k(p',p) = 0. \quad (48)
$$

It must be stressed that this property of the function $\Lambda_k(p',p)$ is a consequence of (41) and thus essentially rests on the assumption that all the renormalization constants are finite quantities.

It is clear from (24) that the function Λ_μ transforms as the matrix γ_μ under a Lorentz transformation. The explicit verification of this from (23) is somewhat involved but can be carried through with the aid of the identity

$$
\left.\begin{aligned}
&\theta(x3)\,\theta(x4)\{f(3),[j_\mu(x),\bar{f}(4)]\}-\theta(x3)\,\theta(34)\,[j_\mu(x),\{f(3),\bar{f}(4)\}]\\
&=\theta(x4)\,\theta(x3)\{\bar{f}(4),[j_\mu(x),f(3)]\}-\theta(x4)\,\theta(43)\,[j_\mu(x),\{\bar{f}(4),f(3)\}]
\end{aligned}\right\}\quad(49)
$$

and the canonical commutators. Eq. (49) can also be used to prove the formula

$$
-C^{-1}\Lambda_\mu(-q',q)\,C = \Lambda_\mu^T(-q,q') \quad (50)
$$

which is, however, also evident from (24) and the charge invariance of the formalism. From the Lorentz invariance it follows that we can write

$$
\Lambda_\mu(p',p) = \sum_{\varrho'=0,1}\sum_{\varrho=0,1}(i\gamma p'+m)^{\varrho'}\,[\gamma_\mu F^{\varrho'\varrho}+p_\mu G^{\varrho'\varrho}+p'_\mu H^{\varrho'\varrho}]\,(i\gamma p+m)^{\varrho} \quad (51)
$$

where the functions F, G and H are uniquely defined and depending only on p^2, p'^2, $(p-p')^2$ and the signs $\varepsilon(p)$, $\varepsilon(p')$ and $\varepsilon(p-p')$. From (50) it then follows

$$F^{\varrho\varrho'}(-p, p') = F^{\varrho'\varrho}(-p', p) \tag{52 a}$$

$$G^{\varrho\varrho'}(-p, p') = H^{\varrho'\varrho}(-p', p). \tag{52 b}$$

Utilizing (51) and (52) we get

$$\left.\begin{aligned} ie\langle 0|\overline{\psi}^{(0)}|q'\rangle \Lambda_\mu(-q', q)\langle 0|\psi^{(0)}|q\rangle &= \langle 0|j_\mu^{(0)}|q, q'\rangle R((q+q')^2) \\ &+\frac{e}{2m} S((q+q')^2)(q_\mu - q'_\mu)\langle 0|\overline{\psi}^{(0)}|q'\rangle\langle 0|\psi^{(0)}|q\rangle \end{aligned}\right\} \tag{53}$$

where, in view of (48),

$$\lim_{-(q+q')^2\to\infty} R((q+q')^2) = \lim_{-(q+q')^2\to\infty} S((q+q')^2) = 0. \tag{54}$$

The equations (53) and (54) are the desired result of this paragraph.

Completion of the Proof.

We are now nearly at the end of our discussion. From the assumptions made about $\Pi(p^2)$ (and its consequences for $\overline{\Pi}(p^2)$, cf. the Appendix), Eqs. (53) and (54), the limit of Eq. (24) reduces to

$$\left.\begin{aligned} \lim_{-(q+q')^2\to\infty}\langle 0|j_\mu|q, q'\rangle &= \langle 0|j_\mu^{(0)}|q, q'\rangle\left[1+\overline{\Pi}(0)+2\frac{N-1}{1-L}\right] \\ &= \langle 0|j_\mu^{(0)}|q, q'\rangle\frac{2N-1}{1-L}. \end{aligned}\right\} \tag{55}$$

Our inequality (4) now gives

$$\left.\begin{aligned} \Pi(p^2) &> \frac{V}{-3p^2}\sum_{q+q'=p}{}' |\langle 0|j_\mu|q, q'\rangle|^2 \\ &\to \frac{V}{-3p^2}\sum_{q+q'=p}{}' |\langle 0|j_\mu^{(0)}|q, q'\rangle|^2\left(\frac{2N-1}{1-L}\right)^2 \\ &= \Pi^{(0)}(p^2)\left(\frac{2N-1}{1-L}\right)^2 \to \frac{e^2}{12\pi^2}\left(\frac{2N-1}{1-L}\right)^2. \end{aligned}\right\} \tag{56}$$

Except for the possibility of N being exactly $\frac{1}{2}$ $\left(\text{independent of } e^2 \text{ and } \frac{m^2}{\mu^2}\right)$ we have then proved that, if all the renormalization constants K, $\frac{1}{N}$ and $\frac{1}{(1-L)}$ are finite, the function $\Pi(p^2)$ cannot approach zero for $-p^2 \to \infty$. This is an obvious contradiction and the only remaining possibility is that at least one (and probably all) of the renormalization constants is infinite.

The case $N = \frac{1}{2}$ is rather too special to be considered seriously. We can note, however, that N must approach 1 for $e \to 0$ and that one of the integrals in I Eq. (75) will diverge at the lower limit for $\mu \to 0$, independent of the value of e. The constant N could thus at the utmost be equal to $\frac{1}{2}$ for some special combination (or combinations) of e^2 and $\frac{m^2}{\mu^2}$. As μ is an arbitrarily small quantity it is hardly possible to ascribe any physical significance to such a solution, even if it does exist.

The proof presented here makes no pretence at being satisfactory from a rigorous, mathematical point of view. It contains, for example, a large number of interchanges of orders of integrations, limiting processes and so on. From a strictly logical point of view we cannot exclude the possibility that a more singular solution exists where such formal operations are not allowed. It would, however, be rather hard to understand how the excellent agreement between experimental results and lowest order perturbation theory calculations could be explained on the basis of such a solution.

Appendix.

It has been stated and used above that: if

$$\bar{f}(x) = P\int_0^\infty \frac{f(y)}{y-x}\,dy \quad (f(0)=0) \qquad \text{(A. 1)}$$

where $f(x)$ is bounded and continuous for all finite values of x and fulfills

$$|f(x+y) - f(x)| < M\,|y| \quad \text{for all } x \qquad \text{(A. 2)}$$

and if the integral converges, both $f(x)$ and $\bar{f}(x)$ will vanish for large values of the argument. This is not strictly true, and in this appendix we will study that point in some detail.

We begin by proving that if the integral in (A. 1) converges absolutely and if

$$\lim_{x\to\infty} \log x\,|f(x)| = 0 \tag{A. 3}$$

it follows that

$$\lim_{x\to\pm\infty} \bar{f}(x) = 0. \tag{A. 4}$$

(Note that the integral $\int^{\infty} \frac{dx}{x\cdot\log x}$ is *not* convergent and that the vanishing of $f(x)$ is already implicit in (A. 3).) To get an upper bound for $\bar{f}(x)$ when $x > 0$ we write

$$\bar{f}(x) = P\int_0^{\infty} \frac{f(y)}{y-x}\,dy = \left(\int_0^{x/2} + P\int_{x/2}^{3x/2} + \int_{3x/2}^{\infty}\right)\frac{f(y)}{y-x}\,dy. \tag{A. 5}$$

(The limit $x \to -\infty$ is simpler and need not be discussed explicitly.) The absolute value of the first term in (A. 5) is obviously less than

$$\frac{2}{x}\int_0^{x/2} |f(y)|\,dy < \text{const.}\,\frac{2}{x}\int_0^{x/2}\frac{dy}{\log y} \to 0. \tag{A. 6}$$

The last term can be treated in a similar way and yields the result

$$\left|\int_{3x/2}^{\infty}\frac{f(y)}{y-x}\,dy\right| \leq \int_{3x/2}^{\infty}\frac{|f(y)|}{y/3}\,dy \to 0. \tag{A. 7}$$

The remaining term can be written

$$\left.\begin{aligned} \left|P\int_{x/2}^{3x/2}\frac{f(y)}{y-x}\,dy\right| &= \left|\int_0^{x/2}\frac{dy}{y}\,[f(x+y)-f(x-y)]\right| \\ < \int_0^{\varepsilon}\frac{dy}{y}\left|f(x+y)-f(x-y)\right| &+ \int_{\varepsilon}^{x/2}\frac{dy}{y}\left|f(x+y)\right| + \int_{\varepsilon}^{x/2}\frac{dy}{y}\left|f(x-y)\right|. \end{aligned}\right\} \tag{A. 8}$$

In view of (A. 2) and (A. 3), the three terms in (A. 8) vanish separately for large values of x. It thus follows

$$\lim_{x\to\infty} \bar{f}(x) = 0$$

q. e. d.

As the function $\Pi(p^2)$ is positive the condition (A. 3) seems rather reasonable from a physical point of view. On the other hand, the functions F_k in (45) are not necessarily positive. It is, however, also possible to construct a more general argument where (A. 3) is not used, and where even the vanishing of $f(x)$ is not needed. Instead, we then require that from

$$\bar{f}(x) = P\int_0^\infty \frac{f(y)}{y-x}\,dy;\; f(y) = 0 \quad \text{for} \quad y \leqq 0 \qquad \text{(A. 9)}$$

will follow

$$f(x) = -\frac{1}{\pi^2} P\int_{-\infty}^{+\infty} \frac{\bar{f}(y)}{y-x}\,dy \qquad \text{(A. 10)}$$

where both $f(x)$ and $\bar{f}(x)$ are finite.

Note that

$$\left.\begin{aligned} &\frac{1}{\pi^2} P\int_{-\infty}^{+\infty} \frac{dz}{(z-x)(z-y)} \\ &= \frac{-1}{4\pi^2}\iint dw_1 dw_2 \int dz e^{i(w_1+w_2)z} \cdot e^{-iw_1 x - iw_2 y} \frac{w_1 w_2}{|w_1 w_2|} \\ &= \frac{1}{2\pi}\int dw_1 e^{iw_1(y-x)} = \delta(y-x). \end{aligned}\right\} \qquad \text{(A. 11)}$$

It then follows that the integral

$$\int^\infty \frac{|1+\bar{f}(x)+i\pi f(x)|^2}{x}\,dx > \int^\infty \frac{|1+2\bar{f}(x)|}{x}\,dx$$

is divergent, because the second term is convergent in view of (A. 10). This is everything that is needed for the proof.

It is, of course, possible to construct functions $f(x)$ where (A. 10) does not follow from (A. 9). In that case we are not allowed to interchange the order of the integrations in (A. 11); but we have already excluded such cases from our discussion.

For simplicity, the statement that the functions "vanish" for large values of the variables has been used in the text. If a more careful argument is wanted the phrase

"the functions have the property that the integral

$$\int^{\infty} \frac{f(x)}{x}\,dx$$

converges" should be substituted for the word "vanish" in many places.

The author wishes to express his gratitude to Professor NIELS BOHR, Professor C. MØLLER, and Professor T. GUSTAFSON for their kind interest. He is also indebted to Professor M. RIESZ for an interesting discussion of the problems treated in the appendix.

CERN (European Council for Nuclear Research)
Theoretical Study Group at the
Institute for Theoretical Physics, University of Copenhagen,
and
Department of Mechanics and Mathematical Physics,
University of Lund.

Indleveret til selskabet den 28. februar 1953.
Færdig fra trykkeriet den 27. maj 1953.

On the Self-Energy of a Bound Electron*

NORMAN M. KROLL** AND WILLIS E. LAMB, JR.
Columbia University, New York, New York

(Received October 7, 1948)

The electromagnetic shift of the energy levels of a bound electron has been calculated on the basis of the usual formulation of relativistic quantum electrodynamics and positron theory. The theory gives a finite result of 1051 megacycles per second for the shift $2^2S_{\frac{1}{2}}-2^2P_{\frac{1}{2}}$ in hydrogen, in close agreement with the non-relativistic calculation by Bethe.

I. INTRODUCTION

BETHE[1] has recently discussed the anomalous fine structure[2] in hydrogen on the basis of non-relativistic quantum electrodynamics. His result for the $2^2S_{\frac{1}{2}}-2^2P_{\frac{1}{2}}$ displacement was

$$\Delta W = W(2^2S_{\frac{1}{2}}) - W(2^2P_{\frac{1}{2}}) = (\alpha^3 Ry/3\pi)\log(K/\bar{\epsilon}), \quad (1)$$

where $\alpha = e^2/\hbar c \sim 1/137$ the fine structure constant, Ry the Rydberg energy $\alpha^2 mc^2/2$, and $\bar{\epsilon}$ an average excitation energy of the atom, calculated to be $17.8Ry$. As Bethe's calculation diverged logarithmically, it was necessary for him to introduce a cut-off energy K for the light quanta which could be emitted and reabsorbed by the atom. On the basis of speculations as to the improved convergence of a relativistic calculation which included positron theoretic effects, Bethe took K equal to mc^2. This led to a value of $\Delta W/h = 1040$ megacycles per second, which was in very good agreement with the then available observation[3] of 1000 Mc/sec.

The purpose of this paper is to show that a relativistic calculation of ΔW does, in fact, give a convergent answer, and to present the results and some details of a calculation based on the 1927–1934 formulation of quantum electrodynamics due to Dirac, Heisenberg, Pauli, and Weisskopf. It will appear from this that the formal relativistic invariance of the present theory is to some degree illusory in that all self-energies diverge logarithmically, so that the difference of two energies such as $W(2^2S_{\frac{1}{2}})$ and $W(2^2P_{\frac{1}{2}})$, although finite, is not necessarily unique. The method we have used has a certain simplicity in its motivation, however, and the results are surprisingly plausible in their mathematical appearance. In any case, the calculations may serve as an illustration of the extent to which physical results may be derived from a divergent field theory.

The calculation is incomplete in several well defined respects. It is only made to order α in the coupling between the electron and the electromagnetic field, and to fourth order in the ratio of the velocity of the atomic electron to the velocity of light. It is expected that these deficiencies will be made up elsewhere. We will make no effort to improve on the low frequency part of the calculation as done by Bethe, for this is essentially a non-relativistic problem.

II. DERIVATION OF EQUATIONS FOR SELF-ENERGY

We start from the Hamiltonian for a system of N electrons moving in an external static electric potential energy field V and interacting with the radiation field. After elimination of the longitudinal and scalar photons in the usual way, we obtain the Hamiltonian

$$H = H_{\text{rad}} + H_{\text{mat}} + H_{\text{int}}, \quad (2)$$

$$H_{\text{rad}} = \int d\mathbf{k} \sum_{\lambda=1}^{2} N_{\mathbf{k}\lambda} \hbar c |\mathbf{k}|, \quad (3)$$

* Work supported by the Signal Corps.

** Now National Research Fellow at The Institute for Advanced Study.

[1] H. A. Bethe, Phys. Rev. 72, 339 (1947). It may be of some interest to observe that if the non-relativistic theory is taken seriously to such an extent that retardation and the recoil energy in the energy denominators are retained, the dynamic self-energy diverges only *logarithmically*, and the $S-P_{\frac{1}{2}}$ level shift *converges*, and, in fact, with K determined to be $K=2mc^2$. The resulting shift of 1134 Mc is in disagreement with the observations.

[2] W. E. Lamb, Jr. and R. C. Retherford, Phys. Rev. 72, 241 (1947).

[3] A later tentative value reported at the April 1948 Washington Physical Society meeting was 1065±20 Mc/sec.

$$H_{\text{mat}}=\sum_{i=1}^{N}\left[c\boldsymbol{\alpha}_i\cdot\mathbf{p}_i+\beta_i mc^2+V(\mathbf{r}_i)\right], \quad (4)$$

$$H_{\text{int}}=-\sum_{i=1}^{N}e\boldsymbol{\alpha}_i\cdot\mathbf{A}(\mathbf{r}_i)+\tfrac{1}{2}\sum_{i=1}^{N}\sum_{j=1}^{N}e^2/r_{ij}. \quad (5)$$

Here e is the (negative) charge on the electron; $\boldsymbol{\alpha}$, β are the Dirac matrices in the form

$$\boldsymbol{\alpha}=\begin{pmatrix}0 & \boldsymbol{\sigma}\\ \boldsymbol{\sigma} & 0\end{pmatrix}, \quad \beta=\begin{pmatrix}I & 0\\ 0 & -I\end{pmatrix}, \quad (6)$$

where $\boldsymbol{\sigma}$ are the usual two-component Pauli matrices. The vector potential of the radiation field is expanded in plane waves normalized in the continuous spectrum as

$$\mathbf{A}(\mathbf{r})=-(i/2\pi)\int d\mathbf{k}\sum_{\lambda=1}^{2}(\hbar c/k)^{\frac{1}{2}}\mathbf{b}_{\mathbf{k}\lambda}\mathbf{e}_{\mathbf{k}\lambda}\times\exp(i\mathbf{k}\cdot\mathbf{r})+\text{conj.}, \quad (7)$$

where $\mathbf{b}_{\mathbf{k}\lambda}^+$, $\mathbf{b}_{\mathbf{k}\lambda}$ are the creation and destruction operators for a light quantum of wave vector $\mathbf{k}$ and polarization type $\lambda=1, 2$.

In positron theory, because of the indefiniteness of the number of electrons, it is convenient to use second quantization for the electrons as well as for the light quanta. Then

$$H_{\text{mat}}\rightarrow\int d\mathbf{x}\psi^+(\mathbf{x})\{c\boldsymbol{\alpha}\cdot\mathbf{p}+\beta mc^2+V(\mathbf{x})\}\psi(\mathbf{x}), \quad (8)$$

$$H_{\text{int}}\rightarrow-\int d\mathbf{x}\psi^+(\mathbf{x})e\boldsymbol{\alpha}\cdot\mathbf{A}(\mathbf{x})\psi(\mathbf{x})+\int\int d\mathbf{x}d\mathbf{x}'\psi^+(\mathbf{x})\psi(\mathbf{x})(e^2/|\mathbf{x}-\mathbf{x}'|)\times\psi^+(\mathbf{x}')\psi(\mathbf{x}')=\mathbf{H}_1+\mathbf{H}_C, \quad (9)$$

where $\psi^+(\mathbf{x})$ and $\psi(\mathbf{x})$ are, respectively, creation and destruction operators for an electron. We will expand $\psi(\mathbf{x})$ in terms of the eigenfunctions $u_n(\mathbf{x})$ of the potential field V

$$\psi(\mathbf{x})=\sum_n \mathbf{a}_n u_n(\mathbf{x}), \quad (10)$$

where the coefficients $\mathbf{a}_n$ are operators corresponding to the destruction of an electron in state n, etc.

We are concerned with the self-energy of a "single" electron bound in some stationary state $u_a(\mathbf{r})$ in the potential field V. In positron theory, this is taken to mean "self-energy of one electron in state a plus the vacuum electrons" minus "self-energy of the vacuum electrons alone." The highly divergent interaction of the extra electron with the infinite charge density of the vacuum electrons must still be removed. This is done by the process of symmetrization[4,5] in which the calculation is also made using the equally justified picture that all the electrons in existence are positively charged, so that the observance of a negatively charged electron in state a corresponds to a vacancy in the sea of negative energy states otherwise filled with positively charged particles. Then the results of the two methods of calculation are averaged. Since in the first picture there is one particle present in addition to the vacuum particles, and in the second picture one particle fewer, the self-term $i=j$ in the electrostatic energy cancels out as does the direct Coulomb interaction between the bound electron and the vacuum electrons, insofar as the latter are not polarized by an external electric field. The result is an avoidance of all singularities worse than logarithmic, and these may be plausibly discarded by renormalization of charge[6] and mass.

The self-energy to order α consists of the first-order Coulomb self-energy W_C and the second-order electrodynamic self-energy W_D. The former will split naturally into a direct or vacuum polarization term W_P and a static exchange term W_S. The static and dynamic terms W_S and W_D were first calculated by Weisskopf[5] in 1934 for the case of a free electron.

To calculate W_C we need the expectation values of the operator

$$H_C=\frac{1}{2}\int\int d\mathbf{x}d\mathbf{x}'\psi^+(\mathbf{x})\psi(\mathbf{x})\times(e^2/|\mathbf{x}-\mathbf{x}'|)\psi^+(\mathbf{x}')\psi(\mathbf{x}') \quad (11)$$
$$=\sum_\alpha\sum_\beta\sum_\gamma\sum_\delta A_{\alpha\beta\gamma\delta}\mathbf{a}_\alpha^+\mathbf{a}_\beta\mathbf{a}_\gamma^+\mathbf{a}_\delta,$$

where

$$A_{\alpha\beta\gamma\delta}=\frac{1}{2}\int\int d\mathbf{x}d\mathbf{x}'u_\alpha^*(\mathbf{x})u_\beta(\mathbf{x})\times(e^2/|\mathbf{x}-\mathbf{x}'|)u_\gamma^*(\mathbf{x}')u_\delta(\mathbf{x}'), \quad (12)$$

[4] W. Heisenberg, Zeits. f. Physik 90, 209 (1934).
[5] V. F. Weisskopf, Zeits. f. Physik 90, 817 (1934).
[6] P. A. M. Dirac, Solvay Congress, 1933.

for the states represented by the Schrödinger functionals

$$\Phi(1_a 0_{r'} 1_\rho),\ \bar{\Phi}(0_r 1_\rho),\ \text{first picture,} \qquad (13a)$$

$$\Phi(0_{(\rho)}0_{(a)}1_{(r')}),\ \Phi(0_{(\rho)}1_{(r)}),\ \text{second picture.} \qquad (13b)$$

Here r denotes any positive energy state, while a prime indicates the exclusion of the state a occupied by the bound electron. The letters ρ, σ denote any negative energy state, while the indices n, α, β, γ, and δ are to be used for a complete set of states of any energy whatever. In the alternate picture, a positive energy state of a positive particle is represented by (ρ) and a negative energy state by (r), (s). The state whose vacancy constitutes our electron is denoted by (a).

Consider the expectation value

$$\Phi^*(1_a 0_{r'} 1_\rho) \sum_{\alpha\beta\gamma\delta} a_\alpha^+ a_\beta a_\gamma^+ a_\delta A_{\alpha\beta\gamma\delta}\, \Phi(1_a 0_{r'} 1_\rho).$$

Using the matrix elements[5] for the destruction and creation operators, we obtain

$$\sum_r A_{arra} + \sum_\rho A_{\rho\rho aa} + \sum_\rho A_{aa\rho\rho} + \sum_\rho \sum_\sigma A_{\sigma\sigma\rho\rho} + \sum_\rho \sum_{r'} A_{\rho r' r' \rho}.$$

Subtracting the vacuum terms

$$\sum_\rho \sum_\sigma A_{\sigma\sigma\rho\rho} + \sum_\rho \sum_r A_{\rho r r \rho},$$

the self-energy of the electron in state a on the basis of the negative particle picture is

$$\sum_r A_{arra} + 2\sum_\rho A_{aa\rho\rho} - \sum_\rho A_{a\rho\rho a} = \sum_n \pm A_{anna} + 2\sum_\rho A_{aa\rho\rho},$$

where the upper or lower sign is to be taken for a positive or negative energy state, respectively. The first term represents an exchange term and diverges only logarithmically. The last term is the direct Coulomb energy of the electron in state a interacting with the sea of negative energy electrons and diverges quadratically. As mentioned above, the worst part of this divergence is removed by the process of symmetrization. On the basis of the alternate picture, we therefore calculate the expectation value

$$\Phi^*(0_\rho 0_a 1_{r'}) \sum_{(\alpha)(\beta)(\gamma)(\delta)} a_{(\alpha)}^+ a_{(\beta)} a_{(\gamma)}^+ a_{(\delta)} A_{(\alpha)(\beta)(\gamma)(\delta)}\, \Phi(0_{(\rho)} 0_{(a)} 1_{(r')}),$$

and obtain

$$\sum_{(r')} \sum_{(s')} A_{(r')(r')(s')(s')} + \sum_{(\rho)} \sum_{(r')} A_{(r')(\rho)(\rho)(r')} + \sum_{(r')} A_{(r')(a)(a)(r')}.$$

The vacuum term is

$$\sum_{(r)} \sum_{(s)} A_{(r)(r)(s)(s)} + \sum_{(r)} \sum_{(\rho)} A_{(r)(\rho)(\rho)(r)},$$

and the difference

$$\sum_{(r)} A_{(a)(r)(r)(a)} - \sum_{(\rho)} A_{(a)(\rho)(\rho)(a)} - 2\sum_{(r)} A_{(a)(a)(r)(r)} = \sum_{(n)} \pm A_{(a)(n)(n)(a)} - 2\sum_{(r)} A_{(a)(a)(r)(r)},$$

represents the self-energy of the electron in state a as calculated on the basis of the positive particle picture. The wave functions u_n and $u_{(n)}$ are identical for physical reasons, so that we may now drop the parentheses. Averaging the two results, we obtain

$$\sum_n \pm A_{anna} + \sum_n \mp A_{aann} = W_S + W_P. \qquad (14)$$

The static term W_S is

$$W_S = \sum_n \pm A_{anna} = \tfrac{1}{2} \sum_n \pm \int\int d\mathbf{x} d\mathbf{x}' u_a^*(\mathbf{x}) u_n(\mathbf{x}) \times (e^2/|\mathbf{x}-\mathbf{x}'|) u_n^*(\mathbf{x}') u_a(\mathbf{x}'), \qquad (15)$$

and by use of a Fourier representation for $1/|\mathbf{x}-\mathbf{x}'|$

$$\frac{1}{|\mathbf{x}-\mathbf{x}'|} = (1/2\pi^2) \int d\mathbf{k}\, \exp(i\mathbf{k}\cdot(\mathbf{x}-\mathbf{x}'))/\mathbf{k}^2, \qquad (16)$$

may be written as

$$W_S = (e^2/4\pi^2) \int \frac{d\mathbf{k}}{k^2} \sum_n \pm \int d\mathbf{x} u_a^*(\mathbf{x}) \times \exp(i\mathbf{k}\cdot\mathbf{x}) u_n(\mathbf{x}) \int dx' u_n^*(\mathbf{x}') \times \exp(-i\mathbf{k}\cdot\mathbf{x}') u_a(\mathbf{x}'). \qquad (17)$$

The polarization term $W_P(a)$

$$W_P(a) = \sum_n \mp A_{aann} = \frac{e^2}{2} \int d\mathbf{x} |u_a(\mathbf{x})|^2 \times \int \frac{d\mathbf{x}'}{|\mathbf{x}-\mathbf{x}'|} \sum_n \mp |u_n(\mathbf{x}')|^2, \qquad (18)$$

may be written as

$$W_P(a)=\int d\mathbf{x}\,|u_a(\mathbf{x})|^2 ev(\mathbf{x}), \qquad (19)$$

where the function $v(\mathbf{x})$ is the potential due to a charge density[7–9]

$$\rho(\mathbf{x})=(e/2)\sum_n \mp|u_n(\mathbf{x})|^2, \qquad (20)$$

induced in the vacuum by the external electrostatic field. The energy W_P vanishes for a free electron.

The second-order electrodynamic self-energy $W_D(a)$ of the electron in state a, according to the electron picture, is given by the difference of the energy $W_D(1_a0_{r'}1_\rho)$ for the electron in state a plus the vacuum electrons and the energy $W_D(0_r1_\rho)$ of the vacuum electrons alone. The vacuum energy $W_D(0_r1_\rho)$ is given by second-order perturbation theory, and involves the virtual emission and re-absorption of a light quantum of wave vector **k** and polarization type λ. There are two types of terms, represented by the following transition schemes:

$$\begin{pmatrix}\rho\rightarrow r+\mathbf{k}\\ r+\mathbf{k}\rightarrow\rho\end{pmatrix} \text{ and } \begin{pmatrix}\rho\rightarrow\rho+\mathbf{k}\\ \sigma+\mathbf{k}\rightarrow\sigma\end{pmatrix}.$$

In the case of the energy $W_D(1_a0_{r'}1_\rho)$, there are some additional transitions which the added electron can make, and some of the previously allowed transitions are prevented by the presence of the atomic electron in state a. One has then the following types of transitions:

$$\begin{pmatrix}\rho\rightarrow r'+\mathbf{k}\\ r'+\mathbf{k}\rightarrow\rho\end{pmatrix},\ \begin{pmatrix}a\rightarrow r+\mathbf{k}\\ r+\mathbf{k}\rightarrow a\end{pmatrix},$$

$$\begin{pmatrix}\rho\rightarrow\rho+\mathbf{k}\\ \sigma+\mathbf{k}\rightarrow\sigma\end{pmatrix},\ \begin{pmatrix}\rho\rightarrow\rho+\mathbf{k}\\ a+\mathbf{k}\rightarrow a\end{pmatrix},$$

$$\begin{pmatrix}a\rightarrow a+\mathbf{k}\\ \rho+\mathbf{k}\rightarrow\rho\end{pmatrix}.$$

The difference of the two corresponding energies is

$$-\int d\mathbf{k}\sum_{\lambda=1}^{2}\Big[\sum_r\big(|(r\mathbf{k}|\mathbf{H}_1|a)|^2/(E_r+\hbar ck-E_a)\big)$$
$$-\sum_\rho\big(|(a\mathbf{k}|\mathbf{H}_1|\rho)|^2/(E_a+\hbar ck+E_a)\big)$$
$$+2\sum_\rho\big((a|\mathbf{H}_1|a\mathbf{k})(\rho\mathbf{k}|\mathbf{H}_1|\rho)/\hbar ck\big)\Big],$$

which may be written as

$$-\int d\mathbf{k}\sum_{\lambda=1}^{2}\Big[\sum_n\big(\pm|(n\mathbf{k}|\mathbf{H}_1|a)|^2/$$
$$(|E_n|+\hbar ck\mp E_a)\big)+2\sum\big((a|\mathbf{H}_1|a\mathbf{k})$$
$$\times(\rho\mathbf{k}|\mathbf{H}_1|\rho)/\hbar ck\big)\Big].$$

Symmetrization affects only the second term, and gives

$$W_D(a)=-\int d\mathbf{k}\sum_{\lambda=1}^{2}\sum_n$$
$$\times\big(\pm|(n\mathbf{k}|\mathbf{H}_1|a)|^2/(|E_n|+\hbar ck\mp E_a)\big)$$
$$+\int d\mathbf{k}\sum_{\lambda=1}^{2}\sum_n$$
$$\times\big(\pm(a|\mathbf{H}_1|a\mathbf{k})(n\mathbf{k}|\mathbf{H}_1|n)/\hbar ck\big). \qquad (21)$$

The last term can be written as

$$\int(d\mathbf{k}/k^2)\sum_{\lambda=1}^{2}\sum_n$$
$$\pm\int\int d\mathbf{x}d\mathbf{x}'u_a^*(\mathbf{x})\alpha\cdot\mathbf{e}_{\mathbf{k}\lambda}u_a(\mathbf{x})$$
$$\times\exp(i\mathbf{k}\cdot(\mathbf{x}-\mathbf{x}'))u_n^*(\mathbf{x}')\alpha\cdot\mathbf{e}_{\mathbf{k}\lambda}u_n(\mathbf{x}'),$$

which will be zero if the polarization current[7–9]

$$\mathbf{j}(\mathbf{x}')=e\sum_n\mp u_n^*(\mathbf{x}')\alpha u_n(\mathbf{x}') \qquad (22)$$

is zero. In the absence of an external vector potential, this current is in fact zero, so that the last term in Eq. (21) will henceforth be dropped. It should be noted that two physically different **k**-spaces are involved in the expressions Eq. (17) and Eq. (21) for W_S and W_D.

---

[7] See reference 4, Eq. (40).
[8] E. A. Uehling, Phys. Rev. **48**, 55 (1935), W. Pauli and M. Rose, Phys. Rev. **49**, 462 (1936), and V. F. Weisskopf, Kgl. Danske Vid. Sels. Math.-Fys. Medd **14**, No. 6 (1936).
[9] R. Serber, Phys. Rev. **48**, 49 (1935).

III. COMPUTATION OF THE SELF-ENERGIES

We turn now to the evaluation of the expressions W_S, W_D, W_P for the self-energy. We shall pay particular attention to the static and dynamic terms W_S and W_D, as the polarization (Uehling) term W_P is directly related to the polarization charge density which has been computed by others.[7–9]

In the calculations to follow, relativistic units will be used throughout, in which $\hbar$, m, and c are taken equal to unity.

Our main interest, of course, is in the case of an electron moving in a Coulomb field for which $V=-e^2/r$. The integrals like

$$(n\mathbf{k}|\mathbf{H}_1|a)=-(ie/2\pi k^{\frac{1}{2}})\int d\mathbf{x}u_n^*(\mathbf{x})\boldsymbol{\alpha}\cdot\mathbf{e}_{\mathbf{k}\lambda}\times\exp(-i\mathbf{k}\cdot\mathbf{x})u_a(\mathbf{x})$$

occur in the theory of the relativistic photoelectric effect and have been studied extensively by Hall.[10] Because of their complexity, it seems hardly likely that we could perform the necessary further operations on them to evaluate such expressions as W_D. Even more must such a direct attack be ruled out for the case of an electron moving in a general potential field $V(\mathbf{x})$, for which the relativistic eigenfunctions $u_n(\mathbf{x})$ are not known. The only remaining method of approach seems to be to make an expansion of some kind. We observe that if the electron is free, the evaluation of the sums is a comparatively simple matter. Thus, if $u_a(\mathbf{x})$ is a plane wave of momentum $\mathbf{p}$, then

$$\int d\mathbf{x}u_n^*\boldsymbol{\alpha}\cdot\mathbf{e}\exp(-i\mathbf{k}\cdot\mathbf{x})u_a$$

is different from zero only if the momentum of the state n is $\mathbf{k}+\mathbf{p}$, and there are only four such states. In the case of a "weakly" bound electron, i.e., for

$$k\gg(a||\mathbf{p}||a),$$

one might expect that the matrix element above would have an appreciable value only when $|E_n|$ is of the order $E_k=+(1+k^2)^{\frac{1}{2}}$. We take advantage of this fact in the method of calculation used.

[10] H. Hall, Rev. Mod. Phys. 8, 358 (1936).

The sums over n can be performed, at least formally, by making use of the completeness of the solutions of the Dirac equation. Thus W_S can be written in the form

$$W_S=(e^2/4\pi^2)\int(d\mathbf{k}/k^2)\sum_n\int d\mathbf{x}u_a^*(\mathbf{x})\times\exp(i\mathbf{k}\cdot\mathbf{x})(H/|H|)u_n(\mathbf{x})\int d\mathbf{x}'u_n^*(\mathbf{x}')\times\exp(-i\mathbf{k}\cdot\mathbf{x}')u_a(\mathbf{x}'),\quad(23)$$

where

$$H=\boldsymbol{\alpha}\cdot\mathbf{p}+\beta+V\quad(24)$$

is the Hamiltonian of the unperturbed electronic motion, and $|H|$ the absolute value of the Hamiltonian, by which we mean an operator having the same eigenstates and spectrum as the Hamiltonian, except that its eigenvalues are taken to be positive. It can most conveniently be computed by representing it as $+(H^2)^{\frac{1}{2}}$. The equivalence of Eqs. (17) and (23) follows from the fact that $H/|H|$ is $+1$ when operating on a positive energy state and -1 when operating on a negative energy state. Using the completeness of the $u_n(\mathbf{x})$, we now find

$$W_S=(e^2/4\pi^2)\int(d\mathbf{k}/k^2)\int d\mathbf{x}u_a^*(\mathbf{x})\times\exp(i\mathbf{k}\cdot\mathbf{x})(H/|H|)\exp(-i\mathbf{k}\cdot\mathbf{x})u_a(\mathbf{x}),\quad(25)$$

so that the problem of computing W_S is reduced to that of finding the expectation value of the operator

$$(e^2/4\pi^2)\int(d\mathbf{k}/k^2)\times\exp(-i\mathbf{k}\cdot\mathbf{x})(H/|H|)\exp(+i\mathbf{k}\cdot\mathbf{x})$$

for the state a. We first note that for any polynomial function $f(\mathbf{p}, V)$

$$f(\mathbf{p},V)\exp(+i\mathbf{k}\cdot\mathbf{x})u_a(\mathbf{x})=\exp(+i\mathbf{k}\cdot\mathbf{x})f(\mathbf{k}+\mathbf{p},V)u_a(\mathbf{x}).$$

This theorem may then be used for any function $f(\mathbf{p}, V)$ such as $H/|H|$, for which a series expansion in $\mathbf{p}$ and V is valid. We therefore write

$$(H/|H|)\exp(i\mathbf{k}\cdot\mathbf{x})u_a(\mathbf{x})=\exp(i\mathbf{k}\cdot\mathbf{x})(H/|H|)_{\mathbf{k}+\mathbf{p}}u_a(\mathbf{x}),$$

where the notation

$$(\ )_{\mathbf{k}+\mathbf{p}}$$

means that the operator $\mathbf{p}$ is to be replaced by $\mathbf{k}+\mathbf{p}$ everywhere it appears within the brackets. One then has

$$W_S=(e^2/4\pi^2)\left(a\left|\int\frac{d\mathbf{k}}{k^2}(H/|H|)_{\mathbf{k}+\mathbf{p}}\right|a\right). \quad (26)$$

In an entirely similar manner, one can show that

$$W_D=-(e^2/4\pi^2)\left(a\left|\int(d\mathbf{k}/k)\sum_{\lambda=1}^{2}\boldsymbol{\alpha}\cdot\mathbf{e}_{\mathbf{k}\lambda}\right.\right.$$
$$\times\left\{\left(\frac{H}{|H|}+1\right)\Big/(|H|+k-E_a)\right.$$
$$\left.\left.\left.+\left(\frac{H}{|H|}-1\right)\Big/(|H|+k+E_a)\right\}_{\mathbf{k}+\mathbf{p}}\boldsymbol{\alpha}\cdot\mathbf{e}_{\mathbf{k}\lambda}\right|a\right). \quad (27)$$

The evaluation of the terms W_S and W_D thus hinges on the expression of the operators

$$(1/|H|)_{\mathbf{k}+\mathbf{p}} \quad\text{and}\quad (1/(|H|+k\mp E_a))_{\mathbf{k}+\mathbf{p}}$$

in terms of operators whose expectation values can be readily obtained.

Turning now to this task, we write

$$\begin{aligned}(|H|)_{\mathbf{k}+\mathbf{p}}&=((H^2)^{\frac{1}{2}})_{\mathbf{k}+\mathbf{p}}=(((\boldsymbol{\alpha}\cdot\mathbf{p}+\beta+V)^2)^{\frac{1}{2}})_{\mathbf{k}+\mathbf{p}}\\&=((1+\mathbf{p}^2+\boldsymbol{\alpha}\cdot\mathbf{p}V+V\boldsymbol{\alpha}\cdot\mathbf{p}+2\beta V+V^2)^{\frac{1}{2}})_{\mathbf{k}+\mathbf{p}}\\&=(1+k^2+2\mathbf{k}\cdot\mathbf{p}+\mathbf{p}^2+2V(\boldsymbol{\alpha}\cdot\mathbf{k}+\boldsymbol{\alpha}\cdot\mathbf{p}+\beta)\\&\qquad+\boldsymbol{\alpha}\cdot\boldsymbol{\pi}V+V^2)^{\frac{1}{2}}, \quad (28)\end{aligned}$$

where $\boldsymbol{\pi}$ denotes an operator $\mathbf{p}$ which operates only on the quantity immediately following it (e.g., $\boldsymbol{\alpha}\cdot\mathbf{p}V=V\boldsymbol{\alpha}\cdot\mathbf{p}+\boldsymbol{\alpha}\cdot\boldsymbol{\pi}V$).

It is clear that the ratio of $(|H|)_{\mathbf{k}+\mathbf{p}}$ to $(1+k^2)^{\frac{1}{2}}$ approaches unity as k becomes large, which corresponds to our previous statement regarding the relationship between the magnitude of the matrix elements and the energy of the state n. Thus we expand

$$(1/|H|)_{\mathbf{k}+\mathbf{p}} \quad\text{and}\quad (1/|H|+k\mp E_a)_{\mathbf{k}+\mathbf{p}}$$

as follows:

$$\begin{aligned}(1/|H|)_{\mathbf{k}+\mathbf{p}}&=1/(E_k+\Delta_k)\\&=1/E_k-\Delta_k/E_k^2+\Delta_k^2/E_k^3-\cdots, \quad (29)\end{aligned}$$

$$\begin{aligned}(1/(|H|+k\mp E_a))_{\mathbf{k}+\mathbf{p}}&\\=1/(D_k^\pm+\Delta_k\mp w_a)&=1/D_k^\pm\\-(\Delta_k\mp w_a)/(D_k^\pm)^2&+\cdots, \quad (30)\end{aligned}$$

where

$$E_k=(1+k^2)^{\frac{1}{2}}, \quad (31)$$

$$\Delta_k=|H|-E_k, \quad (32)$$

$$D_k^\pm=E_k+k\mp1, \quad (33)$$

$$w_a=E_a-1. \quad (34)$$

It is, of course, also necessary to evaluate Δ_k by applying the binomial expansion

$$\begin{aligned}\Delta_k&=|H|_{\mathbf{p}+\mathbf{k}}-E_k=(E_k^2+\delta_{\mathbf{k}})^{\frac{1}{2}}-E_k\\&=\tfrac{1}{2}\delta_{\mathbf{k}}-\tfrac{1}{8}(\delta_{\mathbf{k}}^2/E_k)+\tfrac{1}{16}(\delta_{\mathbf{k}}^3/E_k^2)+\cdots, \quad (35)\end{aligned}$$

where

$$\begin{aligned}\delta_{\mathbf{k}}=2\mathbf{k}\cdot\mathbf{p}+\mathbf{p}^2+2V(\boldsymbol{\alpha}\cdot\mathbf{k}+\boldsymbol{\alpha}\cdot\mathbf{p}+\beta)&\\+\boldsymbol{\alpha}\cdot\boldsymbol{\pi}V+V^2.& \quad (36)\end{aligned}$$

All expansions indicated are to be carried to sufficiently high order so as to include all terms which are effectively of the fourth or lower order in v/c. The operator $\mathbf{p}$ is obviously of first order in v/c, V is of second order because of the virial theorem, while $\mathbf{k}$, β are of zeroth order. Since $\alpha_x^2=\alpha_y^2=\alpha_z^2=1$, $\boldsymbol{\alpha}$ must be regarded as a zero-order quantity until the expansion has been fully worked out.

The expansion of the operators in the manner indicated and the summation over the polarization direction $\lambda=1$, 2 is a straightforward but lengthy matter. This being completed, one is left with a sum of expectation values of various operators

$$1,\ \beta,\ \boldsymbol{\alpha}\cdot\mathbf{p},\ \beta\boldsymbol{\alpha}\cdot\mathbf{p},\ \mathbf{p}^2,\ \beta\mathbf{p}^2,\ V,\ \beta V,$$
$$\boldsymbol{\sigma}\cdot\boldsymbol{\pi}V\times\mathbf{p},\ \boldsymbol{\alpha}\cdot\boldsymbol{\pi}V,\ \beta\boldsymbol{\alpha}\cdot\boldsymbol{\pi}V,\ \boldsymbol{\pi}^2V,\ V\mathbf{p}^2,\ \mathbf{p}^2V,\ \boldsymbol{\pi}V\cdot\mathbf{p},$$
$$\mathbf{p}^4,\ \beta\mathbf{p}^4,\ V^2,\ \beta V^2,\ V\boldsymbol{\alpha}\cdot\mathbf{p},\ \beta V\boldsymbol{\alpha}\cdot\mathbf{p},$$
$$\boldsymbol{\alpha}\cdot\mathbf{p}\mathbf{p}^2,\ w=E-1=\boldsymbol{\alpha}\cdot\mathbf{p}+\beta+V-1,\ w^2,\ \beta w,$$

each multiplied by a combination of some fifty elementary integrals over k. The result of this calculation is given in Eq. (73) below. Before coming to it, we shall first discuss briefly the validity of the expansion used and the form in which the self-energy is expressed.

Assuming for the moment that $\mathbf{p}$ and V can be regarded as numbers less than unity (in relativistic units), then the expansions of

$$(1/|H|)_{\mathbf{k}+\mathbf{p}} \quad\text{and}\quad (1/(|H|+k+E_a))_{\mathbf{k}+\mathbf{p}},$$

(if carried far enough) are valid for all values of k, since E_k approaches unity and D_k^- approaches two as k goes to zero. On the other hand, D_k^+

approaches zero in this limit, so that one should examine the low k behavior for this case. It turns out that that part of

$$\Delta_k - w_a,$$

which does not approach zero as k approaches zero is of the order $(v/c)^2$. Therefore, in the term involving D_k^+ we shall carry our integrals down only to some intermediate wave number k_i, which, for convenience, we take to be of order $\alpha^{\frac{3}{2}}$. This term must then be given a separate treatment for the low k region $0 \leqslant k \leqslant k_i$. One would expect that the result is independent of the precise value of k_i, and this is indeed the case.

The assumption that $\mathbf{p}$ and V are always numbers less than unity is, of course, not valid. For example, in the case of the Coulomb field $V(\mathbf{x})$ becomes infinite at the origin. The region over which V is large, however, is small, so that the contribution to the expectation value from the region in which the expansion is invalid should be small. Again, in the case of the Coulomb field, $\mathbf{p}^2 u_a(\mathbf{x})$ becomes large compared to $u_a(\mathbf{x})$ for small $\mathbf{x}$. It should be observed that in the case of the Coulomb field this circumstance limits the expansion to the power of v/c here used, as the expected values of V^3, $\mathbf{p}^6$, $\boldsymbol{\pi}^4 V$, etc., diverge for S states of the Coulomb field. Although the error introduced by this phenomenon is believed to be small, a numerical estimate would be desirable. We shall not, however, make such an estimate here, as the problem is a purely non-relativistic one.

In order to simplify the appearance and physical interpretation of our results, we have found it convenient to make use of various relationships between the expectation values of the Dirac operators which are valid to the order of v/c required. Thus one can write any solution of the Dirac equation as

$$u = \begin{pmatrix} \phi \\ \omega \end{pmatrix}, \tag{37}$$

where ϕ and ω are two-component wave functions satisfying

$$(\boldsymbol{\sigma}\cdot\mathbf{p}(1/(1+E-V))\boldsymbol{\sigma}\cdot\mathbf{p}+V+1-E)\phi=0, \tag{38}$$

$$\omega=(1/(1+E-V))\boldsymbol{\sigma}\cdot\mathbf{p}\phi, \tag{39}$$

so that for a positive energy state ω is of order v/c with respect to ϕ, which apart from terms of order $(v/c)^2$ is just a non-relativistic two-component Pauli-Schrödinger wave function. One can then see, for example, that

$$\begin{aligned}(a|\mathbf{p}^2-\beta\mathbf{p}^2|a) &= 2\int d\mathbf{x}((1/(1+E-V)) \\ &\quad\times\boldsymbol{\sigma}\cdot\mathbf{p}\phi_a)^*\mathbf{p}^2(1/(1+E-V))\boldsymbol{\sigma}\cdot\mathbf{p}\phi_a \\ &\simeq \frac{1}{2}\int d\mathbf{x}\phi_a^*\mathbf{p}^4\phi_a \\ &\simeq \tfrac{1}{2}(a|\mathbf{p}^4|a),\end{aligned} \tag{40}$$

since $1-E$ and V are of order $(v/c)^2$, and $\int d\mathbf{x}\phi_a\mathbf{p}^4\phi_a$ and $(a|\mathbf{p}^4|a)$ differ only by a quantity of order $(v/c)^6$. One can therefore simplify the final result Eq. (73) by expressing all operators in terms of certain arbitrarily chosen ones which we have taken to be

$$\beta,\ \boldsymbol{\alpha}\cdot\mathbf{p},\ V,\ \boldsymbol{\pi}^2 V,\ \beta\boldsymbol{\alpha}\cdot\boldsymbol{\pi} V,$$
$$V\mathbf{p}^2,\ \mathbf{p}^4,\text{ and } V^2.$$

Our reduction is obtained by using the following relations between expectation values

$$1\rightarrow\beta+\tfrac{1}{2}\boldsymbol{\alpha}\cdot\mathbf{p}-\tfrac{1}{8}\mathbf{p}^4+\tfrac{1}{4}\beta\boldsymbol{\alpha}\cdot\boldsymbol{\pi} V, \tag{41}$$

$$\mathbf{p}^2\rightarrow\boldsymbol{\alpha}\cdot\mathbf{p}+\tfrac{1}{2}\mathbf{p}^4+\tfrac{1}{2}\beta\boldsymbol{\alpha}\cdot\boldsymbol{\pi} V, \tag{42}$$

$$\beta\mathbf{p}^2\rightarrow\boldsymbol{\alpha}\cdot\mathbf{p}+\tfrac{1}{2}\beta\boldsymbol{\alpha}\cdot\boldsymbol{\pi} V, \tag{43}$$

$$\beta V\rightarrow V-\tfrac{1}{2}V\mathbf{p}^2-\tfrac{1}{2}\beta\boldsymbol{\alpha}\cdot\boldsymbol{\pi} V, \tag{44}$$

$$\beta\boldsymbol{\alpha}\cdot\mathbf{p}\rightarrow 0, \tag{45}$$

$$\mathbf{p}^2 V\rightarrow V\mathbf{p}^2, \tag{46}$$

$$\boldsymbol{\pi} V\cdot\mathbf{p}\rightarrow-\tfrac{1}{2}\boldsymbol{\pi}^2 V, \tag{47}$$

$$\boldsymbol{\alpha}\cdot\boldsymbol{\pi} V\rightarrow 0, \tag{48}$$

$$i\boldsymbol{\sigma}\cdot\boldsymbol{\pi} V\times\mathbf{p}\rightarrow\beta\boldsymbol{\alpha}\cdot\boldsymbol{\pi} V+\tfrac{1}{2}\boldsymbol{\pi}^2 V, \tag{49}$$

$$V\boldsymbol{\alpha}\cdot\mathbf{p}\rightarrow V\mathbf{p}^2+\tfrac{1}{2}\beta\boldsymbol{\alpha}\cdot\boldsymbol{\pi} V, \tag{50}$$

$$\beta V\boldsymbol{\alpha}\cdot\mathbf{p}\rightarrow-\tfrac{1}{2}\beta\boldsymbol{\alpha}\cdot\boldsymbol{\pi} V, \tag{51}$$

$$\boldsymbol{\alpha}\cdot\mathbf{p}\mathbf{p}^2\rightarrow\mathbf{p}^4, \tag{52}$$

$$\beta\mathbf{p}^4\rightarrow\mathbf{p}^4, \tag{53}$$

$$\beta V^2\rightarrow V^2, \tag{54}$$

$$w=H-1\rightarrow\tfrac{1}{2}\boldsymbol{\alpha}\cdot\mathbf{p}+V+\tfrac{1}{8}\mathbf{p}^4-\tfrac{1}{4}\beta\boldsymbol{\alpha}\cdot\boldsymbol{\pi} V, \tag{55}$$

$$\beta w\rightarrow\tfrac{1}{2}\boldsymbol{\alpha}\cdot\mathbf{p}+V-\tfrac{1}{8}\mathbf{p}^4-\tfrac{1}{2}V\mathbf{p}^2-\tfrac{1}{4}\beta\boldsymbol{\alpha}\cdot\boldsymbol{\pi} V, \tag{56}$$

$$w^2\rightarrow\tfrac{1}{4}\mathbf{p}^4+V\mathbf{p}^2+V^2. \tag{57}$$

From these relations one finds that the total

contribution of the static and dynamic terms of the self-energy, apart from the low k contribution of the term involving $D_k{}^+$, is

$$(W_S+W_D)'=(\alpha/\pi)\left(a\left|\left(\frac{3}{2}\int_0^\infty (dk/E_k)+\tfrac{1}{4}\right)\beta\right.\right.$$
$$+\tfrac{1}{6}\boldsymbol{\alpha}\cdot\mathbf{p}+\tfrac{2}{3}\boldsymbol{\alpha}\cdot\mathbf{p}k_i+\tfrac{1}{4}\beta\boldsymbol{\alpha}\cdot\boldsymbol{\pi}V-(\tfrac{1}{3}\log(1/k_i)$$
$$\left.\left.-\tfrac{1}{3}\log 2+11/72)\pi^2 V\right|a\right). \quad (58)$$

In order to compute the low k contribution of the term involving $D_k{}^+$, it is convenient to take advantage of the essentially non-relativistic nature of this region and to make use of the previously discussed large and small component reduction Eq. (37). One then readily finds that the resultant expression has, to the order required, just the form of the non-relativistic self-energy, so that Bethe's[1] calculation may be used up to the frequency k_i. The contribution is

$$(\omega_{N.R.})_{k<k_i}$$
$$=(\alpha/\pi)(a|-\tfrac{2}{3}\mathbf{p}^2k_i-\tfrac{1}{3}(\log k_i/\bar{\epsilon})\pi^2 V|a). \quad (59)$$

Adding the two, and observing that $(a|\mathbf{p}^2|a)$ is the same as $(a|\boldsymbol{\alpha}\cdot\mathbf{p}|a)$ to the order required, we find for the total contribution of W_S and W_D

$$W_S+W_D=(\alpha/\pi)\left(a\left|\left(\frac{3}{2}\int_0^\infty (dk/E_k)+\tfrac{1}{4}\right)\beta\right.\right.$$
$$+(\boldsymbol{\alpha}\cdot\mathbf{p}/6)+\tfrac{1}{4}\beta\boldsymbol{\alpha}\cdot\boldsymbol{\pi}V-\left(\tfrac{1}{3}\log\frac{1}{\bar{\epsilon}}\right.$$
$$\left.\left.\left.-\tfrac{1}{3}\log 2+11/72\right)\pi^2 V\right|a\right), \quad (60)$$

and we note that the result is independent of the joining frequency k_i.

As previously mentioned, the direct Coulomb energy term can be expressed in terms of a polarization charge as follows:

$$W_P(a)=e\int d\mathbf{x}|u_a(\mathbf{x})|^2$$
$$\times\int d\mathbf{x}'\rho(\mathbf{x}')/|\mathbf{x}-\mathbf{x}'|, \quad (61)$$

where

$$\rho(\mathbf{x}')=(e/2)\sum_n \mp|u_n(\mathbf{x}')|^2 \quad (62)$$

is the induced charge density calculated by various authors. To the order required, this is found to be

$$\rho(\mathbf{x})=(e/6\pi^2)\nabla^2 V\left(\tfrac{1}{6}+\int(k^2dk/E_k{}^3)\right)$$
$$+(e/60\pi^2)\nabla^4 V, \quad (63)$$

from which one finds

$$W_P(a)=-(2e^2/3\pi)\left(\tfrac{1}{6}+\int(k^2dk/E_k{}^3)\right)$$
$$\times(a|V'|a)-(e^2/15\pi)(a|\nabla^2 V|a). \quad (64)$$

The prime appearing on V' is used to indicate that the gauge of V' has been determined by the fact that it arises from an expression of the form

$$-1/4\pi\int d\mathbf{x}'\nabla^2 V(\mathbf{x}')/|\mathbf{x}-\mathbf{x}'|. \quad (65)$$

We should like to point out that the expression for $\rho(\mathbf{x}')$ can be readily and neatly calculated by methods very similar to those used above in the case of the static and dynamic terms in the self-energy. To show this we first evaluate

$$\rho(\mathbf{x}',\mathbf{x}'')=-(e/2)\sum_n \pm u_n{}^*(\mathbf{x}')u_n(\mathbf{x}'')$$
$$=-(e/2)\sum_n u_n{}^*(\mathbf{x}')(H/|H|)u_n(\mathbf{x}'')$$
$$=-(e/2)\sum_n\sum_{\mu=1}^4\sum_{\nu=1}^4(H/|H|)_{\mu\nu}$$
$$\times u_{n\mu}{}^*(\mathbf{x}')u_{n\nu}(\mathbf{x}''), \quad (66)$$

where the $u_{n\mu}$, $\mu=1, 2, 3, 4$ are the components of u_n, and all operators in $H/|H|$ are to be taken with respect to $\mathbf{x}''$. Making use of the completeness relation

$$\sum_n u_{n\mu}{}^*(\mathbf{x}')u_{n\nu}(\mathbf{x}'')=\delta_{\mu\nu}\delta(\mathbf{x}'-\mathbf{x}''), \quad (67)$$

we obtain

$$\rho(\mathbf{x}',\mathbf{x}'')=-(e/2)\sum_\mu(H/|H|)_{\mu\mu}\delta(\mathbf{x}'-\mathbf{x}'')$$
$$=-(e/2)(\mathrm{Spur}H/|H|)_{\mathbf{x}''}\delta(\mathbf{x}'-\mathbf{x}''). \quad (68)$$

To evaluate this, we Fourier-analyze the delta-function

$$\delta(\mathbf{x}'-\mathbf{x}'')=(1/8\pi^3)\int d\mathbf{k}\,\exp(i\mathbf{k}\cdot(\mathbf{x}''-\mathbf{x}')), \quad (69)$$

TABLE I.

Quantity	$2^2S_{1/2}$	State $2^2P_{1/2}$	$2^2P_{3/2}$
$\nabla^2 V$	1	0	0
$i\beta\boldsymbol{\alpha}\cdot\nabla V$	−1/2	1/6	−1/12

and find

$$\rho(\mathbf{x}', \mathbf{x}'') = (1/8\pi^3)\int d\mathbf{k}\exp(-i\mathbf{k}\cdot\mathbf{x}')\times(\mathrm{Spur}H/|H|)\exp(i\mathbf{k}\cdot\mathbf{x}''),$$

$$= (1/8\pi^3)\int d\mathbf{k}\exp(i\mathbf{k}\cdot(\mathbf{x}''-\mathbf{x}'))\times(\mathrm{Spur}H/|H|)_{\mathbf{k}+\mathbf{p}}\cdot 1(\mathbf{x}''), \quad (70)$$

when $1(\mathbf{x}'')$ is a constant equal to one. Since

$$\rho(\mathbf{x}') = \rho(\mathbf{x}', \mathbf{x}'), \quad (71)$$

we obtain, finally,

$$\rho(\mathbf{x}') = -(e/16\pi^3)\times\int d\mathbf{k}(\mathrm{Spur}H/|H|)_{\mathbf{k}+\mathbf{p}}\cdot 1(\mathbf{x}'). \quad (72)$$

The expression can now be readily reduced to the form Eq. (63) by expanding

$$(\mathrm{Spur}H/|H|)_{\mathbf{k}+\mathbf{p}}$$

in the manner used for the evaluation of the static and dynamic terms.

IV. INTERPRETATION OF RESULTS

The total expression for the self-energy is

$$W(a) = (\alpha/\pi)\Bigg(a\Bigg|\left(\frac{3}{2}\int(dk/E_k)+\tfrac{1}{4}\right)\beta - \left(\frac{2}{3}\int(k^2dk/E_k^3)+\tfrac{1}{9}\right)V' + (\boldsymbol{\alpha}\cdot\mathbf{p}/6) - (i/4)\beta\boldsymbol{\alpha}\cdot\nabla V + \Big(\tfrac{1}{3}\log(1/\bar{\epsilon}) - \tfrac{1}{3}\log 2 + (11/72) - (1/15)\Big)\nabla^2 V\Bigg|a\Bigg), \quad (73)$$

where we have made use of the fact that $\boldsymbol{\pi} = -i\nabla$. We observe first of all that the worst divergence is logarithmic, and that the expression is invariant to the gauge of V.

We next see that the difference of the self-energies of the states $2^2S_{\frac{1}{2}}$ and $2^2P_{\frac{1}{2}}$ of the hydrogen atom does converge, since the expectation values of β, V', (and, therefore, $\boldsymbol{\alpha}\cdot\mathbf{p}$) are, respectively, equal for the two states. (These statements follow from the observation that neither a small change of charge nor mass of the electron will remove the degeneracy.) In order to calculate the numerical difference of $W(a)$ for the two states, we need the values of the expectation values of the remaining operators in Eq. (73). These are given in Table I[11] in units of $\alpha^2 Ry$. The energy difference is then

$$\Delta W = (\alpha^3 Ry/3\pi)[\log(1/\bar{\epsilon}) - \log 2 + (23/24) - \tfrac{1}{5}], \quad (74)$$

which, using Bethe's revised values[12] for the constants $[(\alpha^3 Ry/3\pi) = 135.580$ Mc/sec., $\log(1/\bar{\epsilon}) = 7.7169 - 0.0293]$, gives $\Delta W = 1051$ Mc/sec., and thus differs from the original guess by only a small amount. It should be admitted, however, that one cannot regard this energy difference as uniquely determined, since one is taking the difference of two infinite quantities.

With respect to the determination of the absolute value of the self-energy for a state, it is convenient to attempt a physical interpretation of the terms involved. In this context, it should be observed that even if the coefficients of the V' and β-terms were finite, the effect of these terms would be unobservable. This follows from the fact that they would manifest themselves as a

[11] It should be mentioned here that the expected value of $\nabla^2 V$ really diverges for the S states of the Coulomb field, since it then is equal to the square of the absolute value of the wave function at the origin. Since, however, our evaluation is being carried only to order $(v/c)^4$, one should be able to use the spatial dependence of the Schrödinger wave functions for the evaluation of operators which are themselves of fourth order. Thus, where Dirac wave functions are used, the divergent part is of higher order in v/c, and its neglect is consistent with the neglect of such divergent expressions as the expected value of p^6. If one rounds off the Coulomb potential at a radius considerably smaller than the classical electron radius, the contribution of the divergent part of $(S|\nabla^2 V|S)$ is still quite negligible. (If the charge is assumed to be evenly distributed over a sphere of radius a, then the ratio of the expected value of $\nabla^2 V$ for a Dirac S state to that for a Schrödinger S state is of the order $a^{-\alpha^2} \sim 1 - \alpha^2 \log a$.)

[12] H. A. Bethe, Pocono Conference, 1948.

modification in the real charge and mass of the electron, and thus be included in the observed charge and mass. We shall assume that these terms have been so included in the observed charge and mass and drop them from the self-energy expression. The term in $i\beta\boldsymbol{\alpha}\cdot\nabla V$ is just of the form of the interaction of a Pauli-type intrinsic magnetic moment with a static potential V and can thus be interpreted as implying an additional electronic magnetic moment of $\alpha/2\pi$-Bohr magentons, while the term $\nabla^2 V$ implies a correction to the external potential, or, more specifically, an additional short-range interaction between the electron and a point charge. The term in $\boldsymbol{\alpha}\cdot\mathbf{p}$ is not subject to a direct physical interpretation, and, in fact, must be regarded as having no physical significance. Thus if one applies the self-energy expression (73) (with the β- and V' terms omitted as explained above) to a free electron of momentum $\mathbf{p}$, only the term in $\boldsymbol{\alpha}\cdot\mathbf{p}$ contributes, yielding for the self-energy $(\alpha/6\pi)[\mathbf{p}^2/(1+\mathbf{p}^2)^{\frac{1}{2}}]$. Now if the electron is to be regarded as a particle, the relativistic connection between the momentum and energy of a particle must be retained, so that the self-energy should have the momentum dependence appropriate to a mass correction, that is[13] $\sim 1/(1+\mathbf{p}^2)^{\frac{1}{2}}$ corresponding to the term in β already subtracted. The presence of the non-covariant term $(\alpha/6\pi)[\mathbf{p}^2/(1+\mathbf{p}^2)^{\frac{1}{2}}]$, which is reminiscent of the stress terms in the classical self-energy, can be traced to the fact that the total self-energy is infinite, and can be avoided in the case of the free electron by paying proper attention to the domains of integration in the various $\mathbf{k}$-spaces.[14] That is, in order to keep the total self-energy finite it is necessary to integrate over a finite region of the light quantum space and the electron momentum space. If one integrates over a region which would be spherical for an electron at rest, a covariant result is obtained. One cannot, however, apply this prescription to a bound electron, so that some other means of modifying our self-energy expression must be found to give a covariant expression for the free electron. We proceed by subtracting some free electron operator from the operators contained in our self-energy expression such that the self-energy of a free electron is zero, thereby regarding the total self-energy as contained in the observed mass.

Such a procedure is, of course, not unique: we shall make the simplest subtraction, examine the resultant expression, and then investigate the nature of the lack of uniqueness. Thus if one simply drops the $(\alpha/6\pi)\boldsymbol{\alpha}\cdot\mathbf{p}$ term from the self-energy, one obtains

$$W(a)=(\alpha/\pi)(a|-i\beta\boldsymbol{\alpha}\cdot\nabla V/4+\Big(\tfrac{1}{3}\log(1/2\bar{\epsilon})+(11/72)-(1/15)\Big)\nabla^2 V|a). \quad (75)$$

This expression can be interpreted as arising from an increase in the magnetic moment of the electron of $\alpha/2\pi$-Bohr magnetons and an additional interaction potential given by

$$\delta V_{\text{eff}}=\Big(\tfrac{1}{3}\log(1/2\bar{\epsilon})+(11/72)-(1/15)\Big)\nabla^2 V. \quad (76)$$

These contribute 68 and 983 Mc/sec., respectively, to the level shift.

In accordance with our subtraction prescription we could, however, add any linear combination of free electron operators of order up to $(v/c)^4$ whose expectation value is zero for the free electron. There are seven such operators,[15] *viz.*, 1, β, $\mathbf{p}^2$, $\beta\mathbf{p}^2$, $\boldsymbol{\alpha}\cdot\mathbf{p}$, $\mathbf{p}^4$, $\beta\mathbf{p}^4$. The condition that a linear combination gives zero to order constitutes three constraints, so that there should be four linearly independent combinations giving zero for the free electron. A possible choice for these is the following:

$$\Omega_a=1-\beta-\tfrac{1}{2}\mathbf{p}^2-\boldsymbol{\alpha}\cdot\mathbf{p}+\beta\mathbf{p}^2+\tfrac{3}{8}\mathbf{p}^4, \quad (76a)$$

$$\Omega_b=\mathbf{p}^2-\beta\mathbf{p}^2-\tfrac{1}{2}\mathbf{p}^4, \quad (76b)$$

$$\Omega_c=\mathbf{p}^4-\beta\mathbf{p}^4, \quad (76c)$$

$$\Omega_d=\boldsymbol{\alpha}\cdot\mathbf{p}-\beta\mathbf{p}^2. \quad (76d)$$

The expectation values of the above combinations are all zero for the free electron. Their effect upon the self-energy of a bound electron depends upon their expectation values for a

[13] The energy of a particle of mass m and momentum p is $(m^2+\mathbf{p}^2)^{\frac{1}{2}}$. If m is modified by a quantity δm, then the energy to first order in δm is

$$(m^2+\mathbf{p}^2)^{\frac{1}{2}}+\delta m/(m^2+\mathbf{p}^2)^{\frac{1}{2}},$$

and the correction term with $m=1$, as is appropriate for the electron, is of the form given.

[14] A. Pais, Verh. d. K. Ned. Akad. v. Wet, Section 1, **19**, No. 1 (1947).

[15] Operators of odd order in v/c have been ignored, as these are all zero for the bound electron.

bound electron. Those for Ω_a, Ω_b, and Ω_c are zero, so that their subtraction would have no physical consequences. On the other hand,

$$(a|\Omega_d|a) = -i(a|\beta\boldsymbol{\alpha}\cdot\nabla V|a)/2, \qquad (77)$$

which is precisely the form of interaction of a magnetic moment with a static potential V. Thus, the lack of uniqueness of the subtraction prescription is just such as to make the magnetic moment correction indeterminate, while the correction to the potential is left uniquely determined. Now a purely magnetic measurement of the correction to the magnetic moment of the electron has been made by Kusch and Foley,[16] who obtain a value in good agreement with the value $\alpha/2\pi$-Bohr magnetons theoretically computed by Schwinger.[17] If we adopt this experimental and theoretical result, the $2^2S_{\frac{1}{2}}-2^2P_{\frac{1}{2}}$ separation becomes uniquely determined to be just the value 1051 Mc/sec. obtained above by a direct subtraction (74) of the self-energies for the two states.

---

[16] P. Kusch and H. M. Foley, Phys. Rev. **74**, 250 (1948), also J. E. Nafe and E. B. Nelson, Phys. Rev. **73**, 718 (1948).

[17] J. Schwinger, Phys. Rev. **73**, 416 (1948) and Pocono Conference, 1948.